Monosaccharides

Peter Collins took his BSc degree at Birkbeck College, London University, while employed at Glaxo Laboratories in the team engaged on a synthesis of cortisone. PhD (1963) studies followed, investigating pyranosiduloses in George Overend's monosaccharide group. After a postdoctoral period with Harold Hart at Michigan State University studying cyclohexadiene photochemistry, he took up a teaching post at Birkbeck College, where he has held the Chair of Organic Chemistry since 1987.

His research has centred on monosaccharides, at the forefront being ultraviolet-light-induced transformations of their photo-active derivatives. The syntheses of pseudo-cyclodextrins, oxepan derivatives and carbohydrate ligands for SAP plasma protein have recently featured in his group's activities.

He is Editor of Chapman and Hall's *Carbohydrates—A Source Book* and has published widely in the *Journal of the Chemistry Society, Perkin Transactions 1, Carbohydrate Research* and *Tetrahedron Letters.*

Robin Ferrier did his undergraduate work at the University of Edinburgh and graduate training in the same department under Gerald Aspinall. He was a post-doctoral fellow in Melvin Calvin's group in Berkeley and went from his first teaching position in Birkbeck College, University of London to be Professor of Organic Chemistry in Victoria University of Wellington, New Zealand in 1970.

Research interests have always been near to monosaccharide chemistry with reactions of unsaturated sugars, carbocyclic compounds derived from sugars, free radical reactions and the development of routes to pharmaceutically important products from monosaccharides being the centres of focus for his research students.

He is Senior Reporter for the Royal Society of Chemistry's annual *Specialist Periodical Report on Carbohydrate Chemistry* and has published extensively, mainly in the *Journal of the Chemical Society, Perkin Transactions 1* and *Carbohydrate Research.*

Monosaccharides
Their Chemistry and Their Roles in Natural Products

Peter M. Collins
Birkbeck College, University of London, London, UK

Robert J. Ferrier
Victoria University of Wellington, Wellington, New Zealand

JOHN WILEY & SONS
Chichester • New York • Brisbane • Toronto • Singapore

AFW0541

Other Wiley Editorial Offices

John Wiley & Sons, Inc., 605 Third Avenue,
New York, NY 10158-0012, USA

Jacaranda Wiley Ltd, 33 Park Road, Milton,
Queensland 4064, Australia

John Wiley & Sons (Canada) Ltd, 22 Worcester Road,
Rexdale, Ontario M9W 1L1, Canada

John Wiley & Sons (SEA) Pte Ltd, 37 Jalan Pemimpin #05-04,
Block B, Union Industrial Building, Singapore 2057

Library of Congress Cataloging-in-Publication Data
Collins, Peter.
     Monosaccharides : their chemistry and their roles in natural
     products / Peter Collins, Robin Ferrier.
          p.   cm.
     Includes bibliographical references and index.
     ISBN 0-471-95342-3. — ISBN 0-471-95343-1 (pbk.).
     1. Monosaccharides.   I. Ferrier, Robin.   II. Title.
     QD325.C66   1995
     547.7'813—dc20   94-26989
                                                             CIP

British Library Cataloguing in Publication Data

A catalogue record for this book is available from the British Library

ISBN 0 471 95342 3 (cloth)
ISBN 0 471 95343 1 (paper)

Typeset in 10/12pt Times by Laser Words, Madras, India
Printed and bound in Great Britain by Biddles Ltd, Guildford, Surrey

Contents

Preface

This book is effectively a second edition of one we wrote for different publishers in 1972.[1] In the interim we have often been questioned about a new edition, but a long sequence of duties has seen our good intentions lead in the inevitable direction. To some degree we now perceive the delay as an advantage because in recent years our subject has altered almost out of recognition, and now we have had the opportunity to portray a greatly extended and invigorated branch of modern chemistry. The book has therefore been substantially restructured, almost entirely rewritten and expanded to accommodate these major changes.

As before, 'we have attempted to present a rationalized account using the concepts of current mechanistic organic chemistry and stereochemistry', but whereas in the preface to the first edition we continued, 'in so doing (we) hope to have depicted our subject as a branch of chemistry which is developing in line with general advances', we now aim to depict our subject as a major component of modern organic chemistry with central significance in biology, in particular in the new science of 'glycobiology'. While up to the 1970s carbohydrate chemistry was rather an isolated component of the subject practised by specialists, this is no longer so and many highlights of the newer developments have been contributed by 'mainstream' organic chemists, the subject having been greatly vitalized in consequence. The range of organic chemistry now described is extensive, and parts of the book can be looked on as portrayals of modern general organic chemistry with monosaccharides as model systems.

In appropriate introductory sections we have retained the pedagogic style adopted previously, and we have been mindful of the student and scientist coming to grips with the subject for the first time. In other sections, where we felt the interests of the practising chemist and research students were paramount, we have provided considerable detail and, following suggestions from many colleagues, have included over 1000 references which will aid direct access to the literature. Our references are unusual in not including names of authors, and while this has allowed some space saving, the main reason for the omission is that we do not necessarily identify seminal papers, often preferring reviews or recent publications with useful bibliographies. We use authors' names sparingly in the text when significant work can be attributed appropriately.

We introduce the subject (Chapter 2) by dealing with matters relating to molecular structure and conformation, and we indicate how sugars are biosynthesized and how they may be obtained synthetically from carbohydrate and non-carbohydrate sources. For those unfamiliar with the basis of monosaccharide chemistry the first part of this chapter is essential, and the treatment of the vitally important issue of molecular shape and its consequences will prove useful for the study of conformational analysis at a level beyond that used in most organic chemistry text books. The heart of the book is in Chapters 3, 4 and 5, in which we describe the fundamental organic chemistry of the monosaccharides, and comparison of these sections with the same chapters in the preceding edition will readily reveal how appreciable has been the recent progress made in the understanding of the panorama of the organic chemistry of the sugars. Free radical, carbene, carbanion and carbocation processes are now, for example, all commonly utilized in synthetic work, whereas for the most part only the last of these were recognized as important in the 1970s.

Chapters 6 and 7 cover other main areas of recent advancement: the use of monosaccharides as starting materials in the synthesis of oligosaccharides of biological importance and in the synthesis of enantiomerically pure non-carbohydrate products. Lastly, in Chapter 8 are described many natural products which contain or are related to sugars, the focus often being on compounds that are of interest in medicinal chemistry and of importance in biology. Parts of this chapter could prove useful to chemists approaching glycobiology for the first time or to biologists wishing to examine chemical features of natural materials.

Five appendices are included on topics that we feel add to the book as a data and reference source: the literature and nomenclature of the subject, n.m.r. chemical shift data, polarimetry and the trivial and systematic names of microbiological sugars. Polarimetry has a special place in carbohydrate chemistry, but unlike the other physical methods treated in the last chapter of the first edition (n.m.r. spectroscopy, mass spectrometry, for example) it has not become an everyday technique to all who practise organic chemistry. For this reason it is singled out for extended treatment in Appendix 4 and our earlier chapter on physical methods has been omitted. Treatment of the biochemistry of the sugars has been severely curtailed (Section 2.5.1).

We thank Professor Alistair Stephen for guidance on aspects of Chapter 8, and Dr Howard Carless for valuable suggestions on many topics. Professor Derek Horton kindly allowed us access to the revised version of the IUPAC-IUB rules *The Nomenclature of Carbohydrates* prior to its publication (see Appendix 2). Excellent secretarial service was provided by Yvonne Cuthbert, Janine Doherty, Alison Hetherington and Katherine Prior, and Keri McCombe assisted with the diagrams. We are most grateful to them all. Parts of the book were written when one of us (R.J.F.) was on sabbatical leave in the Department of Chemistry, University of Edinburgh, and Victoria University is thanked for granting the leave and Professor R. Ramage for making available excellent facilities in Edinburgh. Finally, for their encouragement and support, we thank our wives Shirley and Carolyn who, unlike our subject, have appeared steadfast and unchanging for even longer than the time spanned by our two books.

REFERENCE

1. *Monosaccharide Chemistry*, Penguin Books Ltd, Harmondsworth, Middlesex, 1972.

Abbreviations

The following abbreviations have been used in the text.

Ac   acetyl
Ad   adenin-9-yl
AIBN   2,2′-azobisisobutyronitrile
All   allyl
Ar   aryl
Asn   asparagine
Asp   aspartic acid
ATP   adenosine 5′-triphosphate
BBN   9-borabicyclo[3.3.1]nonane
Bn   benzyl
Boc   *t*-butoxycarbonyl
Bu   butyl
Bz   benzoyl
Cbz   benzyloxycarbonyl
c.d.   circular dichroism
c.i.   chemical ionization
DAST   diethylaminosulphur trifluoride
DBU   1,5-diazabicyclo[5.4.0]undec-5-ene
DCC   dicyclohexylcarbodiimide
DDQ   2,3-dichloro-5,6-dicyano-1,4-benzoquinone
DEAD   diethyl azodicarboxylate
DIBALH   diisobutylaluminium hydride
DMAP   4-dimethylaminopyridine
DMF   *N*, *N*-dimethylformamide
DMSO   dimethyl sulphoxide
DMTST   dimethyl (thiomethyl) sulphonium triflate
DNA   deoxyribonucleic acid
ee   enantiomeric excess
Ee   1-ethoxyethyl
e.s.r.   electron spin resonance
Et   ethyl
f.a.b.   fast-atom bombardment
FIAC   1-(2′-deoxy-2′-fluoro-β-D-arabinofuranosyl)-5-iodocytosine
Fru   fructose
Fuc   fucose

Gal	galactose
g.l.c.	gas–liquid chromatography
Glc	glucose
GlcNAc	2-acetamido-2-deoxyglucose
Gly	glycine
HMPIT*	hexamethylphosphoric triamide
h.p.l.c.	high performance liquid chromatography
HSEA	hard sphere *exo*-anomeric effect
IDCP	iodonium dicollidine perchlorate
i.r.	infrared
Kdo	3-deoxy-D-*manno*-2-octulosonic acid
LAH	lithium aluminium hydride
LDA	lithium diisopropylamide
LTBH	lithium triethylborohydride
LUMO	lowest unoccupied molecular orbital
Man	mannose
MCPBA	*m*-chloroperbenzoic acid
Me	methyl
Mem	methoxyethoxymethyl
MNO	4-methylmorpholine *N*-oxide
Mom	methoxymethyl
m.p.	melting point
m.s.	mass spectrometry
Ms	methanesulphonyl
NAD	nicotinamide adenine dinucleotide
NAM	*N*-acetylmuramic acid
NBS	*N*-bromosuccinimide
Neu	neuraminic acid
NIS	*N*-iodosuccinimide
n.m.r.	nuclear magnetic resonance
nOe	nuclear Overhauser effect
Ns	nitrobenzene-*p*-sulphonyl
o.r.d.	optical rotatory dispersion
PCC	pyridinium chlorochromate
PDC	pyridinium dichromate
PET	positron emission tomography
Ph	phenyl
Phth	phthaloyl
Piv	pivaloyl
p.p.m.	parts per million
Pr	propyl

---

* This abbreviation indicates that $(Me_2N)_3PO$ is a phosphoric acid derivative. It is commonly abbreviated either to HMPA or to HMPT, with confusing consequences, particularly since $(Me_2N)_3P$, a phosphorous acid derivative, has been similarly shortened.

p.t.c.	phase transfer catalysis
Py	pyridine
Rha	rhamnose
RNA	ribonucleic acid
Ser	serine
s.i.m.s.	secondary-ion mass spectrometry
SOMO	singly occupied molecular orbital
TASF	tris(dimethylamino)sulphur (trimethylsilyl)difluoride
Tbdms	*t*-butyldimethylsilyl
Tbdps	*t*-butyldiphenylsilyl
Tf	trifluoromethanesulphonyl
TFAA	trifluoroacetic anhydride
Th	theophyllyl
THF	tetrahydrofuran
Thp	tetrahydropyranyl
Thr	threonine
Tips	tetraisopropyldisiloxyl
t.l.c.	thin layer chromatography
Tmb	2,4,6-trimethylbenzoyl
Tms	trimethylsilyl
TPAP	tetra-*n*-propylammonium ruthenate (VII)
TPP	triphenylphosphine
Tps	triisopropylbenzenesulphonyl
Tr	triphenylmethyl (trityl)
Ts	*p*-toluenesulphonyl
U	uracil-3-yl
UDP	uridine diphosphate
UDPG	uridine diphosphate glucose
u.v.	ultraviolet

1 Introduction

1.1 OVERVIEW

Although the primary significance of carbohydrates rests on their major importance in biology, they represent a unique family of polyfunctional compounds which can be chemically manipulated in a multitude of ways. In particular, they can be converted into very many polymeric, oligomeric and monomeric products of importance in, for example, the food, clothing, pharmaceutical and agrochemical industries. Research into the chemistry, biochemistry and biology of carbohydrates has contributed much to these sciences and, in consequence, to medicine; and developments since the 1970s in all these aspects have been phenomenal.

The basis of understanding of the organic chemistry of the carbohydrates was laid in Germany at the end of the 1800s by Emil Fischer, whose work on the parent sugars, which are the building blocks upon which the family is based, with the primitive tools then available seems to grow in stature as more and more reliance is placed on highly sophisticated chromatographic and spectroscopic methods. Succeeding work in European and American research schools, especially those directed approximately between 1925 and 1950 by Haworth and Hudson, respectively, consolidated the foundations of structural carbohydrate chemistry.

The subject developed as a rather specialized branch of organic chemistry possibly because of the (normally) hydrophilic properties of the sugars and the particular techniques that were consequently required for their handling. Also, much of the work with polysaccharides was directed towards their isolation, purification and structure determination, and the research was in consequence repetitive and chemically somewhat limited and it did not command wide attention outside the field. However, from the meticulous initial studies, today's workers inherited a vast body of reliable basic knowledge which has been used to merge the subject with mainstream organic chemistry on the one hand and biology on the other.

For those engaged in the study of carbohydrates, the advent of routine structural analysis by n.m.r. spectroscopy provided a tremendous fillip—especially since derivatives of this class of compounds give spectra that are often extremely well resolved. Consequently, to take a very simple example, the earlier difficulties associated with the determination of the anomeric configurations of simple monosaccharide derivatives disappeared. By the 1960s chromatographic methods and n.m.r. spectroscopy had made the isolation of carbohydrate derivatives and their structural analysis relatively simple, and there was no longer cause to consider them other than orthodox organic compounds. By the early 1970s carbohydrates were recognized as suitable chiral starting materials for the synthesis of a plethora of non-carbohydrate compounds and as chiral auxiliaries in synthesis, and the challenges offered by a multitude of biologically important, complex carbohydrate containing compounds attracted the interest of many leading groups working in synthetic chemistry. Consequently, sugars may be considered to be highly versatile and manageable materials, and, as is demonstrated throughout this book they are now viewed as model compounds on which the whole gamut of standard reactions in the organic chemist's repertoire may be performed — a point well illustrated by the appearance of a recent book devoted to free radical chemistry, the cover of which is adorned with a carbohydrate illustration. Gone indeed are the days when the old adage that carbohydrate chemistry consisted of nothing more than methylation and hydrolysis was sometimes applied.

Since the early 1950s Lemieux has been the one chemist in particular to lead the subject with insight and with continuous reference to both general organic chemistry and biology; his contributions have been uniquely seminal, widespread and illuminating. For example, he was responsible for the first chemical synthesis of sucrose, the introduction of ^{1}H n.m.r. spectroscopy to the field and to stereochemistry in general, the identification of the *endo*- and *exo*-anomeric and reverse anomeric effects, the development of key glycosylation reactions including those of amino sugars, the synthesis of blood group oligosaccharides, and the appreciation of the importance of hydrophobic interactions between carbohydrates and proteins in specific biochemical recognition processes. All of these contributions are described in his lively personalized chemical autobiography,[1] and clearly identify Ray Lemieux as the outstanding contributor to carbohydrate research in modern times.

While it used to be considered that the proteins were the most functionally versatile natural products, it is now clear that carbohydrates rival them in this regard ranging from structural to specifically biochemically active components of plants, animals and micro-organisms. To offset thermal and biochemical combustion of organic matter, photosynthesis alone restores carbon from the inanimate to the animate world, converting atmospheric carbon dioxide primarily to carbohydrate products which are then transformed in plants into structural and food storage materials then to be used by animals as sources of energy. Very many other biological functions are, however, performed by carbohydrate materials; in particular, they play many roles as recognition compounds, for example between cells, as antigens and as blood group substances. They have many key roles, often in association with proteins in, for example, plant and bacterial cell walls, mammalian cartilage tissue and some enzymes. Nucleic acids, the central compounds of molecular biology, contain monomeric carbohydrates as integral parts of their polymeric structures.

Basic to the family are the *monosaccharides* which may be polyhydroxy-aldehydes or -ketones, i.e. *aldoses* or *ketoses*; as phosphate esters they play key roles in biochemical transformations which include the linking of pairs of monosaccharides to give disaccharides. Further inter-unit linking leads to the oligosaccharides and then to the polysaccharides of which starch and cellulose are the best-known members. While these polymers are structurally simple, many polymeric carbohy-drates have highly complex molecular architectures.

It is the purpose of this book to focus in detail on the organic chemistry of the monosaccharides, to illustrate their versatility as synthons and to show how they relate to more complex natural products and to some non-carbohydrate organic compounds.

1.2 REFERENCE

1. R.U. Lemieux, *Exploration with Sugars: How Sweet it Was*, American Chemical Society, Washington, DC, 1990.

2 Preliminary Matters — Structures, Shapes and Sources

Monosaccharides are the chemical units from which all members of the major family of natural products, the carbohydrates, are built. While the simplest are polyhydroxycarbonyl compounds with molecular formulae of the form $C_n(H_2O)_n$ (hence the family name which was derived via the French *hydrate de carbone* and the German *Kohlenhydrat*), many are now known which have somewhat modified formulae; compounds devoid of specific hydroxyl groups or having amino groups in their place are important examples, and many such derivatives play vital roles in biology. The word 'carbohydrate' therefore no longer has an exact definition, nor has the word 'sugar' which is often used as a synonym for 'monosaccharide' but may also be applied to simple compounds containing more than one monosaccharide unit. Indeed, in everyday usage 'sugar' signifies table sugar, which is the disaccharide sucrose comprising one molecule of each of the two monosaccharides D-glucose and D-fructose.

D-Glucose can be considered to be the parent compound of the family which consists mainly (but not exclusively) of five- and six-carbon compounds, i.e. *pentoses* and *hexoses*, 'ose' being the suffix used to denote a sugar. It is not only the most abundant monosaccharide found in nature, but it alone plays a central role in biochemistry (Section 2.5.1). In consequence, it has always been most readily available, and has been studied more than any other member of its family. As 'grape-sugar' it was known to the ancient Persians and Arabians, and it was given its present name by Dumas in 1838, this name having been retained by the pioneering father figure of the subject, Fischer, despite Kekulé's preference for 'dextrose' (1866). In Germany, at the end of the 1800s, Fischer elucidated the structure of glucose and its isomers using chemical and polarimetric methods, the work being recognized then and now as one of the outstanding achievements of early structural work. Fischer was awarded the second Nobel Prize for Chemistry in 1902, and the centenary of the elucidations (1991) was celebrated with the symposium *Emil Fischer — 100 Years of Carbohydrate Chemistry* held in San Francisco in 1992.[1]

The hexoses contain four asymmetric centres in their acyclic modifications, and Fischer's achievement was to determine the relative configurations at the various centres. Since there was no way at that time of determining the absolute configuration of the naturally occurring glucose, i.e. which of the two enantiomers it was, Fischer, in representing the structure, arbitrarily chose one of the possibilities and happily selected correctly. This was established in 1951 by the application of X-ray diffraction analysis to the determination of the absolute configuration of a salt of (+)-tartaric acid,[2] which can be related stereochemically to natural glucose and the other sugars; in carbohydrate terminology it is L-threaric acid. In consequence, the representations of the molecular structures of the monosaccharides now in use, and employed in this book, denote correct absolute as well as relative configurations.

Initially the structure of glucose will be examined, first from the evidence gained by physical methods, and then in relation to some of the chemical properties of the compound. The structural correlation of the other sugars with glucose will then be considered, and an account given of the factors controlling the shapes of molecules and of the isomeric forms which they adopt in solution. Finally in this chapter the different ways of obtaining monosaccharides will be considered.

2.1 THE STRUCTURE OF GLUCOSE

2.1.1 PHYSICAL EVIDENCE

Neutron diffraction analysis has been applied to the most common crystalline modification of glucose (known as the α-D-pyranose form; see later), and has revealed that the compound has a puckered, six-membered, oxygen-containing ring, which is numbered as shown in **2.1** with hydroxyl substituents at C-1–C-4 and a hydroxymethyl group at C-5. It should be noted at this stage that C-1 is unique in having two attached oxygen atoms, i.e. it is a hemiacetal centre, and that all the substituent groups on the ring, except that at C-1, protrude equatorially (Section 2.3.1).[3] For the β-pyranose form, X-ray diffraction studies reveal[4] that all substituents are equatorially bonded to the ring in the same chair conformation. Infrared and ultraviolet spectroscopy confirm that glucose has a cyclic hemiacetal structure, since no absorptions characteristic of a carbonyl group are found in the spectra.

2.1

Much the most revealing physical method for examining glucose in solution is n.m.r. spectroscopy. The ^{1}H spectrum of the most common crystalline form measured immediately after dissolving in D_2O (Figure 2.1a) shows a relatively narrow doublet ($J = 3.6$ Hz) characteristic of a carbon-bonded hemiacetal proton (δ 5.12) and resonances near δ 3.3, 3.4 and 3.6 which show large splittings (*ca* 10 Hz) characteristic of axial protons having axial neighbouring protons on saturated, six-membered rings. After a few hours the spectrum (Figure 2.1b) reveals that partial isomerization has occurred and that now two hemiacetals (α and β) are

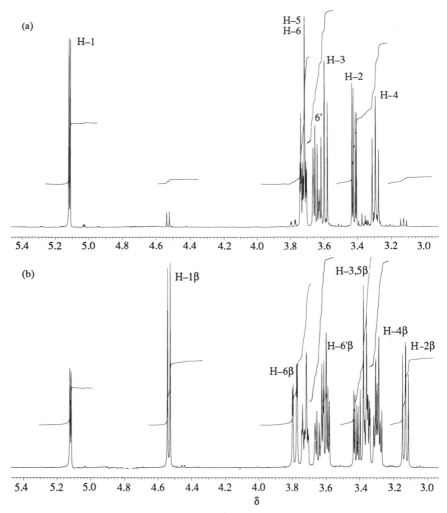

Figure 2.1 ^{1}H n.m.r. spectra (500 MHz) of α-D-glucopyranose. (a) Freshly dissolved in D$_2$O, (b) After equilibration in D$_2$O, referenced with respect to (CD$_3$)$_2$CO (δ 2.12). Note that all δ values in the spectra are +0.03 p.p.m. with respect to the corresponding figures in Table A3.1 (Appendix 3), which were also obtained by use of acetone (δ 2.12) as reference. The cause of the discrepancy is therefore not clear, but it is noted that in *Applications of NMR Spectroscopy in Organic Chemistry*, 2nd Edn, Pergamon Press, Oxford, 1969, the δ value for acetone is 2.09

present, the one newly formed representing about 60% of the mixture and having resonances at, for example, δ 3.13, 3.78 and 4.53 that reveal it to be the 'impurity' present in Figure 2.1(a). It can be noted that the H-1 doublet for the β-isomer is *ca* 0.6 p.p.m. upfield relative to the H-1 resonance of the initial α-compound and that the $J_{1,2}$ value has increased to 7.8 Hz. This figure is typical for an

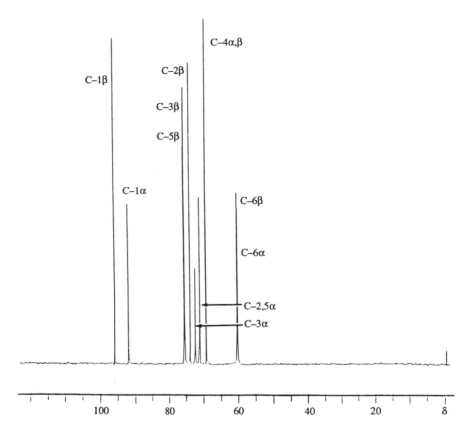

Figure 2.2 ^{13}C n.m.r. spectrum (20 MHz) of equilibrated α,β-D-glucopyranose in D_2O

axial H-1 with a neighbouring axial H-2, and (because of the influence of the ring oxygen atom) is notably smaller than the 10 Hz normally found for diaxial hydrogen pairs.

The ^{13}C spectrum (Figure 2.2) is consistent with these conclusions, and none of the spectra show resonances for an aldehyde group. Furthermore, when α-glucose (containing 20% of the β-isomer) is examined in DMSO, the main component shows a set of five discrete hydroxyl group proton resonances (Figure 2.3) which exhibit vicinal coupling with their neighbouring C-bonded protons. These indicate the presence of four secondary alcohol groups, each giving a doublet, one being a hemiacetal hydroxyl group whose proton is significantly deshielded with respect to the others. A primary alcohol triplet is observed, (δ 4.35) and also a hemiacetal C-bonded proton resonance (δ 4.9), showing vicinal coupling with its O- and C-bonded proton neighbours. The O-1–H, C-1–H and O-6–H resonances for the minor β-isomer are also to be seen. All the features are consistent with the readily available form of glucose having structure **2.1**.

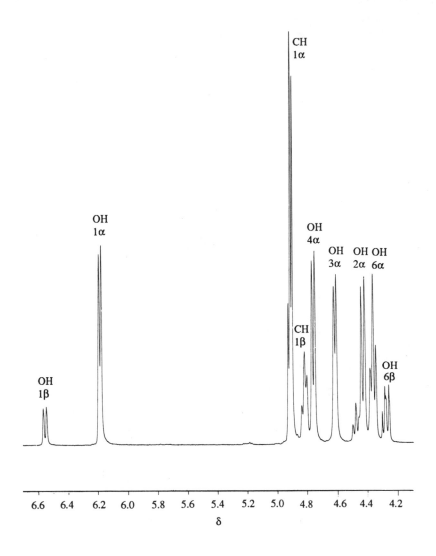

Figure 2.3 Part of the ¹H n.m.r. spectrum (300 MHz) of α-D-glucopyranose (containing 20% of the β-isomer) in DMSO (cf. *Can. J. Chem.*, 1966, **44**, 540)

2.1.2   CHEMICAL EVIDENCE

Elemental analysis and molecular weight determination establish that glucose has the molecular formula $C_6H_{12}O_6$, and the presence of five free hydroxyl groups is readily demonstrated by the preparation of pentasubstituted derivatives. Glucose can also be reduced to a compound (glucitol **3.207**, Section 3.1.5) which has six free hydroxyl groups, can be oxidized to a six-carbon carboxylic acid (gluconic acid **3.214**, Section 3.1.6.a.i) and it readily reduces Fehling's solution and ammoniacal silver nitrate. Furthermore, it may be converted into a wide range of carbonyl derivatives (e.g. an oxime, Section 3.1.3.b). This evidence indicates, therefore, that the compound is a pentahydroxyaldehyde, and since heptanoic acid is formed from it by the reactions shown in Scheme 2.1, it has its six carbon atoms arranged in an unbranched sequence, and is therefore a pentahydroxyhexanal as was established in the initial determination of its structure.

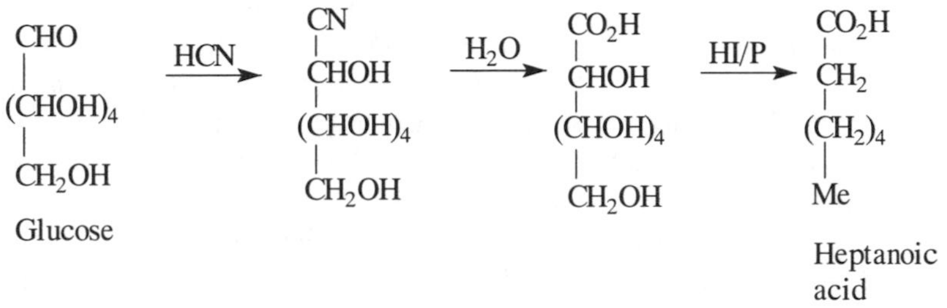

$$
\begin{array}{ccccccc}
\text{CHO} & & \text{CN} & & \text{CO}_2\text{H} & & \text{CO}_2\text{H} \\
| & \xrightarrow{\text{HCN}} & | & \xrightarrow{\text{H}_2\text{O}} & | & \xrightarrow{\text{HI/P}} & | \\
\text{(CHOH)}_4 & & \text{CHOH} & & \text{CHOH} & & \text{CH}_2 \\
| & & | & & | & & | \\
\text{CH}_2\text{OH} & & \text{(CHOH)}_4 & & \text{(CHOH)}_4 & & \text{(CH}_2)_4 \\
& & | & & | & & | \\
\text{Glucose} & & \text{CH}_2\text{OH} & & \text{CH}_2\text{OH} & & \text{Me}
\end{array}
$$

Heptanoic acid

Scheme 2.1

In the 1800s, however, it was recognized that glucose has certain chemical properties which are inconsistent with those to be expected for a simple poly-hydroxyaldehyde. It does not, for example, reduce Schiff's reagent under ordinary conditions, nor does it give the expected dimethyl acetal on treatment with acidified methanol. Instead, isomeric monomethyl derivatives (glycosides, Section 3.1.1) are produced. Furthermore, acetylation of glucose gives two isomeric pentaacetates and, most significantly, two crystalline modifications of the free sugar are known: the α-form (m.p. 146 °C, specific rotation $[\alpha]_D = +111$ in water), which is obtained by crystallization from cold ethanol, and the β-isomer (m.p. 148–150 °C, $[\alpha]_D = +19$ in water), which crystallizes from hot pyridine. All this evidence led to the proposal of a cyclic hemiacetal structure for glucose which, of course, was borne out by the physical evidence obtained much later and which has already been discussed. It does not, however, provide means for ascertaining which oxygen atom is involved in the ring formation, and it should be stressed that it is still a difficult task to establish the size of the rings of free sugars in solution by chemical methods. Delicate equilibria of several modifications are involved (Section 2.4.1), and any chemical reagent added to the systems may disturb the equilibria and so may provide unreliable results. However, glucose reacts in aqueous bromine solution by a mechanism which is believed to involve a dehydrogenation at C-1 of the cyclic hemiacetal form (Section 3.1.6) to give a lactone which has a six-membered ring; consequently, the species undergoing reaction must also have this ring size.

There remains the problem of the relative configurations at the asymmetric carbon atoms in glucose. Fischer determined these by ingenious methods which are outlined in Section 2.2.1, and he assigned structure **2.8** to the acyclic modification of the compound. As has been indicated above, the absolute configuration was not established until 1951.

2.1.3   THE CYCLIZATION OF GLUCOSE

Evidence taken from compounds other than carbohydrates indicates that a polyhydroxyaldehyde would be highly unfavoured relative to a cyclic hemiacetal form, and is in complete agreement with the conclusion that glucose exists in the solid state and in solution in cyclic modifications. 5-Hydroxypentanal **2.2**, for example, exists to the extent of more than 90% in the cyclic form **2.3** in 75% aqueous dioxane, and substitution of large groups (hydroxyls for the sugars) at C-2, C-3 and C-4 would be expected to stabilize the cyclic form with respect to the acyclic in keeping with the finding that substitution generally favours cyclic modifications in ring–chain tautomeric equilibria of this type.[5] One reason for this stems from the lower entropy of substituted acyclic compounds relative to unsubstituted analogues; consequently, decreases in entropy which occur on ring closure are smaller in the substituted cases.[6] Furthermore, α-hydroxyaldehydes hydrate more readily than do non-hydroxylated equivalents consequent upon the electron-withdrawing effect of the substituent.[7] 4-Oxa-5α-cholestan-3α-ol exists in the solid phase and in solution as its cyclic tautomer **2.4**,[5b] and the hemiacetal forms of compounds containing the structural feature **2.5** can be so stable that it is difficult to cause them to react as 4-formylcyclohexanols.[8]

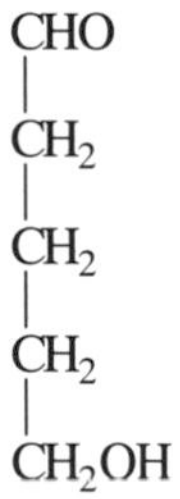

2.2 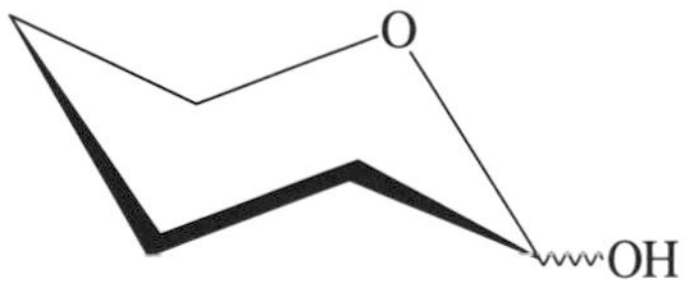   2.3

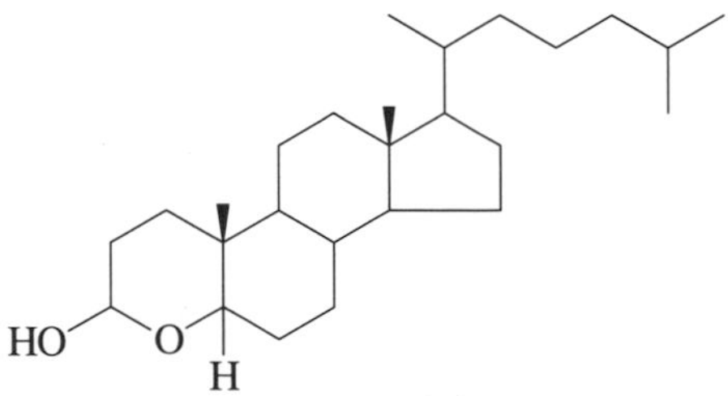

2.4 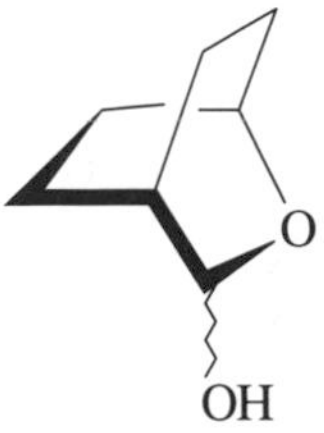   2.5

Cyclization involving the aldehydic function of glucose accounts for its modified reducing properties (Section 2.1.2) and for the existence of isomeric forms of the sugar. The ring closure occurs by nucleophilic attack of the oxygen atom at C-5 on the carbonyl carbon atom after the conformation of the acyclic species is suitably arranged as shown in **2.6**, this procedure resulting in the generation of a new asymmetric centre at C-1 and affording the α- and β-modifications **2.1** and **2.7**. The two acetates formed from glucose (Section 5.3.1) are the pentaesters of these isomers.

2.6   **2.1**   **2.7**

While the five-membered rings which would be formed by a related cyclization involving O-4 are unstable relative to the six-membered rings of compounds **2.1**, and **2.7**, the same does not apply to all sugars, and some have appreciable proportions of five-membered ring forms in solution (Section 2.4.2).

2.1.4   CONVENTIONAL METHODS FOR THE REPRESENTATION OF GLUCOSE — THE D, L AND α, β CONVENTIONS

(a)   The acyclic modification

Although the acyclic modification of glucose is unfavoured relative to the six-membered hemiacetal forms both in the crystal and in solution, it can be an important reactive component in solutions, and methods must be available for the unambiguous representation of it and other acyclic species. The Fischer projection formula **2.8** is traditionally used for this purpose. It is obtained by arranging the carbon chain vertically so that the carbonyl group (C-1) is at the top and the carbon atoms are in a smooth arc. The array of atoms so produced is viewed from the convex side, and therefore **2.8** implies a smooth arc of carbon atoms as indicated by **2.9**, the carbon substituents at C-2–C-5 projecting towards the viewer with C-1 and C-6 receding into the paper.

It follows, therefore, that structures **2.10** and **2.11** represent a completely new compound (D-mannose) since a bond has to be broken and a new bond formed at C-2 to obtain it from D-glucose. Rotation about the C-2–C-3 bond of glucose does not give this isomer but simply a new conformation in which the carbon chain is no longer in the form of an arc. (Beginners should use molecular models to satisfy themselves on this vital point.)

It must be emphasized that these projection formulae are used exclusively for the purpose of portraying structure; they do not imply preferred conformations. On the contrary, they depict relatively high energy states and must not be used in

```
        1  O              CHO                        O              CHO
                     H   C   OH                                HO   C   H
        2    OH     HO    C   H          HO                    HO   C   H
HO      3           H    C   OH          HO
        4    OH                              OH                H    C   OH
                     H    C   OH             OH
        5    OH      H    C   OH                               H    C   OH
        6    OH          CH₂OH             OH                      CH₂OH
```

| **2.8** | **2.9** | **2.10** | **2.11** |

considerations of preferred shape. For such purposes zigzag (or closely related) conformations (Section 2.3.3) must be used for acyclic monosaccharide derivatives.

There are four asymmetric centres but no elements of symmetry in *aldehydo*-glucose **2.8**, and consequently there are 2^4, i.e. 16, aldohexose stereoisomers. The *Rosanoff convention*, introduced in 1906, simplifies the nomenclature problems with these compounds by subdividing them into eight *enantiomeric* (mirror image) pairs and describing all those structures which have the hydroxyl group at the highest-numbered asymmetric centre (C-5 in the glucose case) projecting to the right in the Fischer projection formulae as belonging to the D-series, and the others to the L-series (see illustrations). The formulae **2.1** and **2.8** used up to this point have represented the D-enantiomer of glucose which is the naturally occurring form. The L-isomer has formula **2.12**.

```
                                                        O
                                              HO
                                                        OH
         OH            HO                      HO
         OH                 OH                 HO
                                                        OH
      D-Series         L-Series              L-Glucose
```

2.12

A further point to emphasize is that the Rosanoff convention bears no relation-ship to the sign of the optical rotation of the sugar; many compounds belonging to the D-series are laevorotatory [1 or (−)], and conversely L-compounds can be dextrorotatory [d or (+)].

It would be in keeping with current stereochemical practice in general organic chemistry to adopt the Cahn–Ingold–Prelog conventions and describe D- and L-sugars by use of the (R)- and (S)-designations, but the D, L convention has always been, and still is, used quite generally for carbohydrates, and therefore will be adopted throughout this book. The absolute configurations represented by the Rosanoff–Fischer method are correct.

(b)   The cyclic modifications

As indicated in Section 2.1.1, D-glucose exists in the crystalline state and in solution preferentially in the six-membered, puckered ring forms **2.1** and **2.7**. Historically, however, two other types of formulae have been used to depict the cyclic modifications: Tollens representations and Haworth perspective formulae. The former are based on Fischer projections with very long bonds from O-5 to C-1 to indicate six-membered ring formation, and have serious weaknesses which make them no longer viable; the latter, however, are of appreciable ongoing value and are used extensively in this book. Haworth introduced his formulae to give more realistic pictures of the cyclic forms of the sugars, obtaining **2.14** for the six-membered cyclic form of D-glucose and **2.17** for the five-membered form, the derivatives of which play an important role in the chemistry of the sugar. These rings are depicted as lying perpendicular to the paper with the ring oxygen atoms away from the viewer and are observed obliquely from above. They are derived from the acyclic conformations **2.13** and **2.16**, respectively, and are named *pyranose* and *furanose* after the heterocyclic compounds pyran **2.15** and furan **2.18**. From these, the adjectives *pyranoid* and *furanoid* are derived. Despite their deficiencies in depicting the six-membered rings as planar rather than puckered (e.g. **2.1**), these representations provide unambiguous statements of structure and avoid issues relating to conformational state, and notwithstanding current trends they will be used frequently in this book when the conformation is unknown or not under consideration; puckered rings are mainly reserved for cases in which detailed stereochemical matters are under examination and should always be employed in these circumstances.

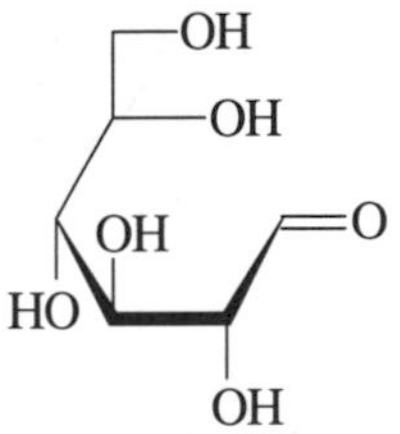

2.13 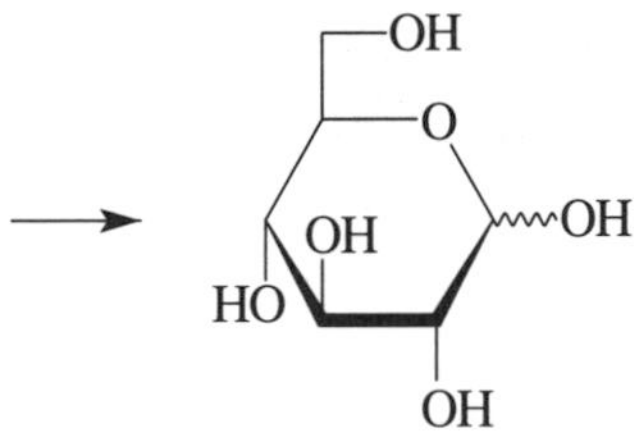   2.14 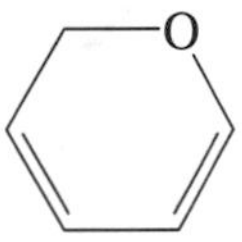   2.15

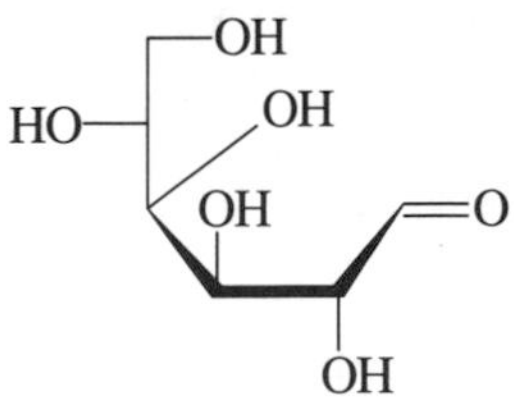

2.16 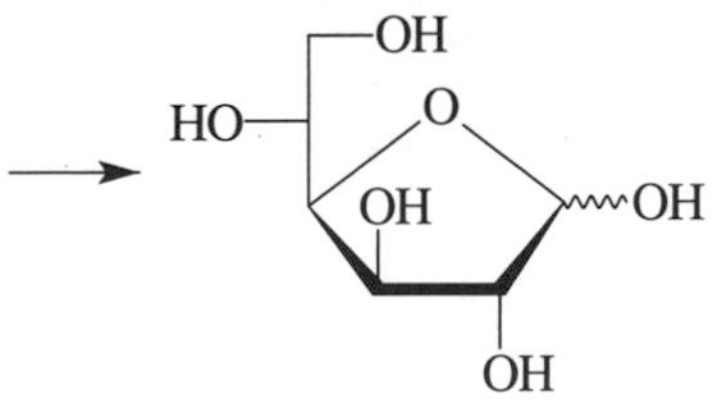   2.17 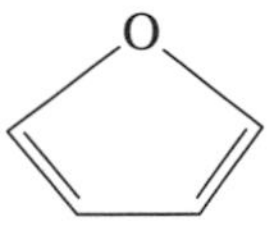   2.18

Hemiacetal ring formation generates a new asymmetric carbon atom at C-1, the *anomeric centre*, thereby giving rise to diastereoisomeric hemiacetals which are called *anomers* and labelled α and β. Freudenberg (1932), using Tollens formulae, developed a convention correlating configurations at the anomeric centre with those at the highest-numbered asymmetric centre of the sugar which is still generally used. Translated to Haworth formulae it becomes: *for D-glucose and all compounds of the D-series, α-anomers have the hydroxyl group at the anomeric centre projecting downwards in Haworth formulae; α-L-compounds have this group projecting upwards as shown in Figure 2.4. The β-anomers have the opposite configurations at the anomeric centre, i.e. the hydroxyl group projects upwards and downwards for β-D- and β-L-compounds, respectively.*

Hudson had earlier (1909) proposed that the α-designation be given to the more dextrorotatory of anomeric pairs of D-sugars (and the more laevorotatory of pairs of L-sugars), and in the great majority of cases this convention and that of Freudenberg concur for free sugars and also their derivatives. Examples are, however, known of α-D-derivatives (Freudenberg) which are less dextrorotatory than their anomers (Appendix 4), and therefore the Hudson system of correlating anomeric configurations with relative optical activity is not reliable and is no longer used.

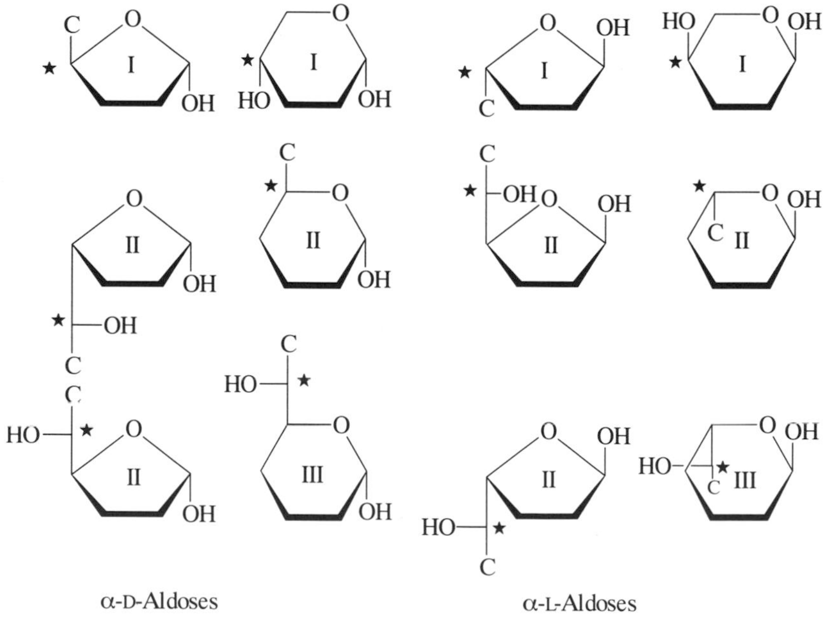

α-D-Aldoses   α-L-Aldoses

Figure 2.4 Anomeric configurations of aldoses: I, aldopentoses; II, aldohexoses; III, aldoheptoses. Asterisks denote the last asymmetric centres the configurations of which determine the D- or L-assignments and serve as references for α- and β-assignments

The anomeric configuration of a sugar or a derivative can be determined by polarimetry or X-ray diffraction analysis, but nowadays n.m.r methods are usually employed, $^1HJ_{1,2}$ values being particularly useful for this purpose. However, configurational assignments were first accomplished with D-glucose by Böeseken (1913) who used the observation that *cis*-1,2-diols on six-membered cyclic compounds preferentially form ionic complexes with borate ions and thus increase the electrical conductivity of boric acid solutions. With D-glucose, the more dextrorotatory isomer gives a positive reaction, whereas its anomer initially does not; the former was therefore allocated the 1,2-*cis*-structure **2.1** and is the α-anomer. With both modifications the conductivities change until finally they have equal values (Section 2.4.1).

Haworth perspective formulae continue to serve carbohydrate chemistry well, although conformational representations are generally favoured by some writers. Additionally, however, together with the recent coming together of the subject with mainstream organic chemistry (see, for example, Chapter 7), has come the tendency for the literature to contain carbohydrate structures represented by the projection methods introduced to carbohydrate chemistry by Mills[9] and favoured in other areas of natural product chemistry. Thus α-D-glucopyranose **2.1** becomes **2.19** or, after rotation, **2.20**, the rings being coplanar with the paper. Although there are advantages in these (especially as problems relating to the representation of typographically bulky groups within rings are removed), the temptation to break with well-established and viable methods will be resisted in this book. In a representation such as **2.19**, three-dimensional perspective is suppressed (although determinable), whereas in the Haworth alternative **2.14** (α-form), three-dimensional features are readily apparent, and we believe firmly that this advantage should not be relinquished.

           2.19   **2.20**

2.2 THE INTERRELATIONSHIPS OF THE MONOSACCHARIDES

2.2.1 ALDOSES

The physical evidence referred to previously provides proof of the relative configurations at the asymmetric atoms of glucose, but no reference has so far been made to the ingenious chemical methods which were devised by Fischer for the initial solution of this problem. Their aesthetic quality and their importance in classical organic chemistry make them worthy of consideration so long after their successful application, their true merit being emphasized by the reminder that they were obtained without any of our modern separatory methods or spectroscopic techniques. Polarimetry, however, played a key role. It comes as no surprise that Fischer reported by letter to his mentor Baeyer on 12th January 1889:

> Unfortunately, the experimental difficulties in this (sugar) group are so great, that a single experiment takes more time in weeks than other classes of compounds take in hours, so only very rarely is a student found who can be used for this work.

Found they were!

Glyceraldehyde ('glycerose' in carbohydrate terms) is the simplest aldose (a *triose* containing an aldehydic group) having one asymmetric centre, and therefore two stereoisomers (enantiomers); there are four *tetroses*, eight *pentoses* and, as has been shown during the discussion of glucose, 16 *aldohexoses*.

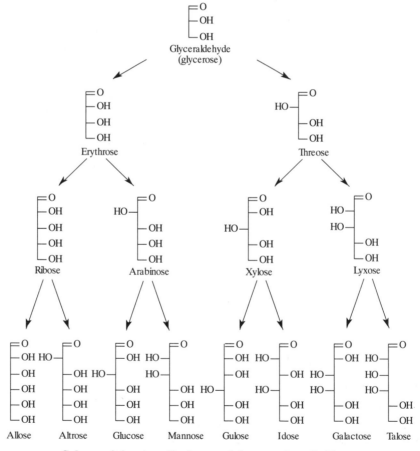

Scheme 2.2   Acyclic forms of the D-series of aldoses

By application of the Kiliani cyanohydrin synthesis (Section 3.1.4.d) it is possible to convert an aldose into two higher diastereoisomeric aldoses named *epimers*,[†] each having one additional hydroxymethylene (CHOH) group. For example, two D-tetroses are obtained from D-glyceraldehyde, and from these four D-pentoses can be prepared as outlined in Scheme 2.2. Here, the D-aldoses (to the hexoses) are named and represented in the acyclic form, and in Scheme 2.3 these same compounds are represented (where possible) in the Haworth α-furanose and α-pyranose ring forms. Since each step in Scheme 2.2 gives rise to two products, a chemical method had to be developed for distinguishing them. This was done by converting each aldose into a derivative with identical terminal groups and determining whether the products were optically inactive (*meso*) or active (asymmetric). Thus, erythrose, ribose, xylose, allose and galactose, on treatment with strong nitric acid, give internally compensated (*meso*) dicarboxylic acids (e.g. tartaric acid from erythrose), whereas from the other aldoses optically active, asymmetric acids are derived. This technique, taken together with the structures of the sugars from which they were derived (Scheme 2.2), therefore allows the five named sugars and their epimers to be structurally characterized.

Glucose, mannose, gulose and idose cannot be assigned structures on this basis. However, Fischer noted that D-glucose and D-gulose give enantiomeric derivatives (**2.21** and **2.22**, respectively) after appropriate oxidation, and therefore D-glucaric

Formula **2.22b** is obtained by rotating **2.22a** through 180° in the plane of the paper as indicated (whereby it remains a Fischer projection of a convex arc)

acid, which is the oxidation product of D-glucose, must have structure **2.21** or **2.22** since no other pair of acids derivable from these four hexoses have this relationship. It was also known that D-glucose is derived from D-arabinose (Scheme 2.2), and consequently D-glucose must have the assigned structure and D-mannose must be its epimer. The structures of D-gulose and D-idose may be similarly deduced.

[†] *Epimer* is used in this book to describe diastereoisomeric compounds which differ only in their stereochemistry at the asymmetric carbon atoms adjacent to the anomeric centres. Isomers which differ at other carbon atoms have the sites of isomerization indicated. For example, D-glucose and D-allose are '3-epimers'.

18

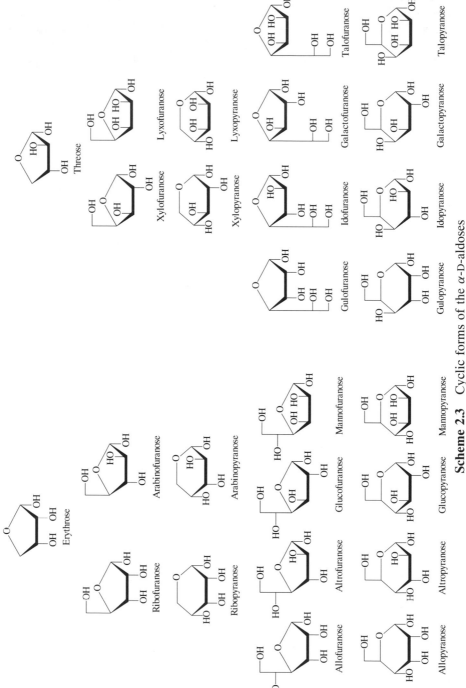

Scheme 2.3 Cyclic forms of the α-D-aldoses

Continuation of the sequences outlined in Scheme 2.2 leads to the D-aldoheptoses of which there are 16 isomers, and then the D-aldooctoses (32 isomers). Fortunately, no new trivial names are used; instead, higher sugars are described systematically, as illustrated for **2.23–2.25**, by use of configurational prefixes derived from the names of the lower members.

D-*glycero*-D-*gluco*-Heptose   L-*threo*-L-*altro*-Octose   D-*xylo*-L-*galacto*-Nonose

2.23   **2.24**   **2.25**

2.2.2   KETOSES

Ketoses, or uloses for purposes of nomenclature, are isomers of the corresponding aldoses with the same number of carbon atoms, and although their carbonyl groups can occur at any secondary position, the naturally occurring compounds, with very few exceptions, are 2-keto derivatives and discussion will be limited to them. Since members of this class contain one fewer asymmetric carbon atom than do the aldoses with the same total number of carbon atoms, there are half the number of isomers, e.g. one D-ketotetrose, two D-ketopentoses and four D-ketohexoses, as shown in Schemes 2.4 and 2.5. As with the aldoses, ring closure can occur to give two anomers which are again designated α and β according to the Freudenberg convention.

In the Haworth perspective formulae, therefore, the hydroxyl group at the anomeric centre, C-2 for the 2-ketoses, projects downwards for the α-D- and β-L-compounds and upwards for the β-D- and α-L-anomers. Again, members of the former classes are generally more dextrorotatory than their anomers.

Higher ketoses are known: D-*altro*-heptulose **2.26** (sedoheptulose), as its phosphate ester, is involved in photosynthesis and in carbohydrate metabolism (Section 2.5.1), and octuloses (e.g. **2.27**) and nonuloses (e.g. **2.28**) have been isolated from plants, avocado pears being the traditional source of such substances.

Because they imply more asymmetry than is present (the *ribo*-configuration, for example, necessarily describing three asymmetric centres), the commonly used names 'ribulose' and 'xylulose' (Scheme 2.4) for the pentuloses are unsatisfactory; '*erythro*-pentulose' and '*threo*-pentulose' are used as preferred systematic alternatives.

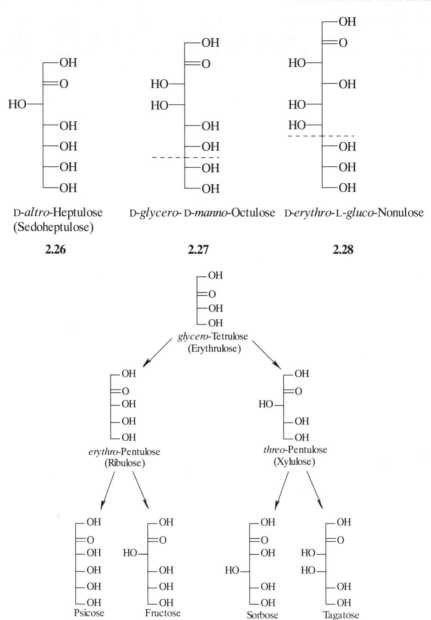

D-*altro*-Heptulose   D-*glycero*- D-*manno*-Octulose   D-*erythro*-L-*gluco*-Nonulose
(Sedoheptulose)

2.26   **2.27**   **2.28**

glycero-Tetrulose
(Erythrulose)

erythro-Pentulose   *threo*-Pentulose
(Ribulose)   (Xylulose)

Psicose   Fructose   Sorbose   Tagatose

Scheme 2.4   Acyclic forms of the D-ketoses

Fischer also used his discovery that aldoses and ketoses which differ only at
C-1 and C-2 all give the same phenylosazones (Section 3.1.3.c.ii) to assist in
establishing the configurational relationships of the sugars; this was of particular
significance in the structural analysis of the ketoses.

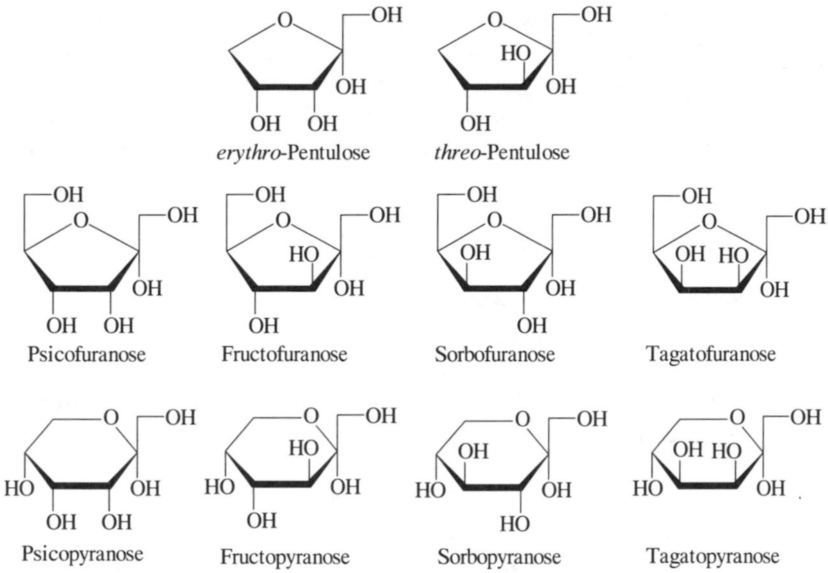

erythro-Pentulose   threo-Pentulose

Psicofuranose   Fructofuranose   Sorbofuranose   Tagatofuranose

Psicopyranose   Fructopyranose   Sorbopyranose   Tagatopyranose

Scheme 2.5   Cyclic forms of the α-D-ketoses

2.3   MOLECULAR SHAPES OF SUGARS AND RELATED COMPOUNDS[10,11]

Frequently, reactions of sugars and their derivatives give rise to several isomeric products, and if such reactions are to be used for syntheses it is desirable that at least the major product can be predicted or, alternatively, that the conditions of a reaction can be controlled to give a required isomer. Before undertaking such a synthesis, therefore, it is necessary to understand whether *kinetic factors* control the formation of products, i.e. whether relative activation energies are most significant, or whether *thermodynamic factors* govern their formation, i.e. whether the thermodynamically most stable compounds are predominantly formed. In the majority of cases, but there are important exceptions (see e.g. Section 3.1.1.a.i), reactions are carried out in the presence of catalysts which allow the equilibration of isomeric products, and thermodynamically controlled compounds are therefore obtained. For this reason and because shape may determine some properties (Section 2.3.5) and biochemical interactions (Section 8.1.2.c), it is necessary to understand the factors governing the relative thermodynamic stabilities of carbohydrates, and to do this the shapes adopted by the molecules and the factors which control them must be appreciated. Although the majority of this book is concerned with the primary structure of molecules, it must be emphasized that their shapes, and the ease with which these shapes can be altered, play vital roles in determining the chemical reactivities as well as the physical and biological properties of sugars and their derivatives and oligomers. Six-membered rings, five-membered rings and acyclic species must be considered.

Conformational analysis was not fully developed until the 1950s,[12] but Hudson and Haworth had 25 years earlier appreciated the potential importance of three-dimensional molecular shapes in sugar chemistry, the latter being credited with the introduction of the word 'conformation' in 1929.[11] It must be noted, however, that structural carbohydrate chemistry was fully developed without the aid of conformational concepts.

2.3.1   SATURATED SIX-MEMBERED RINGS[12]

(a)   Cyclohexane and substituted cyclohexanes

As is readily seen from molecular models (the Dreiding type being particularly suitable for this), cyclohexane can exist basically in two forms which are free from angle strain: (i) the rigid chair form **2.29** and (ii) the flexible form. The former is also free of torsional strain and van der Waals interactions, while the latter suffers on both these counts and is therefore energetically unfavoured. As the name implies, the flexible form can exist in a variety of shapes two of which, the *boat* **2.30** and the *skew* **2.31** are regular (each having four atoms coplanar), and therefore most readily discussed. This does not signify, however, that a molecule in the flexible form must be constrained in one of these conformations. In the boat form of cyclohexane there is eclipsing of hydrogen atoms on the two pairs of coplanar carbon atoms (e.g. A and B) and also steric interaction between the hydrogen atoms at the 'flagpole' positions (C and D). In consequence, the boat conformation is energetically unfavoured relative to the skew form which represents an energy minimum for the flexible modification. Quantitatively, the chair form is 22.2 kJ mol^{-1} (5.3 kcal mol^{-1}) more stable than the skew form, which is again 6.7 kJ mol^{-1} (1.6 kcal mol^{-1}) more stable than the boat.

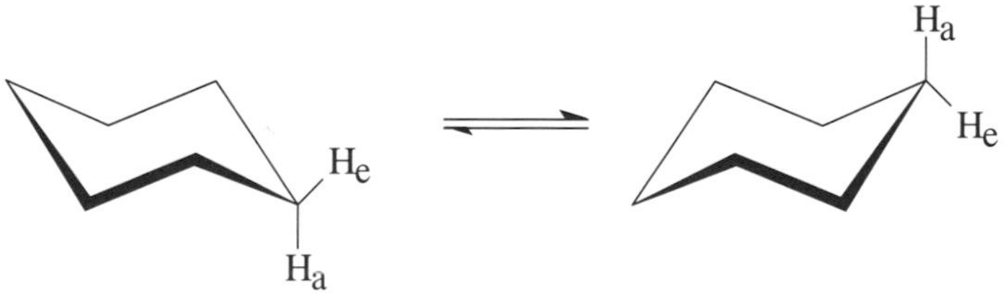

2.29a   2.29b

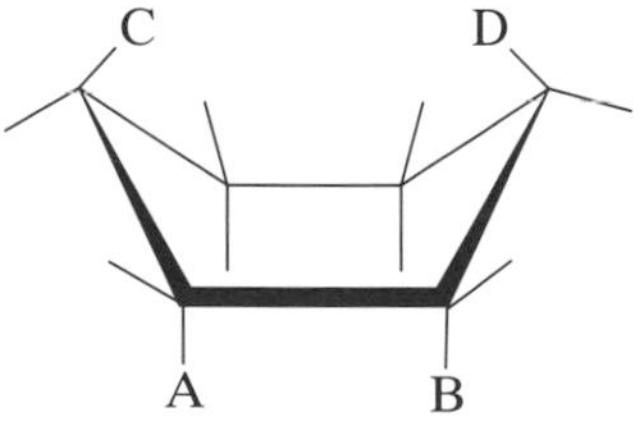

2.30 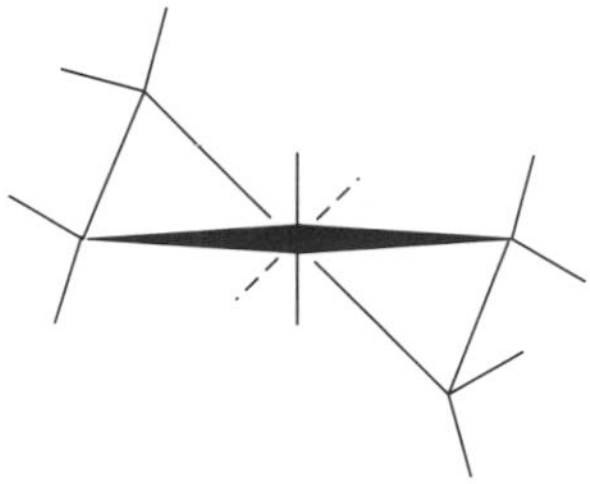   2.31

In the chair conformation **2.29**, the hydrogen atoms of the cyclohexane ring fall into two categories, axial (a) and equatorial (e), which are geometrically (and consequently magnetically) distinct, and therefore might be differentiated by n.m.r. spectroscopy. However, since the ring inversion **2.29a** ⇌ **2.29b** occurs rapidly, the proton spectrum of cyclohexane at room temperature shows only one line, and under these conditions the technique cannot distinguish between the protons. At reduced temperatures the 'ring-flipping' process is retarded sufficiently to allow resolution of the spectral lines, and the signals of the two classes of protons can be distinguished. From a determination of the temperature of coalescence of the signals it has been calculated that the free energy barrier to inversion of the ring is about 40 kJ mol^{-1} (about 10 kcal mol^{-1}).

Monosubstituted cyclohexanes can exist in two non-equivalent chair conformations with the substituents either axial **2.32** or equatorial **2.33**. With the former, van der Waals repulsions occur between the large axial group and the axial protons on the same side of the ring, destabilizing this conformation so that the equilibrium **2.32** ⇌ **2.33** lies to the right, i.e., *large groups attached to the cyclohexane ring prefer the equatorial orientation*. The free energy difference between **2.32** and **2.33** has been established experimentally for many X groups; of particular interest here are the values for the hydroxyl (3.3 kJ mol^{-1}; 0.8 kcal mol^{-1}) and methyl (7.1 kJ mol^{-1}; 1.7 kcal mol^{-1}) groups.

2.32   **2.33**

For 1,2-, 1,3- and 1,4-disubstituted cyclohexanes there are *cis*- and *trans*-isomers, each of which can adopt two chair conformations, the favoured ones being determined to a large extent by the sizes of the substituents. With the 1,4-disubstituted compounds the conformational free energies for the two groups are essentially additive (the bulkier groups will therefore always tend to be equatorial), but as the groups approach each other in their positions of substitution on the ring they can interact in a way which destroys this additivity. For example, two like groups between which there are no attractive interactions on a 1,3-*cis*-substituted ring will repel each other strongly when both are axial, and the molecule will therefore favour the diequatorial conformation even more than would be expected on an additive basis. In 1,2-disubstituted cyclohexanes 'buttressing' effects occur, and, for example, the measured ΔG for the equilibrium **2.34** ⇌ **2.35** is 9.2 kJ mol^{-1} (2.2 kcal mol^{-1}) rather than $7.1 + 3.3 = 10.4$ kJ mol^{-1} (2.50 kcal mol^{-1}). In geminally disubstituted cyclohexanes the deviation from additivity is most marked, and instead of the equilibrium **2.36** ⇌ **2.37** favouring the former by $7.1 - 3.3 = 3.8$ kJ mol^{-1} (0.9 kcal mol^{-1}), the determined

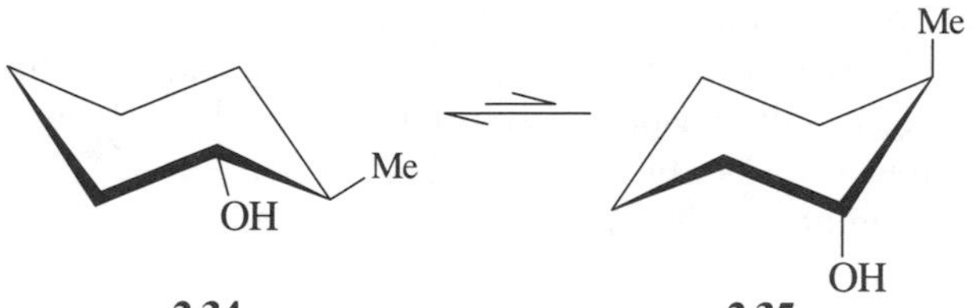

2.34   2.35

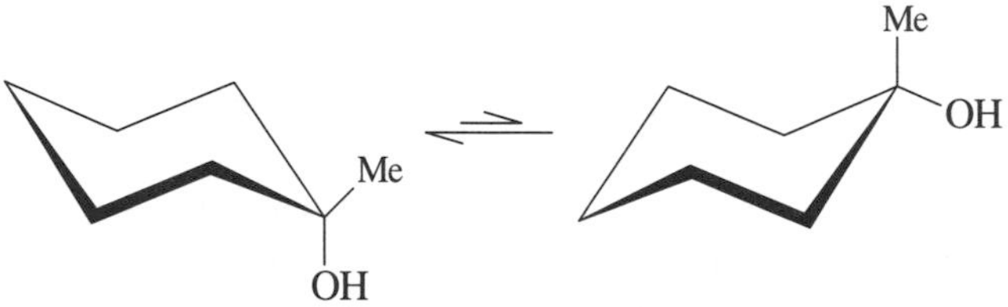

2.36   2.37

value is only 0.84 kJ mol^{-1} (0.2 kcal mol^{-1}). Such factors are of great significance in the calculations of the energies of the pyranoid forms of branched-chain sugars.

Although the causes of the above-mentioned deviations are not fully understood, they are undoubtedly mainly steric in nature. However, for cyclohexanes which carry strongly electronegative substituents in the 1,2-*trans*-relationship, an important dipolar interaction is also in operation. Thus *trans*-1,2-dibromocyclohexane, which would be expected from the above considerations to exist almost entirely in the diequatorial form **2.38**, prefers the alternative form **2.39** to the extent of 68% in carbon tetrachloride solution (dipole moment determination). That this arises from repulsion between the negative ends of the C–Br bond dipoles is suggested by the observation that in more polar solvents (in which dipolar influences are reduced) the equilibrium alters in favour of the diequatorial form **2.38**.

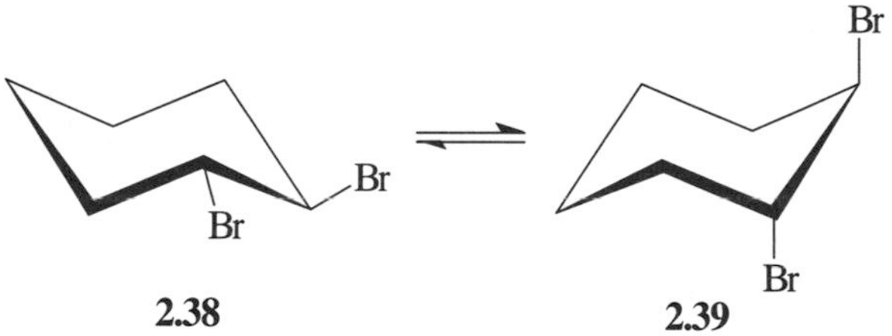

2.38   2.39

In polysubstituted cyclohexanes, calculation of conformational equilibria is complex because of the effects mentioned above combined in some cases with other influences. For example, hydrogen bonding, which in non-hydroxylic solvents stabilizes the diaxial conformation of *cis*-cyclohexane-1,3-diol **2.40**,[13] can complicate the issue. Furthermore, such factors vary when the compounds are in different solvents. Nevertheless, the generalization that the molecules adopt the chair conformation with the largest number of equatorial bulky groups usually applies. For example, *neo*-inositol **2.41** prefers the conformation illustrated.

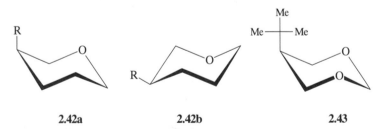

2.40   **2.41**

(b)   Tetrahydropyran and substituted tetrahydropyrans

The introduction of an oxygen atom into a saturated six-membered ring does not alter its basic conformational characteristics: chair forms are still preferred and, with the important exception of electronegative groups attached to the carbon atoms adjacent to the ring oxygen atom, substituents favour the equatorial orientation. This is despite the fact that the ring is irregular since it comprises bonds of unequal length. However, the quantitative preferences of groups on certain ring positions for the equatorial orientation are very different from those on cyclohexane. For example, the conformational free energy of the methyl group in compound **2.42** (R = Me) is only 5.4 kJ mol^{-1} (1.3 kcal mol^{-1}) as opposed to 7.1 kJ mol^{-1} (1.7 kcal mol^{-1}) which is the value for this substituent when it is attached to a cyclohexane ring or to the 2- or 4-position of tetrahydropyran. In related fashion, 3-acetoxytetrahydropyran exists almost equally in the two chair forms **2.42a** and **2.42b** (R = OAc), whereas acetoxycyclohexane exists to the extent of about 75% in the equatorial modification. It is apparent, therefore, that axial groups occupying a β-position relative to the ring oxygen atom of tetrahydropyran suffer fewer repulsive interactions than they do on cyclohexane. Consistent with this is the striking discovery that the 1,3-dioxane derivative **2.43** exists preferentially with the large *t*-butyl group axial.

2.42a   **2.42b**   **2.43**

The exceptional preference shown by electronegative substituents attached to position 2 of tetrahydropyran for the axial orientation was noted above. *This phenomenon, resulting from the operation of the* anomeric effect, *is of the greatest significance in carbohydrate chemistry since it contributes importantly to the free energies of many pyranoid compounds and hence to their preferred conformations and reactivities and also to the compositions of isomeric mixtures at equilibrium.* Compounds **2.44** (X = O-alkyl, O-aryl, O-acyl, halide, etc.) thus prefer the illustrated conformation and the extent of this preference decreases in solvents of increasing polarity and increases with increase in the electronegativity of X. The effect, identified by Edward[14] and particularly by Lemieux,[15] apparently is caused mainly by interaction between the axial lone pairs of electrons on the ring oxygen atoms and the antibonding σ^*-orbitals of the C–X bonds. This leads to shortening of the bond connecting the ring oxygen atom to the anomeric centre and lengthening of the C–X bond in the case of anomers **2.45** having the X group axial relative to the respective bond lengths in the equatorial anomer. Variations of just this kind have been found by X-ray and neutron diffraction analyses of many carbohydrates and have been predicted by theoretical studies applied to model systems such as dimethoxymethane in different conformations.[16]

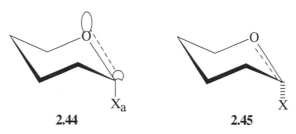

| 2.44 | 2.45 |

While this bonding factor favours axial anomers, there is a cooperative electrostatic factor acting against formation of the equatorial isomers. In the latter, repulsion between the negative end of the C-1–X_e dipole and the unshared electrons of the ring oxygen atom is greater than for the C-1–X_a dipole (see **2.46**), but while this rationalization accounts for the decrease in the anomeric effect with increase in solvent polarity, it is inadequate to explain the bond length changes noted above.

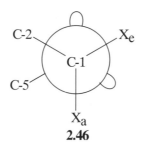

2.46

C-1–O-5 bond projection of a pyranoid ring in the $^4C_1$ conformation

The anomeric effect has profound influences on aspects of the energetics of monosaccharides both in chemistry[16] and biology.[17]

(c) Pyranoid carbohydrates

(i) General

Whilst the two possible chair forms of cyclohexane (Section 2.3.1.a) are indistinguishable, the same does not apply for tetrahydropyran when the ring atoms are

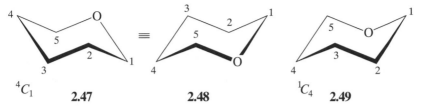

$^4C_1$ **2.47**   **2.48** $^1C_4$ **2.49**

numbered, since the two forms **2.47** and **2.49** are enantiomers, nor does it apply for the pyranoid sugars. It is necessary, therefore, to adopt a nomenclature system which will distinguish between the different chair forms, and it is clearly desirable that such a system should be extendable to cover other conformations, for example boats, skews and five-membered rings.

Conformations are symbolized by the first letter of the general ring shape: chair, C; boat, B; half-chair, H; and skew, S. A reference plane is selected which contains the maximum number of ring atoms, and superscript and subscript numbers describe the out of plane atoms, indicating whether they are above or below the reference plane. Applied to chairs, the convention takes the lowest-numbered carbon atom in the ring as an exoplanar atom. The reference plane, which contains four ring atoms, is thus defined, and **2.47** and **2.49**, when numbered according to the carbohy-drate system for aldoses as shown, become $^4C_1$ and $^1C_4$, respectively. This would appear to follow logically, but were **2.47** to be turned through 180° to give **2.48** (a procedure which leaves the molecule unaffected chemically), this would not be so. It is therefore necessary to define more closely which exoplanar atom is to be superscript and which is to be subscript. The arbitrary choice is made that the exoplanar atom projecting through the side of the plane of the ring from which the ring numbering appears *clockwise* is *superscript*. Thus in **2.47** and **2.48** the numbering appears clockwise when the rings are viewed from above and below, respectively. The atom projecting from this side of the plane of the ring is C-4 in both cases, and the conformation is consequently $^4C_1$. For **2.49** the conformation is now $^1C_4$ unambiguously. The method is readily extended for use with the other regular five- and six-membered ring conformations.[18]

Bulky groups on pyranoid rings, except for electronegative substituents attached at the anomeric centre, favour the equatorial orientation, and therefore preferred chair conformations usually have the largest possible number of such groups equa-torially arranged. On this basis, therefore, the Haworth perspective formula of α-D-glucopyranose **2.14** (α-form) can be converted into the appropriate confor-mational representation by first drawing both possible puckered rings in normal perspectives, then inserting the axial and equatorial bonds and finally entering the ring substituents in 'up' or 'down' positions. The preferred conformation is then selected, **2.1** ($^4C_1$) being clearly favoured over **2.50** ($^1C_4$) (Scheme 2.6). For other compounds the choice between the two chairs may be much less straightforward (see below).

As will be repeatedly stressed, the anomeric effect, which tends to make elec-tronegative substituents at the anomeric position favour the axial orientation, is of the greatest significance in controlling the preferred chair form of pyranoid sugars

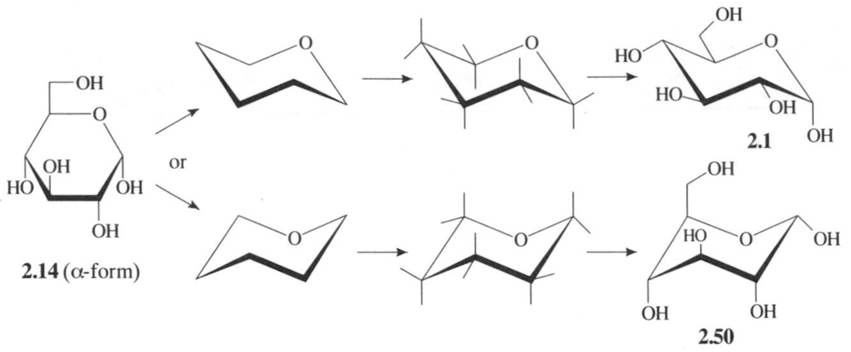

Scheme 2.6

and their derivatives. This is well illustrated by the observation that, largely as a consequence of its influence, compounds **2.51**, where X is Cl or F,[11] and **2.52**[19] exist predominantly in the conformation in which all the bulky groups are axial. On the other hand, as was again first recognized by Lemieux, groups attached to the anomeric centre which reverse the dipole of the bond attaching them to the anomeric carbon atom show a *reverse anomeric effect*. The pyridinium group of compound **2.53**, therefore, has both polar and steric factors combining to cause it to be equatorial, and the additive effect more than offsets the steric interactions between the three axial acetoxy groups. The preferred conformation of the salt in deuterium oxide is therefore as shown.[20]

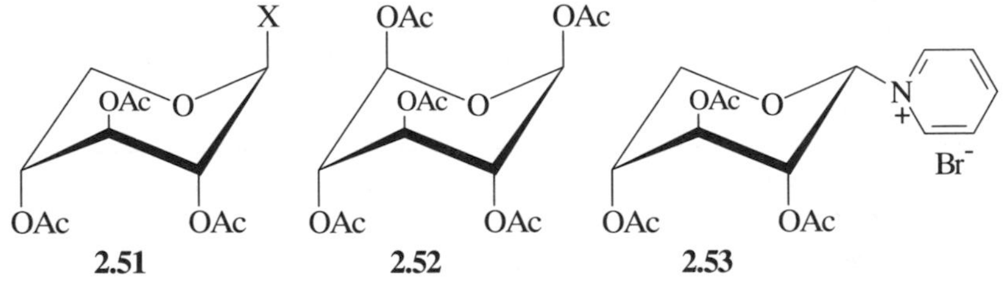

Reference has been made to 'preferred conformations' of pyranoid compounds, implicitly in solution, and this concept will be considered in greater depth in relation to sugars later in this chapter. What is meant by this term? It can signify either that under particular conditions, and at a given time, a compound exists as a mixture of molecules of different and continuously varying conformations (one of which is favoured by representing an energy minimum) or that each molecule has the same shape which lies near to one of the regular conformations. It appears that the former is the case with pyranoid derivatives, and that at ordinary temperatures molecules are continuously inverting. The n.m.r. spectrum of tetra-*O*-acetyl-β-D-ribopyranose, measured at room temperature in deuteriated acetone, indicates that the compound exists in neither the $^4C_1$ nor the $^1C_4$ conformation, and suggests that the molecular shape is intermediate between these. However, at low temperatures the spectrum

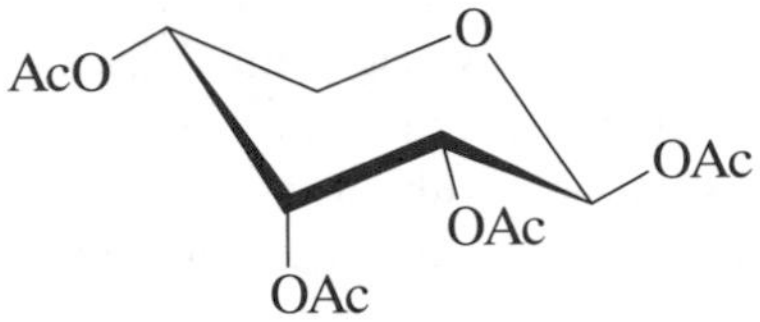

2.54 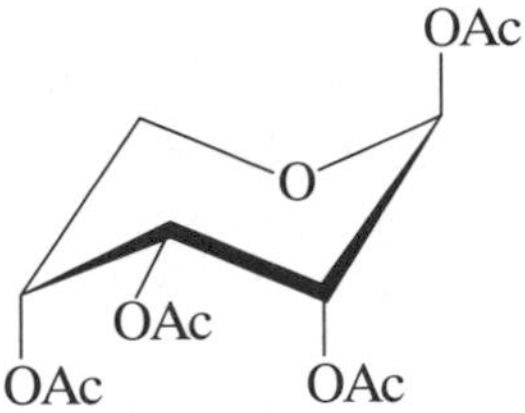   **2.55**

resolves into the sum of the spectra of these two ring forms **2.54** and **2.55** (1:2) and hence the room temperature spectrum is time averaged because the rate of ring inversion is too great to allow independent recognition of the conformers.[21]

Whereas β-D-xylopyranose tetraacetate **2.51** (X = OAc) exists to the extent of 20% in the illustrated all-axial conformation at room temperature in acetone, at 84 °C this form is not detectable.[21] Conformational equilibria are thus temperature dependent.

There is a very important consequence relating to reactivity which arises from this conformational mobility: *compounds, if required, can react in unfavoured but accessible ring shapes*. Thus, for example, alkyl glucopyranosides bearing good leaving groups at C-6, on treatment with base, give high yields of 3,6-anhydro derivatives notwithstanding their favoured $^4C_1$ conformations in which C-6 is quite inaccessible to the O-3 nucleophile. They must therefore react in the alternative $^1C_4$ chairs (Section 4.6.3). When used in reverse, this type of information has led to errors in logic: evidence of reaction such as a ring closure cannot be used to assign preferred conformations to reaction starting materials. All it establishes is that the required transition state shapes can be adopted.

For compounds which have only small energy differences between their two chair conformations, the solvent can play a very large part in controlling the position of equilibrium. Methyl 3-deoxy-β-L-*erythro*-pentopyranoside thus mainly assumes the $^4C_1$ conformation **2.56** in chloroform, while the alternative chair conformation **2.57** is favoured in water. This is partly because the anomeric effect is reduced in the more polar water, and partly because associative hydrogen bonding between the hydroxyl groups in **2.56** is replaced by hydrogen bonding of these groups with the water. On the other hand, the diacetate **2.58** of the diol adopts the $^4C_1$ conformation in all solvents, whereas the analogous trimethyl derivative **2.59**

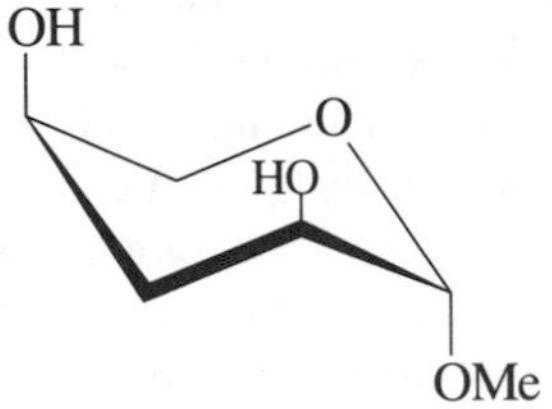

2.56 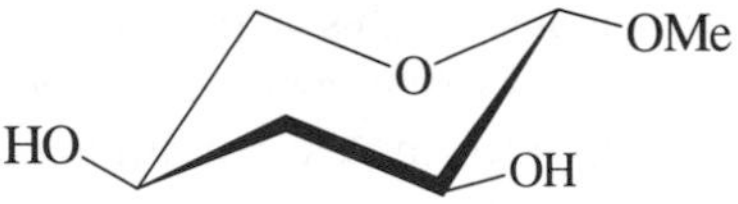   **2.57**

exists mainly in the all-equatorial chair form regardless of solvent.[22] From this the important conclusion can be drawn that the electron densities on the ring oxygen substituents determine the repulsions between them. Whereas methoxy groups have oxygen atoms with relatively high densities and strong repelling inter-actions, the opposite holds for acetoxy groups with their electron-withdrawing acetyl functions — hence the relative ease with which pairs of such groups can be accommodated in *syn*-diaxial relationships on pyranoid rings as, for example, in **2.51–2.53**.

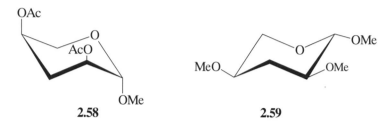

2.58   **2.59**

(ii)   Determination of conformations and their free energies

In theory, n.m.r. spectroscopy provides the ideal means for determining sugar conformations in solution. The majority of the aldopentoses and aldohexoses have been profitably examined, and the determined conformations in water are given in Table 2.2 (p. 33). X-Ray diffraction analysis gives alternative means of deter-mining ring shape but, necessarily, only of crystalline samples. It is noteworthy, however, that the conformations of pyranoid compounds are commonly (but not always) the same in the crystal and in solution.

The first chemical method employed for the determination of conformation in carbohydrate chemistry involved the detection by polarimetric and conductimetric methods of the interaction of the cuprammonium ion with various diols on pyran-oside rings. This pioneering approach, the work of Reeves,[23] paved the way for later research and established the basic factors which control conformational preferences of sugars. As with all other chemical approaches, however, the method is open to the criticism that the reagent may have altered the shapes of the diols during complexing. Physical methods of analysis are therefore much to be preferred.

A different approach, which builds on the work of Reeves, has been refined largely by Angyal and can be used in many cases to determine the favoured conformations of molecules in aqueous solution.[24] The method is deductive rather than experimental, but the agreements between predictions and observations are so good for many compounds (Table 2.2) that it certainly ranks with some of the experimental methods. In this approach the destabilizing influences within each of the two chair conformations of a pyranose sugar are quantified, summed and compared, and provided the energy difference is greater than about 2.9 kJ mol^{-1} (about 0.7 kcal mol^{-1}) (equivalent to an equilibrium ratio of approximately 3:1), the more stable is taken to be the preferred ring shape. In Table 2.1, values are given of the destabilizing effects within a pyranoid sugar ring which contribute

Table 2.1 Destabilizing effects within a pyranoid ring in aqueous solution

	Effect	
Interaction	(kJ mol^{-1})	(kcal mol^{-1})
$O_{axial}-H_a{}^a$	1.9	0.45
$C_a-H_a{}^a$	3.8	0.9
$O_a-O_a{}^a$	6.3	1.5
$C_a-O_a{}^a$	10.5	2.5
$O-O_{gauche}{}^b$	1.5	0.35
$C-O_g{}^b$	1.9	0.45
Anomeric effect in compounds with equatorial hydroxyl group at C-2^c	2.3	0.55
Anomeric effect in compounds with axial hydroxyl group at C-2^c	4.2	1.0
Anomeric effect in compounds with axial hydroxyl groups at C-2 and C-3^c	3.6	0.85
Anomeric effect in compounds with no hydroxyl groups at C-2 and C-3^c	3.6	0.85

^{a}These interactions refer to 1,3-related axial groups on the same side of pyranoid rings.
^{b}These interactions refer to *vicinally* related atoms in the *gauche* relationship; no distinction is made between equatorial–equatorial and equatorial–axial pairs.
^{c}The value for 2-deoxyaldoses is somewhat lower than that determined for 2-alkoxytetrahydropyrans in non-polar solvents, as would be expected. Introduction of an axial electronegative hydroxyl group at C-2 introduces a new dipole which further stabilizes the axial anomer (increased anomeric effect, cf. *trans*-1,2-dibromocyclohexane in Section 2.3.1.a); the opposite applies when an equatorial group is added. When hydroxyl groups at C-2 and C-3 are both axial their influences on the anomeric effect cancel each other.

significantly to its overall energy in aqueous solution. (No appreciable repulsive interaction is thought to arise between an axial oxygen atom and the electrons of the ring oxygen atom.)

As an example of the application of this procedure, the conformations of α- and β-D-altropyranose **2.60** and **2.61** will be calculated. In the $^4C_1$ conformation of the α-anomer **2.60a** the destabilizing features are $O_a-O_a + 3(O_a-H_a) + O-O_g+C-O_g = 6.3+5.7+1.5+1.9 = 15.4$ kJ mol^{-1} (3.66 kcal mol^{-1}). In the $^1C_4$ conformation **2.60b** the corresponding features are $O_a-H_a+3(O-O_g)+2(C_a-H_a)+$ anomeric effect (C-2–OH equatorial) $= 1.9 + 4.5 + 7.6 + 2.3 = 16.3$ kJ mol^{-1} (3.88 kcal mol^{-1}). Since the stabilities of these two conformations as determined by this procedure are so alike, it must be concluded that neither chair will be strongly preferred. Alternatively, the interaction energies are 14.2 kJ mol^{-1} (3.38 kcal mol^{-1}) for structure **2.61a**, and 22.6 kJ mol^{-1} (5.38 kcal mol^{-1}) for structure **2.61b**, and it can safely be concluded that the β-anomer will adopt

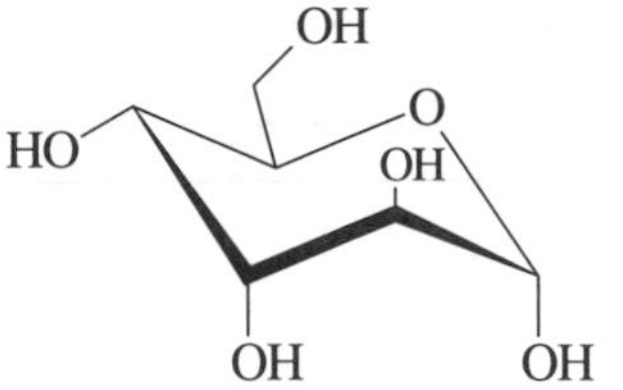

$^4C_1$ **2.60a** 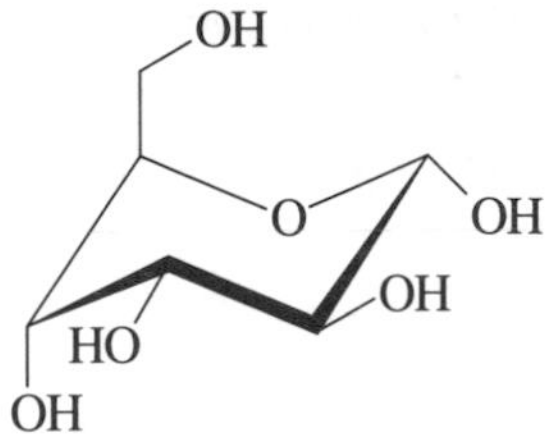 $^1C_4$ **2.60b**

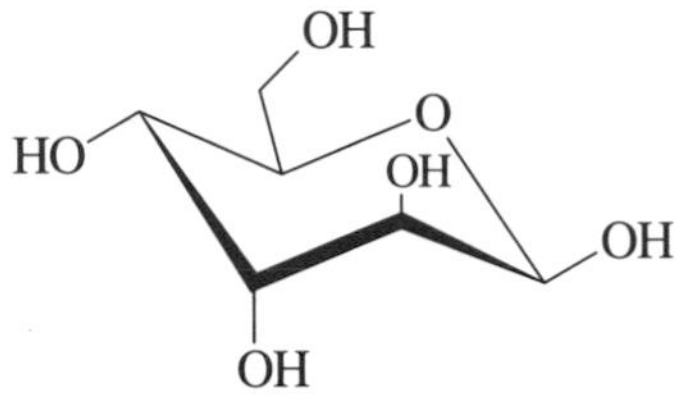

$^4C_1$ **2.61a** 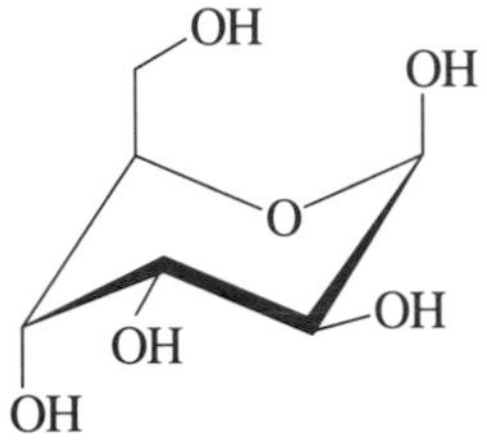 $^1C_4$ **2.61b**

the $^4C_1$ chair form **2.61a** preferentially in solution. These conclusions are borne out by n.m.r. spectral studies. In Table 2.2 the calculated and observed preferred conformations of the pyranose forms of the D-aldopyranoses in aqueous solution are given.

For conformationally flexible six-membered cyclic compounds in solution it is often more appropriate to consider conformational equilibria rather than the preferred conformation. Although, in principle, all possible conformations which contribute to an equilibrium should be considered, attention is frequently confined to the two chairs, and n.m.r. coupling constants and chemical shifts are often used to assess the conformation of a flexible pyranoid compound in terms of the proportions of $^4C_1$ and $^1C_4$ chairs it represents. Table 2.3 provides estimates of this kind for various fully substituted pentopyranose derivatives. From it, for example, the major influence of the anomeric effect on the conformations adopted by β-D-xylopyranosyl compounds is clearly seen. While the methyl glycoside favours the chair conformation with the methoxy grouping in the equatorial orientation ($^4C_1$), the same is so to a lesser extent for the glycosyl acetate, and for the glycosyl chloride the tendency of the even stronger electronegative atom at the anomeric centre to be axial dominates and the molecule favours the inverted chair ($^1C_4$).

2.3.2   SATURATED FIVE-MEMBERED RINGS

(a)   Cyclopentane and substituted cyclopentanes

Although in the planar orientation cyclopentane has a ring largely free of angle strain, all *cis*-adjacent hydrogen atoms are then eclipsed, and the conformation is energetically unfavoured relative to puckered forms. All measurements agree with

Table 2.2 The calculated energy differences between the chair forms of the D-aldopyranoses, and the predicted and observed (n.m.r.) preferred conformations[11,24]

| D-Aldopyranose | $\Delta G(^1C_4 - {}^4C_1)$ | | Preferred conformation | |
	(kJ mol^{-1})	(kcal mol^{-1})	Predicted	Observed
α-Ribose	+0.42	+0.1	—[a]	—[a]
β-Ribose	+2.51	+0.6	—	—
α-Arabinose	−4.81	−1.15	$^1C_4$	$^1C_4$
β-Arabinose	−2.09	−0.5	—	—
α-Xylose	+6.90	+1.65	$^4C_1$	$^4C_1$
β-Xylose	+9.62	+2.3	$^4C_1$	$^4C_1$
α-Lyxose	+2.30	+0.55	—	—
β-Lyxose	+4.39	+1.05	$^4C_1$	$^4C_1$
α-Allose	+6.07	+1.45	$^4C_1$	$^4C_1$
β-Allose	+12.97	+3.10	$^4C_1$	$^4C_1$
α-Altrose	+0.84	+0.2	—	—
β-Altrose	+8.37	+2.0	$^4C_1$	$^4C_1$
α-Glucose	+17.36	+4.15	$^4C_1$	$^4C_1$
β-Glucose	+24.89	+5.95	$^4C_1$	$^4C_1$
α-Mannose	+12.76	+3.05	$^4C_1$	$^4C_1$
β-Mannose	+19.66	+4.7	$^4C_1$	$^4C_1$
α-Gulose	+3.14	+0.75	$^4C_1$	$^4C_1$
β-Gulose	+10.04	+2.4	$^4C_1$	$^4C_1$
α-Idose	−2.09	−0.5	—	—
β-Idose	+5.44	+1.3	$^4C_1$	$^4C_1$
α-Galactose	+14.43	+3.45	$^4C_1$	$^4C_1$
β-Galactose	+21.97	+5.25	$^4C_1$	$^4C_1$
α-Talose	+9.83	+2.35	$^4C_1$	$^4C_1$
β-Talose	+16.74	+4.0	$^4C_1$	$^4C_1$

[a] A dash indicates that the chair forms are determined to be of comparable stability, or that n.m.r. examination indicates that there is no preferred chair conformation.

Table 2.3 Percentages of $^4C_1$ conformations of D-pentopyranose derivatives in solution at 31 °C[11]

Configuration	Methyl tri-O-acetyl pyranosides (%)	Tri-O-acetyl glycosyl acetates (%)	Tri-O-acetyl glycosyl chlorides (%)
α-Ribo	65[a]	77[a]	—
β-Ribo	39[a]	43[a]	6[b]
α-Arabino	17[b]	21[b]	—
β-Arabino	3[a]	4[a]	2[b]
α-Xylo	>98[a]	>98[a]	>98[b]
β-Xylo	81[a]	72[a]	21[b]
α-Lyxo	83[a]	71[a]	91[b]
β-Lyxo	58[a]	39[a]	—

[a] Measured in Me$_2$CO-d$_6$.
[b] Measured in CDCl$_3$.

2.62   2.63

this conclusion, and substituted derivatives are generally found to exist in either an envelope conformation **2.62**, in which one ring atom is out of the plane of the other four, or as the twist form **2.63**, in which three ring atoms are in a plane and the other two protrude above and below it. In the neighbourhood of the exoplanar atoms, the exocyclic bonds correspond approximately to the axial and equatorial bonds of cyclohexane and are described as *quasi*-axial (a′) and *quasi*-equatorial (e′), and substituents at these sites favour the equatorial positions as in cyclohexane. The most stable conformation of a monosubstituted cyclopentane is that which has the substituent on the exoplanar carbon atom and in the *quasi*-equatorial orientation **2.62**. In a 1,2-*cis*-disubstituted ring the interaction energy between the groups is reduced by a deformation at this part of the molecule, and in a 1,3-*cis*-disubstituted compound the groups can occupy *quasi*-equatorial sites at the positions on either side of the exoplanar atom in the envelope conformation. In this way they are subject to minimal interaction energies, and such *cis*-compounds are more stable than the *trans*-isomers for this reason. This would not be interpretable from planar formulae.

For polysubstituted compounds the conformation which permits most bulky groups to be *quasi*-equatorial will be adopted, but although staggering between interacting groups is now minimal it is never as complete as is possible for six-membered cyclic compounds; consequently, substituted five-membered rings are inherently the less stable. Furthermore, since the energy barriers between the various envelope and twist conformations are small relative to those involved in the conformational inversion of six-membered rings, rapid conformational change is a particularly notable characteristic of cyclopentane derivatives.

(b)   Tetrahydrofuran, substituted tetrahydrofurans and furanoid carbohydrates

None of these compounds has a planar ring; all measurements (crystallographic and n.m.r. are most suited) have indicated that, as with cyclopentane derivatives, envelope (E) or twist (T) conformations are preferred. For tetrahydrofuran itself, less strain is removed from the planar molecule by distortions in the region of the oxygen atom than if C-3 is removed from the plane. Conformation **2.64**, therefore, represents an energy minimum, and in substituted derivatives there will be a tendency for such a conformation to be adopted.

2.64   2.65

Tetrahydrofurans with electronegative substitutents at C-2 (anomeric position) are subject to the anomeric effect which favours the ^2E conformation **2.65** (conformations are named as for pyranoid compounds, cf. Section 2.3.1.c.i), and large ring substituents incur fewer steric interactions when they are *quasi*-equatorial; for example, 2-(hydroxymethyl)tetrahydrofuran has a tendency to favour conformation **2.66**. As with cyclopentane derivatives, ring distortion occurs to avoid eclipsing between *cis*-substituents on adjacent ring atoms. These influences acting together result, for example, in methyl α-D-xylofuranoside favouring the ^2E conformation **2.67**, while for methyl α-D-lyxofuranoside the E_4 conformation **2.68** represents an energy minimum.[25] A large number of ribofuranosyl nucleosides (Section 8.2.2.a), which have *cis*-related hydroxyl groups at C-2' and C-3', adopt conformations (e.g. ^3E) which minimize the interactions between these groups.

2.66

2.67   2.68

2.3.3   ACYCLIC CARBOHYDRATES

In linear acyclic compounds group interactions are usually minimized when the carbon chains adopt planar zigzag conformations. X-Ray and n.m.r. evidence indicates that this generalization holds good for carbohydrate derivatives, but distortions occur when bulky groups on 1,3-related carbon atoms have the *cis*-relationship (in either the Fischer projection or the zigzag orientation).[11] Thus D-mannitol **2.69** adopts the conformation **2.70**, but D-glucitol **2.71** in the zigzag conformation **2.72** has *cis*-related hydroxyl groups at C-2 and C-4 (indicated) and, in consequence, exists in the crystal and in solution in the distorted 'sickle' conformation **2.73** which has undergone a 120° rotation about the C-2–C-3 bond to alleviate this interaction.[26]

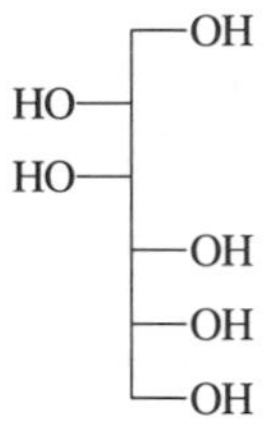

2.69 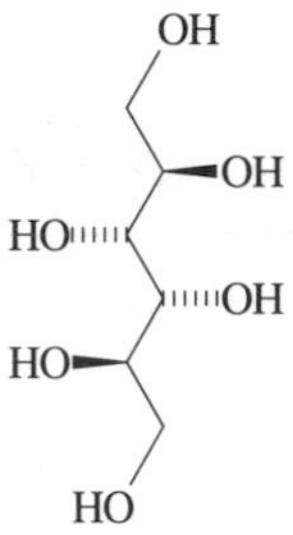   2.70 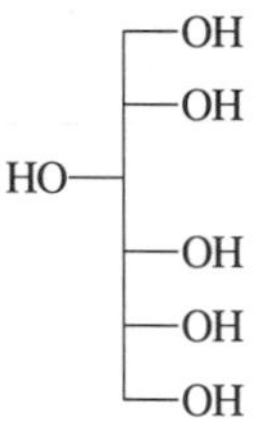   2.71 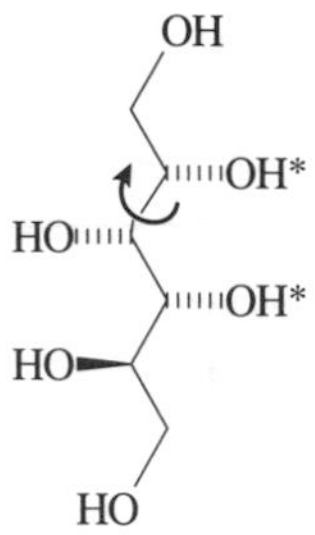   2.72 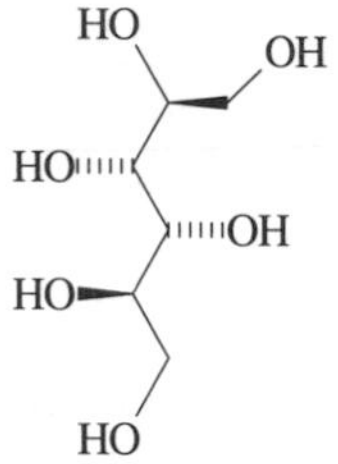   2.73

The 1,3-*syn*-relationships in these compounds are stereochemical equivalents of the 1,3-*syn*-diaxial relationships in pyranoid compounds and thus, in the same way as pairs of *syn*-axial acetoxy groups can be accommodated relatively easily on six-membered rings (Section 2.3.1.c), 1,3-*cis*-related pairs of acyloxy groups would be expected to have less effect in disturbing the regular zigzag shapes of acyclic derivatives than do hydroxyl group pairs.

A very important consequence of the generalization relating to substituted acyclic compounds is that the Fischer projection formulae commonly used for their representation depict conformations which are completely different from those actually favoured by the compounds. This is well demonstrated by the fact that the periodate ion, which is known to cleave *cis*-α-diols preferentially (Section 5.5), attacks *threo*-systems more readily than *erythro*-diols. Thus D-mannitol **2.69** is cleaved at the C-3–C-4 bond faster than at the C-2–C-3 and C-4–C-5 sites. In a similar way, D-*threo*-2,3-butanediol **2.74a**–**2.74c** (Fischer, zigzag and Newman projection, respectively) complexes to give cyclic ions with borate more readily than does the *erythro*-isomer **2.75a**–**2.75c**.

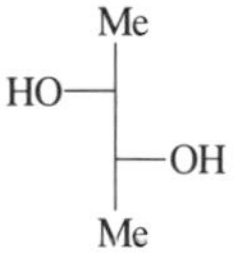

2.74a 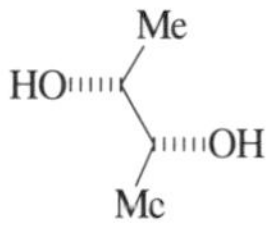   2.74b 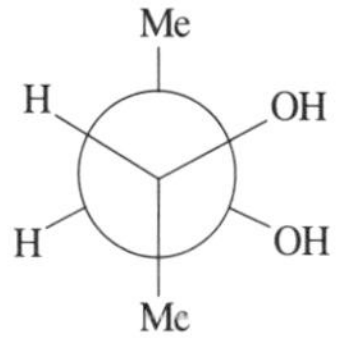   2.74c 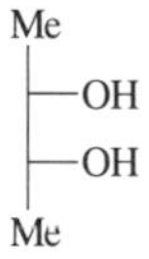   2.75a 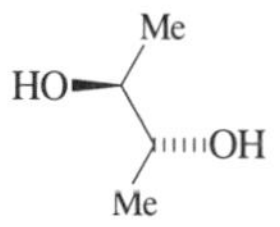   2.75b 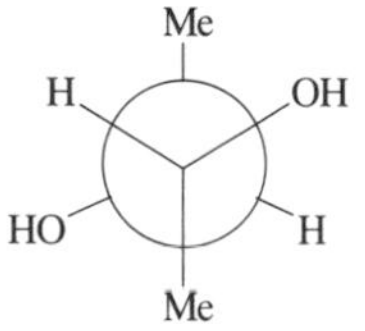   2.75c

2.3.4   THE ORIENTATIONS OF SUBSTITUENT GROUPS ON EXOCYCLIC ATOMS

With advances in analytical methods (in particular X-ray analysis and n.m.r. spectroscopic methods, but molecular modelling calculations are now of appreciable value also) it has been possible to investigate more esoteric features of the preferred conformations of carbohydrate compounds, namely those relating to the orientations adopted by atoms or groups bonded to oxygen and carbon atoms which themselves are attached to sugar rings. It is possible, in some cases, to probe into the orientation of O–H bonds with respect to neighbouring features,[27] but much more significant findings have been made with respect to the important issue of the orientation of groups bonded to the anomeric oxygen atom of pyranosides. Once again Lemieux has made a vital contribution by identifying the *exo-anomeric effect*, which represents a considerable barrier to rotation about the exocyclic C–O bonds at the anomeric centres of such compounds.[28,29] This further stereoelectronic factor causes glycosides with axial and equatorial groups at the anomeric centre to prefer the rotamer states illustrated in **2.76** and **2.77**, respectively, in which the electrons in the shaded orbitals of O-1 (which project out of the paper in the chair representation) are *anti*-periplanar with the C-1–O-5 bonds and thereby undergo $n-\sigma^*$ delocalization. The origins of the effect are analogous to those of the 'anomeric effect' which, for clarity, may be called the '*endo*-anomeric effect'. For axial systems **2.76** the *exo*-anomeric and *endo*-anomeric components oppose each other, thereby making the *exo*-anomeric effect a more important conformational determinant with equatorially substituted derivatives.

C-1–O-1 Projections
2.76   **2.77**

The recognition of this factor has been of the greatest significance in advancing our understanding of the conformations adopted by oligosaccharides in solution, and thence of their biological functioning (Section 8.1.2.c). For the determination of the orientation of one sugar unit with respect to a bonded neighbouring sugar (or other moiety), it is necessary to evaluate the angles denoted by ϕ and ψ in **2.78** (Section 8.1.2.c). While the former is restricted by the *exo*-anomeric effect, it is also found that the limitations on ψ are significant, and hence that many disaccharides and higher saccharides have rather rigid shapes, the knowledge of which has bearing on understanding their biological behaviour.

1,6-Linked disaccharides **2.79** are less easily defined stereochemically because of the further rotation (defined by the angle ω) which is possible about the C-5–C-6 bond. As for all D-glycoses substituted at O-6, the symmetrical orientations around this bond can be represented by **2.80–2.82** — described *tg*, (for *trans/gauche* H–H relationships), *gt* and *gg*, respectively — and an equatorial oxygen atom bonded to C-4 (in glucose components, for example) interacts with O-6 in the first of these, thereby destabilizing it. On the other hand, the C-4 substituent is axial in the case of D-galactose compounds and the *gg* rotamer **2.82** is least favoured.[30]

Nucleosides and nucleotides, important components of nucleic acids, (and thereby the acids themselves; see Section 8.2.2), also present vital questions relating to the rotational status of the groups bonded to the sugar rings. Nucleotides, e.g. adenosine 5′-phosphate **2.83** (R = PO_3H_2), prefer to have the base oriented *anti*

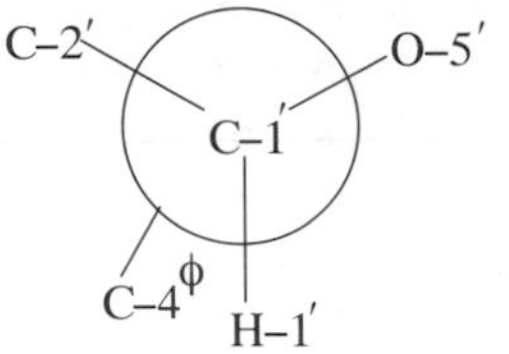

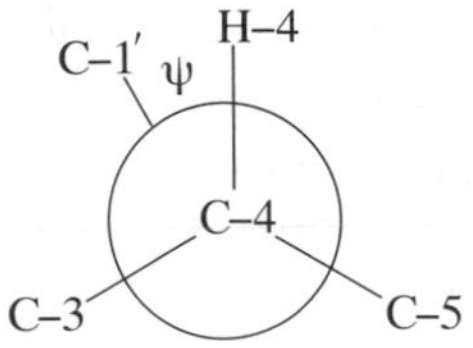

2.78, C-1′−O-1′ Projection **2.78,** C-4−O-1′ Projection

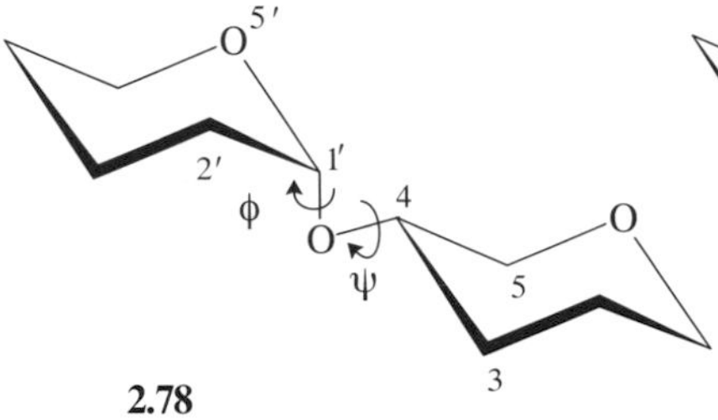

2.78 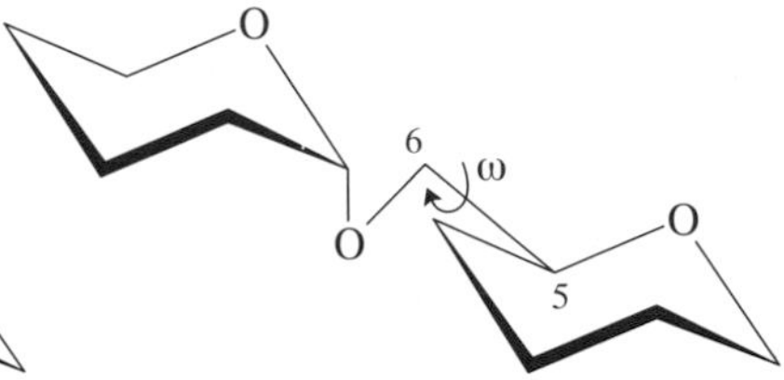   **2.79**

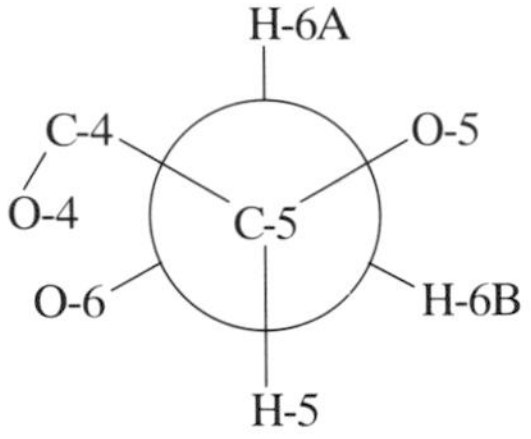

2.80 *tg* 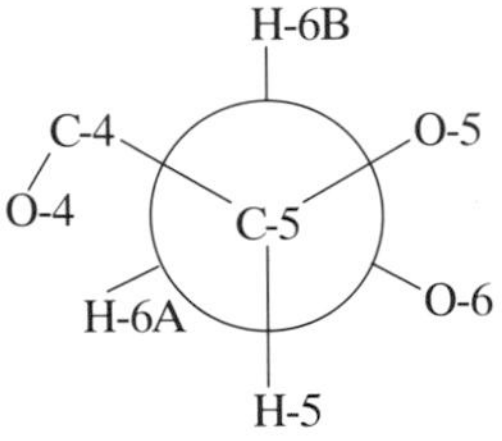   **2.81** *gt* 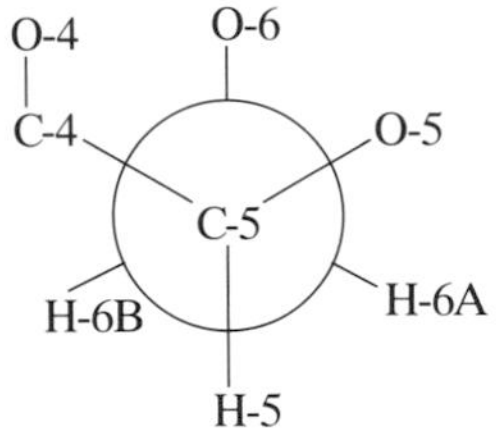   **2.82** *gg*

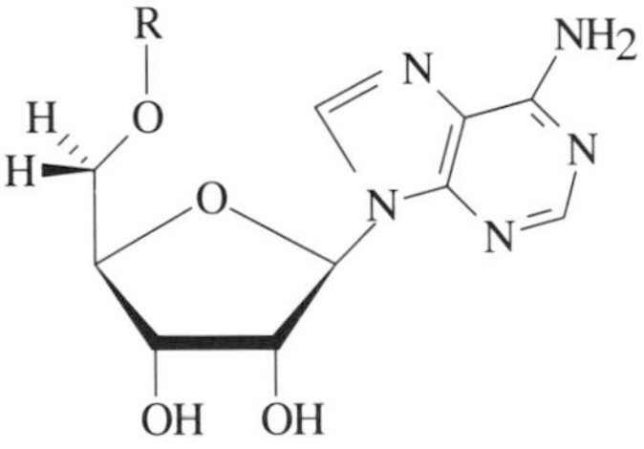

2.83

to the sugar ring (base portion pointing away from C-5′) and the C-4′−C-5′ bond in the *gg* orientation. Purine nucleosides, lacking a substituent at O-5′, are more flexible, adenosine **2.83** (R = H) preferring (60%) the *syn*-rotamer state with the base oriented towards the hydroxymethyl group.[31] On the other hand, pyrimidine analogues prefer the conformation with the carbonyl group at C-2 of the base projecting away from C-5′.

The effect of solvent on features of compounds such as those referred to in this section has been emphasized repeatedly, especially by Lemieux who has shown how the anomeric effects (both *endo* and *exo*[29]), and molecular conformation, including ring substituent orientation, and optical rotation[20b] can be extremely dependent upon solvent. The message is strongly echoed here.

2.3.5   THE PROPERTIES OF MONOSACCHARIDES IN RELATION TO CONFORMATIONAL FACTORS

All physical properties and many chemical and biochemical properties of carbohydrates depend upon molecular shape, and frequently throughout this book reference will be made to phenomena which can only be interpreted in conformational terms. It is not proposed to describe fully the significance of the molecular conformation of sugars at this stage; instead, an indication is given of the types of properties which are conformation dependent.

Equilibria involving pyranoid compounds depend largely upon the axial–equatorial relationships of substituents on the rings. Thus the α/β ratio of pyranoses is governed to some extent by the steric propensity of the solvated anomeric hydroxyl group to acquire the less-hindered equatorial orientation (Section 2.4.2). In a similar fashion, equilibria between different sugars (or their derivatives) are controlled in this way: when α-D-glucopyranose 1-phosphate **2.84** and the *galacto*-isomer **2.85** are brought into equilibrium by an appropriate epimerase enzyme, the *gluco*-compound, having the equatorial substituent at C-4, predominates (75%).[32] Likewise, 2-acetamido-2-deoxy-D-glucose **2.86** exists to the extent of 80% in equilibrium with the *manno*-epimer **2.87** in basic solution.[33]

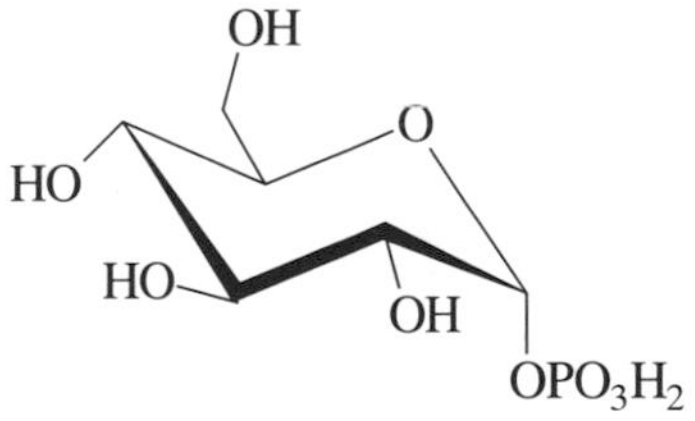

2.84

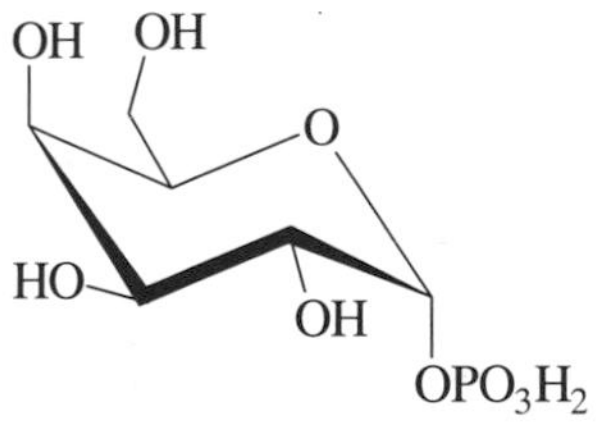

2.85

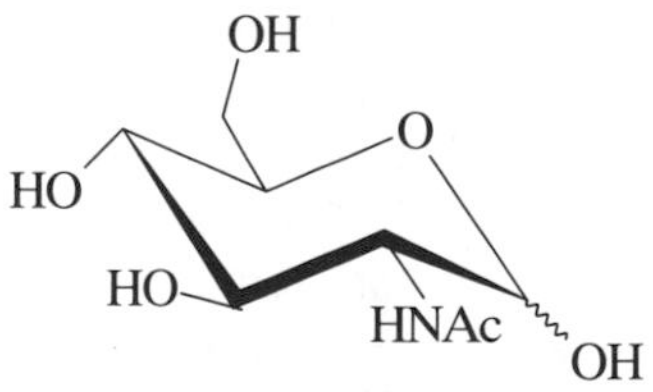

2.86

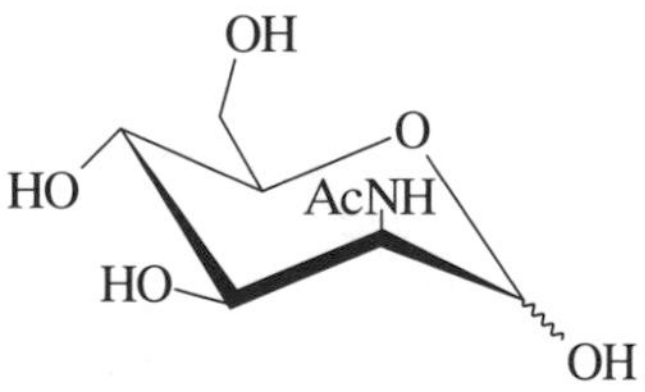

2.87

Functional group reactivity depends largely on conformation. Equatorial anomeric hydroxyl groups of pyranoses are oxidized with aqueous bromine solution more rapidly than when they are axial (Section 3.1.6.a); conversely, axial hydroxyl groups at other positions are oxidized more rapidly with oxygen over a platinum catalyst (Section 4.9.1.a.iii). The relative rates of hydrolysis of anomeric glycosidic bonds are similarly related to their equatorial or axial character (Section 3.1.1.b); methyl glycosides with equatorial aglycons are oxidized selectively with ozone and with chromium trioxide in acetic acid (Section 3.1.1.b.xi), and many other reactions, including free radical reactions (Section 3.3.4), at the anomeric centre are dependent upon the orientation of the group at this position. Equatorial hydroxyl groups are more readily esterified than are their axial counterparts (Section 5.1.3), whereas axial leaving groups are more susceptible to nucleophilic displacement (Section 4.1.2). Cyclic orthoesters involving *cis*-diols on pyranoid rings hydrolyse to give hydroxy esters carrying the ester groups on the axial sites (Section 5.3.4).

Spectral properties and sometimes polarimetric characteristics (appendix 4) of sugars and their derivatives are closely linked to conformational factors. Chromatographic mobilities are also related: those sugars having conformationally stable pyranoid rings generally travel more slowly on paper partition chromatograms, presumably because the equatorial hydroxyl groups can be more fully hydrated than axial groups and consequently increase the hydrophilic character of the compounds. However, galactose and arabinose, with axial hydroxyl groups at C-4, usually migrate more slowly than all the other hexoses and pentoses, respectively, and represent an unexplained exception to this generalization.

There seems also to be a correlation between the natural occurrence of sugars and their ring stabilities. D-Glucose, having the conformationally most stable ring, is, for example, the most abundant of the hexoses. D-Galactose and D-mannose, each having one axial hydroxyl group in the β-modification, are the next most abundant, and so conform with the generalization; D-allose, also with one axial group, is found occasionally, and all the remainder are exceedingly rare if they occur at all. No great significance should, however, be sought in these observations; rather, the natural abundance of sugars should be related to the central role in biochemistry played by D-glucose (Section 2.5.1).

2.4 SUGARS IN SOLUTION

2.4.1 MUTAROTATION

Each crystalline free sugar is a discrete modification, and in dimethyl sulphoxide solution, for example, the integrities of the species are retained. On dissolution in water, however, the hemiacetal ring opens and reforms to give products with different ring sizes and anomeric configurations.[34,35] This equilibration is accompanied by a change in optical rotation known as *mutarotation*, which follows a simple, first-order kinetic law for those sugars (e.g. glucose, mannose and xylose) which give effectively only the two pyranoses at equilibrium (Table 2.4). Alternatively, those sugars (e.g. ribose and altrose) which contain appreciable proportions of furanose forms at equilibrium show complex mutarotations the kinetics of which do not follow a first-order law.

Mechanistically, the isomerizations proceed by way of the acyclic species, as shown in Scheme 2.7 for the interconversions of α- and β-D-glucopyranose, and in keeping with this the mutarotation reaction shows general acid–base catalysis.

Table 2.4 The percentage compositions of sugars in aqueous solution at equilibrium[36,37]

Sugar	Temperature (°C)	Cyclic forms				Acyclic carbonyl form (%)
		α-Pyranose (%)	β-Pyranose (%)	α-Furanose (%)	β-Furanose (%)	
Ribose	31	21.5	58.5	6.5	13.5	0.05
Arabinose	31	60	35.5	2.5	2	0.03
Xylose	31	36.5	63	<1	<1	0.02
Lyxose	31	70	28	1.5	0.5	0.03
erythro-Pentulose	20	—	—	62.8	20.4	16.8
threo-Pentulose	26	—	—	18.1	62.3	19.6
Allose	31	14	77.5	3.5	5	0.01
Altrose	22	27	43	17	13	0.04
Glucose	31	38	62	—	0.14	0.02
Mannose	44	64.9	34.2	0.6	0.3	0.005
Gulose	22	16	81	—	3	—
Idose	31	38.5	36	11.5	14	0.2
Galactose	31	30	64	2.5	3.5	0.02
Talose	22	42	29	16	13	0.03
Psicose	27	22	24	39	15	0.3
Fructose	31	2.5	65	6.5	25	0.8
Sorbose	31	93	2	4	1	0.25
Tagatose	31	71	18	2.5	7.5	0.3

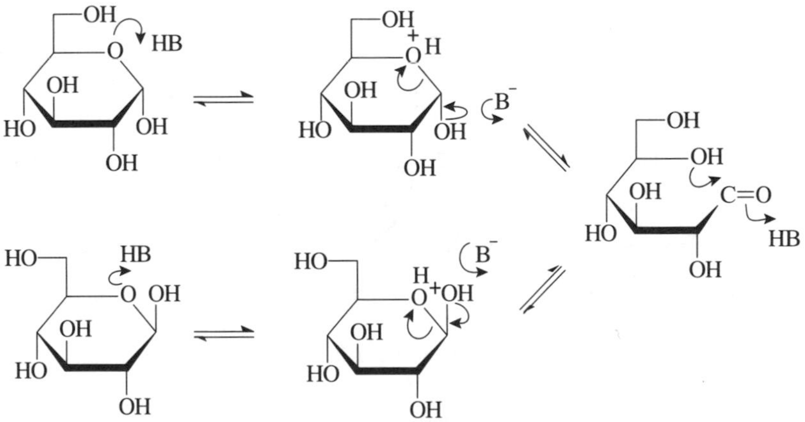

Scheme 2.7

D-Glucose, labelled at O-1 with ^{18}O, mutarotates with only minor loss of this label; that is, O-1 is retained, and consequently a direct displacement mechanism (of the type which occurs during the hydrolysis of glycosides) can be ruled out. Similarly, on this evidence the acyclic hydrated aldehyde is likely to be excluded as a reaction intermediate.

2,3,4,6-Tetra-*O*-methyl-D-glucose mutarotates only slowly in either pyridine or *m*-cresol but exceedingly fast in a mixture of these compounds, which shows that acidic and basic catalysts are required. 2-Hydroxypyridine is an efficient acid–base catalyst.

Several species can be involved in a mutarotation, as shown in Scheme 2.8 for D-glucose. The acyclic *aldehydo*-form, as well as giving the pyranoses, could ring close to give furanoses or, conceivably, oxetanoses or septanoses with four- and seven-membered rings, respectively; furthermore, it could solvate to give the acyclic hydrate. The positions of the equilibria established by the sugars in water must now be considered.

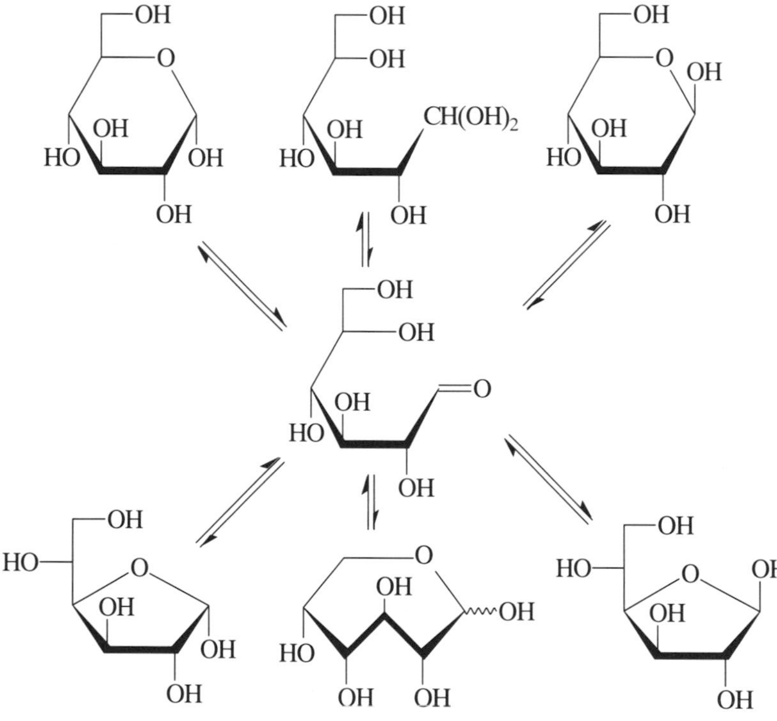

Scheme 2.8

2.4.2 THE SITUATION AT EQUILIBRIUM[36,37]

Considerable information is now available on the compositions of equilibrated mixtures of isomeric forms of the sugars thanks largely to the availability of n.m.r. spectroscopy. Examination of ^{13}C spectra has helped especially with the detection of the minor components, and has been invaluable in analysing solutions of the ketoses which have no anomeric hydrogen atoms. Results for the aldo- and keto-pentoses and -hexoses are given in Table 2.4 (see Figures 2.1b and 2.2 for the ^1H and ^{13}C n.m.r. spectra of equilibrated D-glucose).

Clearly, cyclic forms are heavily favoured over acyclic forms and pyranose rings over furanose forms, in keeping, respectively, with the propensity of comparable hydroxyaldehydes to cyclize (Section 2.1.3) and the relative thermodynamic stabilities of six-membered rings over five-membered analogues. With the latter, eclipsing interactions can never be totally removed, but when substitution patterns are appropriate furanose forms can become very significant, as is the case for idose, talose and especially psicose, all of which also have relatively unstable pyranose forms.

For compounds in aqueous solution, a full appreciation of their state cannot be gained without consideration of their solvation. Findings with the equilibria of sugars in aqueous solution agree well with the hypothesis that pyranoid ring compounds with equatorial hydroxyl groups fit tightly into the water structure and are thereby selectively stabilized. With increase in temperature the proportion of β-D-glucopyranose at equilibrium decreases while that of the α-anomer does not, and the proportions of the furanoses increase (0.6% α and 0.7% β at 82 °C). Similarly, D-fructose comprises 10 and 32% of the α- and β-furanoses at 80 °C and 3% of the sugar is in the form of the acyclic ketone. Further supporting evidence comes from the observations that furanose proportions increase with the addition of organic solvents and with either O-substitution or removal of hydroxyl groups. Removal of the oxygen atom from C-3 of glucose or from C-2 of galactose gives the respective deoxy compound with 20 or 16%, respectively, of the furanose form in D_2O at equilibrium.

Taken together, steric factors and the anomeric effect control the α-pyranose/β-pyranose ratio. It is noticeable from Table 2.4 that when the hydroxyl group at C-2 is equatorial (xylose, glucose, etc.), the latter influence, which is partly polar in origin and weakened in aqueous solution, is overcome by the propensity of the anomeric hydroxyl group to assume the equatorial orientation for steric reasons, and the equatorial anomers are favoured. However, when the hydroxyl group at C-2 is axial (lyxose, mannose, etc.), the anomeric effect increases in significance, perhaps as a consequence of a favourable relationship between the dipoles of the C-1 and C-2 C–OH bonds. In addition, axial anomers have one destabilizing *gauche* interaction fewer than the equatorial analogues, and the combined result is that the former are favoured.

In the majority of cases, the position of the α/β equilibrium for the furanoses lies in favour of the 1,2-*trans*-isomers, as is to be expected since 1,2-*cis*-interactions are important destabilizing factors in five-membered rings. In the case of D-idose, however, the α-furanose **2.88** (1,2-*trans*) is less stable than the β-form, indicating the presence of a destabilizing interaction between the hydroxyl groups at C-1 and C-3 in the former. It is speculated that the surprisingly low concentrations of furanose forms in arabinose and galactose equilibria (these sugars having the *trans–trans* relationship at C-2, C-3 and C-4) can be accounted for by the presence of similar 1,3-repulsions (cf. Section 2.3.2.a).

The aldoses give aldehydic forms in amounts that are too small for detection by ultraviolet absorption spectroscopy, but consistent estimates of the tiny

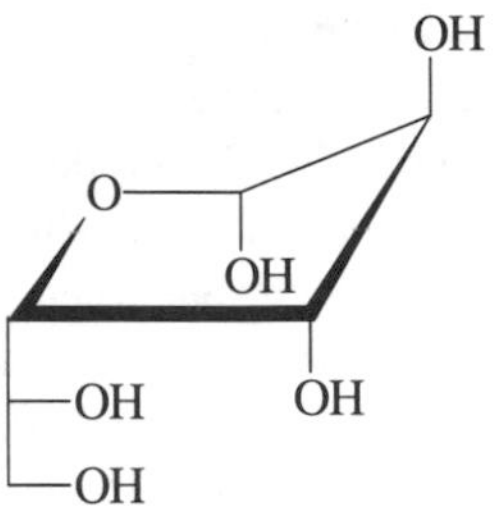

2.88

amounts present have been made by ^{13}C n.m.r., circular dichroism and polaro-graphic methods. Ketoses retain more of the carbonyl forms consistent with the greater difficulty nucleophiles experience in adding to ketones relative to aldehydes, and in the case of the pentuloses, which may cyclize only to relatively unstable furanoses, the acyclic modifications are readily observed.

The hydrated forms of the aldehydic isomers of aldoses have been detected only by highly sensitive methods and then in tiny amounts; the corresponding hydrates of the ketoses seem not to have been detected. Likewise, oxetanose forms, with 4-membered rings, have never been observed and septanoses only in cases in which more stable rings cannot be formed. Thus, 2,3,4,5-tetra-O-methyl-D-glucose exists partly in the acyclic form **2.89** and partly in the ring form **2.90**. (note that this representation does not imply a favoured conformation).

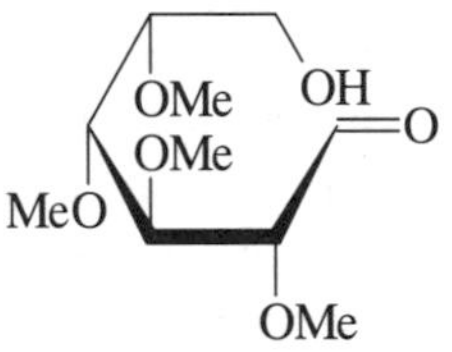

2.89 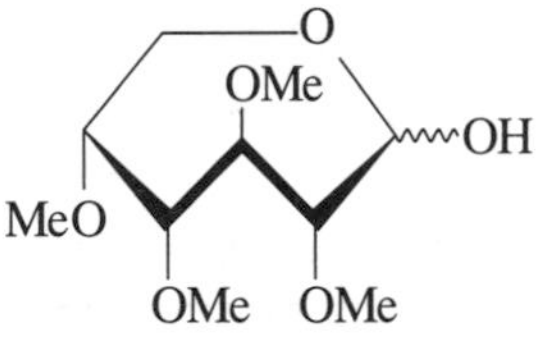   **2.90**

2.5   SOURCES OF MONOSACCHARIDES

Monosaccharides are most readily obtained from natural sources, but the different members of the family are by no means similarly available. For reasons that will be described, D-glucose plays a central role in the biochemistry of the carbohydrates; it is stored in the form of dimers and polymers in very much greater amounts than any other monosaccharide, and is thus much the most readily available. From it, D-galactose, D-mannose and D-fructose of the hexoses and D- and L-arabinose and D-xylose of the pentoses are made naturally by biochemical transformations, and they too are available from polymers. D-Ribose plays a key role as the sugar component of the ribonucleic acids (RNAs), and arabinose is peculiar amongst the simple sugars by occurring in plant products more commonly in the L- than

Table 2.5 Monosaccharide prices

	Price ($US g^{-1})			Price ($US g^{-1})	
	D	L		D	L
Aldopentoses			*Ketopentoses*		
Ribose	0.47	250	*erythro*-Pentulose	430	—
Arabinose	0.33	0.33	*threo*-Pentulose	1350	6800
Xylose	0.13	2			
Lyxose	2.88	30			
Aldohexoses			*Ketohexoses*		
Allose	200	900	Psicose	180	—
Altrose	700	—	Fructose	0.05	—
Glucose	0.02	50	Sorbose	420	0.1
Mannose	0.45	80	Tagatose	—	—
Gulose	1300	1460			
Idose	1000	—			
Galactose	0.12	500			
Talose	340	1500			

the D-configuration. Although several of the other monosaccharides do play roles in biochemical processes, they are not stored in quantity and are therefore not readily available from natural sources. Under these circumstances they and those that do not occur naturally have to be obtained synthetically, and several approaches can be considered: (i) enzymic modification of other carbohydrates or enzymic *in vitro* synthesis; (ii) chemical modification of other sugars or synthesis from non-carbohydrate starting materials.

Table 2.5 gives a list of aldo-pentoses and -hexoses and their ketose isomers together with the approximate costs (Sigma catalogue, 1993) and indicates the commercial availability of both D- and L-enantiomers. The prices are not adjusted for scale.

As will be seen (Chapter 8), many natural products contain sugars in structurally modified forms, the simplest being devoid of specific oxygen atoms. L-Rhamnose, for example, is 6-deoxy-L-mannose, and is widely distributed; 2-deoxy-D-ribose is the sugar of the deoxyribonucleic acids (DNAs). Others have carboxylic acid groups while others have amino groups in place of specific hydroxyl groups or have undergone more substantial change in having branched instead of linear carbon chains as the bases of their structures. Some have combinations of these modifications, micro-organisms and their antibiotics (Appendix 5) offering a particularly rich array of monosaccharides with highly modified structures. Since the natural sources of these important compounds infrequently provide suitable access to them in bulk, chemists must seek alternative means of supply, and this requires that appropriate synthetic routes be available.

In this section a brief indication will be given of how monosaccharides are produced in nature, and examples will be provided of synthetic routes to simple and modified sugars both from readily available carbohydrates and from non-carbohydrates sources.

2.5.1 THE BIOSYNTHESIS OF MONOSACCHARIDES AND DERIVATIVES

Photosynthesis, by which sugars are made from carbon dioxide and water utilizing the energy of sunlight, is the primary natural means for returning carbon to plants and hence animals following respiration, combustion and other degradative processes. Its fundamental significance in biology is therefore readily apparent.

Very briefly, in photosynthesis the carbon dioxide is transferred to D-*erythro*-pentulose 1,5-diphosphate **2.91** to give an unstable β-keto acid which hydrolyses into two molecules of D-glyceric acid 3-phosphate **2.92** (Scheme 2.9). Reduction then gives D-glyceraldehyde 3-phosphate at which point the process forks to afford,

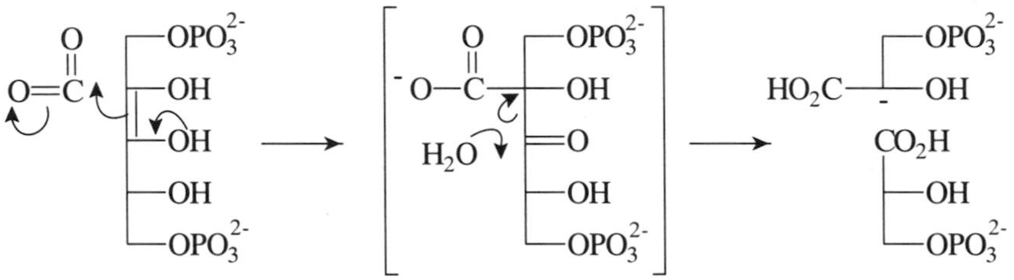

Scheme 2.9

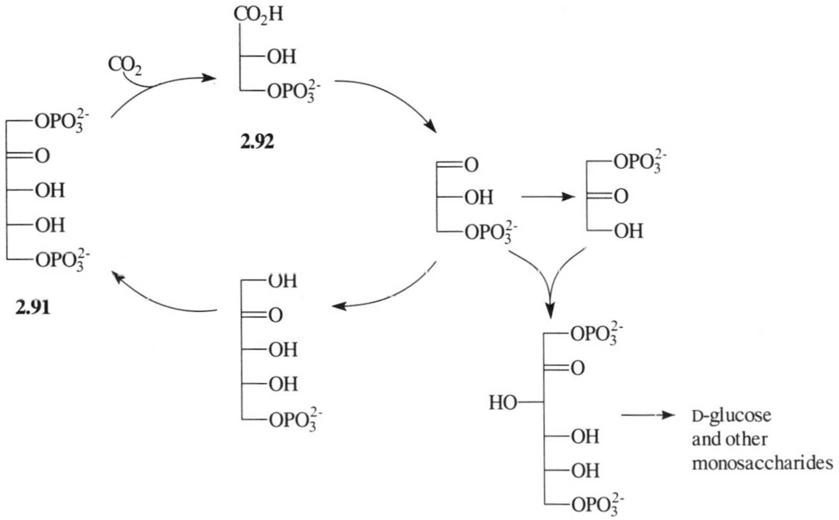

Scheme 2.10

on the one hand, access to the hexoses and, on the other, means of replacing the diphosphate **2.91** utilized initially (Scheme 2.10). The details of these processes were established by Calvin (who was awarded the Nobel Prize in 1961 for the work) and involve several intermediates, including four- and seven-carbon compounds, which are not shown.[38] Each step is catalysed by a specific enzyme and each is recognizable as a well known process in organic chemistry, e.g. aldol condensation. From the D-fructose 1,6-diphosphate **2.93**, D-glucose 1-phosphate and glucose, the most abundant of the monosaccharides, become available.

While the photosynthetic process is employed by plants and some micro-organisms, and is in effect the reverse of the pentose phosphate pathway by which D-glucose phosphate is oxidatively broken down one carbon atom at a time, the reverse of the glycolysis metabolic pathway is used to make glucose by animals and by some micro-organisms. By this process phosphoenolpyruvate is obtained from such feedstock as proteins and fats and biosynthesized into sugars.

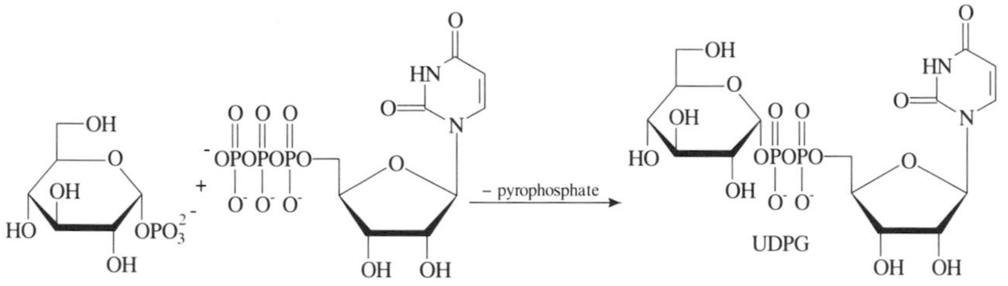

Scheme 2.11

Nature uses a profusion of means for the interconversion of sugars, most of which utilize nucleoside diphosphate sugars, particularly uridine diphosphate glucose (UDPG) formed from α-D-glucopyranose 1-phosphate as shown in Scheme 2.11. In the form of UDPG, D-glucose is enzymically isomerized into D-galactose (epimerization at C-4) and oxidized (at C-6) into D-glucuronic acid and thence converted into D-xylose by decarboxylation. Much more complex changes, however, can be effected as is illustrated by the conversion of D-glucose into L-rhamnose by processes that involve deoxygenation at C-6 as well as configurational changes at C-3, C-4 and C-5 (Scheme 2.12). By such means, therefore, the commonly occurring D-sugars may be converted into members of the much more rare L-series, L-iduronic acid (found in some animal and seaweed polysaccharides, see Section 8.1.3) thus being made from D-galacturonic acid (inversions at C-4 and C-5). L-Arabinose is derived by decarboxylation of the latter acid.

In principle, glycosidic bonds can be formed by the reverse action of glycosidases and phosphorylases, but it is now believed that natural glycosylations also take place primarily by way of the nucleoside diphosphate sugars which act generally as follows

nucleoside DP sugar + ROH $\rightarrow$ sugar–OR + nucleoside DP

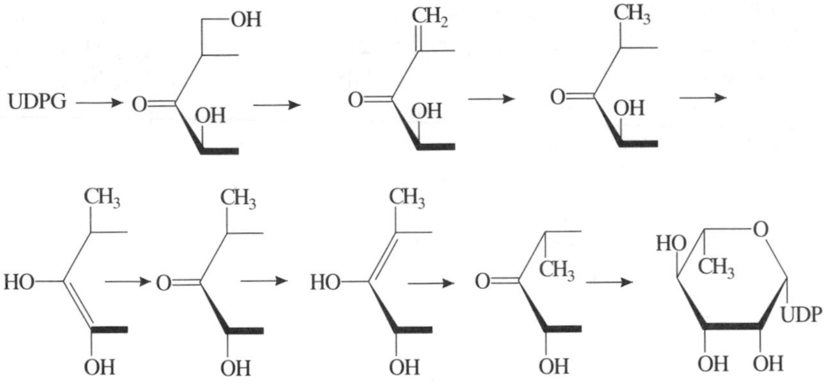

Scheme 2.12

By this general procedure naturally occurring glycosides are produced as are the disaccharides of nature either by glycosylation of free sugars or their phosphates:

$$\text{UDPG} + \text{D-fructose} \rightleftharpoons \text{sucrose} + \text{UDP}$$

$$\text{UDPGal} + \text{D-glucosyl phosphate} \rightleftharpoons \text{lactosyl phosphate} + \text{UDP}$$

Polysaccharide biosynthesis is basically similar but requires an oligomer primer as an acceptor; glycogen synthesis therefore follows the course

$$\text{primer} + \text{UDPG} \rightarrow \text{G}\alpha\text{-primer} + \text{UDP}$$

there being one enzyme present which catalyses the formation of α-(1→4)-bonds and another responsible for glucosylations at position 6. The biosynthesis of cellulose and other polysaccharides is basically similar, UDP being the nucleoside diphosphate used predominantly. However, starch synthesis depends rather on adenosine diphosphate.

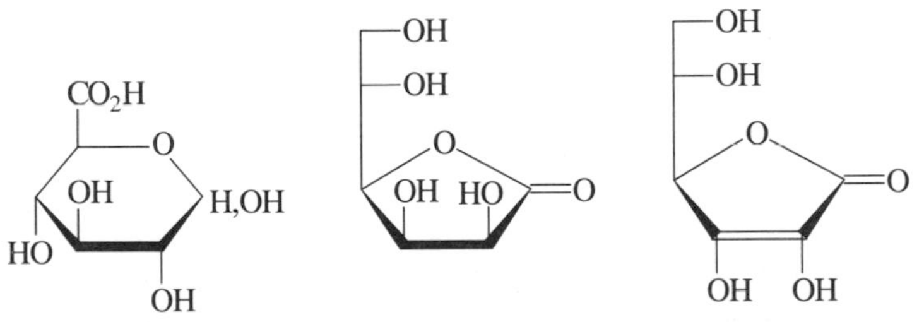

2.94   **2.95**   **2.96** L-Ascorbic acid

Not all modifications of sugars occur at the nucleoside diphosphate sugar level. 2-Acetamido-2-deoxy sugars, for example, are synthesized biochemically from D-fructose 6-phosphate using the amide group of glutamine as a nitrogen source, and the products are acetylated with acetyl coenzyme A. *N*-Acetylneuraminic acid is produced as its 9-phosphate from 2-acetamido-2-deoxy-D-mannose 6-phosphate and phosphoenolpyruvate.

For the biosynthesis of L-ascorbic acid (vitamin C) **2.96** nature again uses D-glucose and then D-glucuronic acid **2.94**. This is converted into the γ-lactone **2.95** of L-gulonic acid following reduction at C-1. Oxidation of **2.95** at C-2 leads to the enediol product.

2.5.2   *EX VIVO* ENZYMIC SYNTHESES OF MONOSACCHARIDES

Although biochemical methods have not been prominent as means of carrying out specific conversions for the purpose of preparing sugars in the laboratory or commercially, it has to be remembered that the enzymic conversions of starch to glucose and of this product to high fructose syrups with enhanced sweetness represent major industrial processes. In the traditional literature of synthetic organic chemistry there is somewhat of a dearth of examples of syntheses of sugars by biochemical procedures, no doubt because enzymes of adequate purity are difficult to obtain and few whole organisms can be employed to effect appropriate conversions. There is, however, a notable exception which is not strictly *ex vivo* in that the bacterium *Acetobacter suboxydans* in aqueous suspension can oxidize the central hydroxyl group of L-*erythro*-triols at the 'head' (of Fischer projections) of alditols (i.e. with D-*erythro*-triols at the 'foot'). By this means, therefore, D-fructose can be obtained from D-mannitol, but of more value since the pentuloses are not readily available by other means are the efficient conversions of D-arabinitol **2.97** and ribitol **2.99** to D-*threo*-pentulose (D-xylulose) **2.98** and L-*erythro*-pentulose (L-ribulose) **2.100**, respectively (Scheme 2.13), on a multigram scale[39] (See also Section 4.9.1.a.iii).

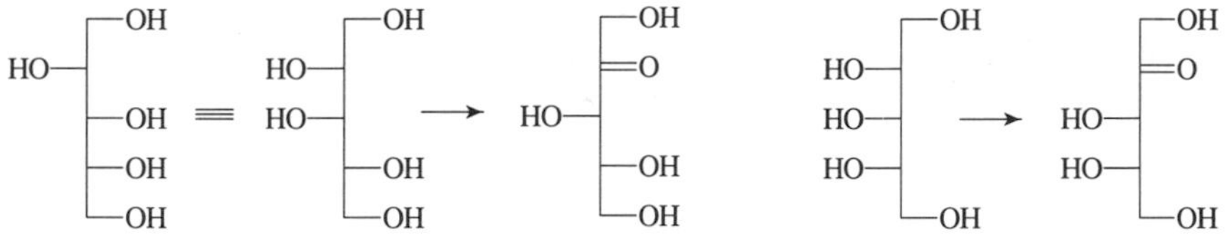

Scheme 2.13

A significant recent development, facilitated greatly by major progress in genetic engineering techniques, has been the *in vitro* application of enzymes of carbohydrate metabolism to the synthesis of monosaccharides.[40] Related work concerns regioselective acylation and deacylation (Section 5.3.3), oxidation and glycosylation processes[41,42] involving sugar derivatives, and it is clear that enzymic and organic chemical methods can now be considered as complementary partners for effecting chemical change in monosaccharide synthetic chemistry.

Free sugars can be synthesized by use of isolated enzymes applied to such simple compounds as dihydroxyacetone **2.101** which can be made by successive additions of hydrogen cyanide to formaldehyde and the derived glycolaldehyde. The carbon chains of the monosaccharides can therefore be assembled effectively atom by atom. Enzymic phosphorylation of dihydroxyacetone can be carried out using a kinase in the presence of a small proportion of adenosine triphosphate, which is the required cofactor, together with phosphoenolpyruvate used in an equimolar amount as the phosphate source. By application of a triose phosphate isomerase the ester **2.102** can be equilibrated with D-glyceraldehyde 3-phosphate **2.103** which is present in the equilibrium in only small amounts; however, catalysed by an aldolase, it reacts stoichiometrically with **2.102** to give D-fructose 1,6-diphosphate **2.104** in high yield. Partial acid-catalysed hydrolysis then affords the 6-phosphate **2.105** from which D-fructose can be obtained with a phosphatase, and also D-glucose 6-phosphate **2.106** and then D-glucose itself by successive uses of an isomerase and a phosphatase (Scheme 2.14).[43]

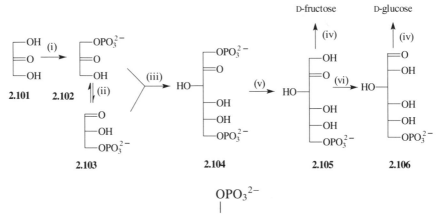

(i) ATP, regenerated *in situ* from $HO_2CC{=}CH_2$, kinase; (ii) isomerase; (iii) aldo-lase; (iv) phosphatase; (v) H_3O^+; (vi) isomerase

Scheme 2.14

While the above syntheses afford sugars which are much more accessible by other means, the methods are extendable to the preparation of rare naturally occurring compounds and also to non-natural analogues. Thus, application of the aldolase step of Scheme 2.14 with D-erythrose as the aldol electrophile gives access to the seven-carbon sedoheptulose **2.107** which is present as an intermediate in the photosynthesis cycle in phosphorylated forms. Structures **2.108–2.110** illustrate the other types of compounds which can be made by use of the respective aldehydes glyoxal, butanal and phenylacetaldehyde.[40]

Furthermore, dihydroxyacetone ^{13}C-labelled at C-2, for example, according to Scheme 2.14 gives the hexoses labelled specifically at C-2 and C-5, and this technology can be adapted to the preparation of an extensive range of specific isotopically labelled sugars.

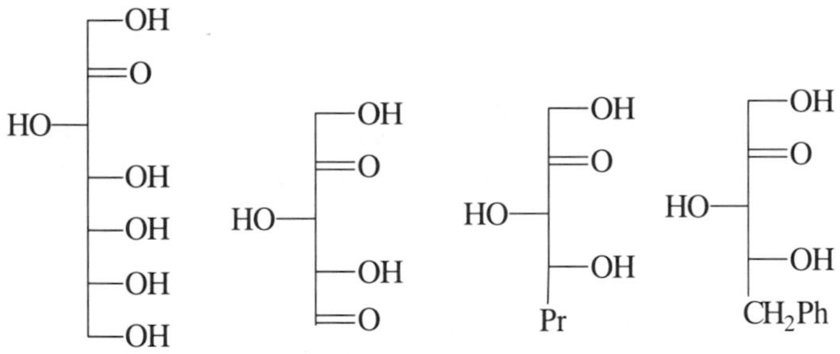

2.107   **2.108**   **2.109**   **2.110**

Transfer of the hydroxyacetyl group ($COCH_2OH$) from hydroxypyruvate to alde-hydes affords an alternative enzyme-catalysed synthesis of ketoses.[44]

In a related manner, other enzymes can be used to transfer the three carbon atoms of phosphoenolpyruvate, which acts as a source of an aldol nucleophile, to aldehydes and thereby provide simple access to 3-deoxyulosonic acids, the synthesis of 3-deoxy-D-*manno*-2-octulosonic acid (Kdo) 8-phosphate **2.111** being illustrated in Scheme 2.15.[45]

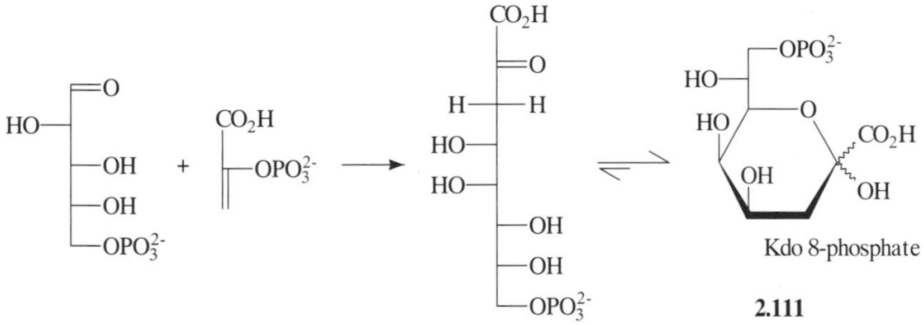

Scheme 2.15

2.5.3   SYNTHESIS OF MONOSACCHARIDES FROM OTHER SUGARS

The abundant naturally occurring sugars serve as obvious potential sources for the rarer members of the monosaccharide family, and many interconversion methods are available. This section, in the main, provides cross references to discussions of relevant reactions.

(a)   Isomerizations

(i)   *Epimerizations*

Under basic conditions aldoses isomerize to their C-2 epimers and the corre-sponding ketoses (Section 3.1.7.a), and specific conditions may be applied for the preparation of particular products.[46] When the hydroxyl groups at C-2 of the initial compounds are O-substituted no ketose can be formed, and alternatively epimerization without ketose involvement can be induced by use of acid molybdate in a remarkable reaction which causes C-1–C-2 interchange (Section 3.1.8.b).

A further acidic method of causing specific inversion of configuration at C-2 of aldoses involves acetolysis of furanose derivatives which have the *cis*-configuration at C-2 and C-3 (Section 3.2.1.a.ii).

(ii) Oxidation and reduction processes

Inversion of specific secondary alcohol centres can, on occasion, be effected very efficiently by oxidation to the ketones followed by reduction (Section 4.9.1.b.ii).[47] β-D-Mannopyranosides are thus obtainable from the more accessible β-D-glucopyranosides by this approach. D-Mannose can be converted to D-fructose by an oxidation strategy by way of 1,3,4,6-tetra-O-acetyl-2,5-O-methylene-D-mannitol and 1,3,4,6-tetra-O-acetyl-5-O-formyl-D-fructose.[48]

(iii) Displacement reactions

Direct configurational inversions at secondary alcohol sites can be effected under Mitsunobu conditions (Section 4.1.2.c), but much more commonly alcohols are substituted by sulphonylation prior to displacement of the sulphonyloxy groups by oxygen nucleophiles (Section 4.1.2.a.ii).

The formation of cyclic acyloxonium ions on sugar rings or chains can provide a means of inverting configuration. Frequently they are produced from vicinal *trans*-related acyl sulphonyl diesters and give access to diol products with *cis*-configurations (Section 4.1.2.b.ii). A related example of this type, which could prove of considerable synthetic value, utilizes carbamoyl group participation and has been used by Kunz[49] to prepare β-mannopyranoside derivatives from the more readily available β-glucopyranosides. With compounds containing acyl esters on several contiguous *trans*-related alcohol groups, acyloxonium ions may rearrange in a relayed manner such that several configurational inversions occur. Conversion of penta-O-acetyl-β-D-glucopyranose to penta-O-acetyl-α-D-idopyranose, which involves inversions at C-1–C-4, is the ultimate example (Section 3.2.1.a.ii).

Formation of epoxides from *trans*-related diols (Section 4.6.1.a) followed by nucleophilic ring opening (Section 4.1.3) frequently affords a way of bringing about inversion at two contiguous centres, D-glucose derivatives being convertible to D-altrose compounds, for example, by this approach.

(iv) Addition reactions

Since most elimination reactions involving monosaccharide derivatives cause some loss of asymmetry, it follows that reciprocal addition processes applied to unsaturated derivatives can result in products which are stereoisomers of the starting materials. Thus, for example, D-talose can be made from D-galactose via D-galactal by peroxidation followed by hydrolysis (Section 4.10.1.b.i). Similarly, D-mannose compounds are the products of *cis*-hydroxylation of 2,3-enes readily obtainable from D-glucose.

Reductions of fully substituted alkenes can also give rise to products with inverted configurations at two centres, D-gulose being obtainable from D-glucose (i.e. inversions at C-3 and C-4) by the sequence[50] 1,2:5,6-di-*O*-isopropylidene-D-glucose → the corresponding 3-ulose → the 3-enol acetate → 1,2:5,6-di-*O*-isopropylidene-D-gulose (sodium borohydride reduction) (see Scheme 4.137 for the first part of this transformation).

(b)   Chain-extending reactions

The carbonyl group offers excellent opportunities for extension of the sugar chains and the formation of 'higher' sugars. Several methods for effecting these chain extensions, the classical one being the Kiliani reaction with cyanide, are covered in Section 3.1.4.d.

(c)   Chain-shortening reactions

Several methods for cleaving the carbon chains of monosaccharides and thus producing 'lower' sugars are covered in later sections: the Wohl degradation of acetylated aldose oximes (Section 3.1.3.b); reactions involving the loss of C-1 from aldonic acids (Section 3.1.6.a.iv); periodate ion cleavage of vicinal diols (Section 5.5); and photodecarbonylation of dicarbonyl sugar derivatives (Section 4.9.1.b.iii).

2.5.4   SYNTHESIS OF MONOSACCHARIDES FROM NON-SUGARS[51]

When the structures of naturally occurring substances are elucidated they become targets for synthesis, and the sugars, notwithstanding their stereochemical complexities, have been no exception. Quite extraordinarily, given these complexities and the paucity of laboratory methods available at the time, Fischer reported the preparation of some of the hexoses in enantiomerically pure form in the late 1800s, just after their structures had been established. By treating 2,3-dibromopropanal with barium hydroxide the German chemists prepared a mixture of racemic hexoses ('acrose', after acrolein used for the starting material), presumably by way of glyceraldehyde which underwent aldol condensation with its tautomer dihydroxyacetone (Scheme 2.16). From the mixed products DL-glucosazone was isolated and from it DL-fructose, the D-enantiomer of which was specifically fermented by yeast, and thus L-fructose, still an extremely rare sugar, was made.[52]

Also from DL-fructose, D-glucose was produced by the sequence DL-mannitol → DL-mannose → DL-mannonic acid (which was resolved by way of strychnine salts) → D-gluconic acid. Carried out with only simple reagents and classical separatory methods, these achievements were prodigious. Interestingly, and of course unknown to Fischer, this approach is almost biomimetic (Section 2.5.1).

Several racemic sugars have been made more recently, many by non-biomimetic methods, the alkene **2.112**, for example, giving access to DL-ribose and -arabinose

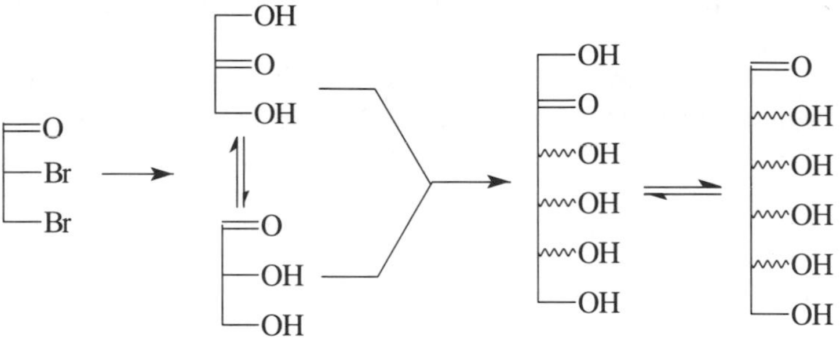

Scheme 2.16

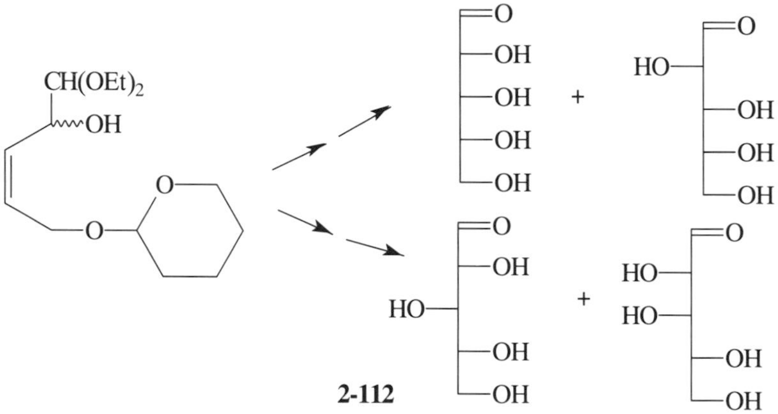

Scheme 2.17

by *cis*-hydroxylation and DL-xylose and -lyxose by epoxidation followed by hydrolysis (Scheme 2.17)[53].

Pyran derivatives have been recognized as attractive starting materials for syntheses of free sugars. For example, the acrolein dimer **2.113** has been converted into the bicyclic **2.114** and hence into DL-glucose.[54] More significantly, the (*S*)-form of the furan derivative **2.115** can be ring expanded to the pyranosides **2.116** by treatment with bromine in methanol, and then, by epoxidation of the double bond, reduction of the carbonyl group, epoxide ring opening and removal of the protecting groups, converted into the natural D-form of glucose.[55]

More extensive use in sugar synthesis has been found for the readily available enantiomers of 7-oxabicyclo[2.2.1]hept-5-enyl derivatives termed 'naked sugars' by Vogel, who has exploited extensively the high stereoselectivity of addition and substitution reactions in this series of bicyclic compounds. For example, the silyl enol ether **2.117** was converted by use of *m*-chloroperbenzoic acid to the *exo*-epoxide which underwent ring opening with the *m*-chlorobenzoic acid by-product. Heating of the resulting silyl hemiacetal gave ketone **2.118** which, with the same

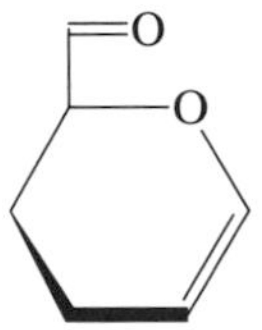

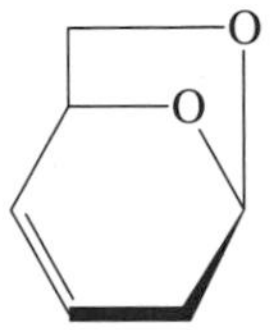

2.113   **2.114**

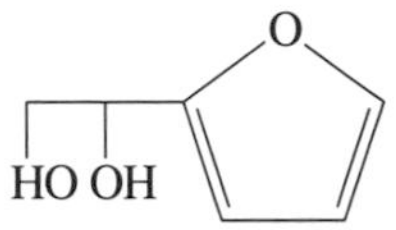

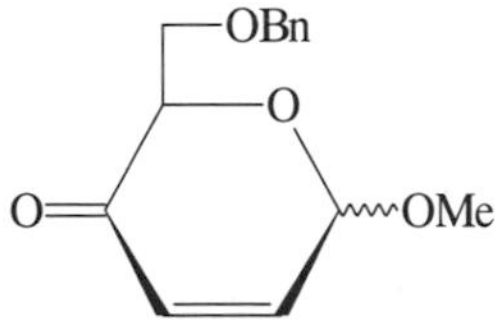

2.115   **2.116**

peracid, afforded the Baeyer–Villiger product **2.119**. Base-catalysed methanolysis of the lactone produced the furanuronate **2.120** unprotected at C-1 which on conversion to the methyl glycoside, reduction of the ester and acidic hydrolysis gave L-allose **2.121** in approximately 35% overall yield from **2.117** (Scheme 2.18).[56] Extension of this work has led to syntheses of optically pure octoses and to a range of sugar derivatives.[57]

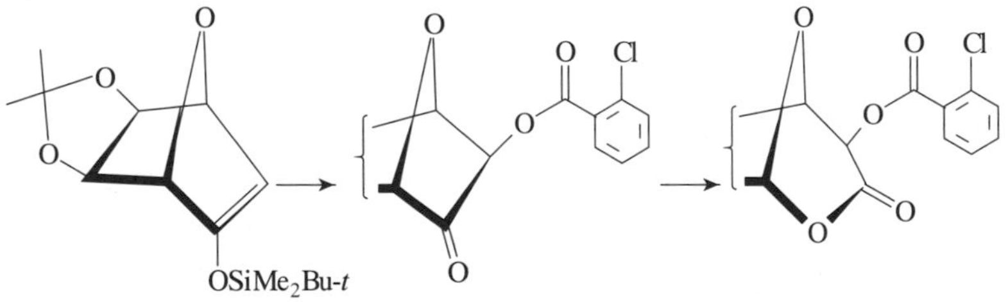

2.117   **2.118**   **2.119**

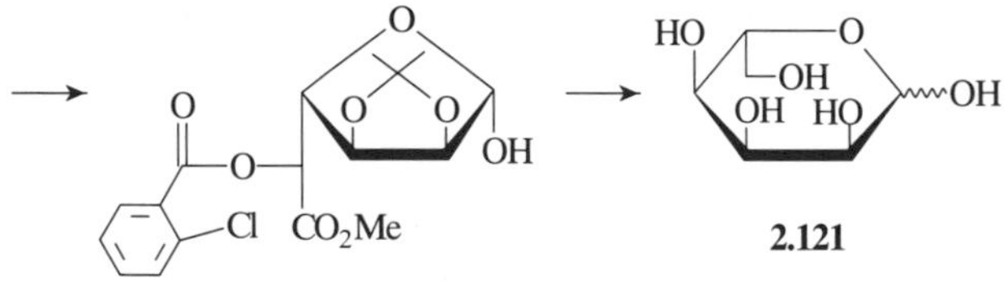

2.120

Scheme 2.18

An outstanding application of asymmetric induction, by which asymmetric reagents are used to introduce asymmetry, and which has led to a general synthesis of enantiomerically pure monosaccharides, involves the Sharpless epoxidation of allylic alcohols by use of titanium(IV) isopropoxide, *t*-butyl hydroperoxide and a single enantiomer of a dialkyl tartrate.[58] This reagent epoxidizes the double bond in

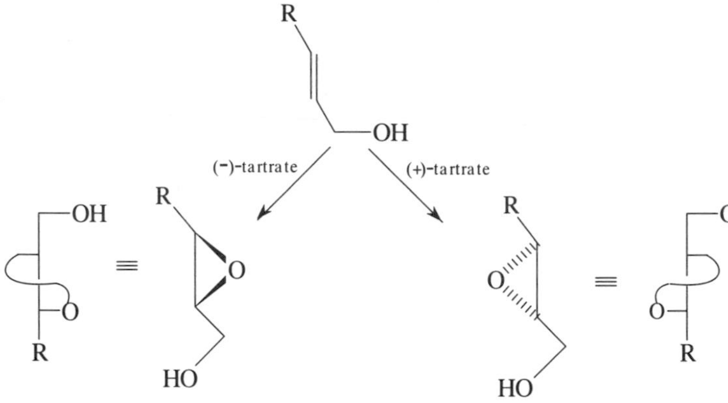

2.122 2.123

Scheme 2.19

an enantioselective fashion which can be predicted by a simple empirical selectivity rule, since when the allyl alcohol is viewed as shown in Scheme 2.19, with the hydroxymethyl group at the lower right-hand side of the diagram, the reagent delivers the epoxide oxygen atom to the upper face of this molecule when the L-(−)-tartrate is employed and to the lower face if the D-(+)-tartrate is used. Thus, in effect, each epoxidation provides a simultaneous means of introducing, at the site of the double bond, two hydroxylated asymmetric centres with the *threo*-configuration. The D- and L-*threo*-systems **2.122** and **2.123** are therefore derived using (−)- and (+)-diethyl tartrate, respectively. By use of an initial substituted 1,2-dihydroxyethyl group for R the synthesis leads to pentitols, and pentoses can be obtained following the oxidation of the alcohol centres of the initial epoxides (e.g. **2.122**). One way to accomplish this is to treat the epoxide products with sodium thiophenate which, for example, reacts with **2.122** to give the 1-thio compounds **2.125** by way of the terminal epoxides **2.124**. Treatment of the derived isopropylidene derivatives **2.126** with *m*-chloroperbenzoic acid in acetic anhydride gives the aldose derivatives **2.127**, and with diisobutylaluminium hydride these result in the 2,3-D-*erythro*-sugars **2.128** (Scheme 2.20).

2.122

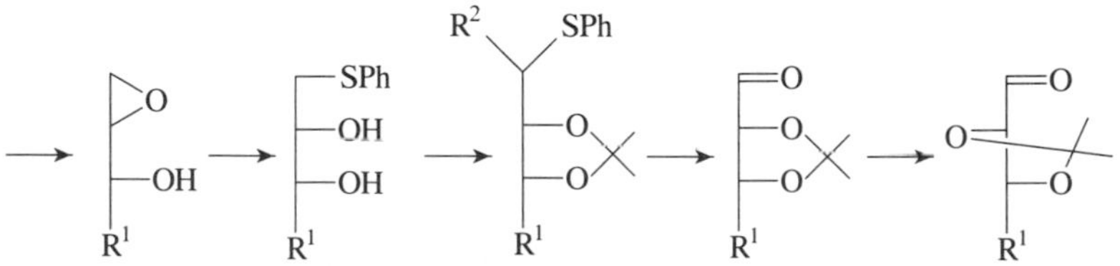

2.124 2.125 2.126 R^2 = H 2.127 R^2 = OAc 2.128 2.129

Scheme 2.20

Since the five-membered rings of compounds **2.128** have the adjacent substituents CHO and R^1 in the unfavourable *cis*-relationship (Section 2.3.2) they epimerize almost completely at C-2 in basic conditions to the *trans*-related D-*threo*-aldoses **2.129** and, from these, sugars with the D-*threo*-configurations at C-2 and C-3 are produced. From the epoxides **2.122**, therefore, aldoses with the structural components **2.130** and **2.131** are derivable, and the alternative epoxides **2.123**, likewise, give access to **2.132** and **2.133** (Scheme 2.21).

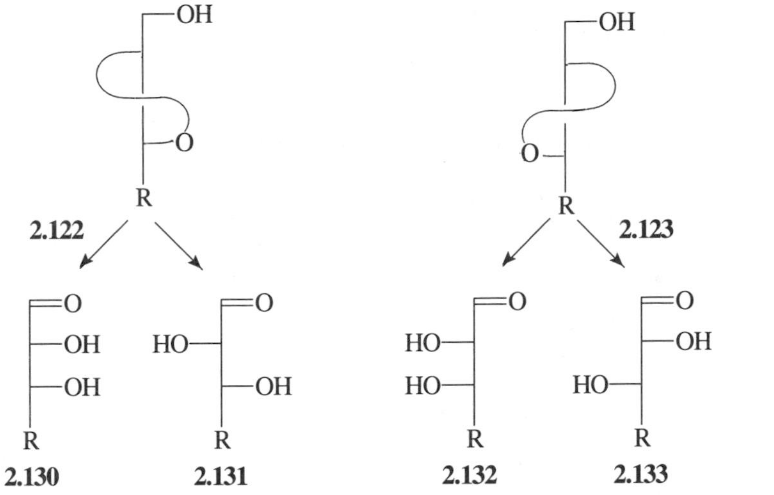

Scheme 2.21

Syntheses of all eight L-aldohexoses have been carried out by Sharpless's group starting from epoxide **2.123** (R = Ph_2CHOCH_2).[59] When the 2,3-*O*-isopropylidene tetroses (acetals of **2.132** and **2.133**, R = Ph_2CHOCH_2) were obtained, two-carbon additions were applied by Wittig procedures to give new allylic alcohols (2,3-dideoxyhex-2-enitol derivatives) which are available for second asymmetric epoxidations. Application of the Sharpless procedure to the synthesis of deoxy sugars is noted in Section 4.2.1.g.

The chemical synthesis of sugars, ranging from Fischer's to the above beautiful achievement with the L-aldohexoses, has therefore produced outstanding chemistry since the late 1800s, and dramatically illustrates the development of methodology in that time. Modified sugars of the kind commonly found in microbiological products can be made by adaptations of the methods illustrated above, but normally are produced by functional manipulation of specific derivatives of the monosaccharides available from natural sources.

2.6   REFERENCES

1. *Angew. Chem., Int. Ed. Engl.*, 1992, **31**, 1541.
2. *Nature*, 1951, **168**, 271.

3. *Science*, 1965, **147**, 1038.
4. *Acta Crystallogr.*, 1968, **B24**, 830.
5. (a) *J. Org. Chem.*, 1960, **25**, 701; (b) *Can. J. Chem.*, 1961, **39**, 2069.
6. *Adv. Carbohydr. Chem.*, 1960, **15**, 11.
7. *Adv. Carbohydr. Chem. Biochem.*, 1984, **42**, 15.
8. *J. Org. Chem.*, 1992, **57**, 573.
9. *Adv. Carbohydr. Chem.*, 1955, **10**, 1.
10. J. F. Stoddart, *Stereochemistry of Carbohydrates*, Wiley–Interscience, New York, 1971.
11. *Adv. Carbohydr. Chem. Biochem.*, 1971, **26**, 49.
12. *Experientia*, 1950, **6**, 316; E. L. Eliel *et al. Conformational Analysis*, Wiley-Interscience, New York, 1965.
13. *J. Chem. Soc., Perkin Trans. 2*, 1993, 1061.
14. *Chem. Ind.*, 1955, 1102.
15. Abstr. Pap. *Am. Chem. Soc.*, 1958, **133**, 31N; *Molecular Rearrangements* (ed. P. de Mayo), Wiley-Interscience, New York, 1964, p. 733; *Can. J. Chem.*, 1987, **65**, 213.
16. *Adv. Carbohydr. Chem.*, 1989, **47**, 45; *Tetrahedron*, 1992, **48**, 5019.
17. *Chem. Rev.*, 1987, **87**, 1047.
18. *J. Chem. Soc., Chem. Commun.*, 1973, 505; *Pure Appl. Chem.*, 1981, **53**, 1901.
19. *J. Chem. Soc., Perkin Trans. 1*, 1980, 2767.
20. (a) *Can. J. Chem.*, 1965, **43**, 2205; (b) *Pure Appl. Chem.*, 1971, **27**, 527; (c) *Chem. Ber.*, 1974, **107**, 2626.
21. *Carbohydr. Res.*, 1969, **10**, 565.
22. *Can. J. Chem.*, 1969, **47**, 4441.
23. *Adv. Carbohydr. Chem.*, 1951, **6**, 107.
24. *Angew. Chem., Int. Ed. Engl.*, 1969, **8**, 157.
25. *Can. J. Chem.*, 1979, **57**, 2504.
26. *Acta Crystallogr.*, 1971, **B27**, 2393.
27. *Can. J. Chem.*, 1977, **55**, 141.
28. *Can. J. Chem.*, 1969, **47**, 4427.
29. *Can. J. Chem.*, 1987, **65**, 213.
30. *J. Carbohydr. Chem.*, 1988, **7**, 239; *J. Carbohydr. Chem.*, 1990, **9**, 287, 601.
31. *J. Org. Chem.*, 1990, **55**, 5784.
32. *J. Biol. Chem.*, 1954, **208**, 293.
33. *J. Am. Chem. Soc.*, 1958, **80**, 3166.
34. *Adv. Carbohydr. Chem.*, 1968, **23**, 11.
35. *Adv. Carbohydr. Chem. Biochem.*, 1969, **24**, 13.
36. *Adv. Carbohydr. Chem. Biochem.*, 1984, **42**, 15.
37. *Adv. Carbohydr. Chem. Biochem.*, 1991, **49**, 19.
38. *J. Chem. Soc.*, 1956, 1895.
39. *Biochem. J.*, 1962, **83**, 8.
40. *Tetrahedron*, 1989, **45**, 5365; C.-H. Wong and G.M. Whitesides, *Enzymes in Synthetic Organic Chemistry*, Elsevier Science Ltd, Oxford, 1994.
41. *Synthesis*, 1991, 499.
42. *Acc. Chem. Res.*, 1992, **25**, 307.
43. *J. Org. Chem.*, 1983, **48**, 3199.
44. *Tetrahedron Lett.*, 1987, **28**, 5525.
45. *Tetrahedron Lett.*, 1988, **29**, 427.
46. *Carbohydrate Chemistry* (ed. J. F. Kennedy), Oxford University Press, Oxford, 1988, p. 382.
47. *Rodd's Chemistry of Carbon Compounds* (ed. M.F. Ansell), Vol. 1F, G Supplement, Elsevier, Amsterdam, 1983, p. 75.
48. *Aust. J. Chem.*, 1971, **24**, 1219.

49. *Carbohydr. Res.*, 1992, **228**, 217.
50. *Carbohydrate Chemistry* (ed. J. F. Kennedy), Oxford University Press, Oxford, 1988, p. 395.
51. (a) *Adv. Carbohydr. Chem. Biochem.*, 1982, **40**, 1; (b) *Rodd's Chemistry of Carbon Compounds* (ed. M. Sainsbury), Vol. 1F, G, H 2nd Supplement, Elsevier, Amsterdam, 1993, p. 213; *Carbohydrate Chemistry* (ed. J. F. Kennedy), Oxford University Press, Oxford, 1988, p. 381.
52. *Chem. Ber.*, 1890, **23**, 370.
53. *Chem. Pharm. Bull.*, 1961, **9**, 316, 492.
54. *Can. J. Chem.*, 1971, **49**, 3342.
55. *Carbohydr. Res.*, 1977, **55**, 165.
56. *Helv. Chim. Acta*, 1989, **72**, 278.
57. *J. Org. Chem.*, 1991, **56**, 1133.
58. *Chem. Br.*, 1986, **22**, 38; *Angew. Chem., Int. Ed. Engl.*, 1985, **24**, 1.
59. *Science*, 1983, **220**, 949.

3 Reactions and Products of Reactions at the Anomeric Centre

As would be expected on the grounds of their hemiacetal character, the anomeric centres, i.e. C-1 of aldoses and C-2 of 2-ketoses, are the most reactive sites within monosaccharide molecules. The extensive range of reactions undergone by sugars at these positions is discussed in this chapter, as are reactions of such derivatives as glycosides, thioglycosides, glycosyl esters and halides, which show specific reactivity at the anomeric centre because cleavage of the anomeric substituents generates carbocations stabilized by mesomeric release of electrons from adjacent oxygen atoms, as exemplified in Scheme 3.1[†] where Z = RO, RS, RCO$_2$ or halogen, respectively.

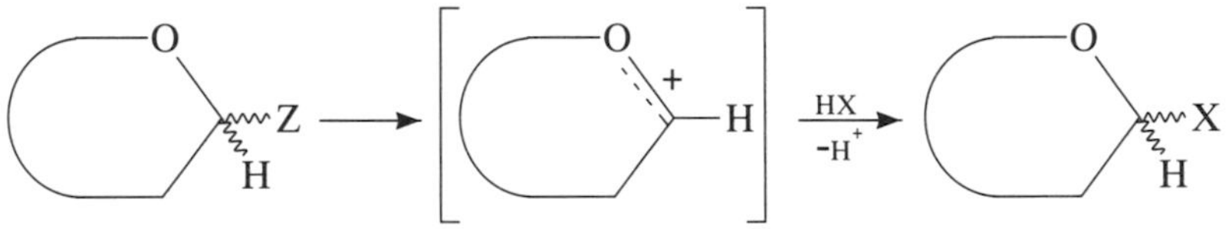

Scheme 3.1

3.1 REACTIONS OF FREE SUGARS

Free sugars can mutarotate in solution to give, ultimately, equilibrium mixtures of five (or more) modifications: the α- and β-furanoses, the corresponding pyranoses and the acyclic forms already discussed (Section 2.4.1). In addition to these tautomeric changes in the presence of acid, intermolecular displacements of the anomeric hydroxyl groups can occur as for the glycosyl derivatives illustrated in Scheme 3.1. However, an alternative route involving stabilized acyclic carbocations may be involved in these displacement reactions, in which case subsequent ring closures can incorporate hydroxyl groups other than those released during ring opening (Scheme 3.2a). In other cases, products appear to arise by reactions involving acyclic carbonyl species, as illustrated in Scheme 3.2(b).

[†] Concerted reactions can take place, although since the ring oxygen atom facilitates the departure of leaving groups from C-1 by its mesomeric influence, it also retards the approach of nucleophiles and, consequently, S_N2 reactivity is not so pronounced as the S_N1 type.

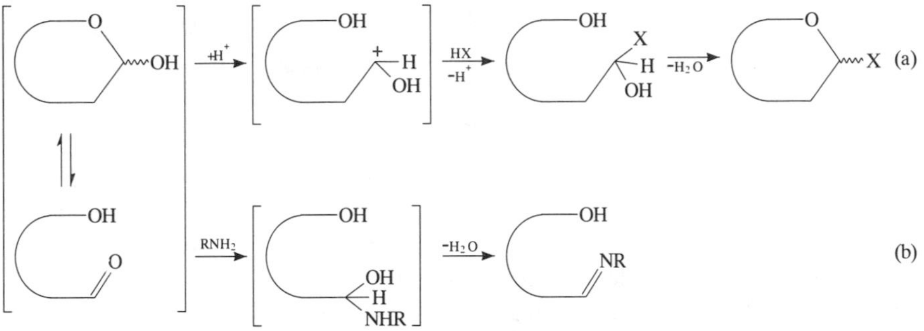

Scheme 3.2

3.1.1   REACTIONS WITH ALCOHOLS — GLYCOSIDES[1-6]

(a)   Preparation of glycosides

(i)   Fischer glycosidation — general features

When sugars are treated with alcohols in the presence of acid catalysts, *glycosides* are formed. This represents one of the easiest means for preparing simple glycosides, but for synthesizing more complex members of this series, of the type widely distributed in nature, the following precursors are usually used: pentenyl glycosides (Section 3.1.1.b.vi), thioglycosides (Section 3.1.2.c.ii), glycosyl sulphoxides (Section 3.1.2.c.iv), glycosyl esters (Section 3.2), glycosyl halides (Section 3.3), glycosyl 1,2-orthoesters (Section 3.2.2), glycosyl 1,2-oxazolines (see **3.360**, Section 3.3.2.a.ii), glycose 1,2-anhydrides (Section 3.1.1.c) and unsaturated derivatives (Section 4.10).

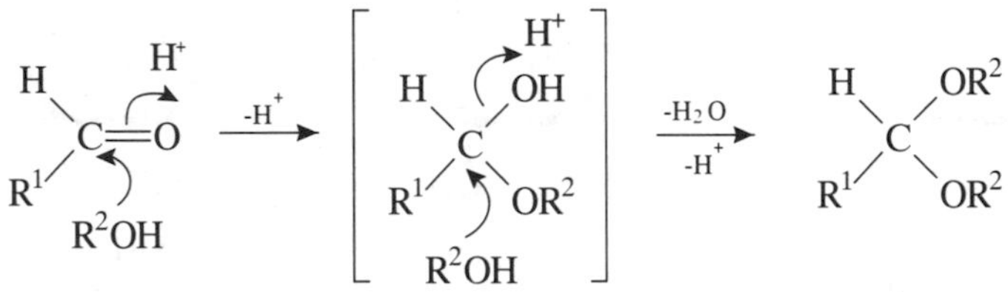

Scheme 3.3

Aldehydes react with alcohols in the presence of acid catalysts to give, initially, hemiacetals and subsequently full acetals (Scheme 3.3). However, certain hydroxyaldehydes with suitably located hydroxyl groups behave differently, forming intramolecular cyclic hemiacetals spontaneously (Section 2.1.3) and yielding mixed acetals upon treatment with acidified alcohols. 5-Hydroxypentanal (Section 2.1.3), for example, on reaction with methanol, gives 2-methoxytetrahydropyran (Scheme 3.4).[7] Sugars behave in a similar but more complex fashion, since more than one hydroxyl group is suitably positioned to participate in hemiacetal formation, and mixtures of acetals (glycosides with

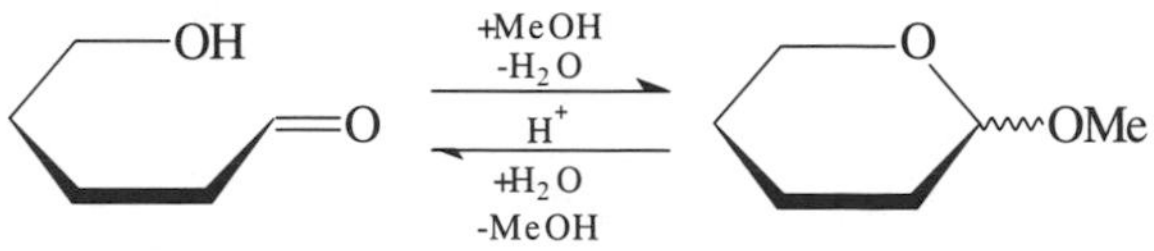

Scheme 3.4

different ring sizes) can result. This reaction is known as the *Fischer glycosidation*; six-membered cyclic products are called *glycopyranosides* and the five-membered ring isomers *glycofuranosides*. The non-carbohydrate parts derived from alcohols in these molecules are called the *aglycons*, and the carbohydrate parts *glycosyl units*, with the two components being linked by oxygen atoms.

When galactose, for example, is heated under reflux in methanol containing anhydrous hydrogen chloride (2%) until equilibrium is attained (12 h), pure crystalline methyl α-D-galactopyranoside **3.1** (α-anomer) can be isolated in 41% yield as its monohydrate by crystallization from water.[8] The mother liquors from this reaction comprise a mixture of the major product **3.1** (α-anomer), the β-pyranoside **3.1** (β-anomer) and the α- and β-furanosides **3.2** (Scheme 3.5). As with the equilibration of free sugars (Section 2.4.1), all monosaccharides do not give the same proportions of glycosides at equilibrium (see below and Table 3.1). However, despite the complexities of the products, the alcoholysis of sugars represents a suitable means of obtaining many simple glycosides. Usually the predominating isomers in the equilibrated mixtures are most easily isolated in crystalline form, but this is not always so, since the methyl D-xylosides, for example, give the crystalline β-pyranoside most readily in spite of the fact that the α-anomer is present to the extent of 65%.

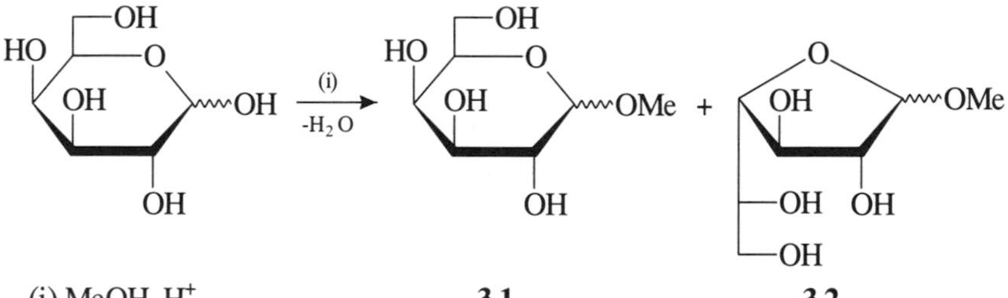

Scheme 3.5

Preparative scale Fischer glycosidations are often carried out in the presence of ion exchange resins, such as Amberlite IR-120 or Dowex 50, which act as the acid catalyst; thereby, troublesome acid neutralizations and resulting salt formation at the end of the reactions are avoided.[9]

Distribution of α- and β-furanosides and -pyranosides at equilibrium in the Fischer glycosidation   The proportions of the various glycosidic forms present in the equilibrium mixtures at the completion of Fischer glycosidations depend upon the relative thermodynamic stabilities of the isomers, and the factors governing these

Table 3.1 The percentage compositions of methyl glycoside mixtures at equilibrium in methanol (35 °C)

Sugar	α-Furano-side (%)	β-Furano-side (%)	Total furano-side (%)	α-Pyrano-side (%)	β-Pyrano-side (%)	Total pyrano-side (%)
D-Ribose	5	17	22	12	66	78
D-Arabinose	22	7	29	24	47	71
2-O-Methyl-D-arabinose	—	—	67	—	—	33
3-O-Methyl-D-arabinose	—	—	51	—	—	49
2,3-Di-O-methyl-D-arabinose	—	—	75	—	—	25
D-Xylose	2	3	5	65	30	95
2-O-Methyl-D-xylose	3	10	13	58	29	87
3-O-Methyl-D-xylose	3	6	9	64	27	91
2,3-Di-O-methyl-D-xylose	4	12	16	54	30	84
D-Lyxose	1	0	1	89	10	99
D-Glucose	0.6	0.9	1.5	66	32.5	98.5
D-Mannose	0.7	0	0.7	94	5.3	99.3
D-Galactose	6	16	22	58	20	78

are similar to those which control the modifications of aldoses present at equilibrium in aqueous solution (Section 2.4.2). Examination of Table 3.1 reveals that for unsubstituted sugars the aldopyranosides are more prevalent than the furanosides and the dialkyl acetals which could have been formed are present in no more than small amounts.[10] As with free sugars, it is possible to rationalize the proportions of glycosides found at equilibrium for different sugars. For example, D-glucopyranosides **3.3** are favoured because these can adopt the $^4C_1$ conformation in which the ring substituents at C-2–C-5 are all equatorially arranged. The furanosides **3.4**, on the other hand, possess an unfavourable *cis*-interaction between the substituents at C-3 and C-4 and are consequently relatively unstable. The D-galactopyranosides **3.5** in the $^4C_1$ conformation have an axial group at C-4 but the furanosides **3.6** have the vicinal *trans*-relationship for the C-2, C-3 and C-4 groups, and therefore more of these glycosidic modifications would be expected to be present than in glucose. Although the D-mannopyranosides **3.7** have an axial C-2 group in the $^4C_1$ conformation, this ring size is still strongly preferred because the furanosides **3.8** are very unstable, having *cis*-interactions between all the substituents at C-2–C-4.

The anomeric equilibrium ratio of the pyranosides can be rationalized by considering the α- and β-anomers in their most stable chair conformations. The methyl D-glucopyranosides **3.3** adopt the $^4C_1$ conformation for the reasons given above and, because of the anomeric effect (see Section 2.3.1.b), the α-anomer with the axial methoxy group predominates.[†] The methyl mannopyranosides **3.7** show an

---

[†] This contrasts with the case of D-glucose in aqueous solution (Table 2.4), for which the β-compound is the more stable, and the difference is attributed mainly to the solvent change. Thus the anomeric effect is more powerful in the less polar methanol than in water, and the effective size of the methoxy group in methanol is less than that of the hydroxyl group in water. On both accounts the axial anomer of the glycoside will be favoured relative to that of the free sugar.

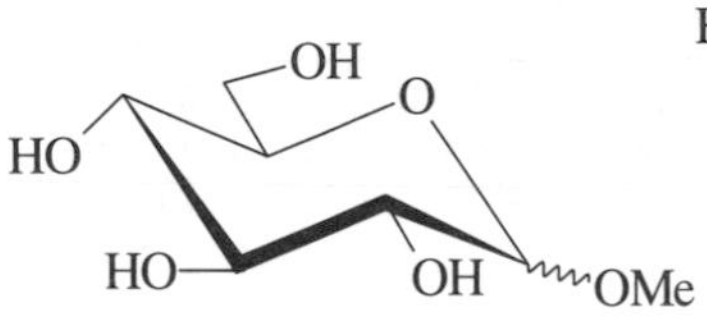

3.3 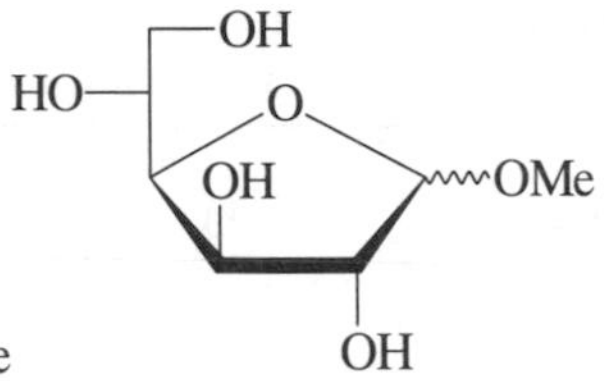   **3.4**

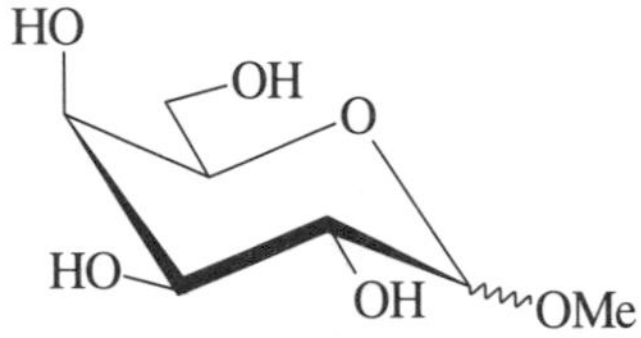

3.5 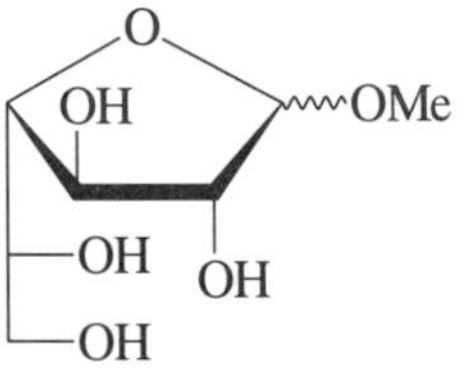   **3.6**

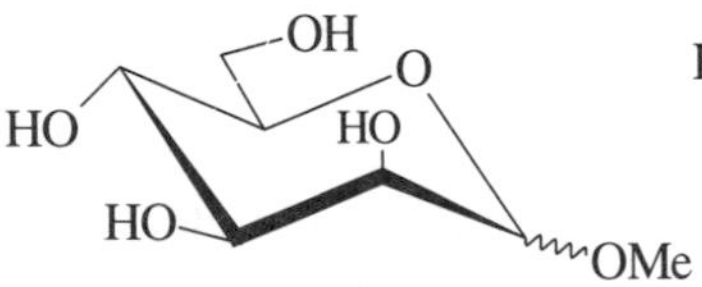

3.7 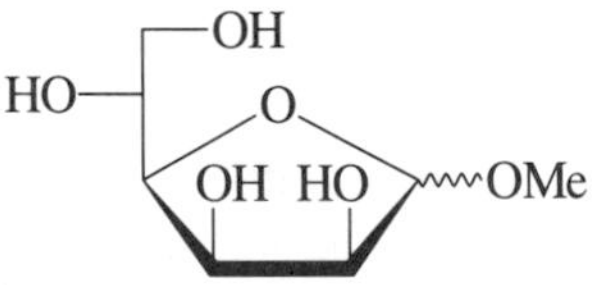   **3.8**

even larger α/β ratio because the substituent at C-2, which is axial in the preferred $^4C_1$ conformation, stabilizes further the α-anomer owing to a stereoelectronic interaction between the bonds at C-1 and C-2, and it further destabilizes the β-anomer (equatorial methoxy group) by additional dipole–dipole interactions. Furthermore, relative to its anomer, the β-compound is subject to an additional *gauche* oxygen–oxygen 1,2-interaction. D-Arabinose, on the other hand, shows a preference for the β-form of the pyranoside which adopts preferentially the $^1C_4$ chair form (Scheme 3.6).

With the furanosides, the favoured anomers have the *trans*-relationship for the substituents at C-1 and C-2, as is also observed with free sugars (see Section 2.4.2). The results in Table 3.1 show that sugars with some of their non-participating hydroxyl groups methylated give glycosides in which the equilibrium is shifted in favour of the furanosides. A similar change is found in the tautomeric equilibria of free sugars that are so substituted (see Section 2.4.2).

Mechanism of the Fischer glycosidation   When Fischer glycosidation reactions are stopped prior to reaching their equilibrium positions the product ratios are different. For example, when D-galactose is heated under reflux with methanolic hydrogen chloride (0.004 mol l^{-1}) for 6 h, 53% of methyl β-D-galactofuranoside **3.6** (β-anomer) can be isolated (cf. the reaction in 2% methanolic hydrogen chloride described above), and the mother liquors contain not only α- and β-pyranosides and the α-furanoside, but also a small amount of the acyclic dimethyl acetal.[11] This simple preparative result suggests that the conversion of the free sugar to

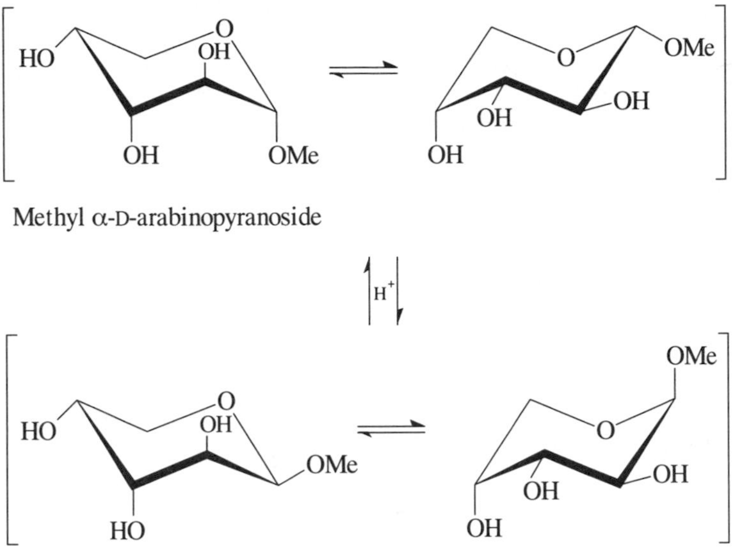

Methyl α-D-arabinopyranoside

Methyl β-D-arabinopyranoside

Scheme 3.6

pyranosides occurs *via* furanosides and reveals that for the preparation of glycoside mixtures rich in furanosides, mild reaction conditions should be used.

The methanolyses of several aldoses have been carefully studied by monitoring (by gas–liquid chromatographic examination of derived volatile ethers) the loss of the aldoses, the formation and reaction of the α- and β-furanosides and the formation of α- and β-pyranosides. Results obtained with D-xylose shown in Figure 3.1 confirm the conclusion reached with D-galactose, since it can be seen that in the course of the reaction a decrease in xylose concentration is accompanied by a rapid, but transient, build-up of furanosides which then isomerize slowly to pyranosides.[12] Knowledge of this reaction has been further increased by studying the behaviour of individual reaction components under methanolysis conditions. It is found that complete glycosidation comprises four individual interdependent reactions: the formation of furanosides from the free sugar, the anomerization of the furanosides, the ring expansion of the furanosides and the anomerization of the resulting pyranosides.

There are two possible routes by which an aldose can be converted into a furanoside, as shown for D-xylose in Scheme 3.7. One involves the reaction of D-xylofuranose which, although present only to a small extent in the reaction medium, could serve as an intermediate since it might be expected to be highly reactive, and because an equilibrium may exist between it and the other sugar modifications (Section 2.4.1). The second route involves formation of the acyclic hemiacetal from the free sugar in the acyclic or any cyclic form (cf. Scheme 3.2a), followed by stereospecific ring closure by participation of O-4 to give the furanosides. Such

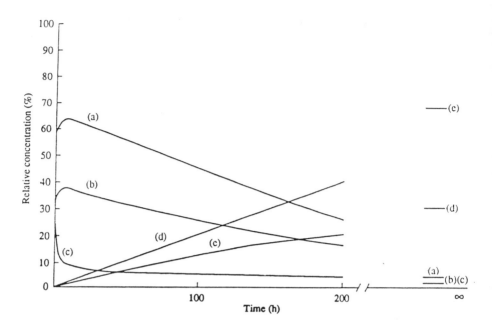

Figure 3.1 Glycosides formed on treatment of D-xylose with methanol containing hydrogen chloride (0.5%) at 25 °C. (a) Methyl β-D-xylofuranoside; (b) methyl α-xylofuranoside; (c) D-xylose; (d) methyl β-D-xylopyranoside; (e) methyl α-D-xylopyranoside (reproduced with permission from *Can. J. Chem.*, 1962, **40**, 224)

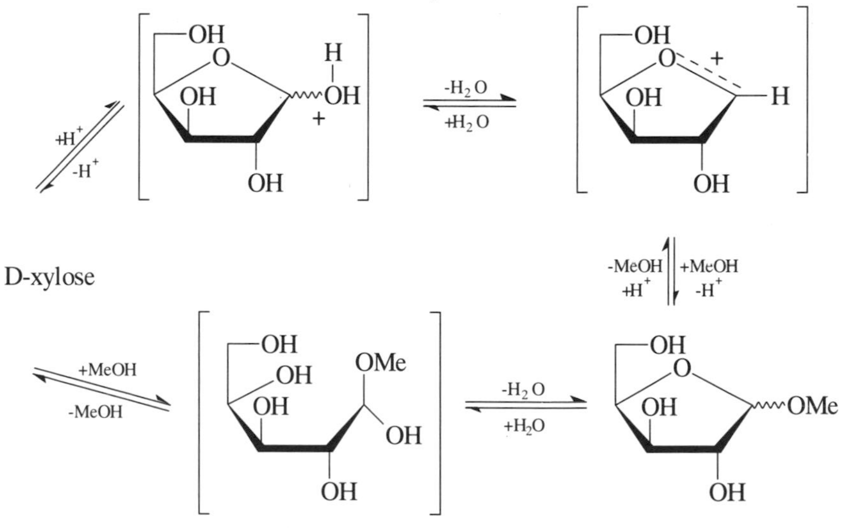

Scheme 3.7

participation to give a five-membered ring would be expected to occur in preference to O-5 participation, even though the latter would give the more stable pyranosides, since it is known from other work that ring closure to give five-membered rings is always faster than analogous ring closure to give six-membered rings.[2,13] There is no specific information available to allow a distinction to be drawn between these possible pathways, but the balance seems to lie in favour of the latter.

Furanoside ring expansion to give pyranosides can occur in two general ways. Cleavage of the C-1–ring oxygen bond can occur accompanied by a concerted ring closure involving O-5 (Scheme 3.8a), or the acyclic ion **3.10** could be formed as a reaction intermediate (Scheme 3.8b). Alternatively, the reaction could proceed by way of the acetal **3.9** which may cyclize by a concerted or two-step process, as shown in Schemes 3.8(c) and 3.8(d), respectively. (Although acetals are known to cyclize to give furanosides preferentially, a small proportion might react as shown and hence lead to pyranosides.) This is the most difficult phase of glycosidation to resolve: some evidence suggests that the routes in Schemes 3.8(a) and 3.8(c) might be favoured.

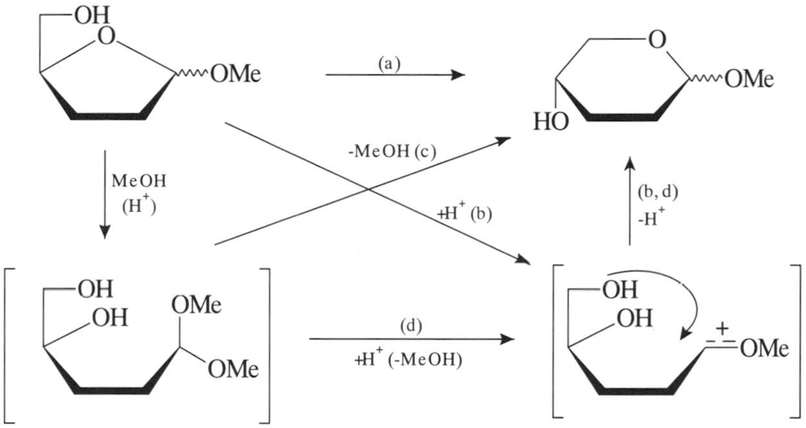

Scheme 3.8

Anomerization of pyranosides almost certainly occurs via cyclic carbocations since it has been shown, using CD_3OH as solvent and n.m.r. spectroscopic monitoring techniques, that the aglycon methyl group in the initially formed products comes from the solvent, as depicted in Scheme 3.9. This result excludes an acyclic ion of the type **3.10** as an intermediate. The anomerization of furanosides, alternatively, may involve acyclic ions such as **3.10** formed after protonation of the ring oxygen atom.

Because of the complex set of competing reactions involved, no simple mechanism can be given for the Fischer glycosidation, but the one put forward by Capon in his seminal review,[1] and shown in Scheme 3.10, most probably represents the key features for the reaction in methanol. It can be confidently stated that

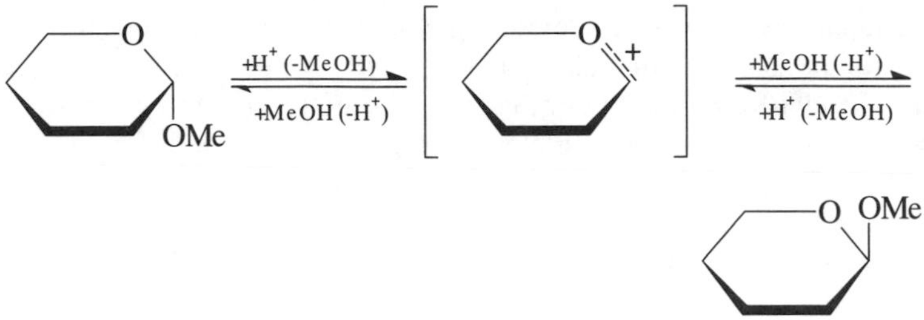

Scheme 3.9

pyranosides are *not* formed by acid-catalysed removal of the anomeric hydroxyl group and replacement by a methoxy group, as is indicated in some texts. Of passing interest is the finding that the intermediate hemiacetals also react inter-molecularly to give the acyclic dimethyl acetals in small proportions in the early stages of the reactions,[13] but, as Fischer originally surmised they might be, they are not, by themselves, the first kinetic products of the reactions.

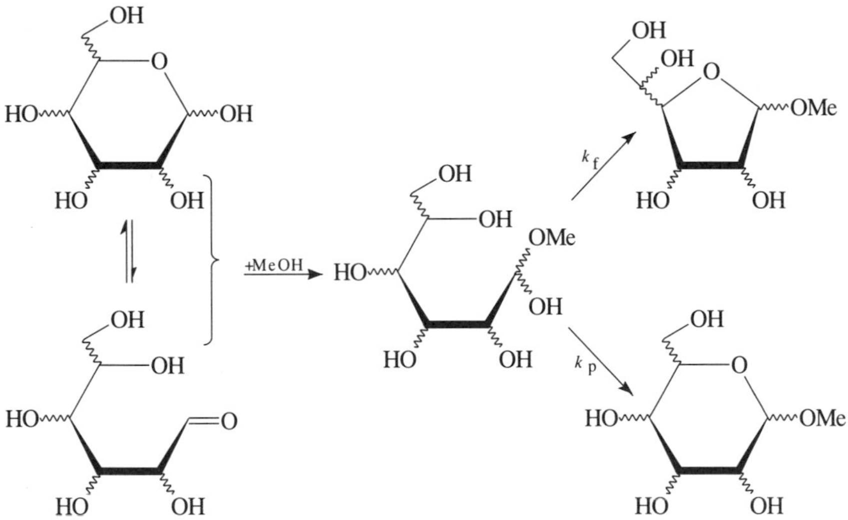

All reactions are reversible; $k_f \gg k_p$

Scheme 3.10   Key features of the Fischer glycosidation

(ii)   Glycoside synthesis by direct displacement of anomeric hydroxyl groups from otherwise fully protected sugars

Displacements involving cleavage of C-1–O-1 bonds at the anomeric centre are usually carried out on suitably reactive glycosyl esters derived from sugars (see Section 3.2), but direct displacements can be achieved by *in situ* activation of the anomeric hydroxyl groups which can be induced to function as leaving groups by means other than protonation. Mukaiyama has introduced a complexing agent which serves this function, but it does require all the other hydroxyl groups in the sugar to be protected.[14] Thus, treatment of the D-ribofuranose ether **3.11** in diethyl ether, containing caesium fluoride as base, with a reagent prepared from diphenyltin sulphide and triflic anhydride, produces the coordinated intermediate **3.12**, from which β-ribosides can be prepared. As illustrated in Scheme 3.11, the β-(1→6)-linked ribosylglucose derivative **3.13** is obtainable by treatment of **3.12** with methyl tri-O-benzyl-6-O-trimethylsilyl-α-D-glucopyranoside. When, however, the reaction is carried out in the presence of lithium perchlorate with the non-silylated glucopyranoside derivative, the α-(1→6)-linked disaccharide **3.15** is formed. The stereochemical outcome of these glycosylations is determined by the structures of the intermediates. In the first reaction, the cyclic intermediate **3.12** has the α-configuration shown because the positively charged tin atom is chelated by the C-2 ether oxygen atom, and is attacked from the β-direction by the nucleophile. In the presence of perchlorate, **3.12** is transformed into the oxonium perchlorate **3.14**, which is attacked by the hydroxyl of the acceptor to give the α-linked disaccharide **3.15** for reasons outlined in Section 3.2.1.c.i.

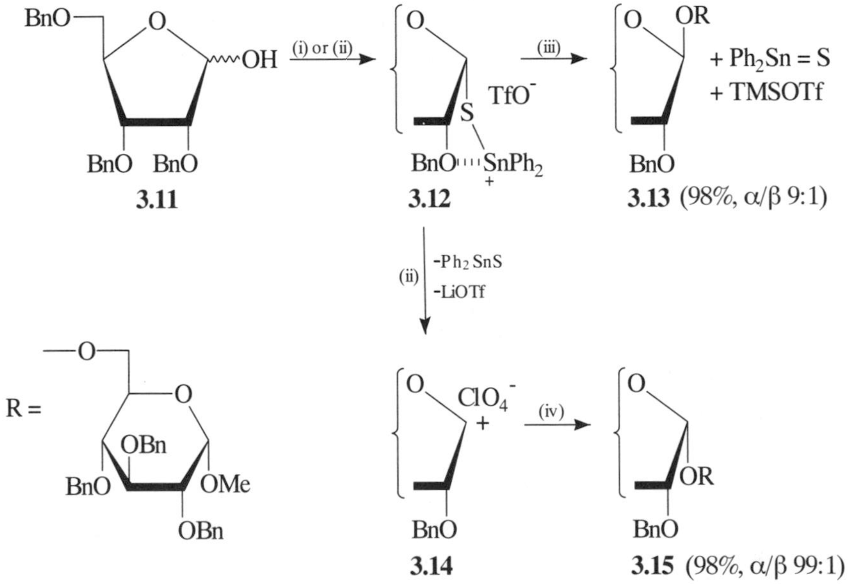

(i) Ph$_2$Sn=S, Tf$_2$O, CsF, Et$_2$O; (ii) Ph$_2$Sn=S, Tf$_2$O, CsF, LiClO$_4$, CH$_2$Cl$_2$;
(iii) ROTms; (iv) ROH

Scheme 3.11

Similar direct displacements of anomeric hydroxyl groups can be achieved under Mitsunobu conditions using phenols, and this has led to a useful method for preparing aryl glycosides. A wide range of partially protected sugars and phenols have been used in these reactions, among which the example given in Scheme 3.12

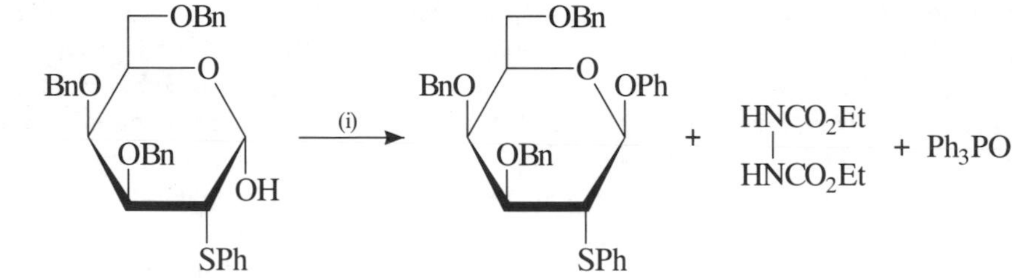

3.16   **3.17** (70%, α/β 7.3:1)

(i) PhOH, Ph₃P, DEAD, PhMe, 0°C

Scheme 3.12

is particularly interesting, since it gives an entry to aryl 2-deoxyglycosides (see Section 4.2). Thus, when 3,4,6-tri-*O*-benzyl-2-thiophenyl-α-D-galactopyranose (**3.16**) is treated with phenol in toluene containing triphenylphosphine and diethyl azodicarboxylate (DEAD), the phenyl galactoside derivative **3.17** is obtained in good yield together with small proportions of its anomer.[15]

Alkyl glycosides can also be produced with alcohols by this method when Lewis acids such as mercury(II) chloride are added to the usual Mitsunobu reagents. Since the glycosides are formed with low stereoselectivity, the method has not been widely used.

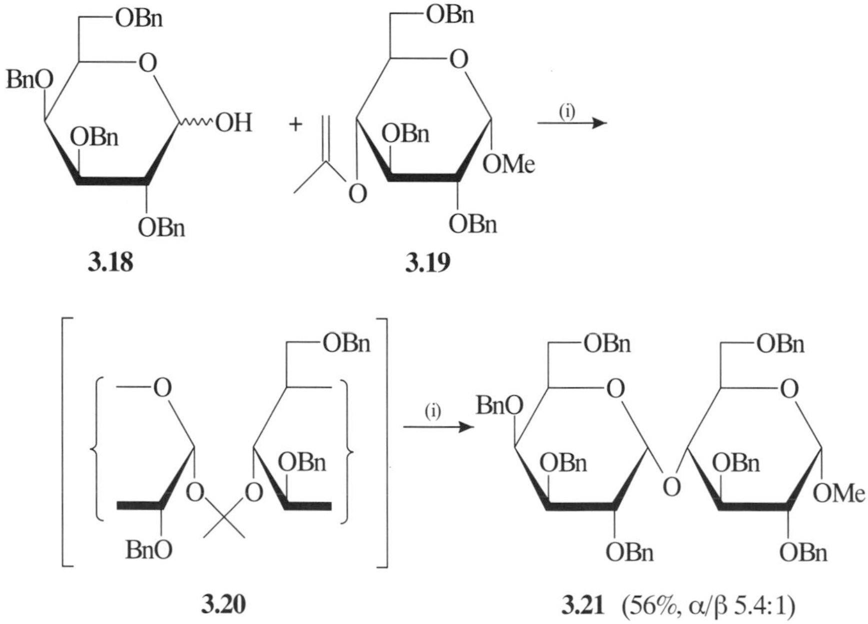

3.18   **3.19**

3.20   **3.21** (56%, α/β 5.4:1)

(i) CH₂Cl₂, TmsOTf

Scheme 3.13

A different method of inducing sugars, fully protected otherwise than at O-1, to function as glycosylating agents in disaccharide syntheses has been introduced by Sinäy and utilizes O-alkenyl ether groups in fully protected acceptor sugars.[16] These react with the anomeric hydroxyls of donor sugars under acid conditions to form transient intermediate mixed sugar acetals, which subsequently yield disaccharides as illustrated by the example in Scheme 3.13. Thus, when the 4-O-isopropenyl glucoside derivative **3.19** and 2,3,4,6-tetra-O-benzyl-D-galactopyranose **3.18** are treated in dichloromethane with trimethylsilyl triflate, silylation of the hydroxyl group in the latter compound occurs. The triflic acid liberated then catalyses the formation of the mixed acetal **3.20** and its subsequent rearrangement with loss of acetone into α/β-(1→4)-linked methyl hepta-O-benzylgalactosylglucoside **3.21**.

The same mixed acetal **3.20** and hence the disaccharide **3.21** can be prepared by a reverse condensation, so called because the 4-hydroxyl group of methyl 2,3,6-tri-O-benzyl-α-D-glucopyranoside reacts with isopropenyl 2,3,4,6-tetra-O-benzyl-α/β-D-galactopyranoside (see Section 3.1.1.b.vi).

(b)   Reactions of glycosides

(i)   Acid-catalysed hydrolysis of glycopyranosides

General features   On treatment with aqueous acid, glycosides are hydrolysed to give free sugars and this reaction is often used for the isolation of monosaccharides from natural products (Chapter 8).[17] In principle, glycoside hydrolysis could be initiated by the cleavage of three different carbon–oxygen bonds: the exocyclic anomeric carbon–oxygen bond, the endocyclic anomeric carbon–oxygen bond or the aglycon–oxygen bond, as depicted in Schemes 3.14(a), 3.14(b) and 3.14(c), respectively.[1] If methyl β-D-glucopyranoside is so treated, methanol and glucose

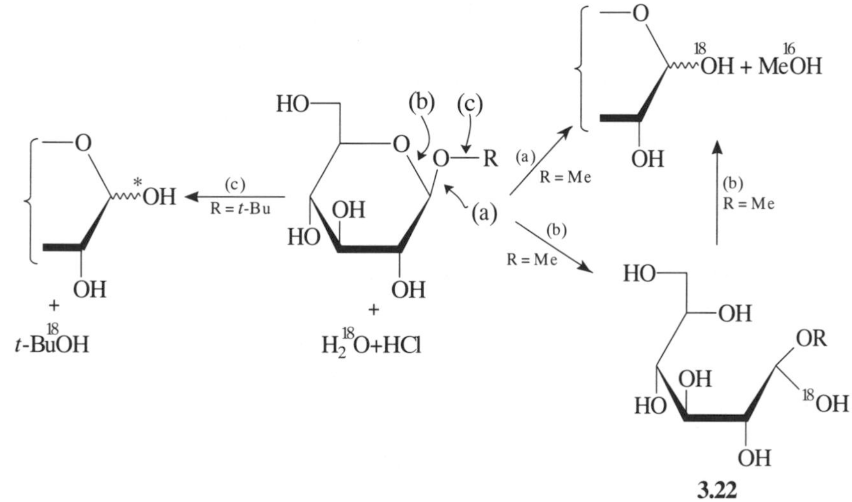

*Initially ^{16}O, but becomes ^{18}O by exchange with the solvent

Scheme 3.14

are liberated, and when carried out in ^{18}O-enriched water this reaction is shown to occur with fission of a glycosyl–oxygen bond either by route (a) or (b) (Scheme 3.14) because the methanol produced is not enriched with [18]O. With *t*-butyl β-D-glucopyranoside, on the other hand, *t*-butanol enriched with [18]O is produced, indicating that route (c) has been followed.[18] This happens only when the aglycon forms more stable carbocations than the glycosyl cations produced by common glycosides (see the conversion of **3.25** into **3.26**, Scheme 3.15).

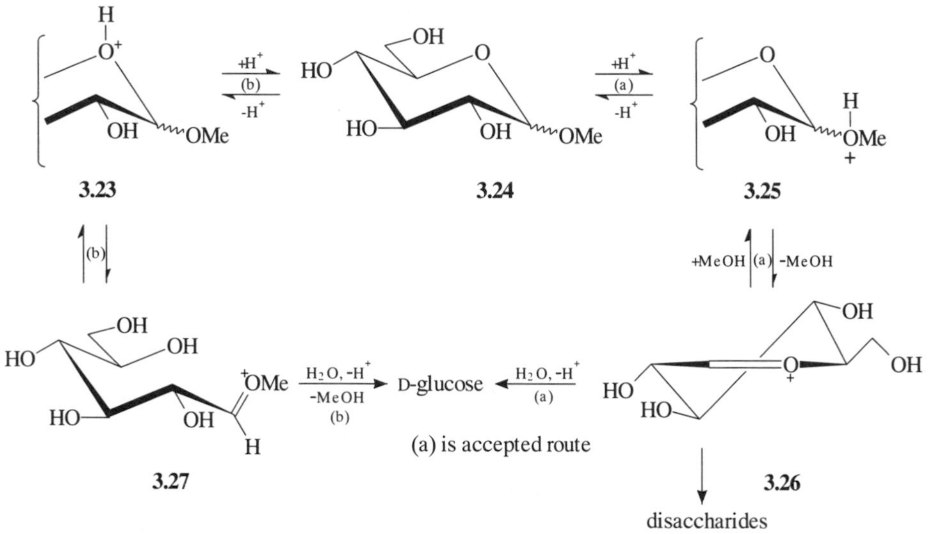

Scheme 3.15

The mechanisms of hydrolysis following routes (**a**) and (**b**) (Scheme 3.14) are depicted in Scheme 3.15 for methyl glucopyranosides **3.24**. The former requires protonation of the aglycon-oxygen atom followed by breakdown of the conjugate acid **3.25** to the cyclic carbocation **3.26** which, when attacked by water, gives the products. Alternatively, protonation of the ring oxygen takes place in route (**b**) and the conjugate acid **3.23** ring opens to the acyclic oxycation **3.27** with the final products arising from a hemiacetal (see **3.22** in Scheme 3.14) produced by water attack on **3.27**. An isotope effect observed in the hydrolysis of methyl α-D-glucopyranoside, and associated with the methoxy oxygen atom, supports initial methoxy–glucosyl bond cleavage since the glucoside in which this oxygen atom is present as [16]O hydrolyses 3% faster than in the case when [18]O is present. A rate difference of this magnitude is compatible with mechanism (**a**) (Scheme 3.15) involving the cyclic oxocarbocation **3.26** which is stabilized by mesomeric electron release from the ring oxygen and thereby adopts a half-chair conformation, as shown, with the C-1 atom sp^2-hybridized.[19] This is the currently accepted mechanism for the hydrolysis of most glycopyranosides, but under conditions of acetolysis (see Section 3.1.1.b.vii) and also methanolysis some compounds appear to react via acyclic ions akin to **3.27** depicted in route (**b**) (Scheme 3.15).[20]

During glycoside hydrolysis, small amounts of dimeric products are sometimes formed even in solutions which are only 0.1 mol l^{-1} in glycoside. They arise by competitive attack on the intermediate oxocarbocations (e.g. **3.26**) from free sugar or glycoside molecules (see reversion, Section 3.1.8.a.).

Effect of anomeric configuration   Of the several anomeric pairs of methyl glycopyranosides which have been studied, anomers with the methoxy groups at C-1 equatorially disposed are usually hydrolysed between two and four times faster than those in which these groups are axial, providing neither of the anomers is conformationally unstable.[†,2,21] It follows, therefore, that with the common hexopyranosides the β-anomers are the more reactive. Protonation of equatorial anomers to give relatively high concentrations of the conjugate acids (unfavourable anomeric effect becoming favourable) probably accounts for this. On the other hand, the ground state energies of the conjugate acids derived from equatorial anomers are lower (reverse anomeric effect) than those of the axial anomers and this factor will increase the reactivity of the latter. Competing factors are therefore in operation which account for the relatively small differences in hydrolysis rates (cf. glycosyl halides) and for the anomalous behaviour of, for example, the phenyl glucopyranosides.

Electronic effects of substituents in the glycosyl ring   The rates of glycopyranoside hydrolyses are very sensitive to the electronic properties of the substituents positioned at C-2, as would be expected from the mechanism given in Scheme 3.15.[1] When the substituent X in the glucoside **3.28** is an electron-withdrawing group, the concentration of the conjugate acid formed in acidic solution by protonation of O-1 is lower than when an electron-releasing group is present; this factor will retard hydrolysis. The cleavage of the glycosyl–oxygen bond is also retarded because the carbocation formed is less stable when the former type of substituent is present. The rate of hydrolysis of glycosides is therefore diminished as the X group increases in its electron-withdrawing power.

Methyl 2-deoxy-β-D-*arabino*-hexopyranoside **3.28** (X = H) is hydrolysed very rapidly, but replacing a hydrogen atom at C-2 by a hydroxyl group (X = OH) causes about a 2000-fold decrease in hydrolysis rate,[‡] whereas replacing it with an amino substituent (X = NH$_2$), which results in a powerful electron-withdrawing group (X = NH$_3{}^+$) in acid media, causes a rate decrease of 3.0×10^5. An *N*-acetamido substituent withdraws electrons much less strongly than does an NH$_3{}^+$ substituent, and consequently the hydrolysis rate of methyl 2-acetamido-2-deoxy-β-D-glucopyranoside **3.28** (X = NHAc) is 800 times faster than that of the unsubstituted 2-amino-2-deoxyglucoside (X = NH$_2$). This explains why *N*-acetylation is an essential first step in the hydrolysis of 2-amino-2-deoxyglycosides (see Section 4.3.2.f.ii). Although part of this rate enhancement is caused by

---

† The anomalous behaviour of the methyl D-gulopyranosides is probably a result of this type of instability.
‡ The steric effect of the hydroxyl group will also reduce the rate of hydrolysis but the contribution from this origin is much smaller than that from the electronic effect (see 3- and 4-deoxyglycosides).

anchimeric assistance from the acetamido groups (see below), the contribution from this factor is relatively small since the rate difference between the corresponding derivatives in the α-glucoside series (where anchimeric assistance is not possible) is of the same order of magnitude.

From the above, it follows that the acetamido compound **3.28** (X = NHAc) is hydrolysed 5.3 times faster than methyl β-D-glucopyranoside (X = OH), a fact which is unexpected on electronic grounds since the acetamido group is slightly more electron withdrawing than the hydroxyl group. This enhanced reactivity arises in this case mainly from direct participation of the carbonyl oxygen atom of the acetamido group at the reaction centre (**3.29**), which gives anchimeric assistance to the departure of the methoxy group.

3.28   **3.29**

3.30

Electronic effects of substituents at other positions in the pyranoid system influence the hydrolysis rates less than do those at C-2, because the distance from the reaction centre is greater.[1] For example, methyl 3-deoxy-α-D-*ribo*-hexopyranoside (3-deoxy-α-D-glucopyranoside) and the 4-deoxy analogue are hydrolysed 20 and 40 times faster, respectively, than is methyl α-D-glucopyranoside (cf. the 2000-fold rate increase for the 2-deoxy compound). The greater hydrolysis rate for the 4-deoxy compared with the 3-deoxy compound indicates that steric factors also exert an influence (see below), but the relatively small size of this rate difference confirms that the dominant effect responsible for the changes produced by the substituents at position 2 is electronic.

Extensive studies have been made on the influence of substituents at C-5 and, in particular, interest has been shown in the hydrolyses of glycopyranosiduronic acids (uronosides), e.g. **3.30** (R = CO$_2$H).[1] Reasons for these studies have stemmed from the need to explain the hydrolytic stability of aldobiouronic acids (Section 8.1.3.a), which are obtained from the partial hydrolysates of uronic acid containing polysaccharides. Methyl α-D-glucopyranosiduronic acid **3.30** (R = CO$_2$H) is hydrolysed more slowly in sulphuric acid (0.5 mol l^{-1}) than is methyl α-D-glucopyranoside

(R = CH$_2$OH) owing to the inductive influence of the carboxylic acid group, and it is this that undoubtedly accounts at least in part for the stability of aldobiouronic acids such as cellobiouronic acid **3.31** towards acids.

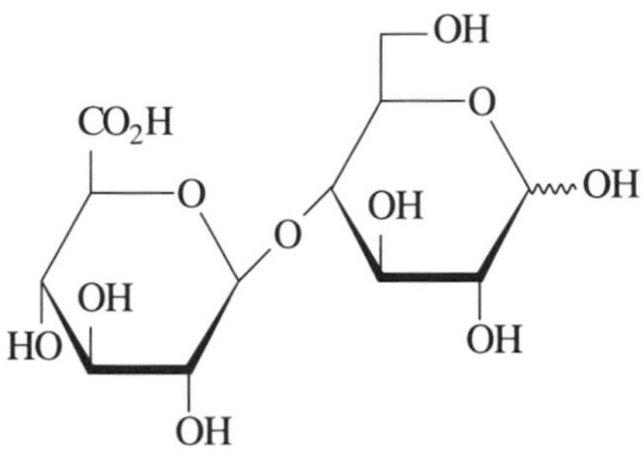

Cellobiouronic acid **3.31** 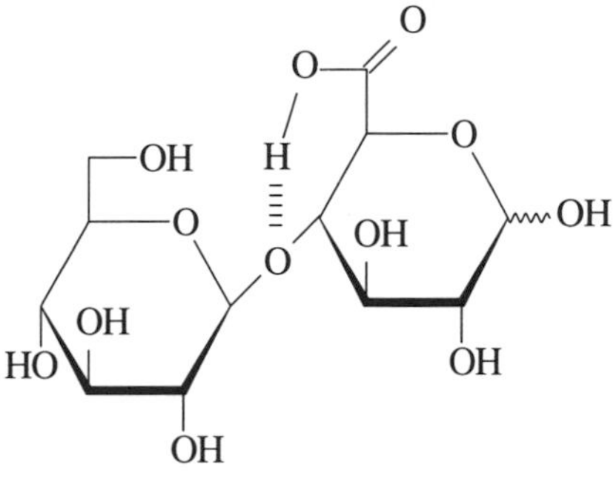   Pseudocellobiouronic acid **3.32**

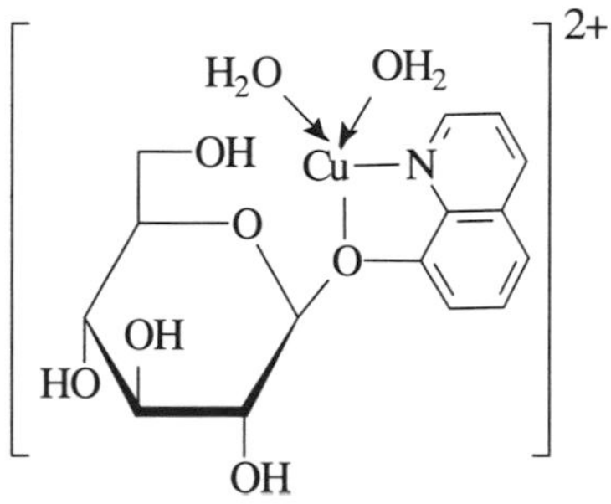

3.33

Structural effects of substituents in the glycosyl rings   Configurational changes in the aldose also affect the rate of hydrolysis.[1] For example, the relative rates of hydrolysis of methyl α-D-hexopyranosides with the *altro-*, *galacto-*, *manno-* and *gluco*-configurations are 18, 5, 3 and 1 respectively. These results can be rationalized by examining the release of steric strain which occurs when the conjugate acids dissociate into methanol and the cyclic oxocarbocations that adopt half-chair conformations (Scheme 3.16). During this transformation, axial groups at C-2 and C-5 in the $^4C_1$ conformation of the conjugate acids move away from the C-4 and C-3 axial substituents, respectively. Hence when these substituents are hydroxyl groups rather than hydrogen atoms, there will be a greater relief of non-bonded interactions during the transformation. Therefore the *manno-* and *galacto*-pyranosides with one axial group at these ring positions (C-2 in the former and C-4 in the latter) react faster than the glucoside, which has all the hydroxyl groups equatorial. For similar reasons the altropyranoside with hydroxyl groups axial at C-2 and C-3 reacts even faster.

Pyranosides having deoxy groups at C-2, C-3, C-4 experience reduced interactions between vicinal groups in undergoing conformational change to give the cyclic

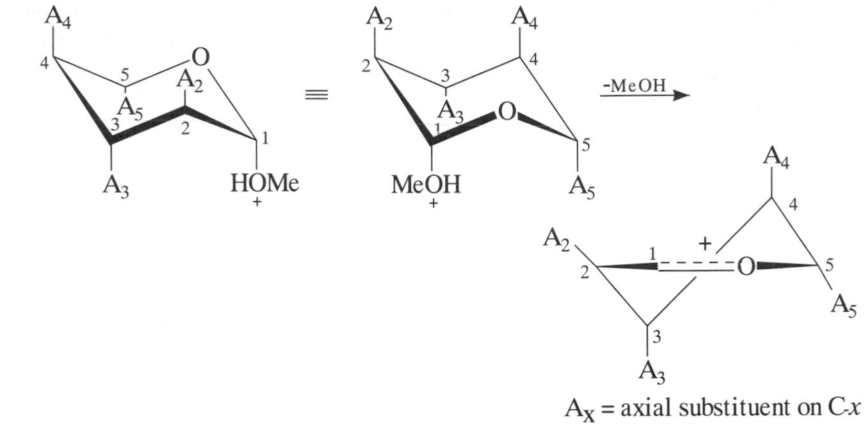

A_X = axial substituent on C-x

Scheme 3.16

oxocarbocations. In consequence their reactions have lower activation energies and higher rates.

Effects of aglycon structure   Substituents in a methyl or phenyl aglycon have a small effect on hydrolysis rates. This is consistent with the composite nature of the proposed reaction mechanism (Scheme 3.15), since an electron-withdrawing group, for example, decreases the rate by reducing the standing concentration of conjugate acid, while it facilitates the rate of cleavage of the glycosyl-oxygen bond.

Two further features of the aglycon structure are important. If it gives rise to a stable oxocarbocation, aglycon–oxygen bond cleavage may occur during hydrolysis (Scheme 3.14c). Furthermore, should the aglycon contain a carboxylic acid residue positioned so that its un-ionized proton can approach the glycosidic oxygen atom in a six-membered ring transition state, easier hydrolysis results because of intramolecular acid catalysis. Pseudoaldobiouronic acids, such as pseudocellobiouronic acid **3.32**, have this structural feature and consequently undergo hydrolysis in dilute acid at a rate faster than expected: such factors affect rates of hydrolysis of glycoside bonds in uronic acid containing polmers.

Glycosides with aglycons which can complex with metal ions have been observed to hydrolyse at increased rates in the presence of these ions. 8-Quinolyl β-D-glucopyranoside, for example, which gives complex **3.33** with copper(II) ions, is hydrolysed 10^5 times faster than the uncomplexed glycoside between pH 5.5 and 6.2.[22] Thus, traces of metal ions can be as effective as strong acid in cleaving this glycoside. This effect is reminiscent of that found with some glycosidase enzymes which only show their full activity in the presence of metal ions, and it is also akin to the so-called *remote activation* used to assist the cleavage of certain substituents at C-1, as described in Sections 3.1.1.b.vi, 3.1.2.c.ii and 3.2.1.c.i.

(ii)   Acid-catalysed hydrolysis of glucofuranosides

Cyclopentyl halides and tosylates undergo unimolecular nucleophilic displacements much faster than do the corresponding cyclohexyl derivatives because reductions in I-strain suffered by the former on passing from their ground states, in which the reacting carbon is tetrahedral, to their transition states, in which it is trigonal, are significantly greater in the five-membered ring compounds than in the six.[23] Consequently, it is not surprising that alkyl glycofuranosides are hydrolysed in aqueous acid at rates several orders of magnitude higher than those for the corresponding

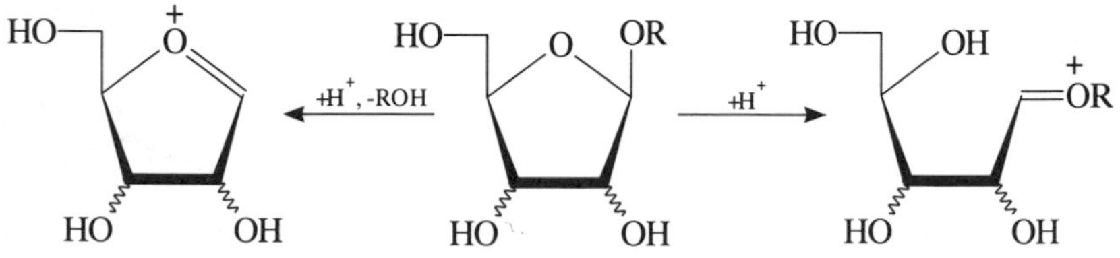

Scheme 3.17

pyranosides. However, the above analogy must be treated with caution since there is a wider range of mechanisms by which furanosides can react, compared to those available to the carbocyclic compounds. Glycosyl–oxygen bond fission has again been found to occur, but kinetic evidence indicates that the mechanism of furanoside hydrolysis is finely balanced, being dependent upon the structure of the aglycon.[24] Those glycosides with electron-withdrawing groups at this site hydrolyse via the furanosyl cation **3.34** (Scheme 3.17), whilst others hydrolyse via the acyclic oxycarbocation **3.35** (cf. Schemes 3.15**a** and 3.15**b**).

(iii)   Alkaline cleavage of glycopyranosides

Alkyl glycosides react with aqueous base only under vigorous conditions. Methyl β-D-glucopyranoside is cleaved slowly at 170 °C in 2.5 mol l^{-1} sodium hydroxide solution, and under such conditions the products formed undergo further decomposition reactions.[25] Aryl glycosides, on the other hand, react under milder conditions particularly when the benzene ring contains electron-withdrawing groups, and when the aldose structure permits participation by a neighbouring hydroxyl group. Phenyl β-D-glucopyranoside **3.36** yields 88% of 1,6-anhydro-β-D-glucopyranose **3.37** when treated with 1.3 mol l^{-1} potassium hydroxide solution at 100 °C for 9 h, as depicted in Scheme 3.18. Support for this reaction pathway comes from the fact that the 2-*O*-methyl derivative of this glucoside does not give the corresponding 1,6-anhydride, nor does phenyl α-D-glucopyranoside under these reaction conditions.

Participation involving *O*-2 of a different type has been observed by Horton with *p*-nitrophenyl α-D-glucopyranoside **3.38**, which is cleaved, via the Meisenheimer

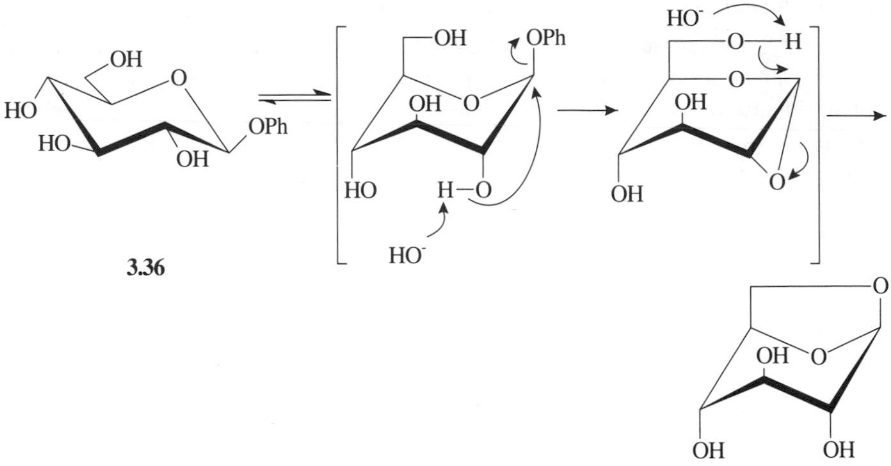

3.36

3.37

Scheme 3.18

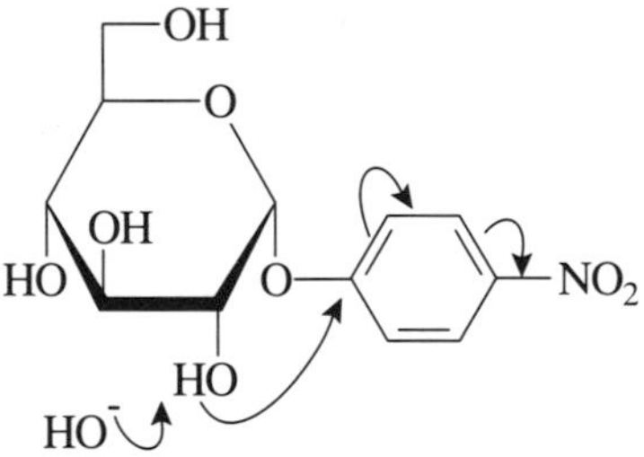

3.38 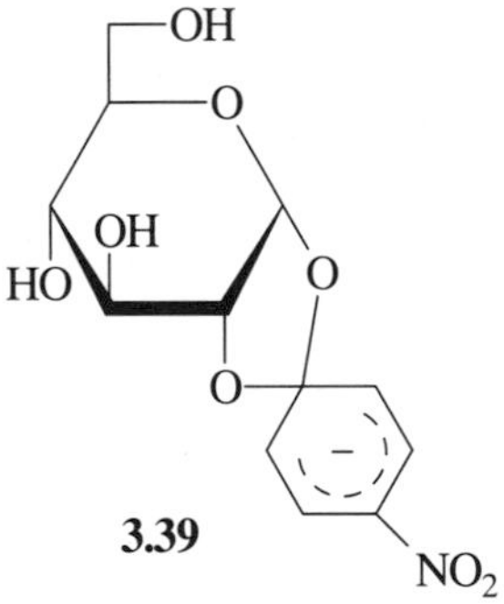   **3.39**

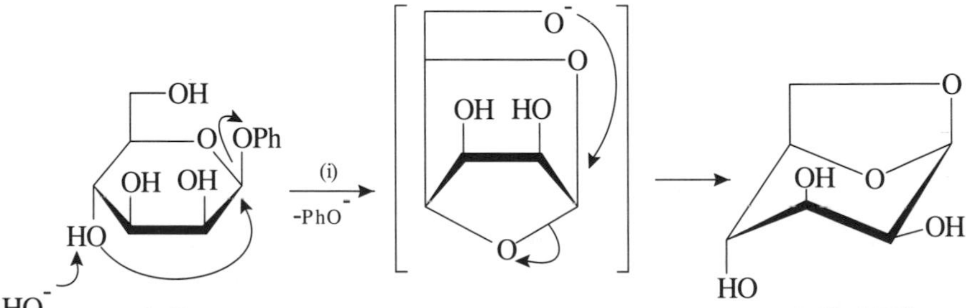

3.40   **3.41** (57%)

(i) 2.6 mol l^{-1} KOH, 100 °C, five days

Scheme 3.19

complex **3.39**, 10^5 times faster than is the phenyl glycoside in 3.9 mol l^{-1} potassium hydroxide solution at 60 °C.[26] Under these conditions the aryl group migrates to O-2 and then to O-3 with subsequent breakdown of the derived 3-O-p-nitrophenyl-D-glucopyranose to saccharinic acids (see Section 3.1.7.b).

Another mechanism for glycoside cleavage, probably involving a 1,4-anhydride, is also indicated (Scheme 3.19) since the 1,6-anhydro product **3.41** is obtained by alkaline treatment of phenyl β-D-mannopyranoside **3.40**, which does not have the 1,2-*trans*-structural feature. Although the intermediate has not been isolated from this reaction, such anhydrides have been prepared (Section 3.1.1.c).

(iv)   Enzymic hydrolysis of glycosides

Many glycosides can also be cleaved by enzymes called glycosidases. These often exhibit a high specificity towards both the sugar and the configuration at the anomeric centre. Maltase, an α-D-glucopyranosidase, is obtained from barley malt and catalyses the hydrolysis of methyl α-D-glucopyranoside, but not that of the β-anomer. On the other hand, almond emulsin is a β-glucosidase which hydro-lyses methyl β-D-glucopyranoside but not the α-anomer. Some glycosidases cleave glycosidic bonds with retention of configuration (β-galactosidases and lysozyme), whereas others do so with inversion (trehalase and β-amylase) by mechanisms thought to be closely related to those involved in acid-catalysed hydrolyses.

Lysozyme, obtained from hen egg white, is a protein containing 129 amino acid residues of known sequence linked in a single polypeptide chain, and crosslinked in four places by disulphide bonds. Its tertiary structure has been determined by X-ray crystallographic methods. This enzyme catalyses the hydrolytic breakdown of certain bacterial cell wall polysaccharides of the type depicted in Scheme 3.20, which contain alternately β-(1→4)-linked 2-acetamido-2-deoxy-D-glucose (abbre-viated to GlcNAc) and 2-acetamido-2-deoxy-3-O-lactyl-D-glucose (N-acetyl-muramic acid, abbreviated to NAM) residues, and gives the tetrasaccharide GlcNAc–NAM–GlcNAc–NAM and the disaccharide GlcNAc–NAM.[1,27] (See Section 8.1.3.d for a discussion of bacterial cell wall structures). Suitable model substrates which have been studied are oligosaccharides containing four, five and six 2-acetamido-2-deoxy-D-glucose units β-(1→4)-linked, i.e. (GlcNAc)$_4$, (GlcNAc)$_5$ and (GlcNAc)$_6$, and these are hydrolysed by the enzyme. In contrast, the catalytic effect of lysozyme upon tri-N-acetyl chitotriose, (GlcNAc)$_3$, is rather weak since a complex is formed between this trisaccharide and the enzyme which inhibits further hydrolytic activity. The structure of this enzyme–inhibitor complex has been determined crystallographically and indicates that the active site is a hydrophobic cleft in the protein. Six hydrogen bonds are involved in the complex, and it is reasonable to assume that similar binding occurs between the enzyme and its substrates. Only two functional groups are correctly positioned for catalytic action in the neighbourhood of a substrate so bound, the carboxylic acid group in the glutamic acid, which is the 35th residue of the polymer, and the carboxylate anion in aspartic acid, which is the 52nd residue. The former acts catalytically as shown in Scheme 3.20, and the latter is believed either to

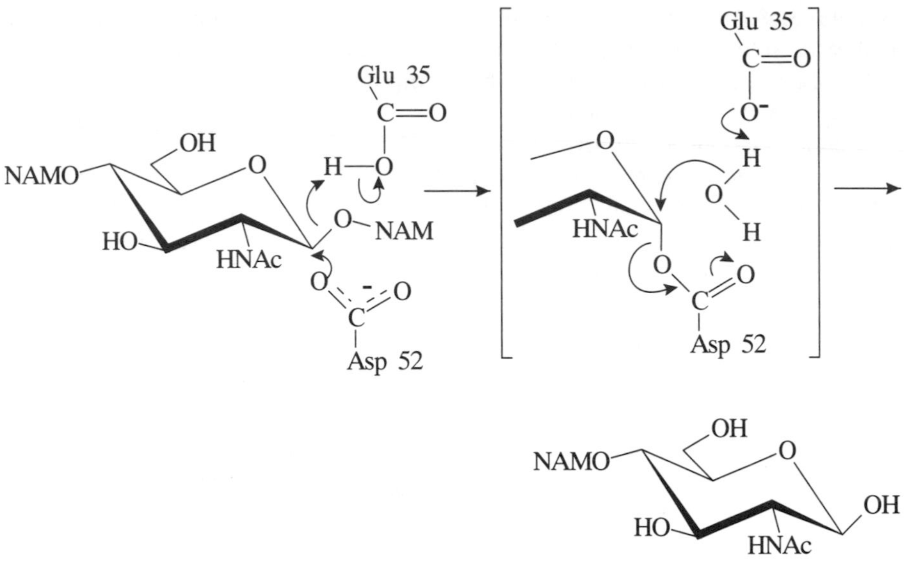

Scheme 3.20

provide direct nucleophilic assistance at C-1 of the glycosyl ring or to catalyse participation at C-1 from the acetamido group.

Hard evidence for the presence of a glycosyl ester intermediate, rather than an ion pair formed between a glycosyl carbocation and an enzyme aspartic acid group, has been found in the hydrolysis of 2-deoxy-2-fluoro-β-D-glucopyranosyl fluoride by a β-glucosidase from an *Agrobacterium* (^{19}F n.m.r. spectroscopy).[28]

(v) Transglycosidation

Treatment of an alcoholic solution of a glycoside with an acid catalyst brings about a substitution closely related to the anomerization of glycosides in which the aglycon derived from the alcoholic solvent replaces the one originally present in the glycoside (cf. Scheme 3.9). Such transglycosidations are particularly useful for preparing simple glycosides from polysaccharides, for example in the preparation of methyl α-D-mannopyranoside from mannans.

Transglycosidation can also be effected enzymically by glycosidases operating in reverse and the method has been used to prepare oligosaccharides. It is usually achieved by treating a glycosyl donor molecule with a large excess of an acceptor in the presence of a glycosidase which is chosen by trial and error so that the glycosylation occurs with the desired anomeric configuration at the required hydroxyl group within the acceptor molecule.[29] Di- or oligo-saccharides and aryl glycosides (also glycosyl fluorides) have been used as glycosyl donors. For example, a reasonable yield of the α-(1→3)-linked galactosylglucose derivative **3.44** is obtained by the highly regioselective glycosylation of methyl α-D-glucopyranoside **3.43** with p-nitrophenyl α-D-galactoside **3.42** as the glycosyl donor in the presence of a

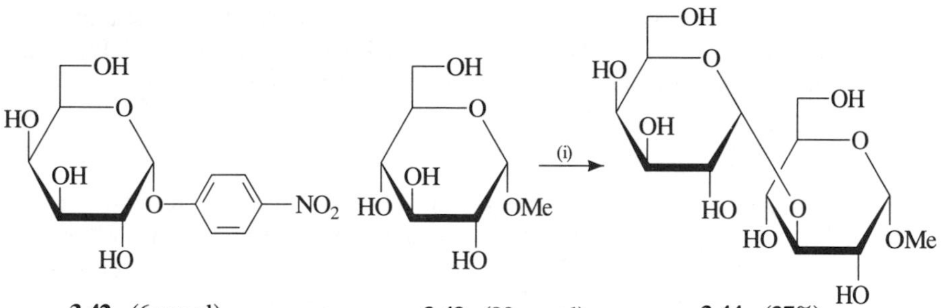

3.42 (6 mmol)   **3.43** (90 mmol)   **3.44** (27%)

$+ \alpha$-(1$\longrightarrow$6)-disaccharide (2%)

(i) coffee bean α-galactosidase, NaH$_2$PO$_4$, H$_2$O/DMF (3:1), 22 °C, seven days

Scheme 3.21

coffee bean α-galactosidase, as shown in Scheme 3.21. The closely related enzymic method of synthesizing oligosaccharides using glycosyl transferases with complex glycosyl phosphates (e.g. UDPG) has also been reported.

Aryl glycosides can be transglycosylated by electrochemical means in appropriate anhydrous polar solvents containing electrolytes, the method requiring only approximately equimolar proportions of the alcohol.[30] Thus, when an electric current is passed through a solution of phenyl 2,3,4,6-tetra-O-benzyl-β-D-gluco-pyranoside **3.45** in acetonitrile containing lithium perchlorate and 1.2 molecular equivalents of methanol, methyl 2,3,4,6-tetra-O-benzyl-D-glucopyranoside **3.48** is obtained as a 1:4 α/β mixture in good yield, as illustrated in Scheme 3.22(a). With participating groups at C-2, β-glycosides are formed almost exclusively as exemplified by phenyl 2,3,4,6-tetra-O-acetyl-β-D-glucopyranoside, which gives, when similarly treated, the methyl glucoside tetraacetate as a 1:32 α/β mixture in 79% yield. Several glycosides have been formed in this way, including disaccharides.

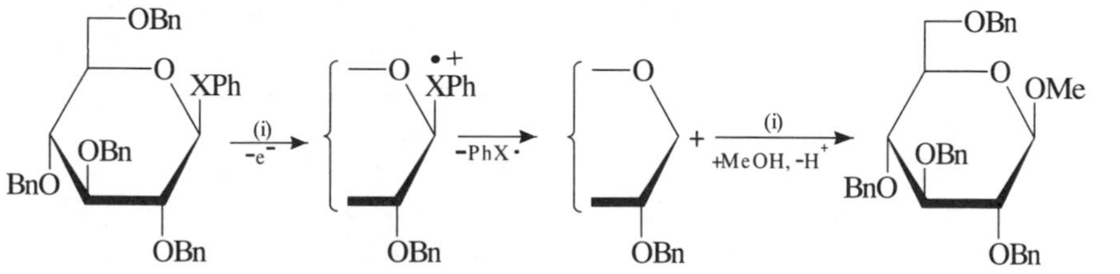

3.45 X = O   **3.46** X = O   **3.47**   (a) **3.48** (84%, α/β 1:4)
3.49 X = S   **3.50** X = S   (b) **3.48** (84%, α/β 1:3)

(a) (i) **3.45**, MeOH, LiClO$_4$, MeCN, electrochemical

(b) (i) **3.49**, MeOH, n-Bu$_4$NBF$_4$, MeCN, electrochemical

Scheme 3.22

The anodic reaction occurs by way of a one-electron transfer from the phenoxy oxygen atom of the aryl glycoside, with the radical cation **3.46** so produced dissociating to yield the pyranosyl cation **3.47** and then the glycoside **3.48** by subsequent alcohol attack. Similar transformations occur even more readily with aryl S-glycosides (Section 3.1.2.c.iv), compound **3.49** again giving the glycosides **3.48**, this time by way of the thio radical cation **3.50** (Scheme 3.22b).

(vi)   Cleavage of alkenyl glycosides

Pentenyl glycosides   Most synthetically useful reactions at the anomeric centre are nucleophilic displacements that occur by way of glycosyl cations, usually generated from electrophilically activated leaving groups in suitable glycosyl derivatives, most frequently halides (Section 3.3), thioglycosides (Section 3.1.2.c.ii) and imidate esters (Section 3.2.3.b). *O*-Glycosides do not often feature in these reactions since their 1-alkoxy substituents are usually poor leaving groups (see, however, Sections 3.1.1.b.vii–3.1.1.b.ix), but Fraser-Reid has found that pent-4-enyl glycosides are an exception since halonium ion additions to the double bonds in their aglycons occur with participation from the anomeric oxygen atoms which transforms them into leaving groups as seen with intermediate **3.53** (Scheme 3.23).[31] Nucleophilic displacement at the anomeric centres is thereby possible. NBS and iodonium dicollidine perchlorate (IDCP) have been most often used as the halonium source together with pentenyl glycosides of a variety of sugars having their non-anomeric hydroxyl groups protected with a wide range of substituents. Displacements on these so-activated glycosides have been achieved in good yields at room temperature with a range of alcohols, particularly those derived from partially protected sugars, thus making this a valuable reaction for oligosaccharide syntheses. For example, pent-4-enyl 2,3,4,6-tetra-*O*-benzyl-α or β-D-glucopyranoside **3.52** with methanol gives methyl glucosides **3.54** as a 1:3 α/β mixture in 85% yield when treated with NBS in acetonitrile, whereas a 75% yield of the α/β-glucosides **3.54** in a 3:1 ratio is obtained when the original reactants are treated with IDCP in an ether/dichloromethane mixture. These transformations occur as shown in Schemes 3.23(a) and 3.23(b) with the furanonium ion **3.53** cleaving to give 2-halomethyltetrahydrofuran and a glucopyranosyl cation which yields α/β mixtures of glycosides, the compositions of which depend on the solvent and the nature of the counteranion. This explains why the glucoside composition is not influenced by the anomeric configuration of the initial pentenyl glycoside.

The α/β-(1→6)-linked disaccharides **3.55** are similarly formed when equimolar proportions of pentenyl 2,3,4,6-tetra-*O*-benzyl-α/β-D-glucoside **3.52** and the partially protected glucoside **3.51** are treated with 1.5 molecular equivalents of IDCP (Scheme 3.23c). Remarkably, and usefully, a disaccharide from self-condensation of the partially protected pentenyl glucoside **3.51** is not formed, which indicates that its pentenyl group is not activated, whereas that of compound **3.52** is. This important difference is attributable to the development of **3.53** as a glycosylating agent, an analogue of which is not concurrently produced from **3.51** because of the deactivating electron-withdrawing influence of its C-2 ester group.

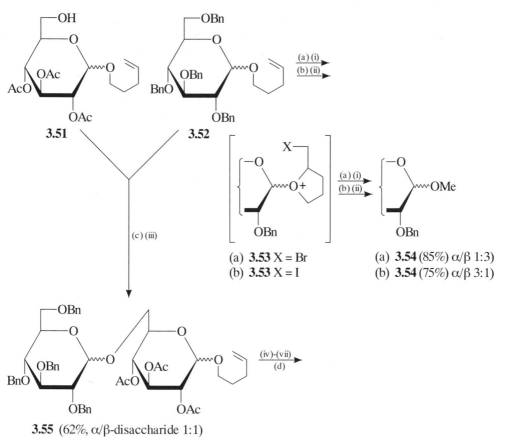

(a) **3.53** X = Br
(b) **3.53** X = I

(a) **3.54** (85%) α/β 1:3)
(b) **3.54** (75%) α/β 3:1)

3.55 (62%, α/β-disaccharide 1:1)

3.56 (60%, α/β 1:1 from **3.55** α-linked disaccharide)

(i) MeOH, NBS, MeCN; (ii) MeOH, IDCP; CH$_2$Cl$_2$, Et$_2$O (1:4); (iii) IDCP, CH$_2$Cl$_2$, molecular sieves; (iv) separate anomers; (v) MeONa, MeOH; (vi) BnBr, NaH, DMF; (vii) 1,2:3,4-di-*O*-isopropylidene-α-D-galactose, IDCP, molecular sieves, CH$_2$Cl$_2$

Scheme 3.23

The observation that electronic and structural features within the pyranose ring can influence the reactivity at the anomeric centre so profoundly that chemoselective reactions can be achieved between different compounds with identical groups at C-1 has led to a new protocol in saccharide-coupling methodology in which the reactive glycosyl donors are said by Fraser-Reid to be *armed* and the unreactive ones *disarmed*.[31] In general, armed compounds have alkoxy or deoxy groups at C-2, whereas disarmed ones have electron-withdrawing substituents such as esters or halides or, alternatively, possess structural features such as 2,3- or 4,6-acetals which deter the formation of the half-chair conformation of the glucopyranosyl cation intermediate (see **3.26**).

The ability to arm or disarm glycosyl donors allows for iterative procedures to be adopted in oligosaccharide syntheses. Thus, for example, the disarmed propenyl α-linked disaccharide glycosides **3.55** can be armed by replacing the acetyl groups with benzyl groups and then subjecting the heptabenzyl disaccharide analogue of **3.55** to the normal coupling protocol. In the presence of 1,2:3,4-di-*O*-isopropylidene-α-D-galactose the trisaccharide **3.56** is produced, as shown in Scheme 3.23(d).

Similar matched pairs of armed and disarmed glycosyl donors derived from thioglycosides,[32] 2-pyridyl thioglycosides[33] (Section 3.1.2.c.ii) and glycals[34] (Section 4.10.1.b.ii) have been found which suggests that this ingenious and potentially valuable method will be of general use in oligosaccharide syntheses (Sections 6.4.1 and 6.4.2). Improvements in means for providing better anomeric selectivities, however, are awaited.

Isopropenyl glycosides   These glycosides function as glycosylating agents under acidic conditions in a process suitable for preparing disaccharides. The reaction is unusual since direct displacement of the anomeric substituent does not take place. Instead, the hydroxyl group of the acceptor sugar adds to the vinyl ether double bond of the donor sugar to give a mixed sugar acetal of acetone, which under the reaction conditions rearranges to give a disaccharide and acetone.[16] (Disaccharide formation from mixed sugar acetals was reported in a synthesis described in Scheme 3.13, Section 3.1.1.a.ii, wherein a reverse condensation was deployed.) For example, when isopropenyl 2,3,4,6-tetra-*O*-benzyl-α/β-D-galactopyranoside and methyl 2,3,6-tri-*O*-benzyl-α-D-glucopyranoside in dichloromethane are treated with trimethylsilyl triflate as a promoter, methyl hepta-*O*-benzylgalactopyranosylglucopyranoside **3.21** as a 4:1 α/β mixture is formed in 70% yield via the transient mixed sugar acetal intermediate **3.20**.

(vii) Acetolysis of glycosides

Acetolysis is frequently used to depolymerize polysaccharides, with cellulose, for example, giving a range of peracetylated oligosaccharides up to heptoses when treated at 30 °C with 5% sulphuric acid in acetic acid/acetic anhydride for 48 h. A reaction time of 100 h gives optimum yields of peracetylated cellobiose

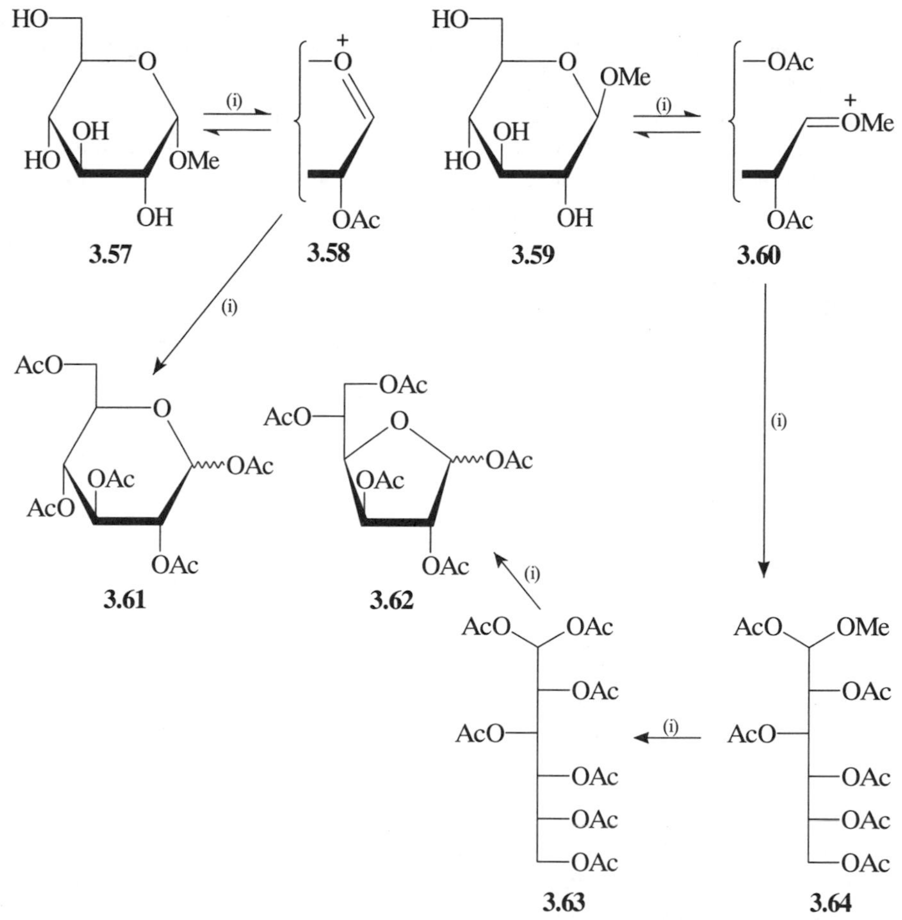

(i) Ac$_2$O, H$_2$SO$_4$, FeCl$_3$

Scheme 3.24

and triose.[35] The reaction is, however, more subtle than these results suggest, as is illustrated by the acetolyses of methyl glucopyranosides **3.57** and **3.59** (Scheme 3.24). Thus, the α-anomer **3.57**, on treatment with acetic anhydride in the presence of sulphuric acid and iron(III) chloride, gives preponderantly (91%) α- and β-glucopyranose acetates **3.61**, whereas the β-glycoside **3.59**, under the same reaction conditions, gives the α- and β-glucofuranose acetates **3.62** and the acyclic heptaacetate **3.63** in 48 and 23% yields, respectively, accompanied by only 24% of the α- and β-glucopyranose acetates **3.61**.[20,36] Consequently, the α-anomer **3.57** can be considered to react exclusively by way of the cyclic ion **3.58**, formed by glycosyl aglycon–oxygen bond cleavage, whereas the acyclic acetyl acetal **3.64**, formed by acetolysis of the acyclic ion **3.60** arising by glycosyl ring oxygen bond cleavage, is the intermediate through which the β-glycoside **3.59** mainly reacts. This is not the exclusive reaction path for the β-glycoside since the acetyl acetal

3.64 affords only furanose acetates **3.62** and heptaacetate **3.63** when subjected to the reaction conditions. The 24% of pyranose acetates **3.61** is presumed to arise by the alternative route involving cyclic ion **3.58**.

All α-pyranosides do not follow the same reaction pathway since acetolysis of methyl α-D-mannopyranoside, for example, gives a mixture of products in the ratio of 3:2 arising from cyclic and acyclic intermediate cations, respectively, and methyl α-D-xylopyranoside gives, in the presence of boron trifluoride, acetolysis products exclusively via the acyclic cation.[37] Thus, in contrast to pyranoside hydrolysis, for which all the evidence suggests that only the cyclic carbocation mechanism operates (see Section 3.1.1.b.i), different pathways are open for pyranoside acetolysis. The factors which determine the courses of these reactions are not yet known and rationalization is not possible by consideration of the basicities of the oxygen atoms at C-1 alone; the influences of different Lewis acid catalysts would merit examination.

(viii) C-Allyl glycosides from glycosides

Treatment of fully benzylated methyl pyranosides in acetonitrile with allyltrimethylsilane and trimethylsilyl triflate gives α-C-glycoside derivatives, as illustrated by the conversion of methyl tetra-O-benzyl-α-D-glucopyranoside into the corresponding α-C-allyl glycoside derivative as the major product (see, Scheme 3.120, Section 3.3.3.d).[38] Closely related reactions have been observed with glycosyl acetates and halides (see Sections 3.2.1.c.iii and 3.3.3.d, respectively). Methods for C-glycoside syntheses have been reviewed[39].

(ix) Thiolysis of glycosides

Transthioglycosidation by acid-catalysed thiolysis of simple O-glycosides can lead to 1-thioglycosides in a process closely related to the Fischer glycosidation, but the reaction has not been studied in detail.[6] Thus, treatment of methyl α-D-mannopyranoside with ethanethiol and concentrated hydrochloric acid gives a low yield of ethyl 1-thio-β-D-mannopyranoside.[40] A more successful protocol involves treating fully etherified O-glycopyranosides with trimethyl(thio-alkyl or aryl) silanes in the presence of zinc iodide and tetrabutylammonium iodide.[41] Under these

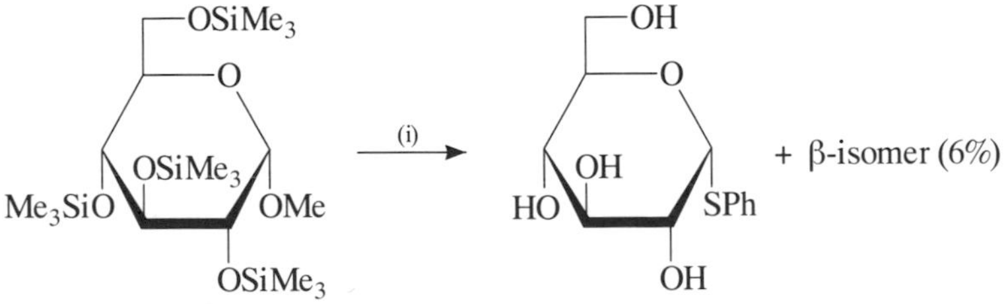

(i) PhSSiMe$_3$, ZnI$_2$, Bu$_4$NI, (CH$_2$Cl)$_2$

Scheme 3.25

conditions, and by use of thiophenyltrimethylsilane with fully trimethylsilylated methyl α-D-glucopyranoside **3.65**, S-phenyl 1-thio-α-D-glucopyranoside **3.66** is formed as the major product, as illustrated in Scheme 3.25. The same procedure is also successful with 2-deoxyglycosides.[42]

(x)   Conversion of glycosides into glycosyl halides

Methyl glycosides of fully protected sugars yield the corresponding glycosyl chlorides or bromides when treated, in refluxing solutions in chloroform, with the relevant α,α-dihalomethyl methyl ether in the presence of a zinc dihalide.[43] Yields for glycosyl chlorides are typically 70%, whereas for the bromides they are only about 50%. A similar transformation can be carried out on glycosyl esters.

Pyranosyl and furanosyl bromides can also be formed by free radical bromination of the corresponding benzyl glycosides. Thus, when carbon tetrachloride solutions of these glycosides are irradiated in the presence of bromotrichloromethane, or heated with NBS, dibromination occurs to give unstable phenyldibromomethyl glycosides which dissociate to give glycosyl bromides typically in 70–80% yield[44]

$$\text{GlcOCH}_2\text{Ph} \longrightarrow \text{GlcOCHBrPh} \longrightarrow \text{GlcOCBr}_2\text{Ph} \longrightarrow \text{GlcBr} + \text{PhCOBr}$$

(xi)   Oxidation of glycosides

Ozone reacts with glycopyranosides that have equatorially oriented aglycons to give the corresponding aldonic esters. Thus, methyl β-D-glucopyranoside tetraacetate **3.67** (Scheme 3.26a) gives methyl gluconate 2,3,4,5,6-pentaacetate **3.71** when so treated. The transformation proceeds under strong stereoelectronic control with the initial abstraction of the axial hydride ion from C-1 by the ozone being favoured

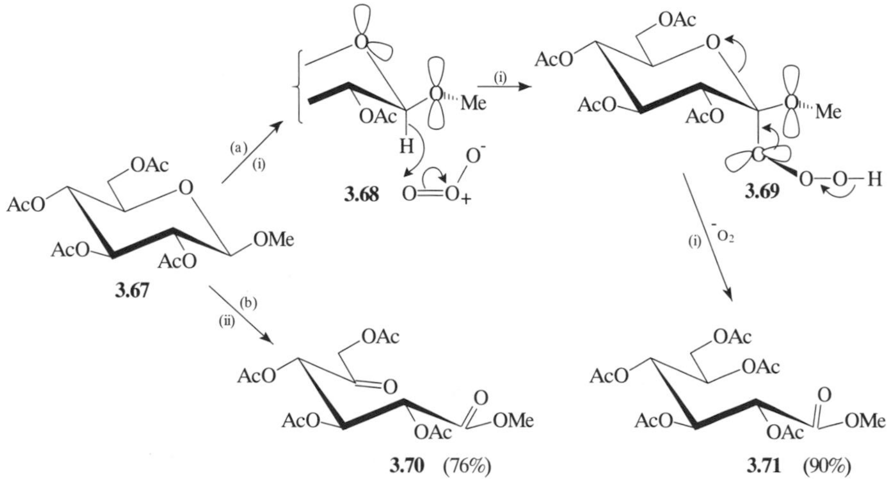

(i) O_3, Ac_2O, NaOAc; (ii) CrO_3, AcOH

Scheme 3.26

because the lone pair orbitals on both oxygen atoms bonded to C-1 can be arranged antiperiplanar to the C–H bond (see **3.68**). The hydrotrioxide **3.69**, which is subsequently formed, then breaks down to give the tetra-*O*-acetyl methyl gluconate which undergoes further acetylation in the reaction medium yielding the ester **3.71**.[45]

Although **3.69** gives no lactone on breakdown, it has been observed that some orthoacids whose structures are closely related to **3.69** decompose less stereoselectively to yield lactones in addition to aldonate esters.[46] *α*-Glycopyranosides in which the H-1 is equatorially disposed are not oxidized by ozone.

The action of chromium(VI) oxide in acetic acid on fully protected glycosides is similarly dependent on glycosidic configuration.[47] The *β*-compound **3.67** is oxidized smoothly by this oxidant which brings about a ring opening analogous to that observed with ozone. However, methyl 2,3,4,6-tetra-*O*-acetyl-D-*xylo*-5-hexulosonate **3.70**, isolated in good yield, is the product that is formed by a subsequent oxidation of the C-5 hydroxyl group in the initial ester (Scheme 3.26b).

(xii)   Spiroketalization of glycosides

Glycoside aglycons in which a radical centre can be generated will, if their structures are correctly designed, abstract anomeric hydrogen atoms intramolecularly to give radicals at this centre that may subsequently react with the aglycon to form spiroketals. The work of Descotes's group on photoinduced hydrogen abstractions by ketones using 3-oxobutyl 3,4,6-tri-*O*-acetyl-*β*-D-glucopyranoside

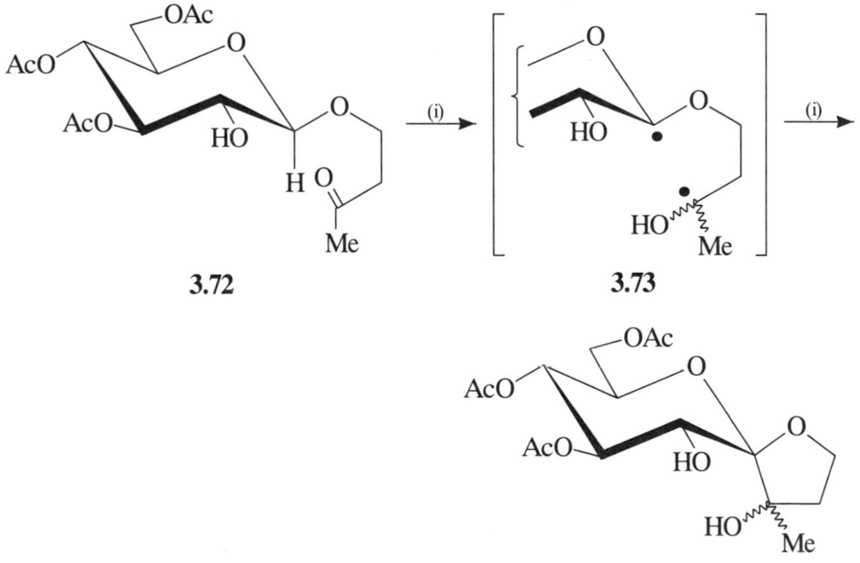

3.72   3.73

3.74   (44%)

(i) *hv*, PhH

Scheme 3.27

3.72 (Scheme 3.27) illustrates this strategy nicely, since u.v. irradiation of this glucoside in benzene gives the β-spiroketal as the *cis/trans* mixture **3.74**.[48,49] This transformation is efficient because the ubiquitous Norrish type II reaction is ruled out for compound **3.72** since there are no γ-hydrogen atoms available to its excited carbonyl group. Of the alternative accessible hydrogens, that at the δ-position is activated by the anomeric oxygen atoms and in consequence is efficiently abstracted under favourable stereoelectronic control to give biradical **3.73** as shown. Ring closure occurs stereospecifically at the anomeric centre by attack at the α-face, thus giving the β-spiroketal **3.74**. The α-anomer of **3.72** reacts only slowly under the same conditions to give unidentified products, probably because abstraction of equatorial anomeric hydrogen atoms is stereoelectronically unfavourable (see Section 3.3.4).

(c)   Intramolecular glycosides — 1,2-, 1,3-, 1,4- and 1,6-anhydroaldoses

In normal glycosides the aglycon and the glycosyl moieties are derived from two separate molecules, but because of the polyhydroxylated nature of sugars it is possible for intramolecular glycosides to be formed by the bonding of an oxygen atom of a hydroxyl group to the anomeric carbon atom of a pyranosyl or furanosyl ring to give anhydropyranoses or anhydrofuranoses. Such compounds are sometimes called *glycosans*. They have all been examined as potential monomer feedstocks for chain growth polymerization and thus as sources of chemically produced specific polysaccharides.[50]

When a sugar is heated in dilute aqueous mineral acid a pseudo-equilibrium is established and among the components are anhydrides formed by reversible intramolecular glycosidation reactions. In principle, any suitably positioned hydroxyl group can take part in this reaction, but the 1,6-anhydrohexopyranoses are the most stable compounds which can be produced in this way. The D-isomers exist with the pyranoid rings in the $^1C_4$ conformation and consequently the common hexoses D-glucose, D-galactose and D-mannose, which are stable in the $^4C_1$ conformation, form these anhydrides only to a minor extent (0.2, 0.8 and 0.8%, respectively). On the other hand, D-aldoses, which are less stable in the $^4C_1$ chair form, i.e., D-talose, D-allose, D-gulose, D-altrose and D-idose, exist in appreciable proportions as their 1,6-anhydrides ($\geq$ 2.8, $\geq$ 14, 65, 65 and $\leqslant$ 86%, respectively).[51]

Inspection of the two extreme examples of the 1,6-anhydrohexopyranoses, **3.75** and **3.76**, formed from D-glucose and D-idose, respectively, shows that the intramolecular steric interactions in the former, which has all its hydroxyl groups axially disposed, are the most severe, whereas in the latter all the hydroxyl groups are equatorially arranged and destabilizing factors are minimal. The other hexoses fall between these two extremes.

In non-protic solvents, sugars behave differently and 1,6-anhydrofuranose isomers are formed under acid conditions. Thus, when allose, galactose, glucose, mannose and talose are heated with azeotropic removal of water in DMF containing tosic acid, 78, 87, 35, 22 and 86% of the respective 1,6-anhydrofuranoses are

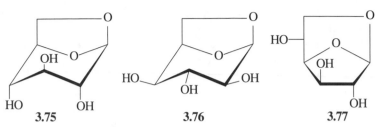

3.75   **3.76**   **3.77**
1,6-Anhydro-D-glucopyranose   1,6-Anhydro-D-idopyranose   1,6-Anhydro-D-glucofuranose

formed, D-glucose giving compound **3.77**. The galactose isomer can be isolated in 33% yield by direct crystallization.[52]

1,6-Anhydro compounds are more common than 1,2-, 1,3- or 1,4-anhydrides because of their greater stability, and they can be synthesized by a wider range of methods.[51] The 1,6-anhydropyranoses of idose, altrose and gulose can be suitably prepared by treatment of the free sugars with acid as outlined above. Alternatively, thermal depolymerization of some polysaccharides can be employed, as for example in the preparation of 1,6-anhydro-D-mannopyranose which is obtained as a distillate when a mannan, found in ivory nuts and composed of β-(1→4)-linked D-mannopyranose units, is pyrolysed under reduced pressure. A plausible mechanism for this reaction involves a 1,4-anhydride intermediate as shown for the alkaline cleavage of phenyl β-D-mannopyranoside (Scheme 3.19). Thus, the pyranose ring must adopt the $_4B_1$ conformation and then the $^1C_4$ conformation, and since this would be more easily achieved by a non-reducing terminal monosaccharide unit, a sequential loss of monomer units appears to be likely. Similar treatment of starch affords a distillate from which 1,6-anhydro-D-glucopyranose can be isolated, but further fractionation of the mother liquor yields a small amount of 1,6-anhydro-D-glucofuranose **3.77** which is also a 1,4-anhydropyranose.

Hexose 1,6-anhydrides are also formed by O-6 attack at the anomeric centre when compounds possessing a good leaving group at C-1 are treated with alkali. (Glucopyranosyl fluoride and 2,3,4-tri-O-acetylglucopyranosyl bromide, for example, both give 1,6-anhydroglucose on treatment with base.) A related reaction is that depicted in Scheme 3.18 and involves O-6 opening of a 1,2-anhydride.

1,6-Anhydrides have similar properties to the alkyl glycosides. They are stable in aqueous base but are susceptible to acid-catalysed ring opening by nucleophiles to give pyranosyl derivatives. For example, 2,3,4-tri-O-acetyl-1,6-anhydro-β-D-glucopyranose yields 2,3,4-tri-O-acetyl-α-D-glucopyranosyl bromide when it is treated with hydrogen bromide in acetic acid, and acid-catalysed hydrolysis of an unsubstituted 1,6-anhydride regenerates the free sugar to an extent which depends upon the structure of the sugar (see above). 1,6-Anhydro-2,3,4-tri-O-benzyl-β-D-mannopyranose upon treatment with the Lewis acid catalyst phosphorus pentafluoride in dichloromethane at $-78\,^\circ$C gives, after debenzylation, a stereoregular 1,6-linked α-mannan containing up to 3000 monomer units.[53] Bimolecular attack on the conjugate acid of the 1,6-anhydride, rather than on an open ion intermediate, is suggested by the stereochemistry at C-1 in the polymer.

1,2-Anhydrides are highly reactive compounds because their already strained oxirane rings are also parts of acetals, and this makes them difficult to isolate and store. Until recently they were synthesized from partially functionalized pyranoses possessing free C-2 hydroxyl groups which could be induced to displace leaving groups from C-1 under appropriate reaction conditions.[54] Although the method has not been widely developed, it is used to prepare Brigl's anhydride **3.80**, the earliest example of this class of compound. Thus, treatment of 3,4,6-tri-*O*-acetyl-2-*O*-trichloroacetyl-β-D-glucopyranosyl chloride **3.78** with ammonia in benzene, as illustrated in Scheme 3.28, selectively removes the trichloroacetyl group and subsequently facilitates attack of O-2 on the anomeric centre to displace the chlorine in **3.79** to give **3.80** in 70% yield.[55] This anhydro sugar was used in the first chemical synthesis of sucrose (see Section 8.1.1.a).

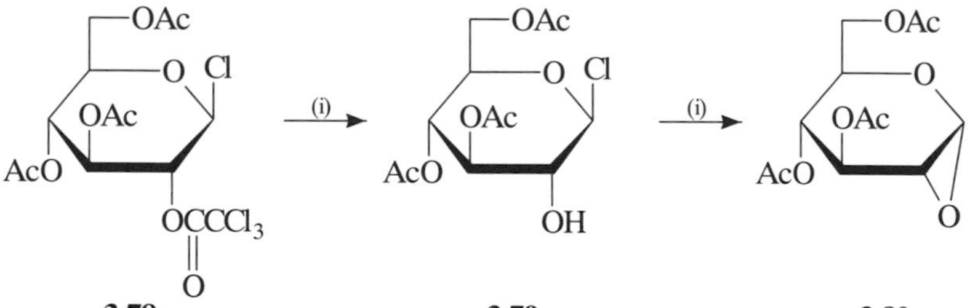

3.78   **3.79**   **3.80**

(i) NH$_3$, PhH

Scheme 3.28

The obvious route to these anhydrides, by direct epoxidation of the double bonds present in suitably protected glycals with peracids, leads to products formed by subsequent opening of the 1,2-anhydro ring by the electrophilic and nucleophilic compounds present in the reaction medium. However, the situation changed with the introduction of dimethyldioxirane which epoxidizes alkenes under very mild conditions, and this has led to the ready availability of a range of 1,2-anhydrides.[56] Thus, for example, tri-*O*-benzyl-D-glucal **3.81** is converted into the 1,2-anhydro-α-D-glucopyranose derivative **3.82** when treated with this reagent, as indicated in Scheme 3.29, and the product is obtained in quantitative yield simply by evaporation of the solvent and the acetone formed as by-product.

The epoxide oxygen–C-1 bonds in these compounds are easily cleaved and consequently compound **3.80**, for example, reduces Fehling's solution and cannot be deacetylated with sodium methoxide without opening of the epoxide ring. They undergo displacements at C-1 with a wide variety of nucleophiles, and if the resident protecting groups are non-participatory, the reactions usually occur with inversion of configuration at C-1. This is illustrated in Scheme 3.29 by a range of trans-formations of 1,2-anhydro-tri-*O*-benzylglucose **3.82**, which can be made and used without isolation.

The formation of products with unprotected C-2 hydroxyl groups could have proved troublesome if these had reacted intermolecularly with the epoxide, but this

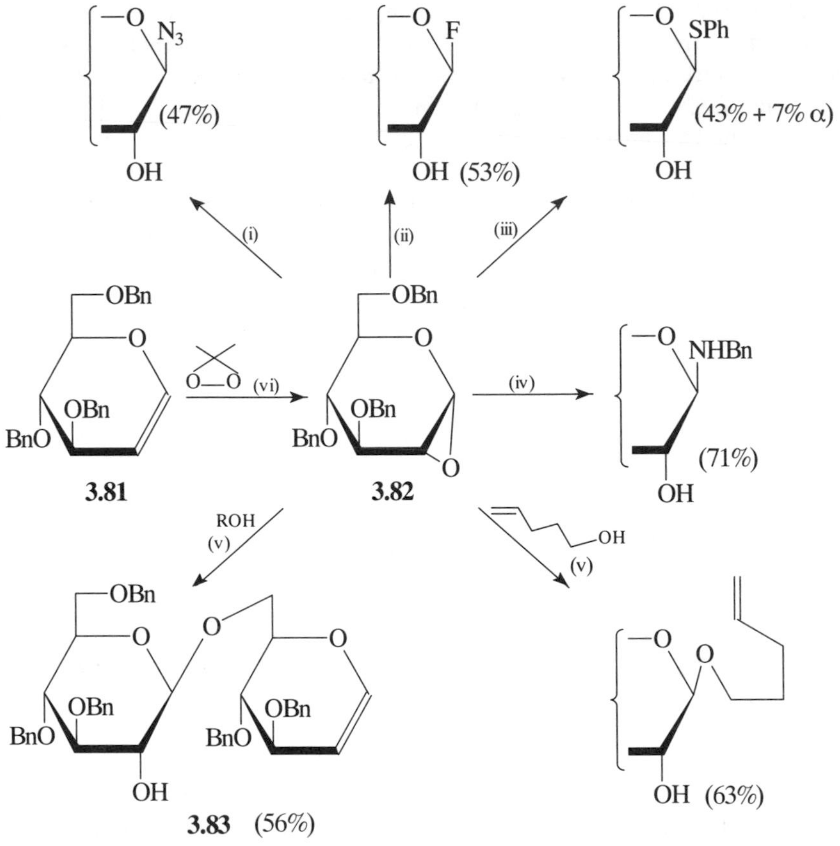

ROH = 3,4-di-O-benzyl-D-glucal

(i) Bu$_4$NN$_3$; (ii) Bu$_4$NF; (iii) Bu$_4$NSPh; (iv) BnNH$_2$, ZnCl$_2$; (v) ZnCl$_2$; (vi) Me$_2$CO, CH$_2$Cl$_2$. All reactions in THF

Scheme 3.29

has not so far been observed. On the contrary, such products can be used with advantage in the construction of complex saccharides and oligosaccharides, since subsequent glycosylations or other transformations may be carried out at the O-2 position.

The reaction of **3.82** with partially protected glycals is of value in the synthesis of oligosaccharides since the unsaturated oligomers formed (see, for example, disaccharide **3.83**) can be epoxidized and made to react with further acceptor glycals.

1,3-Anhydrides are less common than 1,2-anhydrides, although they are prepared in an analogous fashion by displacement of a leaving group from the anomeric centre by O-3. Thus, treatment of 3-O-acetyl-2,4,6-tri-O-benzyl-α-D-mannosyl chloride **3.84** with potassium t-butoxide gives the mannopyranose oxetane derivative **3.85** in high yield, as illustrated in Scheme 3.30a, and the glucose isomer **3.86**

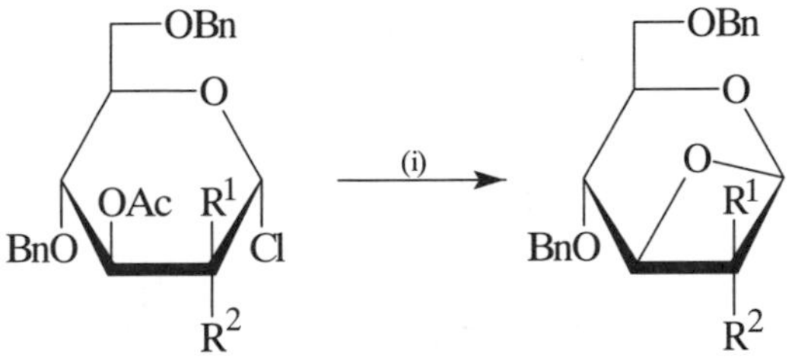

3.84 $R^1 = OBn$, $R^2 = H$ (a) **3.85** $R^1 = OBn$, $R^2 = H$ (95%)

3.86 $R^1 = H$, $R^2 = OBn$ (b) **3.87** $R^1 = H$, $R^2 = OBn$ (95%)

(a) (i) **3.84**, t-BuOK, THF, room temperature, 2 h
(b) (i) **3.86**, MeLi, EtOH, THF, reflux, 24 h

Scheme 3.30

reacts similarly when treated with methyllithium to give **3.87** (Scheme 3.30b), the reactions occurring by sequential deacetylation and intramolecular nucleophilic displacement.[57] These compounds have been prepared for use as polymer precursors but they are also of interest because they possess the 2,6-dioxa[3.1.1]bicycloheptane ring system that is present in the reactive blood platelet aggregator thromboxane A_2 (TXA$_2$), which, being highly susceptible to hydrolysis even under neutral conditions, is very difficult to isolate. The relative stability of compounds **3.85** and **3.87** arises from the presence at C-2 of benzyloxy groups, which would be expected to retard their rates of hydrolysis compared with analogues in which both R^1 and R^2 are hydrogen for the same mechanistic reasons that aldopyranosides hydrolyse slower than do their 2-deoxy analogues (see Section 3.1.1.b.i). Other compounds of this class have been synthesized from non-carbohydrate sources by way of intramolecular nucleophilic displacements, and it is significant that for ring closure electron-withdrawing groups at C-2 are necessary.[58]

A few 1,4-anhydropyranose derivatives (which are also 1,5-anhydrofuranoses) have been synthesized using the method of ring closure brought about by intramolecular nucleophilic displacements. For example, 1,4-anhydro-2,3,6-tri-O-benzyl-α-D-glucopyranose **3.88** is obtained when 2,3,6-tri-O-benzyl-β-D-glucopyranosyl fluoride is treated with sodium hydroxide in methanol.[59] Conversely, O-1 attack at C-4 occurs when 2,3,6-tri-O-methyl-4-O-tosyl-D-glucopyranose is treated with sodium isopropoxide, giving 1,4-anhydro-2,3,6-tri-O-methyl-β-D-galactopyranose **3.89** ($R^1 = OMe$, $R^2 = Me$).[60] Related displacements occur under less basic conditions with 1-O-acyl 4-sulphonates as in the formation of the 1,4-anhydride **3.89** ($R^1 = N_3$, $R^2 = Bz$) from the β-acetate **3.90** by treatment with sodium azide in hot DMF. The anomeric α-acetate also gives the same product, but in this case the first-formed α-C-1 oxyanion must anomerize and then displace the mesyloxy group.[61]

A 1,4-anhydride has been postulated as an intermediate in the alkaline hydrolysis of phenyl β-D-mannopyranoside (Scheme 3.19).

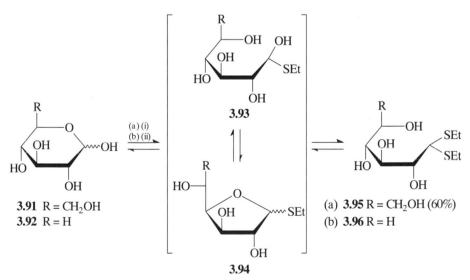

3.88   **3.89**   **3.90**

3.1.2 REACTIONS WITH SULPHUR NUCLEOPHILES — THIOACETALS AND THIOGLYCOSIDES

(a) Preparation of dithioacetals

Sugars react rapidly with alkanethiols in the presence of acid catalysts at room temperature or below to give acyclic dialkyl dithioacetals as the main products, and therefore the reaction is markedly different from the Fischer glycosidation.[62,63] The most common method for preparing aldose dithioacetals involves the use of alkane- or arene-thiols in concentrated hydrochloric acid, as illustrated in Scheme 3.31(a) by the conversion of D-glucose **3.91** into glucose diethyl dithioacetal **3.95**.[64] Typically, the product precipitates from the reaction medium, thus favouring its isolation in high yield and protecting it from hydrolysis. The reaction has been studied under conditions which permit identification of some of the intermediates.[65] D-Xylose,

(a) (i) **3.91**, EtSH, concentrated HCl, 20 °C
(b) (ii) **3.92**, EtSH, DMF (20%), HCl (2.4%)

Scheme 3.31

for example, when treated with ethanethiol containing DMF (20%) and anhydrous hydrogen chloride at 25 °C, yields initially the thiofuranosides **3.94** (R = H) which are then converted into the diethyl dithioacetal **3.96** as the thermodynamically controlled product (Scheme 3.31b). Under these conditions pyranosides do not comprise more than 6% of the reaction products at any stage. The dithioacetal **3.96** is probably formed by the sugar reacting initially through one of its modifications to give the hemithioacetal **3.93** (R = H). This is then protonated on the hemithioacetal oxygen atom to give a conjugate acid which rapidly ring closes with O-4 participation to give the furanosides **3.94** (R = H); alternatively, the furanose form of the free sugar could give the thiofuranosides directly (cf. Scheme 3.7). These are unstable relative to the thioacetal **3.96** because they possess an acetal oxygen atom which gives a higher standing concentration of conjugate acid than is formed from the dithioacetal. Consequently, the thiofuranosides **3.94** (R = H) are more reactive under these reaction conditions than the dithioacetal **3.96**, and the concentration of the latter increases at the expense of the former.

(b)   Reactions of dithioacetals

When D-galactose diethyl dithioacetal **3.97** is treated with dilute aqueous hydrochloric acid, ethyl 1-thio-D-galactofuranosides **3.99** are formed (Scheme 3.32a) via the hemithioacetal **3.98** (R = H), which arises by partial hydrolysis of the starting material.[62] This step is followed by kinetically controlled ring closure (involving O-4) of the type discussed above (cf. the conversion of **3.93** to **3.94** in Scheme 3.31). Prolonged contact with aqueous acid converts these thiofuranosides into thiopyranosides.

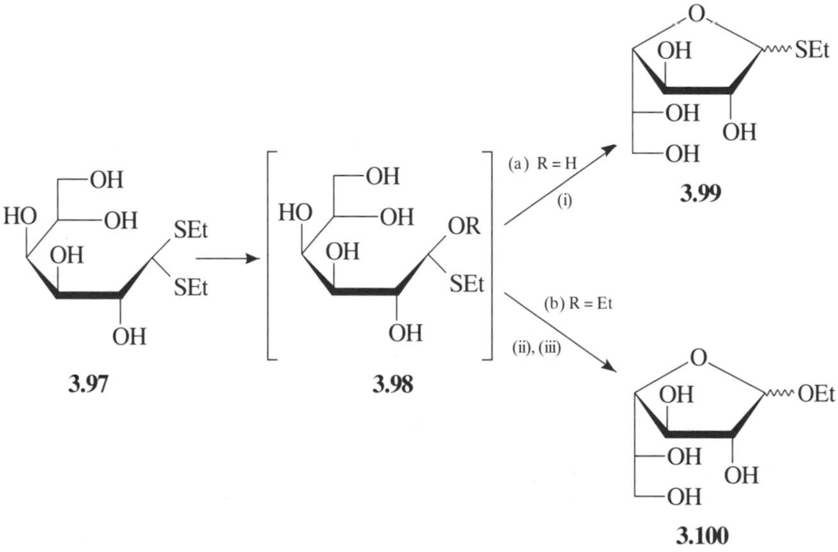

(i) HCl, H_2O ($\sim$ 1%) 20 °C, 20 h; (ii) HgO, 5 h; (iii) EtOH, $HgCl_2$, HgO

Scheme 3.32

The scope of the reaction of dithioacetals is increased by using anhydrous alcohols and mercury(II) chloride in place of aqueous hydrochloric acid.[66] Mercury(II) ions are able to function as selective electrophilic catalysts at sulphur atoms because organosulphur compounds such as dialkyl sulphides, in contrast to their oxygen analogues, complex with mercury(II) salts. Under these conditions, alkyl glycofuranosides are obtained and this represents one of the best general methods for preparing this class of compound. The formation of ethyl D-galactofuranosides by this route is illustrated in Scheme 3.32(b). The initial step in this reaction involves formation of the ethyl thioethyl acetal **3.98** (R = Et) by mercury(II)-catalysed ethanolysis of the diethyl dithioacetal. This is followed by the selective removal of the thioethyl grouping (catalysed by mercury(II) ions) with concomitant ring closure to give the ethyl galactofuranosides **3.100**.

Dithioacetals are the most useful intermediates for preparing acyclic derivatives of sugars since it is possible to protect the hydroxyl groups and then regenerate the aldehydo group either by treating the protected dithioacetal with aqueous mercury(II) salts or, more rapidly, by oxidative hydrolysis with aqueous NBS, as shown in Scheme 3.33(a).[67] The most common protecting groups are acetyl or benzoyl residues, but others have been used.

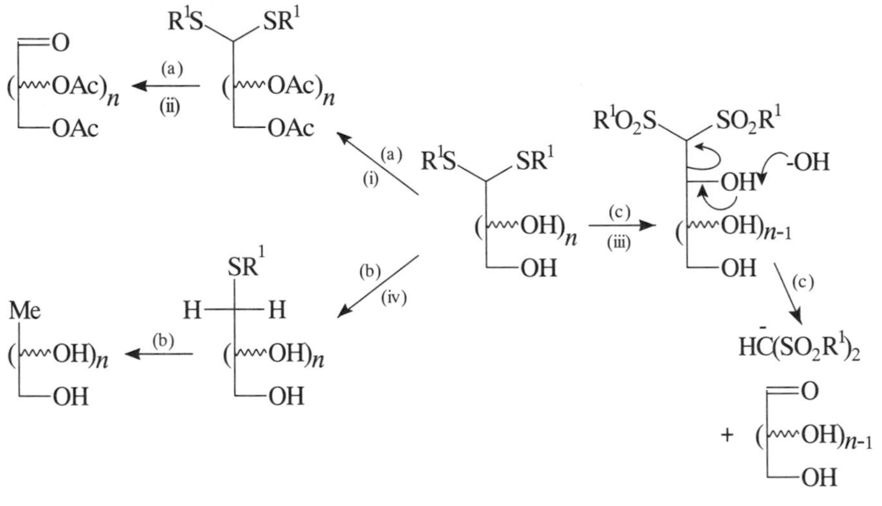

(i) Ac_2O, Py; (ii) $HgCl_2$, $CdCO_3$, H_2O, Me_2CO, 22 °C, 48 h (78% from the *gluco*-compound) or NBS, $CdCO_3$, H_2O, Me_2CO, 0 °C, 3 min (77% from the *gluco*-compound); (iii) R^2CO_3H; (iv) H_2, Ni

Scheme 3.33

Dithioacetals can be converted into the corresponding 1-deoxyalditol compounds by reductive desulphurization with Raney nickel as catalyst (Scheme 3.33b), and under controlled conditions the intermediate 1-S-alkyl-1-thioalditols can be obtained. Oxidation, on the other hand, yields the corresponding disulphones which are useful compounds since mild aqueous base brings about a retroaldol reaction and reduces the carbon chain by one carbon atom (Scheme 3.33c). In this way D-glucose, for example, can be degraded to D-arabinose, and the method compares favourably with other means of descending the aldose series (e.g. see Sections 2.5.3.c, 3.1.3.b and 3.1.6.a.iv).

(c) Preparation, properties and reactions of thioglycosides

(i) Preparation and general properties of thioglycosides

A large number of 1-thioglucosides occur in nature, most commonly in the mustard oil glucosides or glucosinolates; sinigrin **8.79**, for example, has been isolated from the seed of black mustard and the root of horseradish, but the most notable example found in nature is the antibacterial lincomycin **8.82**.

1-Thioglycosides have been prepared in various ways: from glycopyranosides (Section 3.1.1.b.ix), from glycopyranosyl esters (Sections 3.2.1.c.ii and 3.2.3.b.ii), from glycopyranosyl halides (Sections 3.3.3.c and 3.3.5), from 1-thioaldoses (see below) and from dialkyl dithioacetals, which give thiofuranosides as described above in Section 3.1.2.b. In most of their properties these compounds are similar to the O-glycosides, but they show expected differences which result from the different properties of sulphur and oxygen.[63]

1-Thioglycopyranosides are stable in aqueous base at normal temperatures but are hydrolysed by aqueous mineral acid at a slower rate than are the analogous oxygenated glycosides. The difference is small for ethyl β-D-glucopyranoside and its 1-thio analogue but quite large for the phenyl analogues. The slower hydrolysis rate for thioglycosides is to be expected because the extent of conjugate acid formation will be smaller under given conditions than for the oxygen glycosides, since sulphur is the weaker base.

(ii) Displacement reactions of thioglycosides induced by thiophilic reagents

The 1-thioglycosides differ from O-glycosides in their reactions with several electrophiles since they complex with a range of sulphur-specific reagents to form sulphonium cations which, as good leaving groups, are readily displaced. Consequently, thioglycosides have become valuable intermediates in a variety of tranformations, notably glycosylations. Thiophilic metal salts of mercury(II), lead(II), copper(II) and silver(I) were originally the most popular activators for these reactions.[68] Thus, ethyl thioglycosides are converted by silver(I) benzoate into glycosyl benzoates **3.102** *via* the intermediates **3.101** (E = Ag, X = OBz), and phenyl thioglycosides give ethyl glycosides **3.104** when treated with mercury(II) acetate in the presence of ethanol (by way of intermediate **3.101**, E = Hg, X = EtO). Efforts to use these activators in the presence of sugar alcohols, as a route to disaccharides, have met with only limited success, since the glycosylating agents **3.101** they produce are usually of only modest reactivity. However, with the thioglycosides of 2-deoxy sugars, mercury(II) chloride is an effective activator, as demonstrated by the synthesis of the oligosaccharide portion of digitoxin, and lead(II) perchlorate successfully activates a 2-deoxythioglycoside to produce the disaccharide required for the synthesis of avermectin.

Bromination of phenyl thioglycosides under mild conditions with NBS produces an intermediate **3.101** in which the bromosulphonium ion (E = Br) is accompanied by the weakly nucleophilic succinimidyl anion (X). Therefore, if this reaction is carried out in solutions containing alcohols, hydroxyl attack may occur.[68] This procedure has been used to prepare a range of anomeric mixtures of O-glycosides (**3.104**, for example) and a few disaccharides, but has not found wide use in oligosaccharide synthesis. In a recently developed related reaction, glycosyl fluorides **3.103** have been prepared in excellent yields by treating phenyl thioglycosides with NBS and diethylaminosulphur trifluoride (DAST).[42] This constitutes the best route to this class of glycosylating agents (see Section 3.3.5). On the other hand, thioglycosides have long been known as a source of glycosyl bromides and chlorides. The transformation may be brought about by reaction with the corresponding halogen, as shown by the formation of **3.106** via intermediate **3.101**, (E = X = Br).[69] The last two reactions provide good methods for preparing the relatively inaccessible glycofuranosyl halides Section 3.3.2.b.[68]

Recent work on the activation of thioglycosides has centred on finding more potent thiophilic reagents that afford reactive glycosylating agents suitable for general application in oligosaccharide synthesis.[32,70,71] The following have been introduced for this purpose: methyl triflate, dimethyl(methylthio)sulphonium triflate (Me$_2$SSMeOTf or DMTST), nitrosyl tetrafluoroborate (NOBF$_4$), benzeneselenenyl triflate (PhSeOTf), methylsulphenyl triflate (MeSOTf), NIS/TfOH and iodonium dicollidine perchlorate (IDCP), among which DMTST and IDCP seem to be the most promising, and this general approach has become one of the most efficient methods available for the synthesis of oligosaccharides (Chapter 6). Thus, methyl 1-thio-β-D-glucopyranoside tetrabenzoate reacts with 1,2,3,4-tetra-O-benzoyl-β-D-glucopyranose, after treatment with DMTST, to give, in high yield, the β-(1→6)-linked disaccharide represented by **3.105** in Scheme 3.34. IDCP has been an effective promoter for chemospecific glycosidation of partially benzoylated thioglycoside disarmed acceptors by perbenzylated thioglycoside 'armed' donors (cf. Section 3.1.1.b.vi).

An ingenious way of making glycosylating agents from thioglycosides has been found by incorporating into the aglycon a suitably positioned nitrogen atom to supplement the coordination of a Lewis acid activator, as shown in **3.107** for pyridin-2-yl 1-thio-β-D-glucopyranoside coordinated to mercury(II) nitrate.[72] Thus, treatment of the pyridinyl thioglucoside in acetonitrile containing mercury(II) nitrate and ethanol, used in a large excess to compete with the thioglucoside, which is unprotected, gives by way of complex **3.107** the ethyl glucopyranosides in 85% yield as a 2.1:1 α/β mixture. Modest stereoselectivity typifies these reactions. Remote activation of this type which involves coordination at a site distant from the anomeric heteroatom has been exploited in other glycosidation procedures (see **3.33** and Sections 3.1.1.b.iv, 3.1.1.b.vi and 3.2.1.c.i).

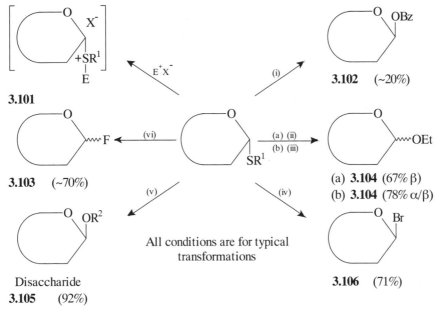

3.101

3.102   (~20%)

3.103   (~70%)

(a) 3.104 (67% β)
(b) 3.104 (78% α/β)

Disaccharide
3.105   (92%)

3.106   (71%)

(i) AgOBz, MeCN, reflux, 3 h; (ii) Hg(OAc)$_2$, EtOH, 25 °C, 0.5 h; (iii) NBS, EtOH, CH$_2$Cl$_2$, molecular sieves, 25 °C, 0.5 h; (iv) Br$_2$, Et$_2$O, 20 °C, 7 min; (v) R^{2}OH, CH$_2$Cl$_2$, DMTST, molecular sieves, 20 °C, 1 h; (vi) NBS, DAST, CH$_2$Cl$_2$, −15 °C

Scheme 3.34

3.107

3.108

Remote activation of a different kind utilizes the ability of a thiophilic reagent to differentiate between single and double carbon–sulphur bonds.[73] Thus, when ethyl 3-*O*-benzyl-4,6-*O*-benzylidene-2-*O*-phenoxythiocarbonyl-1-thio-β-D-glucopyranoside **3.109**, which can be prepared in a straightforward manner, is treated with NIS and triflic acid in an ether/dichloroethane solution containing methyl 2,3,4-tri-*O*-benzyl-β-D-glucopyranoside, α-glycosylation occurs at the free 6-hydroxyl group with concomitant epimerization of the donor to give, in good yield, the α-(1→6)-linked 2-(thiomannosyl)glucoside derivative **3.111**. This glyco-sylation–epimerization occurs as outlined in Scheme 3.35. The iodonium ion from NIS/TfOH reacts chemospecifically with the thiono group to give the intermediate ion which dissociates by intramolecular nucleophilic attack at C-2 from the

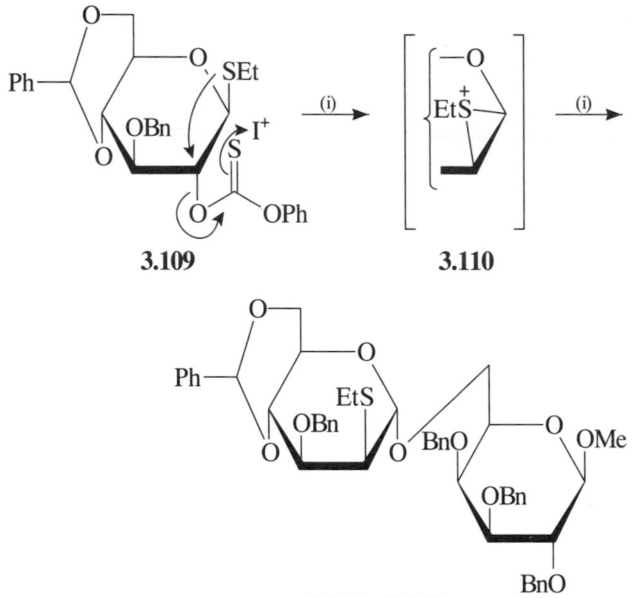

3.109   **3.110**

3.111 (85%)

(i) methyl 2,3,4-tri-*O*-benzyl-β-D-glucopyranoside, NIS, TfOH, Cl(CH$_2$)$_2$Cl, Et$_2$O

Scheme 3.35

trans-disposed thioethyl group at C-1. The episulphonium ion **3.110** so produced then opens by stereospecific nucleophilic attack at C-1 from the alcohol, resulting in migration of the anomeric thioethyl group to C-2 and overall inversion at this site and at the anomeric centre. In accordance with this mechanism α-D-*manno*-analogues of **3.109**, in which the 2-*O*-phenoxythiocarbonyl and the C-1 thioethyl groups are again *trans*-disposed, undergo a similar stereospecific reaction to give 2-ethylthio-β-D-glucosides. Since reductive desulphurization of these products is easily achieved, this method is an ingenious and powerful way of making α- and β-linked disaccharides of 2-deoxy sugars, and this and other methods are discussed in Section 6.4.2.

Glycosyl xanthates, when suitably activated, have been shown by Sinäy and others to be good glycosylating agents particularly valuable for sialylation which cannot be readily achieved with glycosyl halides.[74] Thus, the fully acetylated neuraminic acid xanthate derivative **3.108** has been used with silver triflate and methylsulphenyl bromide in acetonitrile to α-glycosylate suitable partially protected galactose derivatives. By this means *N*-acetylneuraminic acid has been attached to the 3- and 6-positions of galactose, so affording disaccharide elements frequently found at the ends of the oligosaccharide chains of glycoproteins.

Free radical induced displacements applied to phenyl thioglycoside derivatives by use of allylstannanes are mentioned in Section 3.3.4.b as a route to *C*-allyl glycosides.

(iii)   Free radical bromination of thioglycosides

Acylated phenyl thioglycosides react by an alternative free radical route with photoactivated NBS.[75] Thus, when phenyl 2,3,4,6-tetra-O-benzoyl-1-thio-β-D-glucopyranoside **3.114** (R^1 = Bz, R^2 = Ph) is so treated it gives the enone derivative **3.112** (Scheme 3.36a). The reaction is presumably dependent upon an initial hydrogen abstraction from C-1 and consequently β-anomers, with their axial hydrogen atoms at this position, react more readily than the α-forms to produce sulphur-stabilized free radicals which are readily brominated (Section 3.3.4). The bromo derivative **3.113** would be expected to eliminate hydrogen bromide to give a thioglucal, allylic bromination of which would afford a 3-bromo-3-O-benzoylglucal as a precursor of the enone.[76] While the reaction is specific with benzoyl esters, acetates are found also to brominate within the C-2 acetoxy group.

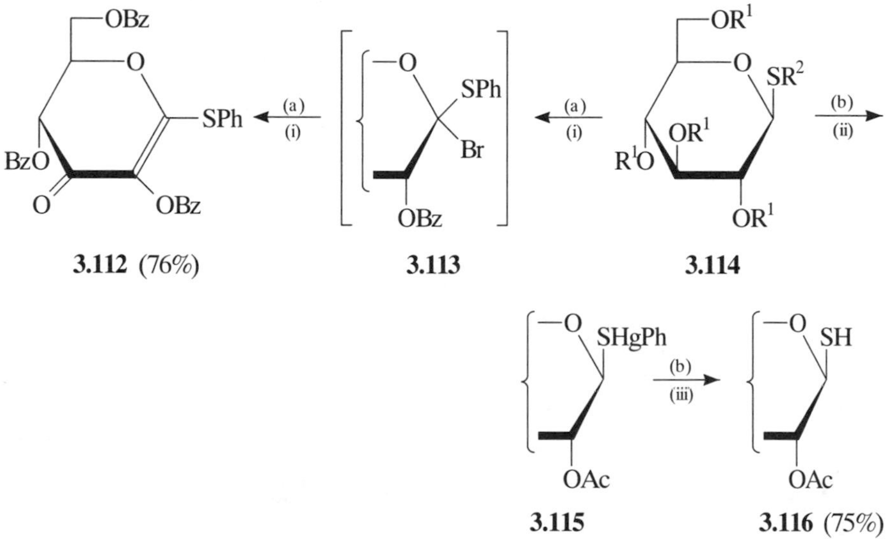

3.112 (76%)   **3.113**   **3.114**

3.115   **3.116** (75%)

(a) (i) **3.114** (R^1 = Bz, R^2 = Ph), NBS, CCl$_4$, reflux, $h\nu$
(b) (ii) **3.114** (R^1 = R^2 = Ac), PhHgOAc, EtOH, reflux, 40 min; (iii) H$_2$S, EtOH
Scheme 3.36

(iv)   Oxidation of thioglycosides — thioglycoside radical cations, sulphoxides and sulphones and their reactions

Under electrolysis conditions, aryl thioglycopyranosides readily undergo anodic oxidation resulting in removal of one electron from the sulphur atom, thereby forming radical cations which readily dissociate to give phenylthiyl radicals and pyranosyl cations; consequently, electrochemical glycosidations are possible, as illustrated in Scheme 3.22(b). O-Phenyl glycosides behave similarly on electrolysis (see Section 3.1.1.b.v), but the reaction of thio compounds benefits from the lower oxidation potential required to generate the sulphur radical cation **3.50** compared with the oxygen equivalent **3.46**. Thus, phenyl 2,3,4,6-tetra-*O*-benzyl-1-thio-β-D-glucopyranoside **3.49** can be used to give the methyl glucoside **3.48** as a 1:3 α/β mixture in 84% yield.[30,77] A range of disaccharides have also been prepared in good yield by this approach. Thus, for example, when an electric current is passed through a solution of phenyl thioglucoside **3.49** in acetonitrile containing lithium tetrafluoroborate and methyl 2,3,6-tri-*O*-benzyl-α-D-glucopyranoside, hepta-*O*-benzyl (1→4)-linked disaccharides are obtained in 73% yield as a 1:3 α/β anomeric mixture.

Oxidation of aryl thioglycosides with *m*-chloroperbenzoic acid in dichloromethane at low temperature gives the corresponding sulphoxides (with chiral sulphur atoms) in good yields, as illustrated in Scheme 3.37, for example, by the conversion of the phenyl thioglucoside **3.117** (R = Bn) into four sulphoxide isomers **3.118** (R = Bn) in 85% yield.[78]

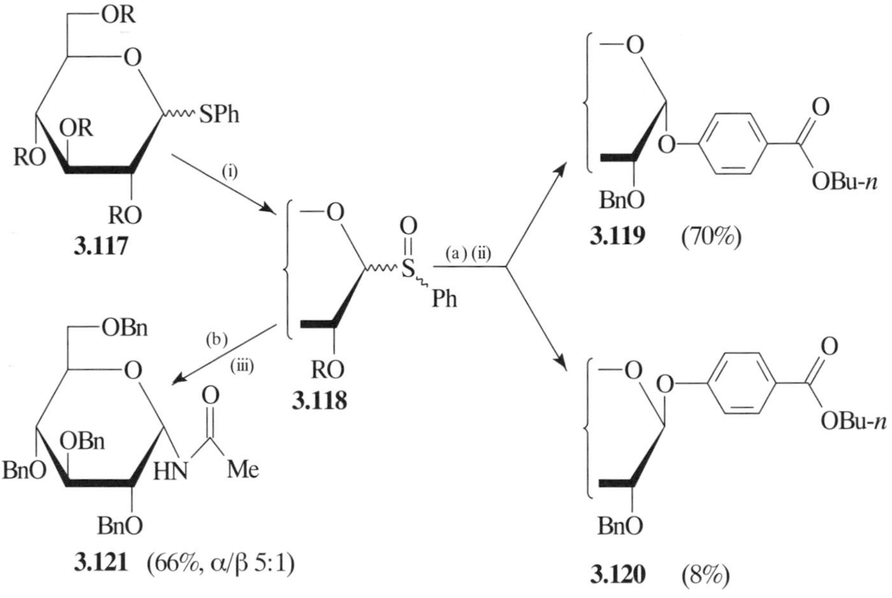

(i) MCPBA, CH$_2$Cl$_2$, −60 to 20 °C; (ii) *n*-butyl 4-hydroxybenzoate, Tf$_2$O, PhMe, −78 to 20 °C; (iii) MeCONHSiMe$_3$, Tf$_2$O, 2,6-di-*t*-butyl-4-methylpyridine, PhMe, −78 to 20 °C

Scheme 3.37

Such aryl sulphoxides, when activated by triflic acid or triflic anhydride in the presence of an acid scavenger, have proved to be extremely effective glycosylating agents for unreactive substrates. Sterically hindered alcohols and phenols and deactivated phenols are rapidly glycosylated at −60 °C in good yields to give preponderantly α-glycosides even from anomeric mixtures of sulphoxides when non-participating groups are present, as illustrated in Scheme 3.37(a) by the conversion of **3.118** (R = Bn) into **3.119** and **3.120** as a 9:1 mixture. If participating groups are present at C-2 (e.g. **3.118**, R = Piv) and dichloromethane is used as solvent, β-glycosides are specifically formed. The efficacy of the method is demonstrated by the direct glycosylation of the poorly reactive nitrogen of N-trimethylsilylacetamide to give the N-glycoside **3.121**, as illustrated in Scheme 3.37(b). The rate-limiting step in these glycosylations is the triflation of the aryl sulphoxide which is dependent upon the substituents present in the phenyl ring, the reactivity increasing in the expected order, i.e. p-nitrophenyl < phenyl < p-anisyl. It is therefore possible to manipulate the reactivity of aryl sulphoxides as glycosyl donors, making them an important class of glycosylating agents for use in iterative oligosaccharide syntheses.[79]

Thioglycosides may be oxidized further to sulphones and this occurs in high yield with peracids between 0° and 20 °C or with potassium manganate(VII).[3]

Whereas most reactions at the anomeric centres described in this chapter are nucleophilic displacements, the polarity of these sites may be reversed in glycosyl sulphones because of the bonded electron-withdrawing sulphonyl group present. Thus, lithiated anomeric species **3.123** may be generated from 2-deoxyglycopyranosyl sulphones **3.122** (R = Tbdms, Me or Bn), as shown in Scheme 3.38. The silylated intermediate **3.123** (R = Tbdms) reacts smoothly with electrophiles to give α-substituted sulphones **3.124** that may be reductively desulphonated with lithium naphthalenide to give the lithiated intermediates **3.125**. Upon protonation, these afford the C-glycosides **3.126** (E = carbon electrophile).

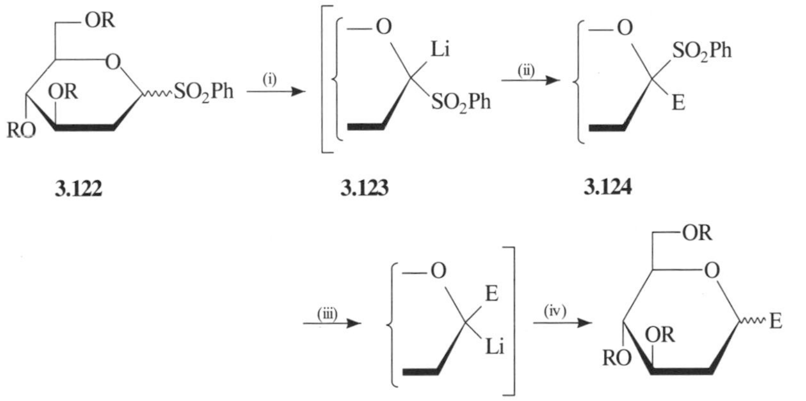

(i) LDA, THF, hexane, $-78\,°C$, 5 min; (ii) E (an electrophilic reagent); (iii) lithium naphthalenide; (iv) MeOH, $-78\,°C$

Scheme 3.38

The stereochemistry of **3.126** depends upon the electrophilic reagent used. With phenyl benzoate (E = Bz), for example, the α-product **3.128** is obtained with good stereoselectivity in 72% yield (α/β 10:1) from **3.122** (R = Tbdms). Since **3.126** (R = Tbdms, E = Bz) is stabilized as enolate **3.127**, preferential protonation at C-1 from the least-hindered equatorial direction gives the product **3.128** with the substituent axial. On the other hand, when benzaldehyde is the electrophile used with **3.122** (R = Tbdms), the intermediate **3.125** (E = CH(OH)Ph) exists as the dilithiated dianion **3.129** which undergoes protonation with retention of configuration to give, after oxidation with pyridinium chlorochromate, the β-anomer of **3.128** in 68% yield.[80]

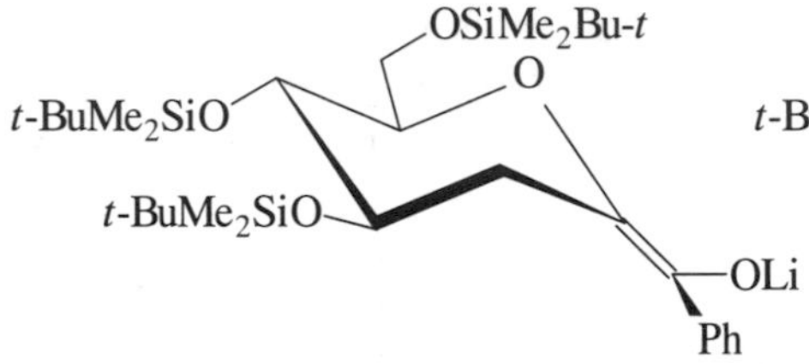

3.127 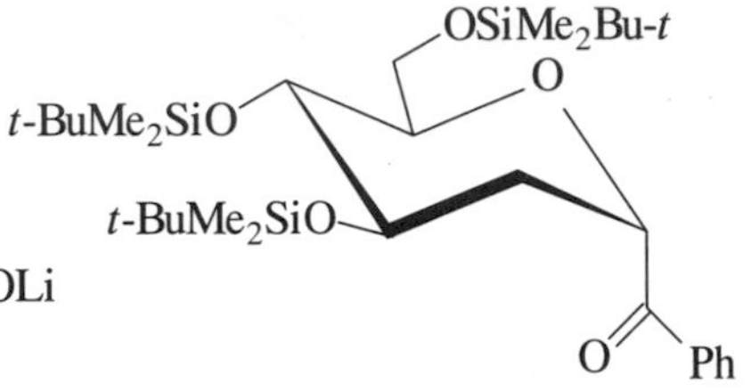   **3.128**

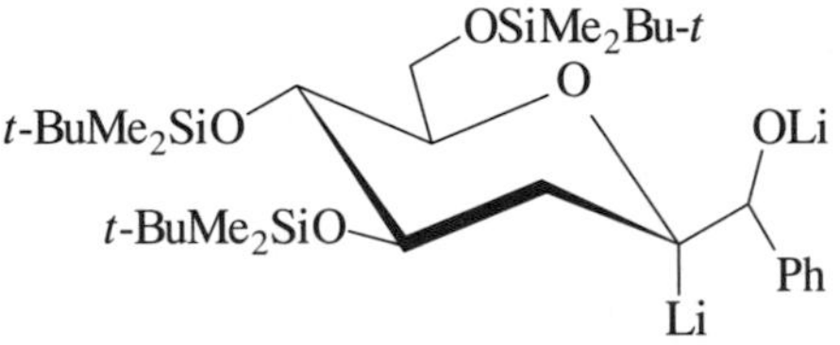

3.129

In contrast, attempts to lithiate glycopyranosyl sulphones result in 1,2-eliminations. Thus, for example, treatment of 2,3,4,6-tetra-*O*-benzyl-β-D-glucopyranosyl

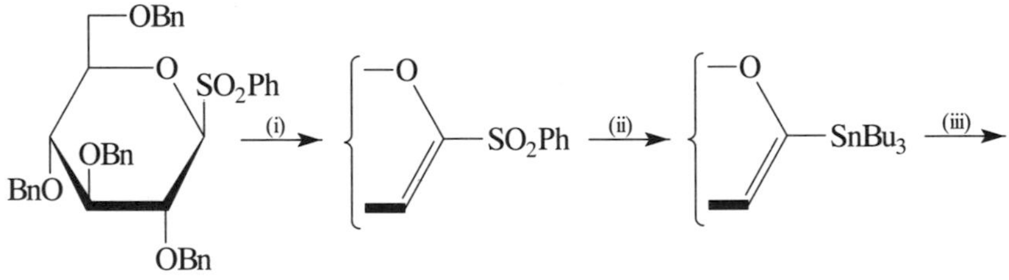

3.130   **3.131** (92%)   **3.132** (77%)

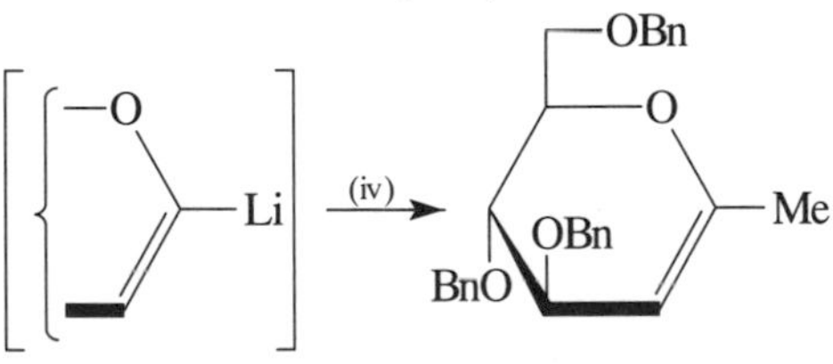

3.133   **3.134** (76%)

(i) *n*-BuLi, THF, −78 °C; (ii) Bu₃SnH, AIBN, PhMe, reflux; (iii) *n*-BuLi, THF, −78 °C; (iv) MeI

Scheme 3.39

phenyl sulphone **3.130** with n-butyllithium in THF at $-78\,^\circ$C causes the elimination of benzyl alcohol and the formation of the tri-O-benzylglucal sulphone **3.131** in almost quantitative yield.[81] From this compound, via the stannane **3.132**[82] and the C-1-lithiated derivate **3.133**, C-1-alkylated glycals (e.g. **3.134**) can be made in good yield (Scheme 3.39).

Ultraviolet light induces carbon-sulphur bond cleavage in sulphones as illustrated by tetra-O-acetyl-β-D-glucopyranosyl phenyl sulphone, which, on exposure to u.v. light in benzene, eliminates sulphur dioxide to form phenyl and glucopyranosyl radicals that lead to a range of carbohydrate products including biglucosyls.[83]

$$\text{GlcSO}_2\text{Ph} \xrightarrow{h\upsilon} (\text{Glc})_2 + \text{GlcPh} + \text{Glc-}p\text{-C}_6\text{H}_4\text{Ph} + \text{GlcH}$$

(v) Reductive desulphurization of thioglycosides

Thiopyranosides can be reductively desulphurized with hydrogen in the presence of Raney nickel to give 1,5-anhydroalditol derivatives (see Scheme 4.105), and similar reactions can be induced by tri-n-butyltin hydride, consequent upon the strong affinity sulphur exhibits for tin radicals.

This latter free radical, reductive desulphurization methodology has been applied to structurally related hemithio orthoesters (e.g. **3.137** and **3.139**) as a means of synthesizing β-glycosides. The synthesis is versatile, since many hemithio orthoesters are available from thiono analogues (e.g. **3.136** and **3.138**) of glycono-1,5-lactones (**3.135**) by methyl iodide induced additions of alcohols (including

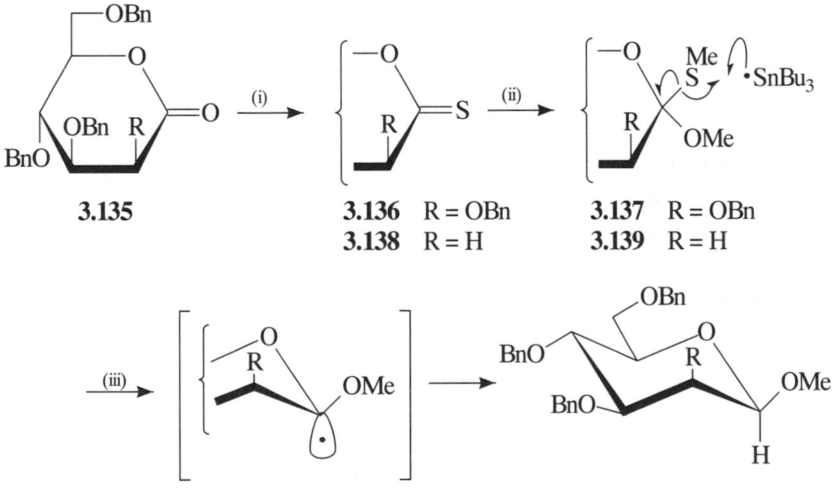

3.135   **3.136** R = OBn   **3.137** R = OBn
                             3.138 R = H   **3.139** R = H

3.140 R = H, OBn   **3.141** R = OBn
                                         3.142 R = H

(i) Lawesson's reagent [Ar(P $=$ S)S]$_2$, PhMe, molecular sieves; (ii) MeI, MeOH, 2,6-di-t-butyl-4-methylpyridine; (iii) Bu$_3$SnH, AIBN, PhMe, $h\upsilon$

Scheme 3.40

partially protected sugars) (Scheme 3.40). These orthoesters can be reductively desulphurized in toluene containing tri-n-butyltin hydride and AIBN by exposure to u.v. light. Thus, for example, from the *manno*-hemithio orthoester **3.137** methyl mannopyranosides **3.141** (α/β 1:18) are produced, and in like fashion the 2-deoxy hemithio orthoester **3.139** gives methyl 2-deoxy-β-D-*arabino*-hexopyranoside **3.142** accompanied by the α-anomer as a 6:1 mixture.[84] The highly stereoselective formation of the β-glycosides of these sugars, which are usually troublesome to prepare by other routes, arises because the reductive desulphurization occurs by way of alkoxy-substituted anomeric radicals (e.g. **3.140**), and these are quenched by axial hydrogen delivery to the α-face, as found for similar glycosyl radicals generated in other ways (see Schemes 3.77 and 3.121 in Sections 3.1.6.a.iv, 3.3.4.a. and 3.3.4.b, respectively).

(vi)   De-S-protection of 1-thiopyranose derivatives — 1-thioaldoses

De-S-alkylation or -arylation of thioglycosides is not a practical route to the parent 1-thioaldoses, but these may be conveniently obtained from the analogous thioacetates as illustrated in Scheme 3.36(b) by the de-S-acetylation of 1-S-acetyl-1-thio-β-D-glucopyranose tetraacetate **3.114**, ($R^1 = R^2 = Ac$). When this compound is treated with phenylmercury(II) acetate, the strong affinity that sulphur exhibits for mercury induces acyl–sulphur bond cleavage to give 1-phenylmercurythio-D-glucose tetraacetate **3.115** in good yield. Treatment of compound **3.115** in ethanol with hydrogen sulphide liberates 2,3,4,6-tetra-O-acetyl-1-thio-β-D-glucose **3.116**.[85] Another method adopted for the preparation of this class of compound is given in Section 3.3.3.c.

1-Thioaldoses are similar to their oxygen counterparts, reducing Fehling's solution and mutarotating in aqueous solution, albeit rather slowly. They differ in forming disulphide-bridged dimers and yielding stable crystalline metal derivatives. One such compound, gold thioglucose, is used as an antiarthritic drug. Their conversion to thioglycosides can be readily achieved by S-alkylation in the presence of alkali. For example, 2,3,4,6-tetra-O-acetyl-1-thio-β-D-glucopyranose **3.116**, on treatment with ethyl bromide in acetone in the presence of potassium carbonate, gives in 82% yield the corresponding ethyl 1-thioglycoside derivative (see **3.306**) with retention of configuration.[63]

(d)   Addition of bisulphite

Aldehydes and ketones often form crystalline adducts with metal or amine bisulphite anions by nucleophilic attack on the carbonyl carbon atom by sulphur. Aldoses behave in a similar way, and this reaction provides a useful method for obtaining acyclic monosaccharide derivatives, as shown for D-glucose in Scheme 3.41.[86] Additions of this type have proved useful for trapping 5-amino-5-deoxyaldoses in their acyclic forms (see Section 4.3.2.g).

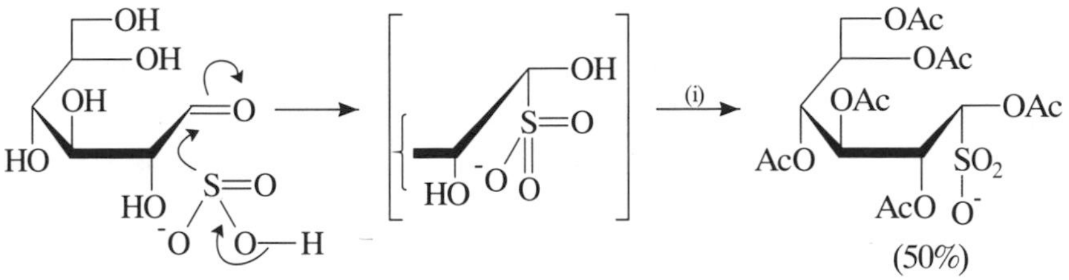

(i) Ac$_2$O, Py

Scheme 3.41

3.1.3   REACTIONS WITH NITROGEN NUCLEOPHILES — GLYCOSYLAMINES, OXIMES, HYDRAZONES AND OSAZONES

(a)   Reactions with ammonia and amines — glycosylamines

(i)   Synthesis of glycosylamines

Aldoses condense with ammonia and with primary and secondary amines with the loss of water in reactions which are analogous to Fischer glycosidations (Section 3.1.1.a). However, whereas glycosidation with alcohols require acid catalysts, sugars react with amines alone, which suggests that the initial condensation involves the acyclic form of the sugars, but details of the mechanism are not known. The products are called *glycosylamines* or *N*-glycosides, and they can occur in five- or six-membered ring forms and in α- or β-modifications, as shown in Scheme 3.42 for products **3.144** and **3.145** from ammonia (R = H) and primary amines.[87] These reagents can also give the uncyclized imines **3.143**, although such structures are usually encountered only with fully protected *aldehydo*-sugars. In solution, isomerization between the various forms (analogous

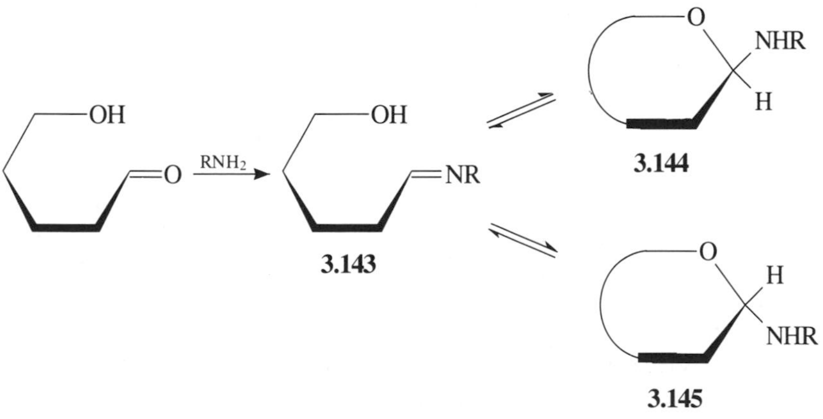

Scheme 3.42

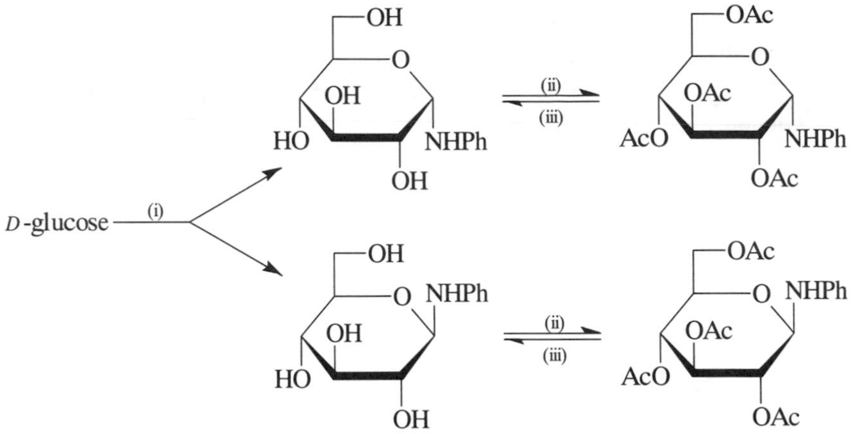

(i) PhNH$_2$, MeOH, reflux, 2 h; (ii) Ac$_2$O, Py; (iii) MeO$^-$, MeOH

Scheme 3.43

to the mutarotation of free sugars) can occur by the route shown in Scheme 3.42, in which the imine form **3.143** is implicated. Consequently, structural analysis of glycosylamines by chemical methods is difficult. However, i.r. and, in particular, n.m.r. spectroscopy have revealed that the crystalline forms of these compounds are usually cyclic.

Glycosylamines derived from primary aromatic amines are frequently crystalline, are less readily hydrolysed than are their alkyl counterparts and consequently have been studied most extensively. They can be prepared, as illustrated in Scheme 3.43, by the formation of the α- and β-anomers of 2,3,4,6-tetra-O-acetyl-N-phenyl-D-glucopyranosylamine following the condensation of glucose with aniline and O-acetylation and separation of the acetates.[88] The ring sizes and anomeric configurations have been established mainly by n.m.r. methods, and it has been shown that they can be de-O-acetylated with methoxide in methanol without loss in configurational integrity at C-1 or other modification in structure. The parent glycosylamines can be made by use of ammonia, but greater stereoselectivity is achieved in the synthesis of glycosylamines by reaction of amines with glycosylating derivatives. Thus, benzyl 3,4,6-tri-O-benzyl-β-D-glucopyranosylamine can be obtained in 70% yield when 1,2-anhydro-3,4,6-tri-O-benzyl-α-D-glucopyranose (see Section 3.1.1.c) is treated with benzylamine.[89] A β-glucosylamine also preponderates when tetra-O-acetyl-α-D-glucopyranosyl bromide is treated with p-toluidine (see Scheme 3.115). However, no anomeric selectivity is observed during amine displacement of fluoride from tetra-O-benzyl-α/β-D-glucosyl fluorides (Section 3.3.5).[90] The ammonolysis of the 1-O-mesylates (prepared and used in situ) provides a high-yielding stereoselective route to glycosylamines unsubstituted on nitrogen, as shown in Scheme 3.44 for tetra-O-benzyl-β-D-glucopyranosylamine,[91] but the possibility of base-catalysed eliminations occurring to give 1,2-unsaturated products in all such instances must

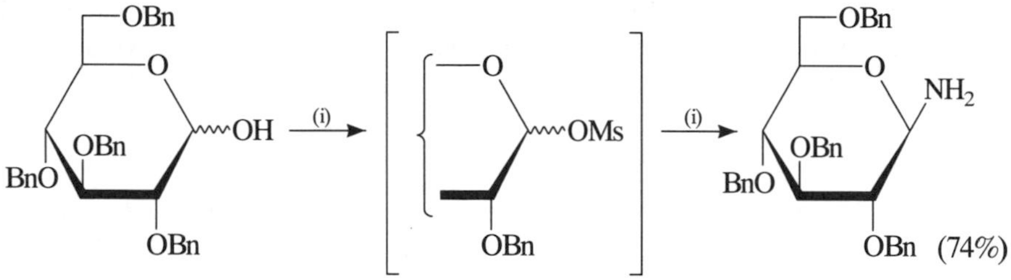

(i) MsCl, CH$_2$Cl$_2$, Et$_3$N, $-20\,°$C, NH$_3$, -20 to $20\,°$C, 20 h

Scheme 3.44

be considered. Another method for preparing glycosylamines involves the reduction of glycosyl azides (Section 3.3.3.b).

The glycosylamine structure occurs in several important classes of natural products, such as *N*-glycosidically linked glycoproteins and nucleosides (see Sections 8.1.4.a and 8.2.2).

(ii)   Reactions of glycosylamines

Glycosylamines can be hydrolysed to free sugars in aqueous acid by a reaction often used in nucleic acid chemistry (Section 8.2.2.a).

Acids also catalyse a transformation called the *Amadori rearrangement*, which often accompanies attempts to prepare glycosylamines from aldoses and amines.[87,92] The products formed are 1-amino-1-deoxyketoses, and a representation of the transformations involved is given in Scheme 3.45. This reaction is related to the Lobry de Bruyn–van Ekenstein reaction of aldoses referred to in Section 3.1.7.a.

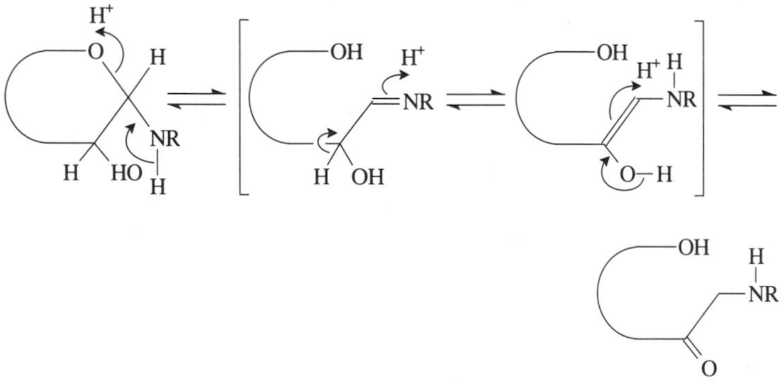

Scheme 3.45

Glycosylamine derivatives are probably implicated in the complex Maillard reaction whereby sugars, amines and amino acids (proteins) condense, rearrange and degrade often during cooking or the preservation of food.[93] The dark-coloured products formed in this reaction are responsible for the non-enzymic browning observed with various foodstuffs.

Glycosylamines are of value in enzymology as active site directed reversible inhibitors of glycosidases,[94] and they have also been used as chiral auxiliaries in a diastereoselective Strecker synthesis of amino acids (Section 7.1.2).[95]

(b)   Reactions with hydroxylamine — oximes

Aldoses show typical carbonyl reactivity when treated with hydroxylamine (Scheme 3.46), but the oximes formed do not make good derivatives with which to characterize sugars because of their high solubility in water. In aqueous solution they exhibit mutarotation because of equilibrations between acyclic and various ring forms (cf. glycosylamines).

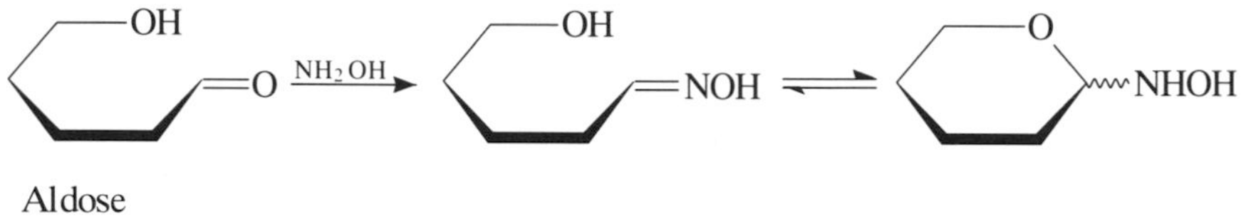

Aldose

Scheme 3.46

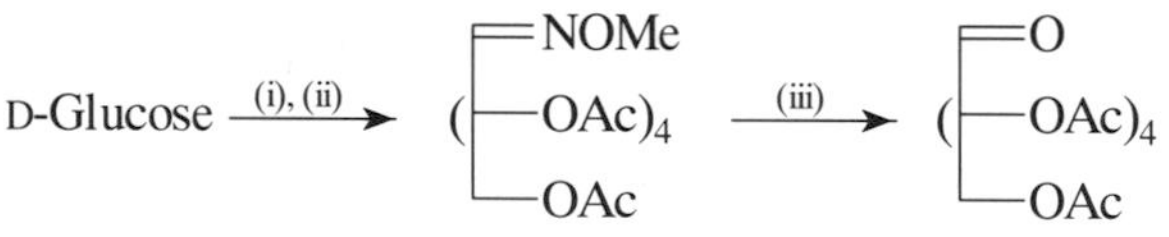

(i) NH_2OMe, Py; (ii) Ac_2O, Py; (iii) O_3, Me_2S

Scheme 3.47

For glucose oxime the α- and β-pyranose and (Z)- and (E)-acyclic forms are present, whereas the arabinose compound exists in aqueous solution in the acyclic forms only.[96] The *aldehydo*-forms of sugars are obtainable in multigram amounts by way of their O-methyl oximes which, after acetylation, may be converted back to aldehydes by ozonolysis of the imino double bond, as shown in Scheme 3.47.[97] In the past, a major interest in sugar oximes has centred around their application in the *Wohl degradation* by which C-1 can be removed from aldoses (Scheme 3.48).[98]

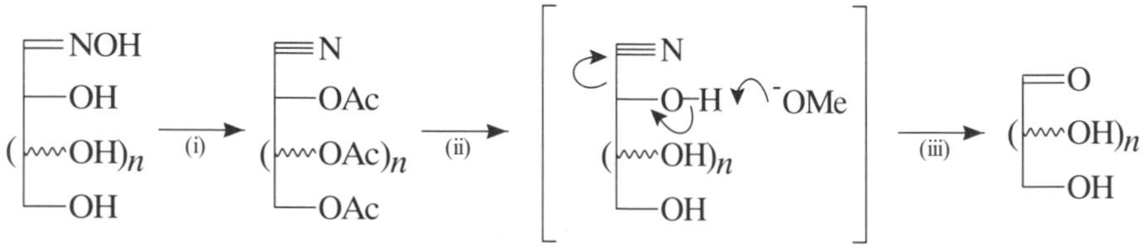

(i) A_2O, Py; (ii) MeO^-; (iii) CN^-

Scheme 3.48

In this way pentoses, for example, can be synthesized from corresponding hexoses in about 20% yield, but the method is now of limited value.

More recently, sugar oximes have been found useful in synthesis, since they can be transformed into a number of compounds with which reactions at the anomeric centre can be carried out. For example, the cyclic form of sugar oximes reacts with aldehydes or ketones to give N-glycosyl nitrones which are useful, readily available intermediates. Thus, the nitrone **3.147** (R = CO_2Bu-t), formed from t-butyl glyoxylate and di-O-isopropylidenemannose oxime **3.146**[99] has been used as a 1,3-dipolar reagent in cycloaddition reactions, as illustrated in Scheme 3.49(a). With ethene it gives the N-mannosyl isoxazolidene derivative **3.148**.[100]

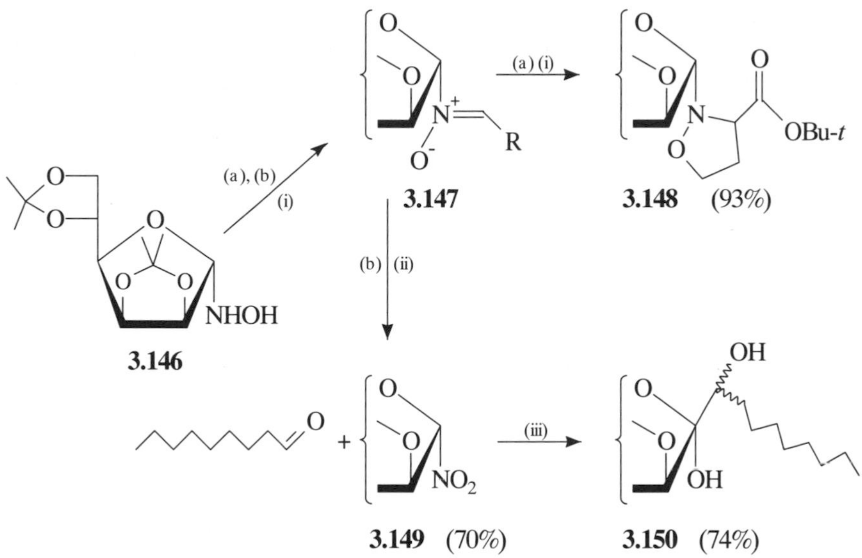

(a) (i) t-$BuCO_2CHO$, CH_2CH_2, $CHCl_3$, 65 atm, 75 °C, 17 h
(b) (i) p-$NO_2C_6H_4CHO$, CH_2Cl_2, 25 °C, 30 h; (ii) O_3, CH_2Cl_2, -78 °C; (iii) Li_2CO_3, $Me(CH_2)_7CHO$, H_2O, Py

Scheme 3.49

Pyranose and furanose glycosyl nitrones can also be oxidized with ozone to give nitro compounds. Thus, for example, the nitrone **3.147** (R = p-$NO_2C_6H_4$), formed from the same sugar oxime and p-nitrobenzaldehyde (Scheme 3.49b), is readily converted in this way into the nitro-*manno*-derivative **3.149**.[101] Carbanions may be generated at the anomeric centres of such compounds, and these can be used in carbon–carbon bond-forming reactions similar to those found for the 2-deoxypyranosyl sulphones (see Section 3.1.2.c.iv). However, milder basic conditions are required such that elimination of 2-alkoxy substituents from pyranose and furanose derivatives is not engendered. In this way the carbanion formed from **3.149** by treatment with lithium carbonate in aqueous pyridine is readily trapped with nonanal to give a water-sensitive 'tertiary' nitro compound, which after spontaneous hydrolysis gives the ketose derivative **3.150**.[102] The nitro group plays a double role at the anomeric centre in these reactions by first facilitating anion formation and then being readily displaced to confer cationic properties at this carbon centre. Acyclic nitrones undergo intramolecular 1,3-dipolar addition reactions in the presence of double bonds, as is illustrated in Section 7.2.2.b.

(c) Reactions with arylhydrazines — arylhydrazones and arylosazones

(i) *Arylhydrazones*

Aldoses and ketoses react with molar proportions of arylhydrazines to give hydrazones[98] (Scheme 3.50) which probably have the acyclic structure **3.151** when first formed, as has been shown for unsubstituted hydrazones.[103] However, after removal of any excess of hydrazine they tautomerize in the aqueous solution to the cyclic glycosylhydrazine forms **3.152**. X-Ray crystallographic methods have shown that the *p*-bromophenylhydrazone of D-ribose, for example, exists in the crystal in the acyclic form, whereas the D-arabinose and D-glucose derivatives have pyranoid ring structures. Hydrazones are sometimes used for characterization purposes, the *p*-nitro- or the 2,5-dichloro-phenylhydrazones being the most suitable because of their low solubilities and good crystallizing properties.

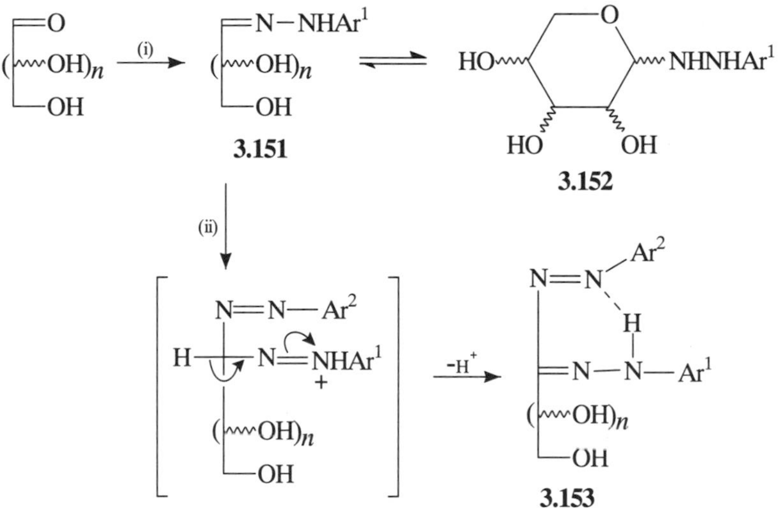

(i) Ar^1NHNH_2, pH 4–5; (ii) Ar^2N_2X, Py, $-5\,^\circ C$
Scheme 3.50

True arylhydrazones yield brilliant-red formazans **3.153** (Scheme 3.50) when treated with benzenediazonium chloride, and this reaction has been developed into a qualitative test for the imine structure of the acyclic form. However, all such tests are unreliable when applied to mobile equilibria of the type often present in such systems.

(ii) Arylosazones—formation

In 1884 Fischer discovered that when free sugars are treated with an excess of phenylhydrazine, highly crystalline, water-insoluble *phenylosazones* are produced.[98] These compounds assumed great importance in his researches which led to the determination of the configurational relationships of the aldoses and ketoses (Section 2.2). However, they are no longer looked upon as satisfactory derivatives. Scheme 3.51 illustrates the general form of the reaction, and shows that the osazones (**3.154**) have no asymmetry at C-1 or C-2, so that sugars which differ structurally only at these positions give the same osazones. In this way Fischer showed, for example, that D-glucose and D-mannose were epimers, and that D-fructose was the corresponding ketose. The exact mechanism of the reaction is still uncertain, but the stoichiometry (Scheme 3.51) shows that one molecule of phenylhydrazine is reduced to aniline and ammonia and that a hydroxyl group in the sugars is effectively oxidized. Two routes have been proposed for this condensation (Scheme 3.52). Both involve the prior formation of phenylhydrazones which undergo further reaction in one of two tautomeric forms, **3.155** and **3.156**.

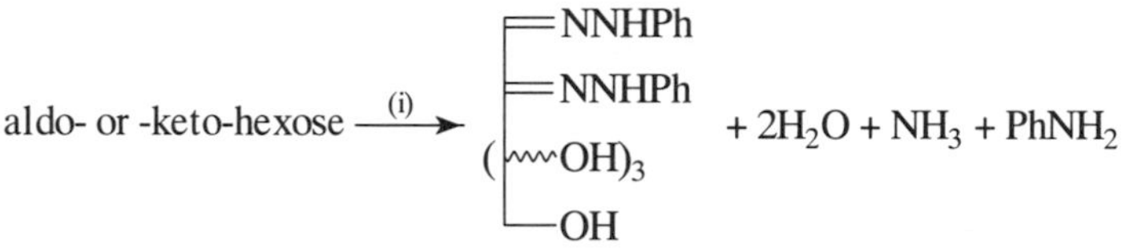

3.154

(i) $PhNHNH_2$ (three molecular equivalents), pH 5–6

Scheme 3.51

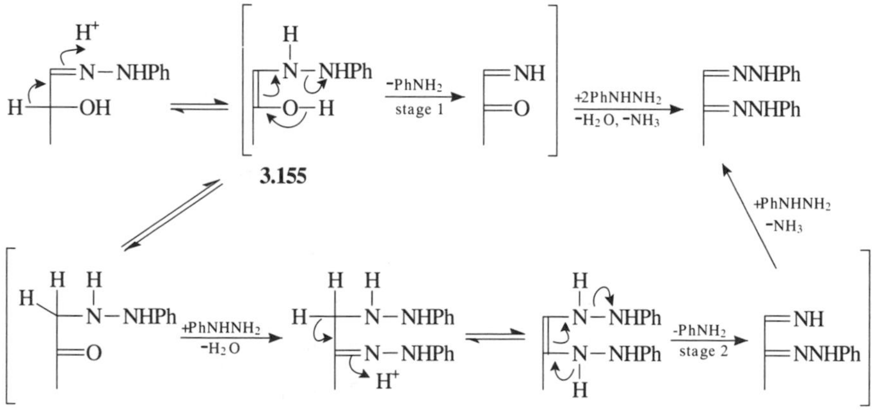

Scheme 3.52

In the first mechanism the oxidation step is thought to occur earlier (stage 1) than in the second (stage 2).

Similarly, the detailed structures of the osazones are not altogether clear and at first sight there appears to be no reason why osazones **3.154** should resist further reaction of the same type. Indeed, it has been shown that when aldoses are treated with an excess of α,α-disubstituted hydrazines, complete reaction does occur as depicted in Scheme 3.53 for the reaction with 1-methylphenylhydrazine. Thus, the α-hydrogen atom in phenylhydrazine appears to be responsible for the normal reaction stopping at C-2. The means by which this hydrogen atom causes this selective reactivity appeared to become apparent when the structures of osazones were studied by X-ray crystallographic and n.m.r. methods. The sugar chains were found to be acyclic with the free hydrogen atoms in the hydrazone residues at C-2 hydrogen bonded to the β-nitrogens in the hydrazone residues at C-1, giving six-membered chelate rings (e.g. **3.157**) which are thought to prevent further reaction. Other properties of phenylosazones can be accounted for by the intramolecularly hydrogen-bonded structure. Thus, methylation would be expected to yield di-N-methyl products; however, only mono-N-methyl derivatives are produced, and the ultraviolet spectra of arylosazones **3.158** and **3.159** are very similar (λ_{max} = 400 nm), whereas the spectrum of the osazone **3.160**, which cannot form a chelated structure, is different (λ_{max} = 355 nm).

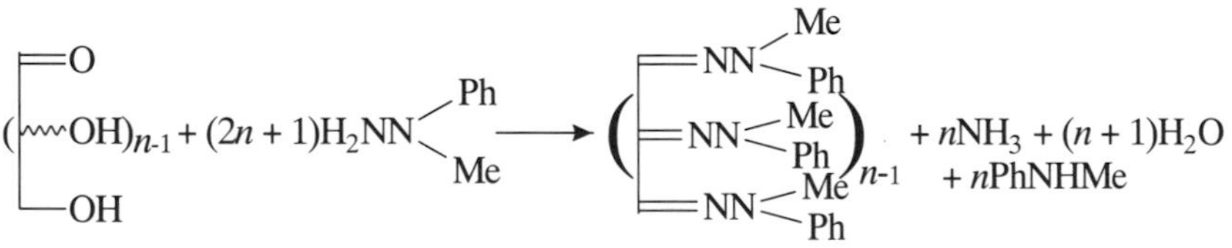

Scheme 3.53

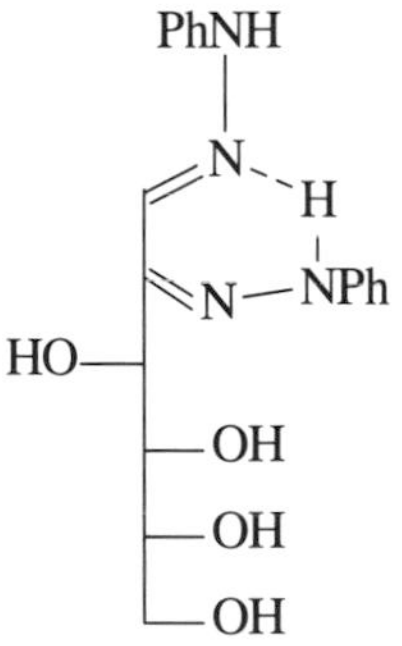

3.157

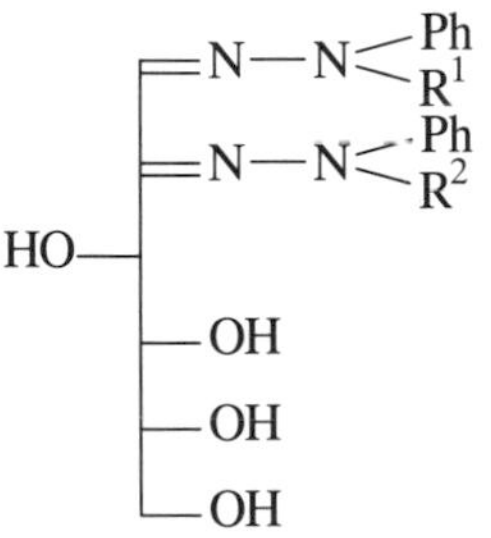

3.158 $R^1 = R^2 = H$

3.159 $R^1 = Me, R^2 = H$

3.160 $R^1 = R^2 = Me$

(iii)   Arylosazone reactions

Removal of the hydrazono groups from these derivatives can be carried out by heating them under reflux with benzaldehyde in the presence of acetic acid, and in this way D-*arabino*-hexosulose (D-glucosone) has been obtained from glucose phenylosazone (Scheme 3.54a). Upon mild oxidation with copper(II) sulphate this phenylosazone gives the osotriazole **3.161** (Scheme 3.54b); such derivatives are useful for characterizing sugars, because unlike the phenylosazones they have sharp melting points and do not mutarotate in solution.

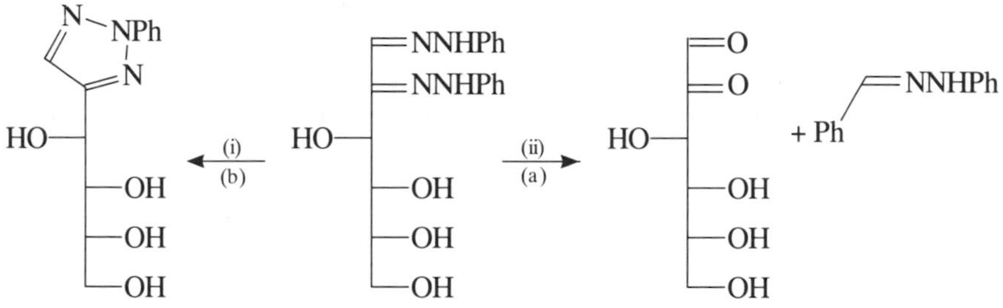

3.161   **3.162** (50%)

(i)$CuSO_4$; (ii) PhCHO, AcOH, reflux 1.5 h

Scheme 3.54

3.1.4   REACTIONS WITH CARBON NUCLEOPHILES

Carbon nucleophiles attack carbonyl groups of aldehydes and ketones to give addition products, usually alcohols, and they also react with sugars, particularly aldoses, to give analogous products. The evidence at present available indicates that these additions occur via the carbonyl forms of the sugars rather than through cyclic modifications, and this view is supported by the fact that glycosides show no reactivity towards these reagents. However, attack at carbonyl groups can only be invoked with confidence when all the probable participating hydroxyl functions are blocked and the free carbonyl group is exposed.

When a non-asymmetric nucleophile $(R^1)^-$ attacks a non-asymmetric aldehyde or ketone of the type R^2R^3CO, a racemic mixture of enantiomeric alcohols $(R^1R^2R^3CHOH)$ is formed. If, however, there is an asymmetric substituent present in either component, nucleophilic addition usually gives rise to two diastereoisomeric alcohols in unequal amounts by the process of *asymmetric induction*. Additions to the carbonyl group in sugars are subject to such asymmetric induction because of the asymmetry in the carbon chain.

(a)   Stereochemistry of additions

Rules have been developed to determine the isomer spread in nucleophilic additions to aldehydes that possess an oxygen-containing substituent on an asymmetric carbon atom adjacent to the carbonyl group, e.g. $R^2HC(OR^1)CHO$. In order to make these predictions the favoured rotamer(s) about the C-1–C-2 bond of the aldehyde must be determined and two models have been developed for this purpose.[104] In the *Cram–Felkin–Anh* model it is reasoned that, of the three bonds between C-2 and its substituents (other than the formyl group), the one attached to the alkoxy group OR^1 would have the lowest σ^*-energy and consequently be able to interact most favourably with the carbonyl π-system. This interaction would be strongest when the overlap is maximized, as in rotamers **3.163** and **3.164** in which the C–OR^1 bonds are orthogonal to the linearly arranged H–C = O bonds. Nucleophilic attack on the carbonyl carbon atom in these rotamers would be expected to be from the side opposite the large OR^1 group, with the trajectory at approximately 100° to the line of the H–C = O bonds as shown in **3.163** and **3.164**. Thus, the incoming nucleophile experiences a smaller steric interaction from H in **3.164** than from R^2 in **3.163**, and consequently the major product arises from the former mode of attack to yield the *erythro*-isomer **3.165**.

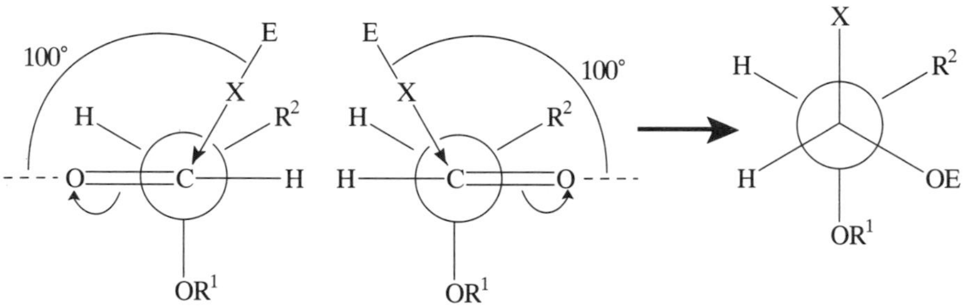

3.163   **3.164**   **3.165**
Cram–Felkin–Anh model   *erythro*-Product

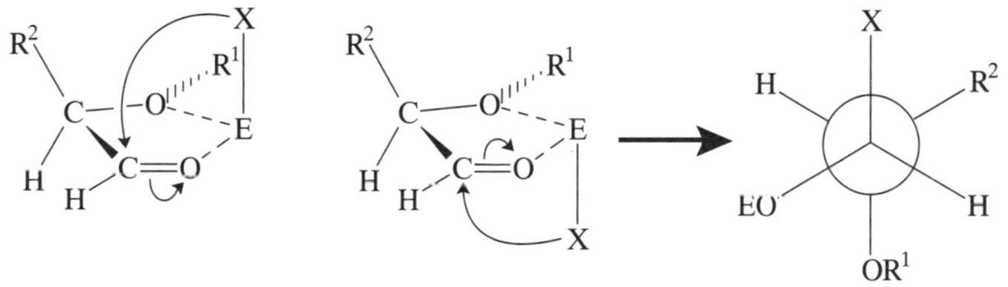

3.166   **3.167**   **3.168**
Cram chelated model   *threo*-Product

In the *Cram chelated* model the conformation in which the aldehyde reacts is easy to predict since it is fixed by chelation with the electrophilic part of the attacking reagent EX. Coordination between EX and the carbonyl and ether oxygen atoms can occur in two ways, as illustrated in **3.166** and **3.167**. The former complex

would yield a product formed by attack from above wherein the nucleophile interacts with the R^2 group on C-2, whereas nucleophilic attack occurs from below in
complex **3.167** and involves the nucleophile in a less sterically demanding interaction with the hydrogen atom on C-2; consequently, the *threo*-product **3.168** would
be expected to be the preponderant isomer.

The two models make contrary predictions, and knowing which to apply
is a problem compounded by the dependence of product stereochemistry on
the many reaction variables. This is highlighted by the results of extensive
studies of additions of carbon nucleophiles to derivatives of relatively simple α-
alkoxyaldehydes and glyceraldehyde.[105] Zinc and tin alkyls, for example, add to
2,3-*O*-isopropylideneglyceraldehyde in a highly stereospecific fashion according
to the Felkin model,[106] whereas alkyl Grignard reagents react selectively with
3-benzyloxy-2-ones to give as the major isomers those predicted by the Cram
chelated model.[107] To compound the difficulties, with many of these reactions only
low stereoselectivities are achieved,[105] and the conclusion has to be drawn that
predictions of the preferred directions of attack of carbanions on aldehydic sugar
derivatives cannot be made with any certainty.

(b)   Reactions with Grignard reagents

The reagents RMgX possess highly polarized carbon–metal bonds and contain
nucleophilic carbon which will bond to the carbonyl group of a fully protected sugar
held in its acyclic form. Various R groups have been used, but those containing a
double or triple bond have been the most versatile, since they can be applied to func-
tionalize the product further. For example, ethynylmagnesium bromide has been
added to 2,3:4,5-di-*O*-isopropylidene-L-arabinose **3.169** to give heptitol derivatives
in high yield with an L-*gluco*-**3.170**/L-*manno*-**3.171** isomer ratio of 3:2. Partial
reduction of the acetylenes to alkenes followed by ozonolysis yields L-glucose and
L-mannose derivatives (Scheme 3.55).[108]

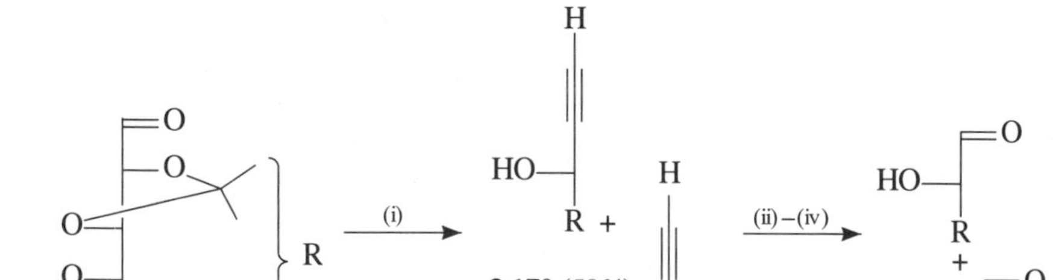

3.169   **3.171** (32%)

(i) HC≡CMgBr, THF; (ii) H$_2$, Lindlar's catalyst, EtOAc, C$_6$H$_{12}$; (iii) O$_3$; (iv) H$_2$,
PtO$_2$

Scheme 3.55

Application of the Cram chelated model to this addition reaction correctly predicts the preferential formation of the L-*gluco*-isomer, since the favoured intermediate would be **3.167** in which R^2 is a sugar chain from C-2 to C-5, X = C≡CH and E = MgBr. On the other hand, **3.164** is favoured by the Felkin model (R^2, X and E as previously defined), which therefore predicts, incorrectly, the L-*manno*-compound as the preponderant product.

Ethynylmagnesium bromide also reacts with aldehydo groups masked as hemiacetals. With compound **3.172**, for example, it gives the chain-extended *allo*-compound **3.173** and a trace of the *altro*-isomer. Synthetically useful C-glycosides can be produced from these diols by treatment with an excess of tosyl chloride, as shown in Scheme 3.56.[109] The propargyl hydroxyl group in **3.173** is tosylated with a surprising degree of selectivity to give the 3-tosylate which spontaneously ring closes with participation from O-6 in an intramolecular displacement reaction with Walden inversion to give the ethynyl α-C-riboside **3.174**.

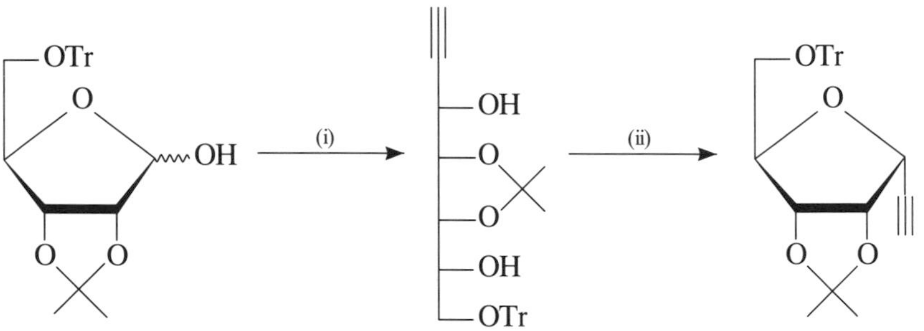

3.172   **3.173**   **3.174**

(i) HC≡CMgBr; (ii) TsCl

Scheme 3.56

Application of the two models for asymmetric induction to Grignard addition to the *aldehydo*-tautomer of **3.172** shows that on this occasion it is the Felkin model **3.164**, in which R^2 is a sugar chain from C-2 to C-5, X = C≡CH and E = MgBr, which is consistent with the finding of the D-*allo*-isomer as the major product. This confirms that the prediction of the major isomers formed in these addition reactions is not straightforward.

(c)   Reactions with allyl-indium and -tin reagents

Important new developments in this area are the tin- and indium-mediated allylations of unprotected carbohydrates in aqueous media which extend the chain by three carbon atoms. The allyl substituent adds to the sugar aldehydo group stereoselectively to give as the major isomer that predicted by the Cram chelated model. A specific application of this methodology to the synthesis of peracetyl neuraminic acid is given in Scheme 4.145, Section 4.9.2.

(d)   Reactions with hydrogen cyanide — cyanohydrins

Cyanide ions represent one of the most readily available forms of nucleophilic carbon, and these react with aldehydes and ketones to give addition compounds called *cyanohydrins*. One of the first methods used (by Kiliani) to ascend the aldose series (Section 2.2.1) involved preparing aldonic acids from aldoses via the derived cyanohydrins.[110] For example, D-arabinose **3.175** yields D-gluconic acid **3.176** and D-mannonic acid **3.177**, as shown in Scheme 3.57. The cyanohydrins are not usually isolated, but are hydrolysed directly to the aldonic acids by aqueous base or acid. From these, aldoses may be obtained by reduction (see Scheme 3.67).

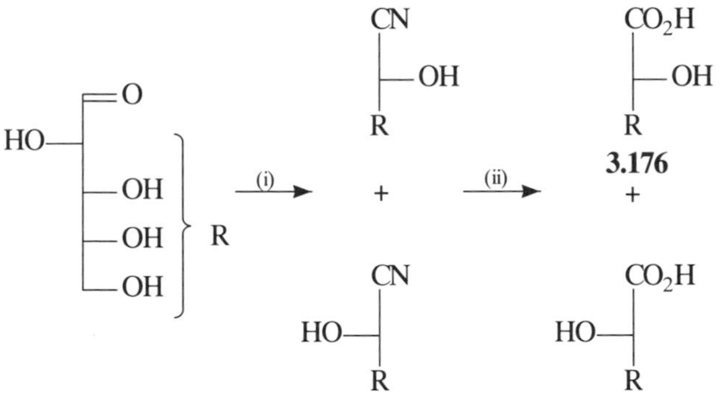

(i) NaCN, H_2O, pH 9; (ii) H_2O

Scheme 3.57

The reaction is subject to asymmetric induction, the *gluco*-isomer in the above case being formed as the major product, as Cram's rule predicts. However, the situation here is more complex than that found with the analagous Grignard reaction since the isomer distribution is affected by the pH of the solution in a way that is not fully understood. It could be that under some conditions the additions are reversible and products are formed under thermodynamic rather than kinetic control. It is not surprising, therefore, that some cyanohydrin preparations obey this rule while others do not.

Ketoses undergo the reaction to give aldonic acids with hydroxymethyl branches at C-2 (Section 4.8.1).

(e)   Reactions with nitroalkanes

In the presence of strong bases nitromethane loses a proton to give a resonance-stabilized carbanion which has been used in a method for extending the carbon chain of aldoses (Scheme 3.58).[110,111] The diastereoisomeric nitroalcohols **3.178** ($R^2 = H$) produced by additions can often be separated by fractional crystallization,

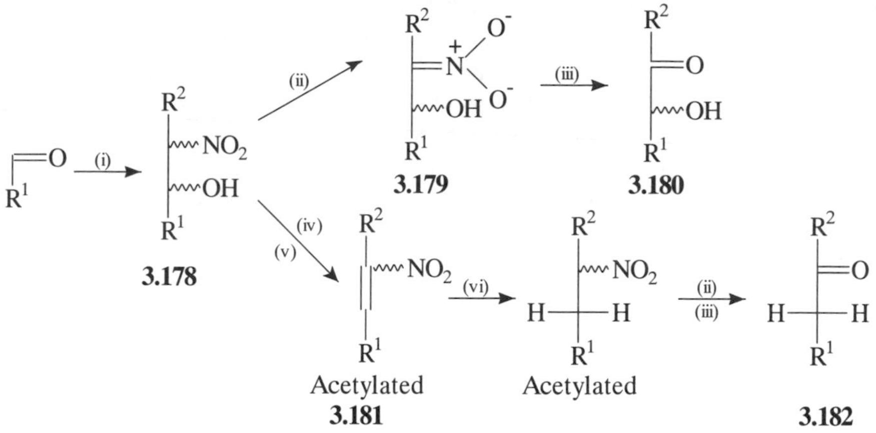

(i) $R^2CH_2NO_2$; (ii) MeO^-; (iii) H_2SO_4; (iv) Ac_2O, H_2SO_4; (v) $NaHCO_3$;
(vi) H_2, Pd

Scheme 3.58

and the conversion of these into new aldoses **3.180** ($R^2 = H$) is achieved by the
Nef degradation applied to the *aci*-sodium salts **3.179** ($R^2 = H$). The scope of this
reaction can be increased and it can be applied to the preparation of 2-ketoses **3.180**
($R^2 = CH_2OH$) by the use of 2-nitroethanol. Additionally, 2-deoxyaldoses **3.182**
($R^2 = H$) can be prepared by reduction of the alkenes **3.181** ($R^2 = H$), obtainable
by dehydration of the nitroalcohols, as shown. Ammonia and alcohols can also be
added to the double bonds of these compounds to give 2-amino-2-deoxy sugars
(Section 4.3.1.e) and 2-*O*-alkylated aldoses, respectively.

(f) Reactions with diazomethane

Diazomethane possesses nucleophilic carbon with which it attacks carbonyl groups,
and this reaction has been used to extend the carbon chains of aldoses and
ketoses. Treatment of 2,3,4,5-tetra-*O*-acetyl-D-arabinose **3.183** with one equivalent
of diazomethane, for example, gives 3,4,5,6-tetra-*O*-acetyl-1-deoxy-*keto*-D-fructose
3.184 (Scheme 3.59). Addition of a second equivalent of this reagent extends
the chain by a further carbon atom to give 4,5,6,7-tetra-*O*-acetyl-1,2-dideoxy-D-
arabino-hept-3-ulose **3.185**.[112]

Diazomethane can also be used with aldonic acid chlorides to extend the carbon
chains of monosaccharides (Section 3.1.6.a.iii). With glycosiduloses it gives rise to
branched-chain sugar derivatives (Section 4.9.1.b.i).

(g) Reactions with Wittig reagents

The Wittig reaction is a useful method for producing carbon–carbon double
bonds from carbon–oxygen double bonds of aldehydes and ketones and is
accomplished by using the nucleophilic carbon available, most commonly, in

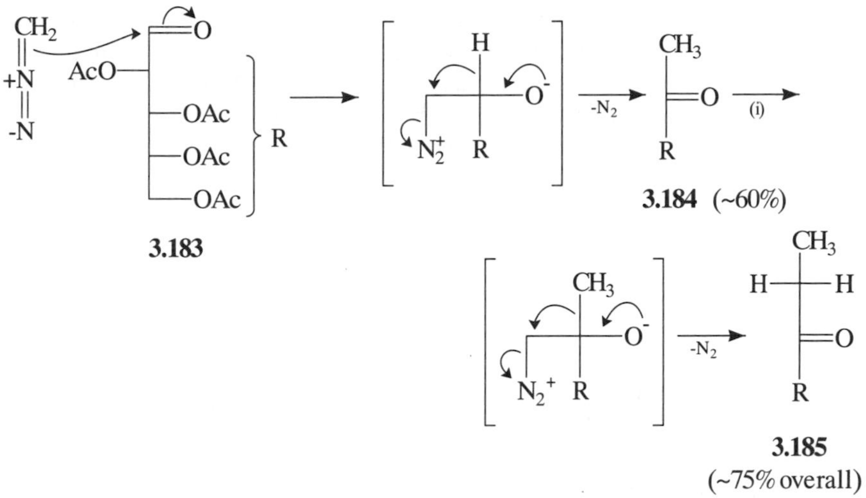

3.183

3.184 (~60%)

3.185
(~75% overall)

(i) CH$_2$N$_2$

Scheme 3.59

triphenylphosphorane ylides **3.186**, called triphenylphosphoranes (Ph$_3$P=CR^{1}R^2). When electron-withdrawing groups are present at the α-position (i.e. R^1 = CO$_2$R^3, CHO, SO$_2$R^3, etc.), the ylides are more stable and easier to work with, but they are really only suitable for reactions with aldehydes. Valuable alternatives to these ylides are reagents prepared from phosphonate esters, which possess a more nucleophilic carbon atom in the derived anion **3.187** because the negative charge cannot be attenuated by delocalization into the d-orbitals of the phosphorus atom, and when these are used to convert aldehydes or ketones to alkenes the reaction is referred to as the Wadsworth–Emmons reaction.[113,114] Some control may be exerted over the (E)- and (Z)-selectivity in these reactions by the choice of ylide, solvent and ylide counterion. As a general rule, stabilized ylides (e.g. **3.186**, R^1 = CO$_2$R^3) and most phosphonate anions **3.187** give (E)-alkenes. On the other hand, non-stabilized ylides (e.g. **3.186**, R^1 and R^2, H or alkyl substituents), under salt-free conditions in polar, non-protic solvents, give (Z)-alkenes in reactions with aldehydes, as do anions (e.g. **3.187**, R^1 = H, R^2 = CO$_2$Me, R^3 = CF$_3$CH$_2$) generated from esters of bis(trifluoroethyl)phosphonoacetate by potassium hexamethyldisilazide in the presence of 18-crown-6 in THF.[113]

$$\overset{+}{Ph_3P} - \overset{-}{C}R^1R^2 \qquad\qquad (R^3O)_2P(O)\overset{-}{C}R^1R^2$$

3.186   **3.187**

Wittig and related reagents have been extensively used with sugar derivatives such as uloses (see Chapter 4), ketoses and particularly aldoses with which they have been successful in producing chain-extended sugars and related compounds.[115] The *aldehydo*-forms of aldoses in which all the hydroxyl groups are protected undergo this reaction in a straightforward fashion, as illustrated by the conversion of the *aldehydo*-D-ribose derivative **3.188** into the heptenoates **3.189** on treatment with *t*-butoxycarbonylmethylenetriphenylphosphorane **3.186** (R^1 = H, R^2 = CO_2Bu-*t*).[116] As would be expected with a stabilized phosphorane, the (*E*)-form of **3.189** is the preponderant isomer (see Scheme 3.60).

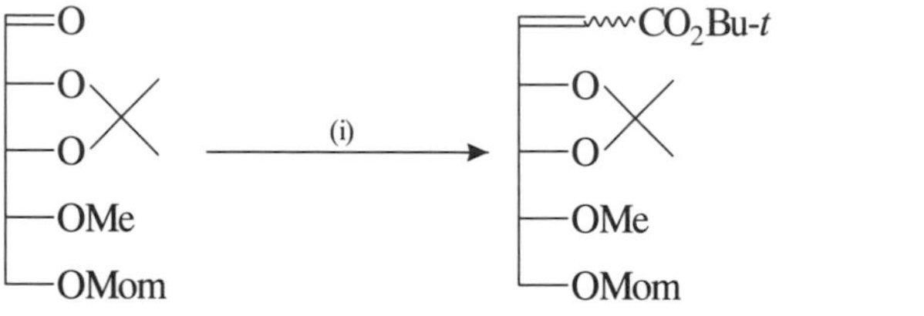

3.188   **3.189** [87%, (*Z*)/(*E*) 1:19]

(i) Ph_3P=$CHCO_2Bu$-*t*, CH_2Cl_2

Scheme 3.60

It is not essential to have the aldoses in the protected acyclic form since even hemiacetals will react as hydroxyaldehydes, as illustrated by 2,3:5,6-di-*O*-isopropylidene-D-mannose **3.190** which, when heated with ethoxycarbonylmethylenetriphenylphosphorane in benzene under reflux, as shown in Scheme 3.61, yields the (*E*)-octenoate **3.191** in good yield. The presence of the

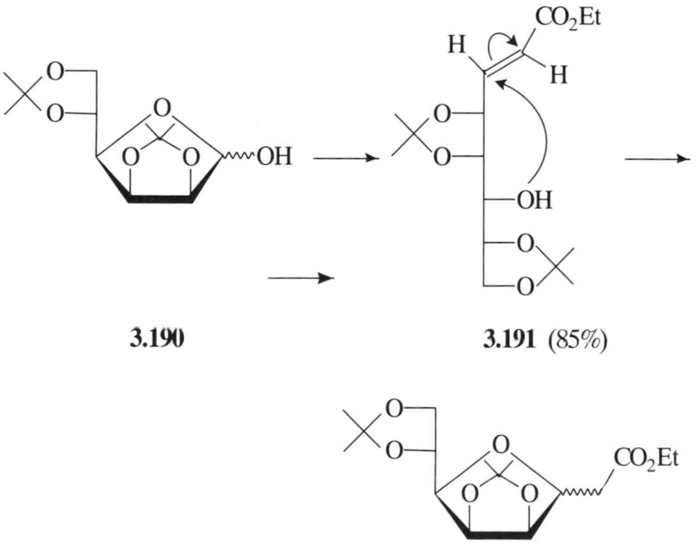

3.190   **3.191** (85%)

3.192

(i) Ph_3P=$CHCO_2Et$, PhH, reflux

Scheme 3.61

unblocked 6-hydroxyl group in the product can, however, influence the outcome of this reaction since use of the more polar acetonitrile as solvent with an excess of ylide yields the C-glycosides **3.192** formed by Michael-like ring closure.[117] The same cyclization may be achieved by treating **3.191** with base.

Unprotected sugars will also undergo Wittig reactions when DMF or pyridine is used as the solvent.[115] Sometimes alkenes are isolated, but often when stabilized ylides are employed the ring closure referred to above occurs, as shown by the reaction of the sulphone-stabilized phosphonate sodium salt **3.196** with D-glucose **3.193**, which affords the α- and β-C-furanosides usually isolated as their acetates **3.194** and **3.195** (see Scheme 3.62).[118]

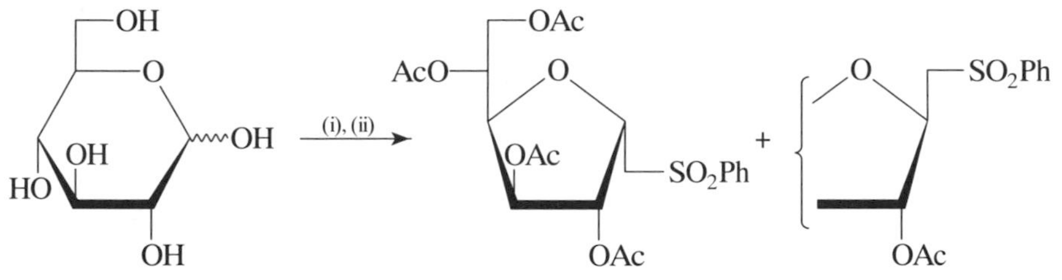

3.193   **3.194** (53%)   **3.195** (17%)

(i) $(EtO)_2P(O)CHNaSO_2Ph$ **3.196** THF; (ii) Ac_2O, Py

Scheme 3.62

(h)   Aldol and related condensations

The Knoevenagel reaction of β-dicarbonyl compounds, such as diethyl malonate, with aldehydes or ketones has been used for extending the length of the carbon chains of aldoses by two carbon atoms. The Verley–Doebner modification of this reaction, which uses malonic acid or its monoalkyl esters, is useful since decarboxylation occurs spontaneously to give unsaturated carboxylic acids in reasonable yields, as illustrated in Scheme 3.63 with 2,3:4,5-di-O-isopropylidene-*aldehydo*-L-arabinose **3.197**. The double bond of the unsaturated acid **3.198** can either be hydrogenated or hydroxylated, and the carboxylic acid residues then converted to *aldehydo* groups with sodium borohydride to give a 2,3-dideoxyheptose **3.201** or two heptoses **3.199** and **3.200**.[119]

It is again not essential to have the *aldehydo* group of the sugar derivative exposed for success with this chain extension reaction since the hemiacetal in an otherwise fully protected aldose reacts in like fashion with monomethyl malonate.

Common sugars can be synthesized from fragments by application of the aldol reaction. Treatment of a mixture of D-glyceraldehyde **3.203** and 1,3-dihydroxyacetone **3.202** with barium hydroxide (0.01 mol l^{-1}) produces a mixture of D-fructose **3.204** and D-sorbose **3.205** in almost quantitative yield, as shown in Scheme 3.64.[120] Thus the *threo*-configuration at the new asymmetric centres is preferred; this is also normally the case when such condensations are effected

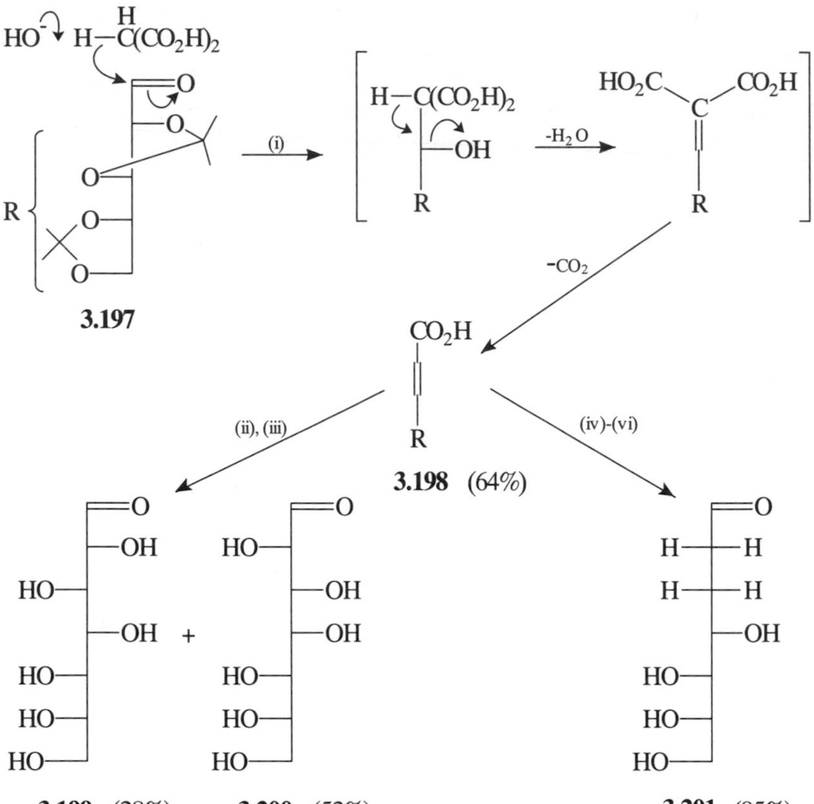

3.197

3.198 (64%)

3.199 (28%)   **3.200** (52%)   **3.201** (95%)

(i) Py, 1% piperidine, 100 °C, 1 h; (ii) OsO$_4$, H$_3$O$^+$; (iii) NaBH$_4$ pH 3; (iv) H$_2$, Pd; (v) NaBH$_4$ pH 3; (vi) H$_3$O$^+$

Scheme 3.63

by the enzyme aldolase (Section 2.5.2). Aldol condensations are also treated in Section 3.1.7.d.

3.1.5   REDUCTIONS TO ALDITOLS

Aldoses and ketoses can be reduced to alditols with the generation of new alcoholic groups from the carbonyl functions. These compounds are named from the aldose from which they can be prepared by replacing the 'ose' suffix with 'itol', e.g. reduction of D-glucose **3.206** gives D-glucitol **3.207**, trivially referred to as 'sorbitol' (Scheme 3.65). Originally, sodium amalgam was the reducing agent most commonly used for these reductions, but now it has been superseded by others, particularly sodium borohydride.[121,122] High pressure hydrogenation of aldoses over nickel is used for the commercial preparation of alditols.

Whereas aldoses give one product on direct reduction, ketoses give two. D-Fructose **3.208**, for example, produces D-mannitol **3.209** and D-glucitol **3.207**

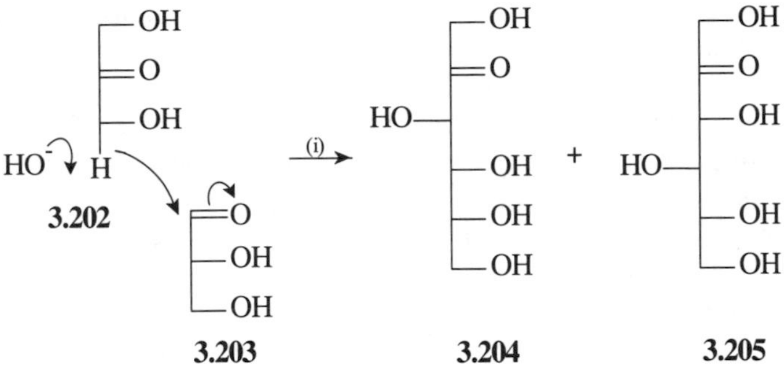

3.202   **3.203**   **3.204**   **3.205**

(i) H_2O, $Ba(OH)_2$ $(0.01 \ mol \ l^{-1})$

Scheme 3.64

because in this case a new asymmetric centre has been generated at C-2 (Scheme 3.65). Although the reactions are depicted as occurring via the open-chain carbonyl forms of the ketose and aldose, it should be borne in mind that, as with other reactions in this section, the carbonyl group can only be implicated with certainty in reactions of compounds in which the hydroxyl groups are protected. It is possible that the reaction with some of these reducing agents could proceed by way of a reductive displacement at the anomeric centres of cyclic forms.

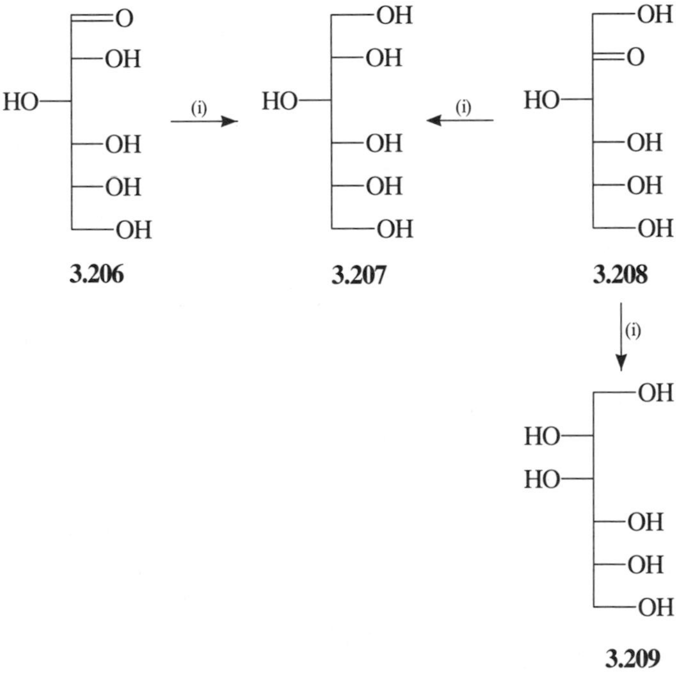

3.206   **3.207**   **3.208**

3.209

(i) $NaBH_4$ or $NaHg$, EtOH

Scheme 3.65

This, however, seems unlikely, since none of the reagents reduce glycosides under normal conditions, and these should be susceptible if displacement mechanisms are involved.

The acyclic alditols (see Section 8.3 for cyclitols), sometimes referred to as *glycitols* or *polyols*, have fewer stereoisomers than the corresponding aldoses because the conversion of the carbonyl into a hydroxymethyl group increases the symmetry of the compounds. There are thus 10 hexitols (cf. 16 aldohexoses), D-glucose and L-gulose giving, for instance, the same reduction product. The added symmetry also causes allitol and galactitol to be *meso*-compounds.

Polyols occur extensively in nature. Glycerol (1,2,3-trihydroxypropane) is widely distributed as the alcoholic component of the long-chain fatty esters of plant and animal oils; D-glucitol is found in many fruits such as apples and plums, and the berries of the mountain ash are also a rich source. D-Mannitol is found as a component of polysaccharides in brown seaweeds, and ribitol as a constituent of teichoic acids. D-Glucitol is non-toxic and slightly hygroscopic with a mildly sweet flavour, and for this reason it is used in large amounts as a humectant and conditioner in pharmaceutical, cosmetic, confectionery and paper products. Xylitol has been used as a sweetening agent that is not bacterially attacked in the mouth. The reactions of these compounds are typical of alcohols and are discussed in Chapter 5.

3.1.6   OXIDATIONS

(a)   Aldonic acids

(i)   Preparation

Aliphatic aldehydes are readily oxidized to carboxylic acids under mild conditions, whereas ketones are resistant. In accordance with this behaviour, D-glucose is converted into D-gluconic acid **3.214** in good yield on oxidation with aqueous bromine solutions over a pH range extending from about 1 to 11. Chlorine and iodine are also effective oxidants and the method is quite general for all aldoses; each gives rise to an acid, named by removing the 'ose' ending from the sugar and replacing it with the 'onic acid' suffix.[123,124] The aldonic acids are usually isolated as their salts. Ketoses, on the other hand, although not inert to oxidation because of their ability to isomerize to aldoses, are moderately stable under the mild oxidation conditions of halogens in water, and therefore it is possible to free ketoses from aldose impurities by converting the latter into acids which can be readily removed. Aldoses and ketoses both give aldonic acids when treated with Tollens reagent, which contains the complex ion $[Ag(NH_3)_2]^+$. The oxidation proceeds with the deposition of a metallic silver mirror on the surface of the reaction vessel and is the basis of a qualitative test for free sugars.[125] Another test uses Fehling's reagent, which contains blue copper(II) ions in a basic solution of sodium potassium tartrate and oxidizes enediol systems derivable from α-hydroxy-ketones or -aldehydes to give red copper(I) oxide.[124,125] This oxidation is rather complex and several organic products are formed in addition to aldonic acids. A test related to this is used clinically to detect glucose in the urine of suspected diabetics.

All sugars under the influence of very strong oxidizing agents such as potassium dichromate and potassium permanganate suffer oxidative degradation; nitric acid converts aldoses into terminal dicarboxylic acids (Section 3.1.6.b). Uronic acids, which are usually prepared by oxidation of the primary alcohol groups in aldoses, are discussed in Section 4.9.3.

Under mildly acidic conditions, aqueous bromine oxidizes D-glucose **3.210** directly through its pyranose ring form to the corresponding lactone **3.211**, as shown in Scheme 3.66(a).[123,124] The initial oxidation of D-glucose occurs almost exclusively via the β-anomer since this reacts about 250 times faster than the α-form at pH 5;[†] in addition, anomerization of the latter to the β-form is slow compared with the oxidation.[1] This difference in oxidation rates can thus be utilized for determining the proportions of anomeric pyranoses present in equilibrated glucose solutions. Initially, the measured concentration of sugar decreases rapidly during the oxidation of the β-form, and this is then followed by a slow reaction, closely dependent on the rate at which the α-anomer mutarotates. Extrapolation of the rate data from the slow part of the reaction back to zero time gives the composition of the initial solution, and the values so obtained for several aldoses agree well with measurements made by other methods (Section 2.4.2).

3.210 R = H
3.212 R = Bn

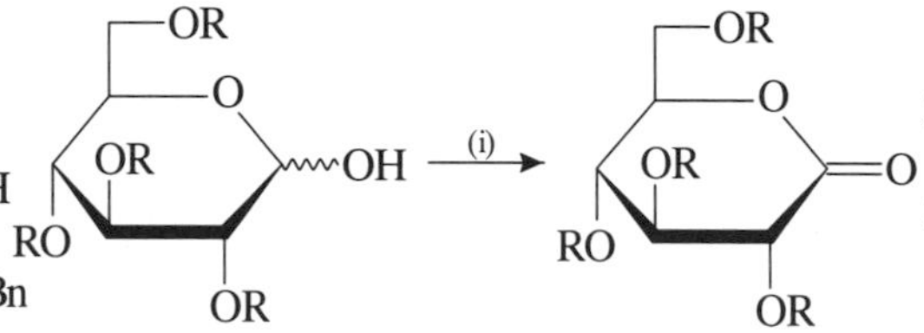

3.211 R = H (87%)
3.213 R = Bn (84%)

(a) (i) **3.210**, Br$_2$, H$_2$O, pH 4–6
(b) (i) **3.212**, DMSO, Ac$_2$O

Scheme 3.66

Partially protected aldoses, e.g. 2,3,4,6-tetra-O-benzyl-D-glucose **3.212**, can also be oxidized to aldonolactone derivatives **3.213** with dimethyl sulphoxide in the presence of acetic anhydride, as illustrated in Scheme 3.66(b),[126] and N-chlorosuccinimide smoothly converts 2,3:5,6-di-O-isopropylidine-D-mannose in the presence of tetrabutylammonium iodide into the corresponding substituted γ-mannonolactone.[127] Aldonic acids can otherwise be produced by the Kiliani cyanohydrin synthesis (Section 3.1.4.d) and by ozone oxidation of β-glycosides, as described in Section 3.1.1.b.xi. Chromium trioxide oxidations of β-glycosides, which are covered in this latter section, afford 5-keto derivatives of aldonic acids.

[†] A mechanism has been suggested for this oxidation by which bromine molecules are initially attacked by anions formed at O-1. Since β-glucose ionizes to give a higher concentration of O-1 anions than the α-anomer and because equatorial groups are better nucleophiles (Section 5.1.1), β-glucose can be expected to be the more reactive form.

(ii) Lactones

Recent developments have promoted compounds of this class into the 'very useful' category of starting materials for synthesis.

The free aldonic acids are often difficult to isolate because evaporation of water from their aqueous solutions usually dehydrates the acids completely and gives rise to lactones. In aqueous solution, unsubstituted aldonic acids are in equilibrium with γ-and δ-lactones, and it is possible by careful selective crystallization to obtain from an aqueous solution of D-gluconic acid, for example, either the free acid **3.214** or the γ-lactone **3.215** or the δ-lactone **3.216** by varying the temperature and pH of the aqueous mother liquor.[128] For different aldonic acids the amounts of each form present at equilibrium vary with structure and with pH of the solution, and, in contrast with the free sugars (Section 2.4.2) and glycosides (Section 3.1.1.a.i), five-membered ring forms of γ-lactones are relatively favoured because their geometries are suited to the bond lengths, bond angles and especially the coplanarity imposed by the partly conjugated $C-C(=O)-O-C$ array of atoms. The δ-lactone rings, on the other hand, adopt half-chair or boat conformations which represent compromises in which normal chairs are distorted in directions which allow only partial flattening in the regions of the lactone groups. In keeping with this, 5-hydroxypentanoic acid exists to the extent of 9% as the δ-lactone at 25 °C in aqueous solution, whereas under these conditions 4-hydroxybutanoic acid is 73% cyclized (γ-lactone); furthermore, aldono-δ-lactones usually hydrolyse faster than do γ-lactones. These differences between the five- and six-membered ring lactones are also reflected in their enthalpies measured from heats of combustion.[129]

3.214   **3.215**   **3.216**

A useful reaction of lactones, which has often been used in combination with the Kiliani synthesis of higher aldonic acids, is their reduction to the corresponding aldoses and alditols. For example, with sodium borohydride or sodium amalgam under carefully controlled mild conditions, D-lyxono-δ-lactone **3.217** may be converted into D-lyxose **3.218** (Scheme 3.67). The applicability of this reaction has been extended by use of the selective reducing property of bis(3-methyl-2-butyl)borane, a reagent which reduces γ-lactones (i.e. internal esters) but leaves substituent acetate or benzoate groups unaffected. It has been used to prepare acetylated furanoses and is of wide application, since several aldonic acids can be obtained as their γ-lactones.[130] The same degree of reduction can also be conveniently achieved with diisobutylaluminium hydride (DIBALH), which is

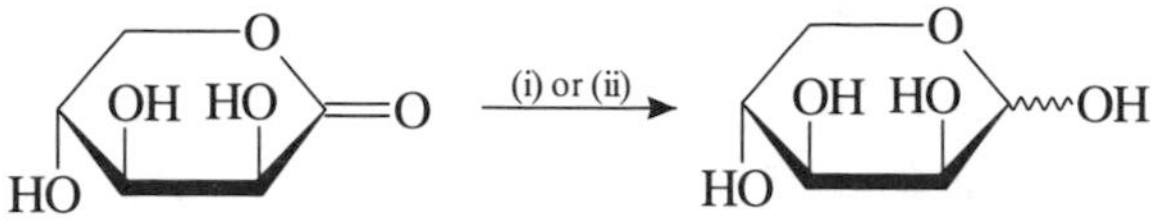

3.217   **3.218**

(i) $NaBH_4$, H_2O, pH 3–4, 0 °C, 1 h; (ii) NaHg (3.5 equivalents of Na), H_2O, pH 3–4, 10 °C, 1 h

Scheme 3.67

commercially available. The reagent is used in toluene solution at low temperature and the aldose products are obtained in high yields.

Under more vigorous reaction conditions than those given in Scheme 3.67, sodium amalgam (10 equivalents of sodium per mole of lactone, pH 7–8) or sodium borohydride (lactone added to sodium borohydride, i.e. reverse addition, pH 8, 40 °C) reduces aldonolactones beyond the aldose stage and gives alditols (Section 3.1.5).

Suitably protected aldonolactones, e.g. tetra-O-benzyl-D-glucono-δ-lactone **3.219**, may be converted into spiro-orthoester derivatives **3.220** by their reaction with diols under the acidic conditions given in Scheme 3.68(a).[131] However, higher yields are obtained when trimethylsilyl derivatives of the diols are used with trimethylsilyl trifluoromethanesulphonate (TMSOTf) as catalyst (Scheme 3.68b), and this method is the only one that gives satisfactory yields when two secondary hydroxyl groups of a sugar are condensed with the lactone, as illustrated by the preparation of **3.223** in Scheme 3.68(c).[132] Such spiroorthoesters of aldonolactones are of interest because the orthosomycin antibiotics, such as everninomicins, flambamycin and hygromycin B, for example, possess this structural feature. A synthesis and reaction of an aldonic acid hemithio orthoester are discussed in Section 3.1.2.c.v.

Fully protected lactones have also been employed as substrates for lithium 1-alkynides in carbon–carbon bond-forming reactions, as illustrated in Scheme 3.68(d) by the synthesis of lactol **3.222** from the benzylated gluconolactone **3.219** and the lithium derivative of 4-trimethylsilyloxy-1-butyne.[133] The lactol **3.222** is of particular value since its triple bond may be partially reduced to give a (Z)-alkene and the protected hydroxyl group in its aglycon may be deblocked and used in an intramolecular glycosidation reaction to give a spiroacetal of the type found in many biologically active natural products.

The Reformatsky reaction, which is usually associated with aldehydes and ketones, has been applied to di-O-isopropylidene-D-gulono-γ-lactone **3.226** to give a chain-extended lactol **3.227** as shown in Scheme 3.69(a),[134] and the same lactone has been converted into the exocyclic dichloromethylene derivative **3.225** upon treatment with hexamethylphosphorous triamide and carbon tetrachloride, as illustrated in Scheme 3.69(b).[39,135] It is thought that the reagents generate a trichloromethylide anion which initiates the reaction by nucleophilic attack on the lactone carbonyl group. The most efficient way of preparing exocyclic methylene

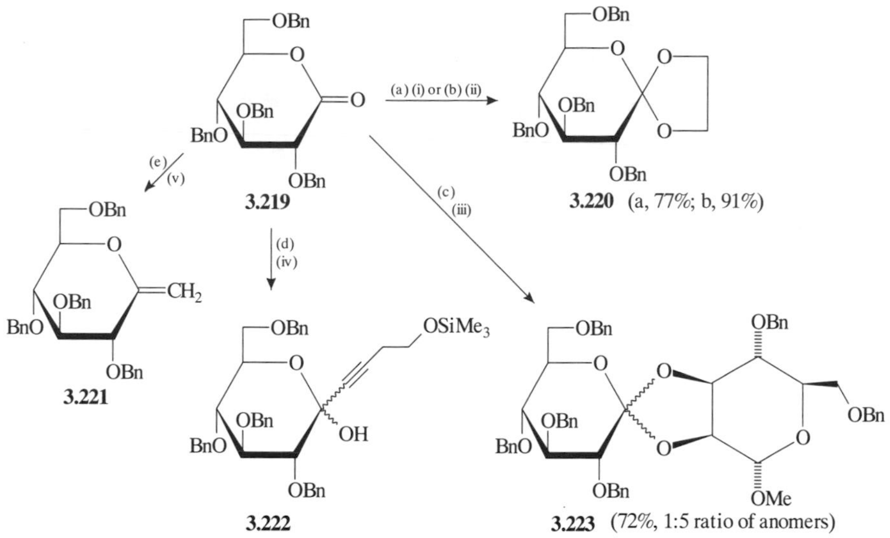

(i) $(CH_2OH)_2$, PhH, H_2SO_4, azeotropic distillation, 10 h;

(ii) $(CH_2OSiMe_3)_2$, CH_2Cl_2, TMSOTf, N_2, 20°C, 2h;

(iii) methyl 4,6-di-O-benzyl-2,3-di-O-trimethylsilyl-α-D-mannoside, CH_2Cl_2,
       TMSOTf, N_2, 0°C, three days;

(iv) LiC≡C$(CH_2)_2$ OSiMe$_3$, Et$_2$O, -15°C, 45 min;

(v) $(C_5H_5)_2$ Ti$\overset{CH_2}{\underset{Cl}{<}}$AlMe$_2$ **3.224**, Py, PhMe

Scheme 3.68

compounds from fully protected lactones is by use of Tebbe's reagent (see **3.224**
in Scheme 3.68e), which provides versatile titanium alkylidene [$(C_5H_5)_2Ti=CH_2$],
a methylenating agent more potent than the corresponding Wittig phosphorus
base compounds and, in consequence, reactive towards lactone (or ester) carbonyl
groups.[39] Thus, for example, tetra-O-benzylglucono-δ-lactone **3.219** is converted
in high yield into the 2,6-anhydroheptenitol derivative **3.221** when treated with
Tebbe's reagent, as shown in Scheme 3.68(e).[136] See also Section 4.10.4.

Carbon–nitrogen bonds can be formed at lactone carbonyl carbon atoms by
attack from nitrogen nucleophiles, as illustrated by, for example, tetra-O-benzyl-
D-glucono-δ-lactone **3.219** which, on treatment with trimethylsilyl azide in the
presence of boron trifluoride etherate, gives the *gem*-diazide **3.228** in a reaction
closely related to the spiroketalization reported in Scheme 3.68(b).[137] Such diazides
can also be prepared in higher yields, but on a small scale, by silver azide double
displacements of halogens from anomeric *gem*-dihalides. Thus, the tetra-O-acetyl
analogue **3.229** ($R^1 = R^2 = N_3$) is obtained in 60% when 1-bromo-tetra-O-acetyl-
β-D-glucopyranosyl chloride **3.229** ($R^1 = Cl$, $R^2 = Br$) is so treated.[137] Related

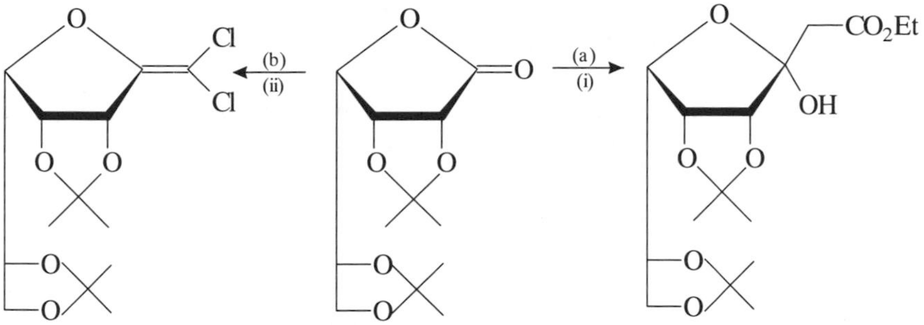

3.225 (92%)   **3.226**   **3.227** (50%)

(i) BrCH$_2$CO$_2$Et, Zn, dioxane, 50°C, 2 h; (ii) P(NMe$_2$)$_3$, CCl$_4$, THF, −30°C

Scheme 3.69

gem-diaza compounds have been prepared by treatment of hydroximolactone derivatives with ammonia. For example, hydroximolactone **3.230** (obtainable in 86% yield from tetra-*O*-benzylglucose oxime by sodium periodate oxidation), after quantitative conversion into its mesylate **3.231**, undergoes addition of ammonia to its carbon–nitrogen double bond when treated as indicated in Scheme 3.70. The intermediate so formed spontaneously suffers mesylate displacement and ring closure to give spirodiaziridine **3.232** as a mixture of *cis*- and *trans*-isomers which are immediately oxidized to give the relatively stable crystalline anomeric diazirine **3.235**.[99,138]

3.228   **3.229**

Recent work has shown that anomeric spirodiazirines **3.235** are precursors of glycosylidene carbene reaction intermediates **3.234**, and this opens the way for new reactions at C-1 of pyranoid compounds that complement carbonium ion, radical and carbanion chemistry at this centre.[99,139] Typical carbene chemistry is exhibited when these diazirine derivatives are decomposed. Thus, alkene addition, the most characteristic of all carbene reactions, is observed when the diazirine **3.235** is thermally decomposed at room temperature in the presence of acrylonitrile to give anomeric spirocyclopropane adducts **3.233** as a mixture of four isomers,[99,139] the low selectivity being expected from a reaction of a nucleophilic alkoxyalkylcarbene with an electron-deficient alkene. When a similar decomposition is carried out in the presence of buckminsterfullerene (C_{60}), the spiro-linked *C*-glucoside of C_{60} **3.236** is obtained, thus giving a new class of compound about which more will undoubtedly be heard.[140]

U.v. irradiation of the diazide **3.228** in acrylonitrile does not give **3.233** via carbene **3.234** as previously reported, but instead, by loss of one molecule of nitrogen, a nitrene is generated which rearranges with concomitant pyranose ring expansion to give a bicyclic tetrazole.[137]

Insertion of carbenes into C–H bonds or other bonds involving hydrogen is another important reaction of carbenes, and with alkoxyalkylcarbenes O–H bonds are particularly vulnerable because insertion can be facilitated by prior protonation of the nucleophilic carbene to give, for example, the familiar anomeric carbonium species **3.237**, as shown in Scheme 3.70. This approach therefore becomes the basis of a new procedure for preparing *O*-glycosides by which solutions of diazirines **3.235** are decomposed either thermally or photochemically in the presence of one

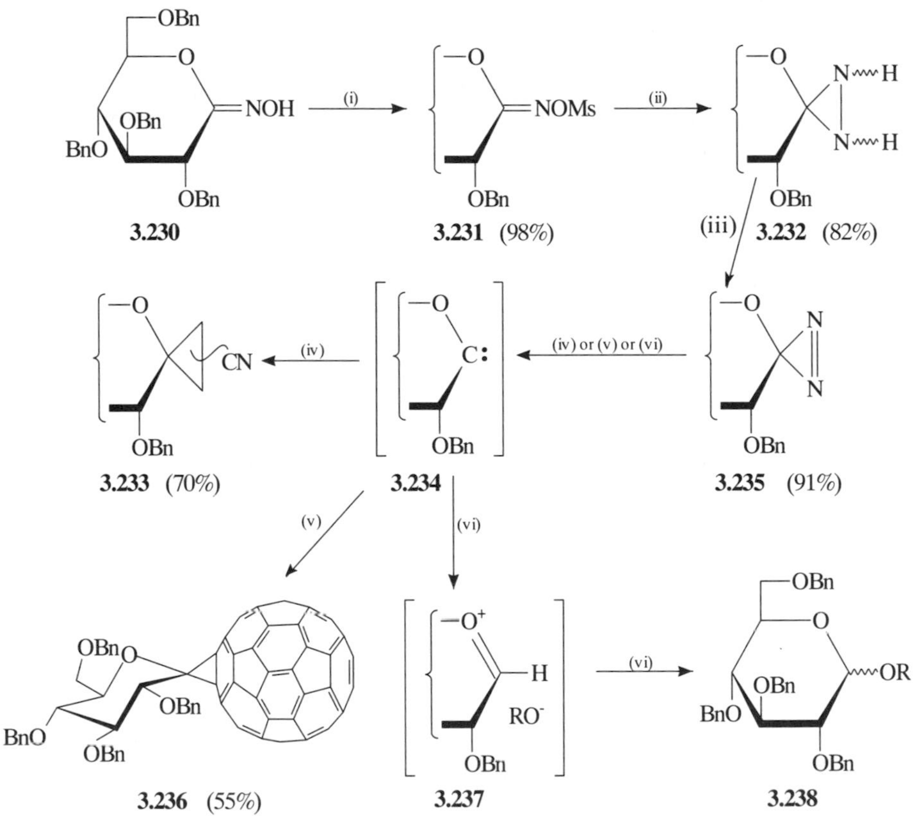

(i) MsCl, Et$_3$N, 0 °C; (ii) NH$_3$, MeOH; (iii) I$_2$, Et$_3$N, MeOH, −25 °C; (iv) CH$_2$=CHCN, 22 °C, 12 h; (v) C$_{60}$, PhMe; (vi) ROH, CH$_2$Cl$_2$ or ROH, THF, *hν*

Scheme 3.70

molecular equivalent of an alcohol or a phenol. The yields and diastereoselectivities achieved in the reactions are dependent upon the pK_a of the alcohol, the reaction temperature, the method of decomposition and the solvent employed.[99] For example, thermolysis of **3.235** in dichloromethane with p-nitrophenol gives, in 75% yield, a 3:2 α/β mixture of p-nitrophenyl glycosides **3.238** (R = $C_6H_4NO_2$), and electron-rich phenols such as p-hydroxyanisole behave likewise. However, some aryl C-glycosides are also formed in the latter case, and are believed to arise following Friedel–Crafts alkylation at the *ortho*-position of the electron-rich phenate ion by the cation of the postulated ion pair **3.237** (R = $MeOC_6H_4$).[99] The lack of steric hindrance commonly associated with reactions which take place via carbene intermediates is nicely exhibited by a similar, easy O-glycosylation of the crowded 2,6-di-t-butyl-4-methylphenol with **3.235**.[99] With common, less acidic alcohols the reaction becomes more sensitive to the conditions, as exemplified by the thermal decomposition of a dichloromethane solution of **3.235** in the presence of isopropanol, which produces equal proportions of the α- and β-isopropyl glycosides **3.238** (R = Pr-i) in a modest (40%) yield improving to 60% and giving an α/β mixture (8:92) when the reaction is photochemically induced at $-65\,°C$.[141] Poor stereoselectivities have so far been observed in disaccharide syntheses. The 1→3-linked disaccharides, obtained in 65% yield when **3.235** is thermally decomposed in dichloromethane in the presence of 1,2:5,6-di-O-isopropylideneglucose, are produced with only modest (67:33) β/α selectivity, which is little improved by the usual changes in reaction conditions.[141]

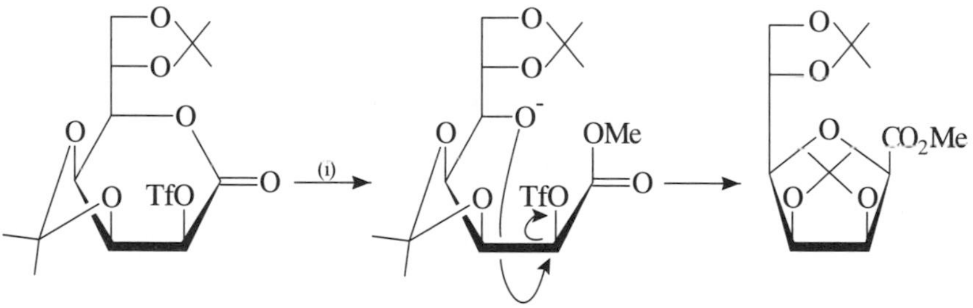

3.239   **3.240**   **3.241** (81%)

(i) K_2CO_3, MeOH, $0\,°C$

Scheme 3.71

Aldonolactones with triflate groups at O-2 undergo a ring contraction reaction when treated with methanol in the presence of potassium carbonate.[142] Thus, for example, the heptono-δ-lactone **3.239** gives the C-glycoside ester **3.241** in good yield when so treated by a reaction which proceeds as shown in Scheme 3.71. The ring-opened intermediate **3.240**, which is formed by methoxide attack at C-1, subsequently recyclizes to yield a product with an inverted configuration at C-2. Some O-unsubstituted lactones ring contract similarly to the corresponding unprotected C-glycoside esters when sequentially treated with triflic anhydride and pyridine in methanol. In this case, it is preferable to avoid the use of potassium carbonate because it produces epoxides from the highly triflated lactone intermediates. Ring contraction with inversion is generally found with all 1 → 5-lactone 2-triflates, and on some occasions with their C-2 mesylate and C-2 iodide analogues. Related reactions are also observed when aldonolactone 2-triflates are treated with methanolic hydrogen chloride.[142]

Somewhat surprisingly, the analogous base-catalysed ring contraction with five-membered ring aldono-γ-lactone 2-triflates has been observed to give four-membered oxetane derivatives, which are of value since they allow access to a range of antiviral oxetane nucleosides (Section 8.2.2.a). The stereochemistry of this transformation is, however, markedly different since oxetanes in which the carboxymethyl groups are *trans* to the adjacent alkoxy substituents are produced with high selectivity.[143]

(iii)   Acyclic derivatives

Aldonic acid nitriles are formed on dehydration of aldose oximes by treatment with acetic anhydride (Scheme 3.48) or by the Kiliani addition of cyanide to aldoses. Aldonamides are usually prepared from lactones by the action of liquid ammonia. These products are easily hydrolysed, but acetylation of the free hydroxyl groups gives stable derivatives (e.g. **3.242**), useful as precursors of aldonic acid chlorides **3.243** (Scheme 3.72).[144]

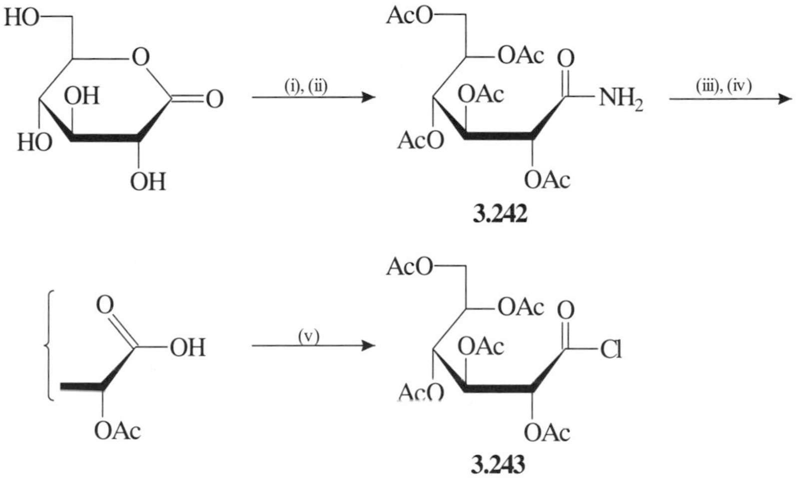

3.242

3.243

(i) NH₃; (ii) Ac₂O, Py; (iii) NOCl; (iv) H₂O; (v) PCl₅

Scheme 3.72

Aldonyl chlorides show two useful reactions: one increases the number of carbon atoms in the aldose by one, whereas the other decreases it by one. Thus, ketose derivatives containing one additional carbon atom are formed by the reaction of diazomethane with poly-*O*-acylaldonyl chlorides. The conversions of tetra-*O*-acetyl-L-arabinonyl chloride **3.244** to *keto*-L-fructose pentaacetate **3.245** and

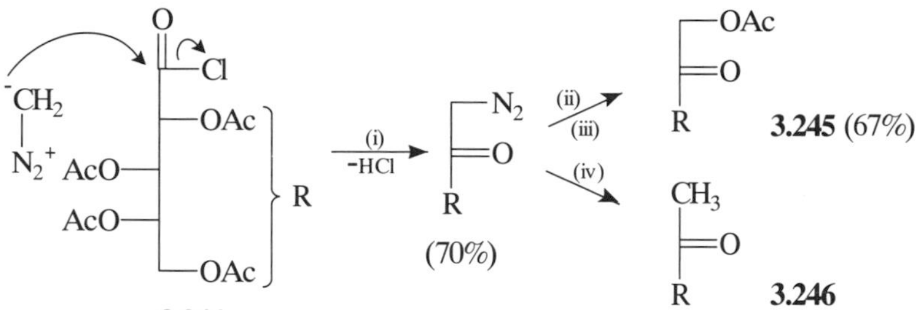

(i) Et_2O, 20 °C, 2 h; (ii) AcOH, $Cu(OAc)_2$; (iii) Ac_2O; (iv) HI, H_2O, $CHCl_3$

Scheme 3.73

l-deoxy-*keto*-L-fructose tetraacetate **3.246** are shown in Scheme 3.73 as examples of this reaction.[145] An example of the chain-shortening reaction is given in Scheme 3.75.

(iv)   Reactions involving loss of C-1

One of the important applications of aldonic acids is in reactions which lead to the loss of C-1 and the formation of lower aldoses. Such applications are illustrated in Schemes 3.74–3.76. Other related reactions by which aldose chain shortening can be effected are shown in Schemes 3.33(c) and 3.48.

The *Ruff degradation*[146] utilizes hydrogen peroxide in the presence of iron(III) ions to convert the calcium salts of hexonic acids into pentoses, as shown for the preparation of D-arabinose **3.250** from calcium D-gluconate **3.249** (X $= \frac{1}{2}Ca^{2+}$, R $-$ II) (Scheme 3.74a).[110] The use of ion exchange resins to remove inorganic materials after completion of the reaction has improved the yields and makes this degradation one of the best methods for aldose chain shortening. The mechanism is not completely understood, but the high specificity obtained under these conditions, compared with the random oxidation usually experienced when sugars are exposed to hydroxyl radicals ($\cdot OH$), has led to the suggestion that specific complexes are formed between the aldonic acids, the iron(III) ions and hydrogen peroxide, which then break down to give the products.

The chain-shortening reaction of aldonyl chlorides referred to above can be achieved by reacting them with peroxides in a variant of the Ruff process. Thus penta-*O*-acetyl-D-gluconyl chloride **3.251** on treatment with *t*-butyl hydroperoxide gives two different intermediate products depending upon the conditions used.[147] In the presence of base, the peroxygluconate **3.252** is formed and appears to result from attack of the hydroperoxide anion on the acyl chloride carbonyl group, whereas without base the hydroperoxide appears to attack the C-1–C-2 acetox-onium ion to yield the orthoperester **3.253** (Scheme 3.75). Both of these compounds give D-arabinose upon treatment with base, but the latter requires less vigorous conditions and gives a better yield. Possible mechanisms for the breakdown of the peroxy compounds are given in the scheme.

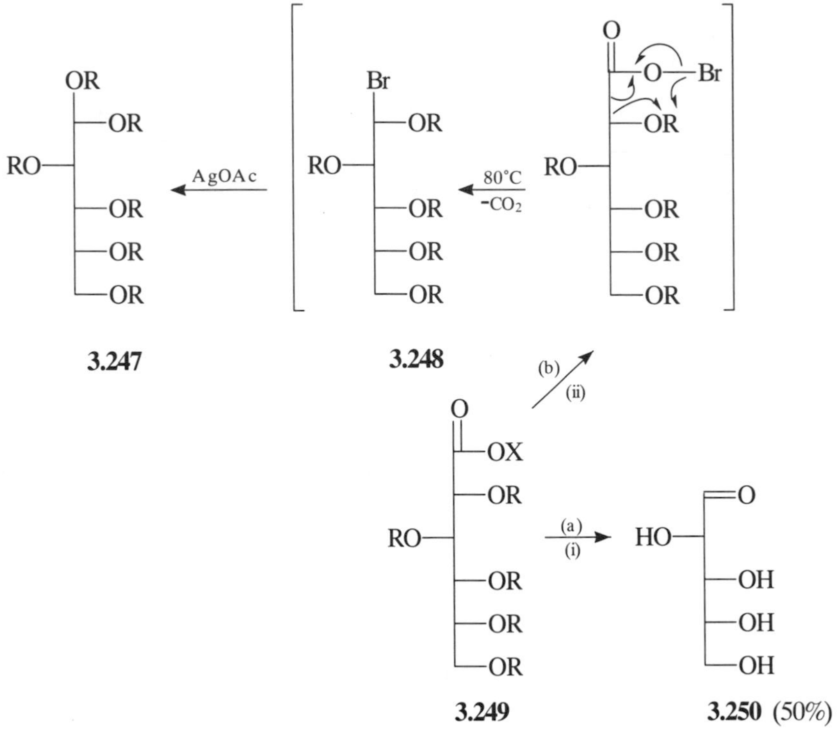

3.247   **3.248**

3.249   **3.250** (50%)

(i) H_2O_2, Fe^{3+} (X = $\frac{1}{2}Ca^{2+}$, R = H); (ii) Br_2, CCl_4, 80 °C, 4 h (R = Ac, X = Ag)
Scheme 3.74

The *Hunsdiecker degradation* is a well-known method for descending homologous series much used in aliphatic chemistry. It involves bromination of the silver salts of carboxylic acids followed by thermal breakdown of the bromo products to give alkyl halides containing one carbon atom fewer than the parent acids. When applied to the fully acetylated derivative of silver D-gluconate **3.249** (X = Ag, R = Ac), the peracetylbromo acyclic derivative of D-arabinose **3.248** (R = Ac) is formed (Scheme 3.74b). This can be converted into the hexaacetyl-*aldehydo*-D-arabinose derivative **3.247** (R = Ac) by treatment with silver acetate.

The *Weerman degradation* applies the Hofmann alkaline conversion (of amides to amines) to aldonamides (Scheme 3.76).[110] The hydroxyl group at C-2 must be free for this degradation to occur, which in practice means that the unprotected aldonamides must be used and these are rather unstable. The method is therefore employed infrequently.

Free radical decarboxylation is a much more satisfactory procedure for removing C-1 from aldose derivatives and can readily be accomplished by application of Barton *O*-acylthiohydroxamate chemistry to aldonic acids. However, the most useful application of this reaction does not involve simple aldonic acids but rather glycosides of ulosonic acids, as illustrated in Scheme 3.77.[148] The 2,6-anhydro-3-deoxyheptonate **3.254** can be prepared in 73% yield by methods described earlier

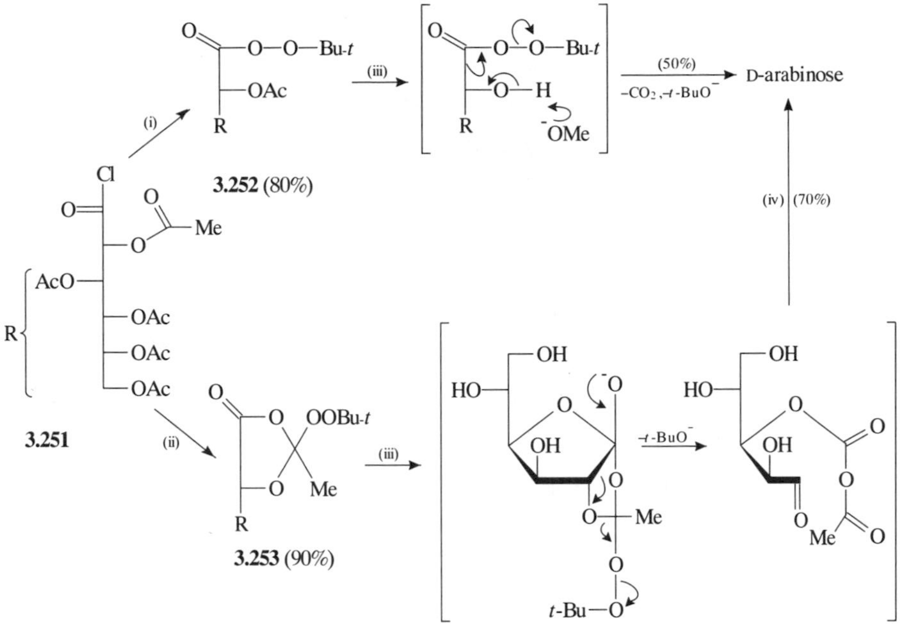

(i) t-BuO$_2$H, Py; (ii) t-BuO$_2$H; (iii) MeO$^-$(deacetylate); (iv) hydrolysis

Scheme 3.75

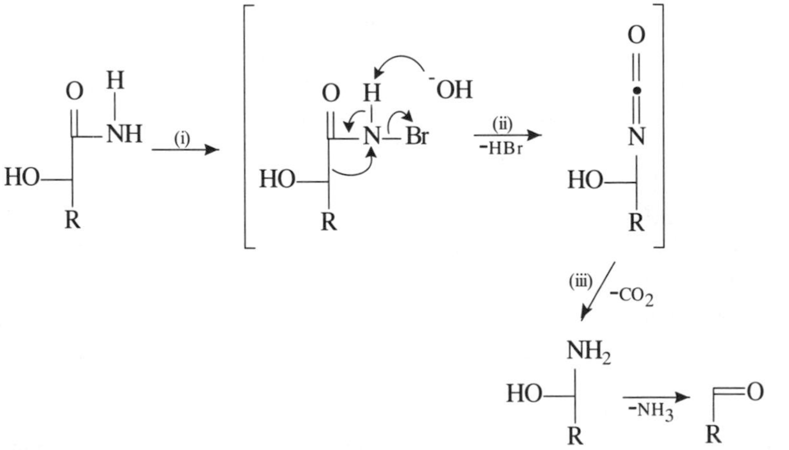

(i)NaOBr; (ii) NaOH; (iii) H$_2$O

Scheme 3.76

(see Scheme 3.38, R = Bn, E = CO$_2$Me). Chlorination of **3.254** at the C-2 position gives the anomeric glycosyl chlorides **3.255** which, with methanol, afford the methyl glycosides **3.256**. These glycosides can be separately saponified and the free acids converted to the thiohydroxamates **3.257** which, after filtration, are decarboxylated by exposure to light. Both anomers of **3.257** give the same

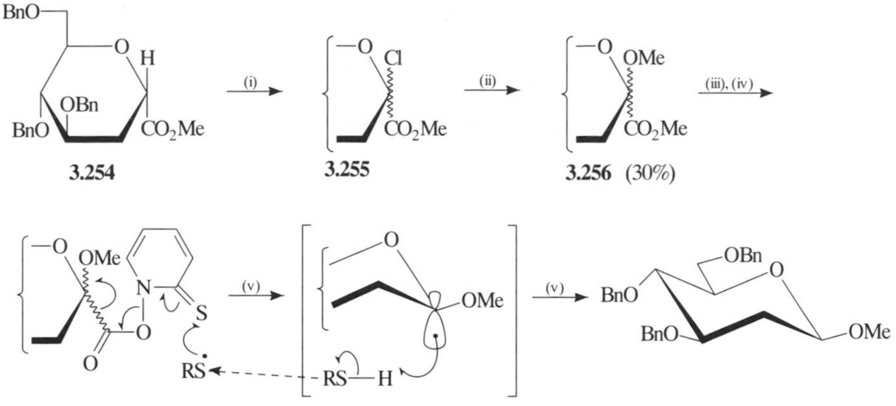

3.254 **3.255** **3.256** (30%)

3.257 **3.258** **3.259** (40%, β/α 10:1)

(i) LDA, C_2Cl_6, THF, -70° C; (ii) MeOH, Ag_2SO_4, $CaSO_4$, I_2, Et_2O; (iii) saponify, KOH; (iv) 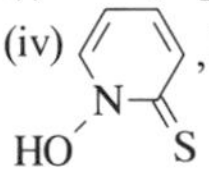, DCC, Et_3N; (v) *hν*, dodecanethiol (RSH)

Scheme 3.77

anomeric mixture of 2-deoxyhexopyranosides **3.259** with the β-form predominating because the common 2-deoxy-D-*arabino*-hexopyranosyl radical **3.258** exhibits a strong preference for hydrogen transfer at its α-face in keeping with the diastereoselectivity observed with some other pyranosyl radicals (Schemes 3.40 and 3.121). The diastereoselective decarboxylation of glycosides of 2-ulosonic acids therefore constitutes another route to β-glycosides related to the thio-orthoester method described earlier.

(b) Aldaric acids

Strong nitric acid appears to be one of the few oxidants which is able to oxidize the terminal groups of aldoses but leave the secondary hydroxyl groups unchanged. D-Glucose **3.260** treated with this reagent gives D-glucaric acid **3.261** (Scheme 3.78).[149] The names of these polyhydroxydicarboxylic acids are derived by replacing the 'ose' ending in the sugar by the 'aric acid' suffix, but formerly these compounds were known as 'saccharic acids'. They can form monolactones or dilactones in which the ring sizes depend on the stereochemistry of the asymmetric carbon atoms. Glucaric acid, for example, can form two monolactones, the 1,4-**3.262** and 6,3-**3.263**, and two dilactones linked 1,4:6,3 and 1,5:6,3.

Galactaric acid (mucic acid) has surprisingly low solubility and can be used in a gravimetric determination of D-galactose.

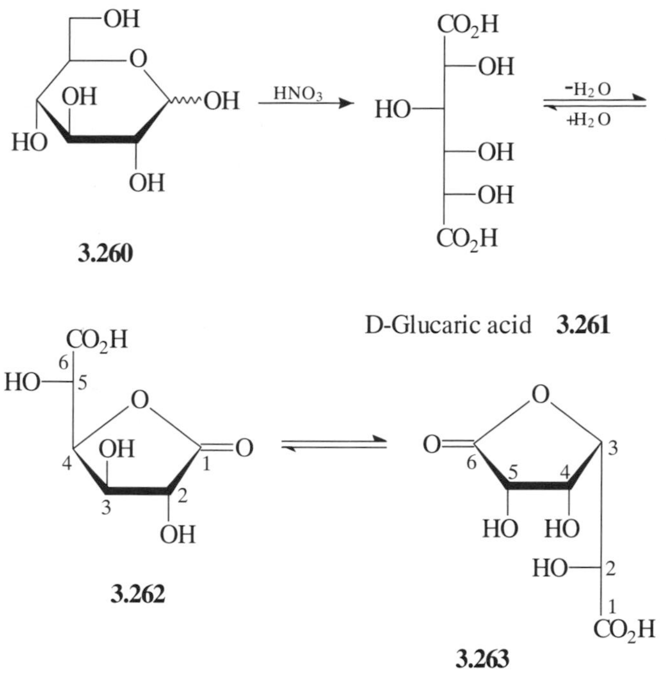

3.260

D-Glucaric acid   **3.261**

3.262

3.263

Scheme 3.78

3.1.7   REACTIONS WITH BASES

(a)   Epimerizations and aldose–ketose isomerizations

Aldoses and ketoses undergo rearrangements in dilute basic solution which depend upon their hydroxyaldehyde and hydroxyketone structures. In 0.035% aqueous sodium hydroxide solution at 35 °C for 100 h, D-glucose **3.269** (Scheme 3.79) gives a mixture containing D-fructose **3.270** (28%), D-mannose **3.264** (3%) and unchanged D-glucose (57%). The glucose–mannose interconversion is an epimerization, and the total rearrangement is known as the Lobry de Bruyn–van Ekenstein reaction which occurs by enolization of the glucose to give the enediol **3.265**, and this can 'ketonize' in three ways to glucose, mannose or fructose.[150]

Because hydroxide ions also catalyse other reactions, pyridine is usually used as the base in preparative work. Thus, heating D-xylose in anhydrous pyridine under reflux for 4.5 h gives D-*threo*-pentulose accompanied by some D-lyxose and other pentoses and pentuloses formed by way of a 2,3-enediol.

Base-catalysed isomerization of aldonic acids also occurs, and pyridine is often used to convert common aldonic acids into their less abundant 2-epimers.

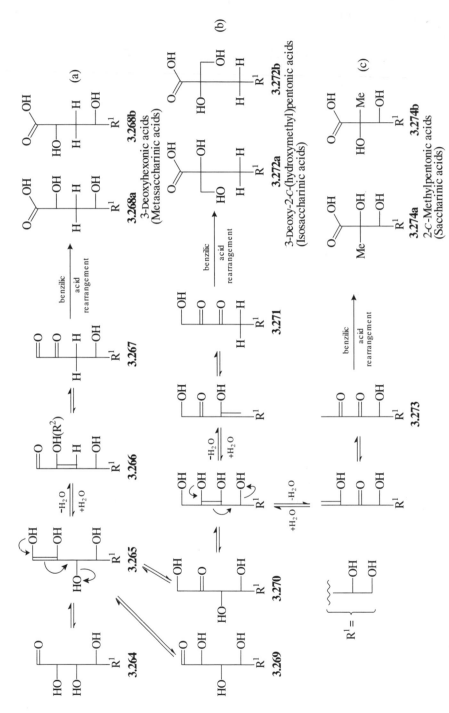

Scheme 3.79

(b)   Saccharinic acid formation[151]

More-concentrated basic solutions (i.e. about 1% calcium hydroxide) convert sugars into three types of isomeric deoxyaldonic acids.[92] Glucose and some of its derivatives, for example, form 3-deoxyhexonic acids (metasaccharinic acids) **3.268** (Scheme 3.79), 3-deoxy-2-*C*-(hydroxymethyl)pentonic acids (isosaccharinic acids) **3.272** and 2-*C*-methylpentonic acids (saccharinic acids) **3.274**. The formation of these products involves the two successive steps illustrated in Scheme 3.80: a direct β-elimination from the free sugar (**a**) or a δ-elimination from an enediol (**b**) and a benzilic acid rearrangement of the α-dicarbonyl compounds outlined in Scheme 3.80(c). When the leaving group in the first step in the reaction is hydroxide (HO$^-$), this elimination is slow because in basic solution such groups are partially ionized. However, when *O*-3 carries an alkyl or acyl substituent, elimination is much more likely. Routes to the various saccharinic acids utilizing these steps are given in Scheme 3.79.

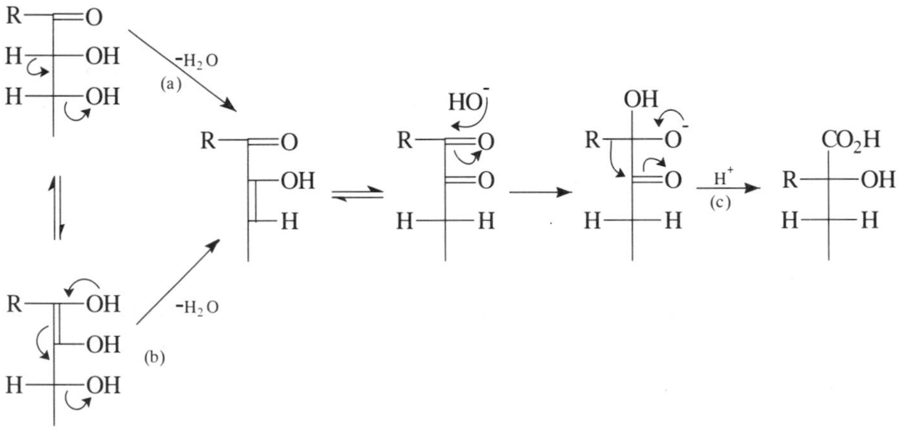

Scheme 3.80

By stopping the reactions at optimum times, before completion, it has been possible to isolate some of the intermediates. This has been facilitated by using selected starting materials with good leaving groups in appropriate positions which increase the rates of intermediate formation relative to the rates of their destruction.

(i)   3-Deoxyhexonic acids (metasaccharinic acids) (Scheme 3.79a)

The intermediate 3-deoxy-D-*erythro*-hexosulose **3.267** is obtainable by treatment of 3-*O*-benzyl-D-glucose with sodium hydroxide since β-elimination of the benzyloxy anion is fast, and thus the concentration of osulose builds up before it reacts further by the benzilic acid rearrangement. Treatment of the isolated intermediate **3.267** with base then gives the metasaccharinic acids **3.268a** and **3.268b** quantitatively, as required by the above-outlined mechanism, and this rearrangement occurs faster with calcium hydroxide than with sodium hydroxide. Thus, the latter is the better base to use if osulose **3.267** is to be isolated, but the former is to be preferred if the metasaccharinic acids are required. Further support for the proposed mechanism (Scheme 3.79a) has been obtained from work with 2,3-di-*O*-methyl-D-glucose which, after elimination, cannot ketonize to a dicarbonyl compound. It therefore finally gives the enal **3.266**, ($R^2 = Me$).

(ii) 3-Deoxy-2-C-(hydroxymethyl)pentonic acids (isosaccharinic acids) (Scheme 3.79b)

4-Deoxy-D-*glycero*-hexo-2,3-diulose **3.271** is the only intermediate given in Scheme 3.79b which has been isolated. This has been achieved by treating 4-*O*-substituted derivatives of D-glucose, e.g. cellobiose, with aqueous potassium hydroxide. The diulose so obtained readily rearranges to isosaccharinic acids **3.272a** and **3.272b** when treated with base.

(iii) 2-C-Methylpentonic acids (saccharinic acids) (Scheme 3.79c)

1-Deoxy-D-*erythro*-hexo-2,3-diulose **3.273**, a proposed intermediate in the formation of the saccharinic acids **3.274a** and **3.274b**, has not been isolated from a basic hexose solution, but it has been prepared by independent synthesis. Treatment of it with base gives saccharinic acids **3.274a** and **3.274b** in high yield.

(iv) Stereochemistry in saccharinic acid formation

A point of interest in the formation of these acids is the product ratio of the C-2 epimers. Two metasaccharinic acids, namely 3-deoxy-D-*ribo*- and 3-deoxy-D-*arabino*-hexonic acids **3.268a** and **3.268b** have been isolated in approximately equal amounts from the products formed by treating D-glucose with sodium hydroxide solution. Careful chromatographic studies have shown that two isomeric isosaccharinic acids are formed in roughly equal amounts when 4-*O*-substituted-D-glucose derivatives are treated with aqueous base. One of these isomers, trivially named '*α*-isosaccharinic acid', is easy to isolate as its calcium salt, and it has been shown to be 3-deoxy-2-*C*-hydroxymethyl-D-*erythro*-pentonic acid **3.272a** by X-ray crystallographic analysis. The C-2 isomer, sometimes called the '*β*-isosaccharinic acid', therefore has the *threo*-structure **3.272b**. In contrast with the transformations leading to iso- and meta-saccharinic acids discussed above, the reaction of D-fructose or 1-*O*-benzyl-D-fructose with calcium hydroxide (Scheme 3.79c) apparently gives only 2-*C*-methyl-D-*ribo*-pentonic acid **3.274a**. Since the isomeric *arabino*-saccharinic acid has been prepared by an independent route, it was determined that this compound was formed only in trace amounts in this reaction. A plausible reason for the high stereospecificity in this particular case is the close proximity of an asymmetric centre to the carbonyl group in the diulose intermediate **3.273**. Thus the alkyl shift in the benzilic acid rearrangement reaction leading to the acid **3.274** will be subject to asymmetric induction to a greater extent than in the corresponding benzilic acid rearrangements with the corresponding *α*-dicarbonyl intermediates **3.267** and **3.271**.

(c)   Base-induced degradations

As well as undergoing the base-catalysed rearrangements referred to above, aldoses and ketoses can be broken down to yield smaller fragments by retroaldol reactions as is typical for β-hydroxycarbonyl compounds (Scheme 3.81). Prolonged treatment of hexoses with moderately concentrated aqueous base yields three-carbon fragments such as 2-hydroxypropionaldehyde, pyruvic acid, methyl glyoxal and lactic acid which are formed from glyceraldehyde and dihydroxyacetone, the first products of the degradations.

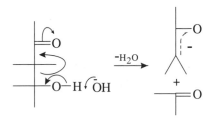

Scheme 3.81

(d)   The formose reaction

The reverse of the process discussed in the above section can be used in the synthesis of monosaccharides (Scheme 3.82). When formaldehyde is treated with mild alkali, a sugar-like syrup called *formose* is produced.[152] This is a complex

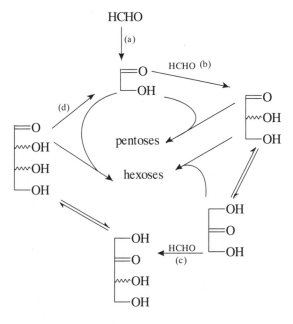

Scheme 3.82

mixture of glycolaldehyde, which under certain conditions can account for 50% of the formaldehyde transformed, together with trioses, tetroses, pentoses and hexoses. The reaction is autocatalytic, exhibiting a long induction period which is followed by a rapid consumption of formaldehyde; glycolaldehyde has been shown to be a key intermediate since the induction period disappears if it is added initially.

The induction period is caused by the very slow formation of glycolaldehyde from two molecules of formaldehyde (Scheme 3.82a), and once the concentration of glycolaldehyde builds up, it can rapidly condense with two more molecules of formaldehyde to give tetroses (Schemes 3.82b and 3.82c). These can then isomerize and fragment by a retroaldol reaction into two glycolaldehyde molecules (Scheme 3.82d), and so the reaction cycle accelerates. Pentoses and hexoses are formed from the C_2, C_3 and C_4 components as indicated.

The nature of the initial condensation is not yet clear, but one suggestion is that it involves the slow formation of a formyl anion which condenses with a formaldehyde molecule.

The formose reaction is considered by many to have been the most likely source of sugars on the primitive earth, i.e. the reaction by which the pentoses necessary for nucleic acid (Section 8.2.2.b) formation were generated. It is significant that paper chromatography indicates that all the pentoses are present in the formose products, although they often constitute only about 1% of the total mixture. Crystalline tosylhydrazones of DL-xylose and DL-arabinose have been isolated. The hexoses are usually present in greater proportions and crystalline phenyl DL-*arabino*-hexosazone and phenyl DL-*xylo*-hexosazone have been isolated in a total yield of 20%, being mainly derived from DL-fructose and DL-sorbose, respectively.

3.1.8   REACTIONS WITH ACIDS[2]

Free sugars undergo changes in acid solutions as a result of their hydroxycarbonyl structures, but they are significantly more stable at low pH values than in basic media. Glucose, for example, is only very slowly degraded under conditions suitable for polysaccharide hydrolysis (e.g. 2 mol l^{-1} mineral acid at $100\,^{\circ}C$).

(a)   Reversion and anhydride formation

When solutions of D-glucose (1%) are heated in dilute sulphuric acid (about 0.15 mol l^{-1}), isomaltose (6-*O*-α-D-glucopyranosyl-D-glucose) is produced (0.1% has been isolated) along with smaller amounts of other disaccharides. The main product is formed by attack of the primary hydroxyl of one glucose molecule upon the anomeric carbon of another in a reaction directly comparable with the methanolysis of the sugar (Section 3.1.1), whereas the other disaccharides are formed by attack from secondary groups which are sterically less favoured. This type of condensation takes place with most free sugars and to some degree it always accompanies glycoside and polysaccharide hydrolysis. It is referred to as *reversion*, and can be troublesome in polysaccharide analysis since disaccharides isolated from the hydrolysates in low yield may be reversion products rather than constituent parts of the original molecules. Problems of this type are reduced by carrying out such hydrolyses in dilute solutions in which intermolecular processes are specifically unfavoured.

Alternatively, attack at the anomeric centres can be effected intramolecularly by suitably disposed hydroxyl groups, and anhydro sugars can be formed. In particular, this occurs with hexoses in which the primary hydroxyl groups participate to yield 1,6-anhydropyranoses in reversible processes. Whereas the more abundant hexoses, e.g. glucose and galactose, give only very small proportions of these anhydrides, the same does not apply to, for example, idose which exists predominantly as the 1,6-anhydride at equilibrium in acid solution. This reaction is treated in Section 3.1.1.c.

(b)   Epimerizations in molybdic acid

Epimerization of aldoses accompanied by ketose formation is a well-established reaction of sugars in mildly-basic solution, as is discussed in Section 3.1.7.a. It proceeds by a series of enolizations and prototropic shifts. A new method has now been discovered by Bilik's group for inducing aldose epimerizations which occur, without ketose involvement, under mildly acidic conditions in the presence of molybdate.[153] Thus, for example, D-mannose in aqueous molybdic acid at pH 4.5 and 90 °C gives a 2.5:1 glucose/mannose mixture, as illustrated in Scheme 3.83. Analysis of the products in the equilibrated solution formed from 1-^{13}C-mannose reveals that the ^{13}C label resides only at C-2 in the glucose and C-1 in the mannose, indicating that this epimerization differs from those involving prototropic shifts and involves a carbon skeleton rearrangement. The transformation is thought to occur within a complex formed between the molybdate and the carbonyl, and the C-2 and C-3 hydroxyl groups of the *aldehydo*-tautomer of the aldose, which induces the bond reorganizations shown schematically in **3.275** → **3.276**. Thus, a 1,2-shift in the carbon skeleton takes place in which C-3 migrates, whilst maintaining its stereochemical integrity, from C-2 to form a new bond by stereospecific attack at the C-1 *aldehydo* group as illustrated. The method can be used to introduce a carbon isotope at C-2 of D-mannose (by use of the C-1-labelled D-glucose analogue as indicated by the asterisks) and thereby at C-2 of D-glucose (or D-fructose) by base-catalysed isomerization.

(c)   Formation of non-carbohydrate products

Another reaction occurs when sugars are treated with concentrated mineral acids and results in the formation of furan derivatives; in general, ketoses react in this way more readily than do aldoses. A plausible reaction pathway is given for hexoses in Scheme 3.84 and, in agreement with this, labelling studies have shown that the aldehydic carbon in the product is derived from C-1 of the sugar.[150] Pentoses yield furfuraldehyde.

A further reaction occurs when acid solutions of hexoses are heated, causing fragmentation of the molecules to take place by way of the furfuraldehyde derivative

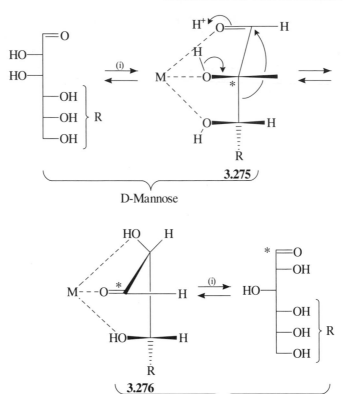

D-Mannose

D-Glucose

(i) molybdic acid, H_2O, pH 4.5, 90 °C, 4 h

Scheme 3.83

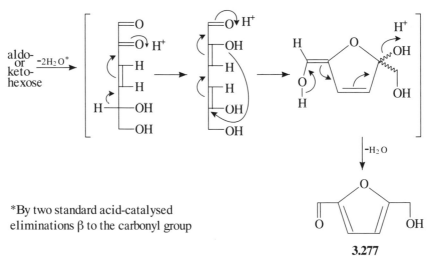

*By two standard acid-catalysed
eliminations β to the carbonyl group

Scheme 3.84

3.277 and yielding formic acid and levulinic acid **3.278**, possibly as shown in Scheme 3.85.

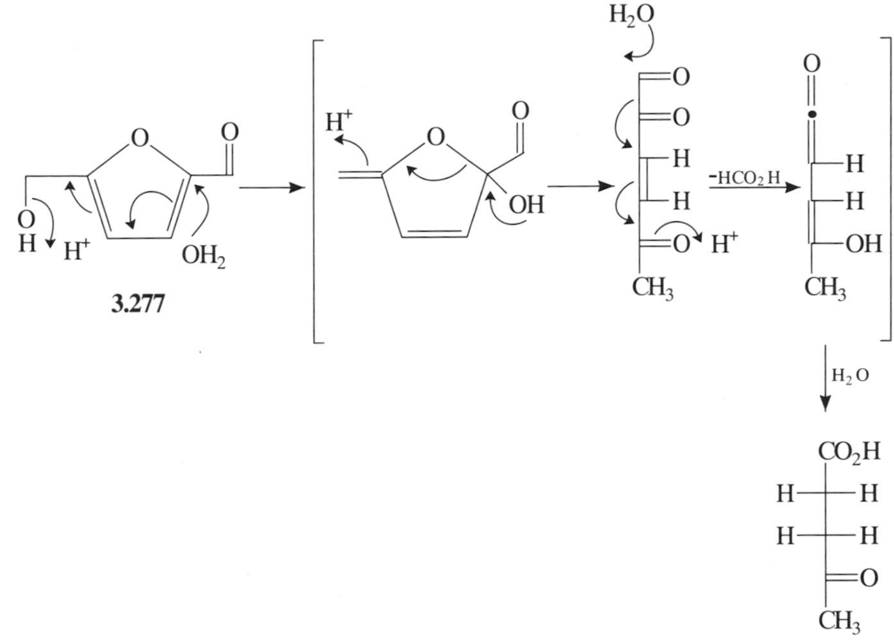

3.277

3.278

Scheme 3.85

Many of the tests used for detecting and analysing for sugars colorimetrically are based upon the production of strongly coloured materials formed from the reactions between acid degradation products and polyaromatic compounds.[154] Some of these reactions allow distinctions to be made between sugars, but the specificities

Table 3.2

Test	Reagents	Carbohydrates	Observations
Molisch	α-Naphthol + concentrated H_2SO_4	Hexoses, pentoses, polymers	Violet colour at H_2O/H_2SO_4 interface
Anthrone	Anthrone + concentrated H_2SO_4	Hexoses Pentoses	Stable blue colour Unstable blue colour changing to red
Tollens	Phloroglucinol + concentrated HCl	Pentoses	Red-violet colour
Seliwanoff	Resorcinol + 1 mol l^{-1} HCl	Ketoses Aldoses	Red colour rapidly formed Red colour slowly formed

have been discovered empirically, and knowledge of the structures of the chromophores is still largely lacking. The Molisch test (Table 3.2) can be applied for any sugar, and other reactions belonging to this category include the anthrone and phloroglucinol tests by which pentoses can be detected and the resorcinol test that distinguishes ketoses from aldoses. All of these tests can be applied with a few milligrams of sugar in aqueous solution, but for many purposes these have been superseded by chromatographic methods, especially t.l.c., h.p.l.c. and g.l.c.

3.2 REACTIONS OF GLYCOSYL ESTERS

3.2.1 GLYCOSYL CARBOXYLATE ESTERS

In acidic media, cyclic peracylated sugars, which can be prepared as discussed in Section 5.3, undergo nucleophilic displacements of the glycosyl acyloxy groups, and therefore a variety of other glycosyl derivatives can be formed from them. The chemospecificity of these displacements arises from the same feature that renders the anomeric centres of free sugars susceptible to specific substitution — the resonance stabilization of the glycosyl carbonium ions formed following coordination of O-1, or of the acyclic analogues formed following coordination of the ring oxygen atom.

Often the displacement of a glycosyl ester group is carried out in the presence of a further ester at C-2 which may participate, thereby leading to a 1,2-acyloxonium ion and a rate-enhanced displacement with subsequent formation of a 1,2-*trans*-related product. Glycosyl esters, the other hydroxyl groups of which are fully protected with non-ester groups, undergo similar reactions but do not show the same anomeric selectivity. They are usually prepared by acylating hydrolysis products obtained from fully protected glycosides.

(a) Anomerization and further isomerizations

(i) Simple anomerizations

Glycosyl acetates anomerize when dissolved in acetic anhydride/acetic acid mixtures containing acid catalysts such as sulphuric acid, perchloric acid, zinc chloride, tin(IV) chloride or tin(IV) trichloride acetate.[1,92] Either α- or β-D-glucopyranose pentaacetate **3.279** or **3.281**, for example, gives a mixture containing 84% of the α-anomer and 16% of the β-anomer at equilibrium, and the increased proportion of the α-compound relative to that found for the free sugar (36%) and methyl glycoside (66%) correlates with the increased influence of the anomeric effect resulting from the greater electronegativity of the glycosyl substituent (Section 2.3.1.b). The rates of exchange of the acetoxy groups at C-1 with the solvent, occurring during anomerization and measured using acetate labelled with ^{14}C in the acetoxy groups, show that the α-anomer exchanges at the same rate as it anomerizes, i.e. every exchange occurs with inversion of configuration to give the β-product (Scheme 3.86a) and every C-1 acetoxy group of the product is derived from the solvent.

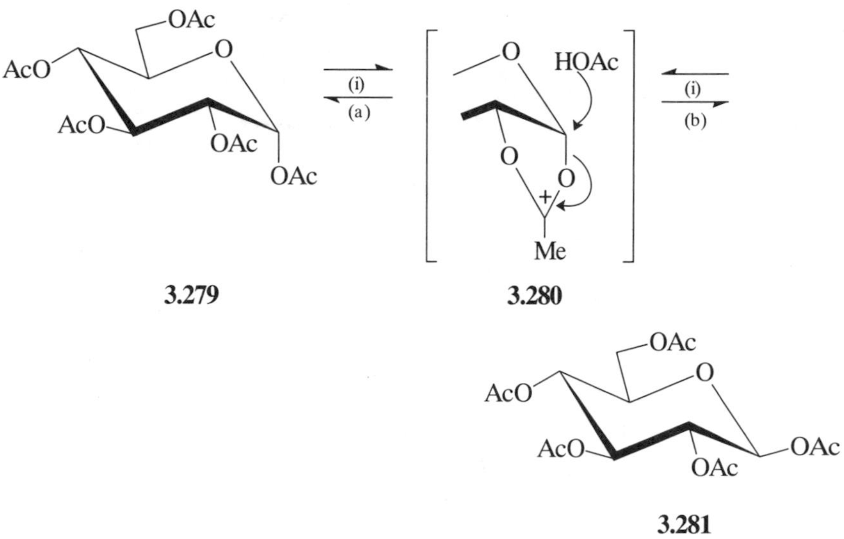

3.279   **3.280**

3.281

(i) AcOH, H$^+$

Scheme 3.86

The β-anomer, on the other hand, exchanges about 15 times faster than it anomer-izes; most of these exchanges therefore occur with retention of the β-configuration and proceed, presumably, via the acetoxonium ion **3.280** (Scheme 3.86b).

The thermodynamically unfavoured anomers of monosaccharide peresters (usually the β-compounds having equatorial anomeric acyloxy groups) are often readily avail-able (Section 5.3.1.a), and therefore anomerizations can provide a useful synthetic route to the more stable forms. In this way, the poly-O-acetyl-α-pyranoses of glucose, galactose, mannose and xylose have been obtained by treating the corresponding β-anomers with sulphuric acid in acetic anhydride/acetic acid mixtures. The reverse situation arises with the D-ribopyranose tetra-acetates, since the α-anomer is the less stable modification and equilibrated mixtures therefore afford mainly the β-form (cf. Section 3.1.1). On the other hand, the difference in stabilities between the α- and β-anomeric modifications of the arabinose acetates is small, and therefore the anomerized mixture is composed of almost equal amounts of each isomer.

(ii)   Complex intramolecular processes

On treatment with antimony pentachloride, fully acetylated aldoses undergo more deep-seated rearrangements involving acetoxonium ion intermediates which cleave to yield products with inverted stereochemistry at one or more of the non-anomeric carbon atoms.[155] Thus, penta-O-acetyl-β-D-glucopyranose **3.282**, on treatment in dichloromethane with antimony pentachloride, gives the D-idopyranose salt **3.283** which crystallizes from solution in 60% yield and provides an ingenious method of obtaining the relatively inaccessible D-idose as its α-pentaacetate **3.284**. The precipitation of **3.283** from solution drives the reaction in the forward direction by

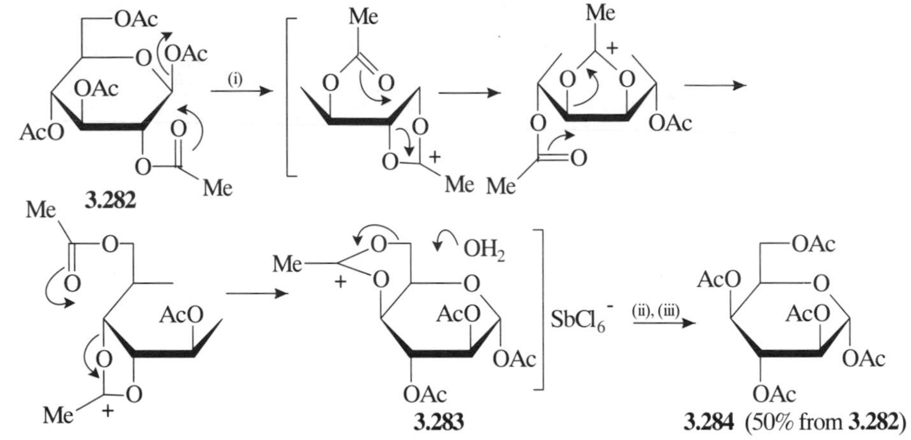
3.282
3.283
3.284 (50% from 3.282)
SbCl6−
(i)
(ii), (iii)

(i) $SbCl_5$, CH_2Cl_2, −10 °C; (ii) NaOAc, H_2O; (iii) Ac_2O, Py

Scheme 3.87

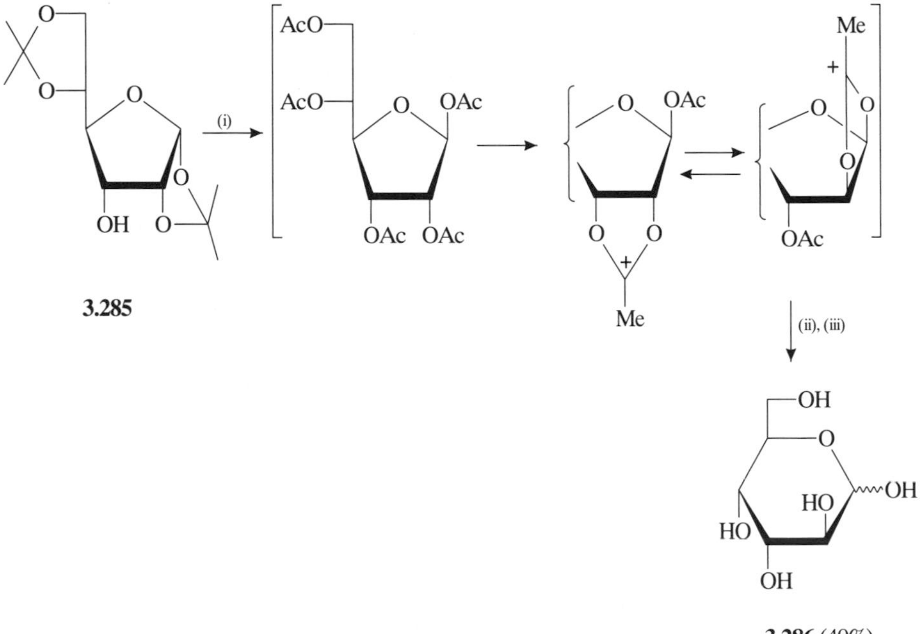
3.285
3.286 (40%)
(i)
(ii), (iii)

(i) AcOH, Ac_2O, H_2SO_4; (ii) H_2O; (iii) Et_3N, MeOH, H_2O

Scheme 3.88

way of the intermediate α-*gluco*-, α-*manno*- and α-*altro*-acetoxonium ion salts, so inversions are brought about at all the secondary ring carbon atoms (Scheme 3.87). With the galactose analogue, five-membered *cis*-acetoxonium ions can only be formed at C-1–C-2 and C-2–C-3 and from the resultant mixture of α-*galacto*- and α-*talo*-salts, penta-*O*-acetyl-α-D-talopyranose has been prepared.

Acyloxonium ions can also be generated during acetolysis of acid-sensitive derivatives and D-altrose **3.286** can be made from D-allose by acetolysis of its 1,2:5,6-di-*O*-isopropylidene derivative **3.285**.[156] The reaction is triggered by formation from the *cis*-diol diacetate system of the 2,3-acetoxonium ion which rearranges under these acetolysis conditions to the 1,2-β-D-*altro*-ion and leads to D-altrose, as illustrated in Scheme 3.88. In this case, cleavage of the anomeric acetoxy group of the allose pentaacetate is not observed because it leads to a relatively unstable α-D-*allo*-1,2-acetoxonium ion which cannot rearrange or react further.

(b) Reactions leading to glycosyl halides

(i) Treatment of acylated glycosyl esters with hydrogen halides

When a per-*O*-acylated glycopyranosyl acetate or benzoate is treated either with hydrogen bromide or hydrogen chloride in acetic acid, or with liquid hydrogen fluoride, the acyloxy group at C-1 is displaced by a halide ion and this reaction represents an important synthetic route to the acylated glycosyl halides.[157] Hydrogen iodide brings about a similar displacement, but side reactions of the unstable glycosyl iodides under these reaction conditions make this a poor method of synthesis. (The best method for preparing glycosyl iodides is given in Section 3.3.3.a.) The most stable anomer (the anomeric effect ensures that the halogen atom is axial) of the glycosyl bromides or chlorides is usually obtained in high yield by this method, irrespective of the anomeric configuration of the precursor. For example, the α-anomer and β-anomer of 1,2,3,4,6-penta-*O*-acetyl-D-glucopyranose both give 2,3,4,6-tetra-*O*-acetyl-α-D-glucopyranosyl bromide **3.289** as a crystalline

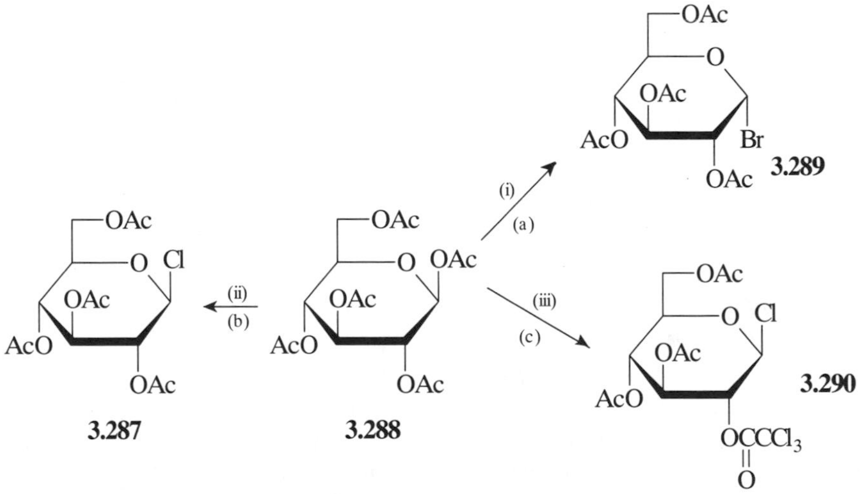

(i) AcOH, HBr, 20 °C, 2 h; (ii) AlCl$_3$, CHCl$_3$, 20 °C, 0.5 h; (iii) PCl$_5$, CCl$_4$, reflux, 5 h

Scheme 3.89

product in 85% yield, as illustrated for the β-anomer in Scheme 3.89a. However, there has been a report that treatment of 2-O-benzyl-1,3,4,6-tetra-O-p-nitrobenzoyl-α-D-glucopyranose with hydrogen bromide in dichloromethane yields mainly the substituted β-glucosyl bromide. The isolation of the thermodynamically less stable anomer from a reaction of this type is unusual and indicates that equilibration of the anomers does not occur under these particular conditions.

(ii)   Treatment of acylated glycosyl acetates with aluminium chloride

A useful method for preparing thermodynamically unstable glycosyl chlorides involves the displacement of the acetoxy group at C-1 of aldosyl acetates possessing 1,2-*trans*-configurational arrangements with chloride generated from aluminium trichloride in chloroform. 1,2,3,4,6-Penta-O-acetyl-β-D-glucose **3.288** thus gives 2,3,4,6-tetra-O-acetyl-β-D-glucopyranosyl chloride **3.287** in high yield when treated under these conditions (Scheme 3.89b).[158] The reagent clearly does not catalyse anomerization reactions efficiently.

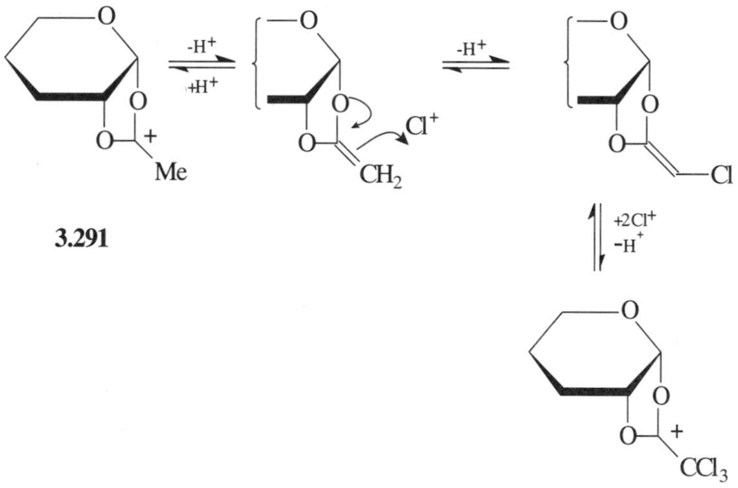

Scheme 3.90

(iii)   Treatment of acetylated glycosyl acetates with phosphorus pentachloride

When 1,2,3,4,6-penta-O-acetyl-β-D-glucopyranose **3.288** is treated with phosphorus pentachloride, a reaction occurs beyond simple exchange at C-1. Not only is the C-1 acetoxy group displaced, but the acetoxy group at C-2 is trichlorinated and the useful β-D-glucosyl chloride derivative **3.290** (Scheme 3.89c) is obtained.[157] (A synthetic application of this compound is given in Scheme 3.28.) Only the β-glucosyl acetate affords this product, and the C-2 acetoxy group is chlorinated preferentially, which suggests the intermediacy of an acetoxonium ion process.[55,159] How this occurs is not known, but an electrophilic mechanism seems most likely as the reaction resembles the well-known halogenation at the α-carbon atom of a ketone. A possible mechanism that starts with the 1,2-acetoxonium ion **3.291** is outlined in Scheme 3.90.

(c)   Reactions leading to *O*-, *S*-, *C*- and *N*-glycosides

(i)   *O*-Glycosides

One of the most convenient methods for preparing aryl pyranosides and furano-sides involves the displacement of the C-1 acetoxy groups of the corresponding glycosyl acetates. The reaction is usually brought about by heating the acetate and a phenol in the presence of an acid catalyst in xylene or chloroform solu-tion, or simply as a melt.[160] The anomeric composition of the products is depen-dent upon the reaction conditions, and it is often considered better to accept a lower overall yield of anomerically pure product rather than a high yield of an anomeric mixture. 1,2,3,4,6-Penta-*O*-acetyl-α-D-glucopyranose **3.292**, for example, gives mainly phenyl 2,3,4,6-tetra-*O*-acetyl-α-D-glucopyranoside **3.293**, the thermo-dynamically more stable anomer, when it is heated in the presence of zinc chloride (Scheme 3.91). The β-anomer **3.294** is obtainable as the main product when tosic acid is used as the catalyst.[160]

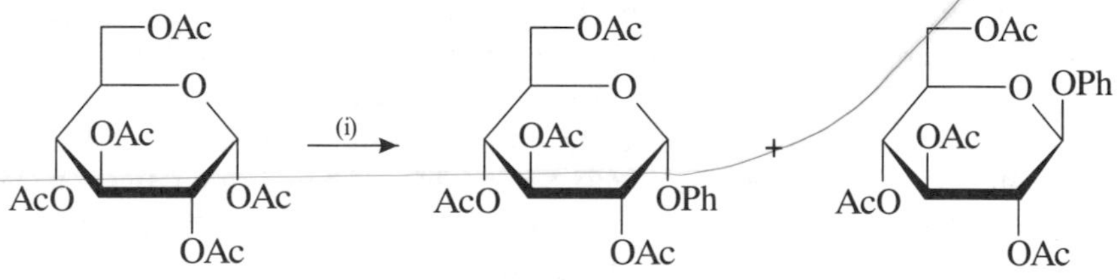

3.292	**3.293** (64%)	**3.294** (26%)

(i) PhOH, ZnCl$_2$

Scheme 3.91

Forcing reaction conditions are unsuitable for *O*-alkyl glycoside synthesis and consequently more efficient Lewis acid catalysts have been introduced to activate the C-1 ester group so that displacements can take place at lower temperatures and with greater stereoselectivities. Thus, glycosyl *p*-nitrobenzoates have been activated with trimethylsilyl trifluoromethanesulphonate[161] and glycosyl acetates with either tin(IV) chloride[162] or trityl perchlorate.[4] For example, when tri-*O*-benzoylribofuranosyl acetate **3.295** is treated with cyclohexanol in dichloromethane containing tin(IV) chloride, as shown in Scheme 3.92(a), the β-cyclohexyl glyco-side **3.296** is formed in good yield at 25 °C, as expected from a compound with a participating ester group at C-2. However, when tri-*O*-benzyl-β-ribosyl acetate **3.297** and cyclohexanol are treated at 0 °C in diethyl ether containing trityl perchlo-rate as the activator and with added lithium perchlorate and molecular sieves to

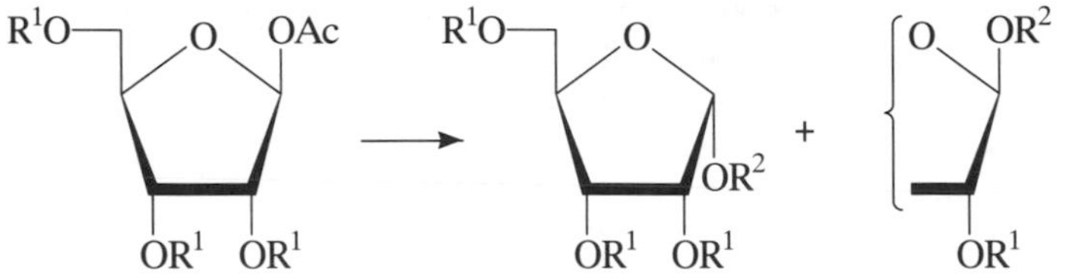

(a) **3.295** R^1 = Bz —(i)→   **3.296** (82%) ⎤
(b) **3.297** R^1 = CH$_2$Ph —(ii)→   **3.298** (60%) + **3.299** (26%) ⎬ R^2 = cyclohexyl
(c) **3.297** R^1 = CH$_2$Ph —(iii)→   **3.299** (83%) ⎦
(d) **3.297** R^1 = CH$_2$Ph —(iv)→   **3.300** (77%) + **3.301** (9%) R^2 = cholestanyl

(i) CH$_2$Cl$_2$, SnCl$_4$, 0 °C, 10 min, cyclohexanol, 25 °C; (ii) Et$_2$O, TrClO$_4$, LiClO$_4$, molecular sieves, cyclohexanol, 0 °C; (iii) Et$_2$O, TrClO$_4$ cyclohexanol, 0 °C; (iv) SnCl$_4$, Sn(OTf)$_2$, LiClO$_4$, CH$_2$Cl$_2$, 22 °C, 1 h, then at -23 °C, cholestanyl-OTms, 5 h

Scheme 3.92

remove the free acid, α-glycosylation preponderates to give furanosides **3.298** and **3.299** as illustrated in Scheme 3.92(b). If, however, the additives are omitted, acid-catalysed anomerization of the α-product occurs to give exclusively the more stable β-furanoside **3.299** (Scheme 3.92c). Greater α-selectivity is achieved in reactions of ribosyl acetate **3.297** with silylated alcohols in dichloromethane at -23 °C when the complex Lewis acid catalyst SnCl$_3$SnCl(OTf)$_2$ (formed from tin(IV) chloride and tin(II) triflate) is used in the presence of a stoichiometric amount of lithium perchlorate.[163] Thus, for example, when cholestanyl trimethylsilyl ether is so treated, the α/β-cholestanyl ribosides **3.300** and **3.301** are produced as a 9:1 mixture, as illustrated in Scheme 3.92(d). The high stereoselectivity arises because the bulky perchlorate anion is believed to be situated on the least sterically hindered β-side of the ribofuranosyl cation intermediate, thereby facilitating silyl alkoxide attack preferentially from the α-face.

Mukaiyama has introduced SnCl$_3$ClO$_4$ and GaCl$_2$ClO$_4$ as catalysts for efficient 1,2-cis-glycosidations.[164] Thus, 2,3,4,6-tetra-O-benzyl-α/β-D-glucopyranosyl acetate reacts, in toluene at room temperature, with methyl 2,3,6-tri-O-benzyl-4-O-trimethylsilyl-α-D-glucopyranoside to give an 85% yield of the α/β-(1→4)-linked glucosylglucopyranosides in the ratio of 9:1 when silver chlorate(I) and tin(IV) chloride or gallium(III) chloride are employed as catalysts.

With specific ester groups at C-1, milder conditions can be used to produce effective glycosylating agents, as is the case with glycosyl 2-pyridinecarboxylates, which have their pyridine nitrogen and carbonyl oxygen atoms well placed to coordinate with Lewis acids to produce good leaving groups as shown, for example, in **3.303**. A range of alcohols, including partially protected sugars, have been glycosylated with either anomer of **3.302** and with copper(II) and tin(II) salts as mild Lewis acid activators, the triflates proving most effective.[165] Treatment of 2,3,4,6-tetra-*O*-benzyl-β-D-glucopyranosyl 2-pyridinecarboxylate **3.302** (β-anomer) with methyl 2,3,6-tri-*O*-benzyl-α-D-glucopyranoside in diethyl ether/dichloromethane containing copper(II) triflate gives mainly the α-(1→4)-linked disaccharide **3.304α** in 70% yield, whereas when the two sugars are coupled in acetonitrile/dichloromethane containing tin(II) triflate a good yield of the β-(1→4)-linked disaccharide **3.304β** is obtained, as illustrated in Scheme 3.93. Similar remote activation to facilitate bond cleavage at C-1 is mentioned in Sections 3.1.1.b.i, 3.1.1.b.vi and 3.1.2.c.ii.

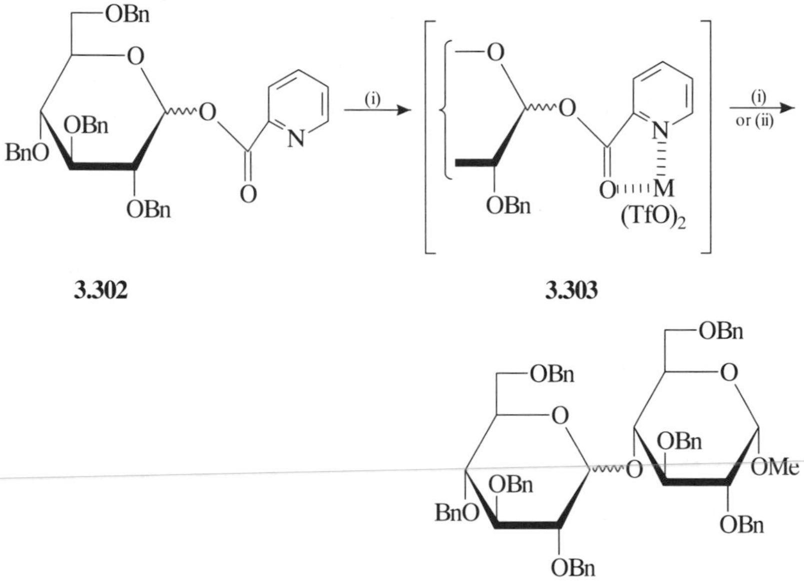

3.302   3.303

(i) **3.304** (70%, α/β 92:8)
(ii) **3.304** (70%, α/β 17:83)

(i) methyl 2,3,6-tri-*O*-benzyl-α-D-glucopyranoside, Cu(OTf)$_2$, Et$_2$O/CH$_2$Cl$_2$;
(ii) methyl 2,3,6-tri-*O*-benzyl-α-D-glucopyranoside, Sn(OTf)$_2$, MeCN/CH$_2$Cl$_2$
Scheme 3.93

(ii) S-Glycosides

Various alkyl and aryl 1-thioglycopyranosides can be prepared from peracylated pyranoses by treatment with thiols in the presence of boron trifluoride etherate. For example, penta-*O*-acetyl-β-D-glucopyranose **3.305** gives ethyl 2,3,4,6-tetra-*O*-acetyl-1-thio-β-D-glucopyranoside **3.306** when so treated with ethanethiol (Scheme 3.94).[166] The same transformation can be brought about more rapidly when zinc chloride is used as illustrated, but the method is of less general applicability.[167] 1-Thio-β-D-glucopyranose pentaacetate can similarly be prepared using thioacetic acid in place of ethanethiol.

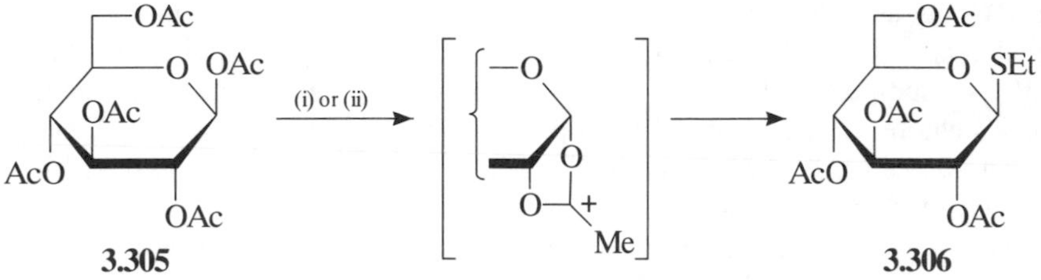

3.305   **3.306**

(i) EtSH, BF$_3$·Et$_2$O, CHCl$_3$, room temperature, 35 h (83%); (ii) EtSH, ZnCl$_2$, 0 °C, 15 h (70%)

Scheme 3.94

(iii)   C-Glycosides

Nucleophilic displacements of C-1 ester groups from suitably protected glycosyl esters by allyltrialkylsilanes give *C*-allyl glycosides under very mild conditions, as illustrated in Scheme 3.95.[39,168] Satisfactory protecting groups are ethers and acetals, and, as the formation of **3.308** from the pyranosyl acetate **3.307** shows, a high degree of stereoselectivity for the α-product can be achieved. This is improved by using as starting material the pyranosyl *p*-nitrobenzoate analogue of **3.307**, which reacts stereospecifically to give 88% yield. Benzoylated pyranosyl and furanosyl esters under similar reaction conditions also afford *C*-allyl glycoside derivatives in good yield, but often with low and unpredictable stereoselectivity.

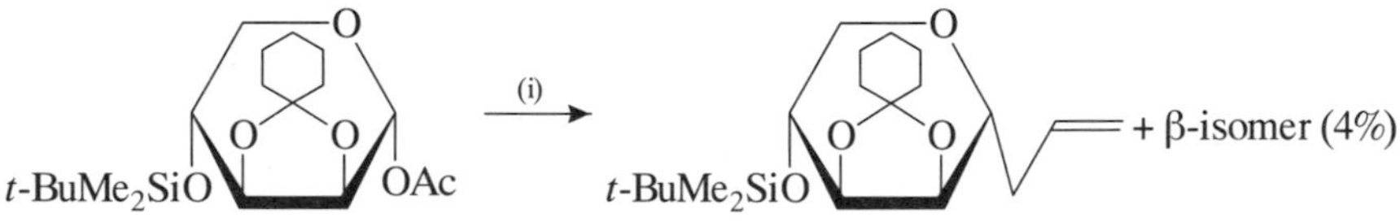

3.307   **3.308** (84%)

(i) CH$_2$=CHCH$_2$SiMe$_3$, BF$_3$·OEt$_2$, MeCN, 0–22 °C, 3 h

Scheme 3.95

(iv)   N-Glycosides

Azide displacement of the anomeric acyloxy group from the readily available penta-*O*-pivaloyl-β-D-galactopyranose **3.309** can be brought about by use of trimethylsilyl azide in the presence of tin(IV) chloride, as shown in Scheme 3.96.[95] Hydrogenation of the product **3.310** readily converts it to the glucosylamine **3.311**, which is of value as a chiral template in diastereoselective syntheses of α-amino acids (Section 7.1.2).[169]

Displacements applied to peracylated furanose precursors assisted by Lewis acid catalysts such as tin(IV) chloride, trimethylsilyl triflate and trimethylsilyl perchlorate are currently one of the more widely used methods for nucleoside synthesis. Thus, treatment of silylated 6-azauracil **3.313** with 1-*O*-acetyl-2,3,5-tri-*O*-benzoyl-β-D-ribofuranose **3.312** and tin(IV) chloride affords 2′,3′,5′-tri-*O*-benzoyl-6-azauridine **3.314** in good yield after desilylation, as illustrated in Scheme 3.97.[170]

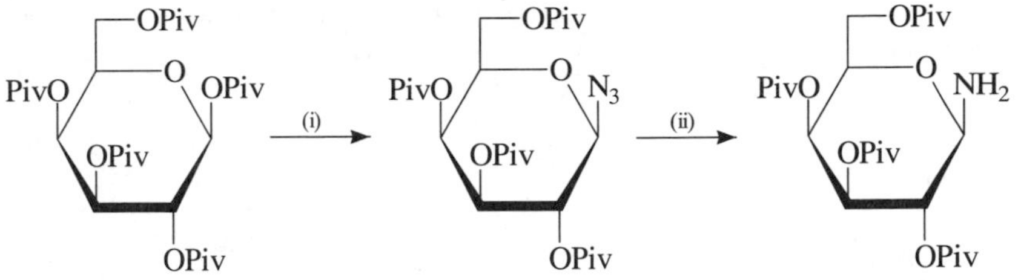

3.309   **3.310**   **3.311**

(i) Me_3SiN_3, $SnCl_4$, CH_2Cl_2; (ii) H_2, Pt

Scheme 3.96

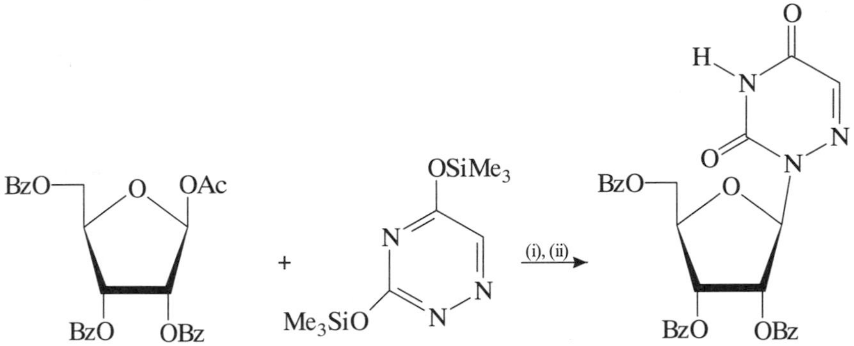

3.312   **3.313**   **3.314** (93%)

(i) $SnCl_4$, $ClCH_2CH_2Cl$, 22 °C, 20 h; (ii) $NaHCO_3$, H_2O

Scheme 3.97

3.2.2   1,2-ORTHOESTERS

Orthoesters, embracing C-1 and C-2 of a sugar ring, are readily formed as outlined in Section 3.3.2.a.iii. Being devoid of carbonyl groups they are stable under basic or neutral conditions but are readily hydrolysed by aqueous acid. Interest in them has centred mainly on their use as glycosylating agents, a role they assume when activated by protic or Lewis acids as is outlined in Scheme 3.98 by the reaction of an orthoacetate **3.318** with an alcohol (R^2OH) in the presence of a proton donor such as tosic acid.[5] Good yields of 1,2-*trans*-glycosides **3.320** have been produced by this reaction which occurs by alcohol (R^2OH) attack on the acetoxonium ion **3.319**, as shown schematically by the arrows (a). When this reaction is applied to disaccharide synthesis yields can be reduced by side reactions. Thus, the alcohol R^2OH can attack the carbocation **3.319** directly to give a new orthoacetate, but since this product eventually rearranges to the required β-glycoside it is not a serious problem. More troublesome is attack by the alcohol R^1OH liberated from the orthoester **3.318** on the acetoxonium ion **3.319**, which affords contaminating glycoside **3.321**. A further problem, however, arises when orthoacetate **3.318** is protonated at the oxygen attached to C-2, since this leads (see arrows (b) in **3.318**) to acetate ester **3.317** and the 2-hydroxypyranosyl ion **3.316**, which is trapped by the alcohol R^{2}OH to give 2-hydroxyglycosides **3.315**.

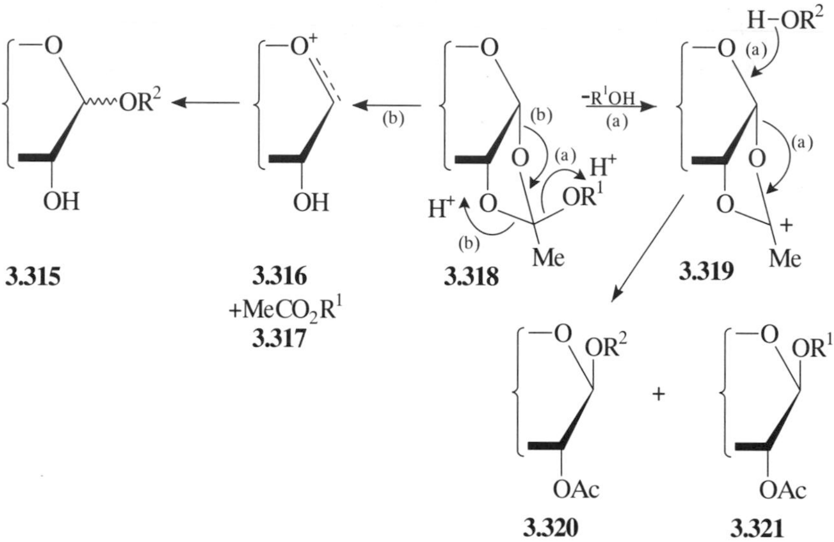

Scheme 3.98

Attempts to minimize these undesirable side reactions have included the replacement of proton acids by Lewis acids such as mercury(II) bromide and cyanide, and the use of orthoesters derived from alcohols that are poor nucleophiles and have low boiling points such that they can be removed from the reaction by distillation.[5] However, the most dramatic improvements have come from the use of cyano- or thio-ethylidene analogues **3.322** as glycosylating agents, trityl perchlorate as catalyst and the acceptor hydroxyl compounds present as their trityl ethers **3.323**. This

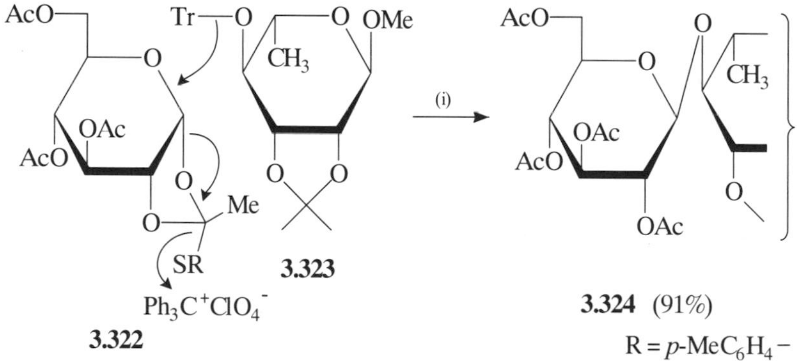

(i) Ph$_3$CClO$_4$, *sym*-collidine, CH$_2$Cl$_2$, 1 h, 22 °C
Scheme 3.99

leads to the preferential cleavage of the cyano or thio group by the trityl cation, as depicted in Scheme 3.99, and the formation of the disaccharide **3.324** by route (a) as defined in Scheme 3.98. A further advantage of this method is that only catalytic amounts of trityl perchlorate are necessary since the reagent is regenerated during the progress of the reaction.[5]

3.2.3 GLYCOSYL IMIDATE ESTERS

(a) Preparation of trichloroacetimidates[4]

Glycosyl imidate esters, particularly the trichloroacetimidates (e.g. **3.327** and **3.330**) introduced by Schmidt, have proved to be more versatile glycosylating agents than the glycosyl carboxylates because of their higher reactivity towards nucleophiles under mild conditions. The *O*-glycopyranosyl trichloroacetimidates have been most studied because they are easy to prepare in both anomeric configurations, and they are conveniently more stable than the corresponding glycosyl bromides.[4] Thus, treatment of tetra-*O*-benzylglucopyranose **3.325** with trichloroacetonitrile in dichloromethane containing potassium carbonate gives the *β*-imidate **3.327** (Scheme 3.100) as the preponderant product, whereas in the presence of the stronger base sodium hydride the *α*-anomer **3.330** is exclusively formed. This selectively arises because the *β*-anion **3.326** reacts more rapidly (see Section 3.1.6.a) with the trichloroacetonitrile than does the *α*-anion **3.329**, and the *β*-imidate **3.327** so formed is sufficiently stable in the presence of potassium carbonate to allow its isolation. With sodium hydride, however, anomerization of

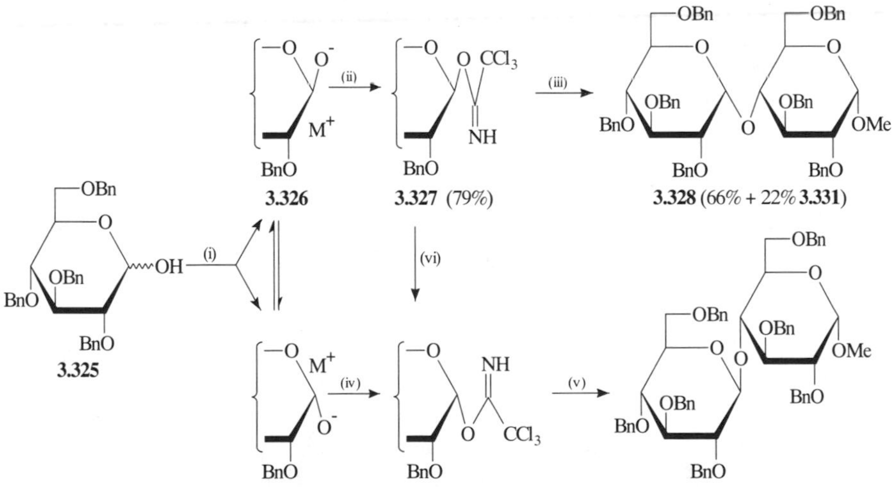

3.325 3.326 3.327 (79%) 3.328 (66% + 22% 3.331) 3.329 3.330 (96%) 3.331 (81% + 4% 3.328)

(i) base (either K_2CO_3 or NaH); (ii) K_2CO_3, Cl_3CCN, CH_2Cl_2, 25 °C, 5 h; (iii) methyl 2,3,6-tri-*O*-benzyl-*α*-D-glucopyranoside, Me_3SiOTf, Et_2O, 22 °C, 1 h; (iv) NaH, Cl_3CCN, CH_2Cl_2, 22 °C, 2.5 h; (v) methyl 2,3,6-tri-*O*-benzyl-*α*-D-glucopyranoside, $BF_3{\cdot}OEt_2$, CH_2Cl_2, −40 °C; (vi) NaH, CH_2Cl_2, 2.5 h

Scheme 3.100

the imidates occurs (probably because imidate formation is reversible), and the thermodynamically more favoured α-anomer **3.330** is produced.[4]

(b) Reactions of glycosyl trichloroacetimidates

Glycosyl trichloroacetimidates undergo nucleophilic displacement reactions with predictable stereochemical outcome under mild conditions. Those with non-participating groups at C-2 react at low temperatures with boron trifluoride etherate catalysis with inversion at C-1, whereas with strong catalysts such as trimethylsilyl triflate and particularly Brönsted acids (e.g. tosic acid) the thermodynamically more stable products are formed. In this way 1,2-*cis*-compounds may be prepared from sugars with the *gluco*- or *galacto*-structure. Sugar trichloroacetimidates with participating groups at C-2 yield 1,2-*trans*-compounds, as would be expected.

(i) Reactions with oxygen nucleophiles

Glycosyl trichloroacetimidates readily yield *O*-glycosides when treated with alcohols, as illustrated in Scheme 3.100 by the formation of disaccharides. Thus, treatment of methyl 2,3,6-tri-*O*-benzyl-α-D-glucoside with the α-imidate **3.330** in dichloromethane containing boron trifluoride etherate gives the β-(1→4)-linked disaccharide **3.331**, whereas when the partially protected sugar is treated with the β-imidate **3.327** in diethyl ether containing trimethylsilyl triflate it gives the α-(1→4)-linked disaccharide **3.328**. Both reactions occur in good yields with inversion of configuration.[4] However, with tosic acid as catalyst, the α-imidate **3.330** and methyl 2,3,6-tri-*O*-benzyl-α-D-glucoside react with retention of configuration to give the thermodynamically more stable α-glycoside **3.328**. In certain circumstances, disaccharide yields are improved by treating the trichloroacetimidate donor with a preformed complex formed between the acceptor sugar and the catalyst (e.g. trimethylsilyltriflate).[171]

The value of glycosyl trichloroacetimidates in oligosaccharide synthesis is extensive, since 2-azido and 2-deoxy sugar derivatives react with inversion of configuration when their reactions are catalysed by boron trifluoride etherate. On the other hand, peracylated 2-phthalimido trichloroimidate derivatives form 1,2-*trans*-linked disaccharides as expected with participating groups. Thus, most constituent sugars found in important natural polymers can be built into synthetic oligosaccharides by way of these intermediates.

In a novel development, glycosyl trichloroacetimidates have been shown to react with hydroxyl compounds in the presence of chloral without activation by acids, as illustrated by the stereospecific formation of the β-(1→6)-linked disaccharide **3.335** (Scheme 3.101).[172] The initial step of the reaction is believed to be the formation of a hemiacetal **3.333** by the reversible addition of the 6-hydroxyl group of methyl 2,3,4-tri-*O*-benzyl-α-D-glucoside **3.332** to chloral, which then delivers the alkoxy group to C-1 and offers a proton to the departing imidate in a highly structured transition state represented in the scheme.

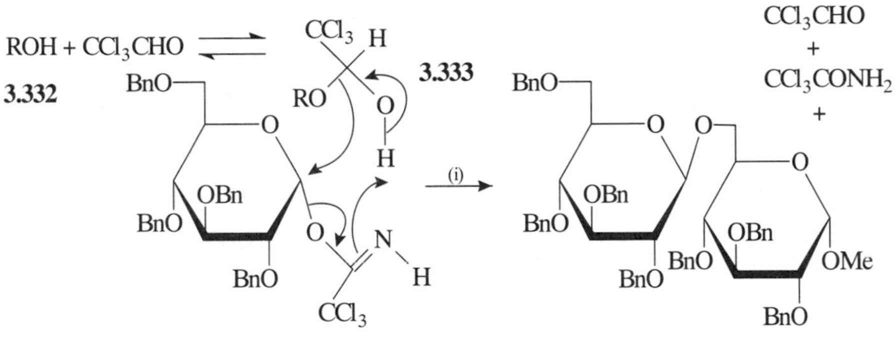

3.334   **3.335** (79%)

ROH = methyl 2,3,4-tri-O-benzyl-α-D-glucopyranoside **3.332** (i) MeCN, $-40\,^{\circ}$C, 18 h

Scheme 3.101

In common with several other analogous glycosylating agents, glucosyl trichloroacetimidates having non-participating groups at C-2 react in acetonitrile with good β-selectivity indicating the intermediacy of the α-acetonitrilium species **3.336**[16] (Scheme 3.102).

Carboxylic acids and mono- and di-esters of phosphoric acid react with glucosyl trichloroacetimidates to give the corresponding glucosyl esters.[4] The reactions occur with inversion of configuration without additional acid catalysis, as shown by the formation of the β-glucosyl dibenzylphosphate **3.337** from **3.334** in Scheme 3.102. The high stereoselectivity of this reaction is lost if the dibenzylphosphate contains traces of additional acid.

(ii)   Reactions with nitrogen, carbon and sulphur nucleophiles

Compounds with nucleophilic nitrogen react with glycosyl trichloroacetimidates, usually in the presence of boron trifluoride etherate, giving products with inverted configurations at C-1. This is illustrated by the preparation of the nucleoside 1-(tetra-O-benzyl-β-D-glucopyranosyl)uracil **3.338** in Scheme 3.102.

C-Glycosides can be formed by glycosyl transfer from sugar trichloroacetimidates to activated aromatic rings in a Friedel–Crafts type of reaction.[39] Thus, as shown in Scheme 3.102, glucosylation occurs at the active 4-position in 1,3-dimethoxybenzene when it is treated with α-imidate **3.334** in the presence of boron trifluoride etherate. As expected, the β-C-glycoside **3.339** is produced.

Thioglycosides are formed from these glycosylating agents on treatment with thiols in the presence of the usual catalyst. Surprisingly, the α-thioglucoside **3.340** is obtained from the benzylated imidate **3.334** with apparent retention of configuration, perhaps following an anomerization step.

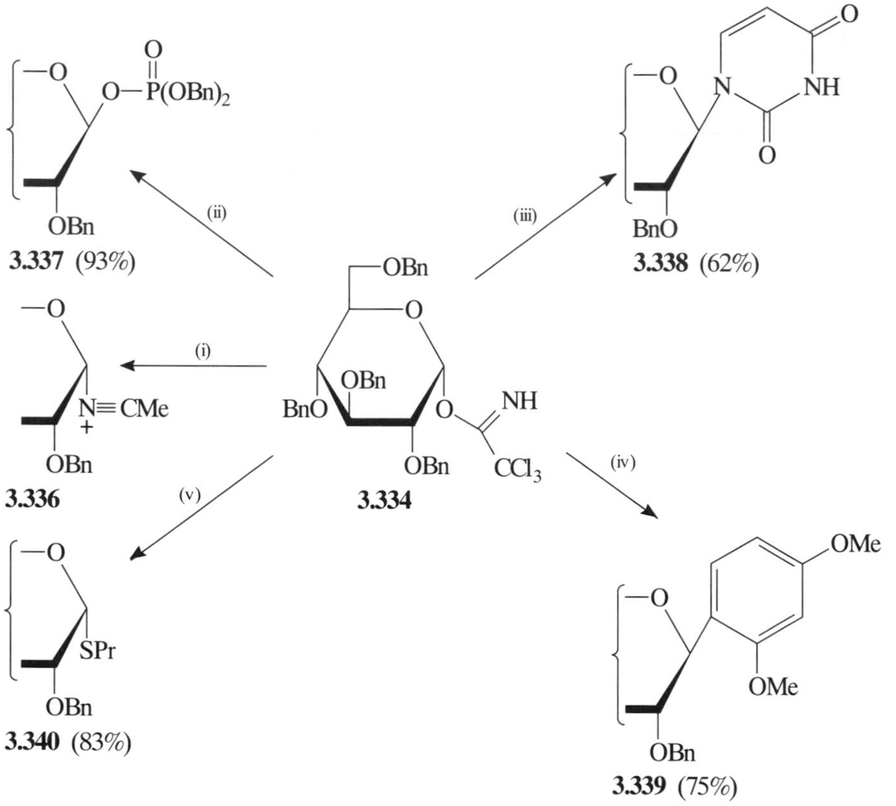

3.337 (93%)

3.338 (62%)

3.336

3.334

3.340 (83%)

3.339 (75%)

(i) MeCN; (ii) (BnO)$_2$P(O)OH, CH$_2$Cl$_2$, 22°C, 1 h;

(iii) 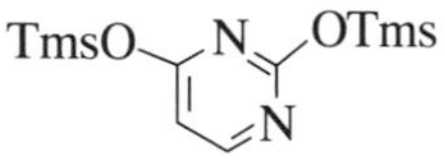 [structure], CH$_2$Cl$_2$, BF$_3$·OEt$_2$, 22°C, 2 h;

(iv) 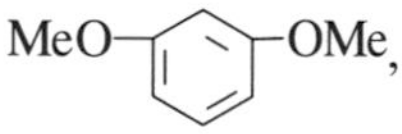 MeO—[ring]—OMe, BF$_3$·OEt$_2$, CH$_2$Cl$_2$;

(v) PrSH, BF$_3$·OEt$_2$, CH$_2$Cl$_2$, molecular sieves

Scheme 3.102

3.2.4   OTHER GLYCOSYL ESTERS

Other methods of generating good leaving groups at C-1 have been tried. Not surprisingly, 1-O-tosyl derivatives, the most obvious choice, have been investigated but glycosyl sulphonates are rather labile and not easily isolated following standard O-sulphonylation reactions. Consequently, they have been prepared by the action of silver tosylate on the corresponding glycosyl halide and used *in situ*, but although they afford glycosides on treatment with alcohols, this method has not gained wide acceptance.[173] However, their involvement is probably significant in the synthesis of glycosides from glycosyl halides when silver sulphonates are used as promoters. A glucosyl mesylate, prepared by direct sulphonation, has proved very useful in the one-pot multigram synthesis of the crystalline tetra-benzyl β-glucosylamine (Scheme 3.44).[91] In this case, the by-products from the initial mesylation do not interfere with the subsequent displacement by ammonia of the mesyloxy group.

In nature, glycosyl transfer occurs via glycosyl pyrophosphate or phosphate derivatives and it is therefore surprising that only a few investigations using glycosyl phosphate derivatives in non-enzymic approaches have been undertaken. Some sugars with phosphorus-containing leaving groups at the anomeric centre have, however, exhibited good glycosylating properties.[174] Glycopyranosyl tetramethyl-phosphoramidates **3.341** appear the most promising since they are easily prepared, have a good shelf life yet, when suitably activated, react well with a range of partially protected sugar acceptors at low temperatures, as shown by the synthesis of the β-(1→6)-linked disaccharide **3.342** in Scheme 3.103.

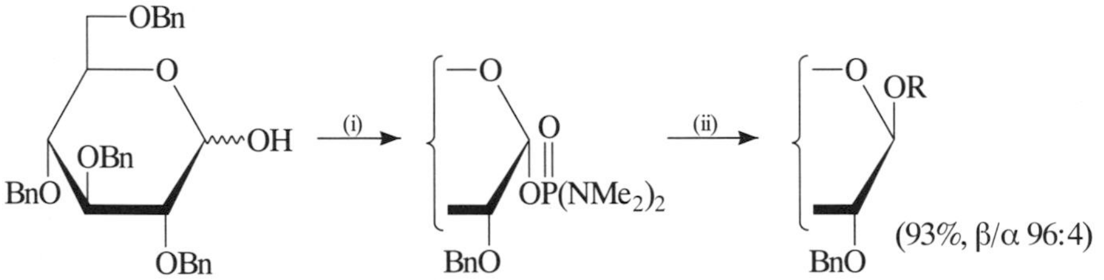

3.341 (74%)   **3.342**, R = methyl 2,3,4-
                               tri-*O*-benzyl-α-D-gluco-
                               pyranosid-6-yl

(i) *n*-BuLi, THF, (Me$_2$N)$_2$P(O)Cl, (Me$_2$N)$_3$P, −78 to −30°C, 4 h; (ii) methyl 2,3,4-tri-*O*-benzyl-α-D-glucopyranoside, TMSOTf, EtCN, −78°C, 0.5 h

Scheme 3.103

3.3   PROPERTIES AND REACTIONS OF *O*-PROTECTED GLYCOSYL HALIDES

The per-*O*-acylglycosyl halides are important synthetic intermediates because they are relatively simple to prepare (Section 3.2.1.b) and because they undergo nucleo-philic displacements of the halides with ease.[151,160] Their reactivity falls in the order iodide > bromide > chloride > fluoride, the last being so stable that deacyla-tion can be achieved with methoxide ions without displacement of the fluoride.[175] On the other hand, the iodides are very unstable and are little used syntheti-cally, while the bromides combine the most useful reactivity with stability and are most often used. The high reactivity of this class of compound towards nucle-ophiles, compared with the low reactivity of halogens at other carbon atoms on the pyranose ring (Section 4.5.2), is to be expected from the α-haloether nature of the glycosyl halides. S_N1 mechanisms are favoured for displacement reactions because the carbocations produced on ionization of the halides are resonance stabilized by the ring oxygen atoms.

Glycosyl halides derived from 2-amino-2-deoxy sugars are discussed in Section 4.3.2.f.iii.

3.3.1 CONFIGURATION AND CONFORMATION

Although it is possible to obtain both the acetylated α- and β-glucopyranosyl chlorides from some aldoses (Section 3.2.1.b), halide equilibration strongly favours one form. Tetra-O-acetyl-β-D-glucopyranosyl chloride **3.343** is stable in pure, inert solvents, but if chloride ions are present anomerization occurs to give an equilibrium mixture containing at least 93% of the α-anomer **3.344** (Scheme 3.104), and with acetylglucosyl bromides the preference for the stable anomer is also great. With these compounds, which have strongly electronegative C-1 substituents, the anomeric effect is particularly significant and consequently there is a marked preference for those anomers which possess conformationally stable rings to which are attached axial halogens. Since the normal preparation conditions allow equilibrations, the more stable anomers are obtainable in high yields. The common aldohexoses thus afford α-anomers, as do D-xylose and D-lyxose; however, from D-arabinose and D-ribose, the β-compounds are obtained. This is mainly because the α-anomers in the conformations in which the halogen atoms are axial (**3.345** for the D-arabinose derivative) suffer a strongly destabilizing 1,3-interaction which is relieved by anomerization and conformational inversion (to give, for example, **3.346**). The powerful influence of the anomeric effect in determining the conformations adopted by acylglycosyl halides is discussed in Section 2.3.1.c.

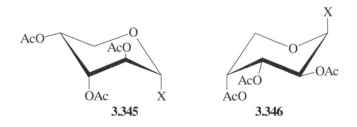

3.343 (7%)   **3.344** (93%)

(i) Me$_4$NCl, MeCN

Scheme 3.104

3.345   **3.346**

3.3.2   REACTIONS OF *O*-PROTECTED GLYCOSYL BROMIDES AND CHLORIDES WITH ALCOHOLS

(a)   *O*-Protected glycopyranosyl bromides and chlorides

(i)   Solvolysis reactions

These glycosyl halides react readily with low molecular weight alcohols alone, but although such solvolytic reactions are not the preferred preparative route to glycosides, they have been used in kinetic investigations (as have hydrolyses).[1] Extensive kinetic studies have shown that the methanolyses of acylglycopyranosyl halides in which the acyloxy group at C-2 and the halogen at C-1 are *cis* proceed by S_N1 mechanisms and that Walden inversions are involved. If there is a *trans*-relationship between these groups, solvolysis occurs with neighbouring group participation, and this not only influences the stereochemistry of the products but also accelerates their rate of formation. Tri-*O*-acetyl-6-deoxy-α-D-mannosyl bromide **3.347**, which has the 1,2-*trans*-configuration, is hydrolysed 20 times faster than is tri-*O*-acetyl-6-deoxy-α-D-glucopyranosyl bromide with the 1,2-*cis*-configuration, and this observation is believed to depend upon participation occurring in the former case (Scheme 3.105).

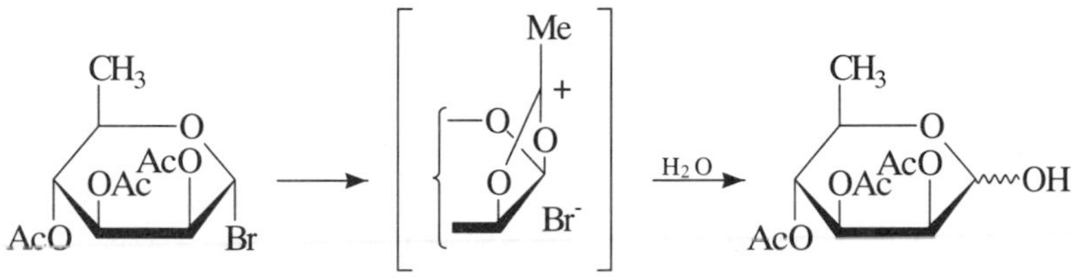

3.347   **3.348**

Scheme 3.105

The unstable anomers of the glycosyl halides (usually the β-forms) are more reactive than the α-forms. For example, tetra-*O*-acetyl-β-D-glucopyranosyl chloride **3.343** is hydrolysed 10^5 times faster than is the α-anomer **3.344**, and although this is in part because of anchimeric assistance from the *trans*-acetoxy group at C-2, the greater instability (i.e. higher free energy content) of the β-anomer is mainly responsible for this large rate difference.[1]

(ii)   Conversions to glycopyranosides

The preparation of *O*-glycosides from these glycosyl halides by their reaction with alcohols is usually carried out in a solvent in the presence of acid scavengers and electrophilic catalysts. However, it is difficult to gain stereochemical control over these displacements because even though anomerically pure halides may be available, it is not easy to bring about specific reactions as they tend to proceed by way of glycosyl carbocations, as mentioned in Section 3.3.2.a.1. Paulsen has pointed out in his authoritative reviews that the anomeric nature of the products is affected by the reactivity of the alcohol acceptor, the class of the glycosyl halide donor, the type of catalyst and/or promoter used and the solvent.[176] The type of substituent at C-2 and its stereochemical relationship to the halide at the anomeric centre are also both highly significant (see Section 3.3.2.a.i).

There exist four structural types of glycosides represented by **3.349**–**3.352** which are progressively difficult to prepare: α-*manno*-**3.349** and β-*gluco/galacto*-**3.350** with 1,2-*trans*-configurations and α-*gluco/galacto*-**3.351** and β-*manno*-**3.352** with 1,2-*cis*-configurations. All pyranosides (with the exception of 2-deoxy compounds) fall into these four categories. The most common hexosides are taken to exemplify them and different approaches are required for the construction of each, as is described below. Routes to type **3.350** hexosides have been most extensively studied, and for this reason they are described first.[4,5,176,177]

3.349   **3.350**   **3.351**

3.352

Conversions to β-glucopyranosides and β-galactopyranosides   Formation of β-glycopyranosides of the *gluco-* or *galacto*-type by halide displacements from acylated α-glycosyl halides with alcohols or phenols in the presence of insoluble silver carbonate or silver oxide is the basis of the well-established Koenigs–Knorr synthesis.[178] All classes of alcohols may be used but the efficiency decreases with increasing complexity of the alcohol, as shown in Scheme 3.106 by the silver carbonate promoted glycosylation of methanol, ethanol, *n*-propanol and *t*-butanol.[179] The yields are often good without being quantitative, and they are improved by removing the water liberated during the reactions by the addition of drying agents, or by use of one of the new promoters such as silver triflate as described later. Kinetic studies have indicated that the displacements proceed by unimolecular mechanisms; but despite this, products formed by Walden inversion processes predominate. At first sight this could be ascribed to the participation of the acetoxy group at C-2 (see **3.280**, Scheme 3.86). However, with heterogeneous catalysts/promoters, it is probable that the cleavage of the carbon–bromine bond and delivery of the alcohol take place synchronously on the surface of the silver carbonate and lead to configurational inversion (see the discussion associated with Scheme 3.112).

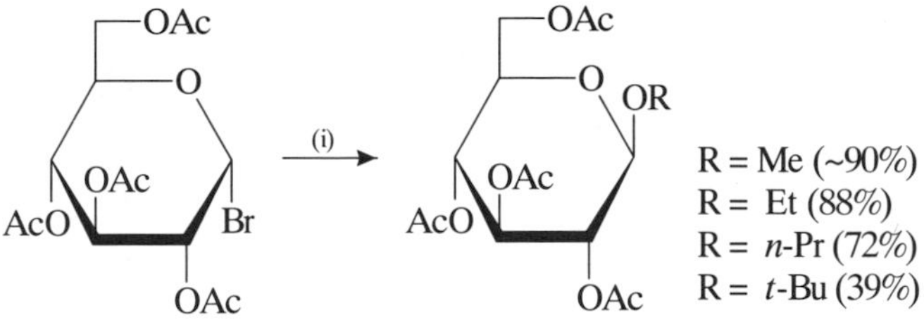

R = Me (~90%)
R = Et (88%)
R = *n*-Pr (72%)
R = *t*-Bu (39%)

(i) ROH, Ag$_2$CO$_3$

Scheme 3.106

Low molecular weight alcohols are often used in excess as the reaction solvents, but more complex alcohols can be taken in equimolar proportions in inert solvents. In this way, disaccharides have often been synthesized. For example, tetra-*O*-acetyl-α-D-glucopyranosyl bromide reacts with the free primary hydroxyl group of 1,2,3,4-tetra-*O*-acetyl-β-D-glucopyranose in chloroform in the presence of silver oxide and Drierite to give the β-(1→6)-linked disaccharide gentiobiose octaacetate in a yield of 74%.[180] Secondary hydroxyl groups on sugar rings, which are less reactive, are glycosylated less efficiently by this method.

Soluble mercury(II) salts have proved to be very effective catalysts in the syntheses of 1,2-*trans*-linked disaccharides by the coupling of acceptor sugar derivatives with free secondary hydroxyl groups and 1,2-*cis*-acylglycosyl halides. The effectiveness of secondary hydroxyl groups in these displacement reactions differs, but for the all-equatorial situation found in glucose derivatives the acceptor reactivity is normally 6-OH ≫ 3- and 2-OH > 4-OH (cf. Section 5.1). Axial hydroxyl groups are usually less reactive than those equatorially disposed; in consequence, reactions of the group at C-4 in galactose derivatives are particularly sluggish.[176] The reactivity of an alcohol group is also affected by the protecting groups present in its molecule, and free hydroxyl groups in sugars partially protected with alkyl and acetal groups are more reactive than those in partially acylated sugars. These points are discussed further in Section 5.1 and are illustrated by the reactions shown in Scheme 3.107. Thus, treatment of **3.353**, which possesses a relatively unreactive 4-hydroxyl group, with tetra-*O*-acetyl-α-D-galactosyl bromide **3.357** in the presence of the Helferich catalyst mercury(II) cyanide gives the fully protected disaccharide **3.355** in 70% yield. Hydrogen cyanide, the by-product, escapes causing little interferance in the course of this reaction. The more potent catalyst mercury(II) bromide with molecular sieves as a hydrogen bromide scavenger brings about the same condensation in even better yield (78%).[176,181] However, the related disaccharide **3.356** is formed in only low yield when the acetylated acceptor **3.354** is treated with the galactosyl bromide **3.357**; the 3-*O*-acetyl group adjacent to the 4-hydroxyl group in **3.354** therefore has a deactivating effect compared with the benzyl residue in **3.353**.[176] On the other hand, the 3-hydroxyl group of **3.358** is readily glycosylated giving a good yield of the β-(1→3)-linked disaccharide **3.359**, even though **3.358** possesses a deactivating acetyl group adjacent to the reacting hydroxyl centre.[182]

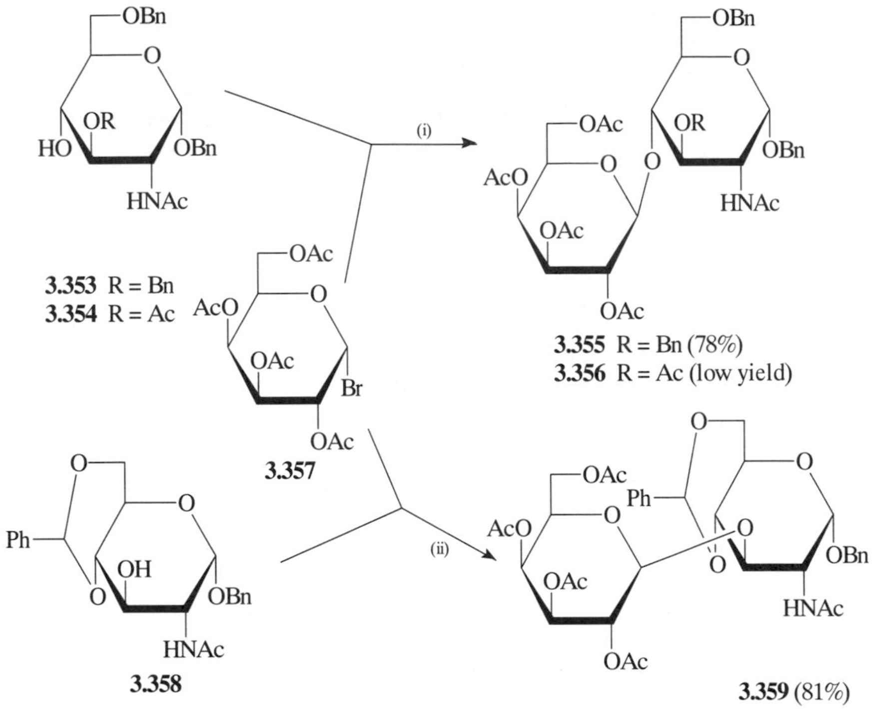

OBn
O
OR
HO
OBn
HNAc
3.353 R = Bn
3.354 R = Ac
OAc
O
AcO
OAc
Br
OAc
3.357
(i)
OAc
O
AcO
OAc
OAc
OR
O
OBn
HNAc
3.355 R = Bn (78%)
3.356 R = Ac (low yield)
O
Ph
OH
OBn
HNAc
3.358
(ii)
Ph
OBn
HNAc
3.359 (81%)

(i) $ClCH_2CH_2Cl$, $HgBr_2$, molecular sieves, 80 °C, 36 h; (ii) PhMe, $MeONO_2$, $HgCN_2$, 60 °C, 8 h

Scheme 3.107

Procedures for the synthesis of β-glycosides from 1,2-*cis*-acylglycosyl halides of 2-amino-2-deoxy-D-glucose and -D-galactose are of necessity somewhat different from those used to prepare their oxygen counterparts because the C-2 acetamido group, for example, has properties not exhibited by the C-2 ester function.[183] Acetamidoglycosyl chlorides, in contrast to their bromides, are sufficiently stable to be usable in disaccharide synthesis in the presence of mercury(II) cyanide. Following participation by the acetamido group, 1,2-*trans*-glycosides are formed via oxazolinium salts. These salts are also very readily produced from the unstable acetamido sugar bromides, and, in contrast with acetoxonium ions, they can lose a proton to give stable oxazolines (see Section 4.3.2.f.iii), and this transformation can lead to loss of glycosylating capability during some glycoside syntheses. Despite this, however, since tosic acid will catalyse the opening of 1,2-oxazoline rings by alcohols to give *trans*-glycosides, as depicted in **3.360**, they have been used as glycosylating agents in oligosaccharide syntheses in which sensitivity towards acid is not a concern. An advantage of these methods is that they offer direct routes to 2-acetamido-2-deoxy-β-D-glucose- and -galactose-containing oligomers commonly found in glycoconjugates (Section 8.1.4).[176]

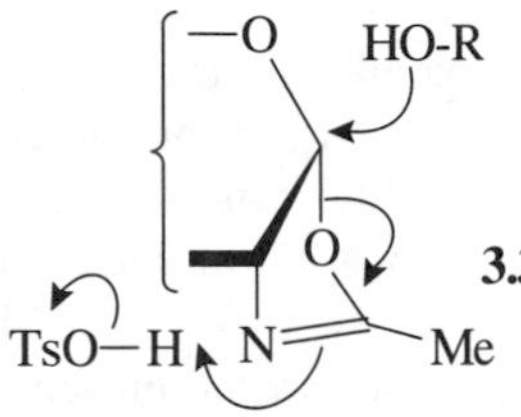

3.360

Phthalimidoglycosyl halides **3.361** have also proved to be useful glycosylating agents when used with silver triflate in the presence of *sym*-collidine, as shown by the synthesis of disaccharide **3.362** in Scheme 3.108.[184] The β-glycoside is formed by the directing influence of the phthalimido group which follows mainly from its size and also from its ability to stabilize the glycosyl carbocation by participation. Phthaloylation of amino groups and methods for their subsequent liberation are discussed in Section 4.3.2.a.

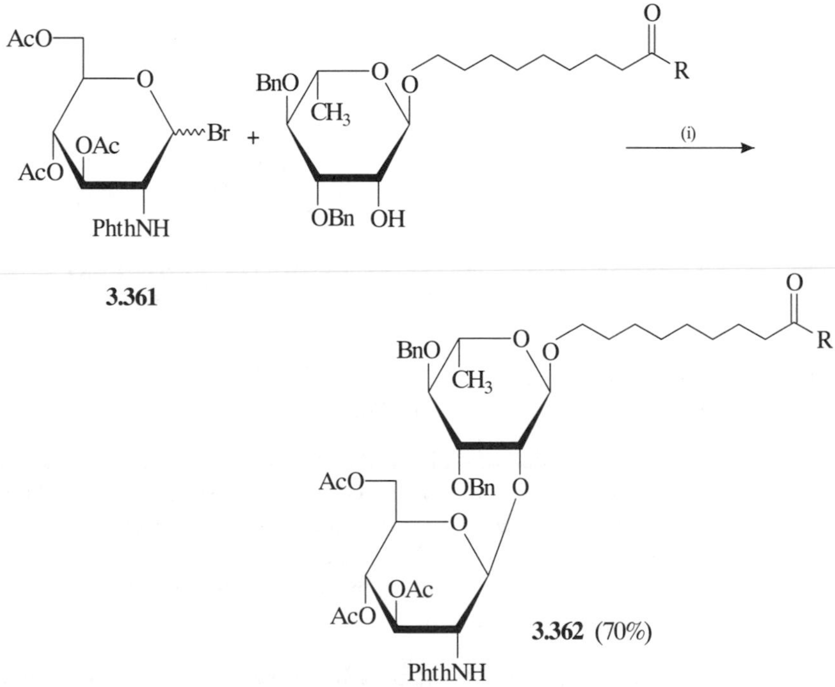

3.361

3.362 (70%)

(i) AgOTf, CH$_2$Cl$_2$, collidine, −40 to 22 °C, 20 h

Scheme 3.108

Conversions to α-mannopyranosides   The synthesis of α-mannopyranosides is easily achieved with acylmannopyranosyl halides in which participation from the C-2 neighbouring group is possible in a process closely related to that found for β-glucoside synthesis reported above. However, because the C-2 acyloxy group is in the alternative configuration, the acyloxonium ion is formed on the β-face of the pyranose ring, and consequently the α-mannopyranoside is produced upon ring opening by an alcohol. Silver triflate or mercury(II) cyanide alone, or in combination with mercury(II) bromide, can be employed as the promoter. Thus, 2,3,4-tri-O-acetyl-6-deoxy-α-D-mannosyl bromide **3.363**, for example, gives the α-$(1\rightarrow3)$-D-linked disaccharide **3.365** by way of **3.364** when treated with methyl 2,4-di-O-benzyl-α-D-rhamnopyranoside in the presence of mercury(II) cyanide, as illustrated in Scheme 3.109.[185]

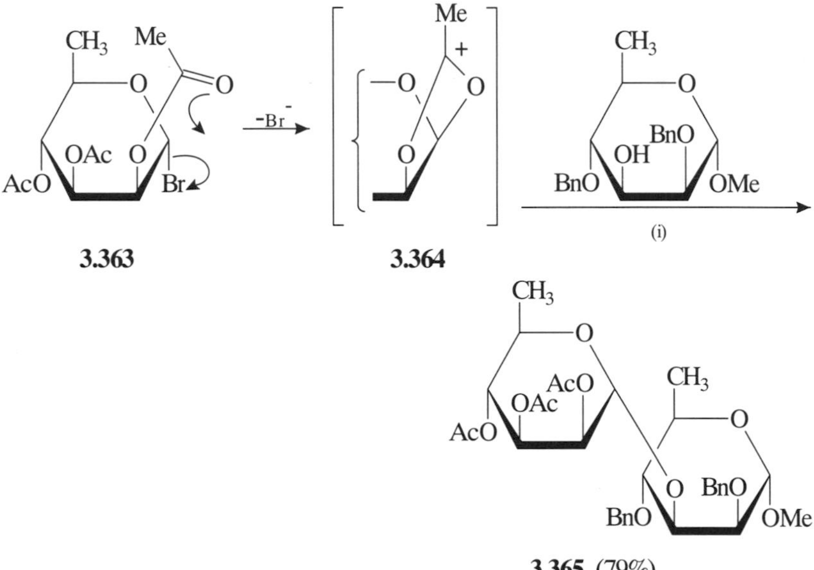

3.363   **3.364**

3.365 (79%)

(i) Hg(CN)$_2$, MeCN, molecular sieves, 20 °C, 2 h

Scheme 3.109

It can generally be concluded that glycosides produced in these reactions have the 1,2-*trans*-configuration irrespective of the configuration of the glycosyl halide, providing the group at C-2 is able to participate in the displacement reaction. It must be noted, however, that each glycosylation reaction, particularly if part of an oligosaccharide synthesis, is an independent challenge sensitive to all the parameters described above.

Conversions to α-glucopyranosides and α-galactopyranosides α-Glycopyranosides of glucose and galactose possess a 1,2-*cis*-arrangement and the rational approach to their preparation involves the use of 1,2-*trans*-glycosyl, i.e. β-glycosyl, halides which must have suitable non-participating groups at C-2 such as benzyloxy, nitrate, trichloroacetate or azide, the last of these groups being particularly useful for subsequent preparation of amino sugar glycosides. When these halides, which, because of their low thermodynamic stability, need particular care for their synthesis, are treated with alcohols in solvents of low polarity in the presence of active catalysts such as silver perchlorate, displacements tend to occur with inversion of configuration. This is illustrated by the examples in Scheme 3.110(a). Thus, 3,4,6-tri-*O*-acetyl-2-*O*-nitro-β-D-glucopyranosyl chloride **3.367** (R = NO$_2$) is converted into the methyl α-D-glucoside derivative **3.366** (R = NO$_2$) on treatment with methanol, silver perchlorate and silver carbonate.[186] α-Linked disaccharides can also be produced as illustrated by the glycosylation shown in Scheme 3.110(b), which gives the maltose derivative **3.368** in 43% yield.[187] This procedure is suitable only when the β-halides are relatively stable and isolable, which is not usually the case because of their inherent thermodynamic instability.

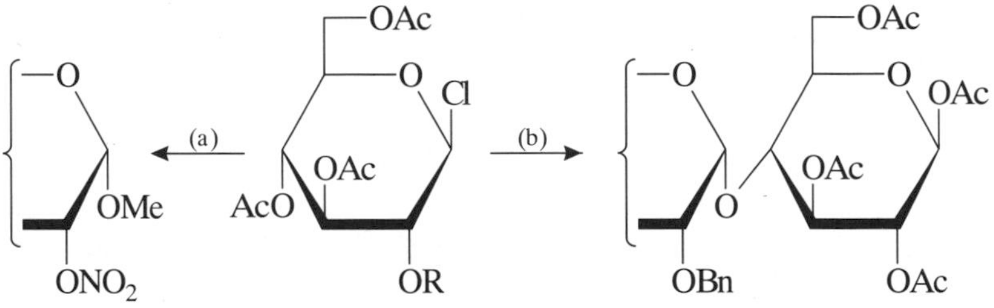

3.366 (80%)   **3.367**   **3.368** (43%)

(a) MeOH, **3.367** (R = NO$_2$), AgClO$_4$, AgCO$_3$, CaSO$_4$, Et$_2$O
(b) 1,2,3,6-tetra-*O*-acetyl-β-D-glucose, **3.367** (R = Bn), AgClO$_4$, *sym*-collidine, Et$_2$O, 0 °C, 4 h

Scheme 3.110

This instability difficulty has been ingeniously circumvented by Lemieux's halide ion catalysed method which permits the use of the abundant, relatively stable α-halo sugars and depends upon the rapid, partial anomerization that α-halides undergo with tetraalkylammonium halides in dichloromethane.[188] Mercury(II) bromide and mercury(II) cyanide, together with silver perchlorate and silver triflate, which are all also soluble in dichloromethane, function likewise and have the advantage of being more active than tetraalkylammonium halides.[176] The method requires that the reaction conditions and the blocking groups in the two sugar moieties be chosen so that the alcohol acceptor reacts only with the low concentration of the more reactive β-halide present to give α-linked disaccharides, as shown by the example in Scheme 3.111. Benzyl protecting groups have been used extensively in these α-glycoside syntheses since they are non-participating and also confer high reactivity to the halo sugars (see Chapter 6).

Conversions to β-mannopyranosides   The 1,2-*cis*-link of the type found in β-mannopyranosides is the most difficult to realize. The thermodynamically more stable α-halides which have non-participating groups at C-2 must be used together with catalysts that minimize anomerization to the more reactive β-halides. Consequently, insoluble silver salts such as silver silicate or silver oxide (see the discussion associated with Scheme 3.106) which favour direct displacement

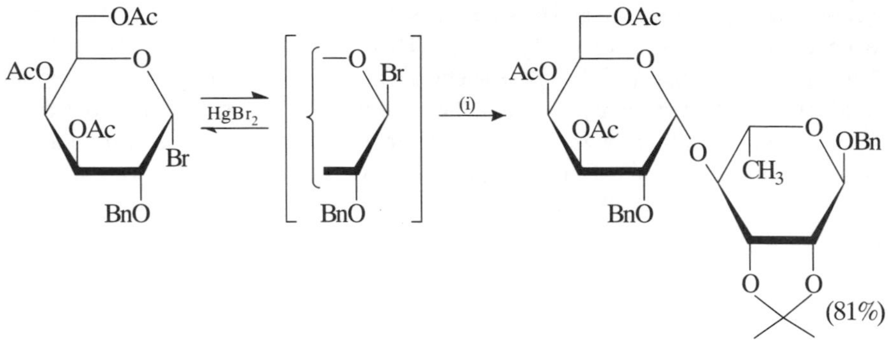

(i) benzyl 2,3-O-isopropylidene-α-L-rhamnoside, HgBr$_2$, CH$_2$Cl$_2$, 4A molecular sieves, 0 °C, 6 h

Scheme 3.111

with Walden inversion are employed,[189] as illustrated by the stereoselective preparation of the β-(1→4)-linked mannosylglucose disaccharide derivative **3.369** (Scheme 3.112).[190]

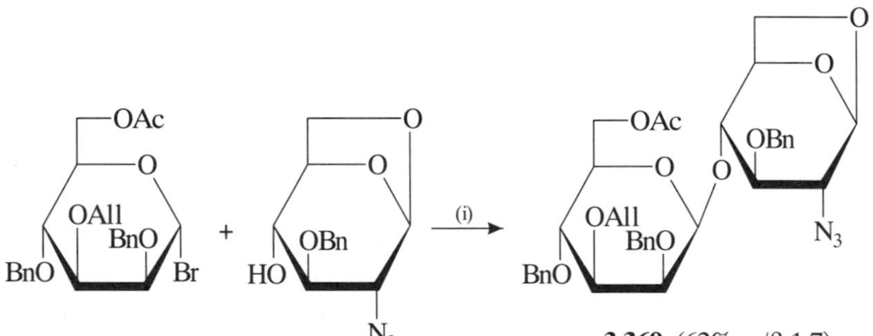

(i) silver silicate, CH$_2$Cl$_2$, 22 °C

Scheme 3.112

A new ingenious intramolecular approach to β-mannopyranoside synthesis involves initial covalent attachment of an aglycon alcohol to the substituent on C-2 in the latent mannosyl donor molecule. Activation of the anomeric centre causes intramolecular delivery of the aglycon alcohol to the β-face of the mannosyl ring (see Section 6.5). This approach can be used to prepare glycosides of other configurations.[191] β-D-Mannopyranosides have recently been prepared by Kunz using epimerizations of β-D-glucopyranosides brought about by displacements on their 3-O-(N-phenylcarbamoyl)-2-O-triflyl derivatives. These reactions occur with neighbouring group participation and inversion at C-2, thereby giving rise to 2,3-O-carbonyl-mannosides *via* the corresponding 2,3-iminocarbonates, as discussed in Section 2.5.3.a.iii. Thus, for example, cholesteryl 4,6-*O*-benzylidene-3-*O*-(*N*-phenylcarbamoyl)-β-D-glucopyranoside is converted in 64% yield by way of its 2-triflate into the corresponding 2,3-*O*-carbonyl-β-D-mannopyranoside derivative when heated at 75 °C for 3.5 h in DMF and pyridine. Other methods for the preparation of these pyranosides are referred to in Section 6.4.1 during treatment of the synthesis of oligosaccharides.

(iii)   Conversions to 1,2-orthoesters

In any displacement reaction at an anomeric centre that occurs with neighbouring group participation from an acyloxy group, the transient acyloxonium ion can be trapped by nucleophiles; if alcohols are present such events give rise to 1,2-orthoesters (Section 3.2.2).[5] With the usual Lewis acid catalysts employed in glycoside syntheses this is a minor side reaction, but in the presence of basic reagents high yields of orthoesters can be formed; with 1,2-*trans*-acylglycosyl halides this is particularly so. For example, when tri-*O*-acetyl-α-L-rhamnosyl bromide (see **3.363** in Scheme 3.109 for the enantiomer) is treated with methanol in the presence of 2,6-lutidine, the methanol attacks the acetoxonium centre in ion **3.364** (enantiomer) to give a 1,2-orthoacetate in approximately 60% yield.[192]

It is also possible to prepare 1,2-orthoesters efficiently from 1,2-*cis*-glycosyl halides using alcohols in the presence of either tetraalkylammonium halides and *sym*-collidine[193] or lead carbonate and ethyl acetate (Schemes 3.113a and 3.113b).[194] These reactions are believed to occur with double inversions at C-1: the initial step in the former is anomerization of the α-bromide **3.370**; in the latter, attack on the bromide **3.370** by the ethyl acetate is thought to produce the ion **3.372**. The orthoesters **3.371** are then produced by attack at C-1 from the participating *trans*-ester group at C-2 with subsequent reaction of the alcohol at the acyloxonium centre.

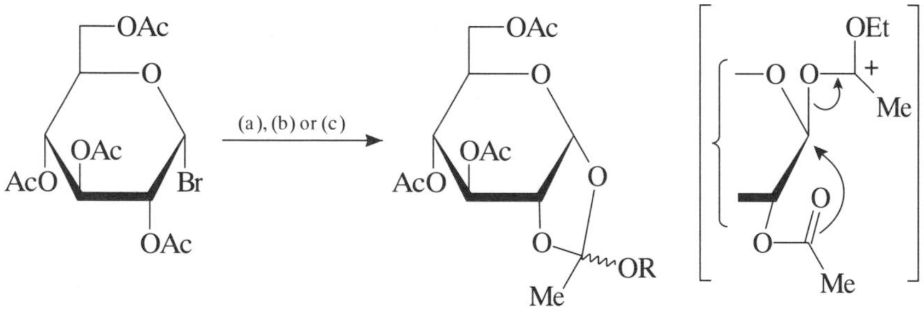

3.370   **3.371**   **3.372**

(a) EtOH, Et_4NBr, *sym*-collidine, 50 °C 12 h (R = Et, 85%)
(b) EtOH, $MeCO_2Et$, $PbCO_3$, $CaSO_4$, reflux, 4 h (R = Et, 62%)
(c) $Me_2NCH(OMe)_2$, AgOTf, CH_2Cl_2, −5 °C, 15 min (R = Me, 87%)

Scheme 3.113

In an alternative preparation of this class of compounds, acylglycosyl halides are treated with dimethylformamide dialkyl acetals under the conditions given in Scheme 3.113(c).[195]

Carbohydrate orthoesters at diols which do not involve the anomeric centre are discussed in Section 5.3.4.

(b) *O*-Protected glycofuranosyl chlorides and bromides

As found in glycoside hydrolysis, furanosyl halides are more reactive than the pyranosyl analogues.[1] *O*-Protected furanosyl halides have been used a great deal to prepare furanosides by their reactions with alcohols under conditions similar to those used for pyranoside syntheses. They have also played a particularly important role in nucleoside synthesis (Section 8.2.2.a). When a participating group is present at C-2, the furanosides with the 1,2-*trans*-configurations are formed, irrespective of the stereochemistry of the furanosyl halides (see above), and even when a non-participatory group is present, 1,2-*cis*-furanosides are not easily obtained. Methods of preparing this class of glycosides are given in Sections 3.2.1.c and 3.3.5.

3.3.3   REACTIONS OF *O*-PROTECTED GLYCOPYRANOSYL BROMIDES AND CHLORIDES WITH OTHER NUCLEOPHILES

(a)   Halides

A well-known reaction of alkyl halides is *Finkelstein's halide exchange*, which also occurs very readily with glycosyl halides and has been used to prepare iodides and fluorides from glycosyl bromides. For example, tetra-*O*-acetyl-α-D-glucopyranosyl bromide **3.374** is converted to the α-iodo compound **3.375** with sodium iodide[196] and to the β-fluoro compound **3.373** with silver fluoride (Scheme 3.114);[157] these results illustrate the different configurations which can be formed under thermo-dynamic and kinetic control. Since fluorine is the most electronegative element it would be expected that the α-glucosyl fluoride would be strongly preferred because of the anomeric effect, but, because of the great strength of the C–F bond anomer-ization of the first-formed β-anomer **3.373** is slow, so the product isolated is that produced under kinetic control. With the iodo analogues the C-1 bond is readily cleaved, thus any β-anomer initially formed in the reaction is anomerized by further attack from iodide ions.

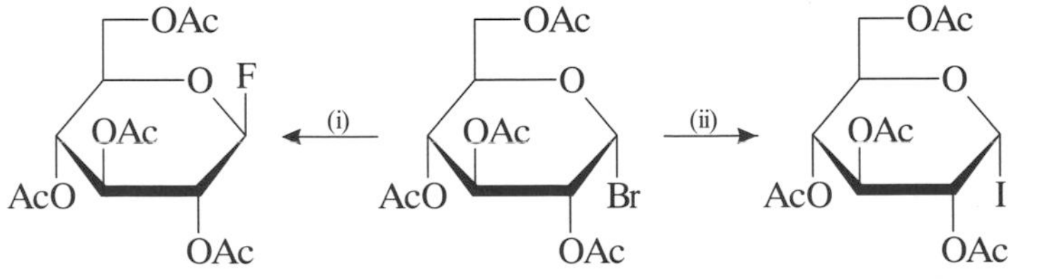

3.373 (50%)   **3.374**   **3.375** (75%)

(i) AgF, MeCN; (ii) NaI, Me₂CO

Scheme 3.114

(b)   Nitrogen nucleophiles

Nucleophilic displacement of the halogen atoms of acylglycopyranosyl halides readily occurs with nitrogenous compounds such as amines. Primary amines usually give glycosylamine derivatives, as shown for the preparation of the N-tolylglucopyranosylamine derivative **3.376** (Scheme 3.115),[197] and tertiary amines such as trimethylamine or pyridine usually give glycosyl quaternary ammonium salts. On the other hand, secondary n-alkylamines often bring about the elimination of the elements of hydrogen halide to give per-O-acetylated 2-hydroxyglycals (Section 4.10.1.a).

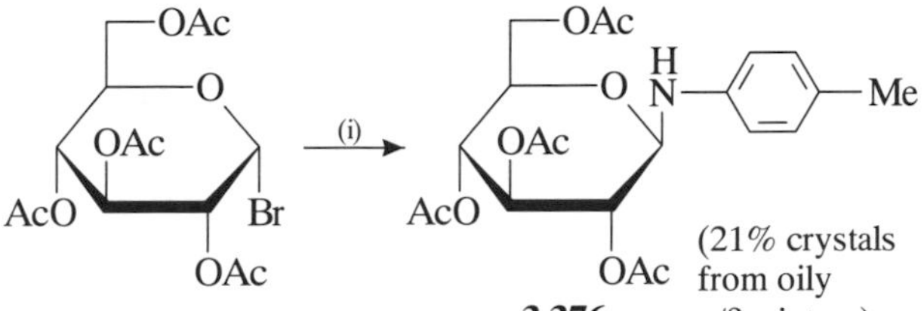

(21% crystals from oily α/β mixture)

3.376

(i) p-MeC$_6$H$_4$NH$_2$, NaOH, Me$_2$CO, 20 °C

Scheme 3.115

The most important application of the displacement reaction is in the Hilbert-Johnson synthesis of nucleosides[198] which are glycosylamines derived from heterocylclic secondary bases (see Section 8.2.2.a). An example of an efficient analogous nucleoside synthesis from a furanosyl acetate is given in Section 3.2.1.c.iv.

Azide ions also react with glycosyl halides to give azido compounds which have been used as precursors of glycosylamines. Scheme 3.116 illustrates how these transformations can be applied with 2-acetamido-3,4,6-tri-O-acetyl-2-deoxy-α-D-glucopyranosyl chloride **3.377**.[199]

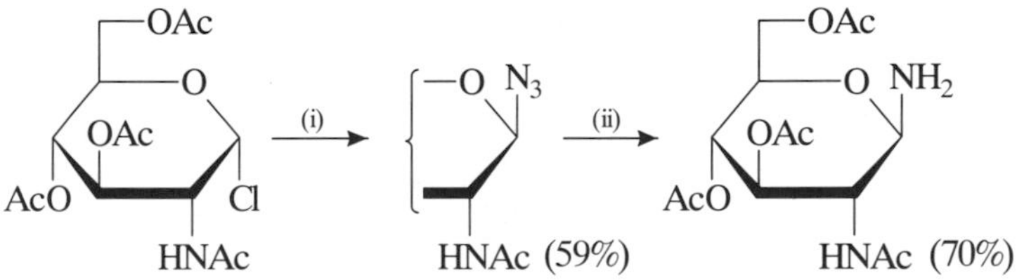

3.377   HNAc (59%)   HNAc (70%)

(i) LiN$_3$, DMF, 80 °C, 3 h; (ii) H$_2$, Pt, EtOAc, 20 °C, 4 h

Scheme 3.116

(c)   Sulphur nucleophiles

Acylglycosyl halides undergo halide displacements with reagents containing nucleophilic sulphur atoms, and these provide a method of preparing 1-thio sugars. For example, thiourea reacting in the enolic form with acetylglucosyl bromide **3.378** yields the adduct **3.379** from which the thiol **3.380** can be obtained with very mild base (Scheme 3.117) (cf. Section 3.1.2.c).[200]

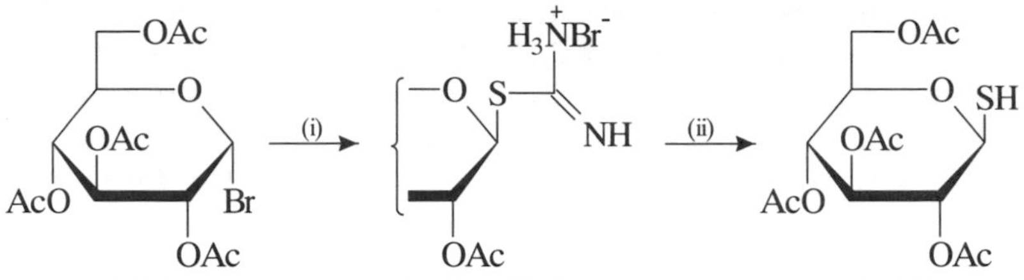

3.378   **3.379** (64%)   **3.380**

(i) $(NH_2)_2C=S$, *i*-PrOH; (ii) $NaHCO_3$, $NaHSO_3$, H_2O, CCl_4

Scheme 3.117

(d)   Carbon nucleophiles[39,201]

Cyanide ion displacement of the halide from acetylglucopyranosyl bromide using mercury(II) cyanide in nitromethane gives a low yield of 2,3,4,6-tetra-*O*-acetyl-β-D-glucopyranosyl cyanide by direct nucleophilic displacement at C-1. This product is accompanied by 3,4,6-tri-*O*-acetyl-1,2-*O*-(1-cyanoethylidene)-α-D-glucopyranose which is formed by cyanide ion attack on an intermediate 1,2-acetoxonium ion.[202] This dichotomy in the mechanistic pathways limits the usefulness of the reaction.

Attempts to utilize acetylglycosyl halides in Friedel–Crafts reactions have been successful. A large excess of catalyst must be used because the acetoxy groups complex with it, and deacetylation accompanies the reaction. Acetylglucosyl chloride **3.381** with moderate amounts of aluminium chloride in benzene affords, after

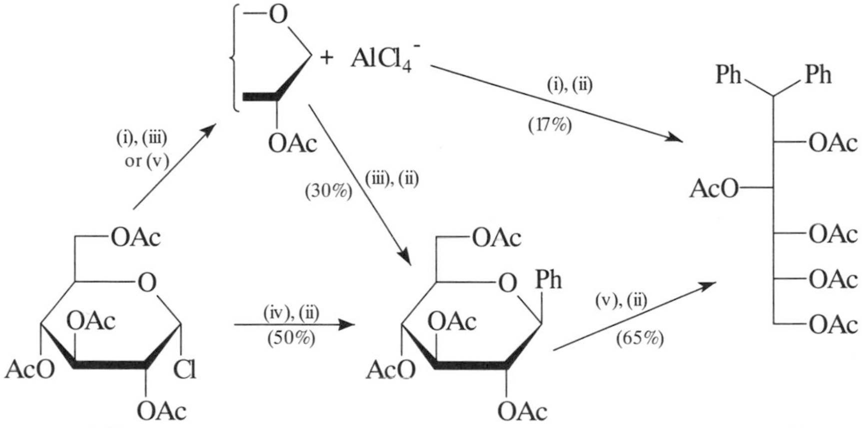

3.381   **3.382**   **3.383**

(i) benzene, $AlCl_3$ (8 molar equivalents); (ii) Ac_2O, Py; (iii) benzene, $AlCl_3$ (5 molar equivalents); (iv) PhMgBr; (v) benzene, $AlCl_3$

Scheme 3.118

acetylation, 2,3,4,6-tetra-*O*-acetyl-β-D-glucopyranosylbenzene **3.382**, whereas with a larger excess of the reagent the pyranose ring is opened to give, again after acetylation, 2,3,4,5,6-penta-*O*-acetyl-1-deoxy-1,1-diphenyl-D-glucitol **3.383** which appears to arise from the former product (Scheme 3.118).[203]

Synthesis of more elaborate *C*-glycosides has been stimulated by their identification as subunits of many natural products (Chapter 8) and by a requirement for them as model substrates, such as enzyme inhibitors, in biological studies and for their use as materials for chiral syntheses (Chapter 7). Condensations using glycosyl halides with reagents possessing nucleophilic carbon, generated from metal alkyls (often Grignard reagents), enolates or alkylsilanes, have been used to this end, giving products which can be employed in further chemical transformations.[39]

Treatment of the β-furanosyl chloride **3.385**, possessing non-participating benzyl protecting groups, with ethynylmagnesium bromide gives mainly the *C*-glycoside **3.386** with the inverted α-configuration (Scheme 3.119a).[204] The enolate anion formed from diethyl malonate, however, reacts with the chloride **3.385** in a non-stereoselective fashion to give similar proportions of the α- and β-*C*-glycosides **3.384** when treated as indicated in Scheme 3.119(b).[201,205]

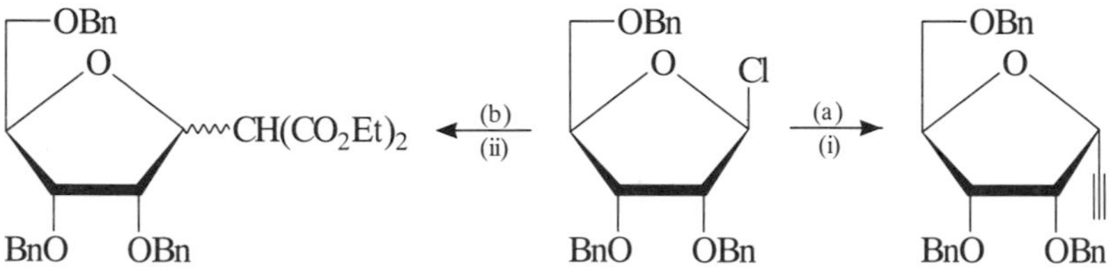

3.384 (85%, α/β 1:1)   **3.385**   **3.386**
                                                    (63% + 8% β)

(i) CH≡CMgBr, THF, C_2H_2, 3.5 h, 22 °C; (ii) $CH_2(CO_2Et)_2$, Na, MeO(CH$_2$)$_2$OMe, 22 °C, 12 h.

Scheme 3.119

C-Glucopyranosides **3.389** and **3.390** are produced in the ratio 10:1 when benzylated glucopyranosyl chloride **3.388** is treated with allyltrimethylsilane as outlined in Scheme 3.120.[38,39] Similar products have been obtained with this reagent in related displacements on glycosides (e.g. **3.387**) as mentioned earlier in Section 3.1.1.b.viii, and also with fluorides (see Section 3.3.5).

C-Nucleosides are a class of *C*-glycosides that involve nitrogen heterocycles similar to those found in nucleosides (Section 8.2.2.a). Many show interesting biological properties such as antiviral, antibacterial, antitumour and antimetabolite activities. Their syntheses, which have been reviewed,[206] are usually accomplished by elaboration of the nitrogen heterocycle from a functionalized *C*-glycoside synthesized by one of the routes described above. However, some direct approaches to their synthesis from furanosyl halides have been developed.[198,206]

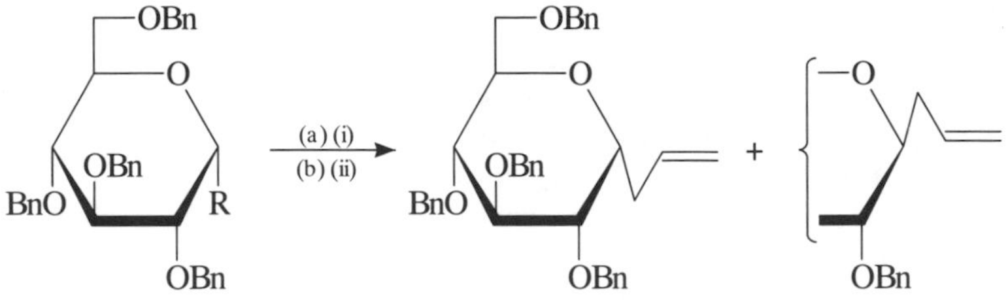

(a) **3.387** R = OMe   **3.389**   **3.390**
(b) **3.388** R = Cl

(i) Me$_3$SiCH$_2$CH=CH$_2$, Me$_3$SiOTf, MeCN, 22 °C, 16 h (**3.389/3.390** 78:8%);
(ii) As for (i) but for 6.5 h   (**3.389/3.390** 74:7%)

Scheme 3.120

(e) Hydrogen and hydride species

Acetohalohexopyranoses can be reduced to 1,5-anhydroglycitol tetraacetates in three ways. Hydrogenolysis catalysed by palladium or platinum in ethyl acetate solution containing amine bases under hydrogen is suitable for glycosyl bromides. The epimerizations that occur to some degree at C-2 probably arise from amine-induced hydrogen bromide eliminations (see Section 4.10.1.a) with subsequent hydrogenation of the 2-acyloxyglycals. Consequently, tetra-O-acetyl-α-D-glucopyranosyl bromide affords tetraacetates of 1,5-anhydro-D-glucitol contaminated with some 1,5-anhydro-D-mannitol. Lithium aluminium hydride induced displacements of the halides from acetoglycosyl chlorides or bromides take place readily with concomitant deacetylation, the anhydroglycitol products usually being isolated after reacetylation.[207] The third method involves free radical processes and is dealt with in Section 3.3.4.

(f) α-Mercuriated carbonyl compounds

Vinyl glycosides cannot be made as described in Section 3.3.2 since 'vinyl alcohol' does not exist; it is the enol form of acetaldehyde. They may be obtained in modest yields and as mixed anomers by transvinylation from, for example, butyl vinyl ether and 2,3,4,6-tetra-O-acetyl-D-glucose in the presence of mercury(II) acetate. Otherwise, eliminations can be applied to glycosides having appropriate leaving groups within the aglycons. The most convenient method utilizes, for example, tetra-O-acetyl-α-D-glucopyranosyl bromide and Hg(CH$_2$CHO)$_2$, made from butyl vinyl ether and mercury(II) oxide and acetate in aqueous ethanol. Rather than reacting as a carbon nucleophile (cf. Scheme 7.27, Section 7.2.2.c), the organomercury reagent acts as an oxygen nucleophile to give vinyl 2,3,4,6-tetra-O-acetyl-β-D-glucopyranoside in 80% yield. 1′-Substituted analogues are available by use of the homologues XHg(CH$_2$COR).[208]

3.3.4   FREE RADICAL REACTIONS OF *O*-PROTECTED GLYCOSYL BROMIDES (AND CHLORIDES)

(a)   Reduction reactions

Tributylstannane free radical reduction of *O*-protected glycosyl bromides and chlorides is claimed to be the best method for preparing 1,5-anhydrides on a large scale (see Section 4.6.5 for an alternative preparation).[209] The transformations can be photochemically or thermally induced and both anomeric forms of tetra-*O*-acetylglucopyranosyl chloride can be used. However, whereas the α-anomers of the bromo and chloro compounds give high yields of product, the β-chloro derivative reacts less favourably, producing poor yields in thermal reactions and reacting only slowly under u.v. light. The reaction has been used to introduce deuterium stereospecifically into 1,5-anhydroglycitols. In keeping with this, treatment of tetra-*O*-acetyl-α-D-glucopyranosyl bromide **3.392** with tributyltin deuteride in toluene gives the 1-α-deuterioanhydroglucitol derivative **3.393** in high yield, as shown in Scheme 3.121(a).[210]

These findings illustrate the strong stereoelectronic influences that govern free radical reactions at anomeric centres.[49,211] Abstraction by tin radicals of axial halogen atoms, as illustrated in **3.392** to give the glucopyranosyl radical **3.395**, is favourable because of stereoelectronic assistance from the axial electrons of the ring oxygen atom (**3.396**), whereas removal of equatorial halogens is not so assisted—hence the less efficient reduction of β-chlorides. Similar diastereoselectivities for axial anomeric hydrogen abstractions by radicals are noted in Sections 3.1.1.b.xii, and 3.1.2.c.iii. The radical intermediates formed by hydrogen abstractions from these β-D-*gluco*-compounds have been shown by e.s.r. spectroscopy to be identical to those produced from the α-halides, and they adopt the boat conformation discussed in the next Section.[211] Radical **3.395** reacts by abstracting a hydrogen/deuterium atom from the tin hydride/deuteride to form a C–H/D axial bond preferentially at its α-face. Similar stereoselective bondings of glycopyranosyl radicals are reported in Section 3.1.2.c.v and Scheme 3.77, Section 3.1.6.a.iv for hydrogen, Section 3.1.2.c.iii for bromine and Sections 3.1.1.b.xii and 3.3.4.b for carbon.

When tetra-*O*-acetyl-α-D-glucopyranosyl bromide **3.392** is treated as described above, but with restricted quantities of tri-*n*-butyltin hydride, the radical **3.395** undergoes an ester rearrangement to **3.394** before being trapped by a hydrogen donor to give the 2-deoxy sugar derivative **3.391** (Scheme 3.121b). To achieve high yields in this transformation, the quantities of tin hydride used must be finely adjusted throughout the course of the reaction such that just sufficient is available to abstract the bromine atom to initiate the process. The same result can be achieved more efficiently by using a radical process involving tris(trimethylsilyl)silane, which possesses an Si–H bond slightly stronger than the Sn–H bond in tri-*n*-butyltin hydride (79 versus 74 kcal mol^{-1}, respectively).[212] Thus, when the bromo sugar

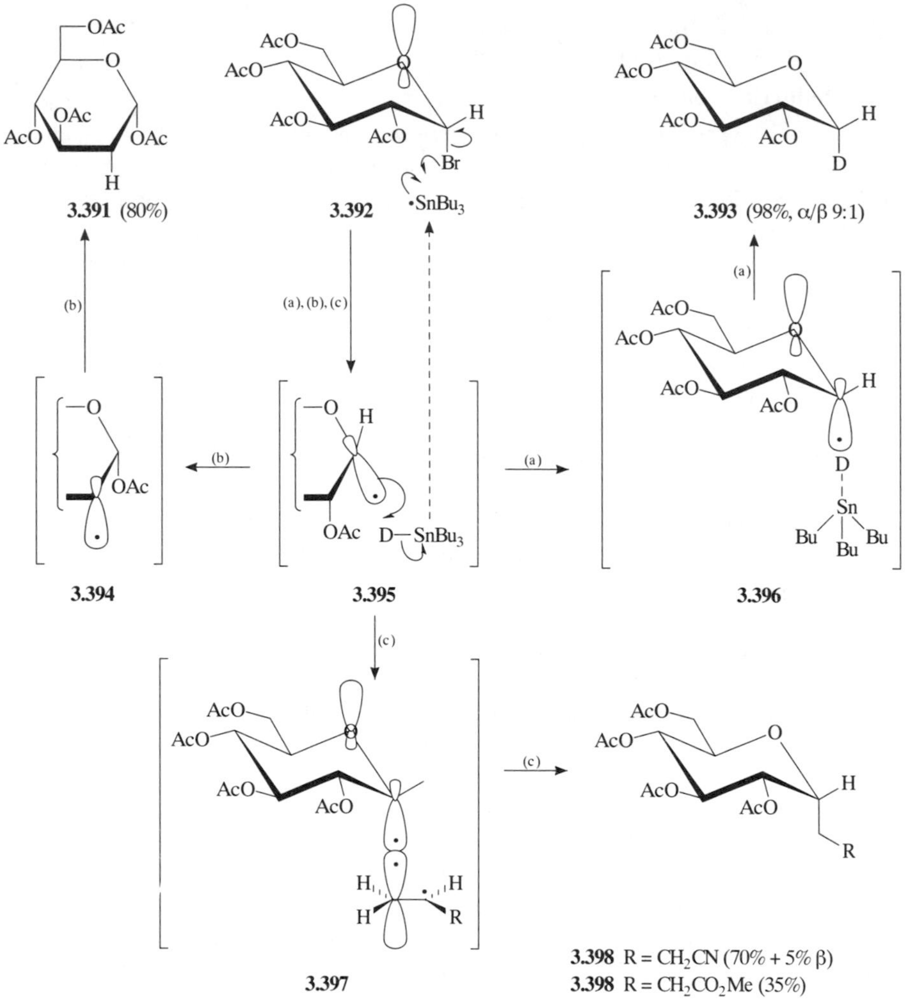

3.391 (80%)   **3.392**   **3.393** (98%, α/β 9:1)

3.394   **3.395**   **3.396**

3.397

3.398 R = CH₂CN (70% + 5% β)
3.398 R = CH₂CO₂Me (35%)

(a) Bu₃SnD or Bu₃SnH, AIBN, PhMe
(b) (Me₃Si)₃SiH, AIBN, PhMe
(c) *n*-Bu₃SnH, CH₂=CHCN (20 molar equivalents), Et₂O, reflux, *hν*: or Ph₃SnH, AIBN, CH₂=CHCO₂Me (10 molar equivalents), PhMe, reflux

Scheme 3.121

3.392 is treated with the silane in toluene with AIBN as initiator, 1,3,4,6-tetra-*O*-acetyl-2-deoxy-α-D-*arabino*-hexopyranose **3.391** is formed in good yield.

(b) Formation of *C*-glycosides

As outlined above in Section 3.3.3.d, most past improvements in *C*-glycoside syntheses have rested on adaptations of carbon nucleophilic displacements on suitable glycosyl donors. This aspect of carbohydrate chemistry has now, however, taken a new direction following the discovery that glycosyl radicals can be generated and efficiently trapped by electron-deficient alkenes.[39,211] Thus, treatment of tetra-*O*-acetyl-α-D-glucopyranosyl bromide **3.392** with the bromine radical abstractor tri-*n*-butyltin hydride or triphenyltin hydride in the presence of several molar proportions of acrylonitrile or methyl acrylate and under light irradiation or in the presence of the radical initiator AIBN (Scheme 3.121c), gives the corresponding cyanoethyl or methoxycarbonylethyl *C*-glucosides **3.398** (R= CH_2CN[213] or CH_2CO_2Me[214]) with a high degree of α-stereoselectivity. Therefore, the α-glucopyranosyl radical **3.395**, formed after bromine abstraction from **3.392** by a tri-*n*-butyltin radical, is trapped stereoselectively from the axial α-direction by its addition to the more electron-deficient methylene carbon of the alkene (see **3.397**). This propensity for axial reaction is exhibited by all glucopyranosyl radicals irrespective of the class of compound or anomeric configuration of the radical precursor, as noted in Sections 3.1.1.b.xii,   3.1.2.c.iii, 3.1.2.c.v, 3.1.6.a.iv and 3.3.4.a.

In a related free radical reaction, the glucosyl halide **3.392** gives, on treatment with allyltri-*n*-butylstannane and AIBN in refluxing toluene, the *C*-allyl glucoside **3.398** (R = $CH=CH_2$) as a mixture of anomers of undetermined composition (but high α-proportions are expected) in 64% yield.[215] Allyl glycosides are also formed from phenyl thioglycosides with this reagent under similar conditions or on exposure to u.v. light.[215]

Whereas most glycopyranosyl cations are believed to exist in half-chair conformations (see **3.26**), e.s.r. spectroscopic studies have revealed that the conformations of glycopyranosyl radicals vary with their structures. Thus, the tri-*O*-acetyl-2-deoxy-2-deuterioglucosyl radical exists in the $^4C_1$ conformation **3.399**, as does the tetra-*O*-acetylmannosyl radical **3.400**, but the tetra-*O*-acetylglucosyl radical adopts the $B_{2,5}$ conformation **3.401**.[211,216] It is significant that the C-2 acetoxy groups have axial orientations in both **3.400** and **3.401**, thereby probably being subject to a stabilizing interaction between the SOMO of the unpaired electron and the LUMO of the neighbouring axial carbon–oxygen bond. However, interactions between the ring oxygen atom and the radical centre can also be expected to play a significant part because analogous conformational changes are not observed with pyranose radicals centred at positions other than C-1.

3.399   **3.400**   **3.401** $B_{2,5}$ Conformation

All of these radicals are quenched from the α-face by both alkenes and trialkyltin hydride so that new substituents are axial in the $^4C_1$ conformations adopted by the final products. This arises because the direction of attack at the radical centre is probably under stereoelectronic control from the ring oxygen atom, and whereas this is easy to see in the case of radicals in the chair conformations **3.399** and **3.400**, it is less clear for those in the boat form. At some stage during the bonding process sterically and stereoelectronically favoured chair transition states are likely to be adopted even from a boat-shaped radical, as depicted in **3.396** and **3.397** for hydrogen and carbon quenching, respectively.[217]

An alternative explanation has been advanced to explain the preponderance of axial products in these radical reactions based on the principle of least motion and steric effects.[218]

3.3.5 REACTIONS OF GLYCOSYL FLUORIDES

Glycosyl fluorides are discussed separately from the chlorides and bromides because, although their reactions are similar, they are significantly less reactive under normal conditions owing to the increased strength of the carbon–fluorine bond. Consequently, they can be used differently. They have been known since the 1920s,[175] but have played a lesser role in carbohydrate chemistry owing in part to their low reactivity and also because the methods available for their preparation were unattractive. The situation changed dramatically with the introduction of some good synthetic routes to these compounds and the discovery of fluorophilic catalysts which convert them into effective glycosylating agents. They can now be prepared in good yields from thioglycosides, by treatment with DAST (see Scheme 3.34),[42] by fluoride ring opening of 1,2-anhydrides (see Scheme 3.29),[56] by DAST-induced 1,2-migrations and subsequent displacements on pyranosyl derivatives

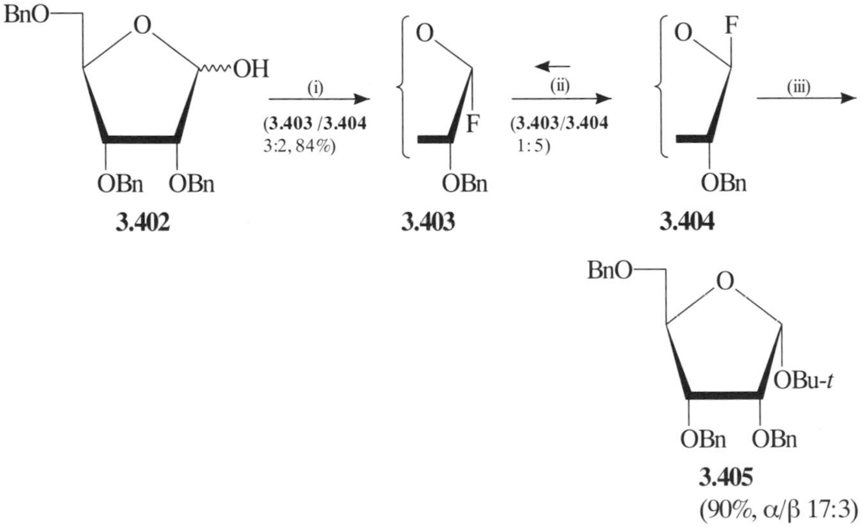

(i) 2-fluoromethylpyridinium triflate, CH_2Cl_2, Et_3N, 22 °C; (ii) $BF_3 \cdot OEt_2$, Et_2O, 22 °C, 10 min; (iii) *t*-BuOH, $SnCl_2$, $TrClO_4$, Et_2O, molecular sieves

Scheme 3.122

unprotected at the O-2 position (see Section 4.5.1.e.iii), and by the action of 2-fluoro-1-methylpyridinium tosylate on anomeric hydroxyl groups in otherwise fully protected sugars. This last approach is illustrated in Scheme 3.122 by the conversion of tri-*O*-benzylribofuranose **3.402** into the ribofuranosyl fluorides **3.403** and **3.404**.[219]

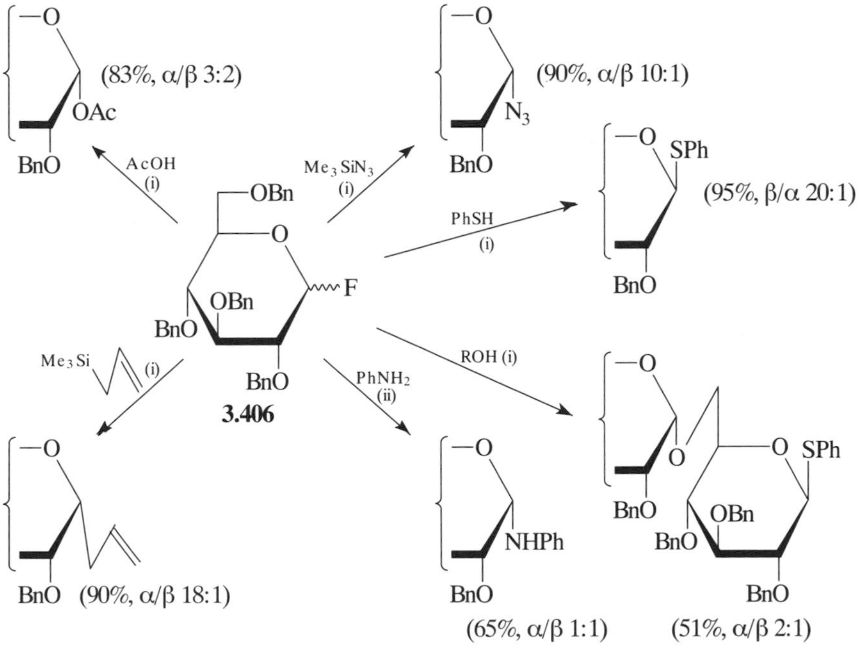

(i) BF$_3$·OEt$_2$, CH$_2$Cl$_2$, molecular sieves; (ii) AlMe$_3$, CH$_2$Cl$_2$ (ROH = phenyl 2,3,4-tri-*O*-benzyl-1-thio-β-D-glucopyranoside)

Scheme 3.123

The value of pyranosyl fluorides as glycosyl donors is demonstrated by the numerous transformations that tetra-*O*-benzyl-α/β-D-glucopyranosyl fluorides **3.406** undergo when activated by boron trifluoride or timethylaluminium. Some useful examples of these reactions are illustrated in Scheme 3.123 by the synthesis of glycosyl esters, glycosyl azides, *S*-glycosides, *O*-glycosides and *N*-glycosides.[90] *C*-Glycosides can also be obtained using similar conditions.[220]

1,2-*trans*-Glycofuranosyl fluorides, protected with non-participating groups, have been found to react with alcohols with Walden inversion in the presence of a tin(II) chloride/trityl perchlorate mixture, thereby affording a good route to 1,2-*cis*-furanosides that are difficult to obtain by other means.[219] Several α-ribofuranosides, including disaccharides, can be synthesized in this way, as illustrated in Scheme 3.122 by the formation of the sterically crowded *t*-butyl riboside **3.405**, which is achieved by treating the chromatographically pure β-fluoride **3.404**, obtained from the mixture of furanosyl fluorides **3.403** and **3.404** by boron trifluoride induced anomerization, as indicated.

3.3.6   ELIMINATIONS OF *O*-PROTECTED GLYCOSYL HALIDES

Acylated glycosyl halides are the most satisfactory starting materials from which glycal and 2-hydroxglycal derivatives may be synthesized. These reactions are described in Section 4.10.1.

3.4   REFERENCES

1. *Chem. Rev.*, 1969, **69**, 407.
2. *Rodd's Chemistry of Carbon Compounds* (ed. S. Coffey), Vol. 1F, Elsevier, Amsterdam, 1967, Chaps 22, 23.
3. *Rodd's Chemistry of Carbon Compounds* (ed. M.F. Ansell), Vol. 1F, G Supplement, Elsevier, Amsterdam, 1983.
4. *Angew. Chem., Int. Ed. Engl.*, 1986, **25**, 212; *Adv. Carbohydr. Chem. Biochem.*, 1994, **50**, 21.
5. A. F. Bochkov and G. E. Zaikov, *The Chemistry of the O-Glycosidic Bond*, Pergamon Press, Oxford, 1979.
6. *The Carbohydrates* (eds W. Pigman and D. Horton), Vol. IA, Academic Press, New York, 1972, p. 279.
7. *Bull. Soc. Chim. Fr.*, 1934, **1**, 1971.
8. *Aust. J. Chem.*, 1965, **18**, 1303.
9. *Methods Carbohydr. Chem.*, 1963, **2**, 328.
10. *Can. J. Chem.*, 1963, **41**, 2743; *Can. J. Chem.*, 1968, **46**, 3085.
11. *J. Am. Chem. Soc.*, 1955, **77**, 1667.
12. *Can. J. Chem.*, 1962, **40**, 224.
13. *Carbohydr. Res.*, 1968, **6**, 75; *J. Org. Chem.*, 1968, **33**, 740.
14. *Chem. Lett.*, 1991, 981
15. *J. Org. Chem.*, 1991, **56**, 5740.
16. *J. Am. Chem. Soc.*, 1992, **114**, 6354.
17. *Adv. Carbohydr. Chem. Biochem.*, 1988, **46**, 251.
18. *J. Chem. Soc.*, 1961, 412.
19. *J. Chem. Soc.*, 1961, 3240.
20. *J. Am. Chem. Soc.*, 1992, **114**, 1905; *J. Am. Chem. Soc.*, 1994, **116**, 2645.
21. *J. Org. Chem.*, 1965, **30**, 153.
22. *J. Chem. Soc., Perkin Trans. 2*, 1973, 1943.
23. E. Eliel, *Stereochemistry of Carbon Compounds*, McGraw-Hill, New York, 1962, p. 266.
24. *Carbohydr. Res.*, 1977, **56**, 277; *J. Chem. Soc., Perkin Trans. 2*, 1985, 1233.
25. *Carbohydr. Res.*, 1972, **24**, 57.
26. *J. Chem. Soc., Chem. Commun.*, 1971, 79.
27. *Chem. Rev.*, 1990, **90**, 1171.
28. *J. Am. Chem. Soc.*, 1990, **112**, 5887.
29. *Synthesis*, 1991, 499.
30. *J. Org. Chem.*, 1986, **51**, 4320.
31. *J. Chem. Soc., Chem. Commun.*, 1988, 823; *J. Am. Chem. Soc.*, 1988, **110**, 5583; *J. Am. Chem. Soc.*, 1991, **113**, 1434; *Synlett*, 1992, 927.
32. *Tetrahedron Lett.*, 1990, **31**, 275, 1331.
33. *Tetrahedron*, 1991, **47**, 9721.
34. *J. Am. Chem. Soc.*, 1989, **111**, 6656.
35. *Methods Carbohydr. Chem.*, 1963, **3**, 143.
36. *Indian J. Chem.*, 1988, **27B**, 527.
37. *Tetrahedron Lett.*, 1971, **12**, 2825.

38. *Tetrahedron Lett.*, 1984, **25**, 2383.
39. *Tetrahedron*, 1992, **48**, 8545.
40. *J. Am. Chem. Soc.*, 1949, **71**, 140.
41. *Carbohydr. Res.*, 1980, **86**, C3.
42. *J. Am. Chem. Soc.*, 1984, **106**, 4189; *J. Org. Chem.*, 1991, **56**, 1649.
43. *J. Org. Chem.*, 1992, **57**, 2455.
44. *Tetrahedron Lett.*, 1989, **30**, 4721.
45. P. Deslongchamps, *Stereoelectronic Effects in Organic Chemistry*, Pergamon Press, Oxford, 1983; *Can. J. Chem.*, 1971, **49**, 2465.
46. *Tetrahedron Lett.*, 1982, **23**, 4823.
47. *Aust. J. Chem.*, 1970, **23**, 1209.
48. *Can. J. Chem.*, 1980, **58**, 2660.
49. *J. Carbohydr. Chem.*, 1988, **7**, 1.
50. *Adv. Carbohydr. Chem. Biochem.*, 1981, **39**, 157.
51. *Adv. Carbohydr. Chem. Biochem.*, 1977, **34**, 23.
52. *Aust. J. Chem.*, 1978, **31**, 1151.
53. *J. Am. Chem. Soc.*, 1969, **91**, 1161.
54. *Carbohydr. Res.*, 1979, **74**, 327; *Carbohydr. Res.*, 1980, **81**, 192.
55. *Methods Carbohydr. Chem.*, 1963, **2**, 400.
56. *J. Am. Chem. Soc.*, 1989, **111**, 6661; *J. Am. Chem. Soc.*, 1992, **114**, 3471, 4518.
57. *Carbohydr. Res.*, 1983, **112**, 141; *Carbohydr. Res.*, 1984, **125**, 165.
58. *J. Am. Chem. Soc.*, 1984, **106**, 3871; *J. Am. Chem. Soc.*, 1985, **107**, 6372.
59. *Tetrahedron Lett.*, 1969, 1459.
60. *J. Org. Chem.*, 1965, **30**, 3951.
61. *J. Am. Chem. Soc.*, 1984, **106**, 450.
62. *Adv. Carbohydr. Chem. Biochem.*, 1976, **32**, 15.
63. *The Carbohydrates* (eds W. Pigman and D. Horton), Vol. IB, Academic Press, New York, 1980, p. 799.
64. *Methods Carbohydr. Chem.*, 1963, **2**, 427.
65. *Carbohydr. Res.*, 1968, **6**, 87.
66. *Methods Carbohydr. Chem.*, 1962, **2**, 354.
67. *Carbohydr. Res.*, 1984, **128**, 11.
68. *Glycoconjugate J.*, 1987, **4**, 97.
69. *J. Org. Chem.*, 1963, **28**, 2986; *Liebigs Ann. Chem.*, 1962, **657**, 179.
70. *Corbohydr. Res.*, 1990, **202**, 225.
71. *Chem. Br.*, 1990, 669.
72. *Carbohydr. Res.*, 1980, **80**, C17.
73. *Tetrahedron Lett.*, 1992, **33**, 2063.
74. *Carbohydr. Res.*, 1990, **195**, 303; *Tetrahedron Lett.*, 1992, **33**, 115; *J. Am. Chem. Soc.*, 1992, **114**, 2256.
75. *Adv. Carbohydr. Chem. Biochem.*, 1991, **49**, 37.
76. *J. Chem. Soc., Perkin Trans. 1*, 1977, 1993.
77. *J. Chem. Soc., Chem. Commun.*, 1990, 718; *Pure Appl. Chem.*, 1991, **63**, 519.
78. *J. Am. Chem. Soc.*, 1989, **111**, 6881.
79. *J. Am. Chem. Soc.*, 1993, **115**, 1580.
80. *Tetrahedron Lett.*, 1985, **26**, 6185, 6189, 6193.
81. *Tetrahedron Lett.*, 1986, **27**, 6201.
82. *Tetrahedron Lett.*, 1986, **27**, 215.
83. *J. Chem. Soc., Perkin Trans. 1*, 1974, 1069.
84. *J. Am. Chem. Soc.*, 1988, **110**, 8716.
85. *Carbohydr. Res.*, 1977, **57**, 73.
86. *Aust. J. Chem.*, 1959, **12**, 97.

87. *The Carbohydrates* (eds W. Pigman and D. Horton), Vol. IB, Academic Press, New York, 1980, p. 881.
88. *J. Chem. Soc.*, 1965, 4497.
89. *Carbohydr. Res.*, 1990, **206**, 361.
90. *J. Chem. Soc., Chem. Commun.*, 1984, 1155.
91. *J. Chem. Soc., Perkin Trans. 1*, 1982, 2139.
92. *Molecular Rearrangements* (ed. P. De Mayo), Part 2, Wiley-Interscience, New York, 1964, p. 709.
93. *Adv. Carbohydr. Chem.*, 1959, **14**, 63; *Angew. Chem., Int. Ed. Engl.*, 1990, **29**, 565.
94. *Carbohydr. Res.*, 1987, **164**, 195.
95. *Angew. Chem., Int. Ed. Engl.*, 1987, **26**, 557; *Synthesis*, 1992, 90.
96. *J. Chem. Soc., Perkin Trans. 2*, 1982, 199.
97. *J. Org. Chem.*, 1989, **54**, 4957.
98. *The Carbohydrates* (eds W. Pigman and D. Horton), Vol. IB, Academic Press, New York, 1980, p. 929.
99. *Pure Appl. Chem.*, 1991, **63**, 507.
100. *J. Chem. Soc., Chem. Commun.*, 1981, 97.
101. *Helv. Chim. Acta*, 1983, **66**, 789.
102. *Helv. Chim. Acta*, 1982, **65**, 2251.
103. *Carbohydr. Res.*, 1983, **117**, 89.
104. *J. Org. Chem.*, 1980, **45**, 3846; *Organic Synthesis Highlights*, VCH, Weinheim, 1991, p. 3.
105. *J. Carbohydr. Chem.*, 1984, **3**, 125.
106. *Tetrahedron Lett.*, 1982, **23**, 4143; *Chem. Lett.*, 1983, 5.
107. *Tetrahedron Lett.*, 1980, **21**, 1031.
108. *Carbohydr. Res.*, 1966, **2**, 315.
109. *J. Chem. Soc., Perkin Trans. 1*, 1975, 1191; *J. Chem. Soc., Perkin Trans. 1*, 1976, 68.
110. *The Carbohydrates* (eds W. Pigman and D. Horton), Vol. IA, Academic Press, New York, 1972, p. 133.
111. *Adv. Carbohydr. Chem.*, 1951, **6**, 291.
112. *J. Am. Chem. Soc.*, 1959, **81**, 243.
113. *Comprehensive Organic Synthesis* (ed. B. Trost), vol. 1, Pergamon Press, Oxford, 1991, p. 125.
114. *Some Modern Methods of Organic Synthesis*, 3rd Edn, Cambridge University Press, Cambridge, 1986, p. 125.
115. *Adv. Carbohydr. Chem. Biochem.*, 1972, **27**, 227.
116. *Tetrahedron Lett.*, 1988, **29**, 6823.
117. *J. Chem. Soc., Chem. Commun.*, 1982, 297.
118. *Tetrahedron Lett.*, 1988, **29**, 693.
119. *Tetrahedron*, 1965, **21**, 803.
120. *Adv. Carbohydr. Chem.*, 1956, **11**, 185.
121. *The Carbohydrates* (eds W. Pigman and D. Horton), Vol. IA, Academic Press, New York, 1972, p. 479.
122. *Methods Carbohydr. Chem.*, 1963, **2**, 65.
123. *Adv. Carbohydr. Chem.*, 1948, **3**, 129.
124. *The Carbohydrates* (eds W. Pigman and D. Horton), Vol. IB, Academic Press, New York, 1980, p. 1101.
125. *The Carbohydrates*, (eds W. Pigman and D. Horton), vol. IIB, Academic Press, New York, 1970, 739.
126. *J. Org. Chem.*, 1967, **32**, 2531.
127. *Synthesis*, 1981, 394.
128. *Methods Carbohydr. Chem.*, 1963, **2**, 16.
129. *J. Chem. Soc., Chem. Commun.*, 1989, 1817.

130. *J. Am. Chem. Soc.*, 1965, **87**, 5475.
131. *Carbohydr. Res.*, 1979, **72**, C9; *Bull. Chem. Soc. Jpn.*, 1980, **53**, 3687. See *Adv. Carbohydr. Chem. Biochem.*, 1994, **50**, 125 for a review of the reactions of aldonolactones.
132. *Carbohydr. Res.*, 1983, **121**, 175.
133. *Carbohydr. Res.*, 1984, **130**, 261.
134. *J. Org. Chem.*, 1979, **44**, 3368.
135. *J. Chem. Soc., Chem. Commun.*, 1984, 449.
136. *J. Org. Chem.*, 1986, **51**, 5458.
137. *J. Chem. Soc., Chem. Commun.*, 1990, 431; *Tetrahedron Lett.*, 1990, **31**, 4441; *Tetrahedron Lett.*, 1994, **35**, 89.
138. *Helv. Chim. Acta*, 1989, **72**, 1371.
139. *Helv. Chim. Acta*, 1991, **74**, 585.
140. *Angew. Chem., Int. Ed. Engl.*, 1992, **31**, 1388.
141. *Helv. Chim. Acta*, 1992, **75**, 621.
142. *J. Chem. Soc., Chem. Commun.*, 1992, 1605; *J. Chem. Soc., Chem. Commun.*, 1993, 1065.
143. *Tetrahedron Lett.*, 1990, **31**, 4787.
144. *Methods Carbohydr. Chem.*, 1963, **2**, 21, 24.
145. *Methods Carbohydr. Chem.*, 1961, **1**, 118.
146. *Methods Carbohydr. Chem.*, 1962, **1**, 77.
147. *Angew. Chem., Int. Ed. Engl.*, 1967, **6**, 950.
148. *J. Chem. Soc., Perkin Trans. 1*, 1990, 945.
149. *Methods Carbohydr. Chem.*, 1963, **2**, 46.
150. *Adv. Carbohydr. Chem.*, 1958, **13**, 63.
151. *Adv. Carbohydr. Chem.*, 1957, **12**, 35.
152. *Adv. Carbohydr. Chem. Biochem.*, 1974, **29**, 173; *Z. Naturforsch., Teil B*, 1977, **32**, 826; *J. Am. Chem. Soc.*, 1989, **111**, 3157.
153. *J. Am. Chem. Soc.*, 1982, **104**, 6764.
154. *Methods Carbohydr. Chem.*, 1962, **1**, 477.
155. *Adv. Carbohydr. Chem. Biochem.*, 1971, **26**, 127.
156. *Can. J. Chem.*, 1972, **50**, 1092.
157. *Adv. Carbohydr. Chem.*, 1955, **10**, 207.
158. *J. Chem. Soc.*, 1959, 636.
159. *Can. J. Chem.*, 1953, **31**, 1040.
160. *Adv. Carbohydr. Chem.*, 1957, **12**, 157.
161. *Chem. Lett.*, 1984, 501.
162. *Methods Carbohydr. Chem.*, 1980, **8**, 243.
163. *Chem. Lett.*, 1989, 145; *Chem. Lett.*, 1991, 533.
164. *Chem. Lett.*, 1991, 985.
165. *Tetrahedron Lett.*, 1991, **32**, 7065.
166. *Carbohydr. Res.*, 1976, **52**, 63.
167. *Can. J. Chem.*, 1955, **33**, 109.
168. *Tetrahedron Lett.*, 1983, **24**, 1563.
169. *Angew. Chem., Int. Ed. Engl.*, 1993, **32**, 336.
170. *J. Org. Chem.*, 1974, **39**, 3654.
171. *Tetrahedron Lett.*, 1991, **32**, 3353.
172. *Tetrahedron Lett.*, 1990, **31**, 327.
173. *Carbohydr. Res.*, 1974, **34**, 79; *Carbohydr. Res.*, 1978, **67**, 147.
174. *Tetrahedron Lett.*, 1992, **33**, 3523.
175. *Adv. Carbohydr. Chem.*, 1961, **16**, 85; *Adv. Carbohydr. Chem. Biochem.*, 1981, **38**, 195; *Adv. Carbohydr. Chem. Biochem.*, 1990, **48**, 91.

176. *Angew. Chem., Int. Ed. Engl.*, 1982, **21**, 155; *Angew. Chem., Int. Ed. Engl.*, 1990, **29**, 823.
177. *Chem. Soc. Rev.*, 1978, **7**, 423.
178. *Adv. Carbohydr. Chem. Biochem.*, 1977, **34**, 243.
179. *Acta Chem. Scand.*, 1949, **3**, 151; *Bull. Soc. Chim. Fr.*, 1936, **3**, 277.
180. *J. Am. Chem. Soc.*, 1938, **60**, 2559.
181. *J. Chem. Soc., Perkin Trans. 1*, 1981, 326.
182. *Chem. Ber.*, 1978, **111**, 2370.
183. *Chem. Rev.*, 1992, **92**, 1167.
184. *Can. J. Chem.*, 1979, **57**, 662.
185. For the L-form see *Tetrahedron*, 1980, **36**, 1261.
186. *Proc. Natl. Acad. Sci. USA*, 1961, **47**, 700.
187. *Carbohydr. Res.*, 1975, **39**, 341.
188. *J. Am. Chem. Soc.*, 1975, **97**, 4056.
189. *Recl. Trav. Chim. Pays-Bas*, 1985, **104**, 171; *Recl. Trav. Chem. Pays-Bas*, 1987, **106**, 596.
190. *Carbohydr. Res.*, 1982, **103**, C7.
191. *Synlett*, 1992, 759; *J. Am. Chem. Soc.*, 1991, **113**, 9376; *Angew. Chem. Int. Ed. Engl.*, 1994, **33**, 1765; for the *gluco*-form see *J. Chem. Soc., Chem. Commun.*, 1992, 913.
192. *Can. J. Chem.*, 1965, **43**, 1918.
193. *Can. J. Chem.*, 1965, **43**, 2199.
194. *Tetrahedron*, 1967, **23**, 693.
195. *Carbohydr. Res.*, 1975, **44**, C14.
196. *Chem. Ind.*, 1954, 20.
197. *J. Chem. Soc.*, 1955, 185.
198. *Carbohydrate Chemistry* (ed. J. F. Kennedy), Oxford University Press, Oxford, 1988, p. 134.
199. *Carbohydr. Res.*, 1978, **67**, 457.
200. *J. Am. Chem. Soc.*, 1951, **73**, 2241.
201. *Adv. Carbohydr. Chem. Biochem.*, 1976, **33**, 111.
202. *Chem. Ber.*, 1961, **94**, 1159.
203. *J. Am. Chem. Soc.*, 1945, **67**, 1759.
204. *J. Chem. Soc., Perkin Trans. 1*, 1974, 1943.
205. *Tetrahedron Lett.*, 1973, 3547.
206. *J. Carbohydr., Nucleosides, Nucleotides*, 1979, **6**, 417.
207. *Methods Carbohydr. Chem.*, 1963, **2**, 197.
208. *J. Chem. Soc., Chem. Commun.*, 1987, 1009; *Carbohydr., Res.*, 1991, **216**, 93.
209. *Carbohydr. Res.*, 1982, **110**, 330.
210. *Tetrahedron Lett.*, 1983, **24**, 3075.
211. *Angew. Chem., Int. Ed. Engl.*, 1989, **28**, 969.
212. *Tetrahedron Lett.*, 1989, **30**, 681.
213. *Carbohydr. Res.*, 1987, **171**, 399.
214. *J. Chem. Soc., Chem. Commun.*, 1983, 944.
215. *Tetrahedron*, 1985, **41**, 4079.
216. *J. Chem. Soc., Perkin Trans. 2*, 1986, 1453.
217. *J. Am. Chem. Soc.*, 1992, **114**, 8375.
218. *Adv. Phys. Org. Chem.*, 1988, **24**, 113.
219. *Chem. Lett.*, 1983, 935.
220. *J. Chem. Soc., Chem. Commun.*, 1984, 1153.

4 Reactions and Products of Reactions at Non-anomeric Carbon Atoms

4.1 INTRODUCTION

4.1.1 GENERAL

A wide variety of reactions can be effected at carbon atoms other than the anomeric, and an extensive range of compounds, often loosely described as *modified sugars*, can thus be prepared. Many such derivatives are closely related to important natural products. Members of the largest category are formally produced by replacement of specific hydroxyl groups of sugars by such substituents as hydrogen, amino, thiol, halogen or alkyl (Scheme 4.1a, e.g. X = H, NH_2, SH, Cl, Me); replacement of carbon-bonded hydrogen atoms only gives stable derivatives when new carbon–carbon bonds are formed (Scheme 4.1b, e.g. X = Me). Hydroxyl groups may, in addition, be formally replaced intramolecularly with the formation of anhydro derivatives (Scheme 4.1c), or, if ring oxygen atoms rather than hydroxyl groups are responsible for the displacements involved, products with altered ring sizes are formed (Scheme 4.1d). Again, if ring carbon atoms displace the hydroxyl groups, ring-contracted branched-chain compounds are obtained (e.g. Scheme 4.1e).

Oxidations at alcoholic centres give carbonyl products (Scheme 4.1f), and eliminations formally give alkenes, or enol or enediol derivatives (Scheme 4.1g). All these products can be subjected to addition reactions which occur by attack at the carbon atoms of carbohydrate chains and give modified, saturated products.

In this chapter methods for effecting the changes illustrated in Scheme 4.1 for each of the above-mentioned classes of products will be discussed. Initially, the important general features of the two most useful reactions applied to obtain modified sugars (especially those formed according to Scheme 4.1a), i.e. nucleophilic displacement reactions and epoxide ring-opening reactions, will be treated.

4.1.2 NUCLEOPHILIC DISPLACEMENT OF LEAVING GROUPS

Leaving groups (L) are displaced by nucleophiles with great ease from the anomeric carbon atoms of pyranoid and furanoid rings, usually by unimolecular mechanisms (Chapter 3). Alternatively, displacement reactions at the other secondary carbon atoms are relatively sluggish, and nearly always occur by S_N2 mechanisms

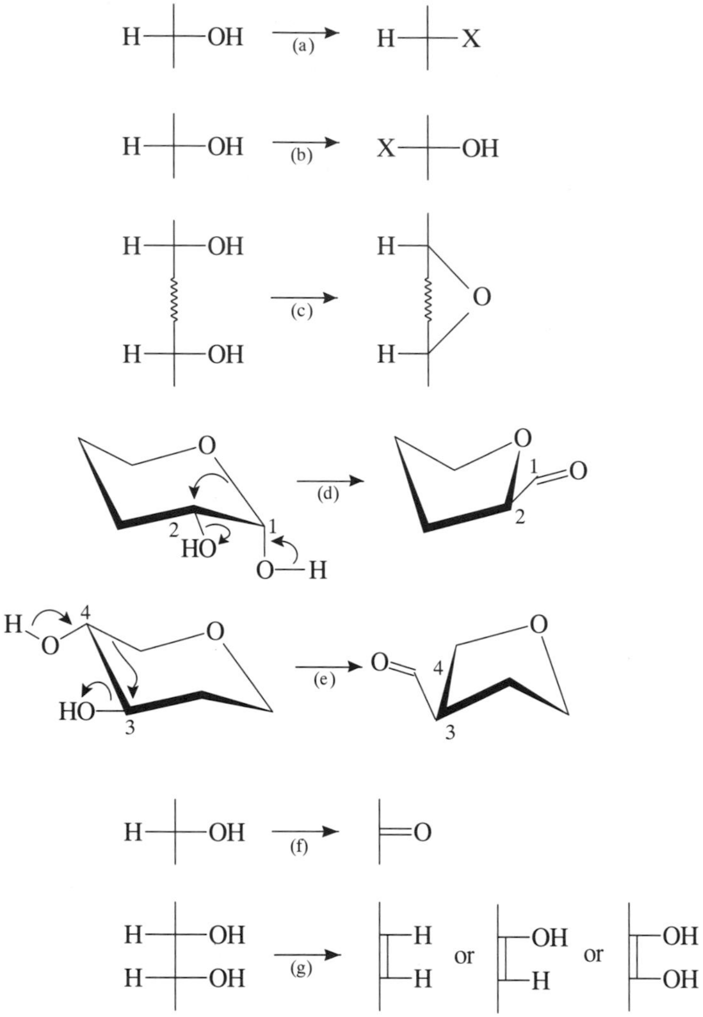

Scheme 4.1

(Scheme 4.2a); S_N1 processes are highly unfavoured because the carbocations which would be produced are destabilized by the electron-withdrawing oxygen-bonded groups usually present on the adjacent carbon atoms (Scheme 4.2b). Dis-placements at primary positions are subject to least steric opposition, and they proceed much more readily than do the corresponding reactions at secondary carbon atoms.

The conditions selected for most displacements are typical of those used for S_N2 reactions. Powerful nucleophiles and high-boiling solvents which will dissolve both substrate and the nucleophilic reagent are preferred. Aprotic solvents are most suit-able, and acetone, butanone and higher-boiling ketones have been extensively used,

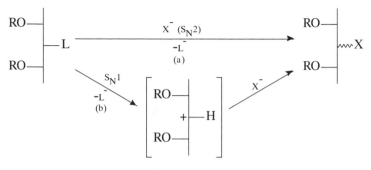

Scheme 4.2

but often with these solvents displacements can only be effected at primary positions. The introduction of N,N-dimethylformamide (DMF), dimethyl sulphoxide (DMSO) and hexamethylphosphoric triamide (HMPIT) has extended the range of displacements possible with carbohydrate derivatives, and many reactions at secondary carbon atoms have been effected. Aprotic solvents facilitate nucleophilic displacement reactions by anions because, unlike protic solvents, they do not hydrogen bond to the anions whose nucleophilic properties are therefore not masked by a solvent cage. Furthermore, the high dielectric constants of these solvents ensure maximum nucleophilicity since ion aggregation is minimized.[1]

Although many leaving groups are used in such reactions in aliphatic organic chemistry, for obvious reasons those whose departure results in cleavage of carbon–oxygen bonds are by far the most important with carbohydrate derivatives. The sulphonate esters have proved particularly valuable in this respect, and, as will be illustrated below, are commonly applied to invert configurations at specific carbon atoms and to replace hydroxyl groups in sugar derivatives by a variety of other functions.[2] Mesylates and tosylates have been used most often but, when displacement on these esters proves to be slow, the corresponding trifluoromethanesulphonates (triflates) can often afford a means of effecting the required reactions (Section 5.3.6).

Displacements can also be induced by *in situ* conversions of sugar alcohols into their alkoxyphosphonium ions using triphenylphosphine with an electrophilic activator, as indicated in the general equation below. Nucleophiles readily attack these salts under very mild conditions with concomitant displacement of triphenylphosphine oxide, thereby providing an important means of forming new bonds to sugar carbon atoms.

$$Ph_3P + EY \underset{+Y^-}{\overset{-Y^-}{\rightleftharpoons}} Ph_3\overset{+}{P}E \xrightarrow[-EH]{ROH} RO\overset{+}{P}Ph_3 \xrightarrow{Z^-} RZ + Ph_3P{=}O$$

Pre-eminent among these reactions is the Mitsunobu procedure,[3] which uses DEAD as the activator to give the betaine **4.1** with which alcohols react to form alkoxytriphenylphosphonium ions. Applications of this method to the synthesis of various modified sugars are discussed in Section 4.1.2.c.

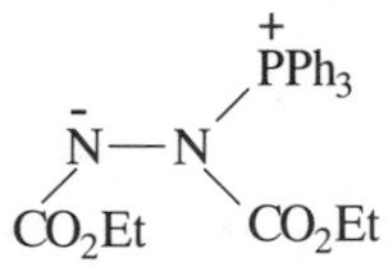

4.1

(a) Displacements of sulphonates without neighbouring-group participation

(i) Displacements of sulphonates from primary centres

It is usually a simple matter to displace selectively a leaving group from the terminal position of a hexopyranoside or a furanoside. For example, only the 6-*O*-mesyloxy residue is displaced from methyl 2,3,4,6-tetra-*O*-mesyl-α-D-glucopyranoside when it is treated with potassium acetate or sodium iodide (Scheme 4.3),[4] and wide use has been made of reactions of this type in synthesis to effect change at the primary sites (see, for example, Section 4.5.1).

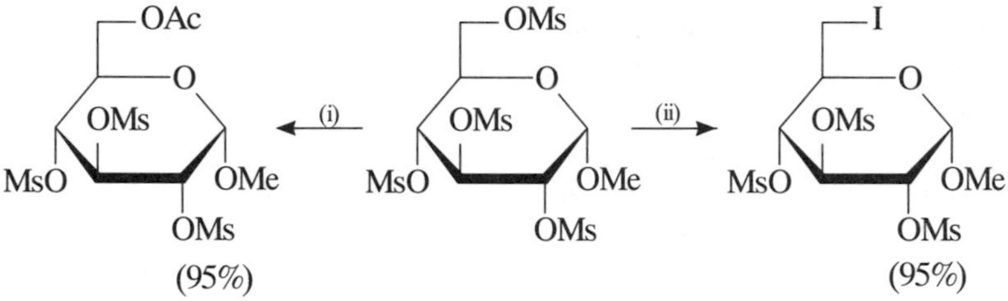

(i) KOAc, AcOH, Ac₂O, reflux, 1 h; (ii) NaI, Me₂CO, reflux, 4 h

Scheme 4.3

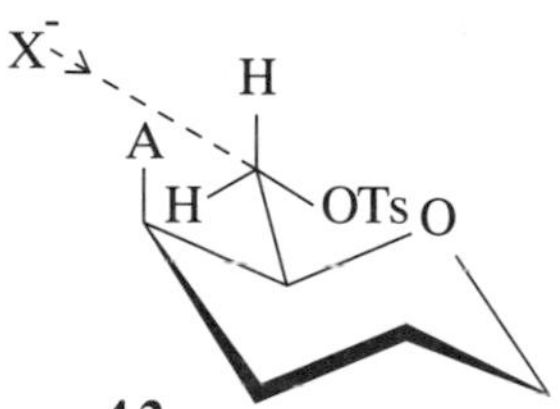

4.2

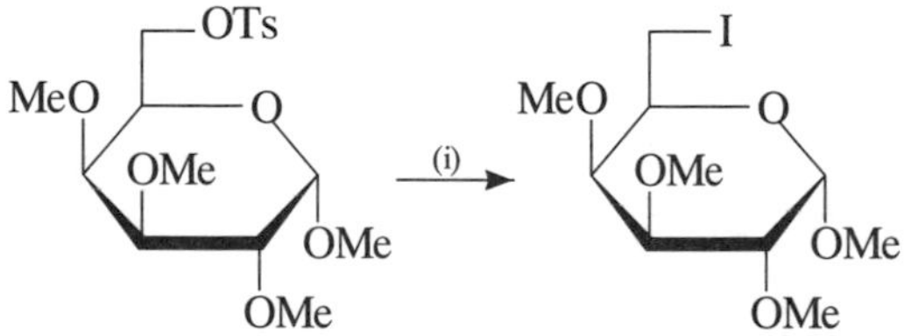

4.3   **4.4** (52% conversion)

(i) NaI, DMF, 120 °C, 5 h

Scheme 4.4

Displacements of primary sulphonate ester residues in hexopyranoside derivatives are, however, not insensitive to structural changes remote from the reaction sites. Thus, an axial substituent at C-4 can exert a rate-retarding effect upon this reaction, as is shown by the slow conversion of methyl 2,3,4-tri-O-methyl-6-O-tosyl-α-D-galactopyranoside **4.3** into its 6-deoxy-6-iodo derivative **4.4** when treated with sodium iodide under the conditions given in Scheme 4.4, which bring about complete conversion of the *gluco*-analogue.[5] Preferred rotamers about the C-5–C-6 bonds in both *galacto*- and *gluco*-isomers would be anticipated to be those depicted in the partial structure **4.2** (see also Section 5.3.6), and the approach path of the nucleophile X^- will be collinear with the C-6–OTs bond and will be subject to interference from group A. Since, in the *galacto*-compounds, this is a methoxy group, its interaction with X^- will be greater than in the *gluco*-series, in which it is a hydrogen atom.[6]

(ii)   Displacements of sulphonates from secondary centres on pyranoid rings
Displacements from positions 3 and 4 are brought about by application of more vigorous conditions than those necessary for primary groups. For example, methyl 2,3-di-O-benzoyl-4,6-di-O-tosyl-α-D-galactopyranoside **4.5** in DMF undergoes displacements by the benzoate ion at the C-6 and C-4 positions to give methyl α-D-glucopyranoside tetrabenzoate **4.6** (Scheme 4.5).[7] Related displacements involving azide ions, thiobenzoate ions and halide ions are discussed in Sections 4.3.1, 4.4.1 and 4.5.1, respectively.

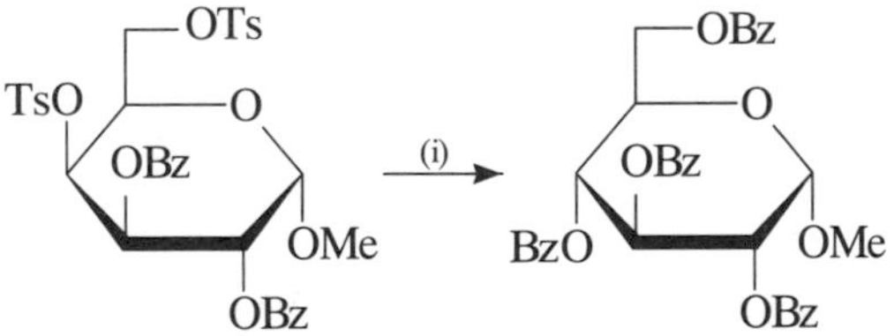

(i) NaOBz, DMF, 140 °C, 24 h

Scheme 4.5

A difference in reactivity dependent upon the orientation of the leaving groups would be anticipated in these reactions, since bimolecular displacements with conformationally locked cyclohexyl tosylates and halides are faster when the leaving groups occupy axial positions. Similar behaviour has been observed with pyranoid systems. The 4-deoxy-4-iodo derivatives of methyl 2,3-di-O-benzyl-6-deoxy-α-D-gluco- and galacto-pyranosides, on being subjected to the Finkelstein reaction ($Na^{131}I$ in acetone was used in these experiments), are interconverted to produce an equilibrated mixture. However, the reaction of the *galacto*-isomer (axial iodine at C-4) to give the *gluco*-compound is approximately 2.5 times faster than the reverse reaction.[8] From an experiment designed in this way, it is possible to conclude that the reaction rate differences depend upon the ground state energies of the two reactants since each bimolecular reaction has the same transition state.[6]

Many displacements at C-3, which provide a suitable strategy for obtaining D-allose derivatives from readily accessible D-glucose compounds (Scheme 4.6), have been brought about and they normally occur almost as readily as those at C-4; however, selective reactions at position 4 have been observed in cases of similarly 3,4-disubstituted glucopyranosides (Section 5.3.6). Displacements of equatorially oriented leaving groups by anionic nucleophilic reagents at either of these positions are greatly retarded if there is a substituent situated in the β-*trans*-axial position. Thus, for example, displacement of the sulphonate ester in 1,2,4,6-tetra-*O*-benzoyl-3-*O*-tosyl-β-D-glucopyranose **4.7** with benzoate ions occurs quite rapidly to give the pentabenzoate **4.8** (Scheme 4.6), but with the α-anomer **4.9** displacement of the sulphonate ester group does not occur (Scheme 4.6) since in this case the approach of the benzoate ion to C-3 is hindered by the axial benzoyloxy group at C-1.[9]

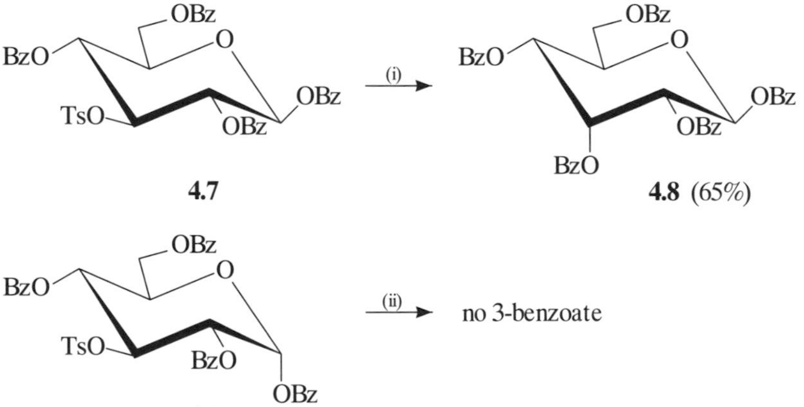

4.7   **4.8** (65%)

4.9

(i) Bu$_4$NOBz, *N*-methylpyrrolidone, 100 °C, 16 h; (ii) *t*-BuNOBz, *N*-methylpyrrolidone, 100 °C, 10 days

Scheme 4.6

Displacements at C-2 with anionic nucleophiles are distinctly difficult, probably because C-1 bears two electron-withdrawing oxygen atoms and retards the expulsions of the leaving groups. The effect is markedly dependent upon the orientation of the anomeric substituents; axially disposed groups make displacement reactions at C-2 impractical when an oxygen substituent is also present at C-3.[10] (The replacement of a 2-chlorosulphonyloxy group by chlorine that has been observed in a methyl α-D-glucopyranoside derivative is an exceptional case.[11]) The general behaviour of 2-*O*-sulphonyl pyranosides is illustrated by the four examples in Scheme 4.7. Thus, even when heated under reflux in DMF with potassium benzoate, methyl 4,6-*O*-benzylidene-3-*O*-methyl-2-*O*-mesylhexopyranosides with the α-*gluco*-**4.10** and α-*manno*-**4.12** configurations yield very little of the respective 2-*O*-benzoyl derivatives **4.11** and **4.13**. This is thought to be because the

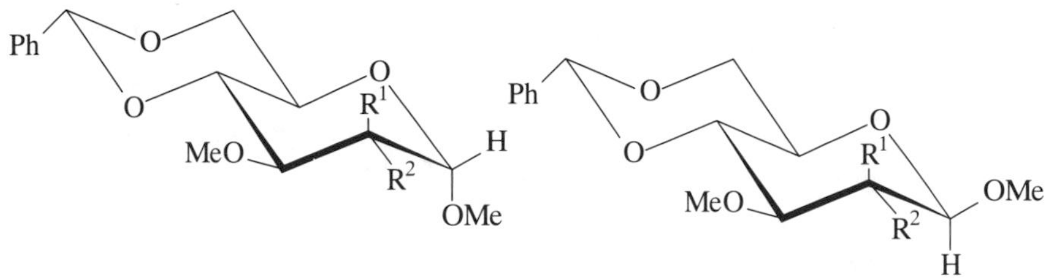

4.10 $R^1 = H, R^2 = OMs$ (i) (<5%)
4.11 $R^1 = OBz, R^2 = H$

4.12 $R^1 = OMs, R^2 = H$ (i) (<1%)
4.13 $R^1 = H, R^2 = OBz$

4.14 $R^1 = H, R^2 = OMs$ (i) (62%)
4.15 $R^1 = OBz, R^2 = H$

4.16 $R^1 = OMs, R^2 = H$ (ii) (70%)
4.17 $R^1 = H, R^2 = OBz$

(i) KOBz, DMF, reflux, 120 h; (ii) as for (i) but for 8 h

Scheme 4.7

transition states that lead to displacement products possess unfavourable dipole interactions between the axial bonds to the anomeric substituents and the trajectories of the nucleophile and leaving group as shown in **4.18**.[10] On the other hand, the 2-*O*-mesyl derivatives with β-*gluco*-**4.14** and β-*manno*-**4.16** configurations give satisfactory yields of the 2-benzoates **4.15** and **4.17**, respectively, when heated under the same, or milder, conditions (Scheme 4.7). This is thought to be because in the transition state **4.19** that leads to these products the dipole of the equatorial bond to the anomeric substituent is not aligned with that of the nucleophile and leaving group, and therefore exerts a weaker rate-retarding effect.[10] With related 3-deoxypyranoside derivatives displacements are more readily achieved under normal conditions, even with α-anomers. Thus, the 3-deoxy-2-*O*-tosyl-α-pyranoside **4.20** gives the 2-azido product **4.21** (and hence the 2-acetamido analogue), as illustrated in Scheme 4.8.[12]

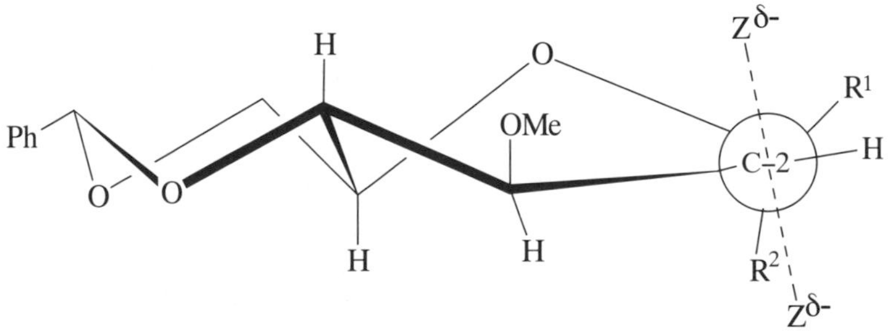

4.18 $R^1 = H, R^2 = OMe$
4.19 $R^1 = OMe, R^2 = H$

Z = nucleophile or leaving group

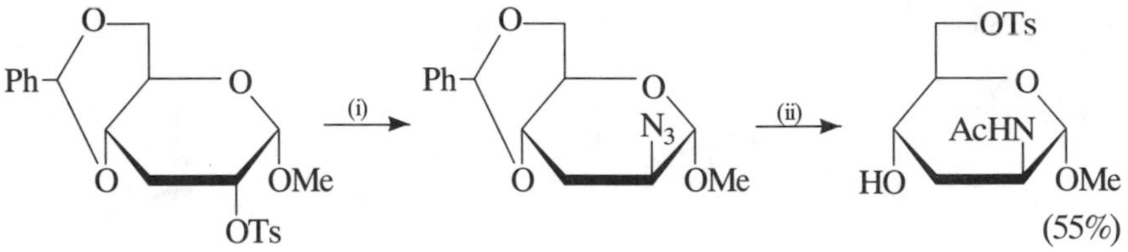

4.20   **4.21**

(i) NaN$_3$, DMF, 140 °C; (ii) reduction, N-acetylation, hydrolysis, selective tosylation

Scheme 4.8

(iii)   Displacements of sulphonates from secondary centres on furanoid rings

S$_N$2 displacement reactions normally occur more readily from five- than from six-membered rings, presumably because of the smaller increase in I-strain involved in attaining the S$_N$2 transition state with the former compounds (see Section 3.1.1.b.ii for further discussion).

Displacements at C-2 and C-3 of furanoid rings have been achieved, and an example of the latter is illustrated by the conversion of 1,2:5,6-di-O-isopropylidene-3-O-tosyl-α-D-allofuranose **4.22** into the 3-azido-3-deoxyglucofuranose derivative **4.24** (Scheme 4.9a).[13] However, with the *gluco*-3-tosylate **4.26** (R = Ts), in which the leaving group occupies an *exo*-position to the bicyclic system formed by the two *cis*-fused five-membered rings, the displacement is only brought about under the more forcing conditions indicated in Scheme 4.10(a) for the synthesis of di-O-isopropylidene-3-azido-3-deoxyallofuranose **4.27** (R = N$_3$).[14] It is possible to displace the *exo*-sulphonate ester residue from **4.26** (R = Ts) more readily with uncharged nucleophiles such as hydrazine under the conditions given in Scheme 4.10(b), probably because such uncharged species encounter a smaller

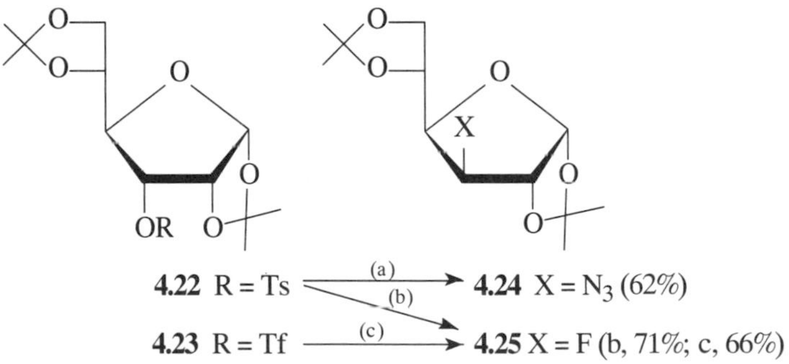

4.22 R = Ts $\xrightarrow[\text{(b)}]{\text{(a)}}$ **4.24** X = N$_3$ (62%)

4.23 R = Tf $\xrightarrow{\text{(c)}}$ **4.25** X = F (b, 71%; c, 66%)

(a) NaN$_3$, DMF, 140 °C, 4 h
(b) Bu$_4$NF, MeCN, reflux, 70 h
(c) TASF, CH$_2$Cl$_2$, ~ 20 °C, 10 min

Scheme 4.9

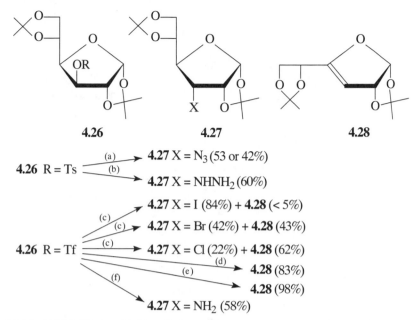

4.26   **4.27**   **4.28**

4.26 R = Ts $\xrightarrow{\text{(a)}}$ **4.27** X = N$_3$ (53 or 42%)

$\xrightarrow{\text{(b)}}$ **4.27** X = NHNH$_2$ (60%)

4.27 X = I (84%) + **4.28** (< 5%)

(c) → **4.27** X = Br (42%) + **4.28** (43%)

4.26 R = Tf $\xrightarrow{\text{(c)}}$ **4.27** X = Cl (22%) + **4.28** (62%)

(d) → **4.28** (83%)

(e) → **4.28** (98%)

(f) → **4.27** X = NH$_2$ (58%)

(a) NaN$_3$, DMF, H$_2$O, 150 °C, 15 days or NaN$_3$, HMPIT, 120 °C, 18 h
(b) (NH$_2$)$_2$, reflux, 36 h
(c) Bu$_4$NX, C$_6$H$_6$, reflux
(d) TASF (see Section 4.5), CH$_2$Cl$_2$, −22 °C, 10 min
(e) DBU, Et$_2$O, 22 °C, 15 h
(f) NH$_3$, ClCH$_2$CH$_2$Cl, 70 °C, two days

Scheme 4.10

electrostatic repulsion from the oxygen atoms at C-1 and C-2 than do charged nucleophiles.[2,14]

The 3-O-triflyloxy residue, which is a much more effective leaving group, is readily displaced even from the glucofuranose derivative **4.26** (R = Tf) to give allo-furanose compounds **4.27** sometimes accompanied by the enose derivative **4.28**, the proportions of which depend upon the relative nucleophilicities and basici-ties of the reagents used, as illustrated in Schemes 4.10(c),[15] 4.10(d),[16] 4.10(e)[17] and 4.10(f).[18] Examples in which the exo-3-hydroxyl group is directly displaced from 1,2:5,6-di-O-isopropylidene-α-D-glucofuranose with and without concomitant rearrangements are given in Section 4.5.1.e.

Displacements at C-2 in furanosyl derivatives are also influenced by the stereo-chemistry of the adjacent substituents, as exemplified by the difference in behaviour between the 2-O-triflyl-α- and -β-furanosides **4.30** and **4.31** when under attack from azide ions. The less sterically crowded α-anomer **4.30** is smoothly converted into the 2-azido compound **4.29** when treated with sodium azide in warm DMF, as illustrated in Scheme 4.11.[19] On the other hand, the β-anomer **4.31**, in which the approach to C-2 is obstructed by the methoxy and benzyloxy groups, is degraded

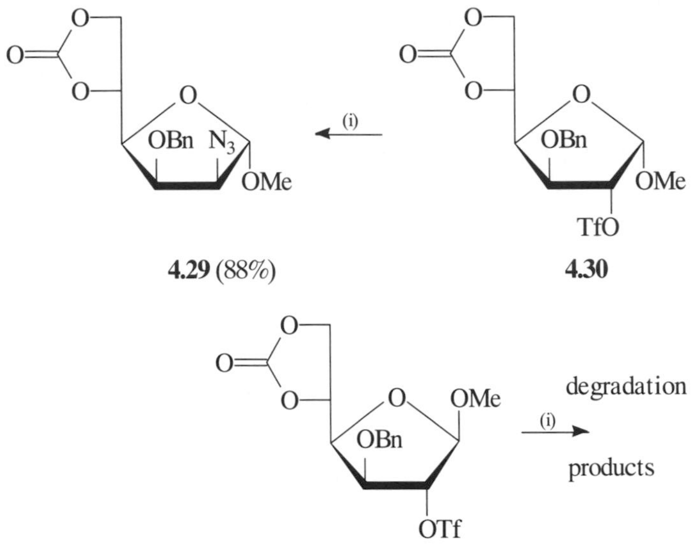

4.29 (88%)   **4.30**

4.31

(i) NaN$_3$, DMF, 50 °C

Scheme 4.11

under these conditions. Halide ion displacement applied to a triflate closely related to **4.30** is reported in Scheme 4.77, Section 4.5.1.a. Non-fluorinated sulphonyl-oxy groups have also been displaced from C-2 by the uncharged hydrazine. (See Section 5.3.6 for further discussions on displacements at particular sites.)

(b) Displacements of sulphonates with neighbouring-group participation

A large number of displacement reactions involve participation by intramolecular nucleophiles and give products other than those formed by direct intermolecular attack.[6,20] Participation by acyl protecting groups is often encountered and will be discussed here, but the structures of sugars are such that a wide variety of other participations are possible, and these are treated under the headings of the products formed. Participation by hydroxyl groups is treated in the sections on anhydrides (Sections 3.1.1.c and 4.6), ring oxygen and ring carbon participations are dealt with in the sections on ring contractions (Sections 4.7.1 and 4.7.2) and examples involving sulphur atoms are given in Section 4.4.2.

(i) Participation by acylamino groups

Neighbouring group participation by the acylamino group in nucleophilic displacement reactions is very common. For example, treatment of methyl 2-benzamido-4,6-O-benzylidene-2-deoxy-3-O-mesyl-α-D-altropyranoside **4.33** with sodium acetate in ethanol gives none of the product in which the acetate ion has displaced the sulphonate ester residue, but, instead, a high yield of the oxazoline **4.35** (Scheme 4.12a) is produced.[21] However, when compound **4.33** is treated with the more basic nucleophilic reagent sodium ethoxide in ethanol, two products are obtained: the oxazoline **4.35** and the *N*-benzoylepimine **4.34** (Scheme 4.12b).[22] Epimines are discussed in Section 4.3.2.e. In the first reaction the oxazoline is produced by attack from the carbonyl oxygen atom of the un-ionized benzamido group, whereas in the second reaction the ambident benzamido anion[†] **4.32** attacks mainly with the nitrogen atom to give predominantly the epimine **4.34**. Intramolecular attack by acylamino nitrogen is usually favoured when basic nucleophiles are used.

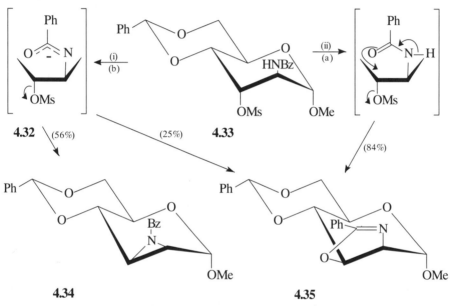

(i) NaOEt, EtOH, reflux, 30 min; (ii) NaOAc, EtOH, reflux
Scheme 4.12

The failure to obtain from compound **4.33** any 3-*O*-acetyl-mannoside derivative by direct displacement with sodium acetate can be ascribed to the *trans*-diaxial stereochemistry of the leaving group at C-3 and the benzamido group at C-2,[20] which represents an ideal arrangement for neighbouring group participation.[6] When the steric arrangement is *trans*-diequatorial, external nucleophiles are able to compete with the benzamido group, as illustrated in Scheme 4.13 by the reaction of methyl 2-benzamido-4,6-*O*-benzylidene-2-deoxy-3-*O*-mesyl-*β*-D-glucopyranoside **4.36** with potassium acetate, in which almost equal amounts of the products **4.37** (R = OAc) and **4.38**, formed respectively by acetate anion and benzamido oxygen attack, are produced.[23] When the glucoside **4.36** is treated with potassium benzylthiolate, not only are the 3-*S*-benzyl-3-thioalloside **4.37** (R = SBn)

[†] An ambident anion is one in which either of two atoms may act as the nucleophilic centre.

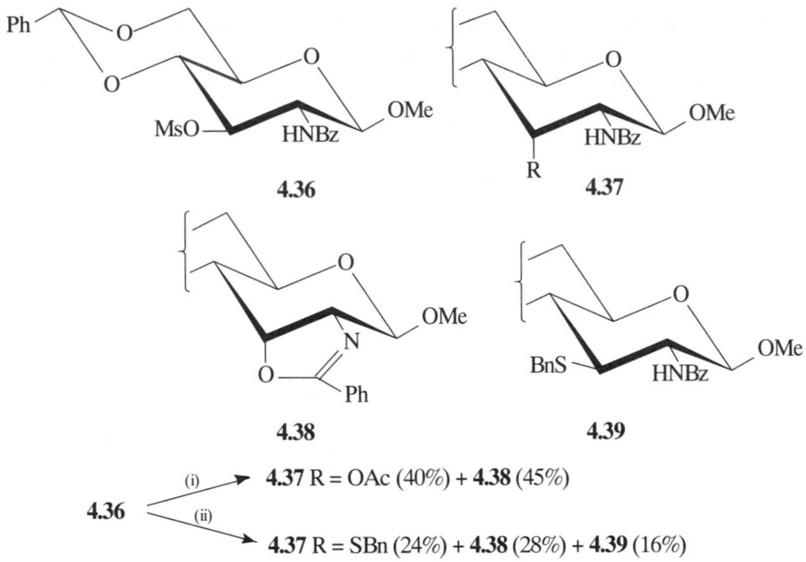

4.36

4.37

4.38

4.39

4.36 $\xrightarrow{(i)}$ **4.37** R = OAc (40%) + **4.38** (45%)

$\xrightarrow{(ii)}$ **4.37** R = SBn (24%) + **4.38** (28%) + **4.39** (16%)

(i) KOAc, DMF, heat; (ii) KSBn, DMF, heat

Scheme 4.13

and the oxazoline **4.38** produced, but the 3-*S*-benzyl-3-thioglucoside **4.39** is also obtained.[23] The thio sugar derivative **4.37** with the *allo*-configuration is formed by direct nucleophilic attack on the mesylate **4.36**, whereas the 3-epimer **4.39** arises by attack of the benzylthiolate either on an oxazolinium ion intermediate or on an *N*-benzoyl-*allo*-epimine intermediate.[20]

Acylamino participation has been of great value for modifying the structures of amino sugars, since by its use it is possible to convert a vicinally related *trans*-amino alcohol via a sulphonate ester into its *cis*-isomer. This is achieved, as shown in Scheme 4.14, by hydrolysing an oxazolinium ion intermediate.[24]

(ii) Participation by acyloxy groups

In contrast with acylamino groups, suitably positioned acyloxy groups often do not participate in nuclcophilic displacements from pyranoid ring systems. For example, 1,2,4,6-tetra-*O*-benzoyl-3-*O*-tosyl-β-D-glucopyranose undergoes exchange with sodium benzoate in what appears to be a direct displacement at C-3, without any participation from either of the adjacent *trans*-benzoyloxy groups (Scheme 4.6). On the other hand, however, treatment of methyl 2-*O*-benzoyl-3-*O*-tosyl-β-L-arabinopyranoside **4.40** with sodium fluoride in moist DMF gives the mixed *cis*-hydroxyl esters **4.42**, produced by hydrolysis of the benzoxonium ion **4.41** (Scheme 4.15).[25] The function of the fluoride ions in reactions of this type is not fully understood; they probably stabilize the acyloxonium ions, but being weak nucleophiles, as mentioned in Section 4.5.1.a, they fail to open the ion **4.41** by attack at one of the secondary carbon atoms.

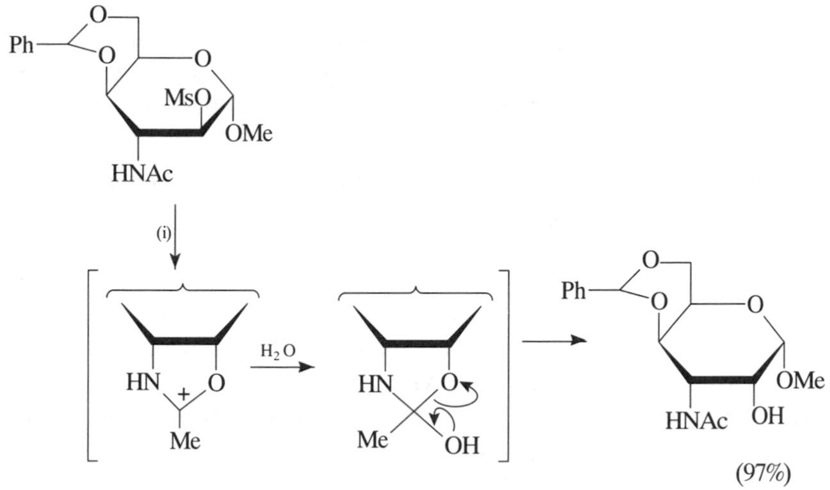

(i) NaOAc, MeOCH$_2$CH$_2$OH, trace H$_2$O, reflux, two days
Scheme 4.14

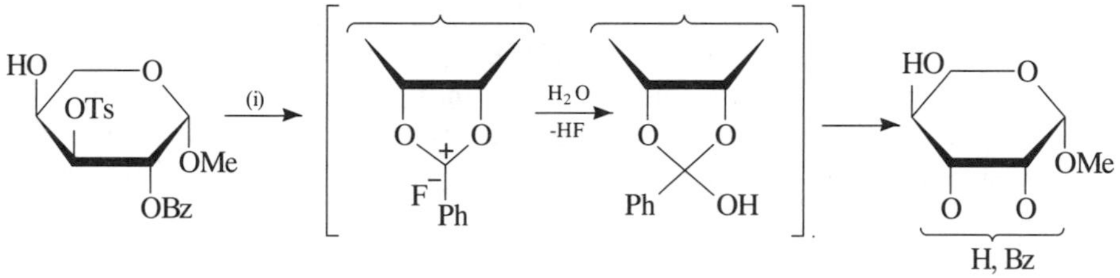

(i) NaF, DMF, trace H$_2$O, reflux, two days
Scheme 4.15

With furanoid derivatives, participation by acyloxy groups occurs during reaction with a large number of nucleophilic reagents. For example, the L-ribose derivatives **4.43** have been prepared from both the L-arabinofuranoside and L-xylofuranoside esters, as illustrated in Schemes 4.16(a) and 4.16(b), respectively.[26]

The lower reactivity of the arabinoside compared with that of the xyloside would be expected, because the displacement is being effected at C-2 in the former. In cases in which specific nucleophilic opening of intermediate acyloxonium ions

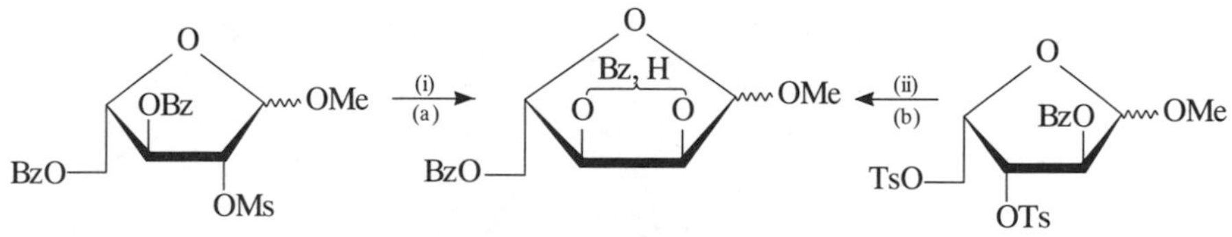

4.43

(i) NaOBz, DMF, trace H$_2$O, reflux, 72 h; (ii) NaOBz, DMF, trace H$_2$O, reflux, 6 h
Scheme 4.16

occurs, single products result. For example, treatment of 3-*O*-acetyl-6-*O*-benzoyl-1,2-*O*-isopropylidene-5-*O*-tosyl-α-D-glucofuranose with acetate ions gives good yields of the 3,6-di-*O*-acetyl-5-*O*-benzoyl analogue of L-idose, the benzoyl group having migrated from O-6 to O-5 via a benzoxonium intermediate.[20]

(c) Displacements of hydroxyl groups following *in situ* activation

Conversions of sugars and their derivatives directly into chlorodeoxy sugars by use of sulphuryl chloride, as discussed in Section 5.3.7.b, represent an early example of this type of chemistry since, in some instances, alcohol groups are initially converted to chlorosulphates which either undergo S_Ni reactions or intermolecular displacements to give the chloro compounds. More recently, similar transformations have been achieved with Vilsmeier's imidoyl chloride reagent (see Section 4.5.1.e.ii), and in related transformations DAST is used as a route to fluorodeoxy sugars (see Section 4.5.1.e.iii).

Chloro-, bromo- and iodo-deoxy sugars have been made from similar sugar precursors using triphenylphosphine and halonium ion electrophilic activators by way of sugar alkoxytriphenylphosphonium ions (see Section 4.1.2, Introduction and specific transformations in Section 4.5.1.e). The generation of these phosphonium ion intermediates by the use of DEAD, as described by the mechanism in Section 4.1.2, was devised by Mitsunobu[3] and has given rise to a versatile reaction for preparing many modified sugars such as thio, and anhydro sugars and *O*-glycosides (see Sections 3.1.1.a.ii, 4.5.1.e.i and 4.6.1.a.i), as well as halogenated derivatives.

In addition to these conversions of hydroxyl-containing sugar derivatives, esters can be prepared from them by treatment with triphenylphosphine, DEAD and a carboxylic acid, and since these esterifications occur by way of a displacement, selectivity for primary positions is exhibited.[3] Thus, methyl 2,3-di-*O*-acetyl-α-D-glucopyranoside on treatment with triphenylphosphine, DEAD and benzoic acid gives methyl 6-*O*-benzoyl-2,3-di-*O*-acetyl-α-D-glucopyranoside in 77% yield, and the S_N2 nature of the reaction is clearly shown by the displacement, with inversion, of the 4-hydroxyl group in the D-*erythro*-hexenoside **4.44** to give, in 92% yield, the 4-*O*-benzoyl analogue **4.45**.[3] 1,2:5,6-Di-*O*-isopropylidene-α-D-glucofuranose, which is known to resist displacement at C-3 (Sections 4.1.2.a.iii and 5.3.6), fails to react in the normal way, and alternatively some benzoylation of the 3-hydroxyl group of the starting material occurs.[3]

               4.44   **4.45**

4.1.3   OPENING OF EPOXIDE RINGS

Although epoxides (Section 4.6.1) are ethers, which usually are rather inert compounds, the strain of the three-membered ring endows them with a very useful
reactivity towards bimolecular nucleophilic attack.[27] Consequently, they can be
used as suitable precursors for many modified sugar derivatives, and they undergo
reactions which are mechanistically related to the displacement processes discussed
above.

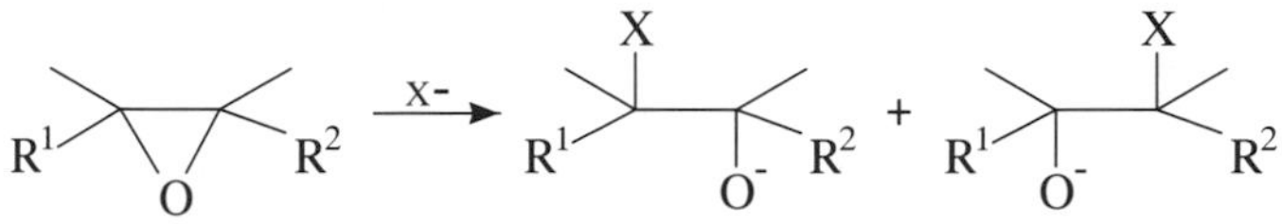

Scheme 4.17

Nucleophiles cause opening of epoxides with Walden inversion at the
carbon atom undergoing attack, as depicted in Scheme 4.17, and give
products with *trans*-configurations as expected for S_N2 reactions. In general,
attack can occur at each of the carbon atoms, and unsymmetrical epoxides
can therefore give rise to mixtures of isomeric products. This does
occur with carbohydrate examples, but on many occasions almost complete
selectivity is observed, especially when the epoxide rings are fused in
conformationally restricted derivatives such as 1,6-anhydropyranoses or *trans*-
fused 4,6-*O*-benzylidenehexopyranosides, as illustrated for the conformationally

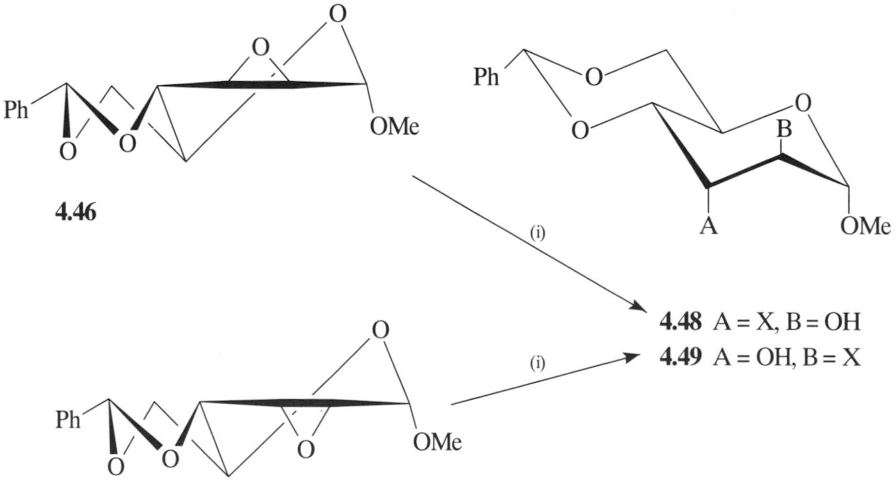

4.46

4.47

(i) HX

4.48 A = X, B = OH
4.49 A = OH, B = X

(i)

Scheme 4.18

bound six-membered ring compounds **4.46** and **4.47** in Scheme 4.18. In these examples, the pyranoid rings adopt half-chair conformations, and the isomers having axial substituents at C-2 and C-3 (**4.48** and **4.49**, respectively) often represent more than 90% of the products. This is in accordance with the Fürst–Plattner rule, formulated from studies of the nucleophilic opening of steroidal epoxides, which can be explained by consideration of the transition states in the ring-opening reactions.[27]

Bimolecular *trans*-opening of the general cyclohexane epoxide **4.50** (taken to be conformationally rigid) can occur in two ways (Scheme 4.19). Attack by the nucleophile at position 1 (Scheme 4.19a) allows the oxygen atom and the carbon atom undergoing reaction and the nucleophile to attain the linear geometry **4.51** required by an S_N2 transition state, and this leads to the ready formation of *trans*-diaxial products. Alternatively, attack at position 2 (Scheme 4.19b) gives a less favourable transition state in which the developing chair conformation **4.52** of the cyclohexane ring causes the three reacting atoms to deviate from the required collinear geometry. Thus, *trans*-diequatorial products, which would arise directly from this mode of attack, are not usually formed. It should be understood, however, that in flexible systems the favoured, initial diaxial products may be conformationally unstable, and may ring invert to give compounds with the new groups in equatorial orientations.

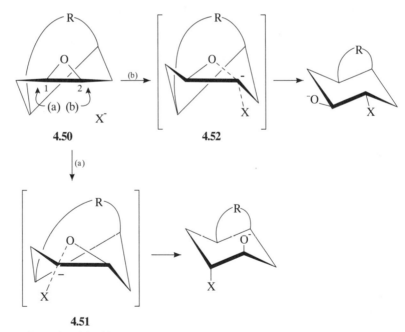

R = conformational locking system

Scheme 4.19

The prediction of products is often difficult with epoxypyranoid derivatives which do not have rigid ring structures because the conformations through which such compounds react can be difficult to forecast. Thus, methyl 2,3-anhydro-6-deoxy-α-L-talopyranoside **4.53** (R = H), which can exist in either the $^5H_0$ **4.53a** or $^0H_5$ **4.53b** conformations (see also Section 2.3.1.c), reacts with methoxide ions to give mainly methyl 6-deoxy-3-O-methyl-α-L-idopyranoside **4.54** (Scheme 4.20), presumably via the $^5H_0$ conformation with *trans*-diaxial opening, while the alternative route is taken by the 4-O-methyl derivative **4.53** (R = Me). With the same reagent, this gives predominantly methyl 6-deoxy-2,4-di-O-methyl-α-L-galactopyranoside **4.55** by *trans*-diaxial opening of the epoxide in the $^0H_5$ conformation.[29] It has been proposed that the $^5H_0$ form is favoured with the 4-hydroxyl compound because a hydrogen bond between the ring oxygen and the hydroxyl group at C-4 can stabilize this otherwise less favoured conformation. In particular, the transition state derived from it would be stabilized in this way.

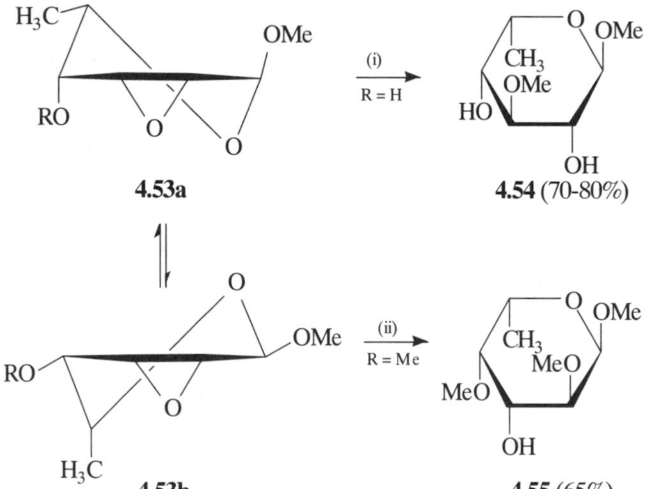

4.53a   **4.54** (70-80%)

4.53b   **4.55** (65%)

(i) MeO$^-$ (R = H); (ii) MeO$^-$ (R = Me)
Scheme 4.20

The opening of epoxides is catalysed by acids, and this fact has been used to devise conditions for opening the epoxide ring of the methyl 2,3-anhydro-β-D-ribopyranoside derivative **4.56** with the poor nucleophile F$^-$. This reaction would not be so successful with normal fluoride salts, but with the acid salt KF.HF it yields the 3-deoxy-3-fluoroxylopyranoside **4.57** (Scheme 4.21).[30]

A neighbouring acetoxy group can have a profound effect upon the rate and stereochemistry of acid-catalysed ring openings of epoxides. Methyl 3,4-anhydro-6-O-trityl-α-D-altropyranoside **4.58** is hydrolysed in 80% acetic acid at 100 °C to a mixture of methyl idoside **4.59** and mannoside **4.60** (Scheme 4.22a), with the former component predominating. (The trityl ether is hydrolytically cleaved under these conditions.) However, on similar treatment, the 2-acetoxy derivative **4.61** is hydrolysed faster than compound **4.58** and the product contains almost exclusively methyl mannoside acetates. The high stereoselectivity and the rate enhancement

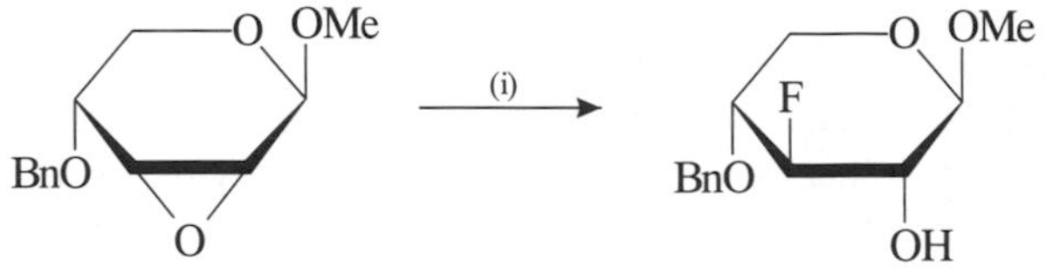

4.56   **4.57** (46%)

(i) KF·HF, (CH$_2$OH)$_2$, reflux, 0.75 h

Scheme 4.21

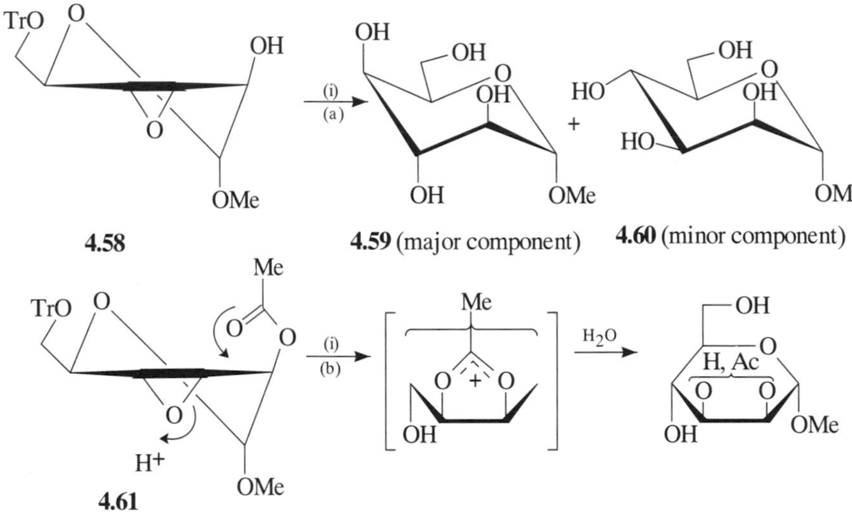

4.58   **4.59** (major component)   **4.60** (minor component)

4.61

(i) AcOH, H$_2$O, 100 °C

Scheme 4.22

in this case are the result of anchimeric assistance from the neighbouring acetoxy group in the opening of the oxirane ring, as illustrated in Scheme 4.22(b).[31]

4.2 DEOXY SUGARS

The deoxy sugars are aldoses or ketoses in which one or more of the hydroxyl groups has been replaced by hydrogen atoms. The accepted nomenclature system for these compounds simply uses an enumerated 'deoxy' prefix and the configurations of the remaining asymmetric carbon atoms are then described by considering them consecutively down the carbon chain. The asymmetric centres need not be contiguous. Deoxyaldoses are more prevalent in natural products than deoxyketoses, those with a terminal deoxy function being the most abundant.[32,33] For example, L-rhamnose (6-deoxy-L-mannose) is found in plant and bacterial polysaccharides, and can be obtained by the hydrolysis of quercetin, a glycoside of certain tree-bark extracts which are used commercially in tanning, whereas L-fucose

(6-deoxy-L-galactose) is found in combined form in animals, plants and micro-organisms. It can be obtained by hydrolysis of the polysaccharide fucoidin, a fucan sulphate found in brown seaweed. 2-Deoxy-D-*erythro*-pentose (2-deoxy-D-ribose) is the exceedingly important sugar component of deoxyribonucleic acids (Section 8.2.2.b). In recent years, mono-, di- and tri-deoxy sugars have been isolated from many antibiotics, bacterial polysaccharides and cardiac glycosides (Chapter 8). See Appendix 5 for the names and structures of deoxy sugars found in microbiological sources and Sections 4.10.1.b.i and 6.4.2 for treatments of the synthesis of 2-deoxyaldose glycosides and oligosaccharides.

4.2.1   SYNTHESES

(a)   Opening of epoxide rings

Epoxide rings can be opened with lithium aluminium hydride, a reagent which can be considered as a source of nucleophilic hydride anions, and the deoxy sugars produced are usually those arising by *trans*-diaxial opening of the rings in accordance with the principles depicted in Schemes 4.18 and 4.19. For example, methyl 2,3-anhydro-4,6-*O*-benzylidene-α-D-allopyranoside **4.47** gives the 2-deoxy-*ribo*-hexoside derivative **4.49** (X = H) in 95% yield,[34] and methyl 2,3-anhydro-4,6-*O*-benzylidene-α-D-mannopyranoside **4.46** gives, in 86% yield, the 3-deoxy-*arabino*-analogue **4.48** (X = H).[35] High stereoselectivity can also be attained with conformationally less rigid compounds, as illustrated by the synthesis of methyl 3-deoxyl-α-D-*xylo*-hexopyranoside **4.63** from the 3,4-anhydrogalactoside **4.62** in Scheme 4.23.[36]

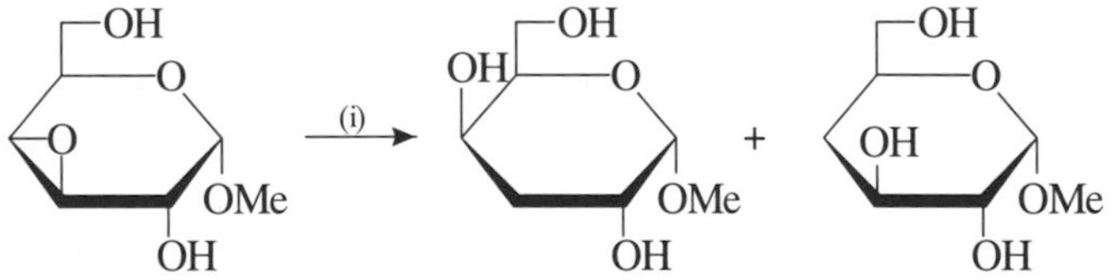

4.62	**4.63**	**4.64**

(i) $\begin{cases} \text{LiAlH}_4, \text{Et}_2\text{O, reflux, 40 h} & (76\%) & (13\%) \\ \text{Ni, H}_2, \text{MeOH, pressure, 105 °C} & (35\%) & (38\%) \end{cases}$

Scheme 4.23

The same products, but usually in very different proportions, are obtained by reductive epoxide ring opening with hydrogen over Raney nickel. The variation arises because the mechanism involved, unlike that operating with the metal hydride reaction, is not bimolecular (S_N2). The *manno*-epoxide **4.46**, for example, again gives predominantly the 3-deoxy-*arabino*-hexoside **4.48** (X = H), whereas the 3-deoxy-*ribo*-hexoside **4.49** (X = H) is the main product of reduction of the *allo*-epoxide **4.47**.[35] See also Scheme 4.23, which illustrates that the epoxygalactoside **4.62** is non-selectively reduced by this reagent to give almost equal proportions of 3-deoxy- and 4-deoxy-glycosides **4.63** and **4.64**, respectively.[36]

(b) Reduction of *C*-halogen derivatives

Bromo- and iodo-deoxy compounds, e.g. **4.65**, **4.66**, can be converted to deoxy sugars e.g. **4.67** by dissolving metal reductions such as with zinc in acetic acid (e.g. Scheme 4.24a) or sodium amalgam in either aqueous diethyl ether or ethanol, by displacements with lithium aluminium hydride, by catalytic reductive cleavage with hydrogen in the presence of Raney nickel (Scheme 4.24b)[37,38] or by free radical initiated hydrogen exchange with tri-*n*-butyltin hydride (see also Section 4.5.2.a).[39] Chloro-sugars can also be reduced by the last two reagents, although often an active form of Raney nickel is required. Scheme 4.25(a) illustrates the efficient conversion of methyl 4,6-dichloro-4,6-dideoxy-α-D-galactopyranoside **4.70** into the methyl 4,6-dideoxyglycoside **4.69**, which occurs with this catalyst in the presence of potassium hydroxide.[38] When triethylamine is used as an acid scavenger, selective hydrogenolysis of the secondary chloride is achieved to give the 4-deoxy-6-chloro-glycoside derivative **4.75** (Scheme 4.25c).[38] Alternatively, only the 6-chloro group is removed when the dimethyl ether of this dichloride (i.e. **4.72**) is heated with lithium aluminium hydride in THF, as illustrated in Scheme 4.25(d) by the formation of 4-chloro-6-deoxyglycoside **4.74**.[38] The specificity for the primary chloride is to be expected with a nucleophilic substitution, whereas the former selectivity indicates a radical reaction, a view borne out by a similar selective cleavage of the secondary carbon–chlorine bond in the 4,6-dichloro-2-acetamido sugar **4.73**, which, on treatment with one molar equivalent of tributyltin hydride in the presence of AIBN as a radical initiator, gives the 6-chloro-4-deoxy compound **4.68** as shown in Scheme 4.25(e).[40] With an excess of this reagent, both primary and secondary carbon–chlorine bonds are reduced, as illustrated in Scheme 4.25b by the conversion of the 4,6-dichloro-2,3-di-*O*-acetylglycoside **4.71** into its 4,6-dideoxy derivative and then into dideoxydiol **4.69** by subsequent deacetylation.[40]

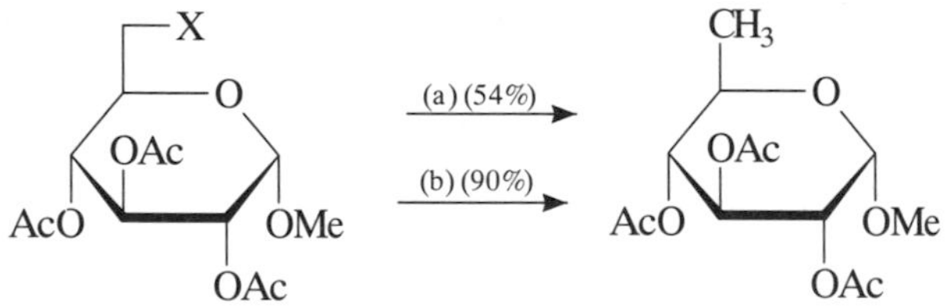

(a) **4.65** X = Br
(b) **4.66** X = I

(a) Zn, AcOH
(b) H₂, Ni, NaOH

Scheme 4.24

Compatibility with ester and amido groups is an advantage tributytin hydride has over lithium aluminium hydride. The mechanism by which the former reduces halides is given in Section 3.3.3.e and accounts for the greater reactivity of the

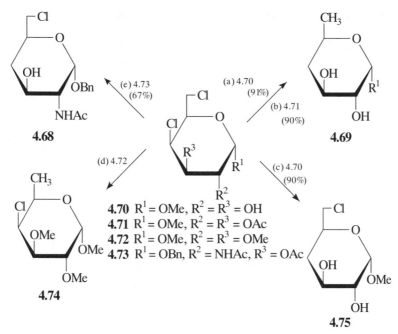

4.68

4.69

4.74

4.70 $R^1 = OMe, R^2 = R^3 = OH$
4.71 $R^1 = OMe, R^2 = R^3 = OAc$
4.72 $R^1 = OMe, R^2 = R^3 = OMe$
4.73 $R^1 = OBn, R^2 = NHAc, R^3 = OAc$

4.75

(a) H_2, Raney NiW4, KOH, EtOH
(b) Bu_3SnH (two equivalents), AIBN, PhMe, N_2, reflux; NaOMe, MeOH
(c) H_2, Raney NiW-4, Et_3N, EtOH
(d) $LiAlH_4$, THF, reflux
(e) Bu_3SnH (one equivalent), AIBN, PhMe, N_2, reflux

Scheme 4.25

chlorine atom at C-4 in compound **4.73**, the secondary radical intermediate formed being more stable than that produced by removal of the chlorine at C-6. For examples of deoxy sugars produced by tributyltin hydride reduction of iodo sugars see Section 4.5.2.a, and for an ingenious synthesis of 2-deoxy sugars by free radical reduction of acetylglycosyl bromides, in which C-2 acetoxy groups participate, see Section 3.3.4.a.

Primary carbon–iodine bonds can be reductively cleaved in methanol solutions containing potassium hydroxide by direct ultraviolet irradiation,[41] and similar results can be obtained when fully protected iodo sugar derivatives are heated at 60 °C under nitrogen for about 20 h in DMF solutions containing *n*-butanethiol and chromium(II) acetate.[42] Some iodo sugars have been reduced by hydrogen over palladium on charcoal in DMF.[43]

(c) Reduction of sulphur-containing derivatives

One of the most efficient ways of preparing deoxy sugars, which is particularly applicable in complex synthetic stategies, is the Barton–McCombie tributyltin hydride radical-induced deoxygenation of suitable thiocarbonyl sugar derivatives (cf. Section 5.3.5.b).[44] Thiobenzoates (RO(C=S)Ph), xanthates (e.g. **4.76**, R = C(=S)SMe, Scheme 4.26) and alkoxythiocarbonylimidazolides (e.g. **4.76**, R = X) were initially employed,[44] but subsequently use has been made of alkoxythiocarbonylpyridones (e.g. **4.76**, R = Y),[45] phenyl thiocarbonates (RO(C=S)OPh)[46] and pentafluorophenyl thiocarbonates (RO(C=S)OC_6F_5).[47]

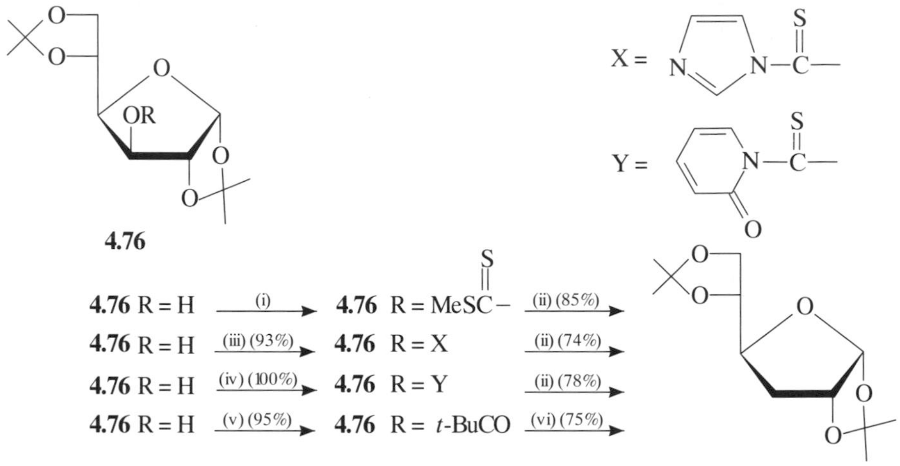
4.76
X =
Y =
4.76 R = H (i) 4.76 R = MeSC(=S)– (ii) (85%)
4.76 R = H (iii) (93%) 4.76 R = X (ii) (74%)
4.76 R = H (iv) (100%) 4.76 R = Y (ii) (78%)
4.76 R = H (v) (95%) 4.76 R = t-BuCO (vi) (75%)
4.77

(i) NaOH, CS_2, MeI, 20 °C; (ii) *n*-Bu_3SnH, AIBN, PhMe, reflux; (iii) thiocarbonyldiimidazole, THF, N_2, reflux; (iv) thiocarbonyldipyridone, PhMe, reflux; (v) *t*-BuCOCl, Py; (vi) HMPIT/H_2O, *hν*

Scheme 4.26

The radical deoxygenation process, the mechanism of which is illustrated in Scheme 4.27, is driven by the affinity of the tin radical for sulphur (stage 1), the formation of a carbonyl group and the concomitant entropy increase (stage 2), and the weakness of the tin–hydrogen bond (stage 3). Excellent yields of deoxy sugars are usually obtained from secondary alcohols, as illustrated in Scheme 4.26 by the conversion of the three different thiocarbonyl derivatives **4.76** (R = C(=S)SMe,[44] X[44,48] and Y[45]) of 1,2:5,6-di-*O*-isopropylidene-*α*-D-glucofuranose into the 3-deoxy sugar derivative **4.77**.

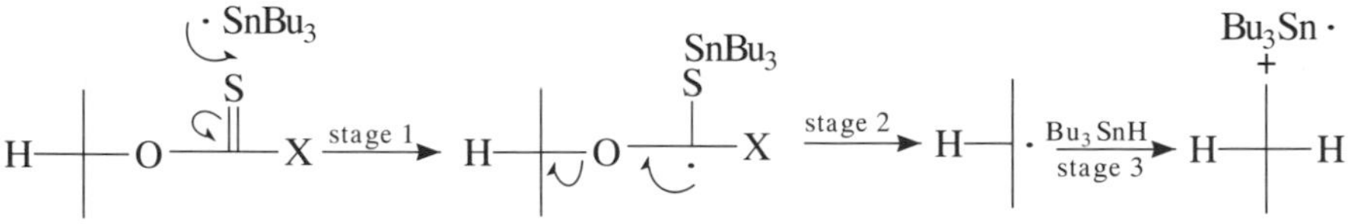
· SnBu3
stage 1
SnBu3
stage 2
Bu3SnH
stage 3
Bu3Sn ·

Scheme 4.27

The deoxygenation of primary thiocarbonyl derivatives is, as expected, less efficient since the intermediate carbon radicals are not formed so readily;[44] in consequence, the parent alcohol is often regenerated under normal reaction conditions. Yields of terminal deoxy sugars are influenced by the thiocarbonyl compound employed, as illustrated in Schemes 4.28(a) and 4.28(**b**) by the different proportions of di-*O*-isopropylidenegalactose **4.82** and its 6-deoxy analogue **4.81** formed from the corresponding thiocarbonylimidazole and thiocarbonylpyridone esters **4.78**[48] and **4.79**,[45] respectively. The efficiency of deoxygenation of primary thiocarbonyl derivatives may be improved by use of higher than normal temperatures.

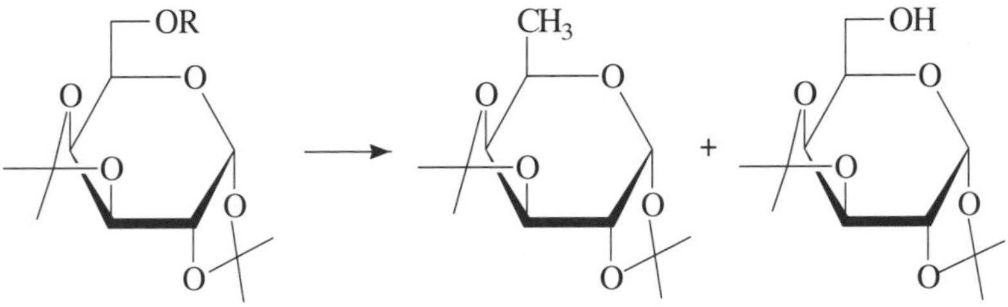

4.78 R = X	$\xrightarrow{\text{(a)(i)}}$	**4.81** (23%)	+	**4.82** (57%)
4.79 R = Y	$\xrightarrow{\text{(b)(i)}}$	**4.81** (62%)	+	**4.82** (10%)
4.80 R = Ts	$\xrightarrow{\text{(c)(ii)}}$	**4.81** (90%)		

(i) *n*-Bu$_3$SnH, AIBN, PhMe, reflux; (ii) NaBH$_4$, DMSO, 110 °C, 6 h (For X and Y see Scheme 4.26)

Scheme 4.28

An extension of this methodology is the monodeoxygenation of cyclic thiocarbonates generated from 1,2-glycols. Regioisomers can be produced in these reactions, but by using the mechanistic principles of radical chemistry, cases in which good selectivity is possible can be identified as with derivatives of terminal diols. For example, 1,2-*O*-isopropylidene-3-*O*-methyl-α-D-glucofuranose 5,6-thiocarbonate **4.83**, which is readily available from the corresponding 5,6-diol by treatment with carbon disulphide and base, is deoxygenated with tributyltin hydride in the presence of an initiator to give the 5-deoxy sugar **4.86** exclusively in good yield (Scheme 4.29).[49] The regiospecificity is accomplished because the first-formed radical **4.84** rearranges as expected to the secondary carbon radical **4.85** rather than the primary one. When the cyclic ester **4.83** is heated in refluxing methyl iodide quantitative conversion to the terminal iodide **4.87** occurs, and since this can be selectively reduced at C-6, **4.83** can act as a source of the 5- or 6-deoxy sugar derivative (Scheme 4.29).[42] With glucopyranoside 2,3-thiocarbonates deoxygenation occurs, but regiospecificity is lost and the 2- and 3-deoxy sugar derivatives are formed in the ratio of 1:2.

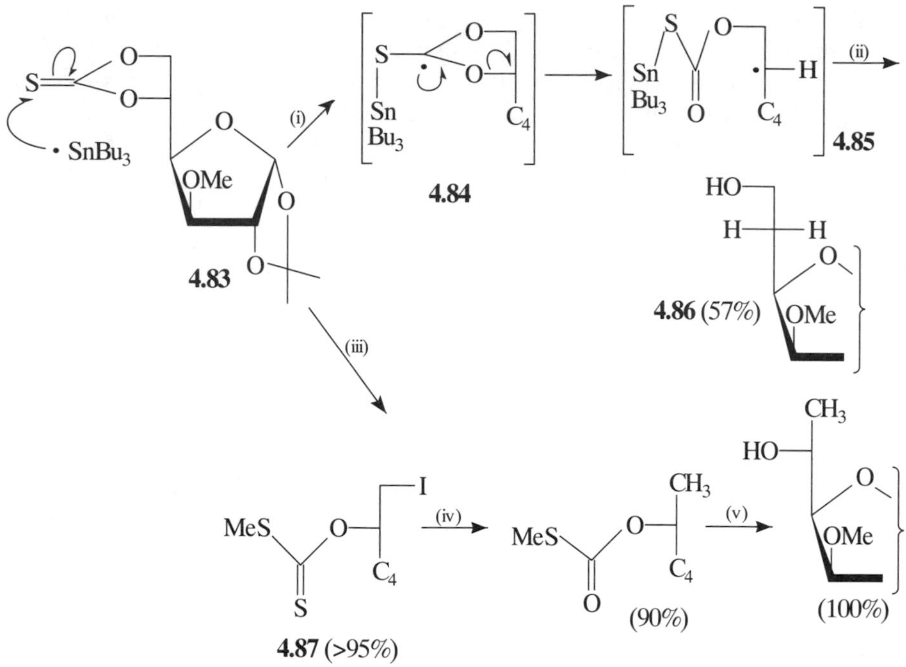

Scheme 4.29

(i) *n*-Bu$_3$SnH, AIBN, PhMe, reflux; (ii) alkaline hydrolysis; (iii) MeI, reflux; (iv) BuSH, Cr(OAc)$_2$, DMF; (v) MeONa, MeOH

Deoxy sugars may also be prepared by the cleavage of carbon–sulphur bonds in thio sugar derivatives (Section 4.4.2).

(d)   Reduction of esters

Reduction of thioesters is discussed in Section 4.2.1.c above.

(i)   *Carboxylate esters*[50]

Solutions of acyl esters, particularly acetates or pivaloates (*t*-BuCO$_2$R, but also triflates; see below), in HMPT containing 5% water may be deoxygenated by irradiation with ultraviolet light. Primary and particularly secondary esters of pyranoid and furanoid rings are efficiently cleaved (see, for example, the conversion of **4.76** (R = *t*-BuCO) into **4.77** illustrated in Scheme 4.26), as are tertiary esters, but with loss of stereochemical integrity. Multiacylated derivatives yield polydeoxy sugars directly; thus, irradiation of methyl 2,3,6-tri-*O*-pivaloyl-α-D-glucopyranoside gives a 20% yield of methyl 2,3,6-trideoxy-α-D-*erythro*-hexopyranoside (methyl α-D-amicetoside, a derivative of amicetose which is a constituent of amicetin) in one step accompanied by 10% of the 2,3-dideoxy 6-pivaloate.[50] The less easy deoxygenation of the primary ester is in keeping with the radical mechanism proposed for this reaction.

(ii)   Sulphonate esters

Terminal deoxy sugars are commonly obtained by lithium aluminium hydride reductions of primary sulphonates. Until recently, the more convenient reagent sodium borohydride proved unsatisfactory for this task, but it has been found to reduce sugar sulphonates in polar non-protic solvents such as DMSO, and this is now one of the methods of choice for preparing large quantities of simple terminal deoxy sugars, as shown in Scheme 4.28(c) with the formation of di-*O*-isopropylidene-D-fucose **4.81** from the corresponding galactose 6-tosylate **4.80** (R = Ts).[51] Lithium aluminium hydride is unsuitable for reducing secondary sulphonates since these are commonly cleaved with sulphur–oxygen bond fission to give the alcohols rather than deoxy derivatives. The sodium borohydride method of deoxygenation has proved satisfactory with some secondary tosylates.[52]

Trifluoromethanesulphonates (triflates) of pyranose and furanose derivatives give deoxy products on reduction with sodium in liquid ammonia, and since the reaction involves free radicals it is most suitably applied to secondary esters. For example, the 3-triflate of methyl 2-acetamido-4,6-*O*-cyclohexylidene-2-deoxy-α-D-glucopyranoside gives the analogously protected 3-deoxy sugar derivative in 80% yield when treated with this reagent for 1 h at −50 °C. The same product is obtained in similar yield when the sugar triflate is irradiated with ultraviolet light in aqueous HMPIT.[53]

(e)   Deoxygenation of carbonyl compounds

The Clemmensen and Wolff–Kishner reactions and the reductive desulphurization of dithioacetals are the three classical methods for reducing ketones to the corresponding methylene compounds. The first of these has not been used with ketones derived from sugars because the strongly acidic conditions required are not compatible with the protecting groups usually present, whereas the other two methods have found limited application in the synthesis of deoxy sugars. The Wolff–Kishner reaction successfully deoxygenates the 3-ulose derivative **4.88** to give compound **4.89** as outlined in Scheme 4.30, but with more complex compounds, such as hexopyranosidulose derivatives (Section 4.9.1), competing base-catalysed elimination reactions occur.[54]

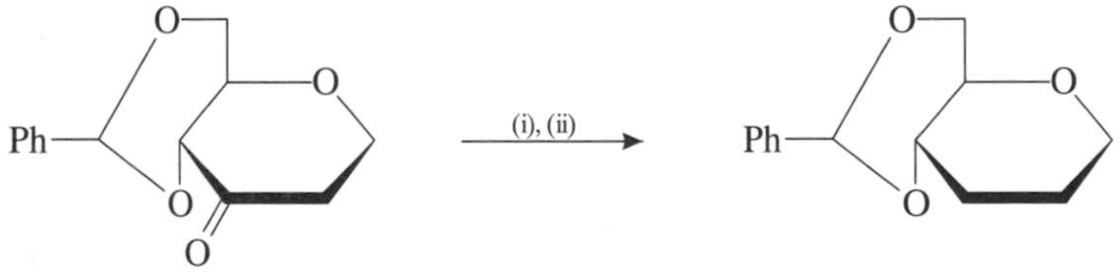

4.88   **4.89**

(i) NH$_2$NH$_2$, O(CH$_2$CH$_2$OH)$_2$, 100 °C, 1 h; (ii) KOH, 100 °C, 2 h
Scheme 4.30

The pyranos-4-ulose **4.90** has been converted into the 4-deoxy sugar **4.92** via the easily formed cyclic polysulphide **4.91**, as illustrated in Scheme 4.31. However,

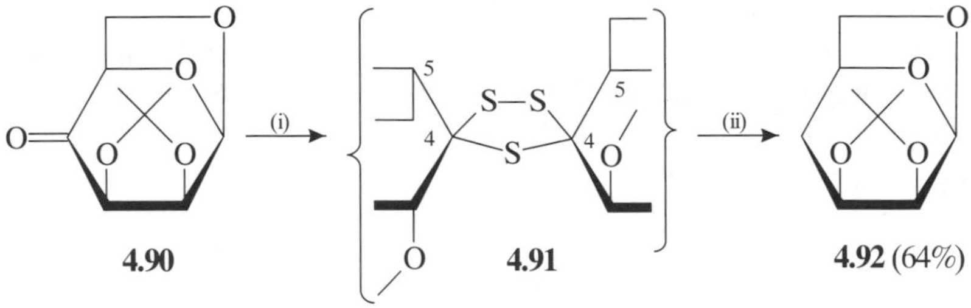

4.90   **4.91**   **4.92** (64%)

(i) P_4S_{10}, Py, 40 °C, 2 h; (ii) H_2, Ni

Scheme 4.31

competing side reactions occur during the thiolation reaction with many other pyranosulose derivatives.[55]

(f)   Additions to unsaturated compounds

Sugars possessing double bonds can serve as useful precursors of deoxy sugar derivatives, the most obvious reduction procedure being catalytic addition of hydrogen (Section 4.10.2.b), and when unoxygenated double bonds are involved, efficient access to dideoxy analogues is provided. Applications of this procedure to unsaturated compounds formed by aldose chain-lengthening reactions are illustrated in Scheme 3.63, and with such examples stereochemical problems do not arise. However, when secondary asymmetric centres are formed, mixtures of stereoisomers arise, the ratios being dependent upon access of the catalyst to the double bonds. While this can depend largely on the stereochemistry at the allylic centres,[56] the stereochemistry at more distant sites can, on occasions, be the dominant determining factor. For example, hydrogenation of the 6-deoxyhex-5-enopyranoside anomers **4.94** and **4.95** gives in the former case the L-*lyxo*-product **4.93** stereospecifically, whereas **4.95** gives a 3:1 mixture of the L-*lyxo*-glycoside and D-*ribo*-glycoside **4.96** and **4.97**, as illustrated in Scheme 4.32.[57]

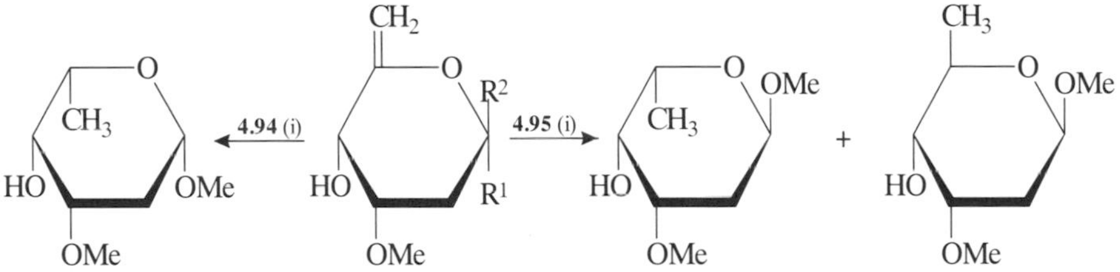

4.93 (99%)   **4.94** R^1 = OMe, R^2 = H   **4.96** (75%)   **4.97** (25%)

4.95 R^1 = H , R^2 = OMe

(i) MeOH, H_2 (1 atm), Pd/BaSO$_4$ (10%), 3 h

Scheme 4.32

The acid-catalysed addition of water or alcohols to glycals is a long established synthetic route to 2-deoxyaldoses and their glycosides, and is illustrated in Section 4.10.1.b.i. Mechanistically analogous additions of monohydroxy sugar derivatives to C-1 of glycals, activated by electrophilic species such as I^+ and $PhSe^+$, have been adapted as highly selective routes to α- or β-linked 2-deoxy disaccharides, the electrophiles being reductively removed from C-2 of the adducts (Sections 6.4.2 and 4.10.1.b.i).[58]

(g) Synthesis from non-carbohydrate allylic alcohols

As an example of the synthesis of deoxy sugars from such alcohols, D-olivose (2,6-dideoxy-D-*arabino*-hexose **4.104**) has been prepared from an aliphatic dienol by application of the Sharpless epoxidation–kinetic resolution procedure (Section 2.5.4). Racemic (*E*)-hepta-1,5-dien-4-ol (rotamers **4.98a** and **4.98b**) (Scheme 4.33) is used as the starting material, and on treatment with the

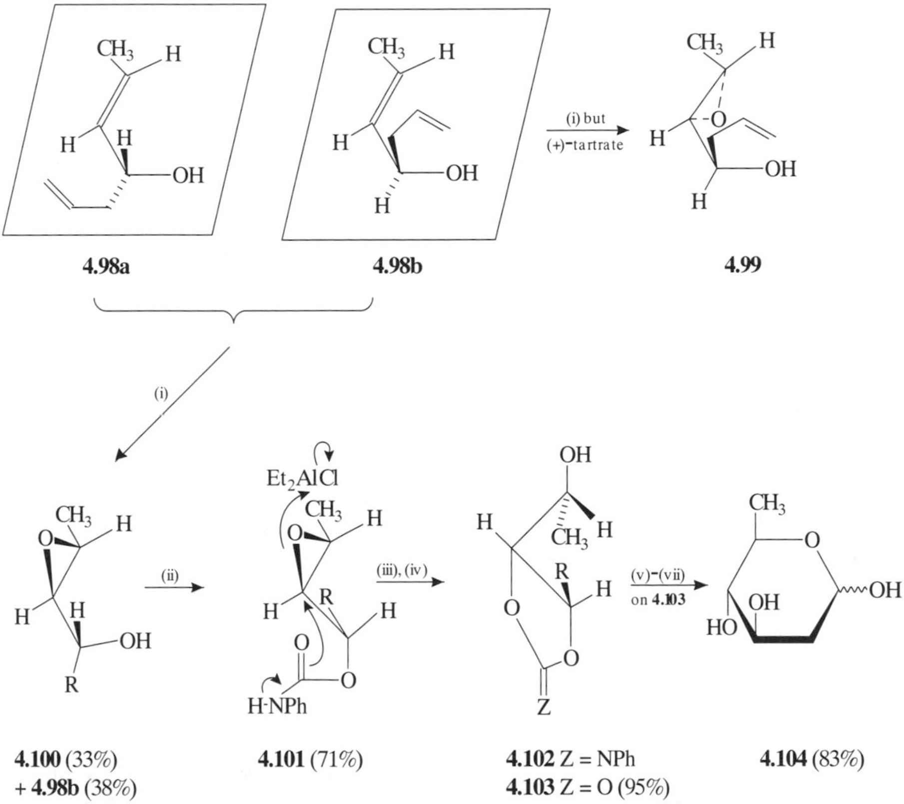

4.98a
4.98b
(i) but
(+)-tartrate
4.99
(i)
(ii)
Et2AlCl
(iii), (iv)
H-NPh
(v)-(vii)
on 4.103
4.100 (33%)
+ 4.98b (38%)
4.101 (71%)
4.102 Z = NPh
4.103 Z = O (95%)
4.104 (83%)

$R = -CH_2CH{=}CH_2$

(i) *t*-BuO_2H (0.4 equivalent), $Ti(OPr\text{-}i)_4$ (one equivalent), D-(−)-diisopropyl tartrate (1.5 equivalents), CH_2Cl_2, −20 °C, 18 h; (ii) PhNCO, Py, CH_2Cl_2, 23 °C, 30 h; (iii) Et_2AlCl, Et_2O, −20 °C, 45 min; (iv) H_2O, H_2SO_4, 23 °C, 2 h; (v) NaOMe, MeOH, 23 °C, 24 h; (vi) O_3, MeOH, −20 °C; (vii) Me_2S

Scheme 4.33

epoxidizing reagents in the presence of D-(−)-diisopropyl tartrate, the (+)-epoxide **4.100** is produced enantiomerically pure[59]. This arises by selective epoxidation of **4.98a, 4.98b** being left unchanged (and now free from its enantiomer) because epoxidation of its upper face is impeded by the allyl substituent and the underface is inaccessible to this enantiomer of the chiral titanium complex which is the oxidizing reagent. The chromatographically separated, resolved alkene **4.98b** may now be expoxidized by use of L-(+)-diisopropyl tartrate to give the (−)-adduct **4.99**, which otherwise may be obtained directly from the racemic starting material by application of the (+)-tartrate. It serves as a precurser of L-olivose.

Stereo- and regio-chemically controlled nucleophilic opening of epoxides in conformationally non-restricted molecules is not as readily achieved as with those that are parts of rigid six-membered ring systems (Section 4.1.3). 2,3-Epoxyalcohols and their ether derivatives often show a preference for nucleophilic attack at the epoxide ring carbon atom furthest from the hydroxyl or alkoxy group, but this selectivity is lost with highly substituted epoxyalcohols. However, a method using α-acyloxyepoxides, which depends upon neighbouring group participation, gives predictable products regiospecifically. The carbanilino (PhNHCO) substituent present in phenyl urethanes is the most successful participating group and this is used in the olivose synthesis[59]. Thus, treatment of the phenyl urethane **4.101,** derived from the (+)-epoxyalcohol **4.100,** with diethylaluminium chloride in ether at −20 °C gives after acid hydrolysis the D-*arabino*-carbonate **4.103**, presumably *via* the iminocarbonate intermediate **4.102** (Scheme 4.33). Ozonolysis of the trihydroxyalkene obtained after saponification of the cyclic carbonate gives the 2,6-dideoxy sugar D-olivose **4.104** in good yield. Application of the Sharpless epoxidation to the synthesis of aldoses is illustrated in Section 2.5.4.

(h) Other methods

Metasaccharinic acids (3-deoxyaldonic acids) are readily available from free sugars (Section 3.1.7.b.i), and can be converted to 2-deoxyaldoses by application of, for example, the Ruff degradation (Section 3.1.6.a.iv). In this way, 2-deoxy-D-*erythro*-pentose is available from D-glucose, and although the yield in the reaction is low, the required materials are inexpensive. Conversely, a method for preparing 3-deoxyaldoses applies a chain-extending reaction, such as the Kiliani synthesis, to 2-deoxyaldoses (Section 3.1.4.d).

4.2.2 REACTIONS

The chemical reactions of the deoxy sugars are, as would be expected, essentially the same as those of normal sugars but their reactions are often much faster. The modifications in the chemistry of this class of compound usually result from the deoxy group having a special relationship to the anomeric centre. For example, 4-deoxyaldoses cannot form furanose rings, and 5-deoxyaldoses cannot form pyranose rings. They therefore are forced to adopt the alternative ring forms and their appropriate reactions, such as glycoside formation, are modified. Furthermore,

2-deoxyaldoses yield hydrazones in the usual way, but they cannot form phenylosazones when treated with an excess of phenylhydrazine (Section 3.1.3.c).

A less obvious consequence of the presence of the deoxy function in 2-deoxy-aldoses is its effect on glycoside formation and hydrolysis. These reactions both occur with marked ease, as is illustrated by the fact that methyl 2-deoxy-α-D-*arabino*-hexopyranoside is hydrolysed in aqueous acid about 2000 times faster than is methyl α-D-glucopyranoside (Section 3.1.1.b.i). This feature makes it virtually impossible to remove acid-labile protecting groups such as benzylidene and isopropylidene acetals by selective hydrolysis from 2-deoxyaldosides — a device often employed in syntheses involving the corresponding 2-hydroxyl compounds. For methyl glycosides this problem can be overcome by acid-catalysed methanolytic removal of the acetals, which leaves the methyl glycosidic bonds intact. However, anomerization is likely to occur under these conditions. These examples of enhanced reactivity induced by removal of hydroxyl groups from C-2 are typical of many reactions of deoxy sugars in general, as discussed in Section 3.1.1.b.i.

4.3   AMINO SUGARS

Amino sugars are aldoses or ketoses which have a hydroxyl group replaced by an amino group at any position other than the anomeric carbon.[60] (Glycosylamines

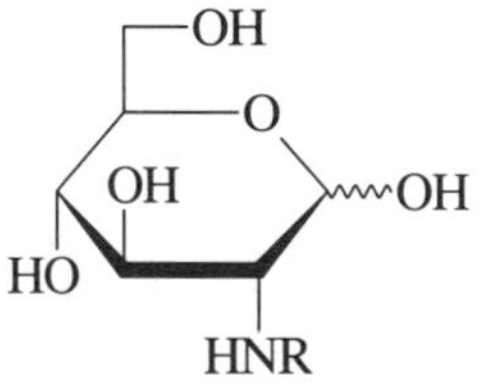

4.105

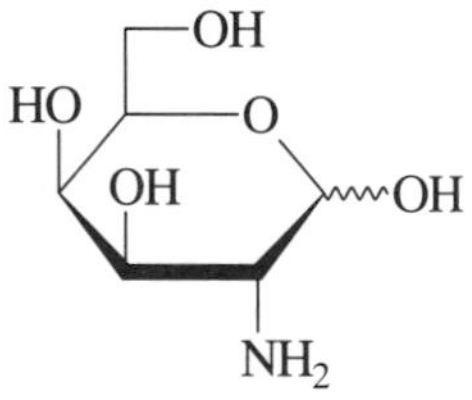

4.106

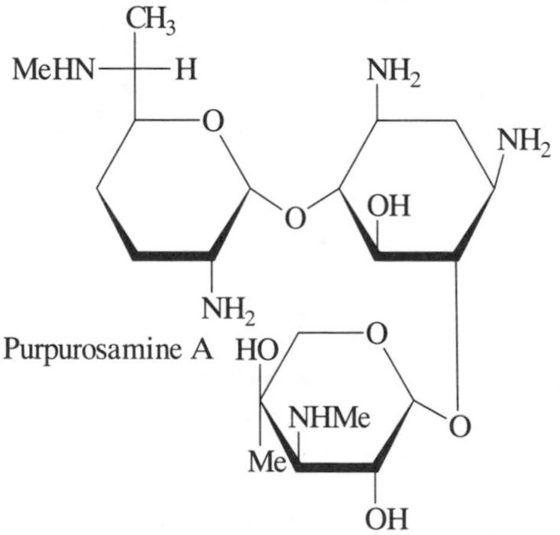

Purpurosamine A

Garosamine

4.107 Gentamicin C$_1$

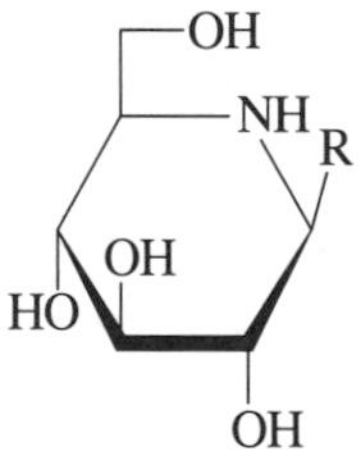

4.108 R = OH
4.109 R = SO$_3$H

are discussed in Section 3.1.3.) These substances are named after the sugar from which they are derived by use of the enumerated prefix 'aminodeoxy'.

2-Amino-2-deoxy-D-glucose (D-glucosamine or chitosamine **4.105**, R = H) is abundant in nature, appearing in particular in the polysaccharide chitin as its *N*-acetyl derivative **4.105** (R = Ac). Several other 2-amino-2-deoxy sugars also occur, 2-amino-2-deoxy-D-galactose **4.106** being one of the more common since it is a constituent monosaccharide unit of dermatan and chondroitin sulphate, polysaccharides found in mammalian tissue and cartilage (Section 8.1.4.a). Because these polymers are difficult to purify and the sugar is difficult to isolate, galactosamine derivatives, in contrast to their *gluco*-epimers, are expensive. Amino sugars also occur naturally with amino groups at positions 3,4,5 or 6. Nonulosaminic acids (neuraminic acids, Sections 3.1.4.c and 4.9.2), for example, are 5-amino-5-deoxynonose derivatives usually found in combined form in mucolipids or mucopolysaccharides isolated from animal sources. Several clinically important antibiotics, e.g. streptomycin **8.60**, erythromycin **8.59** and gentamicin C **4.107**, have amino sugars as constituents and at least one amino sugar, nojirimycin **4.108**, itself shows antibiotic activity.

4.3.1   SYNTHESES

(a)   Displacement reactions

This method of synthesis, which is one of the most widely used, involves displacements of leaving groups in monosaccharide derivatives with reagents containing nucleophilic nitrogen atoms such as azide ions, hydrazine or ammonia. Compounds with free amino groups can then, where necessary, be generated from the products. Azides, for example, can be converted into amines by a wide range of reducing agents which include lithium aluminium hydride, sodium borohydride, hydrogen over metal catalysts, phosphines, low valence metal ions, hydrogen sulphide, thiols and tin(II)–sulphur complexes.[61]

Formation of terminal amino compounds by sequential use of the displacement method and reduction is relatively easy, as illustrated by the preparation of methyl 6-amino-6-deoxy-α-D-glucopyranoside **4.110** in Scheme 4.34.[62] By use of DMF or HMPIT as solvent, displacements at certain secondary positions can

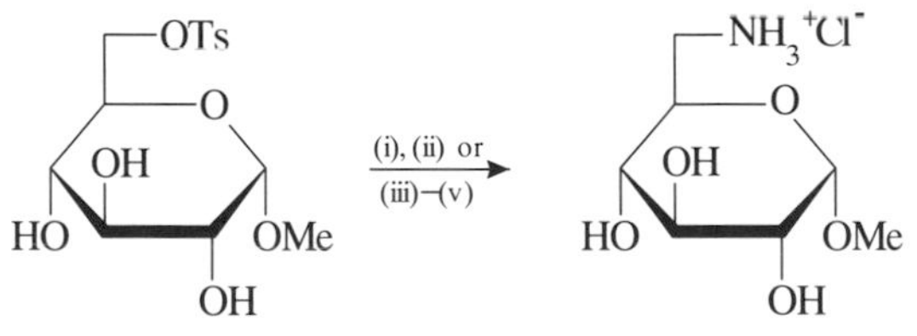

4.110

(i) NH₃, MeOH, 120 °C, high pressure (64% yield); (ii) HCl; (iii) NaN₃, Me₂CO, H₂O, reflux, 72 h; (iv) H₂, Pd/C, EtOH, 60 °C; (v) HCl (48% overall)

Scheme 4.34

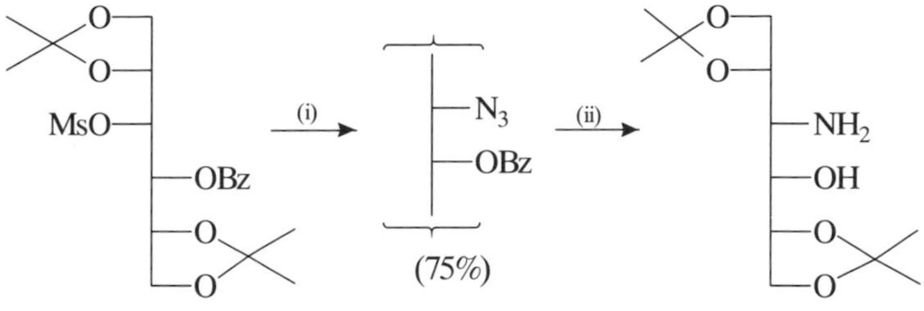

4.111   **4.112** (70%)

(i) NaN$_3$, HMPIT, 80 °C 19 h; (ii) LiAlH$_4$, Et$_2$O
Scheme 4.35

be achieved. For example, with the acyclic derivative 4-O-benzoyl-1,2:5,6-di-O-isopropylidene-3-O-mesyl-D-mannitol **4.111**, the sulphonate group undergoes displacement by azide ion with Walden inversion (Scheme 4.35) to give, after reduction, the 3-amino-3-deoxyaltritol derivative **4.112**.[63] Similar results are obtained by displacing sulphonate ester residues from pyranoid rings.[64] In the example given in Scheme 4.36, the conversion of the 6-deoxyallopyranoside 3-triflate **4.113** into the 3-azido-*gluco*-compound **4.114** is facilitated by the addition of a crown ether to the solution of sodium azide in DMF, thereby activating the nucleophile by complexing with the sodium ions and inducing an efficient displacement at a low temperature.[65]

Polyamino sugars have been prepared by displacement reactions as illustrated in Scheme 4.37 for the synthesis of benzyl 2,3,4,6-tetraaminotetradeoxy-α-D-glucopyranoside hydrochloride **4.117**. A double displacement with azide ions on the 4,6-dimesylate of benzyl 2,3-diacetamido-2,3-dideoxy-α-D-galactopyranoside **4.115** gives the diazide **4.116** which can be selectively reduced with hydrogen over palladized carbon, subsequently deacetylated and neutralized with hydrochloric acid to give the product **4.117** in good yield.[66]

Related displacements have been carried out on furanoid 2- and 3-sulphonates as illustrated in Schemes 4.9, 4.10 and 4.11, which offer routes to 2- and 3-amino furanose derivatives.

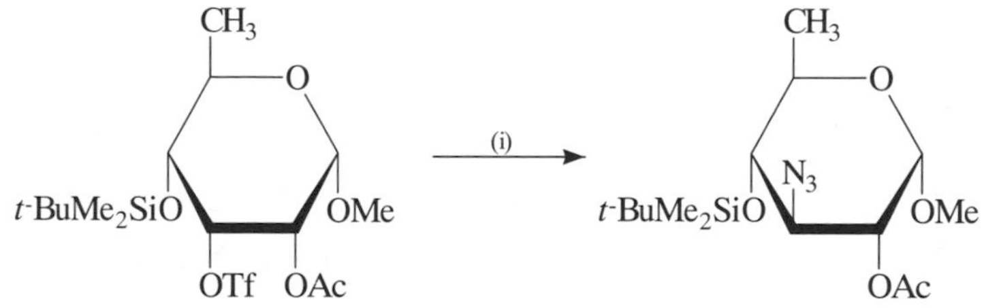

4.113   **4.114** (83%)

(i) NaN$_3$, 15-crown-5, DMF, 25 °C, 0.5 h
Scheme 4.36

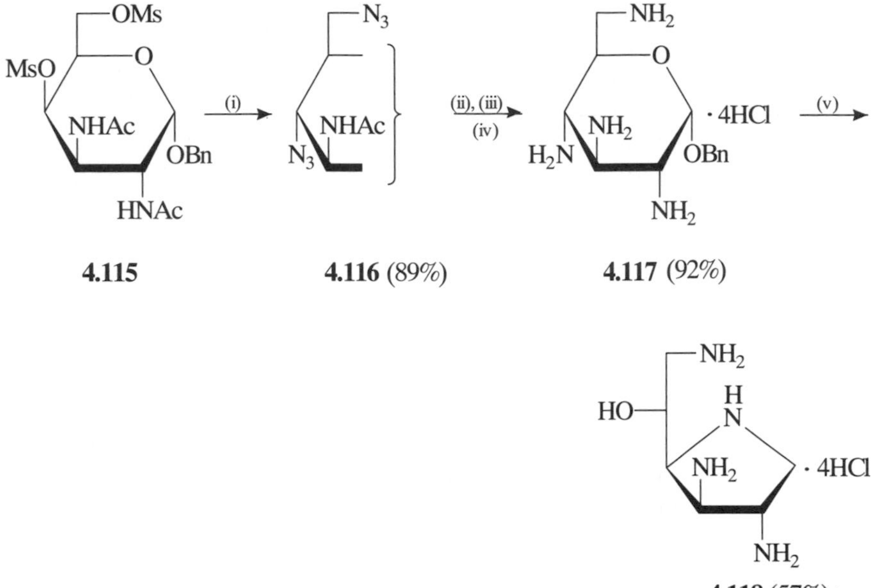

4.115   **4.116** (89%)   **4.117** (92%)

4.118 (57%)+
free sugar (12%)

(i) NaN₃, DMSO, 100 °C, 10 h; (ii) H₂, Pd/C, MeOH, 1 h; (iii) NaOH, H₂O, 20 h, 100 °C; (iv) HCl, H₂O; (v) H₂, Pd, H₂O, HCl, 20 h

Scheme 4.37

Intramolecular delivery of a nitrogen nucleophile by participation of neighbouring carbamoyl groups in sulphonate displacements has been utilized in synthetic routes to amino sugars.[67] Thus, as shown in outline in Scheme 4.38, carbamoyl groups displace sulphonates on treatment with base to give oxazolidinones, e.g. **4.119**, that can be subsequently hydrolytically ring opened and deprotected to give *cis-vic*-aminoalcohols, e.g. **4.120**. Related strategies have been employed with epoxides having trichloroacetimidates on adjacent oxygen atoms.[68]

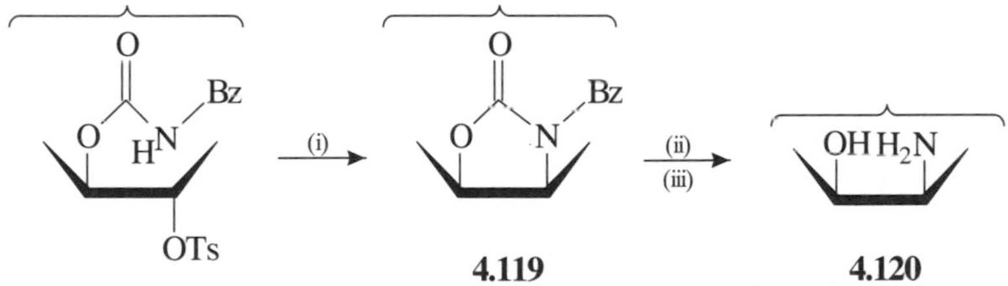

4.119   **4.120**

(i) NaH, THF, 0 °C; (ii) LiOH, THF, H₂O, 20 °C, 2 h; (iii) NaOH, H₂O, reflux 5 h

Scheme 4.38

2-Azido-2-deoxypyranosyl fluorides, e.g. **4.121**, which are of value in the syntheses of amino sugar containing oligosaccharides, are available in good yield from glycosyl azides unprotected at O-2 (e.g. **4.122**) by DAST-induced displacements involving neighbouring group participation and migration of the azide group. See Section 4.5.1.e.iii and Scheme 4.88 for related examples.[69]

4.121   **4.122**

(b)   Opening of epoxide rings

Nucleophilic opening of epoxides is very similar to the displacement method in that the same nucleophiles and similar reaction conditions are commonly used. Methyl 2,3-anhydro-4,6-O-benzylidene-α-D-allopyranoside **4.47** upon heating under reflux for 22 h with sodium azide in aqueous ethanol in the presence of a small amount of ammonium chloride gives, in 85% yield, the 2-azido-2-deoxy-D-altropyranoside derivative **4.49** (X = N$_3$) in accordance with the general Scheme 4.18.[13] Alternative solvents and other nitrogen nucleophiles have also been used.[70]

(c)   Cyclizations of dialdehydes with nitromethane

Nitromethane, in basic solution, condenses with two carbonyl groups of the readily available dialdehydes obtained by periodate oxidation of appropriate cyclic poly-hydroxy precursors (Section 5.5) to give recyclized *aci*-nitro salt isomers, which, on treatment with acid, give deoxynitro compounds.[71] The nitro groups can then be reduced to amino functions and this method has been developed into a useful synthesis of amino sugars. Thus, the dialdehyde **4.123**, formed by oxidation of methyl α-D-glucopyranoside, on treatment in methanol with nitromethane in the presence of sodium methoxide produces a mixture of four isomeric sodium salts of methyl 3-*aci*-nitro-3-deoxy-α-D-hexopyranosides which differ in their stereo-chemistries at C-2 and C4 (Scheme 4.39). Compounds **4.128** and **4.129** are the major components, and immediate protonation gives a mixture of nitro compounds **4.130** and **4.131** (X = NO$_2$) which can be catalytically reduced to give the corre-sponding 3-amino-3-deoxyhexopyranosides **4.130** and **4.131** (X = NH$_2$). In this way, the first of these, 3-amino-3-deoxymannoside **4.130**, (X = NH$_2$), can be produced free of the other isomers in 30% yield simply by fractional crystalliza-tion of its hydrochloride. The *gluco*-compound **4.131**, (X = NH$_2$) can then be isolated as its tetraacetate in similar amounts, although this requires chromato-graphic separation to free it from the other products and represents only a partial isolation of this isomer.

If the above-mentioned protonation step is delayed and the *aci*-nitro compounds, after isolation, are allowed to isomerize in aqueous solutions, the concentrations

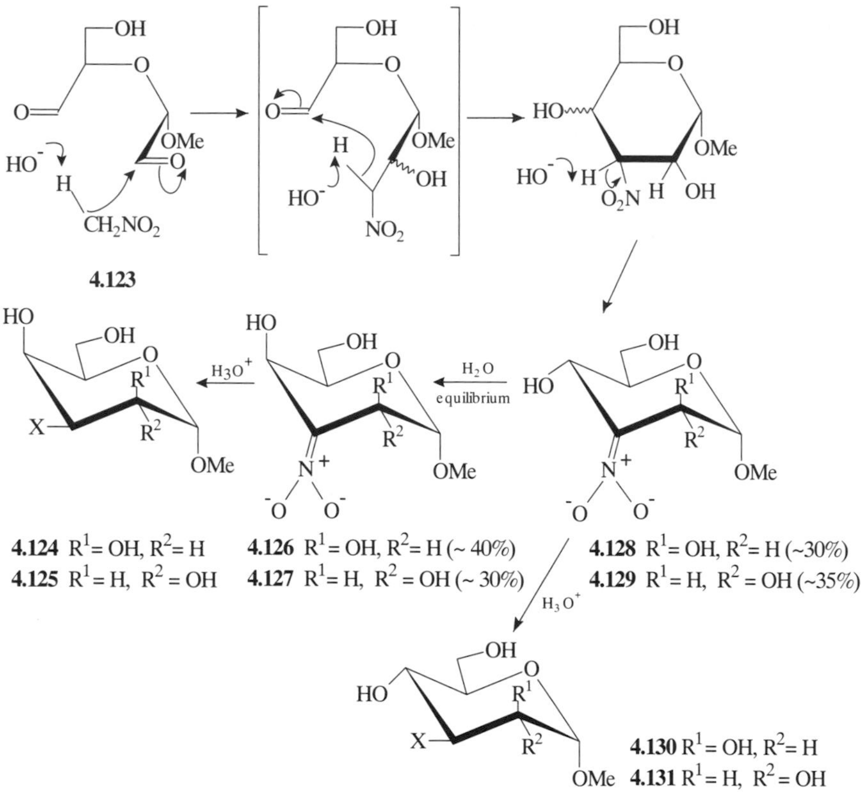

4.123

4.124 R^1= OH, R^2= H   **4.126** R^1= OH, R^2= H (~ 40%)   **4.128** R^1= OH, R^2= H (~30%)
4.125 R^1= H, R^2 = OH   **4.127** R^1= H, R^2 = OH (~ 30%)   **4.129** R^1= H, R^2 = OH (~35%)

4.130 R^1= OH, R^2= H
4.131 R^1= H, R^2= OH

Scheme 4.39

of the major salts **4.128** and **4.129** decrease, while the D-*lyxo*-isomer and D-*xylo*-isomer **4.126** and **4.127** accumulate as shown in Scheme 4.39. In this way the scope of this reaction can be extended, since protonation and reduction of this new mixture produces methyl 3-amino-3-deoxy-α-D-talopyranoside **4.124** (X = NH$_2$) and methyl 3-amino-3-deoxy-α-D-galactopyranoside **4.125** (X = NH$_2$) which can be isolated in 40% and 30% yields, respectively, as their hydrochlorides.

The reaction may be extended to the synthesis of more complex amino-sugar derivatives, since whereas the dialdehyde produced from 1,6-anhydro-D-glucose yields the *ido*-compound **4.132** mixed with its C-2 and C-4 isomers when treated with nitromethane in the usual way, compound **4.133** is formed when the condensation is carried out in the presence of benzylamine. The formation of **4.133** probably arises by attack of nitromethane on initial benzylamine imines to give nitrodiamines, subsequent reduction of which gives triaminotrideoxy analogues.[72]

Application of the reaction sequence to dialdehydes obtained from nucleosides offers a route to 3-amino-3-deoxyhexopyranosyl-purines and -pyrimidines,

4.132 R = OH   **4.134** R = H
4.133 R = NHBn   **4.135** R = Me

as illustrated by the conversion of uridine into the 3-amino-3-deoxyglucose derivative **4.134**.[73]

The scope of this reaction has been further enhanced by the condensation of dialdehydes with homologues of nitromethane, which provides a simple route for the simultaneous introduction of a nitro function and a carbon branch at C-3 in a pyranoid ring. Thus, treatment of the dialdehyde from uridine with nitroethane under the usual conditions gives four nitro adducts isolated in 28, 5, 5 and 1% yields. The major product, which has the *gluco*-configuration, is readily transformed upon reduction into the branched-chain amino sugar **4.135** (see Section 4.8).[74] The value of these nitro derivatives in synthesis is further augmented by the formation of nitroalkenes, as outlined in Scheme 4.40 for the conversion of methyl 3-deoxy-3-nitro-β-D-glucopyranoside **4.136** into the α, β-unsaturated nitro sugar derivative **4.137**. A variety of nucleophiles readily add to this compound under basic conditions leading to a wide range of 3-nitro-3-deoxy sugar derivatives, e.g. **4.138**, which bear newly introduced groups at position 2 and upon reduction yield the corresponding 3-amino-3-deoxy analogues. Direct catalytic reduction of compound **4.137** yields the 2,3-dideoxy-3-nitro derivative **4.139** which can be further reduced to the corresponding amino sugar.

(d)   Cyclizations of dialdehydes with phenylhydrazine

The dialdehyde **4.140**, obtained by periodate oxidation of methyl 4,6-*O*-benzylidene-α-D-glucopyranoside, condenses with two equivalents of *N*-methyl-*N*-phenylhydrazine to give a bis(methylphenylhydrazone) as expected. If, however, the dialdehyde is treated with phenylhydrazine under similar conditions, a cyclization occurs, probably via the monophenylhydrazone **4.141**, as illustrated in Scheme 4.41, to give methyl 4,6-*O*-benzylidene-3-deoxy-3-phenylazo-α-D-glucopyranoside **4.142**, which can be catalytically hydrogenolysed, in good yield, without alteration of the stereochemistry at C-3, to give the 3-amino-3-deoxy-α-D-glucopyranoside **4.143**.[75] A 2,3-unsaturated phenylhydrazono analogue of **4.137** can be obtained from compound **4.142**, and by additions such as those illustrated

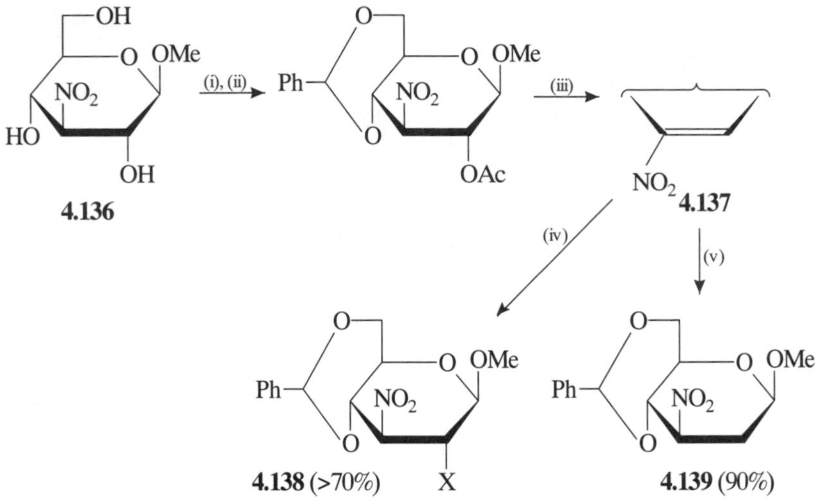

4.136 ... **4.137**

4.138 (>70%) X   **4.139** (90%)

X = MeO, EtO, PhS, R$_2$N, EtOCOCH$_2$ (EtOCO)$_2$CH, RCHNO$_2$
(i) PhCHO, H$^+$; (ii) Ac$_2$O,Py; (iii) NaHCO$_3$, PhH, reflux, four days; (iv) KX;
(v) H$_2$, Pd

Scheme 4.40

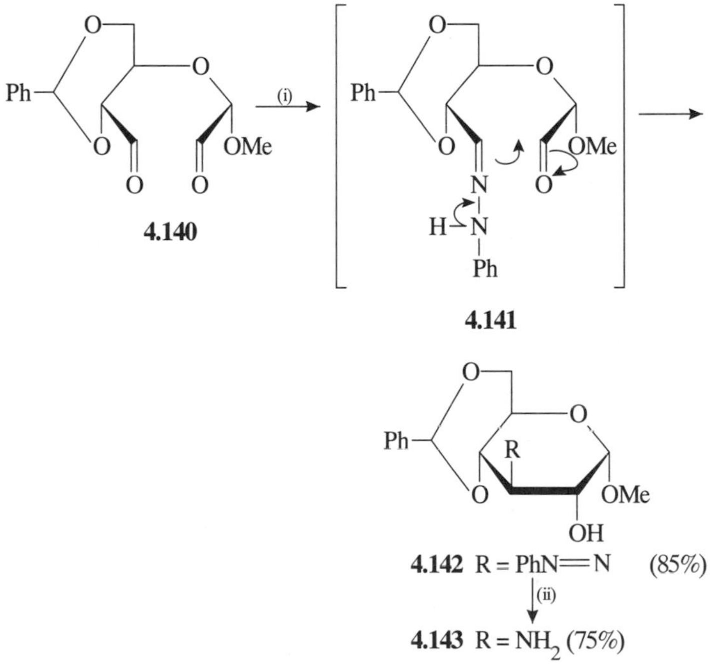

4.140

4.141

4.142 R = PhN=N   (85%)

4.143 R = NH$_2$ (75%)

(i) H$_2$NNHPh, Py, H$_2$O; (ii) H$_2$, Ni, EtOH, 80 °C, 8 h

Scheme 4.41

in Scheme 4.40 saturated amino-sugar derivatives with a range of specifically introduced groups at C-2 can be prepared.

(e)   Addition of nitrogenous reagents to double bonds

(i)   Chloronitroxylation

The electron-rich double bonds of glycals (Section 4.10.1.b.i) are highly susceptible to attack by electrophiles and the addition of electrophilic nitrosyl chloride has been used as a means of synthesizing 2-amino-2-deoxy sugar derivatives, as illustrated in Scheme 4.42 for tri-O-acetyl-D-glucal **4.144**. The nitrosyl chloride adduct is isolated as dimer **4.145** which can be converted, as illustrated, into α-glycosidulose oximes (e.g. **4.146**) in good yield in highly stereoselective reactions with alcohols.[76] Alternatively, the adduct **4.145** gives, in high yield, oximated glycosylulose acetates related to **4.146** on acetolysis.[76] Reduction of these products provides a means of obtaining amino sugar derivatives (see Section 4.3.1.f).

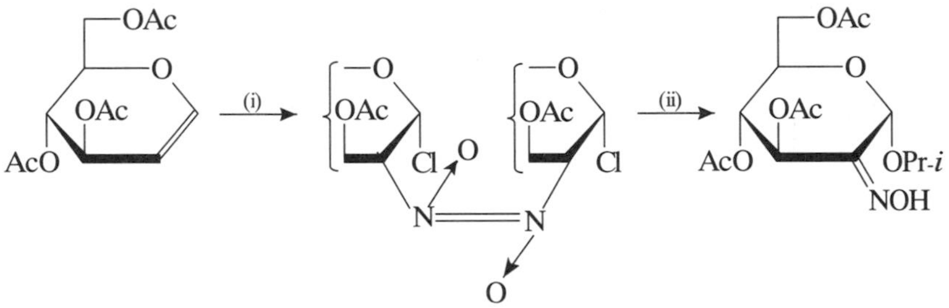

(i) NOCl, CH_2Cl_2, $-80\,°C$; (ii) i-PrOH, CH_2Cl_2

Scheme 4.42

(ii)   Azidonitration

Although full details of the mechanism are unclear, azidonitration of double bonds by sodium azide and cerium(IV) ammonium nitrate is thought to be initiated by azido radical attack. Consequently, 2-azido-2-deoxyglycosyl nitrates are obtained regiospecifically from all glycals, as illustrated in Scheme 4.43 by the conversion of tri-O-acetylgalactal **4.147** into the 2-azido-2-deoxygalactose derivatives **4.148** and a small amount of the talo-isomer **4.149**, a transformation which also indicates that a high degree of stereoselectivity is possible with these reactions.[77] Reasonable stereoselectivities have been reported for additions to the acetates of lactal[78] and xylal,[79] but with glucal triacetate the stereoselectivity is lower, a small excess of 2-azidomannose being formed over the gluco-isomer.[80]

Simple glycosides, glycosyl acetates and, most usefully, glycosyl halides can be readily obtained from the nitrates, as indicated in Scheme 4.43 by the synthesis of the 2-azido-2-deoxygalactosyl chloride **4.150**. Thus, this method makes available on a multigram scale 2-azido-2-deoxyglycosyl donors that are of value in oligosaccharide synthesis in which α-linked 2-amino sugars are required. The

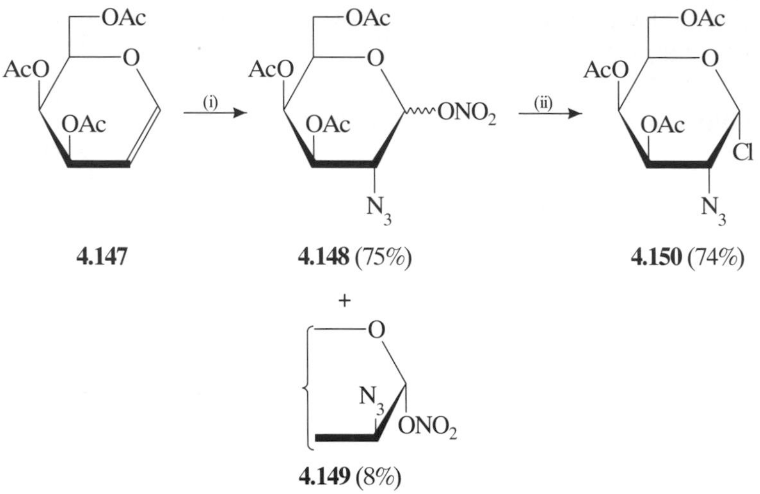

4.147   **4.148** (75%)   **4.150** (74%)

+

4.149 (8%)

(i) (NH$_4$)$_2$Ce(NO$_3$)$_6$, NaN$_3$, MeCN, −20 °C, 5 h; (ii) **4.148**, Et$_4$NCl, 22 °C, 5 h

Scheme 4.43

2-azido-2-deoxygalactose derivative is particularly useful since galactosamine compounds, otherwise difficult to prepare, can be obtained by its use.

(iii)   Oxyamination

The *p*-toluenesulphonylimidoosmium reagent **4.155**, which may be generated from Chloramine T and osmium tetroxide, brings about *cis*-addition of hydroxyl and

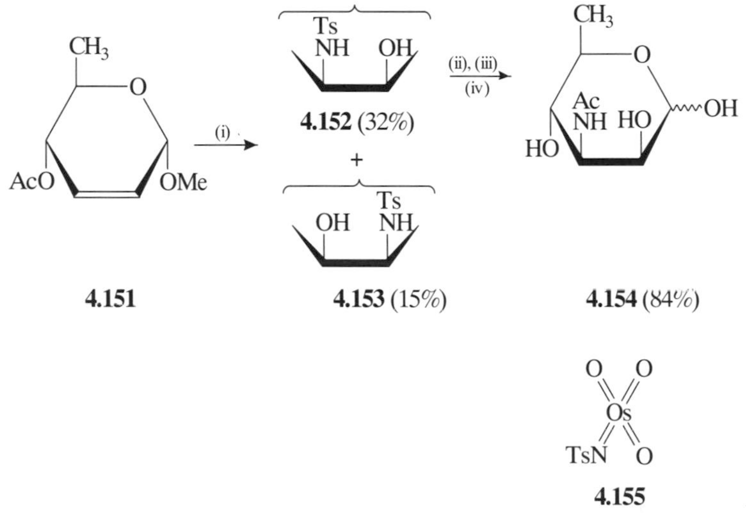

4.151   **4.152** (32%)   **4.154** (84%)

+

4.153 (15%)

4.155

(i) Chloramine T·3H$_2$O, OsO$_4$ (0.01 equivalent), AgNO$_3$, *t*-BuOH, C$_6$H$_{14}$, 40 °C, 20 h; (ii) Na, NH$_3$, 15 min; (iii) Ac$_2$O, Et$_3$N; (iv)H$_3$O$^+$

Scheme 4.44

N-tosylamino groups to the least-hindered face of double bonds, as illustrated in Scheme 4.44. With the hex-2-enopyranoside **4.151**, the addition occurs exclusively to the top face of the double bond to give mannose derivatives, presumably *via* cyclic intermediates which break down to give either the 3-tosylamino derivative **4.152** or the 2-tosylamino isomer **4.153**. With an excess of Chloramine T the osmium tetroxide is effectively recovered after the addition reaction and therefore only catalytic quantities are required.[81] In contrast to the additions to glycals described in Section 4.3.1.e.i and 4.3.1.e.ii above, the regioselectivity with 2- and 3-enopyranosides is often not high,[82] and in the illustrated example the products **4.152** and **4.153** are formed in a 2:1 ratio.[81] The N-tosyl group may be removed with sodium in liquid ammonia and by such treatment, and subsequent N-acetylation and hydrolysis, the major product **4.152** can be converted into N-acetylmycosamine **4.154**.

(iv) Amination of nitroalkenes

Nucleophilic reagents do not normally add to isolated double bonds, but if the bonds have electron-withdrawing neighbouring groups with which they can conjugatively interact, Michael-like addition reactions become possible. An example of additions of amines to nitroalkenes has already been given in Scheme 4.40. Nitromethane aldose adducts (Section 3.1.4.e) can also be used in this type of reaction. Addition of ammonia to such a system, e.g. **4.156**, provides a route to 2-amino-2-deoxy sugars, as illustrated for the conversion of D-xylose into 2-acetamido-2-deoxy-D-glucose **4.157** in Scheme 4.45.[83]

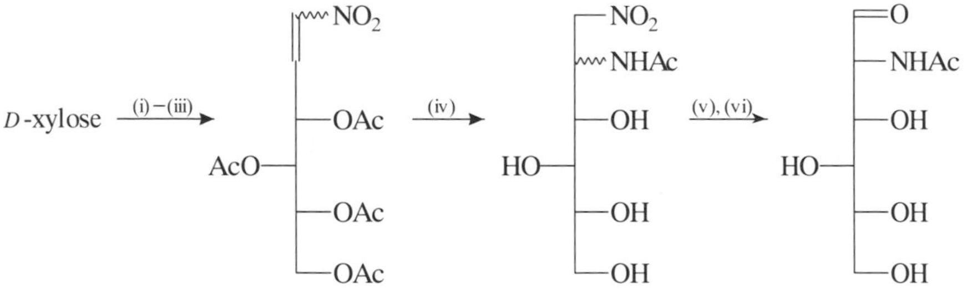

4.156 (26%)   **4.157** (35%)

(i) $MeNO_2$, MeO^-; (ii) Ac_2O, H_2SO_4; (iii) $NaHCO_3$, PhH, reflux; (iv) NH_3, MeOH, 22 °C, 20 h; (v) $Ba(OH)_2$, H_2O; (vi) H_2SO_4, 22 °C, 18 h

Scheme 4.45

(v) Neighbouring group mediated iodoamination

Regio- and stereo-controlled iodocyclizations of allylic trichloroacetimidates offer a route to *cis*-hydroxyamino sugars from hexenopyranosides, e.g. **4.158**, as illustrated by the synthesis of a daunosamine derivative in Scheme 4.46. Thus, **4.159** is converted regio- and stereo-specifically by the action of iodonium dicollidine perchlorate (IDCP) into the trichloromethyloxazoline **4.160**, from which all the carbon–halogen bonds are removed by radical tin hydride hydrogenolyses. Subsequent acid hydrolysis of the reduced product gives N-acetyldaunosamine methyl glycoside **4.161**.[84]

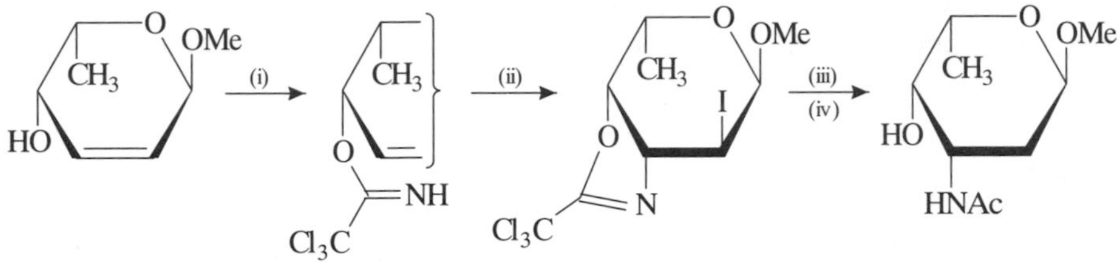

4.158   **4.159**   **4.160**   **4.161** (75%)

(i) NaH, CCl$_3$CN, CH$_2$Cl$_2$; (ii) IDCP, MeCN, three days; (iii) PhH, AIBN, Bu$_3$SnH, reflux, 4 h; (iv) Py, H$_2$O, TsOH, 100 °C, 1.5 h

Scheme 4.46

(f) Reduction of ulose oximes

A standard method for preparing amines is the reduction of oximes or their O-alkyl or -acyl derivatives by hydrogenation or the use of reagents such as lithium aluminimum hydride, borane, or a zinc–copper couple in acetic acid. Since the reductions can be highly stereoselective, this has become a very useful method for preparing amino sugar derivatives from oximes derived from protected ulose derivatives.

That the stereochemical outcome can be dependent upon the reducing condition is illustrated in Schemes 4.47 and 4.48. Thus, borane reduction of acylated oxime **4.162** gives mainly compound **4.163** with its 3-acetamido group equatorial (i.e. axial attack by hydrogen at C-3, Scheme 4.47a), whereas catalytic reduction of its unacetylated analogue **4.164** gives the isomeric 3-amino sugar **4.165** with an axial

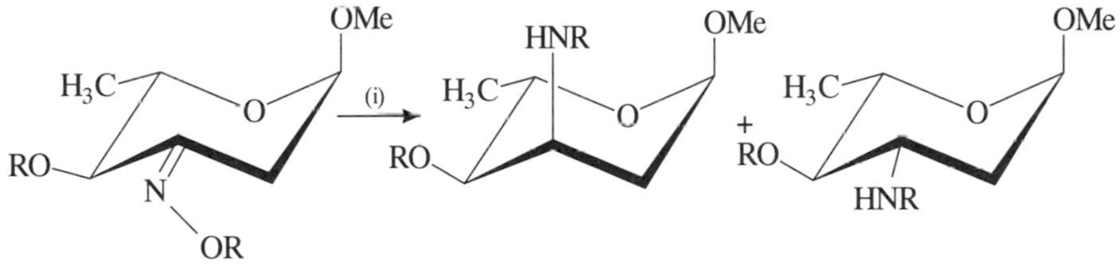

| 4.162 R = Ac | $\xrightarrow{\text{(a) (i)}}$ | R = Ac (3%) | 4.163 R = Ac (90%) |
| 4.164 R = H | $\xrightarrow{\text{(b) (ii)}}$ | 4.165 R = H (79%) | R = H (14%) |

(i) borane, THF; (ii) H$_2$, Ni, MeOH

Scheme 4.47

amino group as the preponderant product (Scheme 4.47b).[85] This is believed to arise because the oximino portion of molecule **4.164** is absorbed onto the hydrogen-rich catalyst surface at its lower (equatorial) face, facilitating preferential delivery of an equatorial hydrogen atom to C-3. The disposition of substituents adjacent to the oximino group also influences the stereochemistry of reduction. Thus, on catalytic hydrogenation, the β-methyl glycoside oxime **4.167** gives exclusively the β-glycoside of mannosamine **4.166** with an axial amino group at C-2, in line with the reasoning used above to explain the catalytic reduction of compound **4.164.** On the other hand, the α-anomer **4.168**, which has an axial methoxy group at C-1 able to deter the approach of the catalyst to the lower (equatorial) face of the C-2 oximino group, gives preponderantly the glucosamine derivative **4.169**, albeit in modest yield, as recorded in Scheme 4.48.[86]

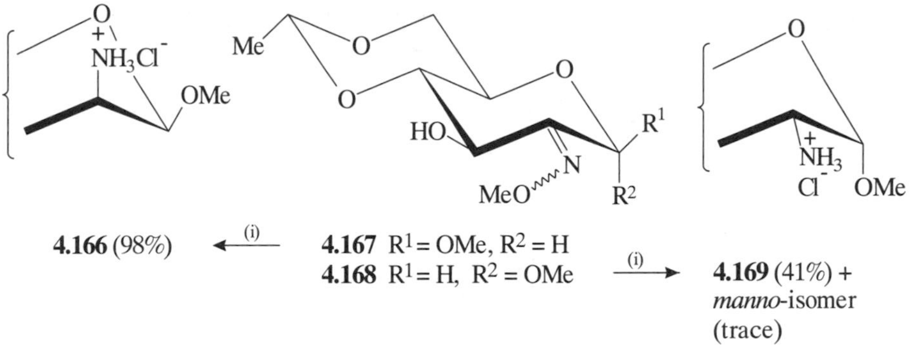

4.166 (98%)   ←—(i)—   **4.167** R¹ = OMe, R² = H
                             4.168 R¹ = H, R² = OMe   —(i)→   **4.169** (41%) +
                                                                     manno-isomer
                                                                     (trace)

(i) H₂, Pd/C, MeOH, HCl

Scheme 4.48

The reductions of the 2-oximino-α-glycoside **4.168** with a range of reagents have been compared: with lithium aluminium hydride it yields the 2-amino-2-deoxymannose derivative as the major isomer, whereas with borane in THF or hydrogen over palladium under acid conditions it gives the glucosamine derivative with high selectivity.[87]

(g) Reactions of aldoses with arylamines and hydrogen cyanide

This method represents a modification of the Kiliani synthesis of aldoses. The free sugars are condensed with aromatic amines to give imines which are not isolated, but are treated directly with anhydrous hydrogen cyanide. Hemihydrogenation of the resulting aminonitriles accompanied by hydrogenolytic cleavage of the aromatic residues, followed by hydrolysis, gives, in high yield, mixtures of epimeric 2-amino-2-deoxyaldoses in which one isomer usually preponderates (Scheme 4.49). Several 2-amino-2-deoxy-pentoses and -hexoses have been prepared by this route and the crude yields are ordinarily very high, but considerable losses are often incurred during the fractional crystallizations involved in isolating the pure isomers.[60]

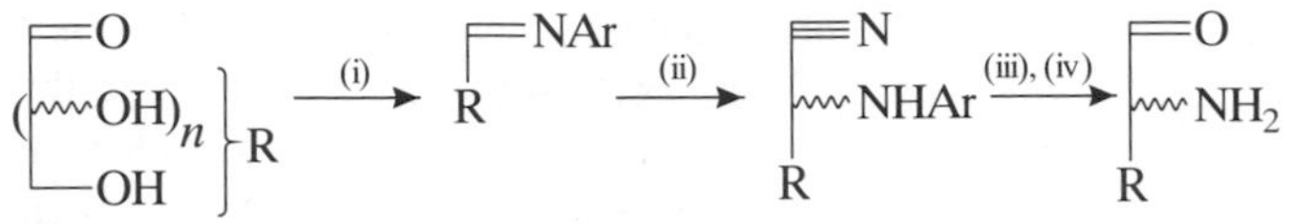

$$\left(\substack{=O \\ \text{\tiny$\sim\sim$}OH \\ -OH}\right)_n R \xrightarrow{\text{(i)}} \substack{=NAr \\ R} \xrightarrow{\text{(ii)}} \substack{\equiv N \\ \text{\tiny$\sim\sim$}NHAr \\ R} \xrightarrow{\text{(iii), (iv)}} \substack{=O \\ \text{\tiny$\sim\sim$}NH_2 \\ R}$$

(i) $ArNH_2$; (ii) HCN; (iii) H_2, Pd; (iv) H_2O

Scheme 4.49

(h) Epimerizations of aminodeoxy sugars

Solvolyses of compounds containing acylated amino residues *vicinal* and *trans* to good leaving groups lead to displacement reactions in which the acylamino groups participate. This provides a method for inverting the stereochemistry at carbon atoms adjacent to those bearing amino groups, as, for example, in the conversion of methyl 3-acetamido-4,6-O-benzylidene-3-deoxy-2-O-mesyl-α-D-idopyranoside to the 2-hydroxyl analogue with the *gulo*-configuration (Scheme 4.14).

Conventional epimerization, specific for sites adjacent to the anomeric centre and usually induced under basic conditions (Section 3.1.7.a), has been applied to 2-acetamido-2-deoxyaldoses but, since equilibration processes are involved, separations are usually required for the isolation of the epimeric products.

4.3.2 REACTIONS

Derivatives of aminodeoxyaldoses protected at C-1 by, for example, glycosidation undergo the reactions expected for alcohols and amines.

(a) Acylation, de-N-acylation and acyl migration

Peracetylation is straightforward, but preferential substitution of the hydroxyl or amino groups can be achieved if the conditions are carefully chosen. N-Acetylation occurs selectively when free amino groups are liberated from their salts in ethanolic or aqueous solutions containing acetic anhydride (Scheme 4.50a). On the other hand, selective O-acetylation results with acyl halides in acid media in which the amino groups are protonated and therefore devoid of their nucleophilic character (Scheme 4.50b). An alternative route to these types of derivative involves selective deacetylation of peracetylated amino sugars. Treatment of such compounds with methanolic ammonia removes the O-acetyl groups, as does a solution of 1 mol l^{-1} aqueous sodium hydroxide at 0 °C.[60] Methoxide in methanol reacts similarly leaving the acetamido derivatives (Scheme 4.51a) which usually only cleave under strongly acidic or basic conditions.[60] Conversely, triethyloxonium

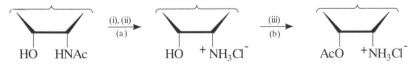

(i) basic resin; (ii) Ac_2O, H_2O; (iii) AcCl, AcOH

Scheme 4.50

tetrafluoroborate removes *N*-acetyl groups selectively giving *O*-acetylated amino products via ethyl acetamidium tetrafluoroborates, e.g. **4.170**, which can often be isolated (Scheme 4.51b).[88] De-*N*-acetylation has also been achieved by the action of calcium in liquid ammonia.[89]

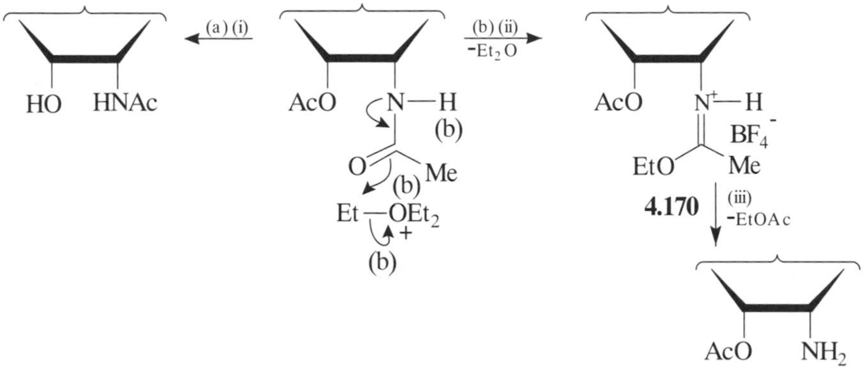

(i) MeOH, NH$_3$, or MeO$^-$, MeOH; (ii) Et$_3$OBF$_4$; (iii) H$_2$O

Scheme 4.51

Partially acylated aminodeoxy sugars can undergo acyl migrations which cause difficulties in structure determinations. For example, in basic media, migration occurs from oxygen to nitrogen and *N*-acetyl derivatives predominate, whereas in acid solutions the esters are favoured because the amino groups are trapped as alkylammonium ions (Scheme 4.52).

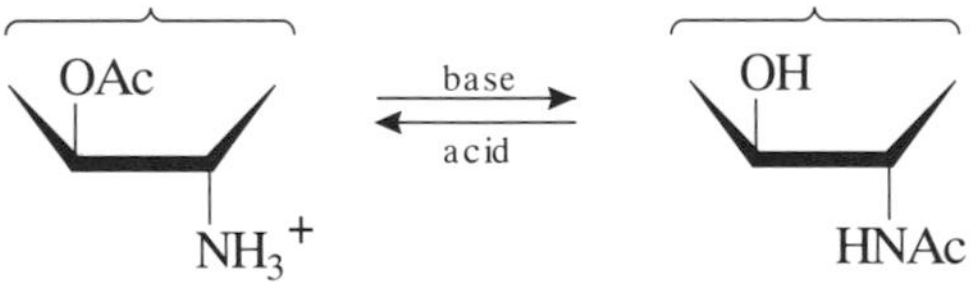

Scheme 4.52

2-*N*-Phthaloyl-protected glycosamine derivatives are frequently used in the syntheses of oligosaccharides which contain 2-amino-2-deoxy sugars (see Sections 3.3.2.a.ii, 4.3.2.f.iii and 6.6.2). They are prepared from amino sugar derivatives by reaction with phthalic anhydride at pH 9, as illustrated in Scheme 4.53(a).[78,90] Conversely, amino groups are usually liberated from phthalimido sugars by hydrazinolysis (Scheme 4.53b),[90] but unfortunately this reaction is not general.[91] Garegg and Ogawa have found that efficient deprotection can be achieved by reductive hydrolysis, as illustrated in Scheme 4.53(c).[91,92]

(b)   Alkylation

N-Methylation can be conveniently accomplished by selective *N*-formylation followed by reduction of the formamido groups (Scheme 4.54a). One of the

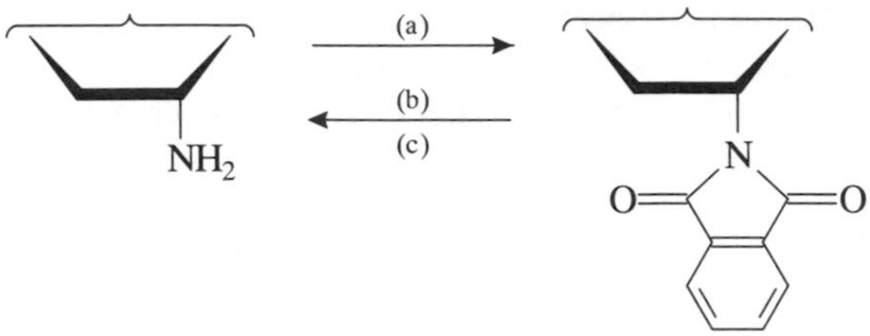

(a) phthalic anhydride, EtOH, H_2O, Na_2CO_3, pH 9
(b) NH_2NH_2, EtOH, H_2O, reflux
(c) $NaBH_4$, i-PrOH, H_2O followed by AcOH

Scheme 4.53

most satisfactory methods for selective N,N-dimethylation is hydrogenation over palladium catalysts in the presence of aqueous formaldehyde (Scheme 4.54b). Reductive amination methods have been used to prepare alkylamino sugar derivatives by treating the corresponding amino sugar and ketones or aldehydes together with sodium cyanoborohydride.[93] Direct selective O-methylation of hydroxylated amino sugar derivatives has not been found possible, and therefore methyl ethers are prepared by selective N-acetylation followed by methylation of the free hydroxyl groups and hydrolysis of the acetamido residues.

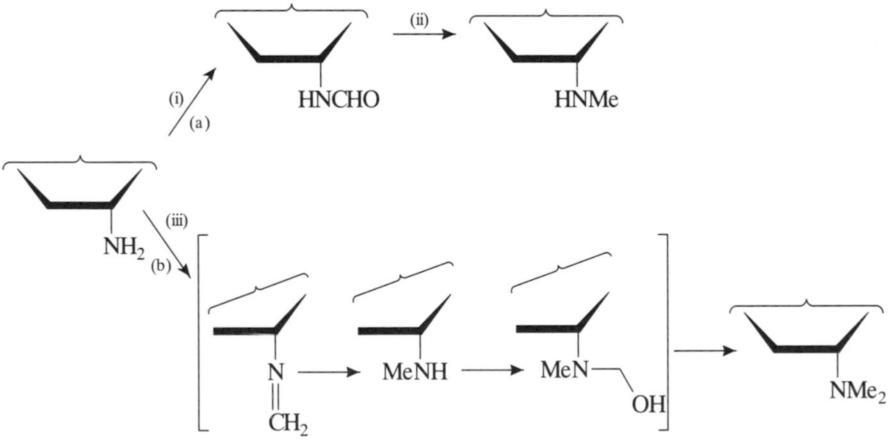

(i) HCO_2Et, Et_3N; (ii) $LiAlH_4$; (iii) 40% HCHO, H_2, Pd

Scheme 4.54

(c) Nitrous acid deaminations[94]

When aliphatic amines are treated with nitrous acid they form unstable diazonium salts which rapidly break down to give nitrogen and carbocations. The latter can then either react directly with the solvent or with available intramolecular nucleophiles, or can undergo Wagner–Meerwein rearrangements. Sugar derivatives

with free amino groups behave similarly. 1,2,3,4-Tetra-*O*-acetyl-6-amino-6-deoxy-
D-glucopyranose, upon diazotization, reacts directly with water to give the 6-
hydroxyl tetraacetate, whereas the diazo derivatives of methyl 2-amino-2-deoxy-[95]
and 3-amino-3-deoxy-glycopyranosides yield less accessible carbocations and give
ring-contracted furanoid products by 1,2-shifts of pyranoid ring bonds (see, for
example, the transformation **4.367** → **4.366**). These rearrangements are, respectively, identical to those in Schemes 4.109 and 4.110 which accompany displacements of the corresponding 2- and 3-sulphonate esters. Nitrous acid deaminations
have been used to cleave heteropolysaccharides at linkages involving 2-amino-2-
deoxy sugars as illustrated in Scheme 8.8 with heparin.

(d)   Reactions with acids and bases

All aminodeoxy sugars are basic, and many of them form crystalline salts with
mineral acids. Their basic strength varies with the location of the amino groups on
the carbon chain, but they are all weaker than cyclohexylamine, probably because
of the inductive effects of the oxygen-bearing substituents on the adjacent carbon
atoms. When the anomeric positions are unprotected, aldohexoses with amino
groups at positions 2,3 or 6 are stable in acid, but in basic solution the 2- and
3-aminodeoxy sugars readily lose their amino groups. The reaction of the latter
compounds occurs particularly quickly, by a mechanism similar to that proposed
for the base-catalysed degradation of the aldoses given in Section 3.1.7.c, and
the aldose–ketose isomerizations that occur under those conditions give, from the
former compounds, 2-imines which deaminate on hydrolysis. Aldoses possessing
amino groups at C-4 or C-5 behave differently, being more stable in base than in
acid (see Section 4.3.2.g).

(e)   Displacement reactions with participation from
nitrogen — epimines

Displacements of sulphonate esters from acylamido sugar sulphonate derivatives
can occur with neighbouring group participation from the nitrogen atom of
trans-vic-acylamido residues to give epimines (aziridines), which are nitrogenous
analogues of epoxides. For example, as shown in Scheme 4.55(a), the bulky reagent
sodium isopropoxide is a suitable catalyst for producing the 2,3-dideoxy-2,3-
epiminoallopyranoside **4.173**, (R = H) from the 3-*O*-tosyl glucoside **4.171** without
any direct displacement of the tosylate by alkoxide, although debenzoylation
does also take place. Sodium hydride brings about a similar conversion but
debenzoylation does not occur with this reagent and **4.173** (R = Bz) is formed in
95% yield.[96] Another route to epimines utilizes azido compounds, as demonstrated
by the conversion of **4.172** into **4.173** (R = H) (Scheme 4.55b).[97] A further
example is given in Scheme 4.12, which illustrates that acylamino groups can
act as ambident nucleophiles in displacements of this kind.

   Although structurally similar to the epoxides, epimines do not react with bases or
with lithium aluminium hydride, and with acids their behaviour is also different (cf.

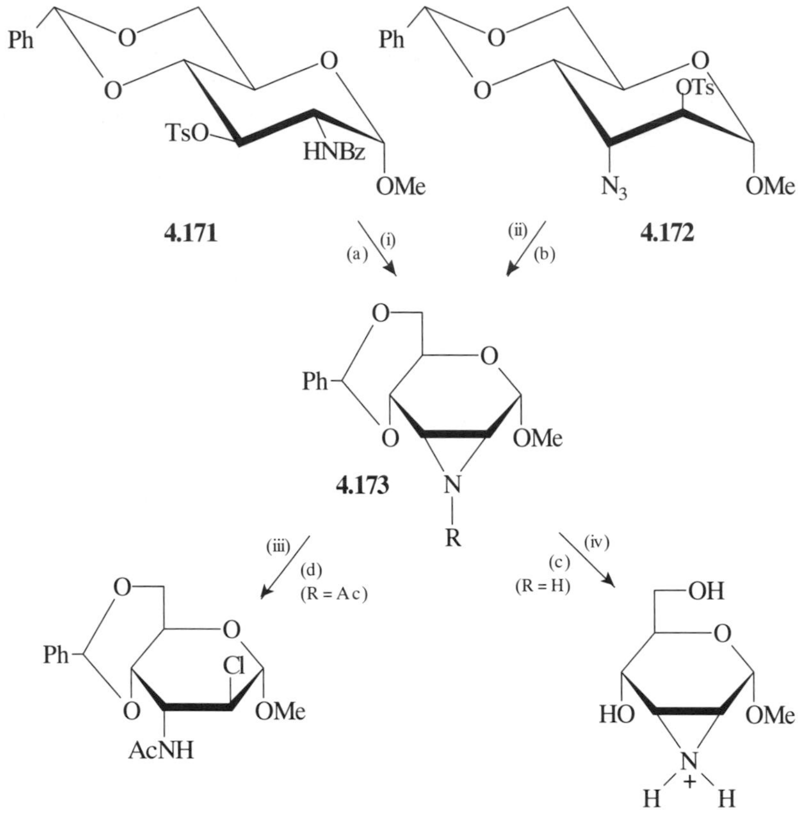

(i) Me$_2$CHONa (**4.173**, R=H, 90%), or NaH, THF (**4.173**, R=Bz, 95%); (ii) (NH$_2$)$_2$, Ni, MeOH, reflux, 2 h (**4.173**, R=H, 66%), or LiAlH$_4$, Et$_2$O; (iii) HCl, Me$_2$CO, −25 °C, 0.5 h; (iv) HCl, Me$_2$CO, 20 °C, 0.5 h

Scheme 4.55

Section 4.1.3) since they are able to form alkylammonium salts. It is possible, for example, to remove benzylidene groups from protected sugar derivatives to leave epimine rings intact (Scheme 4.55c), but N-acetylated derivatives are much more susceptible to ring opening, and mild treatment of compound **4.173** (R = Ac) with hydrogen chloride yields the *trans*-diaxial ring-opened product with the benzylidene residue intact, as illustrated in Scheme 4.55(d).

Sugar epimine rings have been opened with a variety of nucleophilic reagents, among them azide, thiolacetate and halide ions, and, as with epoxides, *trans*-diaxial ring opening usually occurs. For example, sodium azide attacks C-2 in the *allo*-epimine **4.173** (R = H) and its N-benzoyl derivative **4.173** (R = Bz) to give *altro*-products in 85 and 56% yields, respectively. The α-D-*manno*-isomer of **4.173**, (R = H) undergoes attack from azide or thiocyanate ions at C-3 to give the 3-azido- and 3-thiocyanato-D-*altro*-products, in keeping with the Fürst–Plattner rule (Section 4.1.3).

There are some reactions of specific aminodeoxy sugars that are best discussed separately, particularly those which involve interaction between the amino functions and the anomeric centres.

(f)   Reactions of 2-amino-2-deoxyaldoses

(i)   Stability

In neutral aqueous solution, the unsubstituted 2-amino-2-deoxyaldoses are unstable and are consequently stored as salts or in amido forms. Decomposition of the free compounds produces many substances, some being dimers, which are readily oxidized to pyrazines (Scheme 4.56).

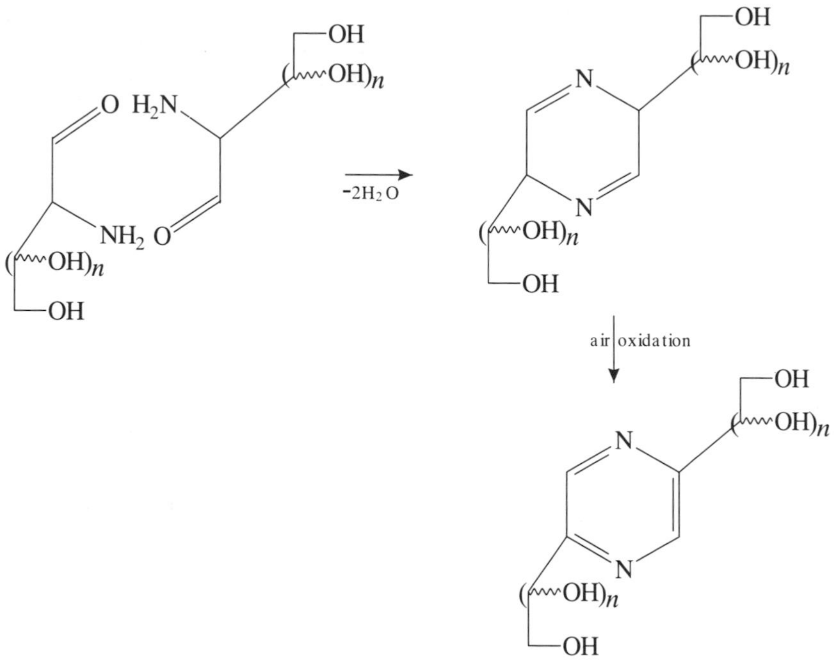

Scheme 4.56

2-Amino-2-deoxyaldoses are oxidatively degraded by one carbon atom under very mild conditions with ninhydrin (1,2,3-triketoindane hydrate). Hexose compounds are thus degraded to pentoses which are easily identified by paper chromatography, and the reaction can thus be utilized for the partial identification of amino sugars.

(ii)   Glycosidation

Fischer glycosidations of 2-amino-2-deoxy sugars (and hydrolyses of their glycosides) are precluded by the protonation of the basic amino groups to give ammonium ions which render the next step in the reaction unfavourable (see Sections 3.1.1.a and 3.1.1.b). This problem can be overcome by N-acetylation since the acetamido function, being less basic, does not retard the reaction in this way. Other methods for preparing methyl glycosides of 2-amino-2-deoxyaldoses are given below.

(iii) Acylglycosyl halide formation

Because of the proximity of N-acyl groups to the reactive centres at C-1 in glycosyl halides of 2-acylamino-2-deoxy sugars, rearrangements can occur (Scheme 4.57) which are dependent upon the nature of the N-acyl groups.[98] N-Acylglucosyl bromides, for example, in acetic acid/hydrogen bromide, which is the usual brominating reagent, equilibrate with two other compounds and the one most readily isolated also depends on the conditions, as illustrated in Scheme 4.57(a). The N-acetyl derivative **4.174** (X = Br, R = Me) readily forms the oxazolidinium bromide **4.176** (R = Me), whereas the N-p-methoxybenzoyl compound **4.174** (X = Br, R = p-MeOC$_6$H$_4$) forms the oxazolinium bromide **4.175** (R = p-MeOC$_6$H$_4$). The N-benzoyl derivative **4.174** (X = Br, R = Ph), although unstable, may be isolated in the unrearranged form from acid media, but rearranges to **4.175** (R = Ph) when dissolved in anhydrous ether. If solutions containing these equilibrium mixtures are neutralized, oxazolines **4.179** can be isolated.

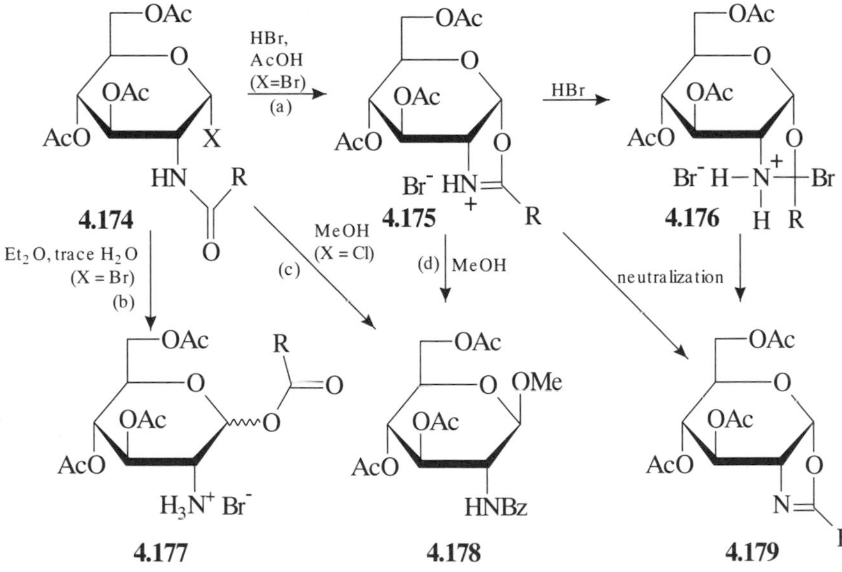

Scheme 4.57

Chlorides are less prone to these changes and consequently 2-acetamido-3,4,6-tri-O-acetyl-2-deoxy-α-D-glucopyranosyl chloride **4.174** (X = Cl, R = Me), for example, may be prepared in 79% yield by treating 2-acetamido-2-deoxyglucose with acetyl chloride.[99]

In the presence of water, N-acylglycosyl halides form 1-O-acyl-2-amino hydrogen halide salts, as shown for the *gluco*-derivative **4.177** (R = Me) in Scheme 4.57b, in keeping with the known behaviour of monoacylhydroxyamines in acid media (see Section 4.3.2.a).

Compounds in this series are useful for preparing 1,2-*trans*-glycosides. For example, either the l-halo derivative **4.174** (X = Cl, R = Ph) or the oxazoline hydrobromide **4.175** (R = Ph) gives rise to the same methyl β-glucopyranoside derivative **4.178** when treated with methanol (Schemes 4.57c and 4.57d). However, the readily isolated oxazolines are the most useful glycosylating reagents in situations in which catalysis by acids at elevated temperatures can be tolerated, and they have been successfully employed in some oligosaccharide syntheses and in the preparation of glycosides with functionalized aglycons, e.g. **4.181**, as illustrated in Scheme 4.58.[100] Oxazolines may be conveniently synthesized from peracylated 2-aminoaldoses by tin(IV) chloride or trimethylsilyl triflate catalysed transformations. Thus, for example, 2-acetamido-1,3,4,6-tetra-O-acetyl-2-deoxy-α-(or β-)D-glucopyranose gives in good yield 2-methyl-(3,4,6-tri-O-acetyl-1,2-dideoxy-α-D-glucopyrano)-[2,1-d]-2-oxazoline **4.180** when so treated, as shown in Scheme 4.58.[101,102]

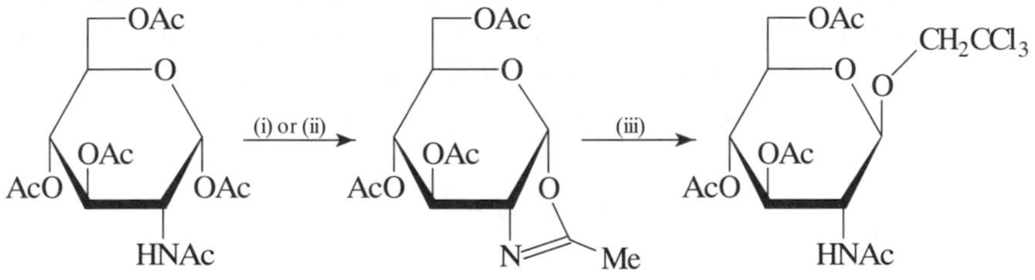

4.180 (82%)   **4.181** (46%) + α - isomer (6%)

(i) SnCl$_4$, CH$_2$Cl$_2$, ~ 22 °C, 3 h (82%); (ii) Cl(CH$_2$)$_2$Cl, Me$_3$SiOTf, 50 °C, 16 h (95%); (iii) Cl$_3$CCH$_2$OH, TfOH, 100 °C, 2 h

Scheme 4.58

2-Phthalimido sugars, however, yield usefully stable bromides which do not form oxazoline derivatives. Consequently, they are often used in oligosaccharide synthesis, yielding 1,2-*trans*-intersaccharide glycosidic links more efficiently than oxazolines (see Scheme 3.108, Section 3.3.2.a.ii and Section 6.6.2).[91] Methods of synthesis of amino sugar containing oligosaccharides have been reviewed.[103]

(g)   Reactions of 4-amino-4-deoxy- and 5-amino-5-deoxy-aldoses

The amino groups in these sugars occupy special positions in relation to the anomeric carbon atoms since if these latter centres are not protected, it is possible for the amino groups to participate in ring formation; 5-amino-5-deoxyaldoses thus form piperidinoses (e.g. **4.182**) and the 4-amino-4-deoxy analogues form substituted pyrrolidine compounds (e.g. **4.184**).[104] These heterocyclic systems are very unstable towards acids and are readily dehydrated to give final products which are often substituted heterocyclic aromatic compounds such as 3-hydroxypyridine **4.183** formed from 5-amino-5-deoxy-D-xylose **4.182** (Scheme 4.59).

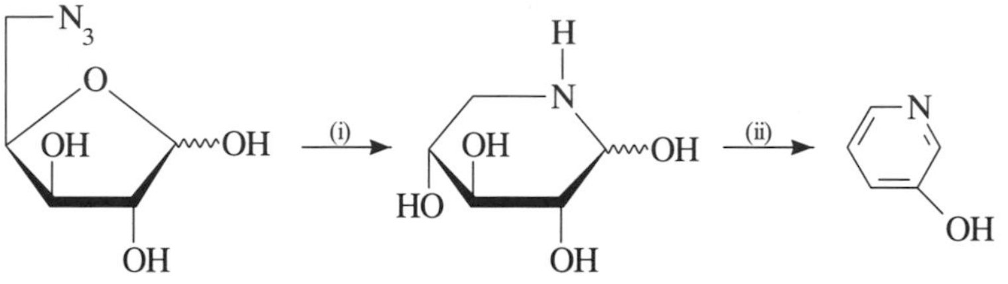

4.182   **4.183**

(i) H_2, Pd, pH 7; (ii) H_3O^+

Scheme 4.59

Because of this behaviour, special procedures have been adopted to obtain these amino sugars. One method, used when azido derivatives are the precursors, involves removal of all the protecting groups, followed by reductive generation of the amines under neutral conditions (e.g. Scheme 4.59). Alternatively, when the free amino groups are present in sugar derivatives protected with acid-labile residues, hydrolysis with aqueous sulphurous acid can be used since the aldoses are then trapped as soon as they are produced (Scheme 4.60). The aldoses can then be liberated from these bisulphite adducts with aqueous base.

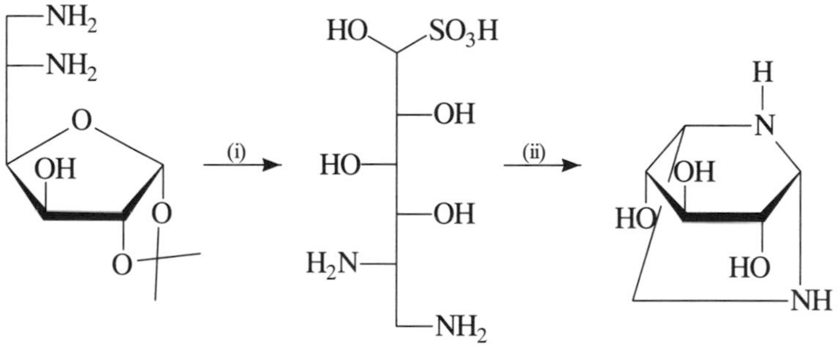

(i) H_2SO_3, H_2O; (ii) H_2O, OH^-

Scheme 4.60

4-Amino-4-deoxyaldoses exist either as five-membered nitrogen heterocycles or pyranoid compounds; alternatively, the 5-amino analogues can form furanoid or piperidine rings. 4-Amino-4-deoxy-L-xylose, for example, exists in the five-membered ring form **4.184** in equilibrium with a dimeric modification. Thus, a 4-amino group is sufficiently nucleophilic to form a five-membered ring even in an aldose with a structure that would otherwise strongly prefer the six-membered form.

A further illustration of this tautomeric preference is found during the hydrogenoly-sis of the benzyl glycoside of 2,3,4,6-tetraaminotetradeoxy-α-D-glucopyranose **4.117** (Scheme 4.37), which gives a large amount of the triaminopyrrolidine **4.118** presumably by reduction of an intramolecular Schiff base which arises by dehydration of the five-membered nitrogen heterocycle of the tetraamino sugar.

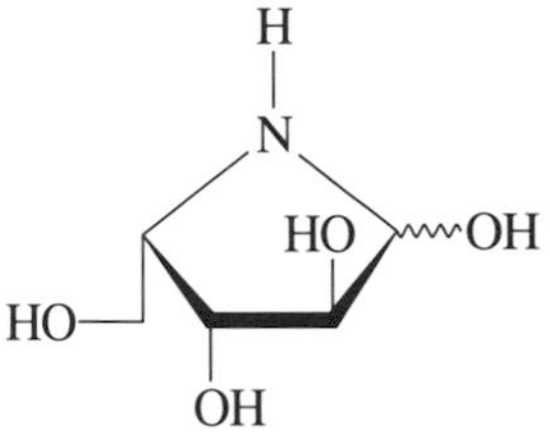

4.184

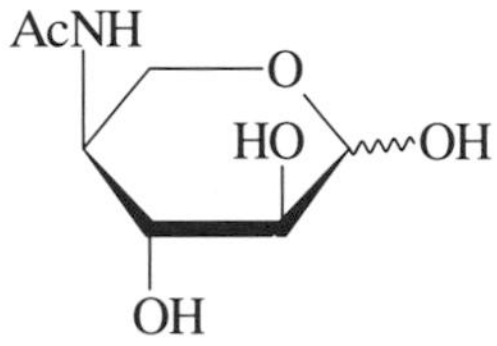

4.185

N-Acetylation of the 4-amino group in the aminoxylose **4.184** reduces its nucleo-philicity and in consequence the acetamido compound **4.185** exists in the pyranose form. A further demonstration of the reduced nucleophilicity of the nitrogen atom in an acetamido residue is given by the *N*-acetyl derivative of 5-amino-5-deoxy-D-xylose, which, although found to exist predominantly in the sterically preferred six-membered ring, is in equilibrium with significant amounts of the furanose form (Scheme 4.61). Thus, the acetamido group at C-5 is unable to hold the molecule exclusively in the six-membered ring form even though ring closure through oxygen produces the less common furanose ring modification. Nojirimycin (5-amino-5-deoxy-D-glucose) **4.108**, which is an antibiotic, has been synthesized by methods similar to those described above[105] and from nonsugar precursors.[106] It exists, as expected, in the six-membered ring form, and as such it can be trapped as a bisulphite adduct **4.109**. Many sugar analogues having phosphorus in the hemiacetal ring have also been prepared,[107] and compounds with sulphur in the ring are referred to in Section 4.4.2.

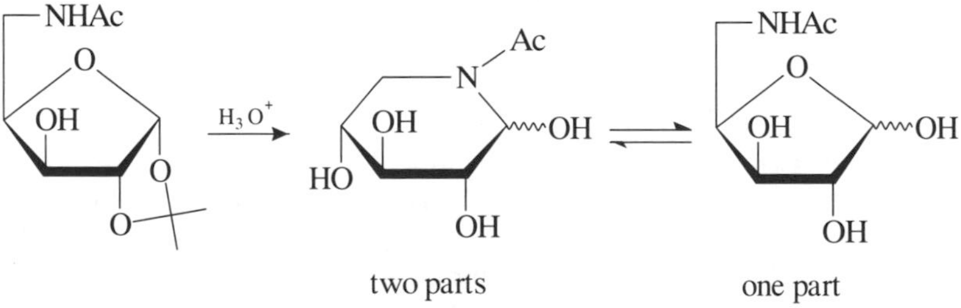

two parts   one part

Scheme 4.61

(h)   Colour tests

There are two sensitive colour tests in common use for amino sugars:[60] the *Morgan–Elson* test, which gives a red coloration when 2-acetamido-2-deoxyaldoses are treated as shown in Scheme 4.62(a), and the Elson–Morgan test, which gives a similar colour when 2-amino-2-deoxyaldoses are treated as outlined in Scheme 4.62(b).

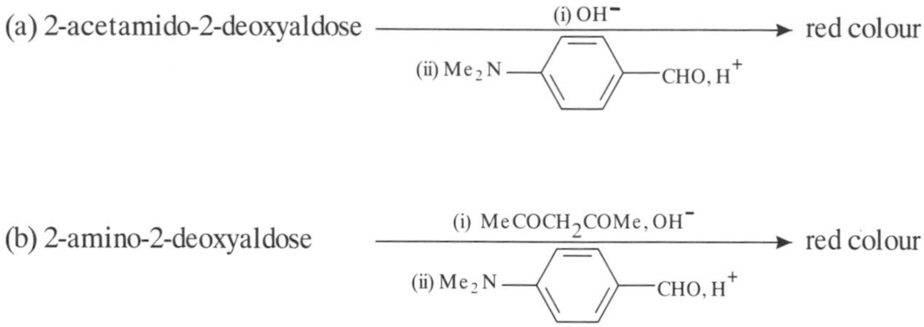

(a) 2-acetamido-2-deoxyaldose ⟶ (i) OH⁻ / (ii) Me₂N⟨⟩CHO, H⁺ ⟶ red colour

(b) 2-amino-2-deoxyaldose ⟶ (i) MeCOCH₂COMe, OH⁻ / (ii) Me₂N⟨⟩CHO, H⁺ ⟶ red colour

Scheme 4.62

4.4 THIO SUGARS

Replacement of oxygen atoms by sulphur in carbohydrates gives a class of compounds called thio sugars.[108] The special cases of thioglycosides and thioacetals in which the exocyclic oxygen atoms at the anomeric centres are replaced were considered in Section 3.1.2; compounds containing sulphur at positions other than this will be discussed here. These derivatives are named by adding an enumerated 'thio' prefix to the name of the sugar to denote the replacement of oxygen by sulphur. ('Deoxy' is therefore not required.) The only thio sugars that occur to a significant extent in nature are the thioglycosides (Section 8.2.3); however, there are exceptions: 5'-methylthioadenosine has, for example, been detected in yeast extracts, and its *arabino*-isomer has been isolated from the glands and mantle of the sea slug *Doris verucoca*.[109] 5-Thiomannose is present in a marine sponge[110] and 2,6-dideoxy-4-thio-D-*ribo*-hexopyranose is present in S-aroylated form in the antitumour compound calicheamicin.[111]

4.4.1 SYNTHESES

(a) Displacement reactions

One of the most common methods for introducing sulphur into sugar molecules involves the displacement of suitably positioned leaving groups by use of reagents containing nucleophilic sulphur such as the thiocyanate ion $^-$SCN, thiolacetate Me(C=O)S$^-$, benzylthiolate PhCH$_2$S$^-$ and ethyl xanthate EtO(C=S)S$^-$,[112] which can be used to give sulphur-containing products and thence thio sugars by deprotection. As expected, these displacement reactions occur readily at primary positions, as illustrated by the thiolacetate displacement applied to the 5-*O*-tosylxylose derivative **4.186** as the key step in the preparation of 5-thio-D-xylose **4.187** (Scheme 4.63). A more sluggish displacement by thiocyanate ion occurs at C-4 in methyl 2,3,6-tri-*O*-benzoyl-4-*O*-tosyl-α-D-glucopyranoside **4.188** to give the 4-thiocyanogalactose

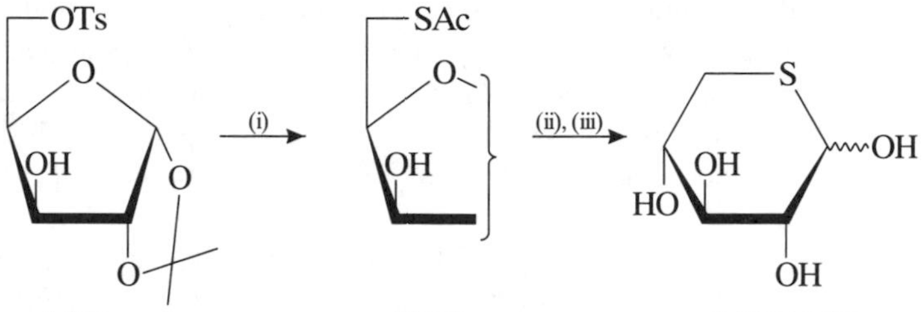

4.186   (68%)   **4.187** (43%)

(i) KSAc, DMF, reflux, N$_2$, 4 h; (ii) MeONa, MeOH; (iii) H$_2$O, H$_2$SO$_4$

Scheme 4.63

derivative **4.189**, as illustrated in Scheme 4.64.[113] Subsequent reduction and deprotection gives 4-thiogalactoside **4.190** which, on acetolysis, is trapped as a 4:2:1
mixture of acetates **4.191, 4.192** and **4.193**, which clearly shows that this sugar has
a strong preference for the five-membered form in which sulphur is in the ring.

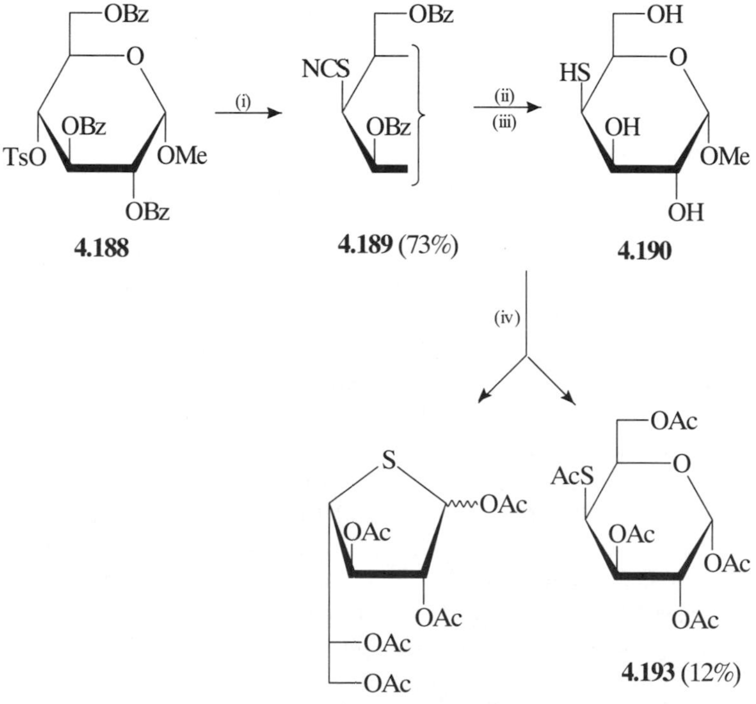

4.188   **4.189** (73%)   **4.190**

4.191 α–form (45%)
4.192 β-form (25%)

4.193 (12%)

(i) KSCN, DMF, 110 °C, 48 h; (ii) Zn, AcOH, reflux, 24 h; (iii) MeONa, MeOH,
20 °C, 2 h; (iv) AcOH, Ac$_2$O, H$_2$SO$_4$

Scheme 4.64

A thermally induced intramolecular isomerization undergone by methyl xanthates can be used to prepare thio sugars (see Section 5.3.5.b). For example, 1,2:5,6-di-*O*-isopropylidene-3-*O*-[(methylthio)thiocarbonyl]-α-D-glucofuranose **4.26** (R = (C=S)SMe), upon pyrolysis, gives the 3-*S*-[(methylthio)carbonyl]thio isomer (RS(C=O)SMe),[114] which on ammonolysis yields 1,2:5,6-di-*O*-isopropylidene-3-thio-α-D-glucofuranose.

A Mitsunobu reaction has been used to convert primary sugar alcohols into derivatives of the corresponding thio sugars in one-pot reactions (Section 4.1.2.c). Thus, di-*O*-isopropylidene-α-D-galactopyranose **4.194** on treatment with thiolacetic acid, diisopropyl azodicarboxylate and a triarylphosphine in THF is converted into the 6-*S*-acetyl-6-thiogalactose derivative **4.195** in 80% yield.[115]

4.194 X = OH
4.195 X = SAc

(b) Opening of epoxide rings

A reaction related to the above is nucleophilic opening of epoxides, and by this means, for example, benzylthiol converts methyl 2,3-anhydro-4,6-*O*-benzylidene-α-D-allopyranoside **4.47** to the 2-thio derivative **4.49** (X = SBn), which yields methyl 2-thio-α-D-altropyranoside **4.49** (X = SH) after reductive removal of the protecting groups.[116]

Sometimes with nucleophiles such as thiocyanate or thiourea a more complex reaction occurs with epoxides giving episulphides directly, as illustrated in Scheme 4.65 by the conversion of the 5,6-anhydro-L-*ido*-compound **4.196** to the D-*gluco*-5,6-episulphide **4.197**.[117] A plausible mechanism for this displacement is illustrated in the scheme. Treatment of the *allo*-epoxide **4.47**, mentioned above, with thiourea gives methyl 4,6-*O*-benzylidene-2,3-thioanhydro-α-D-mannopyranoside in 63% yield.[118] Presumably, a change in conformation of the ring-opened intermediate **4.49** (X = S(C=NH)NH₂) from a chair to a boat form is required for the *manno*-episulphide to be formed by the mechanism outlined above. Episulphides can be ring opened with nucleophiles as illustrated in Scheme 4.65.[117]

(c) Additions to unsaturated compounds

The addition of sulphur-containing reagents to unsaturated sugar derivatives offers a useful route to thio sugars, many of the known reactions of this type involving free radical processes. When readily available glycal derivatives (Section 4.10.1) are treated with reagents possessing SH groups, 1,5-anhydro-2-thioalditol products

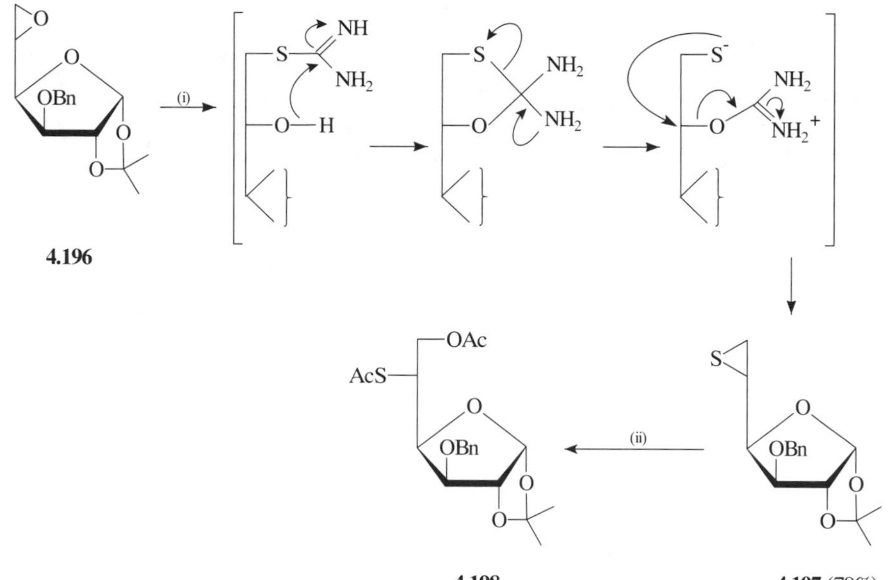

4.196

4.198   **4.197** (79%)

(i) $(NH_2)_2C=S$, MeOH, 25 °C, 60 h; (ii) KOAc, Ac_2O, AcOH, reflux, 16 h

Scheme 4.65

are formed in keeping with the high regiospecificity shown by these compounds in radical addition reactions. For example, triacetyl-D-glucal **4.199** reacts with thiolacetic acid in the presence of free radical initiators to give 3,4,6-tri-*O*-acetyl-2-*S*-acetyl-1,5-anhydro-2-thio-D-mannitol **4.200** and -glucitol **4.201**, as illustrated in Scheme 4.66.[119]

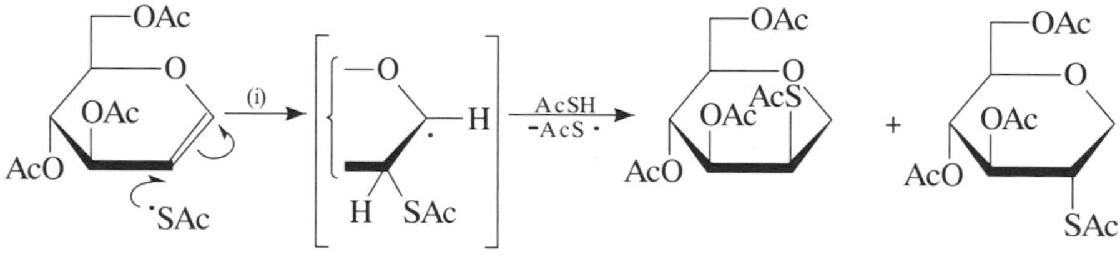

4.199   **4.200** (61%)   **4.201** (25%)

(i) AcSH, cumene hydroperoxide

Scheme 4.66

In similar fashion, benzylthio or thiolacetate radicals, generated by ultraviolet irradiation of the corresponding thiols, add to methyl 5-deoxy-2,3-*O*-isopropylidene-β-D-*erythro*-pent-4-enofuranoside **4.202** regiospecifically (Scheme 4.67). Stereospecificity at *C*-4 is also high, with the bulky thiomethyl substituent placed *trans* to the 2,3-ketal ring to give the 5-thio-D-riboside derivative **4.204**.[120] As this is the thermodynamically preferred product and not that expected from direct

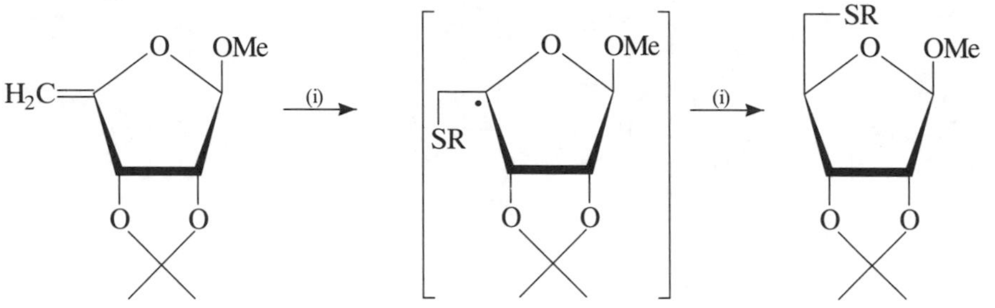

4.202   **4.203**   **4.204** R = Ac (61%)
                                                                    4.204 R = Bn (69%)

(i) AcSH, or BnSH, $h\nu$, N_2, 64 h

Scheme 4.67

hydrogen abstraction by the intermediate radical **4.203,** it seems probable that, under the conditions of the addition, the final hydrogen abstraction step is reversible. Hydrogen radical removal from ether sites like C-4 is known in nucleoside chemistry (Section 4.5.1.d).[121]

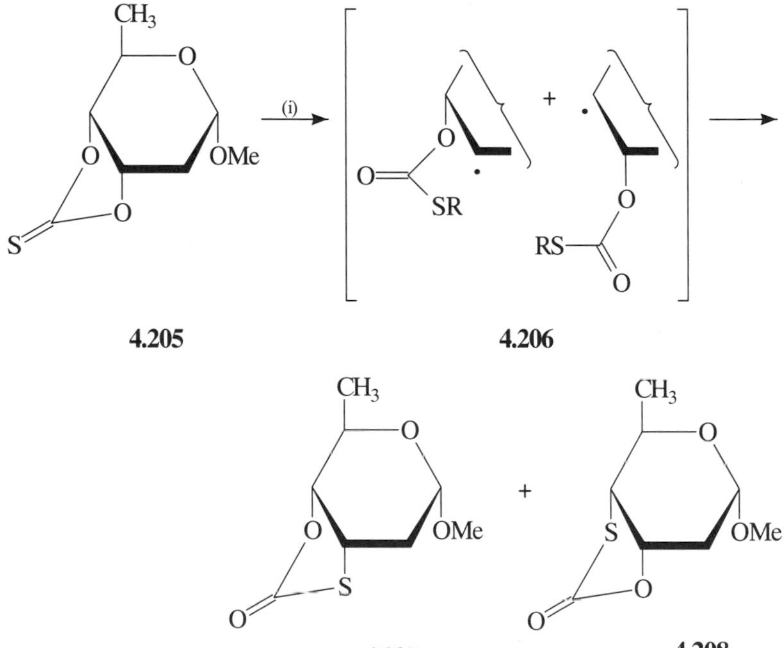

4.205   **4.206**

4.207   **4.208**

R = $SnBu_3$
(i) AIBN, n-Bu_3SnH (0.5 equivalent), PhH, reflux

Scheme 4.68

(d) Radical-induced replacements of oxygen by sulphur

A novel route to thio sugars involves a radical-induced rearrangement of thionocarbonates. Thus, treatment of 3,4-thionocarbonate **4.205**, as illustrated in Scheme 4.68, with AIBN and a restricted amount of tributyltin hydride gives the 3- and 4-thio sugar derivatives **4.207** and **4.208** with retention of configuration. From these, O-methyl glycosides of thio sugars can be liberated by subsequent saponification. The reaction mechanism is similar to that given in Scheme 4.29 for the conversion of thionocarbonates to deoxy sugars. Instead of the intermediate radicals **4.206** being reduced by the reagent, however, they are trapped by the sulphur atoms of tributyltin thioester groups.[122]

4.4.2 REACTIONS

Many reactions of thio sugars and their derivatives differ from those of the oxygen analogues even though sulphur and oxygen are such closely related elements. A most important feature is that thioalkyl groups can be easily converted into *sulphonium cations* by treatment with suitable electrophilic reagents, and this produces derivatives with good leaving groups which can be used in nucleophilic displacement reactions (cf. Section 3.1.2.c.ii). A related characteristic is the high nucleophilicity of sulphur.

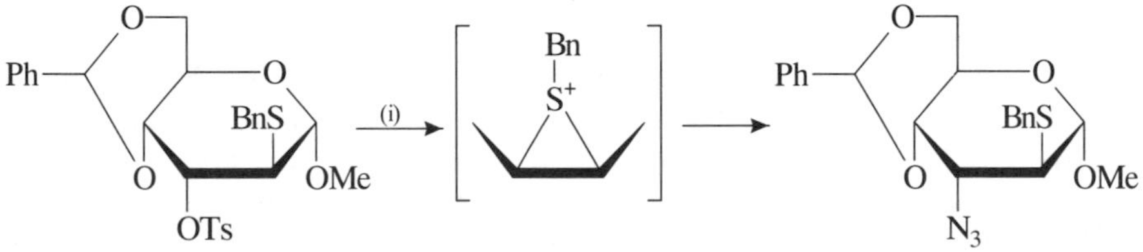

(i) NaN_3, $MeOCH_2CH_2OH$, $100\,°C$, 2 h

Scheme 4.69

This high nucleophilicity is indicated by the participation of the thiobenzyl group when methyl 4,6-O-benzylidene-2-benzylthio-3-O-tosyl-α-D-altropyranoside **4.209** is treated with sodium azide to give the 3-azido-3-deoxy sugar derivative **4.211** with the *altro*-configuration (Scheme 4.69). The overall retention of configuration indicates that this reaction occurs via the episulphonium ion **4.210**.[123] Another example of this type of participation is in the formation of a sulphur bridge in methyl 2,6-anhydro-2-thio-α-D-altropyranoside **4.212** (Scheme 4.70).[124]

Many examples of sulphur group migration are known, particularly with dithio-acetal derivatives from which thioalkyl groups may leave C-1 with particular ease. When 5-O-tosyl-L-arabinose diethyl dithioacetal **4.213**, for example, is heated in aqueous acetone, one of the sulphur atoms displaces the tosyloxy group and the cyclic sulphonium ion **4.214** is formed. Attack by O-4 at C-1 then leads to the ethyl 5-thioethyl-1-thiofuranosides **4.215** as the main products (Scheme 4.71).[125]

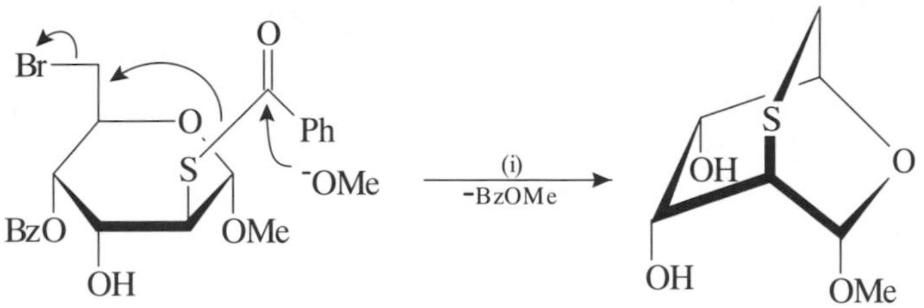

4.212

(i) MeO⁻, MeOH

Scheme 4.70

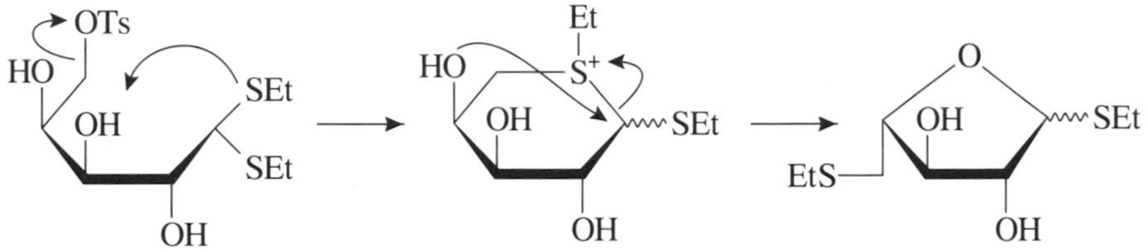

4.213   **4.214**   **4.215**

Scheme 4.71

In related fashion, acid-catalysed thiolyses of aldose dithioacetals in which an acyloxonium ion **4.216** may be established at C-2–C-3 lead to thioalkyl migrations from C-1 to C-2 with inversion of configuration.[126] In situations in which the ester group may then form a further acyloxonium ion with, for example, O-4, the process can repeat and thio groups may be relayed from C-1 to C-2 to C-3, etc., until products containing several thio groups are produced (Scheme 4.72). Ethanethiolysis of 3-*O*-benzoyl-1,2:5,6-di-*O*-isopropylidene-α-D-glucofuranose gives, directly, 40% of ethyl 4-*O*-benzoyl-2,3,6-tri-*S*-ethyl-1,2,3,6-tetrathio-α-D-mannopyranoside, the thioethyl groups all being introduced at C-1 and then relayed to C-2, C-3 and C-6 of acyclic dithiocetals, the last step involving a six-membered acyloxonium ion.[127]

Again, the high nucleophilicity of sulphur is responsible for the ease with which suitably positioned thiols add intramolecularly to the carbonyl carbon atoms of free sugars. Thus, aldoses with a thiol group at C-4 or C-5 favour five- and six-membered sulphur-containing rings, respectively.[104] 4-Thio-D-galactose derivatives (e.g. **4.190**) strongly prefer the form with sulphur in the ring, as shown by the spread of tautomeric isomers on acetylation (e.g. **4.191–4.193** in Scheme 4.64), and likewise the 5-thio analogues of D-xylose **4.187** and D-glucose exist, predictably, in the six-membered form.

5-Thio-D-glucose, one of the closest analogues of D-glucose available, has been synthesized in several ways, but one of the shortest routes which gives a reasonable overall yield is by way of 3-*O*-benzyl-1,2-*O*-isopropylidene-5,6-epithio-α-D-glucofuranose **4.197**. Acetolysis of this compound opens the episulphide by

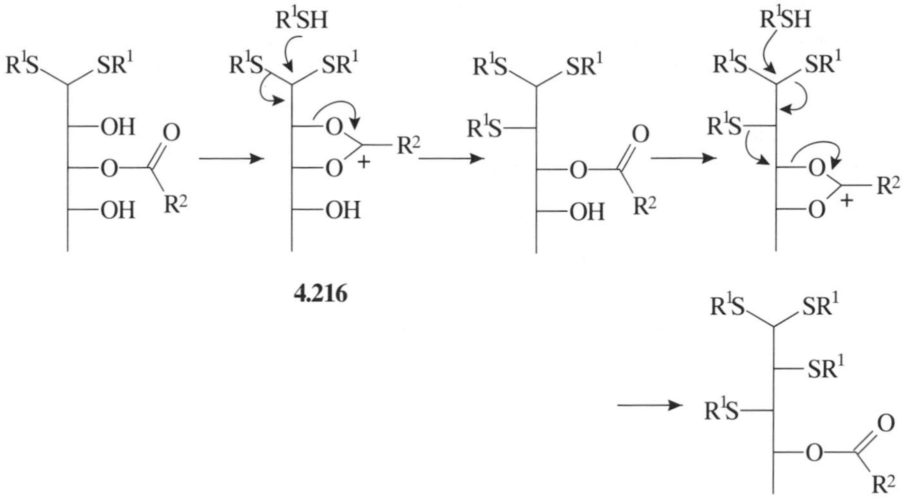

Scheme 4.72

acetate attack at the primary position to give the 5-thio-D-glucose derivative **4.198** (Scheme 4.65), deprotection of which gives the thio sugar.[117] The α-form of this sugar mutarotates in water faster than α-D-glucopyranose and gives an equilibrium mixture comprising 85% and 15% of the α-anomer and β-anomer of the six-membered ring form, in contrast to equilibrated solutions of D-glucose wherein the β-form preponderates (Section 2.4.2).[128] Glycopyranosides of such 5-thio sugars undergo acid-catalysed hydrolysis about an order of magnitude faster than do those of the parent sugars.[104] The physiological activity of 5-thioglucose is different from that of D-glucose: it inhibits the transport and cellular uptake of D-glucose, it functions as a growth inhibitor for parasites which have a high glucose requirement, and it can be used as a non-toxic agent to inhibit spermatogenesis in mice and thereby temporarily control their fertility. It is of consequence in cancer research since it is highly effective in destroying hypoxic cells and protects oxic cells from radiation damage.

Sugars containing free thiol substituents readily form disulphides when treated with oxidizing agents such as iodine, hydrogen peroxide or air, but the reaction is easily reversed by reducing agents (Scheme 4.73). Many carbohydrate thiols are frequently isolated in the oxidized state. Most thiols form stable salts with heavy metals and thio sugars are often stored as their mercury derivatives.

$$2RSH \underset{\text{reduction}}{\overset{\text{oxidation}}{\rightleftharpoons}} RSSR$$

Scheme 4.73

Sulphur is able to form compounds with a valence greater than two by use of its unfilled d-orbitals, so oxidation of sulphides (R^1SR^2) gives sulphoxides (R^1SOR^2)

and finally sulphones ($R^1SO_2R^2$). Sulphoxide groups do not have planar structures and therefore the sulphur is potentially asymmetric. Accordingly, when the di-*O*-acetyl derivative of **4.212** is oxidized with sodium periodate, it yields two separable diastereoisomeric sulphoxides. Further oxidation of either sulphoxide with peracid gives the same sulphone.

The most important reaction of non-anomeric thio sugars from a preparative standpoint is reductive cleavage of the sugar–sulphur bonds, and this is much used in preparing deoxy sugar derivatives. The reaction can often be effected at atmospheric pressure with Raney nickel using the hydrogen adsorbed on the catalyst, as illustrated in Scheme 4.74 by the preparation of methyl 4,6-*O*-benzylidene-2,3-dideoxy-α-D-*erythro*-hexopyranoside **4.218** from the corresponding 2,3-episulphide **4.217**.

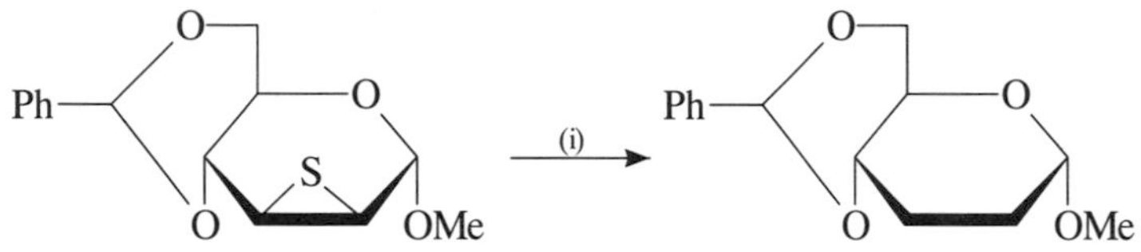

4.217   **4.218**

(i) H$_2$, Ni, dioxane

Scheme 4.74

4.5   DEOXYHALO SUGARS

The deoxyhalo sugars are compounds in which hydroxyl groups at positions other than the anomeric centre are replaced by fluorine, chlorine, bromine or iodine atoms. They are named by using enumerated '*deoxy*' and '*halo*' prefixes; 2-chloro-2-deoxy-D-glucose is an example.[37,129] Compounds having halogens attached to the anomeric centre are named glycosyl halides and are discussed in Section 3.3.

4.5.1   SYNTHESES

(a)   Displacements of sulphonyloxy groups

One of the best-established methods for replacing hydroxyl groups by halogen substituents is the application of the nucleophilic displacement of sulphonate esters discussed in Section 4.1.2.a. Sugar tosylates and mesylates are often used but the introduction of triflate esters (Section 5.3.6) has made possible displacements at secondary carbon atoms that are impractical with the older derivatives. Alkali metal halides or the more highly soluble tetraalkylammonium salts are frequently used as the halide source. The former are of particular value for carrying out displacements at secondary positions usually when used in aprotic polar solvents such as DMF, DMSO or HMPIT. Sodium iodide, however, has a high solubility in several organic solvents, and therefore displacements can often be easily accomplished with this highly nucleophilic halide in refluxing acetone or butanone. On the other hand, fluoride displacements at secondary positions are the most difficult to achieve on account of the weakly nucleophilic character of this ion. Caesium fluoride, tetrabutylammonium fluoride, tetrabutylammonium difluoride (Bu_4NHF_2) and tris(dimethylamino)sulphur (trimethylsilyl)difluoride [$(Me_2N)_3S(Me_3SiF_2)$] (TASF) are the most successful salts for the synthesis of secondary deoxyfluoro sugars.[130]

6-Deoxy-6-halohexopyranose derivatives are easily prepared often by selective displacement of the primary sulphonate ester groups, as shown by the formation of the 6-deoxy-6-iodoglucose derivative **4.220** from the corresponding mesylate **4.219** (Scheme 4.75),[131] and the formation of the 6-deoxy-6-fluoroglucose derivative **4.224** from the trimesylglucofuranose derivative **4.223** (Scheme 4.76a).[132]

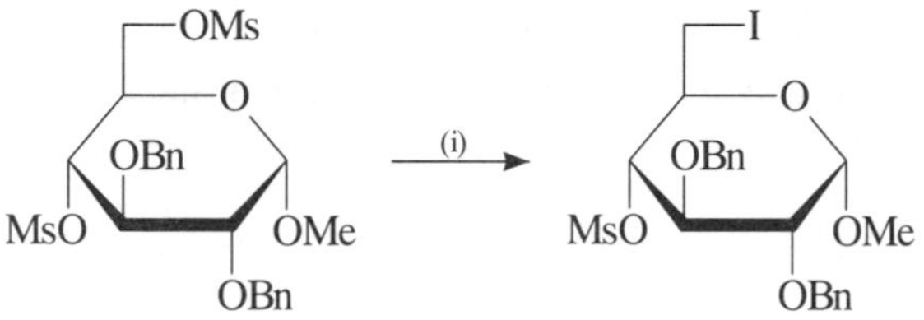

4.219   **4.220** (85%)

(i) NaI, MeCOEt, reflux, 6 h

Scheme 4.75

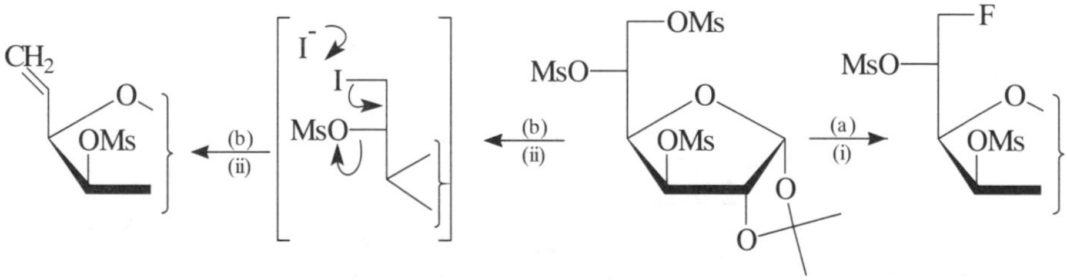

4.221 (87%)   **4.222**   **4.223**   **4.224** (60%)

(i) KF·$2H_2O$, MeOH, reflux, 20 h; (ii) NaI, MeCOEt, reflux, 1.5 h

Scheme 4.76

Alkene products can be formed when iodide displacements are carried out on *vicinal* disulphonate esters, as illustrated by the conversion of **4.223** to the enose derivative **4.221** by way of the primary iodide intermediate **4.222** (Scheme 4.76b; see also Section 4.10.2.a).[133] In such reactions the secondary sulphonate groups do not undergo intermolecular displacement. On the other hand, direct displacements at C-3 in the 3-*O*-triflyl derivative of di-*O*-isopropylidene-α-D-glucofuranose **4.26** (R = Tf) can be induced, as illustrated in Scheme 4.10(c).[15] However, only the 3-deoxy-3-iodoallofuranose derivative **4.27** (X = I) is formed in a satisfactory yield; the 3-bromo-and 3-chloro-allose derivatives **4.27** (X = Br, or Cl) are accompanied by substantial proportions of the 4-enose **4.28**, and this is the exclusive product obtained by attempted displacement with fluoride ions (Scheme 4.10d).[134] Reasons for this product spread are given in Section 4.1.2.a.iii.

In marked contrast, all the 3-deoxy-3-halo-1,2-O-isopropylidene-α-D-gluco-furanoses are readily available from the corresponding 1,2-di-O-isopropylidene-3-O-sulphonylallofuranose derivatives. Even intractable fluoride displacements occur efficiently, not only with the 3-O-triflate **4.23** but also with the less reactive 3-O-tosylate **4.22**, thereby giving ready access to the 3-fluoro derivative **4.25**, as illustrated in Schemes 4.9(c) and 4.9(b).[134,135]

2-Deoxy-2-halo sugars are prepared only with some difficulty by the displacement method, as explained in Section 4.1.2.a.ii, and in keeping with the importance of anomeric configuration in these reactions, 2-deoxy-2-fluoro-gluco- and -manno-pyranose derivatives are obtained efficiently only from the β-anomers of the 2-triflates of the corresponding manno- or gluco-pyranosides. The α-anomers give 2-fluoro derivatives, at best, in low yields.[130]

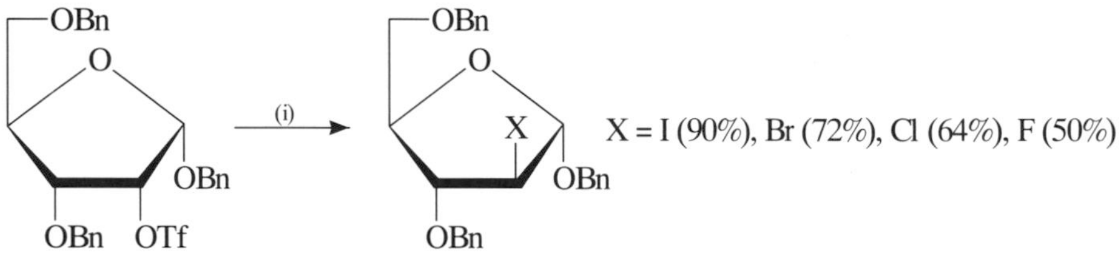

X = I (90%), Br (72%), Cl (64%), F (50%)

4.225   **4.226**

(i) NaI or LiBr or LiCl in DMSO, HMPIT, 22 °C, or n-Bu₄NF, THF, 0 °C

Scheme 4.77

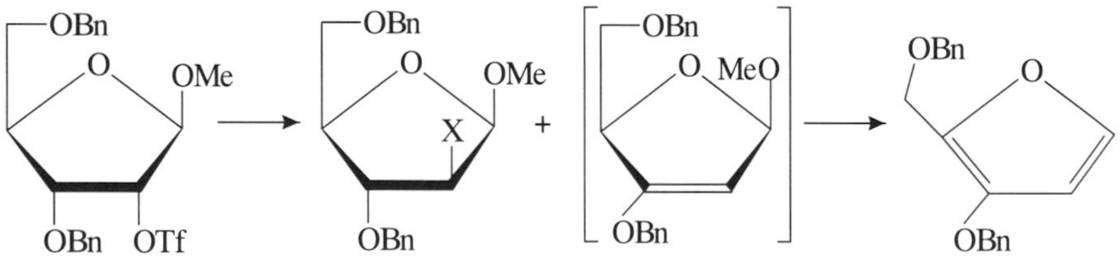

4.227   **4.228** X = I, Br, Cl   **4.229**   **4.230**
                      (14–19%)   (major product)

(i) NaI or LiBr or LiCl in DMSO, HMPIT, 22 °C

Scheme 4.78

Likewise, anomeric configuration plays a role in the synthesis of 2-deoxy-2-halo-furanose derivatives. All four 2-deoxy-2-halo-α-D-arabinofuranosides **4.226** (X = F, Cl, Br, I) have been obtained from the corresponding 2-O-triflyl-α-D-ribofuranoside derivative **4.225**, as illustrated in Scheme 4.77, and the yields of halo sugars show that the efficiency of the displacement is dependent upon the nucleophilicity of the halide.[136] In marked contrast, β-ribofuranoside **4.227**, in which C-2 is more sterically hindered, gives only modest yields of the 2-deoxy-2-halo-β-arabinofuranosides **4.228** (X = I, Br, Cl) under these conditions (Scheme 4.78). Instead, the furan derivative **4.230**, formed by way of **4.229**, preponderates in all cases.[137]

(b)   Opening of epoxide rings

Because of the inherent difficulty in displacing secondary mesyloxy and tosyloxy ester groups from pyranoid and furanoid rings (Section 4.1.2.a), the opening of epoxides by halide ions has often been used for introducing halogens, especially fluorine, into sugar molecules.[130] For example, methyl 2,3-anhydro-4,6-*O*-benzylidene-α-D-allopyranoside **4.47**, when treated with sodium iodide in the presence of acetic acid/sodium acetate or magnesium bromide in tetrahydrofuran, gives the 2-deoxy-2-iodo- and 2-bromo-2-deoxy-altroside derivatives **4.49** (X = I, Br) in high yield. Epoxides can also be opened with hydrogen halides, as illustrated by the conversion of methyl 3,4-anhydro-α-D-galactopyranoside with aqueous hydrochloric acid in acetone to a mixture of methyl 3-chloro-3-deoxy-α-D-gulopyranoside (45%) and methyl 4-chloro-4-deoxy-α-D-glucopyranoside (24%). Examples of the synthesis of deoxyfluoro[30] and bromodeoxy[138] sugar derivatives by this approach are given in Schemes 4.79 and 4.21.

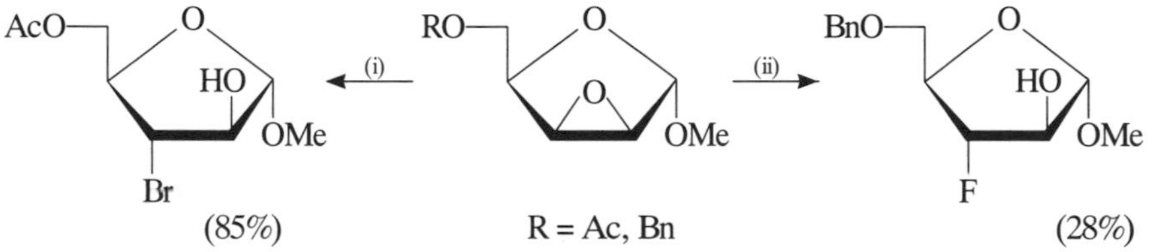

(85%)   R = Ac, Bn   (28%)

(i) (CH₂OMe)₂, MgBr₂, reflux, 3.5 h; (ii) KHF₂, (CH₂OH)₂, reflux 1 h

Scheme 4.79

A novel method of opening epoxides involves the use of (chloromethylene)dimethyliminium chloride, which yields either monochloro or dichloro derivatives depending upon the reaction conditions. For example, methyl 2,3-anhydro-4,6-*O*-benzylidene-α-D-allopyranoside **4.231**, when treated with this reagent, gives the 2-chloro-2-deoxy-3-*O*-formylaltroside **4.233** upon hydrolysis of the first-formed adduct **4.232**, produced as shown in Scheme 4.80(a). Alternatively, if the reaction mixture is heated, methyl 3,4-*O*-bezylidene-2,6-dichloro-2,6-dideoxy-α-D-altroside **4.235** is obtained by way of a rearrangement of the benzylidene group shown in **4.234** (Scheme 4.80b).[139]

(c)   Brominolysis of benzylidene acetals

N-Bromosuccinimide converts benzylidene acetals into bromodeoxy benzoates, and offers an excellent method of preparing 6-bromodeoxyhexoses (See also Section 5.4.4).[140] Methyl 4,6-*O*-benzylidene-α-D-galactopyranoside **4.236**, for

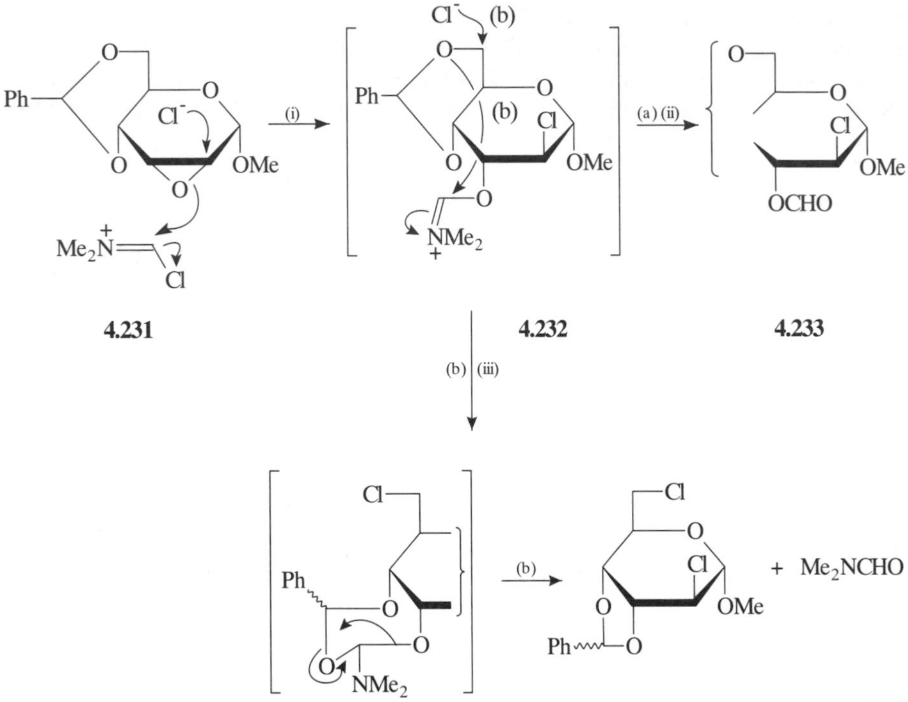

4.231   **4.232**   **4.233**

4.234   **4.235** (~90%)

(i) $Me_2\overset{+}{N}$=CHCl Cl$^-$, (HCCl$_2$)$_2$, ~ 23 °C, 15 h; (ii) NaHCO$_3$, H$_2$O; (iii) 110 °C, 25 h

Scheme 4.80

example, gives methyl 4-O-benzoyl-6-bromo-6-deoxy-α-D-galactopyranoside **4.239** almost quantitatively when treated with this reagent in carbon tetrachloride (Scheme 4.81), and the yield from the *gluco*-analogue is almost as good.[141] The reaction is simple to execute, and gives excellent yields of bromomethyl secondary benzoates when applied to benzylidene derivatives formed from primary and secondary hydroxyl groups. When the attachment is to two secondary positions and there are no stereochemical constraints to effect regioselectivity, mixtures of isomers are produced as illustrated in Scheme 4.82. Similar transformations can be brought about photochemically with bromotrichloromethane as the halogen source.[142]

The mechanism of the ring-opening reaction probably involves free radical bromination of the benzylidene acetal carbon atoms to give, for example, **4.237** followed by breakdown of these bromo derivatives to the stabilized carbocations **4.238** which are then attacked by bromide at the least-hindered position, as shown in Scheme 4.81.

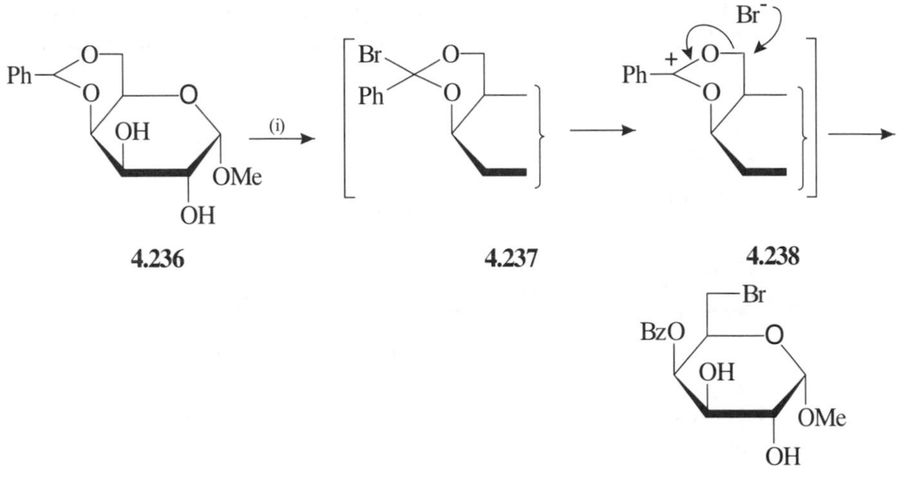

4.236   **4.237**   **4.238**

4.239 (90%)

(i) NBS, CCl$_4$, C$_2$H$_2$Cl$_4$, BaCO$_3$, reflux, 24 h
Scheme 4.81

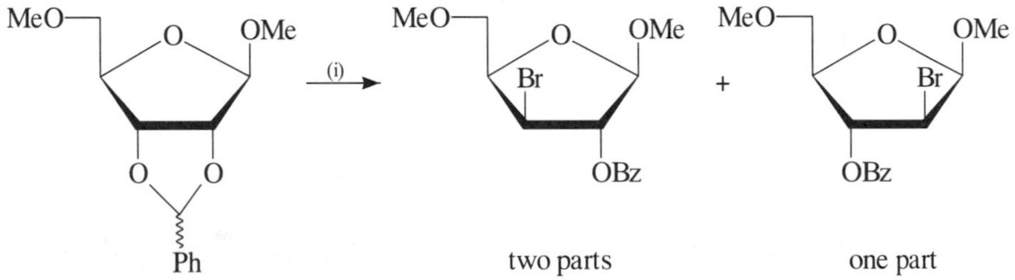

two parts   one part

(i) NBS, CCl$_4$, reflux

Scheme 4.82

(d)   Photobromination of monosaccharide derivatives[121]

Bromine can be introduced directly into particular ring positions of certain sugar derivatives by free radical processes which can show high selectivity and efficiency; the reaction is effectively the only means of causing direct hydrogen substitution on the carbon frameworks of monosaccharide derivatives. N-Bromosuccinimide or bromine can be used as the halogen source in carbon tetrachloride or bromotri-chloromethane, and reactions are initiated by irradiation under a tungsten or heat lamp. The hydrogen bromide generated as by-product when bromine is used can be removed by a scavenger such as potassium carbonate.

Hexopyranuronic acid derivatives undergo substitution at the captodative C-5 centres, compound **4.240** giving 47% yield of crystalline **4.241** which is a derivative of L-*xylo*-hex-2-ulosonic acid **4.242**, a precursor of L-ascorbic acid **4.243** (vitamic C). Hydrolysis of compound **4.241** therefore represents a novel route to the vitamin.

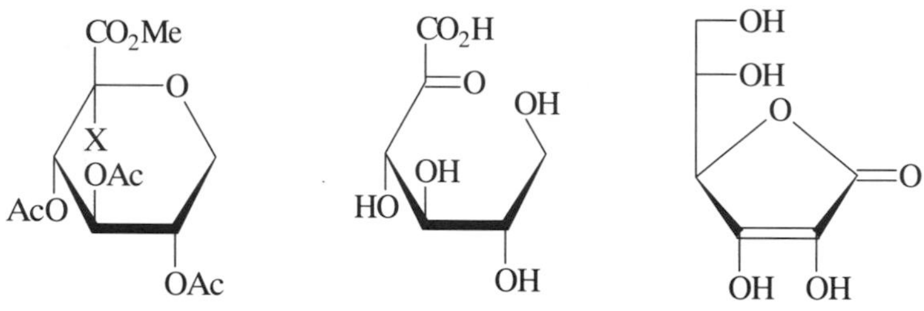

4.240 X = H
4.241 X = Br

4.242

4.243

It is not necessary, however, to have overtly radical-stabilizing groups such as carboxylates in the substrates, penta-*O*-acetyl-β-D-glucopyranose **4.244** giving the 5-bromide **4.247** in high yield, as does the 5-epimer penta-*O*-acetyl-α-L-idopyranose **4.246**. This establishes that each of the acetates reacts by hydrogen abstraction from C-5 to give the common radical **4.245** which combines with bromine in the axial orientation (Scheme 4.83) (cf. Section 3.3.4.b).

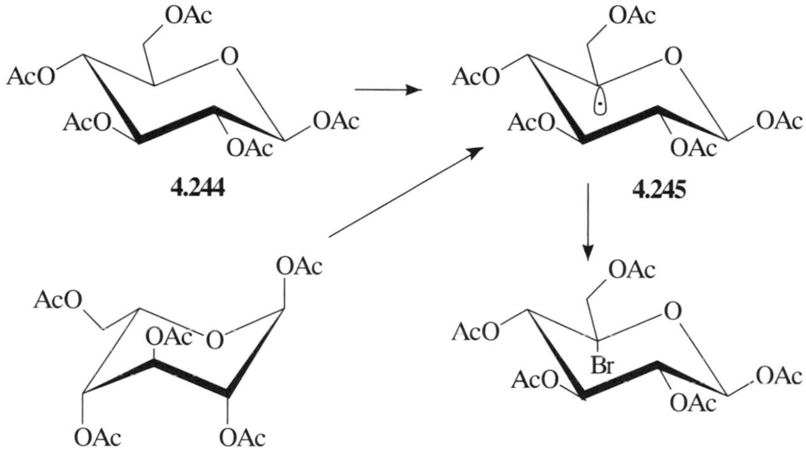

4.244

4.245

4.246

4.247

Scheme 4.83

For this reaction to occur at C-5 it is necessary that there be a strongly electron-withdrawing group at the anomeric centre so that competitive radical stabilization, and hence bromination, does not occur at this acetal-like position. This is well illustrated in Scheme 4.84 by the case of the glycosyl halides: tetra-O-acetyl-β-D-glucopyranosyl fluoride **4.248** gives mainly the 5-bromide **4.249** with only small amounts of **4.250**. On the other hand, the chloride **4.251** affords analogous products **4.252** and **4.253** with the reverse selectivity.

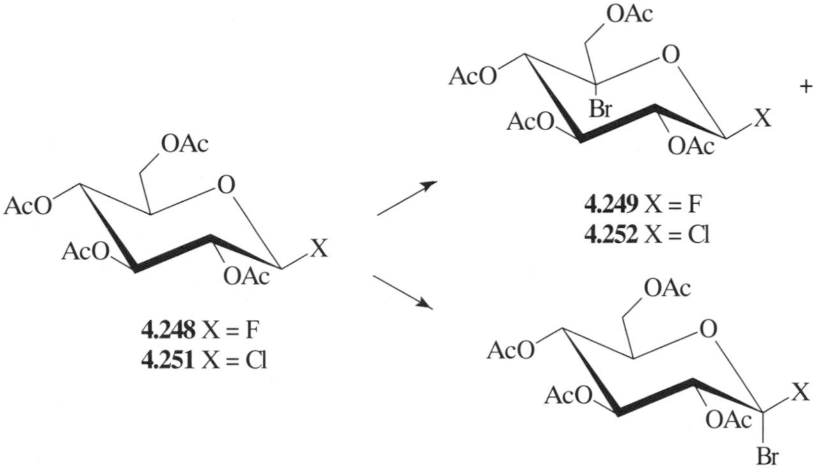

4.249 X = F
4.252 X = Cl

4.248 X = F
4.251 X = Cl

4.250 X = F
4.253 X = Cl

Scheme 4.84

In all brominations of pyranoid compounds of the above kind, substrates with axial groups at the anomeric centre react more slowly: these groups impede substitution at C-5 and also hydrogen abstraction at C-1 (cf. Section 3.3.4).

Compounds **4.254–4.257** are others which are subject to effective photobromination, the hydrogen atoms undergoing substitution being indicated.

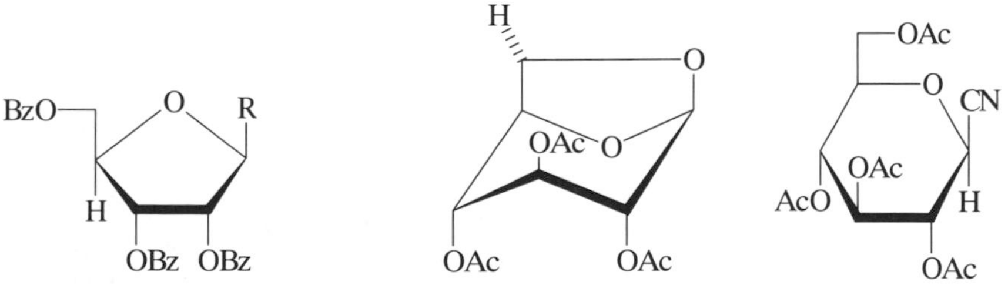

4.254 R = OAc
4.255 R =

4.256

4.257

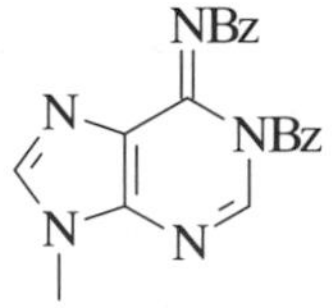

(e)   Direct replacement of hydroxyl groups

*(i)   Application of reagents generated from triphenylphosphine and
   a halogen source*

Several reagents of this type are now available which react by the mechanistic pathway outlined in Sections 4.1.2.a and 4.1.2.c. Some may be used with unprotected sugars, in which case selective halogenation of primary sites is often achieved, whereas others are used to halogenate compounds with unprotected hydroxyl groups in otherwise fully protected sugars. Displacement with inversion of configuration at secondary positions is usually observed but, with sterically hindered alcohols, cases involving protecting group migration and halogenation at sites other than those initially hydroxylated are known.

Mixtures of triphenylphosphine and tetrahalomethanes produce the reagents $Ph_3P^+CX_3X^-$ which, when used in pyridine with unprotected pyranoid and furanoid sugars, react with the hydroxymethyl groups to give primary halo sugars.[143] Thus, treatment of methyl β-D-glucopyranoside **4.258** with triphenylphosphine and carbon tetrachloride, tetrabromide or tetraiodide at room temperature for 18 h produces the corresponding 6-deoxy-6-haloglucosides **4.260** in almost quantitative yields by way of the phosphonium adduct **4.259**. In DMF, partially protected nucleosides can be brominated and chlorinated at the primary positions; for example, **4.261** gives **4.262** (X = Br, Cl) in 55 and 70% yields, respectively.[144] These reagents have also been used to prepare secondary deoxyhalo sugars from partially protected derivatives, as illustrated in Scheme 4.85(a) by the smooth conversion of di-*O*-isopropylidene allose **4.263** into the 3-chloro-3-deoxyglucose derivative **4.265** (X = Cl).[145] The di-*O*-isopropylideneglucose derivative **4.264**, however, in which C-3 is sterically hindered, gives the rearranged 6-chloro-6-deoxyglucose derivative **4.267** (X = Cl) when so treated, as illustrated in Scheme 4.85(f).

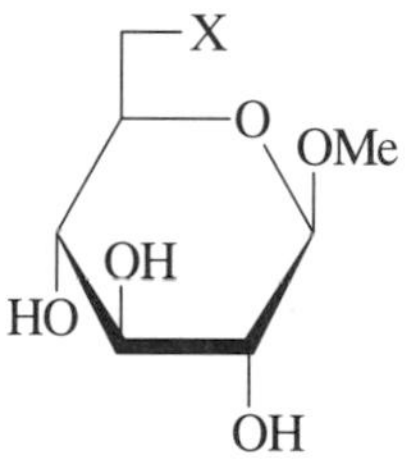

4.258 X = OH
4.259 X = OP⁺Ph₃X⁻   **4.261** X = OH
4.260 X = Cl, Br, I   **4.262** X = I, Cl

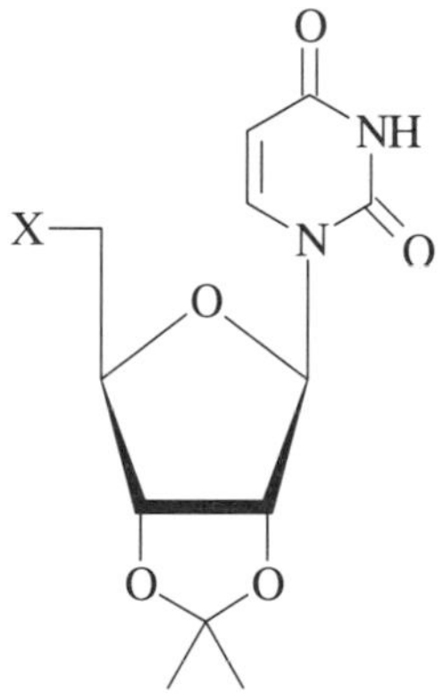

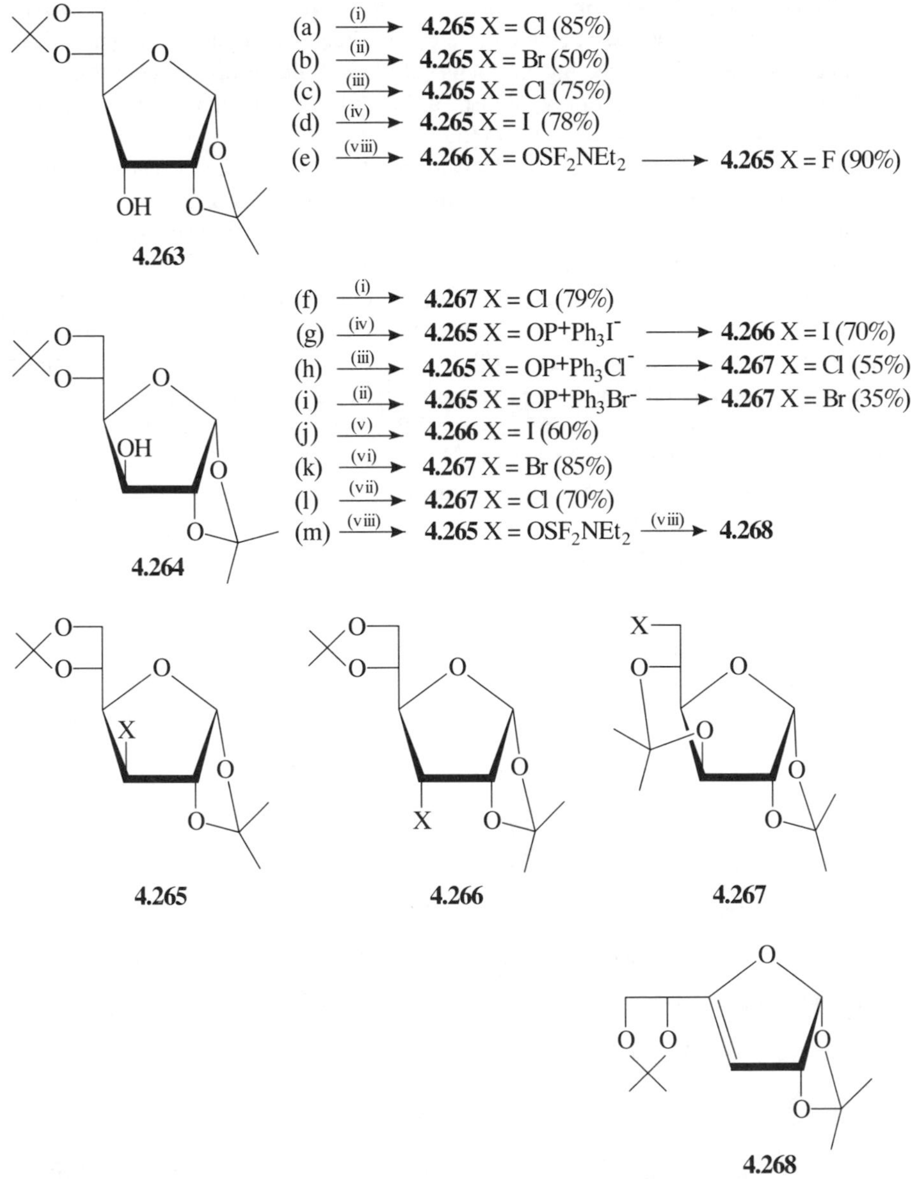

4.263

(a) $\xrightarrow{(i)}$ **4.265** X = Cl (85%)
(b) $\xrightarrow{(ii)}$ **4.265** X = Br (50%)
(c) $\xrightarrow{(iii)}$ **4.265** X = Cl (75%)
(d) $\xrightarrow{(iv)}$ **4.265** X = I (78%)
(e) $\xrightarrow{(viii)}$ **4.266** X = OSF$_2$NEt$_2$ $\longrightarrow$ **4.265** X = F (90%)

4.264

(f) $\xrightarrow{(i)}$ **4.267** X = Cl (79%)
(g) $\xrightarrow{(iv)}$ **4.265** X = OP$^+$Ph$_3$I$^-$ $\longrightarrow$ **4.266** X = I (70%)
(h) $\xrightarrow{(iii)}$ **4.265** X = OP$^+$Ph$_3$Cl$^-$ $\longrightarrow$ **4.267** X = Cl (55%)
(i) $\xrightarrow{(ii)}$ **4.265** X = OP$^+$Ph$_3$Br$^-$ $\longrightarrow$ **4.267** X = Br (35%)
(j) $\xrightarrow{(v)}$ **4.266** X = I (60%)
(k) $\xrightarrow{(vi)}$ **4.267** X = Br (85%)
(l) $\xrightarrow{(vii)}$ **4.267** X = Cl (70%)
(m) $\xrightarrow{(viii)}$ **4.265** X = OSF$_2$NEt$_2$ $\xrightarrow{(viii)}$ **4.268**

4.265   **4.266**   **4.267**

4.268

(i) TPP, CCl$_4$, reflux 96 h; (ii) TPP, BnBr, DEAD, PhMe, reflux; (iii) TPP, PhCOCl, DEAD, PhMe, reflux; (iv) TPP, MeI, DEAD, PhMe, reflux; (v) TPP, imidazole, I$_2$, PhMe, reflux; (vi) TPP, NBS, DMF, 50 °C, 1 h; (vii) Me$_2\overset{+}{N}$=CHCl Cl$^-$, CHCl$_2$CHCl$_2$, 20 °C, 3 h; (viii) Et$_2$NSF$_3$, Py, distil

Scheme 4.85

Mitsunobu conditions (triphenylphosphine/diethyl azodicarboxylate (TPP/DEAD) with alkyl or acyl halides; see Section 4.1.2.c) have been employed in the synthesis of halodeoxy sugars.[3] Thus, heating a toluene solution of diisopropylideneallose **4.263** under reflux with the reagent prepared from benzyl bromide gives the 3-bromo-3-deoxyglucose derivative **4.265** (X = Br) (Scheme 4.85b), and the corresponding chloro and iodo derivatives are obtained in good yields in related reactions using organic halides as the halogen source, as outlined in Schemes 4.85(c) and 4.85(d).[146]

Direct displacement also occurs to give the 3-deoxy-3-iodoallose derivative **4.266** (X = I) when the C-3-*gluco*-epimer **4.264** is treated with TPP/DEAD and methyl iodide, as illustrated in Scheme 4.85(g), but rearrangements take place when **4.264** is caused to react with chlorinating or brominating agents, the 6-deoxy-6-halo-1,2:3,5-di-*O*-isopropylideneglucofuranoses **4.267** (X = Cl, Br) being the main products formed, as illustrated in Schemes 4.85(h) and 4.85(i). The relevant intermediate phosphonium salts **4.265**, (X = OP⁺Ph₃) may be isolated when these reactions are conducted without heating.[146]

A degree of selectivity has been achieved with partially protected glucopyranosides possessing two free secondary hydroxyl groups. Thus, methyl 2,6-di-*O*-*t*-butyldimethylsilyl-β-D-glucopyranoside **4.271** can be converted mainly to 3-deoxy-3-iodoalloside **4.269** or the 3-bromo-3-deoxyalloside **4.270** as illustrated in Schemes 4.86(a) and 4.86(b). With the α-glucoside **4.272**, reactions occur preferentially at C-4 and **4.273** and **4.274** are formed (Schemes 4.86c and 4.86d).[147]

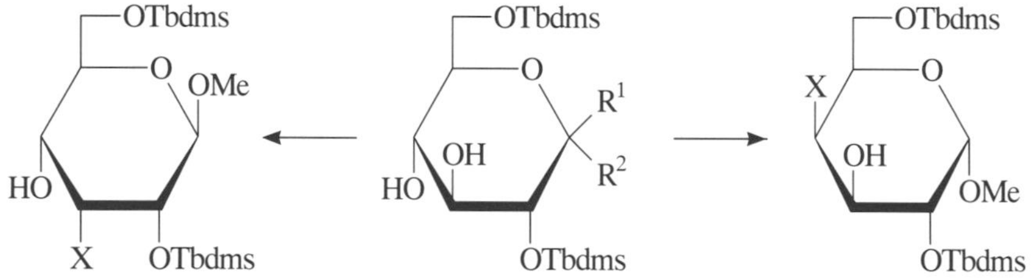

4.269 X = I (74%) ⟵ (a)(i) **4.271** R¹ = OMe, R² = H (c)(i) ⟶ **4.273** X = I (29%)
4.270 X = Br (67%) ⟵ (b)(ii) **4.272** R¹ = H, R² = OMe (d)(ii) ⟶ **4.274** X = Br (42%)

(i) TPP, MeI, DEAD, PhMe, molecular sieves; (ii) TPP, HBr, DEAD, PhMe, molecular sieves

Scheme 4.86

A triphenylphosphine/imidazole/iodine mixture reacts with unprotected glycosides in toluene selectively at primary positions to give terminal deoxyiodoglycosides. For example, methyl α-D-glucopyranoside gives, when heated with this mixture in toluene, methyl 6-deoxy-6-iodo-α-D-glucopyranoside in 80% yield, but in toluene/acetonitrile, a solvent of increased polarity in which the reactants are more soluble, displacements occur at both the 4- position and 6-position to give methyl 4,6-dideoxy-4,6-diiodo-α-D-galactopyranoside in 70% yield.[148] Reactions also occur at single hydroxylic centres with sugars partially protected with acetyl, benzyl or isopropylidene residues. Thus, treatment of diisopropylideneglucose **4.264** with this reagent, as illustrated in Scheme 4.85(j), affords the 3-deoxy-3-iodoallose derivative **4.266** (X = I) without rearrangement. Improved yields of iodo sugars have been achieved in some cases in shorter reaction times by replacing the imidazole and iodine with triiodoimidazole.

Triphenylphosphine and N-halosuccinimides in dimethylformamide have been used to halogenate sugars and nucleosides, both unprotected and partially protected with isopropylidene or acetyl groups.[149] For example, the free hydroxymethyl group reacts when alcohol **4.275** is heated for 2 h at 50 °C, with the N-bromo- or N-chloro-succinimide reagent to give the 6-bromo derivative **4.276** or the 6-chloro derivative **4.277** in yields of 70 and 68%, respectively. On the other hand, the secondary hydroxyl group in diisopropylideneglucose **4.264** reacts with the reagent based on N-bromosuccinimide to give the rearranged 6-bromo-6-deoxyglucose derivative **4.267** (X = Br) when treated under the conditions described in Scheme 4.85(k).[149]

Triphenylphosphite methiodide [$(PhO)_3P^+MeI^-$] and dihalides [$(PhO)_3P^+XX^-$], which are closely related to the reagents described above, have been used to introduce halo groups into partially protected sugars and nucleosides,[150] 2′, 3′-O-isopropylideneuridine **4.261** giving 96% of the 5′-deoxy-5′-iodo derivative **4.262** (X = I) upon treatment for 15 min with the former reagent in DMF at 25 °C. Under similar conditions, 5′-deoxy-5′-iodouridine is selectively formed in 65% yield from the unprotected nucleoside.[151]

(ii) Application of Vilsmeier's imidoyl chloride reagents ($Me_2N^+=CR_2Cl^-$)

Various imidoyl chlorides convert alcohols and diols into alkyl chlorides or formyl esters depending on the reaction conditions. A useful route to chlorodeoxy sugars from carbohydrates possessing one unprotected hydroxyl group is provided by chloromethylene-N,N-dimethyliminium chloride. For example, with this reagent 1,2:3,4-di-O-isopropylidene-α-D-galactopyranose **4.275** in trichloroethane at 25 °C yields the adduct **4.278** which on reflux undergoes nucleophilic displacement by chloride ions at position 6 with loss of dimethylformamide to give the 6-chloro-6-deoxy derivative **4.277** in excellent yield.[152] Alternatively, the adduct **4.278** can be

4.275 X = CH_2OH
4.276 X = CH_2Br
4.277 X = CH_2Cl
4.278 X = $CH_2OCH=\overset{+}{N}Me_2Cl^-$
4.279 X = CH_2OCHO
4.280 X = CH_2I
4.281 X = $CH_2P^+Ph_3I^-$
4.282 X = $^-CHP^+Ph_3$
4.283 X = HC=CHR

hydrolysed to give the 6-*O*-formyl ester **4.279** in better than 70% yield. With certain derivatives, rearrangements of the type illustrated in Scheme 4.85(l) have also been encountered during work with this reagent. A further use of this compound is given in Section 4.5.1.b, where its reaction with epoxides is discussed.

With partially protected sugars with two free hydroxyl groups this reagent produces diformyl esters, 1,2-*O*-isopropylidene-3-*O*-methyl-α-D-glucofuranose **4.284** giving the diester **4.286** (Scheme 4.87a R = H), whereas with α-chlorobenzylidene-*N*,*N*-dimethyliminium chloride the same sugar diol is converted into the 5-*O*-benzoyl-6-chlorodeoxyglucose derivative **4.287** (Scheme 4.87b R = Ph).[153] The stability of the benzoxonium ion **4.285**, (R = Ph) intermediate is responsible for the change in reaction pathway.

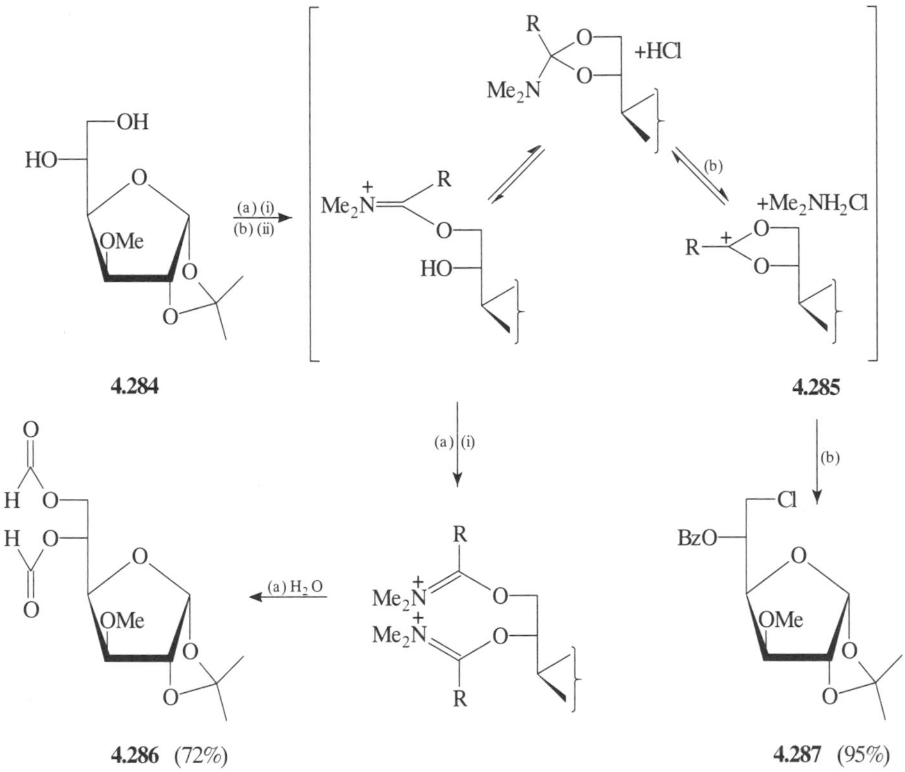
4.284
4.285
4.286 (72%)
4.287 (95%)
(a) (i)
(b) (ii)
(a) H₂O
(b)
+HCl
+Me₂NH₂Cl
OMe
BzO
Cl

(i) $Me_2\overset{+}{N}{=}CHCl\ Cl^-$, Py, ~ 20 °C, 22 h; (ii) $Me_2\overset{+}{N}{=}CPh\ Cl^-$, Py, ~ 20 °C, 22 h

Scheme 4.87

(iii) Application of diethylaminosulphur trifluoride (Et_2NSF_3, DAST)

Alkyl fluorides can be obtained directly from the corresponding alcohols by treatment with DAST, 1,2:5,6-di-*O*-isopropylidene-α-D-allose **4.263** giving the intermediate **4.266** (X = OSF_2NEt_2), which is transformed by nucleophilic fluoride attack into the deoxy-3-fluoroglucose derivative **4.265** (X = F) (Scheme 4.85e). The corresponding di-O-isopropylideneglucose **4.264**, however, forms the intermediate **4.265** (X = OSF$_2$NEt$_2$) when similarly treated (Scheme 4.85m), but steric crowding prevents fluoride attack at C-3 and elimination occurs to give the 3-enose derivative **4.268** as the main product.

Some selectivity has been observed with this reagent; thus, methyl α-D-glucopyranoside gives methyl 6-deoxy-6-fluoro-α-D-glucopyranoside after a short reaction period, whereas methyl 4,6-dideoxy-4,6-difluoro-α-D-galactoside is formed after prolonged treatment.[154]

Attempts to prepare 2-deoxy-2-fluoro sugars by DAST treatment of glycoses which are fully protected other than at O-2 are thwarted if certain substituents, such as azido, thiophenyl, acyloxy or alkoxy, which are prone to migrate, are present in a *trans*-disposition at the anomeric carbon atom. In such cases, the aglycon group migrates to C-2 with inversion of configuration whilst fluorination takes place at C-1, as illustrated in Scheme 4.88.[69] From glycosyl azides and acetates and from benzyl glycosides, which have hydroxyl groups at C-2 *trans*-disposed to the glycosyl substituents, the corresponding glycosyl fluorides with azido, acetyl and benzyl groups at C-2, possessing the alternative *trans*-stereochemisty, can be produced. Since glycosyl fluorides have appreciable value as glycosylating agents, this approach offers powerful opportunities for the syntheses of O-, S-, N- and C-glycosides and oligosaccharides (Section 3.3.5).

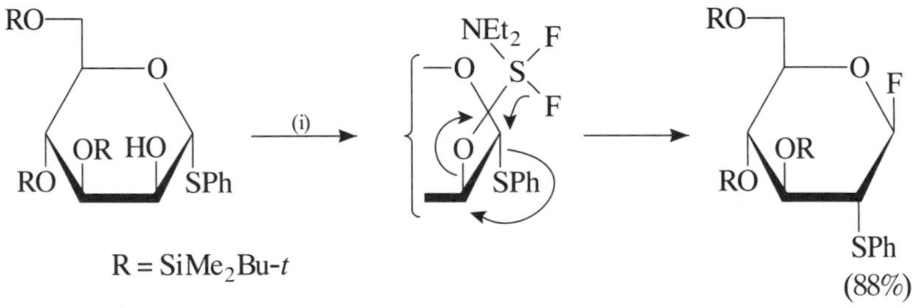

R = SiMe$_2$Bu-t

(i) DAST, CH$_2$Cl$_2$

Scheme 4.88

A similar aglycon migration has been observed during an azide displacement reaction applied to a 2-deoxy-2-iodo-N-sulphonyl-α-D-mannosylamine derivative, the product being a 2-deoxy-2-sulphonylamino-β-D-glucopyranosyl azide which has been used to make an asparagine-linked glycopeptide unit.[155]

(f)   Additions to unsaturated derivatives

One of the most common ways of introducing halogens into organic compounds is by electrophilic addition to alkenes, and this approach can be applied to carbohydrates. Tri-O-acetyl-D-glucal in solution reacts with halogens to give mixtures of 2-deoxy-2-haloglycosyl halides (Section 4.10.1.b.i). An efficient variant of this reaction for fluoro sugars utilizes xenon difluoride additions catalysed by boron trifluoride etherate, as illustrated in Scheme 4.89.[156] Separation of the 2-deoxy-2-fluoroglucosyl fluoride anomers **4.288** and **4.289** from the small amount of the mannosyl isomer **4.290** and subsequent hydrolysis with aqueous sulphuric acid gives 2-deoxy-2-fluoroglucose.

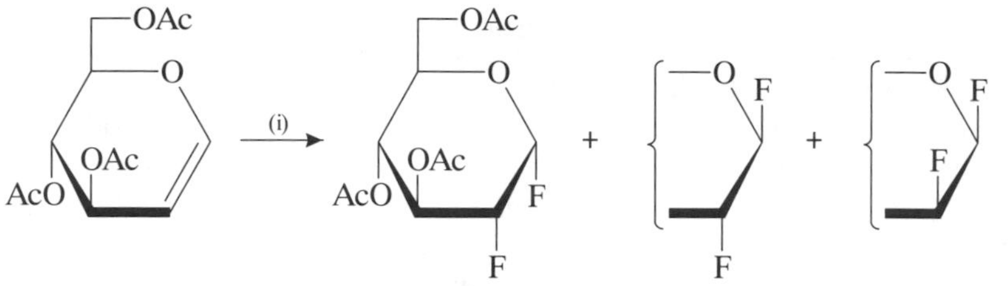

4.288 (61%)   **4.289** (12%)   **4.290** (5%)

(i) XeF$_2$, Et$_2$O·BF$_3$, Et$_2$O, PhH, 22 °C

Scheme 4.89

(g) Direct replacement of carbonyl oxygen atoms

A standard reaction of the carbonyl groups of aldehydes and ketones is their conversion into *gem*-dihalides (chlorides and fluorides) upon treatment with poly-halogen derivatives of phosphorus and sulphur. This approach has been most extensively used in carbohydrate chemistry to prepare *gem*-difluorides, which are of interest because of their potential value in biological studies. Treatment of methyl 3,4-*O*-isopropylidene-β-L-*erythro*-pentopyranosid-2-ulose **4.291**, for example, with DAST in benzene gives methyl 2-deoxy-2,2-difluoro-3,4-*O*-isopropylidene-α/β-L -*erythro*-pentopyranoside **4.292** with good efficiency (Scheme 4.90).[157]

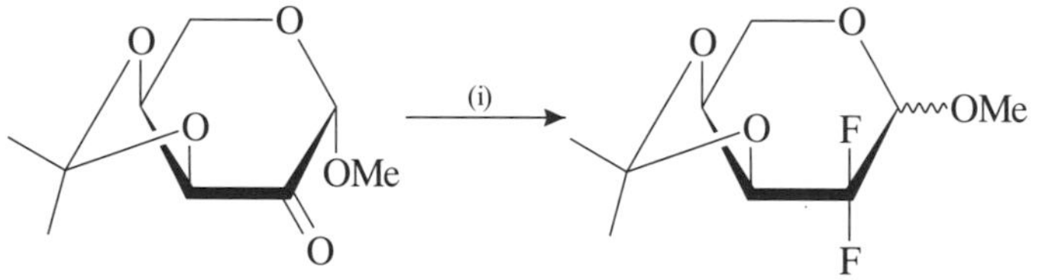

4.291   **4.292**   (78%)

(i) DAST, PhH, 22 °C

Scheme 4.90

4.5.2   REACTIONS

(a)   Chemical reactions[37,129]

The halogen atoms of halogenated sugar derivatives can function as leaving groups in nucleophilic substitution reactions, particularly when the halides are at primary centres; with glycosyl halides (Section 3.3), the ease of the displacements falls in the order I > Br > Cl > F. Iodo and bromo compounds are most suitable for such reactions; secondary fluoro analogues are exceedingly inert. The products of nucleophilic displacements are similar to those discussed in Section 4.1.2

for sulphonates; for example, deoxy sugars are obtained with lithium aluminium hydride, as illustrated in Scheme 4.25(d).

Triphenylphosphine displacements on alkyl halides give phosphonium salts which strong bases convert into phosphorus ylides (phosphoranes) that are useful in Wittig alkene syntheses. A limited number of reactions of this type have been carried out on primary sugar halides. Thus, for example, the 6-deoxy-6-iodogalactose derivative **4.280**, when treated at 110 °C with triphenylphosphine in tetramethylene sulphone for 40 h, gives the phosphonium iodide **4.281**. On treatment with *n*-butyllithium in THF/HMPIT at −60 °C, this salt gives the ylide **4.282** which at −10 °C reacts smoothly with a range of aromatic and aliphatic aldehydes, including dialdose derivatives, to give mainly the (Z)-isomers of chain-extended alkenes (e.g. **4.283**).[158]

When 6-bromo- or 6-iodo-6-deoxypyranose derivatives are treated with weakly nucleophilic reagents such as silver fluoride in pyridine[159] or DBU,[160] eliminations to give 5,6-enose derivatives occur rather than displacements (Section 4.10.3). Similar transformations also occur with basic reagents such as sodium methoxide.[161] Other elimination reactions involving iodo derivatives have already been referred to in Section 4.5.1.a.

Zinc in acetic acid (as well as several other reagents noted in Section 4.2.1.b) reduces carbon–halogen bonds of halo sugars. Under different conditions, zinc causes dealkoxyhalogenation at appropriate vicinal structural units [C(X)C(OR)] within sugar molecules, and Vasella has shown that methyl 2,3,4-tri-*O*-benzyl-6-bromo-6-deoxy-α-(or β-)D-glucopyranoside **4.293**, on treatment with zinc in refluxing ethanol, reacts in this way to give the ring-opened unsaturated aldehyde **4.296** (see Scheme 4.91).[162] The process may be initiated by two single-electron transfers from the zinc to the carbon–halogen bond, followed by a Grub-like fragmentation illustrated in **4.295**. Such transformations can also be induced at room temperature in THF when zinc/silver/graphite is the two-electron donor, as illustrated by the conversion of **4.294** into **4.297** in Scheme 4.91. Other deoxyhalo sugars have been dealkoxyhalogenated with this reagent to give a variety of alkenes.[163]

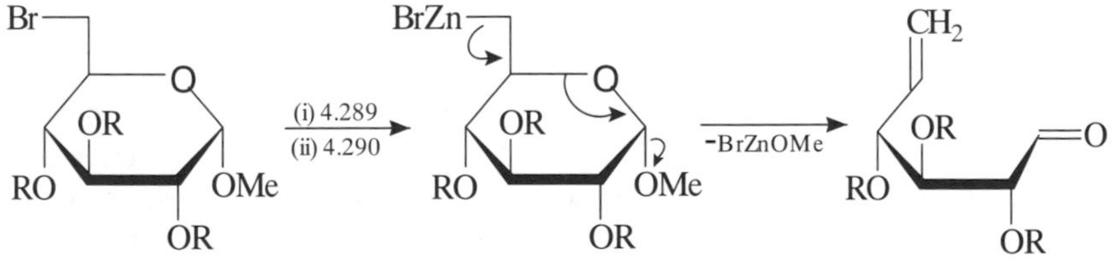

4.293 R = Bn
4.294 R = Me

4.295

4.296 R= Bn
4.297 R = Me (90%)

(i) Zn, EtOH, reflux;   (ii) Zn/Ag/graphite, THF, 20 °C, 45 min

Scheme 4.91

Free radical reduction of primary and secondary sugar halides with tributyltin hydride in the presence of AIBN is a reliable route to deoxy sugars. Thus, for example, ethyl 2-bromo-2-deoxy-α-D-mannopyranoside triacetate **4.298** gives ethyl 2-deoxy-*xylo*-hexopyranoside **4.299** in high yield when so treated, by way of the C-2 radical intermediate **4.303** (R^2 = Et) (cf. Section 3.3.4.a).[164] Examples of similar transformations with chlorodeoxy sugars are given in Schemes 4.25(b) and 4.25(e), Section 4.2.1.b.

4.298 X = Br, R^1 = Me   **4.303**   **4.304** R^3 = Me, R^4 = H
4.299 X = H, R^1 = Me   **4.305** R^3 = H, R^4 = Me
4.300 X = I, R^1 = CH=CH$_2$ **4.306** R^3, R^4 = CH$_2$
4.301 X = I, R^1 = C≡CH
4.302 X = CH$_2$CH$_2$CN, R^1 = H

Radicals generated from halo sugars can be trapped by alkenes prior to final bonding to hydrogen, and this has been developed as a good method for elaborating the carbon frameworks of sugars (see also Section 4.8.2.b).[165] The intramolecular version of this reaction is particularly efficient, as demonstrated by the 5-*exo-trig* cyclization of the radical **4.303** (R^2 = CH$_2$CH=CH$_2$) generated from the allyl 2-deoxy-2-iodo-α-D-mannopyranoside **4.300**, which gives in 60% yield the bicyclic compounds **4.304** and **4.305** in a 1:1 ratio.[166] The propargyl 2-iodomannoside analogue **4.301** undergoes a similar transformation to give the bicyclic exo-alkene **4.306** in 64% yield *via* a 5-*exo-dig* cyclization of the radical intermediate **4.303** (R^2 = CH$_2$C≡CH).[167]

Carbon radicals in pyranose and furanose rings can also be trapped inter-molecularly by electron-deficient alkenes, but the stereoselectivity is normally lower than in intramolecular cases. Thus, acrylonitrile reacts with the pyranose radical **4.303** (R^2=Me) to give methyl 3,4,6-tri-*O*-acetyl-2-cyanoethyl-2-deoxy-α-D-glucoside and -mannoside **4.302** in the ratio of 11:9.[165] In cases in which both substituents adjacent to radical centres on pyranoid rings are equatorially disposed, there is a significant preference for the formation of equatorial bonds, whereas with axial substituents at the adjacent carbon atoms axial bond forma-tion preponderates.[165] Likewise, axial C–C bonds are formed with appreciable selectivity at the anomeric centre (Section 3.3.4.b).

Carbon–iodine bonds are homolytically cleaved on absorption of u.v. light, and the radicals produced readily abstract hydrogen atoms from alcohols. Thus, u.v. irradiation of sugar iodides provides a route to terminal deoxy sugars.[41]

(b)   Reactions in biological systems[37]

Some unsubstituted deoxyhalo sugars are relatively stable and can be used to probe the importance of specific hydroxyl groups in biological systems. For example, D-galactose (i), the 6-fluoro (ii), 6-chloro (iii) and 6-iodo (iv) derivatives of 6-deoxy-D-galactose and 6-deoxy-D-galactose itself (v) are transported across hamster intestinal membrane at rates which decrease in the order (i) > (ii) > (iii) > (v) > (iv), and it has been concluded that hydrogen bonding is important in the transport mechanism since OH and F form strong intermolecular hydrogen bonds, whereas the others cannot. The sharp decrease in transport rate for the iodo compound is attributable to steric effects. An interesting consequence of the ability of fluorine to hydrogen bond was observed in studies of the transport of D-glucose across mammalian intestinal membrane: 6-deoxy-6-fluoro-D-glucose competes with D-glucose in the transport mechanism and thus reduces the rate of glucose assimilation.

3-Deoxy-3-fluoro-D-glucose is incorporated into yeast cells where it blocks glucose and galactose metabolism and stimulates respiration.[130]

2-Deoxy-2-fluoro-D-glucose has proved to be useful for studying various human disorders since it enters metabolic cycles in a way similar to D-glucose, but does not complete the cycle. Consequently, its ^{18}F-labelled analogue, which is a positron emitter, has been employed in nuclear medicine using positron emission tomography (PET) for the study of brain disorders.[130]

Halogenated nucleosides often show enhanced biological activity. For example, 1-(2'-deoxy-2'-fluoro-β-D-arabinofuranosyl)-5-iodocytosine (FIAC) **4.307** shows inhibitory activity against herpes simplex virus type 1 and type 2, varicella zoster virus and cytomegalovirus.[168] An extraordinary naturally occurring 4'-fluoronucleoside, nucleocidin, is noted in Section 8.2.2.a.

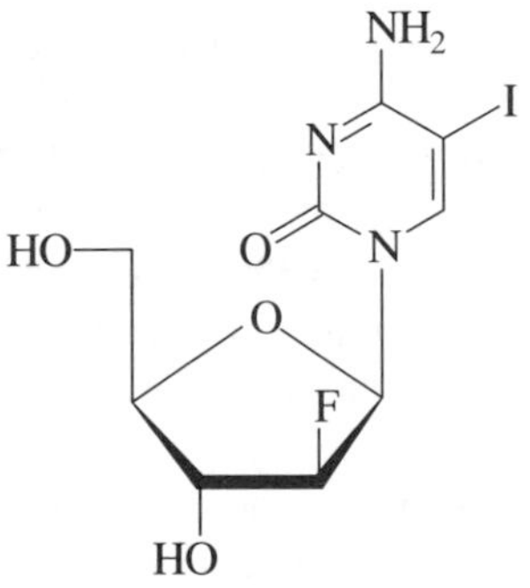

4.307

The trichloro sugar 'sucralose', 4,1',6'-trichloro-4,1',6'-trideoxy-*galacto*-sucrose **5.86** (R = OH), which is a stable, non-toxic, non-nutritive compound 650 times sweeter than sucrose with a similar taste profile, is mentioned in Section 5.3.7.b.

4.6 ANHYDRO SUGARS (COMPOUNDS FORMED VIA INTRAMOLECULAR DISPLACEMENTS WITH HYDROXYL GROUPS)

Formal loss of water between two of the hydroxyl groups in sugar derivatives gives 'anhydro sugars' which can be considered to be intramolecular ethers. Compounds derived from furanoid and pyranoid precursors have bicyclic ring systems and many of the examples taken in this section are of this type. They are named by use of the enumerated 'anhydro' prefix and they will be considered in order of anhydro ring size. Anhydrides involving the anomeric centres are intramolecular glycosides and are discussed in Section 3.1.1.c.

4.6.1. SYNTHESES AND REACTIONS OF OXIRANES (EPOXIDES)

Sugar derivatives which possess three-membered oxygen-containing rings are often described as epoxides and are named as anhydro sugars; the more systematic term, oxirane, is not often used since its introduction leads to nomenclature difficulties. Epoxides are frequently obtainable as stable compounds but are important synthetic intermediates because the three-membered rings are readily opened by various nucleophiles to yield a wide variety of modified sugars.[27]

(a) Syntheses

(i) Intramolecular nucleophilic displacement reactions

Sugar epoxides are usually prepared by intramolecular displacement reactions involving participation from α-hydroxyl groups. Most frequently, α-hydroxysulphonates are used and the formation of the three-membered anhydro rings is most readily achieved when the leaving groups and the hydroxyl oxygen atoms can adopt orientations in which they are coplanar and *trans* to each other. In pyranoid rings this requires the groups to be *trans*-diaxial. This is nicely demonstrated by the formation of various aldose 2,3-epoxides.

   1,6-Anhydro-2-mesyl-β-D-galactopyranose **4.308** (Scheme 4.92a), which possesses a *trans*-diaxial arrangement of the groups at C-2 and C-3, readily forms 1,6:2,3-dianhydro-β-D-talopyranose **4.309** when treated with base,[169] whereas 1,6-anhydro-2-*O*-tosyl-β-D-altropyranose **4.310** does not react because the *trans*-diequatorial disposition of the vicinal hydroxyl and sulphonate ester groups is not favourable for ring closure (Scheme 4.92b).[27] Alternatively, in the preferred $^4C_1$ conformation of methyl 4,6-*O*-benzylidene-2-*O*-tosyl-α-D-glucopyranoside **4.311a**, the sulphonate ester and hydroxyl groups at C-2 and C-3, respectively, are also *trans*-diequatorially arranged, but when this compound is treated with base the 2,3-anhydromannoside derivative **4.312** is produced (Scheme 4.93).[170,171] This reaction is thought to occur via the skew $^3S_5$ conformation **4.311b** in which the substituents at C-2 and C-3 adopt the required *trans*-diaxial orientations. With the altrose derivative **4.310**, the rigidity imposed by the 1,6-anhydro ring prevents the assumption of a related conformation.

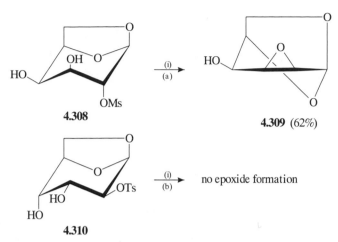

4.308

4.309 (62%)

no epoxide formation

4.310

(i) MeO⁻, MeOH

Scheme 4.92

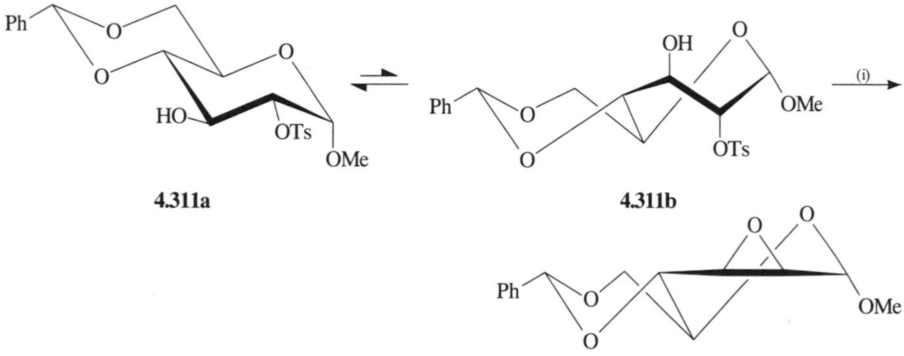

4.311a

4.311b

4.312   (91%)

(i) NaOH, polyethylene glycol, PhH, reflux, 0.5 h

Scheme 4.93

The synthesis of some 2,3-anhydro-4,6-*O*-benzylidenehexopyranosides can be very conveniently achieved by preferential base-catalysed hydrolysis of 4,6-*O*-benzylidene-2,3-di-*O*-tosyl-gluco- and -galacto-pyranosides, as illustrated in Scheme 4.94. The α-glucoside ditosylate **4.313** efficiently yields the α-*allo*-epoxide **4.314** with a range of bases in various solvents,[170] whereas epoxide formation from the β-glucoside **4.315** and the α-galactoside and β-galactoside **4.317** and **4.320**, respectively, is sensitive to the reaction conditions.[172] The phase transfer catalytic conditions given in Scheme 4.94 most effectively convert all four ditosylates into epoxides in good yields.[173]

These reactions presumably proceed by initial selective hydrolysis of one of the tosyl groups, namely that at the O-2 of the α- and β-*gluco*- and α-*galacto*-compounds and at the O-3 of the β-*galacto*-ditosylate. While it is difficult to

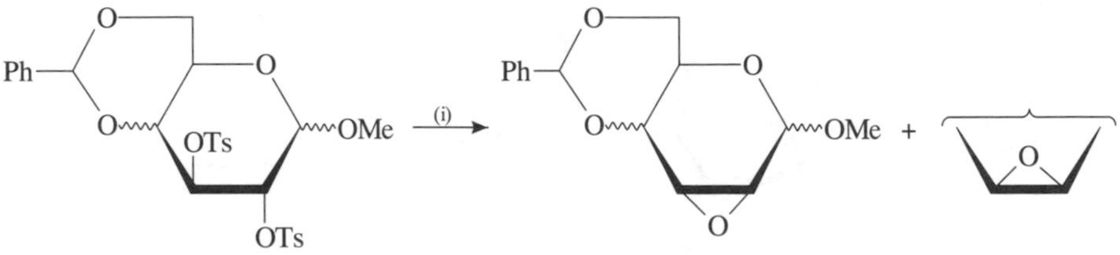

α-gluco-**4.313** $\xrightarrow{\text{(i)}}$ α-allo-**4.314** (90%)
β-gluco-**4.315** $\xrightarrow{\text{(i)}}$ β-allo-**4.316** (84%)
α-galacto-**4.317** $\xrightarrow{\text{(i)}}$ α-gulo-**4.318** (64%) + α-talo-**4.319** (25%)
β-galacto-**4.320** $\xrightarrow{\text{(i)}}$ β-talo-**4.321** (79%)

(i) KOH, Bu$_4$NHSO$_4$, PhH, H$_2$O, DMSO (5%), MeOCH$_2$CH$_2$OH (2%)

Scheme 4.94

rationalize these results, it is notable that in the skew conformations required for epoxide formation (see **4.311b**), compounds **4.313**, **4.317** and **4.320** undergo selective hydrolysis at ester groups which have axial neighbouring oxygen atoms, whereas in **4.315** neither ester has such a neighbour and a predisposition for the 2-ester group to undergo selective hydrolysis is revealed.

Methyl 2,3-di-O-benzoyl-6-deoxy-4-O-tosyl-α-D-glucopyranoside **4.322**, under very mildly basic conditions, gives methyl 3,4-anhydro-6-deoxy-α-D-galactoside **4.323** in good yield. This compound, however, is very sensitive to base and is easily partially converted, following attack at C-3 by the C-2 oxygen nucleophile, to the 2,3-anhydroguloside **4.324** (Scheme 4.95). The equilibrium mixture contains 40% of the 3,4-anhydrogalactoside and 60% of the 2,3-anhydro-gulo-isomer, both of which can be isolated as their acetyl derivatives.[174]

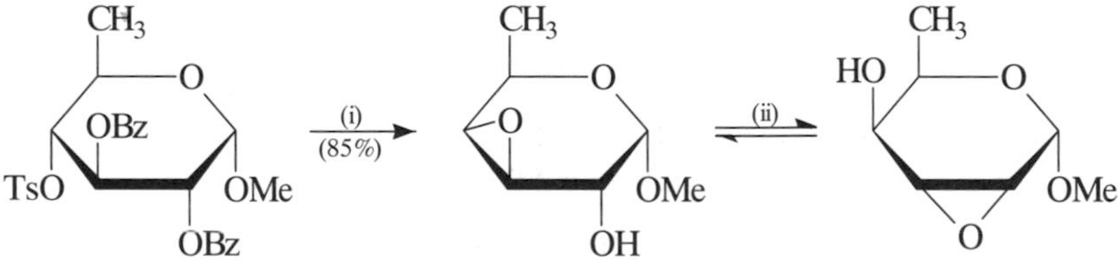

4.322   **4.323** (two parts)   **4.324** (three parts)

(i) MeO$^-$, MeOH, CHCl$_3$, 0 °C; (ii) 0.01 mol l^{-1} NaOH, 42 °C, 80 min

Scheme 4.95

With conformationally mobile systems in which the required antiperiplanar arrangement is readily achieved (even though it may not be the favoured conformation), epoxide ring formation is rapidly accomplished, as illustrated in Scheme 4.96(**a**) by the formation of 5,6-anhydro-1,2-O-isopropylidene-α-D-glucofuranose **4.326** from **4.325**.[170,171]

A very useful, high-yielding and direct method of synthesis of epoxides involves treating trans-disposed vic-diols with DEAD/TPP, a reagent combination which

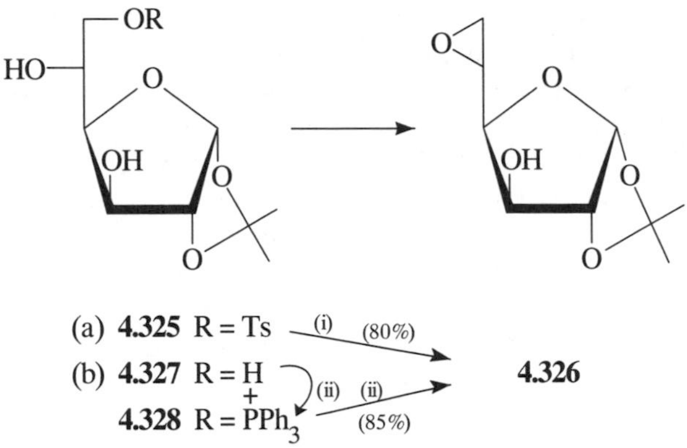

(a) **4.325** R = Ts (i) (80%)
(b) **4.327** R = H ⎫ (ii) (ii)   **4.326**
    4.328 R = PPh₃ ⎭   (85%)

(i) MeONa, MeOH, CHCl₃; (ii) TPP, DEAD, PhH, molecular sieves.
Scheme 4.96

initially forms the betaine adduct (see **4.1**). On treatment with this reagent, the conformationally mobile trihydroxyglucofuranose derivative **4.327** again readily gives the 5,6-epoxy-glucofuranose derivative **4.326** in good yield, this time by way of the 6-oxyphosphonium intermediate **4.328** (Scheme 4.96b; see also Section 4.1.2.c).[175]

Methyl 4,6-*O*-benzylidene-α-D-altropyranoside (see **4.48**, X = OH), which has the C-2 and C-3 hydroxyl groups diaxial in its favoured conformation, as expected, also rapidly forms an epoxide on similar treatment with DEAD/TPP, and since the 2,3-epoxymannose derivative (see **4.46**) is produced, the betaine seemingly derivatizes the 3-hydroxyl group preferentially (to give **4.48**, X = OP⁺Ph₃).[176] Methyl 4,6-*O*-benzylidene-α-D-glucopyranoside, on the other hand, slowly forms the 2,3-epoxyallose derivative by way of the 3-oxyphosphonium intermediate, again indicating preferential derivatization at O-3. The lower reactivity of the *gluco*-isomer is caused by the conformational change necessary to facilitate the epoxide ring closure (cf. Scheme 4.94).[176]

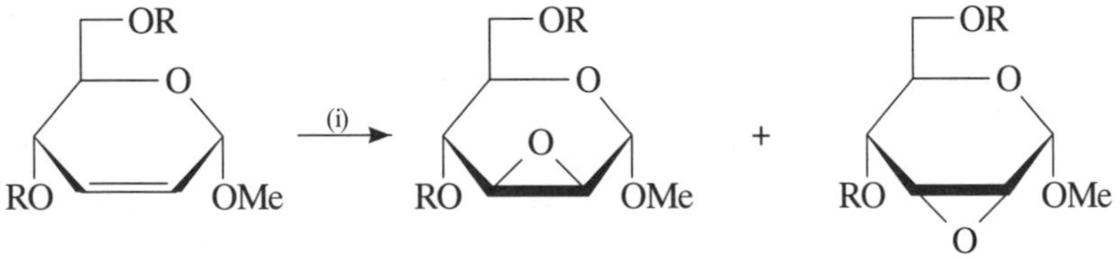

    4.329   **4.330**   **4.331**

(i) H₂O₂, PhCN, MeOH

Scheme 4.97

(ii)   Epoxidation of double bonds

Oxidation of double bonds with peracids, the most general method in organic chemistry for producing epoxides, can be used in carbohydrate work, but diols are the more usual starting materials because they give easier access. When alkene epoxidation is used, mixtures of isomers usually result. Thus, treatment of methyl 4,6-di-*O*-acetyl-2,3-dideoxy-α-D-*erythro*-hex-2-enopyranoside **4.329** (R = Ac) in methanol with hydrogen peroxide and benzonitrile gives the *manno*-epoxide **4.330** (R = Ac) and *allo*-epoxide **4.331** (R = Ac) in a ratio of 3:2 (Scheme 4.97), whereas the isomer distribution is reversed to 1:3 when the unacetylated analogue **4.329** (R = H) is similarly treated. This follows the known generalization that allylic hydroxyl groups direct the oxygen atoms into *cis*-positions, probably by hydrogen-bonding stabilization of the transition states leading to these isomers.[27,177] See Section 3.1.1.c for the stereospecific epoxidation of a glycal with dimethyldioxirane, and Sections 2.5.4 and 4.2.1.g for enantioselective epoxidations of achiral allyl alcohols.

(b) Reactions

Reactions of oxirenes have already been discussed under the headings of the modified sugars which are formed, e.g. deoxy (Section 4.2), amino (Section 4.3), thio (Section 4.4) and halo (Section 4.5) derivatives, and the general principles which control these reactions have been considered in Section 4.1.3.

4.6.2 SYNTHESES AND REACTIONS OF OXETANES IN 3,5-ANHYDROALDOFURANOSES

Several sugars containing four-membered anhydro rings (i.e. oxetanes) are known; most are 3,5-anhydroaldofuranose derivatives.[178] Upon treatment with methoxide ions, 1,2-*O*-isopropylidene-5-*O*-tosyl-α-D-xylofuranose **4.332** undergoes displacement at C-5 with participation from the oxygen atom at C-3 and gives the 3,5-anhydroxylose derivative **4.333** in high yield (Scheme 4.98).[179]

Another approach to 3,5-anhydrides is by application of the ring contraction that α-triflated urono-γ-lactones undergo on treatment with base. Thus, reaction of the readily available 1,2-*O*-isopropylidene-5-*O*-triflyl-α-D-glucofuranurono-6→3-lactone **4.335** with potassium carbonate in methanol gives the methyl 3,5-anhydrouronate derivative **4.336** in good yield (Scheme 4.99).[180] 2,4-Anhydrides of aldonic acids also contain an oxetane ring, and these can be prepared similarly from aldono-γ-lactone sulphonates as discussed in Section 3.1.6.a.ii.

Oxetanes and epoxides undergo similar reactions, but the former have lower reactivity as would be expected for less-strained ring systems. Thus, the oxetane ring in the 3,5-anhydroxylose compound **4.333** is opened by sodium methoxide in refluxing methanol with attack occurring at the least-hindered primary carbon atom to give 1,2-*O*-isopropylidene-5-*O*-methyl-α-D-xylose **4.334**. Aqueous acid, on the other hand, hydrolyses the isopropylidene residue faster than the oxetane ring, as illustrated in Scheme 4.98.[181] The syntheses of oxetanes from glycals are mentioned in Section 4.10.1.b.i and the formation and reaction of an oxetane derived from a pyranosulose are referred to in Section 4.9.1.b.iii.

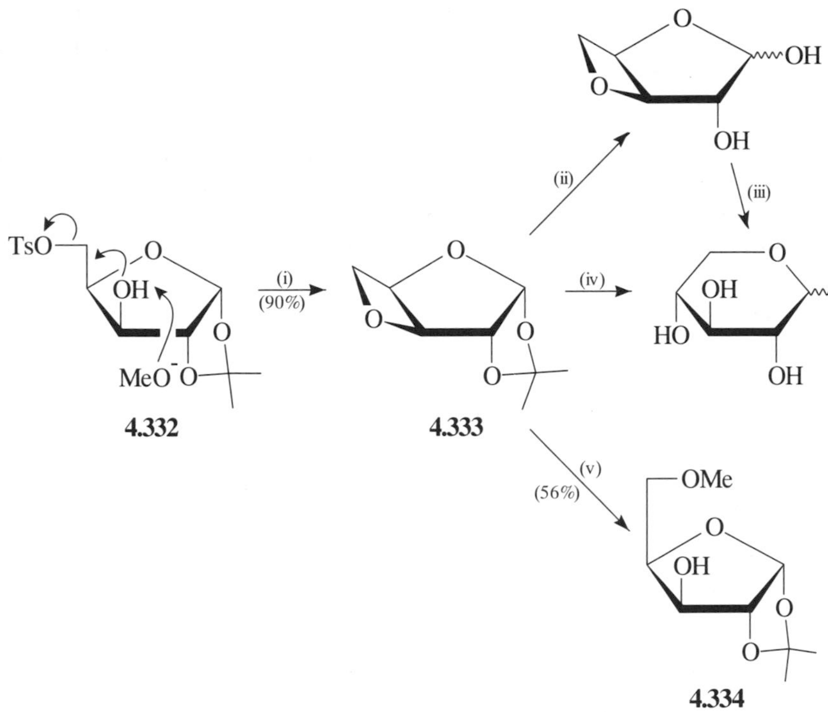

(i) MeO$^-$, MeOH, 25 °C 18 h; (ii) 0.5 mol l^{-1} H$_2$SO$_4$, 100 °C, 2 min; (iii) H$_3$O$^+$;
(iv) 0.5 mol l^{-1} H$_2$SO$_4$, 100 °C, 10 min; (v) as for (i) under reflux

Scheme 4.98

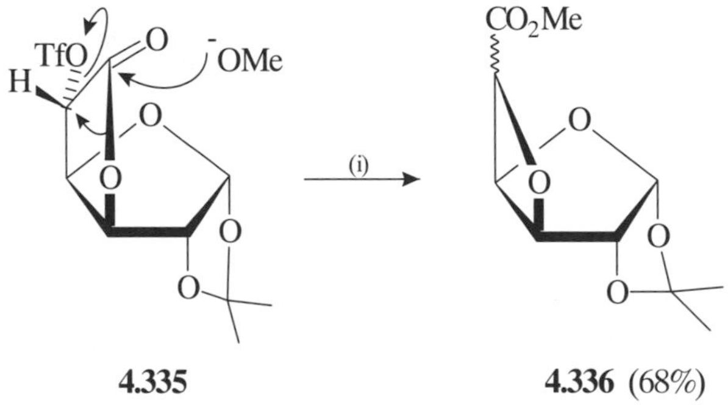

(i) MeOH, K$_2$CO$_3$

Scheme 4.99

4.6.3 SYNTHESES AND REACTIONS OF TETRAHYDROFURANS IN 3,6-ANHYDROPYRANOSES, 3,6-ANHYDROFURANOSES AND 2,5-ANHYDROFURANOSES[182,183]

When cyclic hexose derivatives possessing good leaving groups such as sulphonates, halides or nitrates at C-6 and *cis*-related hydroxyl groups at C-3 are treated with base, intramolecular nucleophilic attack of O-3 upon C-6 occurs and tetrahydrofuran rings are formed. Examples of this type of reaction have been reported for hexoses locked in either the pyranose or furanose ring form, as shown, for example, in the preparation of methyl 3,6-anhydro-β-D-glucopyranoside **4.338** (Scheme 4.100a).[183]

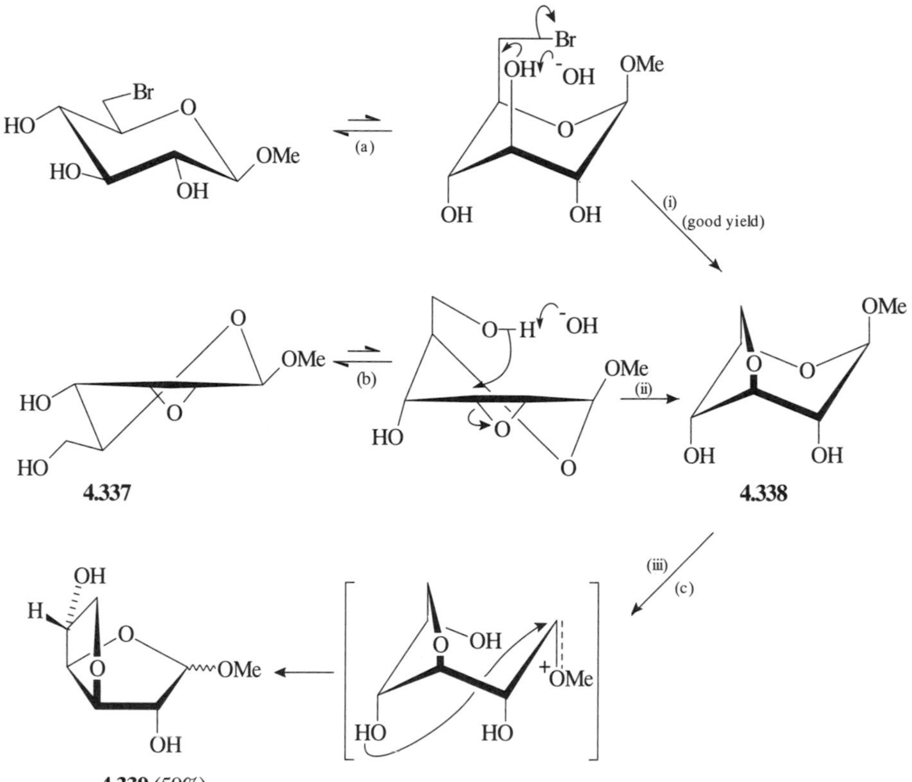

(i) NaOH, EtOH; (ii) NaOH, H$_2$O; (iii) MeOH, H$^+$

Scheme 4.100

A related ring closure can even occur when the C-3 hydroxyl group is benzylated. Thus, attempts to prepare a 1,6-difluoro sugar by treatment of 2,3,4-tri-*O*-benzyl-glucopyranose with DAST are only partially successful. Fluoride ion displacement takes place readily at the reactive anomeric centre to give the glucopyranosyl fluoride, but displacement at the primary centre incurs participation from O-3 with concomitant attack of fluoride on the benzylic carbon atom, as illustrated in **4.340**.[184] An alternative route to the 3,6-anhydroglucoside **4.338** involves base-catalysed opening of the epoxide ring in methyl 2,3-anhydro-β-D-allopyranoside **4.337** by intramolecular attack by O-6, as depicted in Scheme 4.100(b).[185]

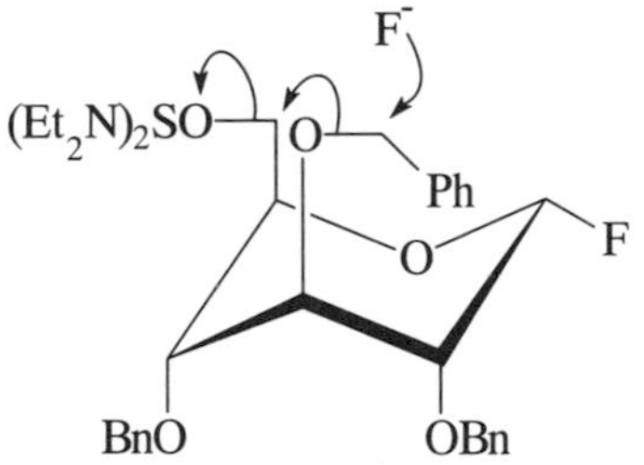

4.340

Compound **4.338** and its furanosyl analogue **4.339** both have tetrahydrofuran rings as parts of their bicyclic structures. However, their dissimilar ring systems impart different properties; of the two, the 3,6-anhydropyranoside **4.338** is more strained, and upon treatment with acidified methanol it gives its less-strained isomers **4.339** (Scheme 4.100c).[183] When methyl 3,6-anhydro-α-D-galactopyranoside **4.341** is similarly treated, the ring strain is relieved by the formation of the dimethyl acetal **4.342** (Scheme 4.101) because in this case it is not sterically possible for the hydroxyl group at C-4 to attack the anomeric centre.[186]

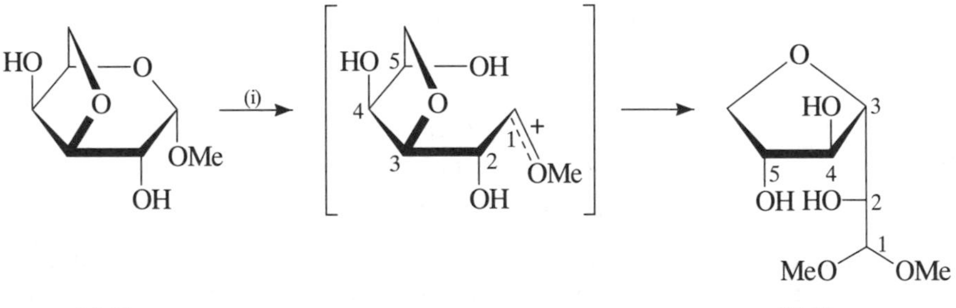

4.341   **4.342**

(i) MeOH, HCl, 22 °C

Scheme 4.101

For steric reasons the hydroxyl group at C-3 in methyl 5-*O*-tosyl-α-L-arabino-furanoside **4.343** cannot participate in ring closure, as was the case with the xylose derivative **4.332** (Scheme 4.98). However, when the former is treated with methoxide ions, the hydroxyl group at C-2, which is *cis* to the carbon atom bearing the leaving group, participates to form the 2,5-anhydride **4.344** (Scheme 4.102a).[182] So strained is the furanoside ring in this [2.2.1]bicyclo system, however, that glycoside hydrolysis occurs in distilled water (Scheme 4.102b).

2,5-Anhydrides are also formed when 1,2-*O*-isopropylidene-5-*O*-tosylaldo-furanose derivatives are treated with methanol and hydrochloric acid, and the transformation offers a good way of preparing anhydrides with *xylo-*, *lyxo-* and

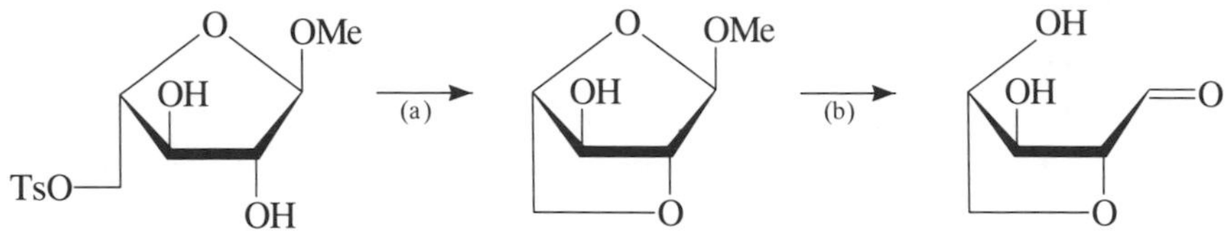

4.343   **4.344** (30%)

(a) MeO⁻, MeOH

(a) MeO$^-$, MeOH

(b) H$_2$O

Scheme 4.102

ido-configurations.[187] Thus, for example, when 3-*O*-benzyl-1,2-*O*-isopropylidene-5,6-di-*O*-tosyl-α-D-glucose **4.345** is so treated (Scheme 4.103), 2,5-anhydro-3-*O*-benzyl-6-*O*-tosyl-L-idose dimethyl acetal **4.347** is formed in good yield. Anhydro ring closure occurs in this reaction after methanolysis of the isopropylidene residue to give dimethyl acetal, as depicted in **4.346.**

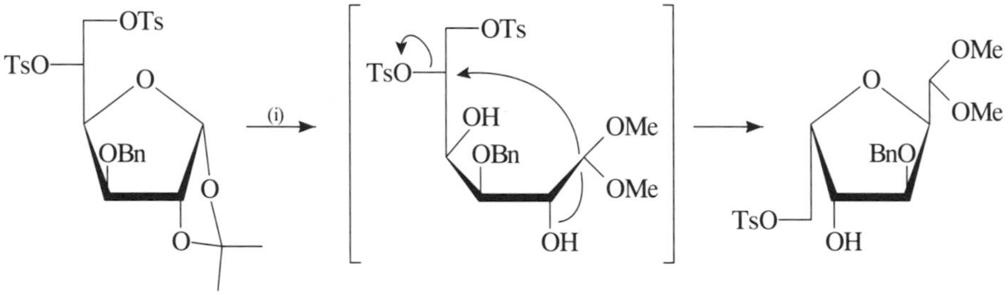

4.345   **4.346**   **4.347** (80%)

(i) MeOH, HCl($\sim$ 1%), H$_2$O ($\sim$ 1%), reflux, 40 h

Scheme 4.103

A synthesis of 2,5-anhydro derivatives from aldono-δ-lactones is discussed in Section 3.1.6.a.ii, and the formation of these compounds as ring-contracted products in displacements from C-2 of pyranose derivatives is described in Section 4.7.1.b.

4.6.4   SYNTHESES AND REACTIONS OF TETRAHYDROPYRANS IN 2,6-ANHYDROPYRANOSES

Attempts to displace the axial sulphonate ester from C-2 in the 1,6-anhydrogalacto-pyranose derivative **4.348** with fluoride ions in methanol fail, but instead the C-1–O-1 bond migrates to give the 2,6-anhydroglycosyl carbocation **4.349** which reacts with the solvent (Scheme 4.104).[188] The 2,6-anhydro-3,4-*O*-isopropylidene-D-talopyranosides **4.350** produced can be hydrolysed in aqueous acid to leave the tetrahydropyran ring intact and give 2,6-anhydro-D-talose **4,351**.

Methyl 2,6-anhydro-α-D-mannopyranoside has been obtained in 20% yield by base-induced displacement, with participation from the 6-hydroxyl group, of a 2-*O*-mesyloxy group in a glucopyranose derivative.[189]

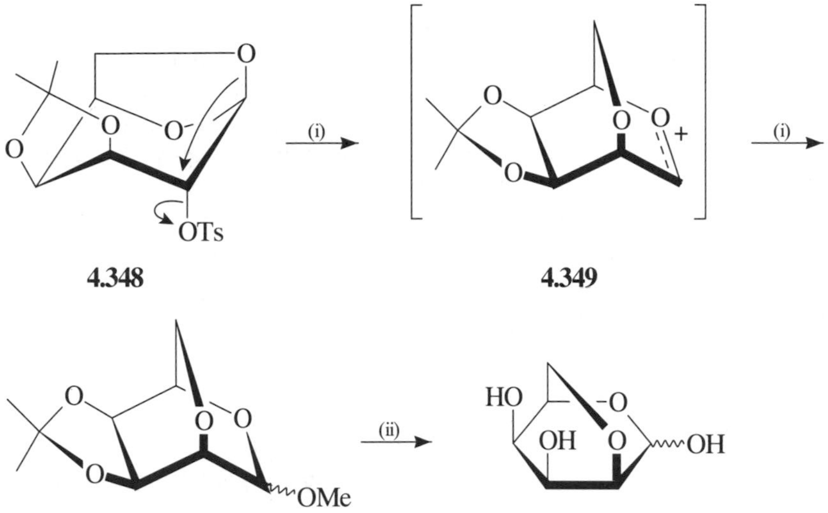

4.348   **4.349**

4.350 (76%)   **4.351**

(i) MeOH, KF·2H$_2$O, 150 °C, 16 h; (ii) H$_3$O$^+$

Scheme 4.104

4.6.5 SYNTHESES AND REACTIONS OF ANHYDROALDITOLS[190,191]

The 1,4- and 1,5-anhydroalditols are formed, respectively, by reduction of 1-halo-or l-thioalkyl-furanosyl or -pyranosyl derivatives. For example, ethyl 2,3,4,6-tetra-O-acetyl-1-thio-β-D-glucopyranoside **4.352** is converted to tetra-O-acetyl-1,5-anhydro-D-glucitol **4.353** with Raney nickel (Scheme 4.105),[192] and 2,3,5,6-tetra-O-acetyl-D-galactofuranosyl chloride yields 1,4-anhydro-D-galactitol in 41% yield when reduced with lithium aluminium hydride (see Sections 3.3.3.e and 3.3.4.a).[193] Reactions of these anhydrides are typical of those of polyhydroxyl compounds, which are discussed in Chapter 5; all functionality deriving from the presence of acetal centres is absent.

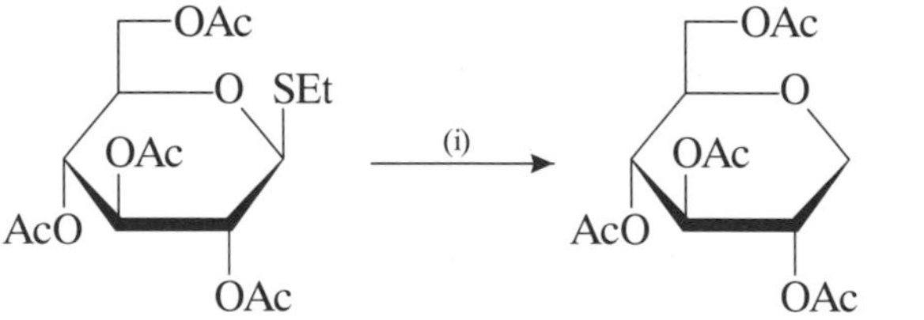

4.352   **4.353** (63%)

(i) H$_2$, Ni, MeOH

Scheme 4.105

1,4:3,6-Dianhydro-D-hexitols are formed by heating those D-hexitols with *threo*-configurations at C-3 and C-4 in hydrochloric acid,[194] compounds **4.354–4.356** being derived from D-glucitol, D-mannitol and D-iditol, respectively. (1,4:3,6-Dianhydro-L-glucitol is produced from D-gulitol.) In addition to the systematic carbohydrate names, the trivial names given under the formulae are also in current use.[194] These anhydrides contain two five-membered rings fused to give bicyclic 'V-shaped' molecules having the free hydroxyl groups projecting inwards or outwards (*endo* or *exo*), and such ring systems can only be formed by *cis*-fusion. The ring closure processes occurring during their formation involve O-3 and O-4 participation either in acid-catalysed displacements of the terminal hydroxyl groups (e.g. Scheme 4.106)[195] or displacements of chloride from 1,6-dichloro-1,6-dideoxy intermediates formed by preferential chlorination of the primary sites. Alternatively, hexitol derivatives possessing good leaving groups at particular positions can be utilized. For example, 3,4-di-*O*-mesyl-D-mannitol, which is obtainable from the 1,2:5,6-di-*O*-isopropylidene derivative after acid hydrolysis for a brief period, is converted into 1,4:3,6-dianhydro-D-iditol **4.356** after prolonged (4 h) treatment by participation from the distal primary hydroxyl groups. With base, however, the dimesylate forms 2,3:4,5-dianhydro-D-iditol by proximal hydroxyl group involvement.[196]

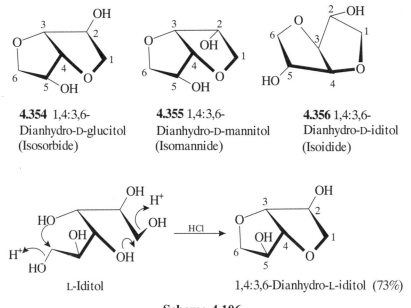

4.354 1,4:3,6-Dianhydro-D-glucitol (Isosorbide)

4.355 1,4:3,6-Dianhydro-D-mannitol (Isomannide)

4.356 1,4:3,6-Dianhydro-D-iditol (Isoidide)

L-Iditol

1,4:3,6-Dianhydro-L-iditol (73%)

Scheme 4.106

The orientation of the hydroxyl groups at C-2 and C-5 has a considerable effect on the chemical reactivities of these compounds. Cyclic acetals can be formed from the *manno*-isomer **4.355** which possesses two *endo*-hydroxyl groups sufficiently close for condensation with carbonyl compounds (Section 5.4.3).[197] The *endo*-hydroxyl group in the *gluco*-isomer **4.354** can be preferentially oxidized by oxygen in the presence of platinum (Section 4.9.1.a.iii)[198] and selectively tosylated

(Section 5.1.3),[199] intramolecular hydrogen bonding being probably responsible for the enhanced reactivity of this hydroxyl in the latter case.

Nucleophilic displacements of sulphonyloxy groups from esters of these compounds also show marked dependence on stereochemistry, the *endo*-sulphonyloxy groups being more easily displaced (cf. Section 4.1.2.a.iii). For example, treatment of 1,4:3,6-dianhydro-2,5-di-*O*-mesyl-D-mannitol with sodium benzoate in DMF gives, in 85% yield the 2,5-di-*O*-benzoyl derivative with the L-*ido*-structure by nucleophilic displacement with inversion at C-2 and C-5,[200] whereas similar treatment of the *gluco*-isomer displaces only the *endo*-sulphonate ester grouping at C-5 to give the monobenzoate also with the L-*ido*-configuration.

With uncharged nucleophiles such as amines or ammonia, *exo*-sulphonate ester groups can be displaced (cf. Section 4.1.2.a.iii), as illustrated by the conversion of 1,4:3,6-dianhydro-2,5-di-*O*-tosyl-L-iditol **4.357** into the tricyclic product **4.358** (Scheme 4.107).[197,201]

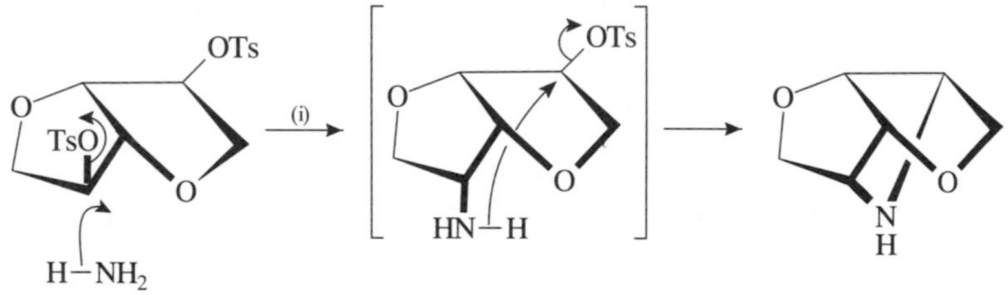

4.357   **4.358**

(i) MeOH, NH$_3$ 160 °C, pressure, 30 h

Scheme 4.107

Several derivatives of these dianhydrides have commercial uses in food processing, polymer manufacture and pharmaceuticals, noteworthy examples in the last category being the 5-mono- and 2,5-di-nitrates of 1,4:3,6-dianhydroglucitol which are used as vasodilators in patients with heart conditions.[194]

4.7 COMPOUNDS FORMED VIA INTRAMOLECULAR DISPLACEMENTS BY NUCLEOPHILES OTHER THAN HYDROXYL GROUPS (RING CONTRACTION REACTIONS)

In some nucleophilic displacements of leaving groups from aldopyranoid derivatives, direct reactions do not take place; instead, 1,2-shifts of pyranoid ring bonds occur and lead to furanoid products.[20,202] The bonds which migrate are always those which involve the most nucleophilic atoms. For example, when the leaving group is at position 2 or 4, the ring oxygen atom participates as the displacing nucleophile in preference to similarly situated carbon atoms. With a leaving group at C-3, participation occurs from C-5 rather than from the electron-deficient C-1.

4.7.1 DISPLACEMENTS WITH RING OXYGEN ATOM PARTICIPATION

(a) Displacements involving leaving groups at C-4

Treatment of methyl 6-deoxy-2,3-O-isopropylidene-4-O-mesyl-α-D-mannopyran-oside **4.359** with sodium acetate in DMF at reflux temperature gives no talopyranoside derivative which would be formed by direct nucleophilic displacement at C-4, partly because of interference exerted by the axial substituent at C-2 on the approaching nucleophile (Section 4.1.2.a). Instead, the 5-O-acetyl-6-deoxy-2,3-O-isopropylidene-α-D-talofuranoside **4.360** (seven parts) and the β-L-allofuranoside isomer **4.361** (one part) can be obtained in 65% yield.[203] This ring contraction proceeds by participation from the ring oxygen atom and migration of the C-5–O bonding electrons to form a C-4–O bond with inversion of configuration.[†] The retention of configuration observed at C-5 in the major product is believed to result from shielding of one side of this carbon atom by the departing mesyloxy group from attack by the acetate anion, as illustrated in Scheme 4.108.

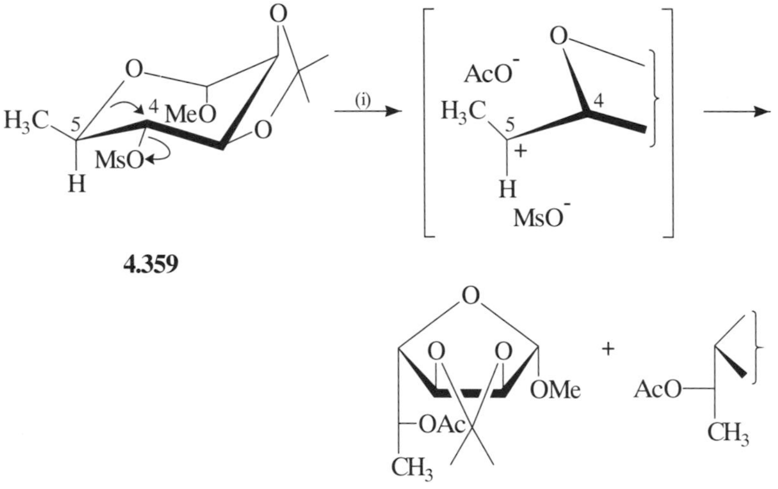

4.359

4.360 (57%)   **4.361** (8%)

(i) NaOAc, DMF, reflux

Scheme 4.108

The paths followed by displacement reactions of this type can be dependent upon the nucleophile used. Compound **4.359** does not undergo ring contraction when treated with hydrazine; instead, the product of direct displacement is obtained, which, after hydrogenolysis, gives methyl 4-amino-4,6-dideoxy-2,3-O-isopropyl-idene-α-D-talopyranoside. From this observation it can be speculated that because hydrazine is uncharged it is more able to align with the axial O-2 than is the naked acetate anion, and it can therefore take part in an S_N2 transition state.

[†] Participations of this kind are most favoured when the migrating bonds are coplanar and *trans* (anti periplanar) to the bonds undergoing rupture.

(b)   Displacements involving leaving groups at C-2

Since C-2 has the same relationship to the ring oxygen atom in pyranoid compounds as has C-4, displacements of certain equatorially oriented leaving groups at C-2 can also proceed with the above-mentioned kind of ring contraction. In this case, migration of the bond between the ring oxygen atom and C-1 is involved. Thus, hydrolysis of methyl 2-*O*-nitrobenzene-*p*-sulphonyl-α-D-glucopyranoside **4.362** yields 2,5-anhydro-D-mannose **4.364**, which arises by attack of water on the stabilized ion **4.363** as shown in Scheme 4.109.[204]

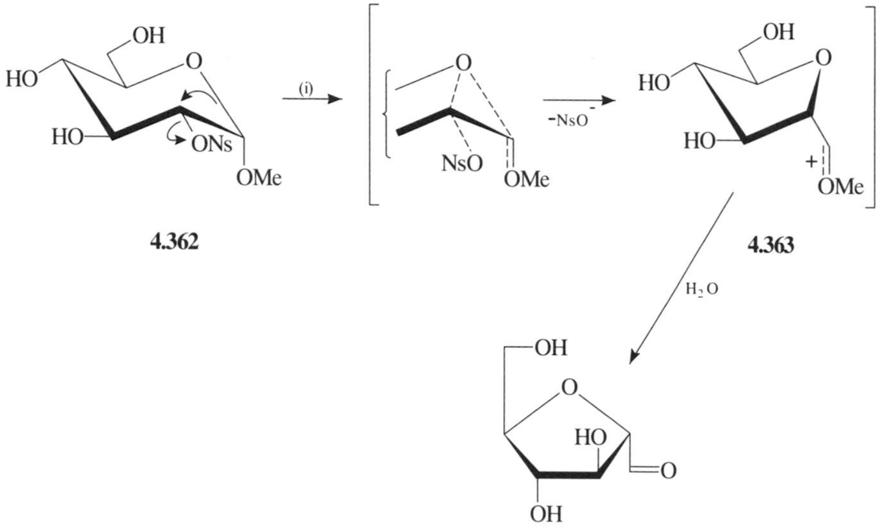

4.362

4.363

H₂O

4.364   (26% 2,5-anhydro-
D-mannitol after reduction)

(i) H₂O, NaOAc, 100 °C, 6 h

Scheme 4.109

Another closely related reaction is the deamination with nitrous acid of 2-amino-2-deoxyaldoses (Section 4.3.2.c). The *gluco*-compound, for example, gives an unstable C-2 diazonium salt which breaks down by a route similar to that shown in Scheme 4.109 to give 2,5-anhydro-D-mannose.[94] The reaction is of value for the specific cleavage of the glucosaminide bonds of heparin (Section 8.1.3.a).

4.7.2   DISPLACEMENTS WITH CARBON ATOM PARTICIPATION

Pyranoid systems, possessing good leaving groups equatorially disposed at C-3 and free hydroxyl groups at C-4, react with carbon participation and migration of the C-4–C-5 bond electrons to give aldehydo groups from the C-4 functions. This is illustrated in Scheme 4.110 by the solvolysis of the glucopyranoside 3-sulphonate ester **4.365** and by the breakdown of a diazonium salt, formed during the deamination of a 3-amino-3-deoxyglycopyranoside **4.367**, to give the 3-deoxy-3-*C*-formyl-*xylo*-hexofuranoside **4.366**.[204] A related 2-deoxy-2-*C*-formyl-*arabino*-hexofuranoside,

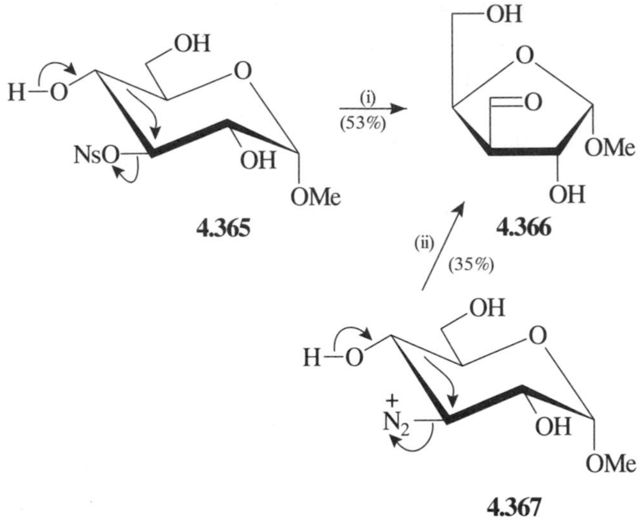

(i) NaOAc, AcOH, H₂O, pH 5, 100 °C; (ii) NaNO₂, H₂O, HCl
Scheme 4.110

which would arise by analogous rebonding of C-1 to C-3, is not formed in these reactions probably because the migratory aptitude of the C-5–C-4 bond is greater than that of the electron-deficient C-1–C-2 bond. For other manifestations of electron deficiency at this site see displacements at C-2 (Sections 4.1.2.a.ii and 4.1.2.a.iii), gem-diol formation with pyranosid-2-ulose derivatives (Section 4.9.1.b.i) and substitution reactions at 2-hydroxyl groups (Section 5.1.3).

4.8   BRANCHED-CHAIN SUGARS

Branched-chain sugars[32,205] have carbon substituents at the non-terminal carbon atoms of the chains. There are two main ways in which such substitutions can be made: (i) C-bonded hydrogen atoms can be replaced as in **4.369**, in which case the compounds are C-substituted derivatives of the normal straight-chain compounds and are classified here as members of the 'dehydro' group; or (ii) hydroxyl functions can be replaced as in **4.368**. In the naming of the latter class the 'deoxy' prefix is included to denote the absence of the hydroxyl substituent at the branching carbon atom, and members can be described as belonging to the 'deoxy' group of branched-chain sugars. Compounds **4.370**, having nitro and amino groups as well as C-methyl groups bonded to C-3 of hexose derivatives, can be considered to be analogues of the dehydro members.

Branched-chain sugars have been found as components of many natural products: D-apiose **4.371** (3-C-hydroxymethyl-D-*glycero*-tetrose) occurs widely in plant polysaccharides and glycosides, duckweed being a particularly rich source,[206] and D-hamamelose **4.372** (2-C-hydroxymethyl-D-ribose) is found as the digalloyl ester **4.373** in the bark of witch-hazel.

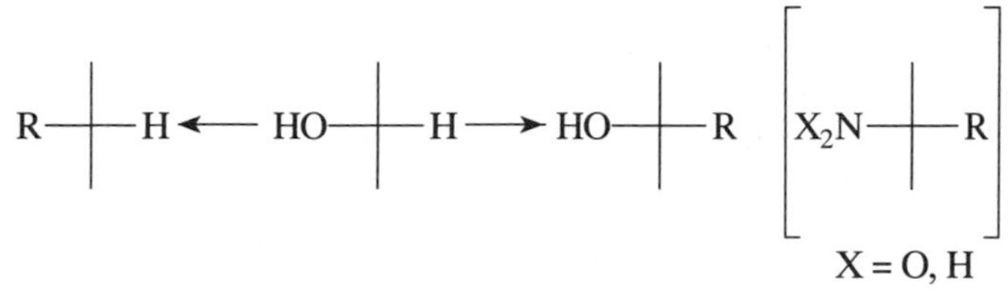

4.368   4.369   4.370

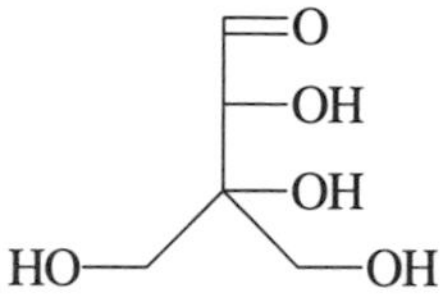

D-Apiose   D-Hamamelose
4.371   4.372

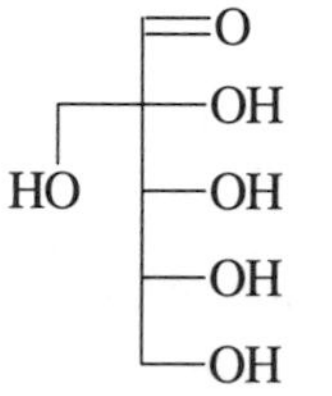

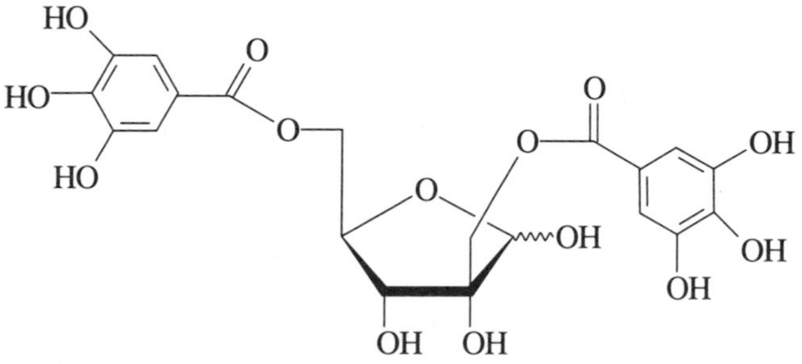

4.373

Antibiotics derived from micro-organisms, mainly of the various strains of *Streptomyces* (Section 8.2.1.c), contain many other members of this class. So far, methyl, formyl, hydroxymethyl, 1-hydroxyethyl, glycolyl, acetyl and butyl side chains have been found in these sugars, and almost all of them have terminal deoxy groups as well. With the exception of blastmycinone (2-*C*-butyl-2,5-dideoxy-3-*O*-(3-methylbutanoyl)-DL-arabinono-1,4-lactone), they are of the type **4.369** with the tertiary hydroxyl group sometimes derivatized. Further examples are L-streptose (5-deoxy-3-*C*-formyl-L-lyxose), which is a component of streptomycin (Section 8.2.1.c), L-mycarose (2,6-dideoxy-3-*C*-methyl-L-*ribo*-hexose) and its 3-*O*-methyl ether L-cladinose, which are found in carbomycin (magnamycin) and erythromycin, respectively, and aldgarose (4,6-dideoxy-3-*C*-[1-(*S*)-hydroxyethyl]-D-*ribo*-hexose 3,1′-carbonate, see Scheme 4.112). Vancosamine (3-amino-2,3,6-trideoxy-3-*C*-methyl-L-lyxo-hexose) and evernitrose (2,3,6-trideoxy-3-*C*-4-*O*-dimethyl-3-nitro-L-*arabino*-hexose) are exceptions in being respectively amino and nitro branched-chain sugars with branch points of the type illustrated in

4.370. Trivial and systematic names of branched-chain sugars which occur in microbiological sources are given in Appendix 5.

Probably the most significant branched-chain sugar derivative is the unstable carboxylic acid 2-*C*-carboxypent-3-ulose 1,5-diphosphate, which is the putative product initially formed during the photosynthetic 'fixing' of carbon dioxide by addition to D-*erythro*-pentulose 1,5-diphosphate, and being a β-keto acid it hydrolyses rapidly and gives two molecules of D-glyceric acid (see Scheme 2.9).

4.8.1 STRUCTURAL ANALYSIS

Early work on the gross structural features of branched-chain sugars was done by classical methods. For example, D-hamamelose **4.372** was converted to 2-methylpentanoic acid, which indicates the degree and site of branching in the carbon chain, and the difficulty experienced in acylating one of its hydroxyl groups indicates that a tertiary alcohol is present. The assignment of configuration, particularly at the branch point, for members of the dehydro series is more difficult. Some of the methods which have been used for this purpose depend upon stereoselective reactions, and in this category periodate oxidations of *vic*-diols (Section 5.5) have proved the most reliable. Pentono-γ-lactones of the D-*ribo*-form **4.374** (X = H), D-*arabino*-form **4.375** (X = H), D-*xylo*-form **4.376** (X = H) and D-*lyxo*-form **4.377** (X = H) oxidize with periodate at the relative rates 32:1:3:50. This is in keeping with expectations since the isomers which possess *vic-cis*-diols reduce periodate faster than those isomers with *trans*-diol units. Applied to the four branched-chain lactones **4.374**–**4.377** (X = CH$_2$OH), which can be synthesized by hydrogen cyanide additions to D-*erythro*- and D-*threo*-pentulose, the same oxidation method gives the relative rates 100:1:3:100, and thus the isomers with *cis*- and *trans*-2,3-diols can be readily identified. Since D-hamamelose can be converted to hamamelonic acid, which gives the lactone **4.374** (X = CH$_2$OH), the D-*ribo*-structure can be assigned to this branched-chain sugar.[207]

4.374   **4.375**   **4.376**   **4.377**

Various other chemical methods for stereochemical assignments have been employed, but often final proof has had to await formal chemical synthesis. Physical methods have also been used, but because branched-chain sugars of the dehydro type do not possess hydrogen atoms at the branch points, the most valuable ^{1}H n.m.r. parameter for solving these stereochemical problems, the vicinal proton–proton coupling constant, is not available. Chemical shift analysis in both the ^{1}H and ^{13}C n.m.r. spectra can, however, be used profitably when pairs of isomers are available.[205] One technique which has been successfully applied utilizes the deshielding effect of oxygen atoms on spacially close protons. For example, with stereoisomers of the type **4.378** and **4.379** the positions of resonance of the ring hydrogen atoms (H_a) are usually found to be displaced about 0.2 p.p.m. to lower field in the former. See the discussion of compounds in Scheme 4.112 for a further example of this method and the application of the nuclear Overhauser effect (nOe), which is another n.m.r. technique useful, in particular, for detecting methyl groups that are in close proximity to certain protons.

Several signals in the ^{13}C n.m.r. spectra of dehydro branched-chain sugars can be of value in stereochemical assignments. For example, the α-carbon atom of a branch group which is axially oriented on a pyranoid ring resonates upfield relative to the same carbon atom in the isomer in which the branch chain is equatorially disposed. The stereochemistry at the secondary carbon atoms in deoxy branched-chain compounds can be determined in the usual way from ^1H-^1H coupling constants.

In particular cases, infrared spectroscopy has proved to be very useful for stereochemical analyses of pyranoid compounds of this series.[208] The method is capable of detecting structural features which have hydroxyl groups and substituted hydroxyl groups in the 1,3-diaxial relationship to each other. In such systems, the free hydroxyl functions can hydrogen bond to the substituted oxygen atoms as shown in **4.380**, and this can be easily detected by measurement of the stretching frequency of the O–H bonds in dilute carbon tetrachloride solution. Compounds which allow such bonding have stretching frequencies near 3500 cm^{-1}, which is more than 100 cm^{-1} lower than the value for the stereoisomers with unbonded hydroxyl groups.

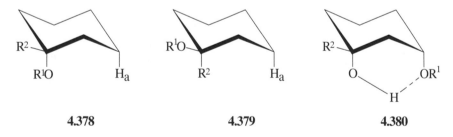

4.378   **4.379**   **4.380**

4.8.2   SYNTHESES

(a)   Dehydro branched-chain sugars

The addition of carbon nucleophiles to ketones derived from carbohydrates is the method most frequently used to prepare this class of compound (Section 4.9.1.b.i). D-Apiose **4.371**, for example, has been synthesized in modest yield using 3-O-benzyl-D-fructose **4.381** as the ketonic precursor and cyanide ion as the carbon nucleophile, as shown in Scheme 4.111(a).[209] The subsequent carbon chain degradation relies on preferential attack by lead tetraacetate at primary, secondary α-diol groups. Another route to the same branched-chain sugar

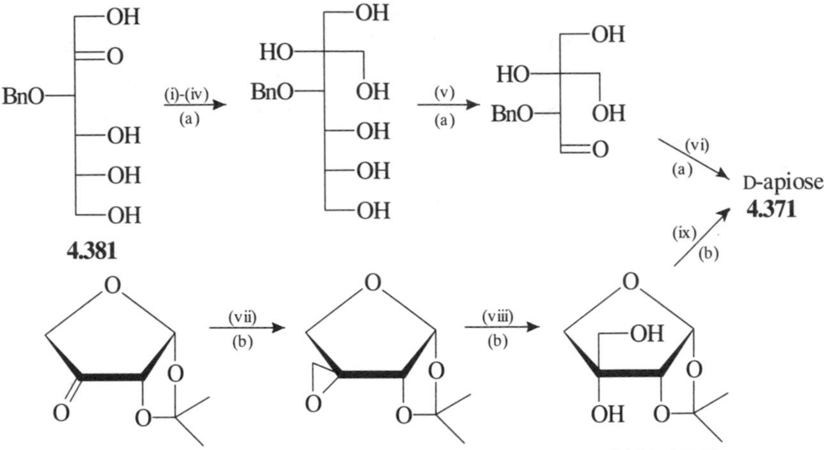

4.381   **4.371**   D-apiose

4.382   **4.383** (49%)   **4.384** (55%)

(i) NaCN, pH 9.3; (ii) hydrolysis of cyanohydrin; (iii) lactonization; (iv) NaBH$_4$;
(v) PbOAc$_4$; (vi) H$_2$ Pt; (vii) CH$_2$N$_2$, MeOH, Et$_2$O; (viii) NaOH, EtOH; (ix) H$_3$O$^+$

Scheme 4.111

involves reaction of diazomethane with the carbonyl group in 1,2-*O*-isopropylidene-
α-D-*glycero*-tetros-3-ulose **4.382** to give the epoxide **4.383** as the major product.
(See Section 4.9.1.b.i for a discussion of the mechanism of this addition and
for a closely related addition initiated by dimethylsulphonium methylide.) The
three-membered ring of this compound can be opened with hydroxide ions
by exclusive attack at the least-hindered methylene carbon atom to give 3-
C-hydroxymethyl-1,2-*O*-isopropylidene -α-D-*erythro*-tetrose **4.384**, from which
D-apiose **4.371** is obtained upon mild acid hydrolysis (Scheme 4.111b).[210]
The chemistry and biochemistry of this sugar have been the subject of a
review.[206] Related syntheses of D-mycarose and its C-3 epimer and other
carbonyl addition reactions leading to branched-chain sugars are mentioned in
Section 4.9.1.b.i.

Aldgarose, which possesses a more elaborate asymmetric 1-hydroxyethyl
side chain, has been synthesized as outlined in Scheme 4.112 by addition of
the 'umpoled' nucleophile 2-lithio-2-methyl-1,3-dithiane to the pyranosidulose
4.385.[211] The resulting mixture of compounds **4.386** and **4.387** with the D-*ribo* and
D-*xylo*-configurations respectively, can be separated and the products sequentially
demercaptalated, reduced, esterified with diphenyl carbonate and debenzylated to
give the (1′S)-isomer and (1′R)-isomer **4.389**, as outlined in Scheme 4.112 for the
ribo-isomer. Assignments of configurations to the addition products **4.386** and **4.387**
were facilitated by the bulk of the methyldithianyl groups, which requires them to
assume the equatorial orientation in both epimers. Thus, compound **4.386**, which
exhibits large (7.0 Hz) $J_{1ax.2ax}$ and large (10.6 Hz) $J_{4ax.5ax}$ coupling values, exists in
the $^4C_1$ conformation which, with the methyldithianyl group equatorially disposed,

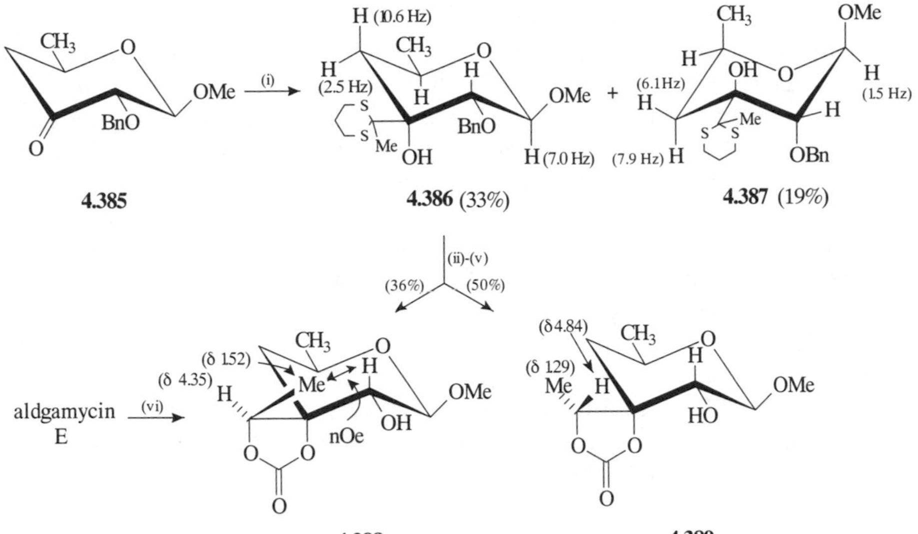

4.385   **4.386** (33%)   **4.387** (19%)

4.388   **4.389**

(i) 2-lithio-2-methyl-1,3-dithiane, C_6H_{14}, THF, N_2, $-78\,°C$, 2.5 h; (ii) $HgCl_2$, HgO, MeOH, reflux, 1.5 h; (iii) $NaBH_4$; (iv) $(PhO)_2CO$, DMF, $NaHCO_3$, $150\,°C$, 30 h; (v) Pd/H_2, MeOH, H_2O; (vi) MeOH, H_2SO_4, $22\,°C$, 15 h

Scheme 4.112

must be the *ribo*-compound **4.386**. The compound with the small (1.5 Hz) $J_{1eq,2eq}$ constant and $J_{4,5}$ values of 6.1 and 7.9 Hz is the *xylo*-isomer **4.387** having the $^1C_4$ chair and the bulky group at C-3 again equatorial. The configurations within the branch chain were ascertained by comparison of the chemical shifts of the l'-hydrogen atoms and the l'-methyl groups of the side chains of the two isomers. As illustrated for the *ribo*-compounds in Scheme 4.112, the isomer in which the l'-methyl group was deshielded by its proximity to the C-2 hydroxyl was assigned the (l'S)-configuration **4.388**, whereas the isomer in which the l'-hydrogen was so deshielded had the alternative (l'R)-configuration **4.389**. A nuclear Overhauser effect (nOe) observed only between H-2 and the l'-methyl group in the former compound confirmed its (l'S)-*ribo*–structure **4.388**, and this material was found to be identical to the methyl β-aldgaroside obtained from the naturally occurring macrolide antibiotic aldgamycin E.

One of the most popular synthetic routes, ideally suited to the formation of simple alkyl branched-chain sugars, involves the addition of Grignard reagents, as illustrated by the formation of the 3-*C*-methyl derivative **4.450** (see Scheme 4.133a, Section 4.9.1.b.i). The method can be adapted to give funtionalized branch substituents since alkene and alkyne groups, similarly introduced, may be subsequently elaborated by chemical means.[205] Functionalized groups have also been introduced directly by the closely related Reformatsky reaction (see Scheme 4.133b) and also by the base-catalysed addition of nitromethane or acetonitrile.[205]

(b)   Deoxy branched-chain sugars

Recently developed free radical chemistry with carbohydrates has opened new routes to deoxy branched-chain sugars which rest on organotin hydride induced deoxygenation of sugar thiocarbonyl derivatives and subsequent trapping by electron-deficient alkenes of the sugar carbon radical intermediates.[165,212] The reaction of the methyl xanthate **4.390** of di-*O*-isopropylideneglucose with acrylonitrile is typical, and on heating in the presence of tributyltin hydride and AIBN these two reactants give an 8:3 isomeric mixture of the 3-*C*-(2′-cyanoethyl)-3-deoxy-*gluco*- and -*allo*-derivatives **4.391** and **4.392** (Scheme 4.113).[213] The reaction occurs by way of a C-3 radical intermediate which is trapped with the acrylonitrile preferentially from the sterically least-hindered *exo*-direction.[213] A related reaction with the *galacto*-isomer of **4.390** (inverted configuration at C-4) gives in 35% yield 3-*C*-cyanoethyl-3-deoxy-1,2:5,6-di-*O*-isopropylidene-α-D-galactose stereospecifically, because in this case the *endo*-face of the furanose ring is even more severely blocked.[213]

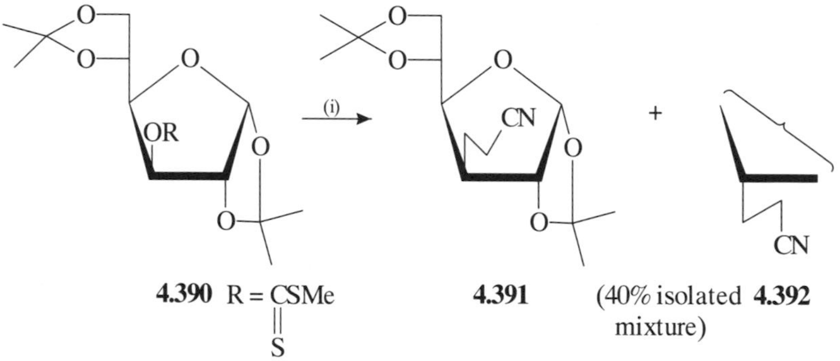

4.390 R = CSMe   **4.391**   (40% isolated **4.392**
          ‖   mixture)
          S

(i) *n*-Bu₃SnH, CH₂=CHCN, AIBN, PhMe, reflux

Scheme 4.113

A very powerful alternative to the tin hydride method that is suitable for *C*-allylation of sugars utilizes allyltri-*n*-butylstannane, which serves as both the precursor of the radical chain transfer agent and as the radical trap.[214] The reagent has been used with a range of thiocarbonyl sugar derivatives and initiation with u.v. light has been shown to be more effective than chemical initiators such as AIBN. By these means benzyl 2,3-*O*-isopropylidene-L-lyxopyranoside 4-phenylthiocarbonates **4.393** are efficiently transformed into the corresponding 4-*C*-allyl-4-deoxy derivatives **4.395**, as indicated by **4.394** (Scheme 4.114).[215]

Free radical methodologies have also been applied to deoxyhalo sugars for making branched-chain deoxy sugars, and although the method is less direct than the deoxygenation route, in cases in which sugar halides are readily available it is very useful.[216] In Section 4.5.2.a, for example, the conversion of 2-deoxy-2-halo sugars into a variety of C-2 branched sugar derivatives was described, the approach being particularly suitable when applied intramolecularly. Applications

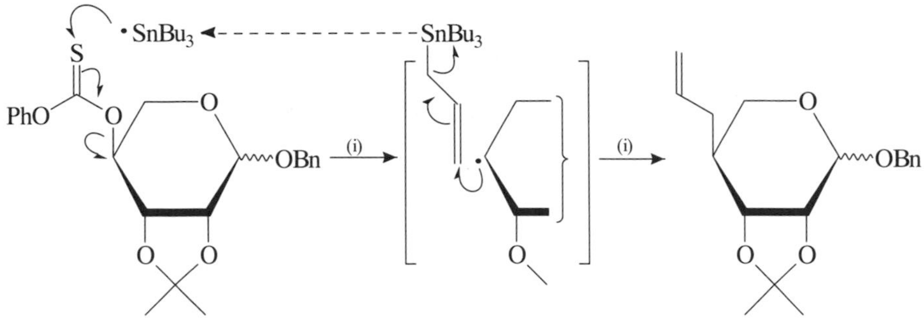

4.393   **4.394**   **4.395** (62%)

(i) $CH_2 = CHCH_2SnBn_3$, hv, AIBN, PhMe

Scheme 4.114

of this methodology in the reverse sense, i.e. by the trapping of radicals generated in substituents of unsaturated monosaccharide derivatives, e.g. **4.396**, can be used to produce mono- or di-branched products in good yield and with high stereoselectivity, as illustrated in Scheme 4.115.[217]

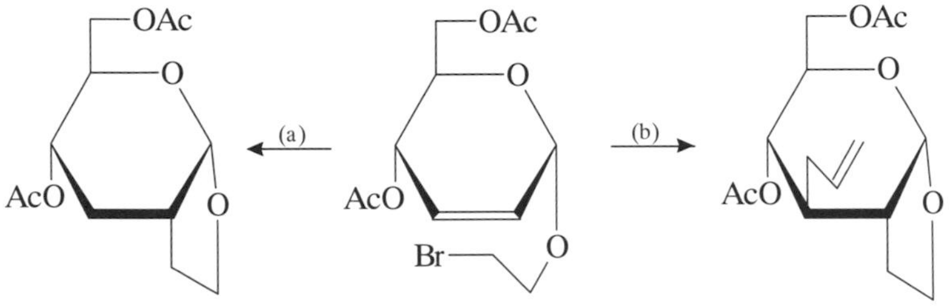

(64%)   **4.396**   (56%)

(a) Bn_3SnH, AIBN, PhH, reflux
(b) $CH_2=CHCH_2SnBn_3$, AIBN, PhH, reflux

Scheme 4.115

The deoxy class of branched-chain sugars can also be formed by the hydrogenation of suitably substituted pyranoid or furanoid derivatives which contain exocyclic double bonds of the type found in **4.398**, obtained from ketone **4.397** (Section 4.9.1.b.i). In this way, the branched-chain sugar **4.399**, which can be further reduced to the 3-deoxy-3-*C*-hydroxyethyl ribofuranose derivative **4.400**, is obtainable (Scheme 4.116). The catalytic hydrogenation of the double bond in **4.398** is highly stereoselective, the hydrogen adding from the less-hindered β-face of the furanoid ring.[218]

The formation of a 4-*C*-cyano-4-deoxypyranoside by S_N2 displacement with tetrabutylammonium cyanide on a relevant 4-*O*-triflyl derivative is one of the few examples of a direct displacement reaction being used as a route to branched-chain sugar derivatives.[219] Epoxide ring opening with carbon nucleophiles has

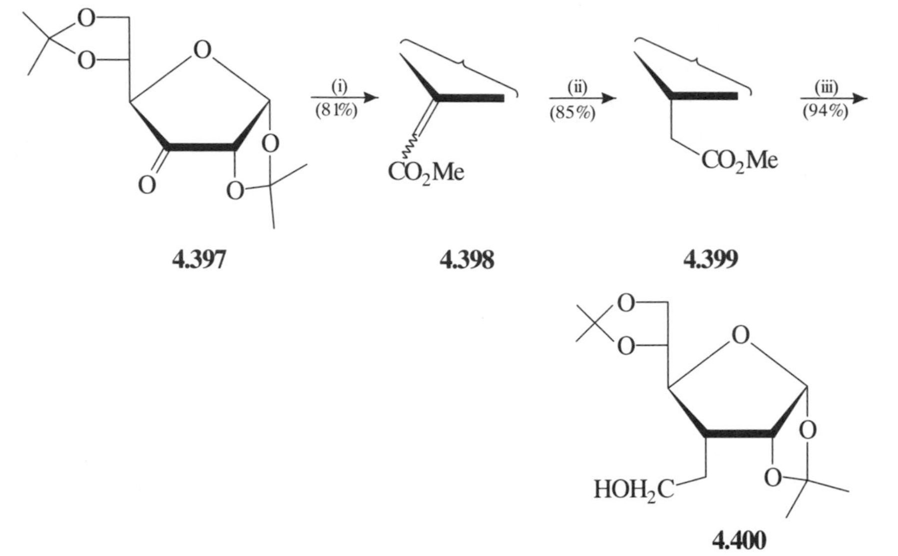

4.397   **4.398**   **4.399**

4.400

(i) $(MeO)_2POCH_2CO_2Me$, DMF, t-BuOK; (ii) H_2, Pd, EtOH; (iii) $LiAlH_4$, THF

Scheme 4.116

proved more successful. Organomagnesium and organolithium reagents have been used, as have lithium dimethyl cuprate, diethyl sodiomalonate and 2-lithio-1,3-dithiane.[205] Care must be taken in these reactions to remove halide ion impurities from reagents prepared from halo compounds, since they can interfere with the desired ring-opening reactions; furthermore, the basic nature of these reagents can produce unwanted unsaturated products. Nevertheless, good yields of branched-chain products can be achieved, as is demonstrated by the reaction of methyl 2,3-anhydro-4,6-O-benzylidene-α-D-mannopyranoside **4.46** with lithium dimethyl-cuprate or ethylmagnesium chloride, which gives 3-deoxy-3-C-methyl or -3-C-ethyl altrosides **4.48** (X = Me[220] or Et[221]) in yields of 70 and 54%, respectively. Dimethylmagnesium in large excess is, however, claimed to be the best reagent for the conversion of **4.46** into **4.48** (X = Me).[222]

   Geminally substituted branched-chain sugar derivatives, in which both the hydrogen atoms and the hydroxyl groups at the same secondary positions of monosaccharides are replaced by carbon-bonded substituents, are also known. These compounds are usually encountered only as intermediates, particularly when sugars are used as starting materials in syntheses of enantiomerically pure non-carbohydrate compounds, as described in Section 7.2. One of the promising methods for constructing such quaternary carbon atoms utilizes a Claisen rearrangement of allyl vinyl ethers derived from sugars, as set out in Scheme 4.117. The unsaturated ester **4.401**, which is readily available from the corresponding pyranosidulose (see Section 4.9.1.b.i), is an excellent precursor for this reaction. Thus, selective reduction of enoate **4.401** gives the allyl alcohol **4.402**, which on vinylation by mercury(II)-induced exchange with ethoxyethene provides the

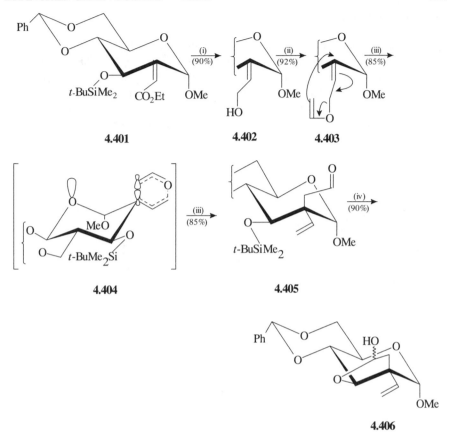

4.401   **4.402**   **4.403**

4.404   **4.405**

4.406

(i) LiAlH$_4$, Et$_2$O, 0 °C, 0.5 h; (ii) EtOCH=CH$_2$, (CF$_3$CO$_2$)$_2$Hg; (iii) PhCN, reflux; (iv) n-Bu$_4$NF

Scheme 4.117

thermally sensitive allyl vinyl ether **4.403**. Upon heating, this dienyl ether undergoes a Claisen transformation by way of transition state **4.404** to give the C-2 geminally branched vinyl *aldehydo*-sugar **4.405** in high yield.[223] The single isomer formed has the new carbon bond in an axial disposition such that the aldehydic substituent is *cis* to the C-3 silyloxy group, as indicated by the ready formation of lactol **4.406** that occurs upon desilylation.

The observed stereoselectivity leading to the new bond at C-2 being axial can be ascribed to through-space delocalization from the axial lone pair of the ring oxygen atom acting in the transition state **4.404**. A similar effect applies in related rearrangements at C-4[224] but not at C-3, and in consequence axial bonding occurs at the former position; however, with Claisen rearrangements at C-3 equatorial bonding takes place, as is found in these rearrangements in carbocyclic systems where stereoelectronic assistance from ring oxygen atoms is not possible.[225]

4.9 DICARBONYL COMPOUNDS

Second carbonyl functions can be introduced into monosaccharides by oxidation of secondary hydroxyl groups, thus yielding aldosuloses **4.407** from aldoses, or diuloses **4.408** from ketoses. If, on the other hand, the primary hydroxyl group in an aldose is oxidized to the aldehydo function, a dialdose **4.409** (R = H) is produced. A uronic acid **4.409** (R = OH) (Section 4.9.3) is formed if the oxidation proceeds to the carboxylic acid stage. Keto derivatives of aldonic acids, aldulosonic acids, are covered in Section 4.9.2.

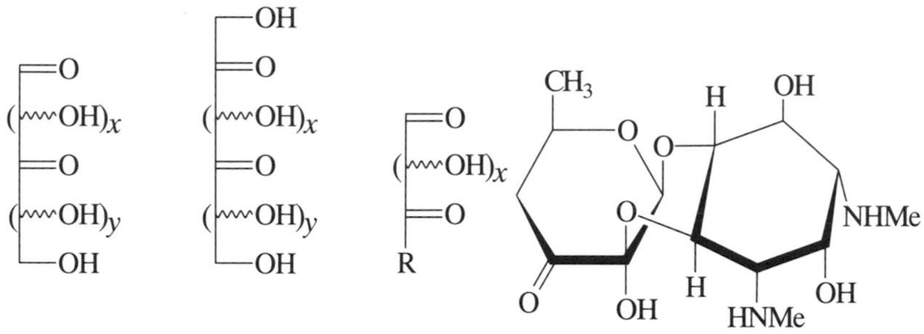

4.407   **4.408**   **4.409**   **4.410**

4.9.1   DIALDOSE, ALDOSULOSE AND DIULOSE DERIVATIVES

Only a few derivatives of this class of compounds have been found in nature, and these occur in antibiotics and plant glycosides which have yielded examples of various forms of hexopyranosides having carbonyl groups at C-2, C-3 or C-4.[226] For example, the antileukaemic principle isolated from the root of *Datisca glomerata* is a glycoside which contains a 2-*O*-acetyl-6-deoxy-α-L-*ribo*-hexopyranosid-3-ulose unit.[227] Very occasionally, tricarbonyl sugar derivatives have been found; spectinomycin **4.410** is based on the aldosdiulose 4,6-dideoxy-D-*glycero*-hexopyranosid-2,3-diulose.[228]

(a)   Syntheses

There are two direct ways of approaching the synthesis of derivatives of these compounds: one involves the partial protection of a monosaccharide in the desired ring form, leaving only a specific hydroxyl group to be oxidized, whereas the other method uses less highly substituted derivatives, often glycosides, and reagents which cause selective oxidation. Some less general methods for obtaining these derivatives are also noted in this section.

(i)   Oxidation of hydroxyl groups in monohydroxy compounds

The conversion of partially protected monosaccharides into dicarbonyl derivatives in good yields requires reagents which oxidize alcohols to aldehydes or ketones under mild, neutral conditions, and these have been available since the mid-1960s.[229-231] Oxidants based upon dimethyl sulphoxide, ruthenium tetroxide and chromium (VI) reagents, usually complexed with pyridine, possess the necessary characteristics, and are now most favoured for the conversion of primary and secondary alcohols into aldehydes and ketones while leaving intact a large number of blocking substituents used in carbohydrate chemistry.

Oxidations with dimethyl sulphoxide based reagents   Dimethyl sulphoxide (DMSO) alone does not oxidize alcohols to carbonyl compounds, but in

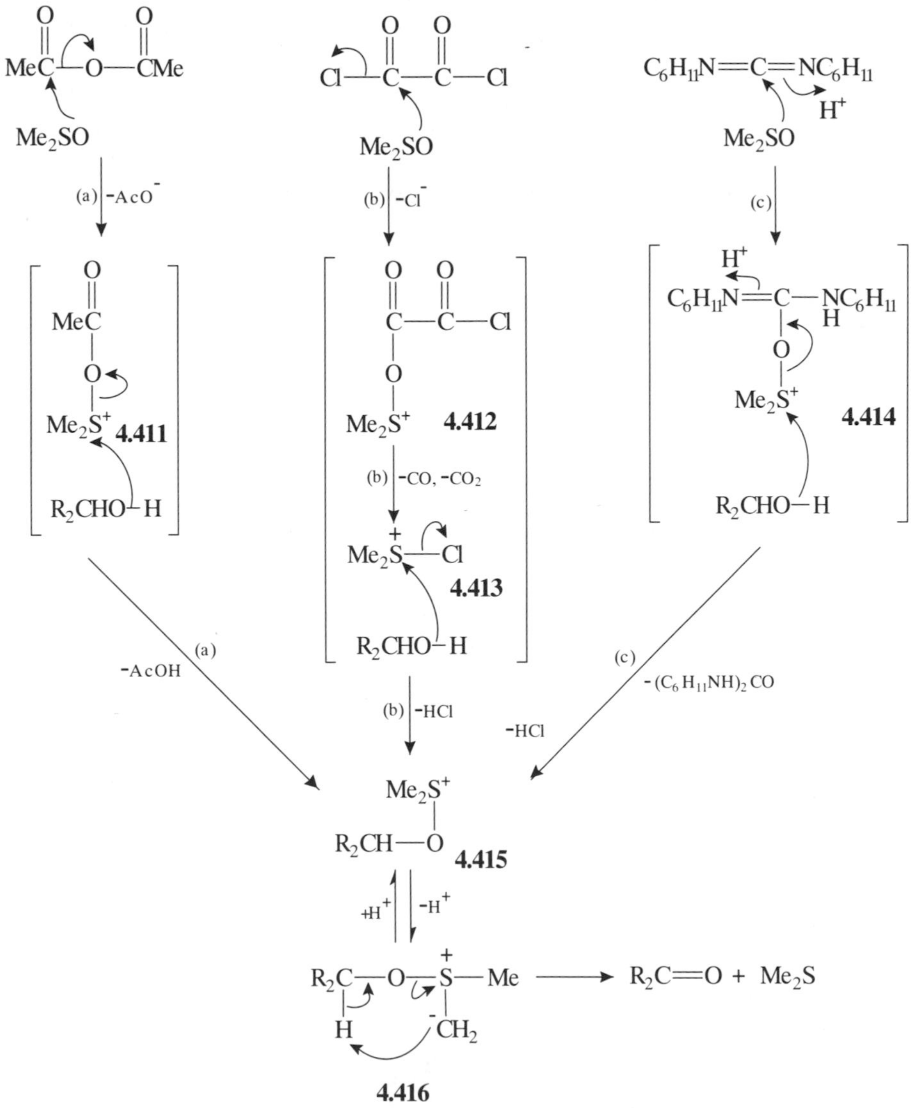

Scheme 4.118

the presence of an electrophilic activating agent such as acetic anhydride, trifluoroacetic anhydride (TFAA), oxalyl chloride, chlorine, phosphorus pentoxide or dicyclohexylcarbodiimide (DCC) with a proton donor it effects this transformation smoothly.[232] Acetic anhydride/dimethyl sulphoxide probably oxidizes by way of the active intermediate **4.411** shown in Scheme 4.118(a), and TFAA forms an analogous trifluoro intermediate. Oxalyl chloride probably activates the DMSO in a similar fashion except that the first-formed intermediate **4.412** is believed to break down with loss of carbon monoxide and carbon dioxide to give the dimethylchlorosulphonium chloride **4.413** oxidizing agent, as illustrated in Scheme 4.118(b). This same intermediate is produced directly in DMSO/chlorine mixtures. With dicyclohexylcarbodiimide, together with phosphoric acid or pyridinium trifluoroacetate as the proton source, the reaction probably relies on **4.414** as the oxidant, as shown in Scheme 4.118(c). The alkoxydimethylsulphonium salt **4.415** is produced in all these reactions, with the final proton abstraction from the carbon undergoing oxidation probably taking place intramolecularly *via* the ylide **4.416**. Among the large number of protecting groups unaffected by these reagents are acetamido, azido and sulphonyloxy groups, cyclic acetals, phenylboronate esters and methyl, triphenylmethyl and *t*-butyldimethylsilyl ethers, and some of these are used in the examples given below. With acylated derivatives, eliminations β to the newly formed carbonyl groups tend to occur giving enones as the isolated products.[233]

A particular merit of the dimethyl sulphoxide based reagents is their ability to oxidize primary alcohols to aldehydes without further oxidation to the carboxylic acids. 1,2:3,4-Di-*O*-isopropylidene-α-D-galactopyranose **4.417**, for example, is converted in good yield to the *galacto*-hexodialdose derivative **4.418** with dimethyl sulphoxide activated with either DCC[229] or oxalyl chloride,[232] (Scheme 4.119a). Virtually none of the dialdose derivative is obtained with the reagent activated by acetic anhydride; instead, some acylation to give di-*O*-isopropylidenegalactose 6-acetate occurs, but mainly the 6-*O*-(thiomethyl)methyl ether is formed by alkylation of the 6-hydroxyl group with the methylenesulphonium cation, as

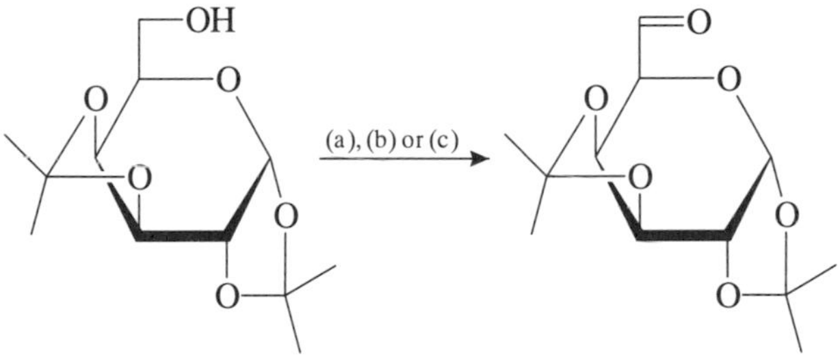

                    4.417   **4.418**

(a) DMSO/DCC or DMSO/ClCOCOCl (80%)
(b) CrO$_3$2Py, CH$_2$Cl$_2$, Ac$_2$O, 25 °C, 2 h (93%)
(c) PDC, Ac$_2$O, DMF, CH$_2$Cl$_2$, reflux, 2 h (71%)
                    Scheme 4.119

outlined in Scheme 4.120 (see Section 5.2.1).[229] An advantage with reactions in which the DMSO is activated with oxalyl chloride or TFAA is that they occur rapidly at $-65\,°C$, at which temperature these competing reactions are suppressed.

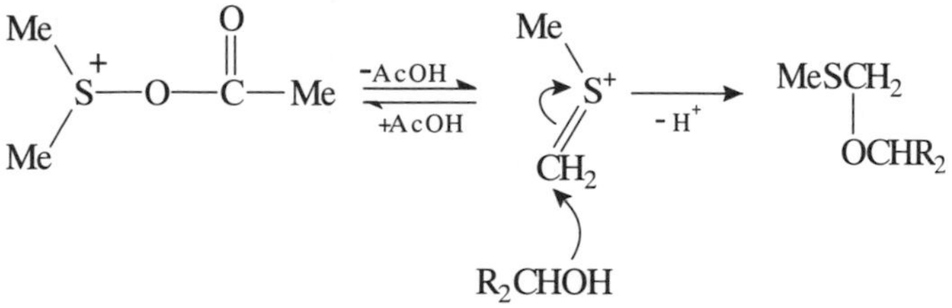

Scheme 4.120

The unprotected secondary equatorial hydroxyl group in methyl 4,6-O-benzylidene-2-O-tosyl-α-D-glucopyranoside **4.419** is efficiently oxidized by DMSO activated with trifluoroacetic anhydride,[232] DCC[234] or phosphorus pentoxide[235] to give methyl 4,6-O-benzylidene-2-O-tosyl-α-D-$ribo$-hexopyranosid-3-ulose **4.420** in good yield, as illustrated in Scheme 4.121(a); again, a lower yield is obtained with the reagent activated with acetic anhydride.[236]

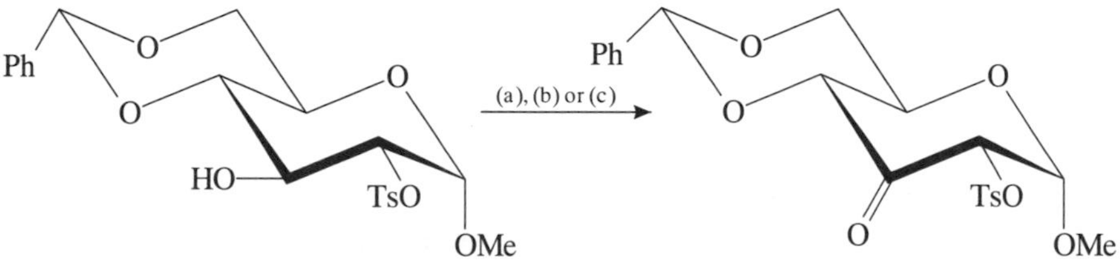

(a) TFAA (89% yield) or DCC (80%) or P_2O_5 (90%) or Ac_2O (50%), all used with DMSO
(b) $CrO_3 2Py$, CH_2Cl_2, Ac_2O, 25 °C, 15 min (95%)
(c) PDC, CH_2Cl_2, Ac_2O, DMF, reflux 16 h (93%)

Scheme 4.121

Epimerization at the carbon atoms adjacent to the newly formed carbonyl groups can occur during some of these oxidations, as is illustrated in Scheme 4.122 by the formation of the same aldosulose derivative **4.422** from either methyl 3-benzamido-4,6-O-benzylidene-3-deoxy-α-D-glucopyranoside **4.421** or its $altro$-isomer **4.423** on treatment with the DCC-activated reagent.[229]

Hydroxyl groups attached to furanoid rings can also be oxidized by most of these reagents, as exemplified by the conversion of di-O-isopropylideneglucofuranose **4.424** into 1,2:5,6-di-O-isopropylidene-α-$ribo$-hexofuranos-3-ulose **4.425** under the various conditions indicated in Scheme 4.123(a).[229,235,237]

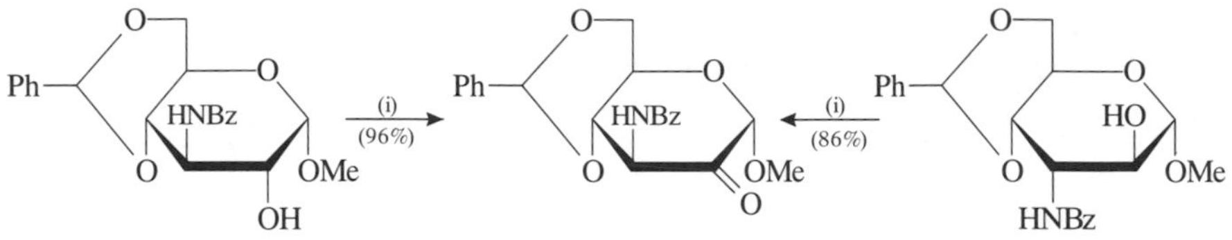

4.421   **4.422**   **4.423**

(i) DMSO, DCC

Scheme 4.122

Oxidations with ruthenium tetroxide   Ruthenium tetroxide is a very powerful oxidant conveniently prepared from hydrated ruthenium dioxide by treatment with sodium periodate.[238] The oxidizing power of RuO_4 is attenuated by using it in low concentration in inert solvents such as carbon tetrachloride, and when used in this way it rapidly oxidizes isolated secondary hydroxyl groups in partially protected sugars, as illustrated in Scheme 4.124(a) by the straightforward oxidation of the partially protected L-rhamnoside **4.426**, (R = OMe) to give the pyranosid-4-ulose **4.427** (R = OMe).[239]

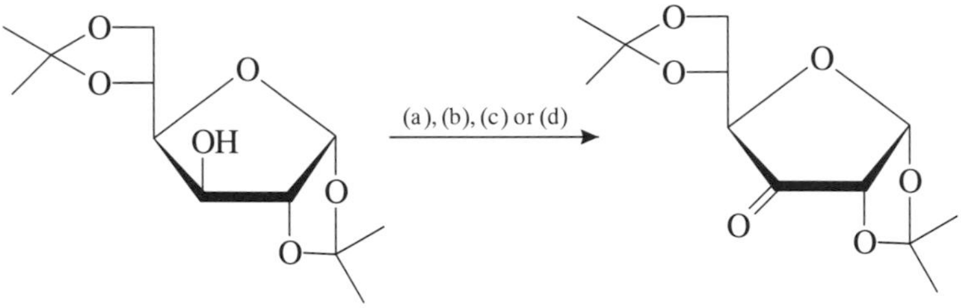

4.424   **4.425**

(a) P_2O_5 (65% yield) or Ac_2O (60%) or DCC, H_3PO_4 (70%) or TFAA (84%). All with DMSO

(b) RuO_4 catalyst, KIO_4, H_2O, $CHCl_3$ (95%)

(c) $CrO_3 2Py$, CH_2Cl_2, Ac_2O, 25 °C, 15 min (90%)

(d) PDC, CH_2Cl_2, Ac_2O, reflux, 2 h (96%)

Scheme 4.123

RuO_4 appears to be able to oxidize axial and equatorial hydroxyl groups on six-membered rings with equal ease, and without isomerization at the adjacent carbon atoms. *Exo-* and *endo*-hydroxyl groups, which are part of bicyclic systems containing two *cis*-fused five-membered rings, are also both oxidized by this reagent, as is shown by the formation of the dicarbonyl compound **4.428** from 1,4:3,6-dianhydro-D-glucitol **4.354** (Section 4.6.5).[240] The RuO_2 is re-formed during these reactions and it may be recovered and reused. Alternatively, the cyclic process can be made continuous by using catalytic amounts of RuO_2 and an excess of aqueous sodium periodate.[241] Although the method is less satisfactory when applied to **4.426** (R = OMe),[242] it efficiently oxidizes the 3-hydroxyl group of 1,2:5,6-di-O-isopropylidene-α-D-glucofuranose **4.424** giving the 3-ulose derivative **4.425** in good

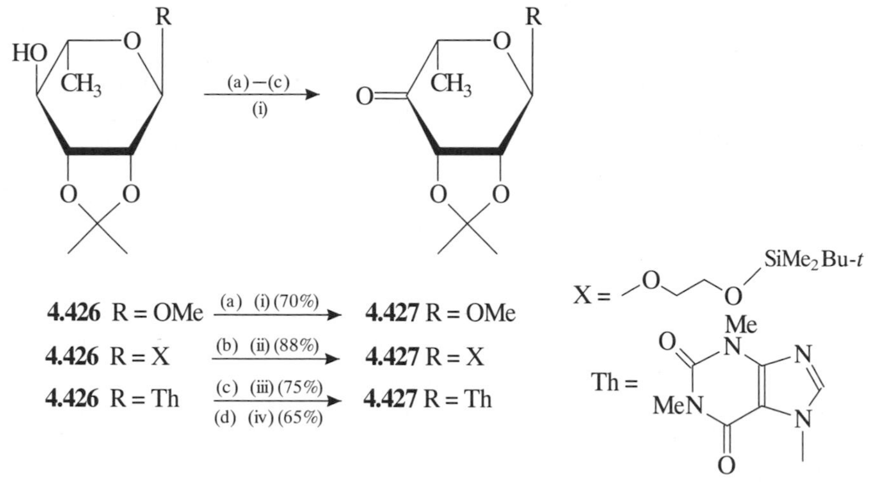

4.426 R = OMe $\xrightarrow[\text{(i)}]{\text{(a)}\ \text{(i)}\,(70\%)}$ **4.427** R = OMe

4.426 R = X $\xrightarrow[\text{}]{\text{(b)}\ \text{(ii)}\,(88\%)}$ **4.427** R = X

4.426 R = Th $\xrightarrow[\text{(d)}\ \text{(iv)}\,(65\%)]{\text{(c)}\ \text{(iii)}\,(75\%)}$ **4.427** R = Th

$X = $

$Th = $

(i) RuO_4, CCl_4; (ii) TPAP, MNO, molecular sieves, CH_2Cl_2; (iii) PCC, molecular sieves, CH_2Cl_2; (iv) PDC, molecular sieves, CH_2Cl_2

Scheme 4.124

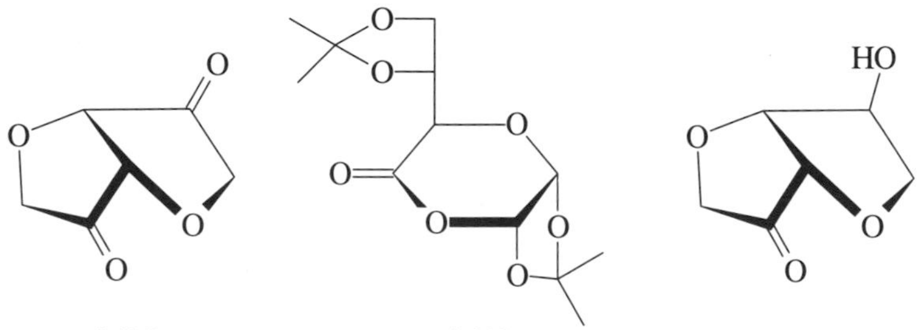

4.428   **4.429**   **4.430**

yield (Scheme 4.123b), and circumventing further Baeyer–Villiger-like oxidation of the ketone to the lactone **4.429** that occurs when an excess of RuO_4 is used.[238]

A highly efficient but less aggressive ruthenium oxidant has been prepared from this metal in its heptavalent state by Ley's group.[243] They have shown that tetra-n-propylammonium ruthenate(VII), n-Pr_4NRuO_4(TPAP), functions well as a catalytic oxidant which can be used to prepare aldehydes and ketones from a wide range of primary and secondary alcohols. The reaction is carried out at room temperature on the alcohol dissolved in dichloromethane or acetonitrile in the presence of a catalytic amount of TPAP and with 4-methylmorpholine N-oxide (MNO) as the inexpensive co-oxidant, the mild conditions being well suited to most partially protected sugar derivatives. Thus, the partially protected silyloxyethyl rhamnopyranoside **4.426** (R = X) gives the pyranosid-4-ulose **4.427** (R = X) when treated with this reagent, as illustrated in Scheme 4.124(b), a transformation that could not be efficiently accomplished with RuO_4.[242]

Oxidations with chromium(VI) compounds   Chromium(VI) oxides and related compounds complexed with pyridine have found considerable application in the oxidation of primary and secondary alcohols which possess acid-sensitive groups. Indeed, the first successful, generally applicable method for preparing glycosulose derivatives employed chromium trioxide in pyridine solution. However, yields were often modest and freeing the products from the very large excess of oxidant used was troublesome. Methyl 2,3-*O*-isopropylidene-α-L-rhamnopyranoside **4.426** (R = OMe), for example, gives the derivative **4.427** (R = OMe) in only 38% yield when treated with this reagent.[244] The oxidant in these reactions is the $CrO_3 \cdot 2C_5H_5N$ complex which is better prepared separately and used in dichloromethane solutions.[245] It is particularly effective in the presence of acetic anhydride, with which both primary and secondary alcohols are oxidized under the conditions indicated in Schemes 4.119(b) and 4.121(b). It is proposed that the acetic anhydride assists by acetylating the first-formed intermediate **4.431** to give an acetate **4.432** that breaks down more readily, as outlined in Scheme 4.125.[246] The value of the acetic anhydride is further highlighted in Scheme 4.123(c) by the efficient conversion of di-*O*-isopropylideneglucose **4.424** into the corresponding furanos-3-ulose derivative **4.425**,[246] which is not formed in the absence of the anhydride.[247]

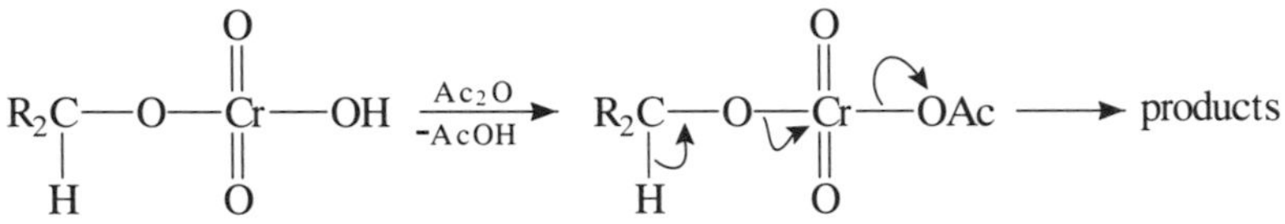

4.431   **4.432**

Scheme 4.125

Further-improved oxidations requiring only three molecular equivalents, or less, of the oxidant have been achieved with pyridinium dichromate(VI), $(C_5H_5NH)_2Cr_2O_7$(PDC),[230] and the mildly acidic pyridinium chlorochromate(VI), $C_5H_5NHCrO_3$(PCC).[248] The effectiveness of both reagents with carbohydrate derivatives is claimed to be improved in the presence of molecular sieves, as illustrated in Schemes 4.124(c) and 4.124(d) by the oxidation of the nucleoside derivative **4.426** (R = Th) to give 7-(6-deoxy-α-L-*lyxo*-hexopyranosyl-4-ulose)theophylline **4.427** (R = Th).[249] A review of ketonucleosides has been published.[250] Efficient oxidations of carbohydrate derivatives have also been achieved with PDC activated by acetic anhydride in dichloromethane under reflux. For example, under these conditions, high yields of the furanos-3-ulose derivative **4.425** and the pyranos-3-ulose derivative **4.420** are obtained, as illustrated in Schemes 4.123(d)[251] and 4.121(c),[251] respectively. The primary hydroxyl of the di-*O*-isopropylidenegalactose **4.417** is similarly oxidized to give the aldehydo product **4.418** in good yield, overoxidation to the uronic acid being suppressed by the presence of DMF, as illustrated in Scheme 4.119(c).[251]

(ii)   Oxidation of hydroxyl groups following their derivatization

Certain derivatives of alcohols and α-glycols can be oxidized to give monoketo compounds, and in this section three reactions of this type are discussed: photo-oxidation of pyruvates, bromine oxidation of O-stannylene derivatives and butyl-lithium-induced transformations of O-benzylidene derivatives.

Photorearrangement of sugar pyruvate esters   Primary and secondary alcohols are converted into aldehydes and ketones by photoinduced Norrish type II rearrange-ments of their pyruvate esters.[41] (See Section 4.9.1.b.iii for other examples.) These reactions are carried out by exposing benzene solutions of sugar monopyruvates, otherwise fully protected with photostable groups, to light from a medium pressure mercury lamp. A range of sugar derivatives can be oxidized in this way, but the method is of particular value for preparing peracylated glycosuloses which readily undergo β-elimination reactions in the presence of many chemical oxidants. Thus, for example, 1,3,4,6-tetra-O-acetyl-α-D-glucopyranose 2-pyruvate **4.333** upon irra-diation in benzene solution gives tetra-O-acetyl-α-D-*arabino*-hexos-2-ulose **4.434** in good yield by the Norrish type II rearrangement depicted in Scheme 4.126.[41] The product is predictably unstable and readily undergoes a series of eliminations and rearrangements to give kojic acid diacetate **4.435**.

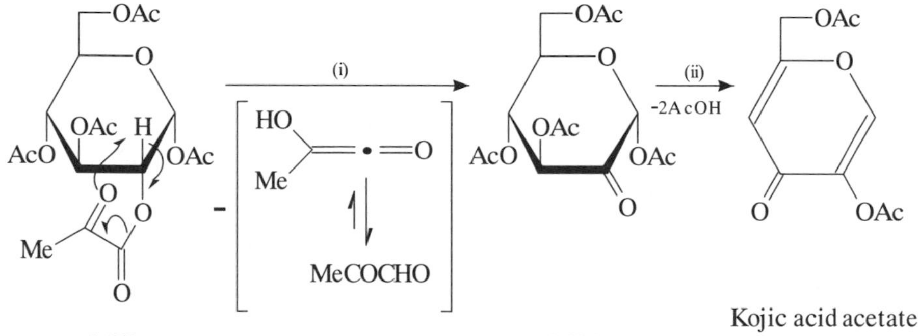

4.433   **4.434** (70%)   Kojic acid acetate
                                                                    4.435

(i) *hv*, PhH; (ii) Et₃N or heat at 80 °C

Scheme 4.126

Photolysis of secondary azides, which are available from secondary alcohols, can also be used to prepare such ketonic derivatives, but the method has not been widely adopted.[252]

Bromine oxidation of O-stannylene derivatives   O-Stannylene derivatives of diols have been shown by David to be readily oxidizable to hydroxyketones with molecular bromine. The reaction is particularly useful because many sugar derivatives with several free hydroxyl groups react selectively with dibutyltin oxide at *cis*-α-glycol sites (see Section 5.1.4), and the crude stannylenes so formed can be oxidized. When O-stannylenes are formed at *cis-vic*-hydroxyl

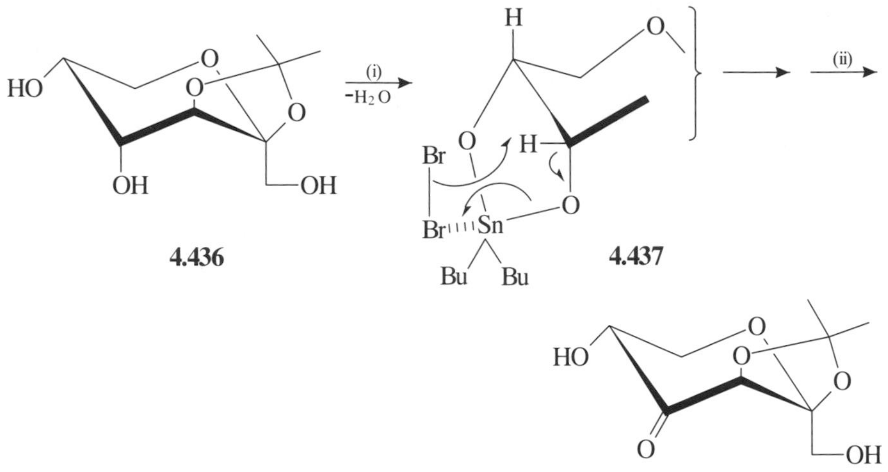

4.436   **4.437**

4.438 (56%)

(i) Bu$_2$SnO, PhH, azeotropic distillation, Br$_2$, MeOCH$_2$CH$_2$OMe, molecular sieves;
(ii) H$_2$O

Scheme 4.127

groups on pyranose rings, the axial carbon–oxygen bonds are selectively
converted into carbonyl groups. The value of the reaction is illustrated by
2,3-*O*-isopropylidene-β-D-fructopyranose **4.436**, which has three unprotected
hydroxyl groups but undergoes oxidation at only one when treated sequentially
with dibutyltin oxide and bromine followed by water under the conditions given
in Scheme 4.127 to give 2,3-*O*-isopropylidene-β-D-*threo*-hexo-2,4-diulopyranose
4.438.[253] The regioselective oxidation has been explained in terms of a cyclic
transition state **4.437** in which the bromine is associated with the tin atom in such
a way that the equatorial hydrogen atom is more accessible for abstraction.

Butyllithium-induced rearrangements of O-benzylidene derivatives   Some acetal
derivatives of sugars undergo an eliminative rearrangement when treated with
strong bases: 2,3- or 3,4-*O*-benzylidenepyranoses, otherwise fully protected with
stable groups, for example, eliminate benzaldehyde to give deoxypyranosuloses
when treated with *n*-butyllithium. This can be an attractive route to specific 'sugar
ketones' if the *O*-benzylidene derivatives are readily available, as is the case
with methyl 2,3:4,6-di-*O*-benzylidene-α-D-mannopyranoside **4.439**. On treatment
with *n*-butyllithium in THF, as illustrated in Scheme 4.128, this compound
regioselectively eliminates benzaldehyde from the 2,3-positions to give the 2-
deoxypyranosid-3-ulose **4.441** via the enolate **4.440**.[254] The elimination occurs by
preferential abstraction of the axial C-3 hydrogen, as illustrated in **4.439**.

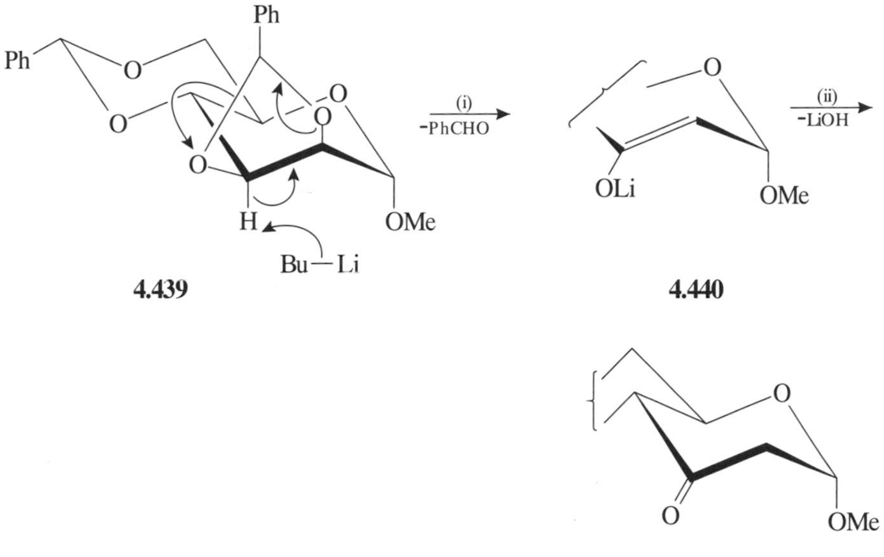

4.439   **4.440**

4.441 (91%)

(i) *n*-BuLi, THF; (ii) H$_2$O

Scheme 4.128

(iii)   Oxidation of specific hydroxyl groups in polyhydroxy-compounds

Selective oxidations with platinum and oxygen   Alcohols in the presence of oxygen and finely divided platinum catalysts undergo dehydrogenation reactions which have been shown to be applicable to a wide variety of sugar derivatives.[255] Since aldoses are easily oxidized to aldonic acids under these conditions, it is customary to protect the anomeric centres before oxidation of secondary hydroxyl groups, and unless uronic acid derivatives are required, it is also usual to use compounds which possess no primary alcohols.

Axial secondary hydroxyl groups attached to pyranoid rings are usually oxidized preferentially, as is shown by the conversion of methyl 6-deoxy-β-D-allpyranoside **4.442** to its 3-ulose derivative **4.443** (Scheme 4.129).[256] The specificity is thought to be dependent upon the greater accessibility of the equatorial carbon–hydrogen bonds which undergo cleavage during the reactions. A further selective oxidation occurs with 1,4:3,6-dianhydro-D-glucitol **4.354**, in which only the *endo*-hydroxyl group (*exo*-hydrogen atom at the reaction site) reacts to give the hydroxyketone **4.430** in 39% yield.[198]

Oxidation with Acetobacter suboxydans   *Acetobacter suboxydans* is a bacterium isolated from stale beer which in aqueous suspension shows specific dehydrogenase activity. It generally oxidizes compounds containing the structural unit **4.444** to

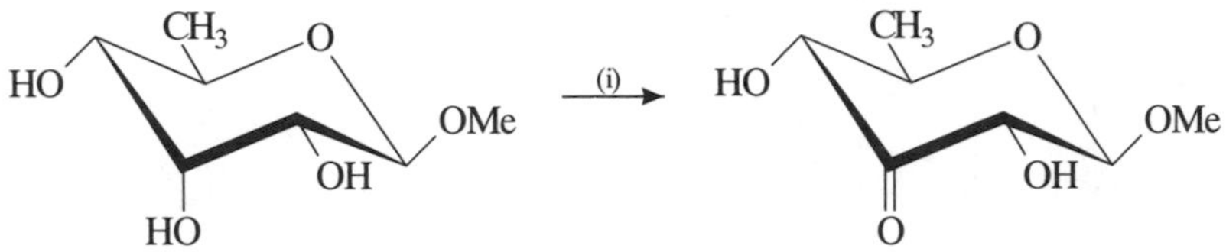

4.442   **4.443** (20%)

(i) Pt, O_2, 25 °C, 4 h

Scheme 4.129

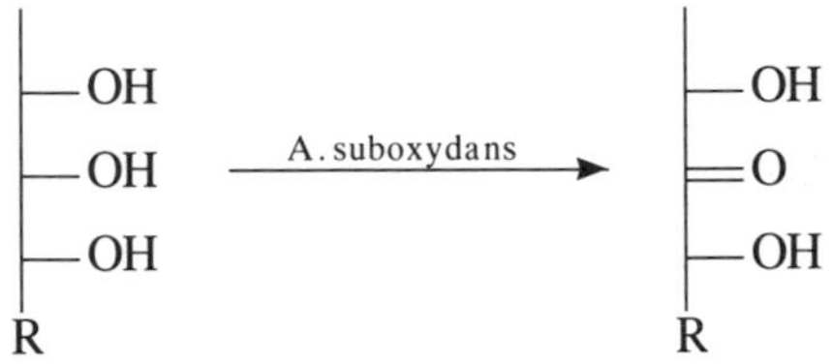

4.444

Scheme 4.130

keto derivatives according to 'the rule of Bertrand and Hudson', as shown in Scheme 4.130. When applied to free aldoses the organism usually produces aldonic acids and ketoaldonic acids, but, with alditols or carbohydrates held in acyclic modifications such as aldose dithioacetals and dialkyl acetals, aldosulose derivatives have been obtained. Thus, inoculation of neutral aqueous solutions of D-glucose diethyl dithioacetal **4.445** brings about oxidation at C-5 to give a good yield of D-*xylo*-hexos-5-ulose diethyl dithioacetal **4.446**, which exists in the cyclic furanose form **4.447** (Scheme 4.131).[257]

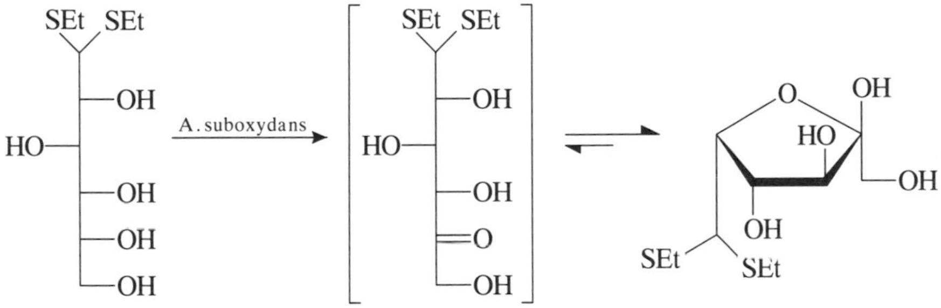

4.445   **4.446**   **4.447**

Scheme 4.131

Alditols with the appropriate stereochemistry are converted to ketoses. For example, D-arabinitol and D-ribitol are oxidized by these bacterial cells to D-*threo*- and L-*erythro*-pentulose, respectively (see Section 2.5.2), and the oxidation of D-glucitol is a key step in the commercial preparation of ascorbic acid outlined in Scheme 4.143.

(iv)   Other syntheses of dicarbonyl compounds

There are several other less direct reactions by which dicarbonyl compounds can be prepared. The formation of phenylosazones from aldoses or ketoses involves oxidation at the carbon atoms adjacent to the original carbonyl functions, and aldos-2-uloses (osones) can be liberated from the phenylosazones by transferring the phenylhydrazine residues to benzaldehyde in acid-catalysed exchange reactions (Section 3.1.3.c.iii). Unsaturated sugar derivatives with double bonds forming parts of enolic systems often give deoxydicarbonyl products upon hydrolysis (Section 4.10).

Dialdose derivatives can be formed by glycol cleavage of suitably protected aldose derivatives possessing vicinal unprotected diol groups on side chains. For example, 3-*O*-benzyl-1,2-*O*-isopropylidene-α-D-glucofuranose **4.448** gives the D-*xylo*-dialdofuranose derivative **4.449** when treated with lead tetraacetate, as illustrated in Scheme 4.132;[258] alternatively, sodium periodate brings about the same transformation.

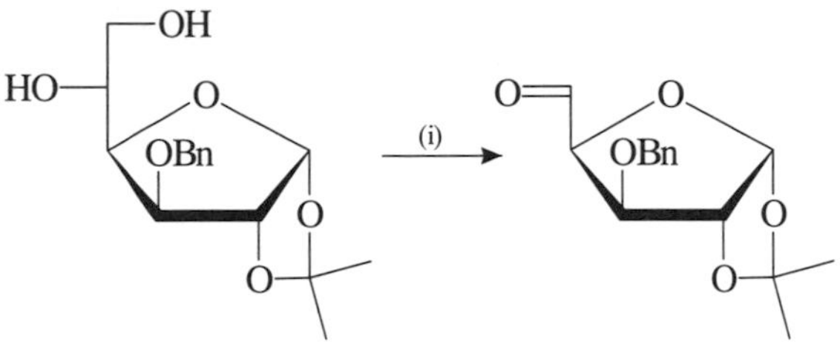

4.448   **4.449**

(i) Pb(OAc)$_4$, PhH, warm

Scheme 4.132

(b)   Reactions

The chemical reactions associated with the free carbonyl groups in fully protected dialdose, aldosulose and diulose derivatives are very similar to those of aldehydes and ketones, and most may be classified as reactions at the carbonyl groups, reactions at the carbon atoms adjacent to the carbonyl groups or photochemical reactions. A review of the role of glycosuloses in the chemical synthesis of branched-chain sugars has been published.[205]

(i)   Reactions at the carbonyl groups in fully protected glycosulose derivatives

Reduction   The carbonyl groups in these compounds may be reduced with many reagents, and these reactions can provide methods for obtaining rare sugar derivatives in good yield since the oxidation–reduction sequence often occurs with high stereoselectivity and can cause inversion of configuration at the carbon atom involved. For example, methyl 4,6-*O*-benzylidene-2-*O*-tosyl-α-D-*ribo*-hexopyranoside-3-ulose **4.420** gives only the allopyranoside derivative in high yield when reduced with sodium borohydride.[234] The hydride attack occurs almost exclusively from the equatorial direction because axial attack is prevented by the presence of axial methoxy group at C-1. But when axial attack is unhindered as in the corresponding 2-O-benzoyl-β-analogue of **4.420**, a 1.0:1.1 mixture of the *allo*-product and *gluco*-product is formed on reduction.[259] A similar reduction of 1,2:5,6-di-O-isopropylidene-α-D-*ribo*-hexofuran-3-ulose **4.425** gives 1,2:5,6-di-O-isopropylidene-α-D-allose (see **4.263**) in high yield.[260] This is produced by stereoselective hydride attack at C-3 from the sterically unhindered *exo*-direction relative to the bicyclic system formed by the two five-membered rings. Similar steric hindrance arises in the nucleophilic displacement reactions discussed in Section 4.1.2.

Addition of Grignard reagents   A wide variety of Grignard reagents have been added to the carbonyl groups of several of these compounds as a means of synthesizing dehydro branched-chain sugars (Section 4.8.2.a). Methylmagnesium iodide, for example, reacts with the keto group of the methyl 4,6-O-benzylidene-2-deoxy-α-D-*erythro*-hexopyranosid-3-ulose **4.451** in a highly stereoselective way to give the branched-chain sugar derivative **4.450** (Scheme 4.133a), from which the branched-chain sugar D-mycarose (Section 4.8) has been synthesized.[261]

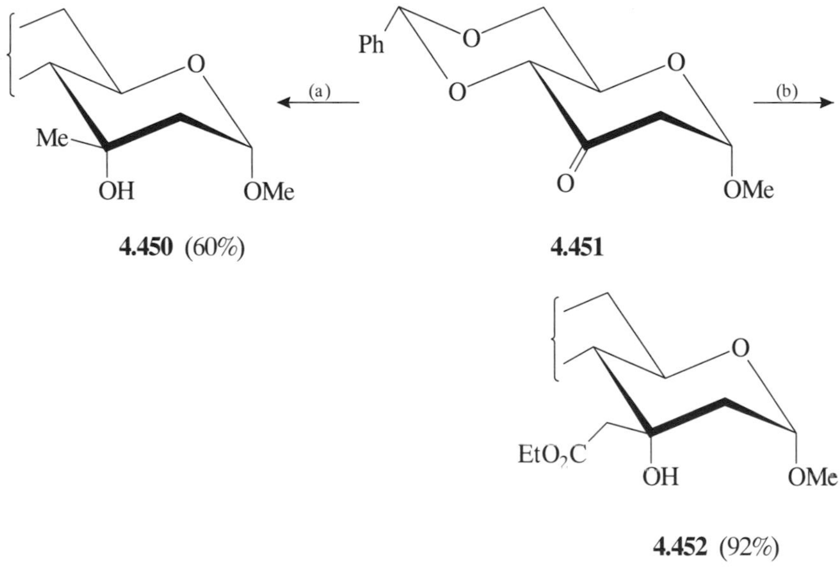

4.450 (60%)   **4.451**

4.452 (92%)

(a) MeMgI, (MeOCH₂)₂, 22 °C, 20 h
(b) Zn/Ag/graphite, BrCH₂CO₂Et, THF, −78 °C, 10 min
Scheme 4.133

Reformatsky additions   The recent introduction of new methods for activating zinc has given a new lease of life to this classical reaction, which offers an efficient method of introducing functionalized branches into sugars by way of zinc-mediated additions of α-haloesters to glycosulose derivatives. Thus, **4.451** reacts with ethyl bromoacetate when treated with an activated zinc reagent (prepared from potassium graphite, zinc chloride and some silver acetate activator), as illustrated in Scheme 4.133(b), to give methyl 2-deoxy-3-*C*-(ethoxycarbonyl)methyl-α-D-*ribo*-hexopyranoside **4.452**.[262]

Additions of other carbon nucleophiles   As well as the organomagnesium and organozinc reagents previously referred to, the reactions of organolithium compounds with glycosulose derivatives have been studied as routes to dehydro branched-chain sugar derivatives (see Section 4.8.2.a).[205] Among the reagents that have been investigated[205] are alkyllithiums, 1-methoxyvinyllithium, 1,1-dimethoxy-2-lithio-2-propene and 2-lithio-2-methyl-1,3-dithiane, the last of these being used in the synthesis of D-aldgarose depicted in Scheme 4.112. Nitromethane, acetonitrile and hydrogen cyanide in basic media have also been added to glycosulose derivatives.[205] A cyanohydrin formed by the last reagent on reaction with 3-*O*-benzyl-D-fructose **4.381** was used in an early synthesis of D-apiose (Scheme 4.111a).

Addition of sulphur ylides   Dimethylsulphoxonium methylide reacts with the carbonyl group of the 3-ulose derivative **4.451**, as illustrated in Scheme 4.134(a), to give the *spiro*epoxide **4.454** which, with lithium aluminium hydride, gives the branched-chain sugar derivative **4.450**, thus showing that the carbanion of the ylide, like Grignard reagents, attacks the ulose from the equatorial direction to give initially the intermediate **4.453**.[263]

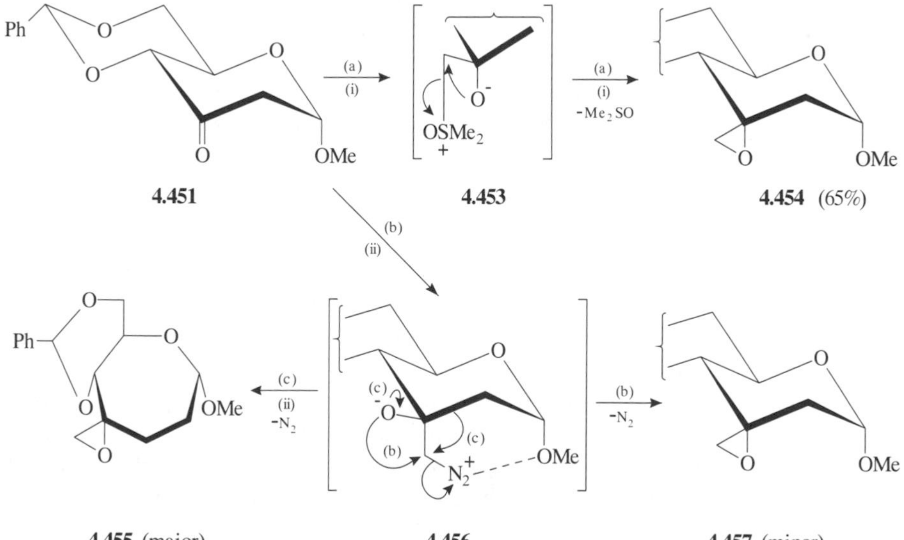

(i) $Me_2S^+OCH_2^-$; (ii) CH_2N_2, MeOH, Et_2O

Scheme 4.134

Addition of diazomethane   Diazomethane can react with ketones in two ways, as exemplified by cyclohexanone which gives a 4:1 mixture of cycloheptanone and the *spiro*-compound methylenecyclohexane oxide. The 3-ulose in Scheme 4.134 behaves similarly, giving the two epoxides **4.455** and **4.457**, the former being derived from a corresponding ring expanded ketone.[264] Intermediate **4.456**, formed by diazomethane attack on the carbonyl group, is the precursor of both products since it can lose nitrogen either with participation from the oxygen atom to give epoxide **4.457** (Scheme 4.134b) or with participation by C-2 in the ring to give a homologous ketone (Scheme 4.134c), which reacts further with diazomethane to give the second spiroepoxide **4.455**.

The stereochemistry of the addition of diazomethane to the pyranosid-3-ulose **4.451** is different from that of sulphur ylides and methyl Grignard reagents which approach the ketone from the equatorial, least-hindered direction. Axial attack by diazomethane is thought to be controlled mainly by the attractive electrostatic forces between the diazomethyl cation and the aglycon oxygen atom, as depicted in **4.456**. Subsequent reduction of epoxide **4.457** thus affords a branched-chain sugar derivative which is epimeric at C-3 with compound **4.450**.

Addition of phosphorus ylides   These carbonyl compounds undergo condensations with phosphorus ylides in the Wittig and related reactions and give rise to sugar derivatives with exocyclic double bonds that can be converted into branched-chain sugars. In principle, two geometric isomers can be produced, as is the case in the Wadsworth–Emmons condensation with the furanos-3-ulose **4.397** (Scheme 4.116). There are, however, examples in which only one isomer is formed, as with methyl 4,6-*O*-benzylidene-3-*O*-*t*-butyldimethylsilyl-α-D-*arabino*-hexopyranosid-2-ulose, which, when treated with the phosphorane $Ph_3P=CHCO_2Et$ in refluxing acetonitrile, gives the unsaturated ester in 85% yield exclusively as its *E*-isomer (see **4.401**). This reagent also gives single geometric isomers with some other pyranosid-3- and -4-uloses, but the reason for this irregular stereospecificity is not clear.[265]

Addition of hydroxyl-compounds   3,4-*O*-Isopropylidenepyranosid-2-ulose derivatives form hemiacetal and *gem*-diol solvent adducts, respectively, at neutral pH in alcoholic and aqueous solutions in readily established equilibrium processes which can be reversed simply by solvent evaporation, thereby giving back the ulose derivative unchanged.[266] Related pyranos-3- and -4-ulose derivatives are not prone to this behaviour. Additions of this type are typical of many ketones possessing electronegative substituents, and with these aldosuloses they may be attributed to the electron-deficient nature of C-2 which has been referred to previously in Section 4.1.2.a.ii. In keeping with this proposition, pyranos-2-uloses, fully protected with electron-withdrawing groups such as acetyl esters, form C-2 *gem*-diols which are sufficiently stable to be isolated.[267]

Furanosuloses, on the other hand, should hydrate less readily because on additions to their carbonyl groups the I-strain incurred is unfavourable compared with that experienced with the six-membered ring analogues (see Section 3.1.1.b.ii). It is therefore surprising that 1,2:5,6-di-*O*-isopropylidene-α-D-*ribo*-hexofuranos-3-ulose **4.425** forms a stable crystalline hydrate particularly as C-3 is not recognized as an excessively electronegative site (as borne out by the behaviour of pyranos-3-uloses). Dehydration of this compound is not spontaneous but requires heating in benzene with azeotropic removal of water or the use of molecular sieves. Favourable hydrogen bonding in the *gem*-diol structure is thought to be responsible for its stability.[268] Some 1,2-*O*-isopropylidenefuranos-3-uloses derived from tetroses and pentoses also form crystalline hydrates.[210]

Condensation with amine derivatives   The carbonyl compounds under consideration form oximes, hydrazones and semicarbazones, as expected for ketones or aldehydes. For example, the oximes **4.164**, **4.167** and **4.168** in Schemes 4.47 and 4.48 are readily prepared from the corresponding pyranosidulose derivatives and they can be used as a means of preparing amino sugars.

(ii)   Reactions at carbon atoms adjacent to the carbonyl groups in otherwise fully protected derivatives

Ketones that possess hydrogen atoms at carbon atoms adjacent to the carbonyl groups can readily lose protons in the presence of bases to give enolate anions which behave as ambient nucleophiles in reactions with electrophilic reagents. Whether the carbon or oxygen nucleophiles engage in substitution reactions with an electrophilic reagent, such as an alkyl halide, is difficult to predict because of the large number of reaction variables involved, but in general oxygen–carbon bonds tend to form with electrophilic reagents that react by an S_N1 route, whereas carbon–carbon bonding takes place with reagents that tend to the alternative S_N2 reactivity.[269]

Fully protected glycosuloses behave likewise, and because of their polyoxygenated structures it would be anticipated that they should be more susceptible to base-catalysed deprotonation. This has been verified with methyl 4,6-*O*-benzylidene-2-deoxy-α-D-*erythro*-hexopyranosid-3-ulose **4.458**, which undergoes deuterium exchange at the C-2 axial position to give **4.460** by way of enolate **4.459** 1000 times faster than 4-*t*-butylcyclohexanone undergoes analogous exchange, as illustrated in Scheme 4.135(a).[270] This reaction is also in keeping with the general view that, in the kinetically controlled steps, electrophiles, in this case D$^+$, react with carbon from a direction orthogonal to the plane of the enolates to give six-membered products with the electrophiles axial.[271]

Alkylation of the same enolate **4.459** with chloromethyl methyl ether gives the enol ether **4.461**, in good yield, whereas methyl bromide reacts at carbon, thereby giving the 2-deoxy-2-*C*-methyl branched-chain sugar **4.462** under similar conditions, which are noted in Schemes 4.135(b)[272] and 4.135(c),[273] respectively. The regioselectivity exhibited in these reactions is in agreement with the prediction based on the reactivity of the two halides employed. At first sight, the formation of the branched-chain-*ribo*-sugar **4.462** with an equatorial C-2 methyl group appears to contravene the rule of orthogonal attack at the enolate carbon. However, this isomer is probably derived by base-catalysed epimerization of the first-formed

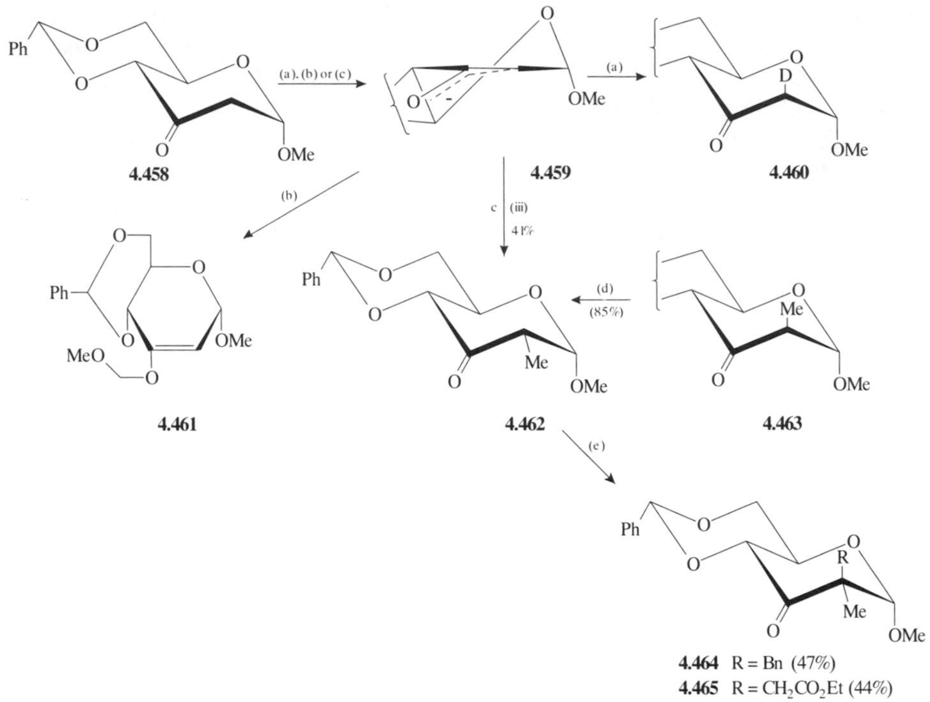

Scheme 4.135

(a) $D_2O/MeOD$, K_2CO_3, THF
(b) n-BuLi, $ClCH_2OMe$
(c) n-BuLi, THF, MeBr, HMPIT
(d) NaOMe, MeOH, $0°C$, 15 min
(e) LDA, RBr, HMPIT, THF, $-70°C$

axial *arabino*-isomer **4.463**, a compound that has been independently prepared by oxidation of the 3-hydroxyl group in methyl 4,6-*O*-benzylidene-2-deoxy-2-*C*-methyl-α-D-altropyranoside[220] and found to isomerize rapidly to **4.462** in high yield, as illustrated in Scheme 4.135(d).[274] Isomerizations of this type are possible with most α-monosubstituted ketones because of the availability of the base-sensitive α-hydrogen atom, but when such hydrogen atoms are not present products of axial *C*-alkylation arise. Thus, treatment of **4.462** with benzyl bromide or ethyl bromoacetate under the conditions given in Scheme 4.135(e) gives the branched-chain sugar derivative **4.464** or **4.465**, respectively, in which the incoming alkyl group occupies the axial position.[273] Similarly, methyl 2-*O*-benzoyl-4,6-*O*-benzylidene-α-D-*ribo*-hexopyranosid-3-ulose (OBz in place of Me at C-2 in **4.462**) is axially methylated at C-2 in 64% yield when treated with methyl iodide in DMF containing potassium *t*-butoxide.[275]

Isomerizations of glycosulose derivatives have been used on many occasions as a means of obtaining rare sugar derivatives. When, for example, methyl 6-deoxy-2,3-*O*-isopropylidene-α-L-*lyxo*-hexopyranosid-4-ulose **4.466** is treated with aqueous pyridine under reflux, it equilibrates *via* the enolate **4.467** with its C-5 epimer **4.468** and gives a final isomeric mixture composed of four parts of the more stable **4.466** and one part of **4.468** (Scheme 4.136).[276] Presumably an enolization of this type accompanies the oxidation of the altroside **4.423** discussed earlier (Scheme 4.122), and is an analogue of the conversion of **4.463** to **4.462**.

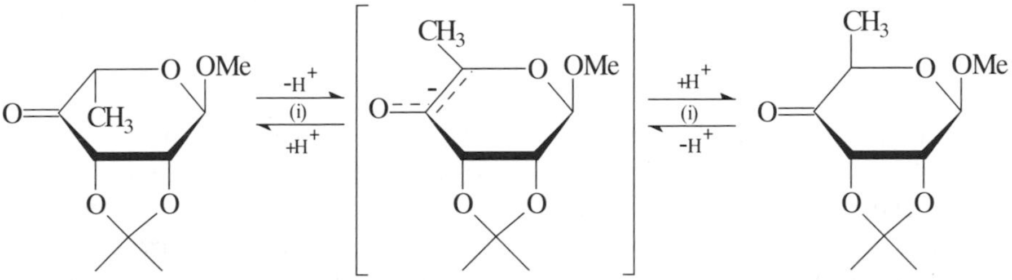

4.466   **4.467**   **4.468**

(i) Py, H$_2$O, 100 °C, 2 h

Scheme 4.136

On acetylation, one enolic form of the *ribo*-hexofuran-3-ulose derivative **4.469** is trapped as the enolacetate **4.470** (Scheme 4.137), reduction of which with sodium borohydride or hydrogen over palladium provides a suitable means of access to the rare sugar D-gulose.[277]

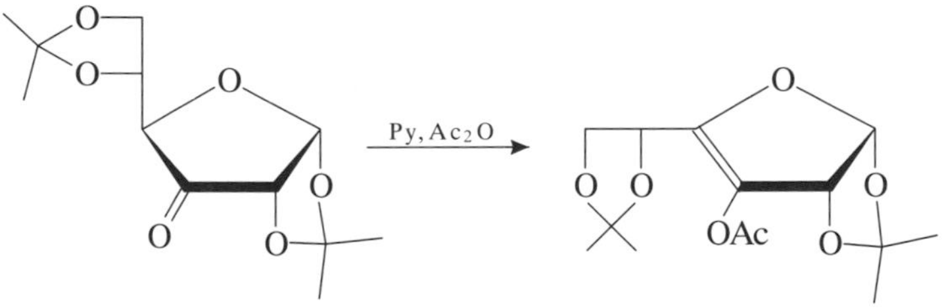

4.469   **4.470**

Scheme 4.137

Glycosidulose compounds often possess alkoxy or acyloxy substituents that may act as leaving groups in a β-position to the carbonyl function, and consequently such systems undergo elimination reactions under the influence of either acid or base catalysis to yield enones. Glycosid-3-uloses which have leaving groups situated at the anomeric centres undergo this reaction particularly readily, as illustrated in Scheme 4.138.[229,233] For a further example see Scheme 4.126.

(iii)   Photochemical reactions of otherwise fully protected glycosulose ketones

Glycosidulose ketones undergo a diverse range of photoreactions in solution.[41] When ring substituents possessing γ-hydrogen atoms accessible to the carbonyl

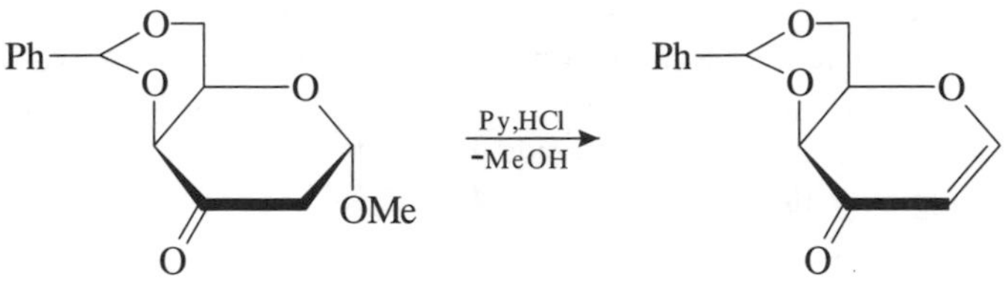

Scheme 4.138

group are present, Norrish type II reactions occur on irradiation with u.v. light,[278] as illustrated, for example, by methyl 4,6-*O*-benzylidene-2-*O*-methyl-α-D-*ribo*-hexopyranosid-3-ulose **4.471**, which undergoes a hydrogen atom transfer from the C-2 methoxy group to the photoexcited carbonyl oxygen atom, thereby giving the 1,4-biradical **4.472** (Scheme 4.139). Cyclization of this intermediate gives the hydroxyoxetane derivative **4.473**, whereas its fragmentation produces formaldehyde and the enol **4.475**, ketonization of which yields the 2-deoxypyranosid-3-ulose derivative **4.474** or the enone derivative **4.476** when accompanied by methanol elimination, as shown in Scheme 4.139.[278]

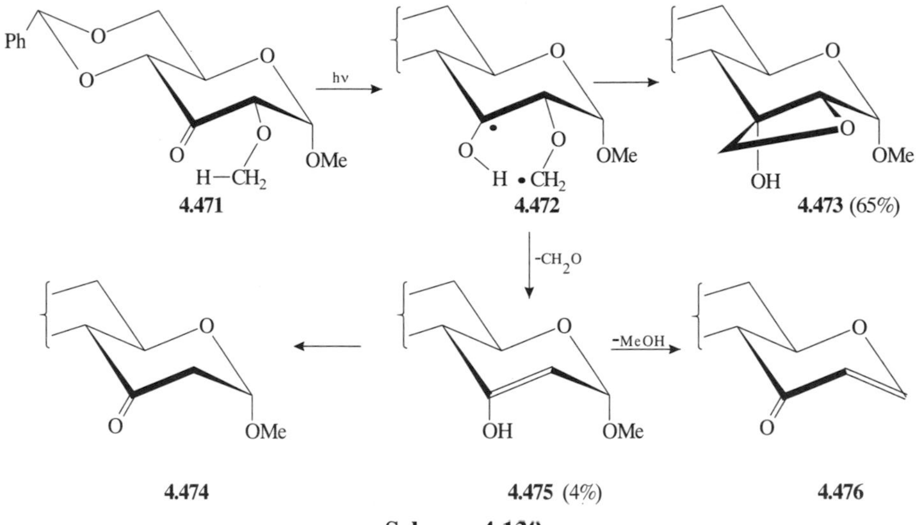

Scheme 4.139

The product distribution in these Norrish type II reactions is sensitive to the stereochemistry of the substituent adjacent to the carbonyl group. Thus, whereas the *ribo*-compound **4.471** gives the oxetanol **4.473** in good yield and only trace amounts of 2-deoxy-3-ulose **4.474** and enose **4.476** (Scheme 4.139), its *arabino*-isomer, which reacts *via* the biradical **4.477**, produces **4.474** and **4.476** in high yield (85% conversion) and no oxetanol.[278] These products are formed because fragmentation of the biradical intermediate **4.477** is stereoelectronically favoured relative to cyclization, which is not the case with intermediate **4.472**.

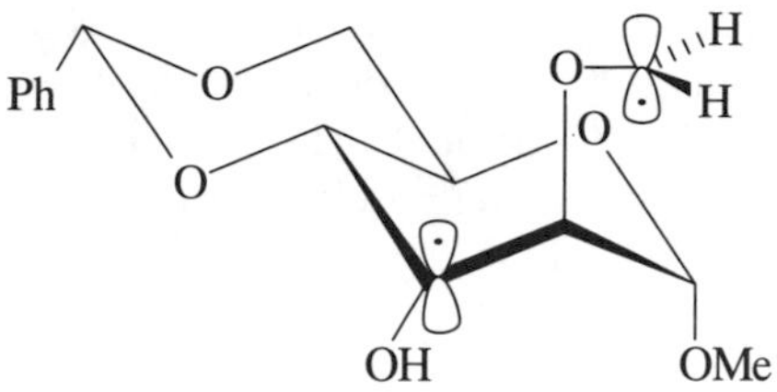

4.477

In the absence of accessible hydrogen atoms, excited pyranosid-2- and -4-ulose derivatives, which have tetrahydropyran-3-one structures, undergo Norrish type I cleavage to give alkyl acyl biradicals. These intermediates undergo many reactions, the most significant being decarbonylation to further biradicals or rearrangement to cyclic carbenes. On exposure to u.v. light methyl 2,3-*O*-isopropylidene-α-L-*lyxo*-hexopyranosid-4-ulose **4.478**, in a range of solvents including benzene, *n*-pentane, acetonitrile, *t*-butanol and even methanol, undergoes α-cleavage to give the biradical **4.480** which decarbonylates to **4.481** and subsequently ring closes, thereby giving the furanosides **4.482** as shown in Scheme 4.140.[279] The unstable, sterically strained isomeric ketone **4.479** is also produced (up to 20%) in the early stages of the photolysis and arises by ring closure of biradical **4.480**, so verifying that the initial α-cleavage is at the C-3–C-4 bond as depicted in the scheme.[280]

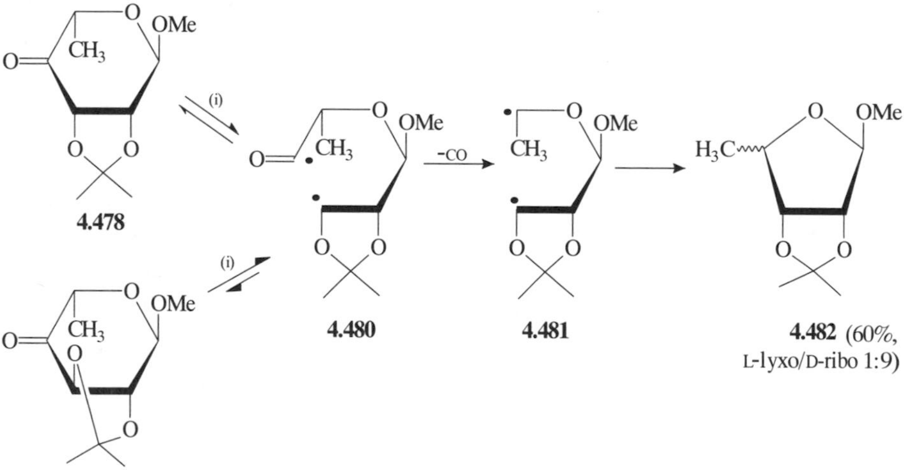

4.478

4.479

4.480

4.481

4.482 (60%, L-lyxo/D-ribo 1:9)

(i) *hv*, PhH

Scheme 4.140

In like fashion, 1,2:4,5-di-*O*-isopropylidene-β-D-*erythro*-hex-2,3-diulose-2,6-pyranose **4.483**, which is structurally related to a pyranos-2-ulose, also gives a decarbonylated biradical related to **4.481**, but this produces alkene **4.485** by intramolecular hydrogen atom transfer in addition to the α-anomer and β-anomer of di-O-isopropylidene-D-*erythro*-pentulose **4.484**,[281] from which D-*erythro*-pentulose (D-ribulose) can be obtained in 40% yield by subsequent distillation and deprotection (Scheme 4.141). Because ketone **4.483** is readily available from D-fructose, the reaction provides a sugar chain shortening procedure of practical value.[281]

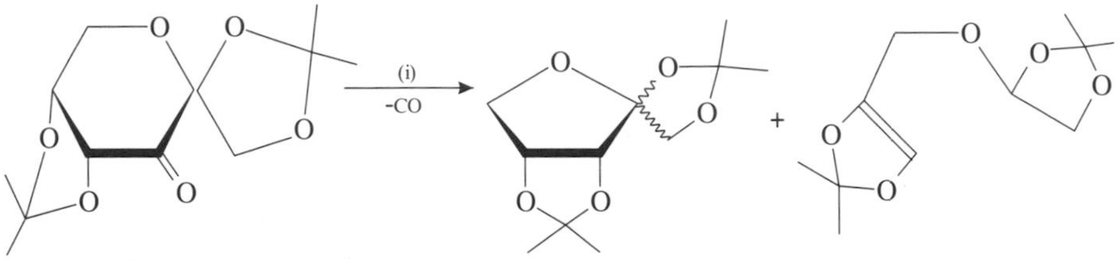

4.483   **4.484** (35–45%)   **4.485**

(i) *hν*, C_6H_{14} (or C_6H_6)

Scheme 4.141

Most pyranosid-2-ulose derivatives undergo type II photochemistry, but if the aglycons are derived from tertiary alcohols type I reactions and decarbonylation occur: the major products are alkenes rather than furanosides.[282]

Solutions of ketone **4.486** at ambient temperatures also decarbonylate on exposure to u.v. light, but at −70 °C in diethyl ether/ethanol mixtures the biradical intermediate **4.487** rearranges to the carbene **4.488** which is trapped by ethanol to give the dioxepan derivatives **4.489** in good yield (Scheme 4.142).[283] A related ring expansion is also observed with furanos-3-ulose derivatives.[284]

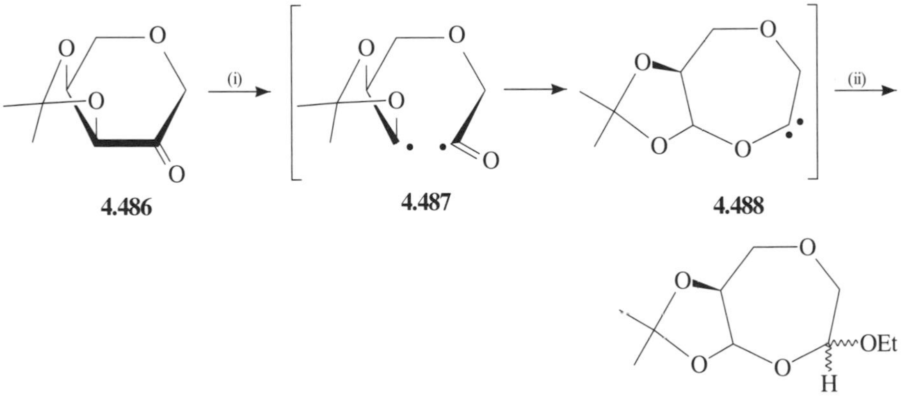

4.486   **4.487**   **4.488**

4.489

(i) *hν*, EtOH/Et$_2$O, −70 °C; (ii) EtOH

Scheme 4.142

The photoreactions of pyranos-3-uloses, which have been studied with a range of 1,2:4,6-diacetal derivatives of α-D-*ribo*-hexos-3-ulose, are different from those of the pyranos-2-uloses and pyranos-4-uloses. Decarbonylation does not occur in any solvent. Instead, in the relatively polar acetonitrile a novel rearrangement into the corresponding 2,3:4,6-diacetal derivatives of mannono-1,5-lactone takes place in yields of about 50%,[285] which can be explained in terms of an initial type I cleavage of the C-2–C-3 bond. See the cover of this book for an illustration of this process which was discovered during the authors' researches. In methanol, on the other hand, the 3-*C*-hydroxymethyl-*α*-D-gluco- and -*α*-D-allo-pyranose branched-chain sugar analogues of the starting materials are each produced in similar (20%) yields in addition to the lactone derivatives.[286]

(iv)   Reactions of unprotected glycosiduloses

Glycopyranosiduloses, which are the most studied derivatives of this group, have properties related to those of the fully protected analogues. Since tautomerization leads to enediols which can ketonize to different pyranosiduloses, isomerization is readily possible. However, this does not seem to occur spontaneously, as is supported by paper chromatography at low pH values which indicates that these compounds are homogeneous and not isomeric mixtures; reduction with sodium borohydride yields only products formed by direct reduction of the discrete carbonyl groups.[231]

The action of bases on glycosiduloses causes enolizations and eliminations which lead eventually to carbon–carbon bond cleavage reactions (cf. Section 3.1.7.c) and the formation of acidic products.

4.9.2   ALDULOSONIC ACIDS (ASCORBIC ACID, Kdo AND N-ACETYLNEURAMINIC ACID)

The best-known acids in this class are those which have the keto group at a penultimate carbon atom, i.e. the 2- and 5-hexulosonic acids and the 2- and 4-pentulosonic acids. The most important of these is L-ascorbic acid, or vitamin C, which is L-*threo*-hexulosono-1,4-lactone 2,3-enediol **4.492**.[287] The biochemical activity of this compound is specific, and is not exhibited by any of its three stereoisomers. L-Ascorbic acid is prepared commercially from D-glucitol **4.490** (Scheme 4.143), the key step in the synthesis being the highly stereospecific biochemical oxidation to L-sorbose **4.491** with *Acetobacter suboxydans* (Section 4.9.1.a.iii).

Ascorbic acid is a powerful reducing agent and is easily oxidized by air to dehydroascorbic acid **4.493** (Scheme 4.144a). For this reason, antioxidant stabilizers are added to commercially processed fruit preparations to preserve the vitamin C content. The hydroxyl groups of ascorbic acid form the usual derivatives, but that at C-3 possesses unusually high reactivity because of the acidity conferred by the electron-withdrawing effect of the carbonyl group. For example, it can be selectively methylated with diazomethane to give the 3-*O*-methyl ether **4.494** (Scheme 4.144b)[288] and selectively acetylated with ketene.[289]

The two other notable compounds in this class are both 3-deoxy-2-ulosonic acids. One is 3-deoxy-D-*manno*-2-octulosonic acid (Kdo) **8.24**, which is found in bacterial polysaccharides (see Section 8.1.3.d).[290] and has been synthesized in

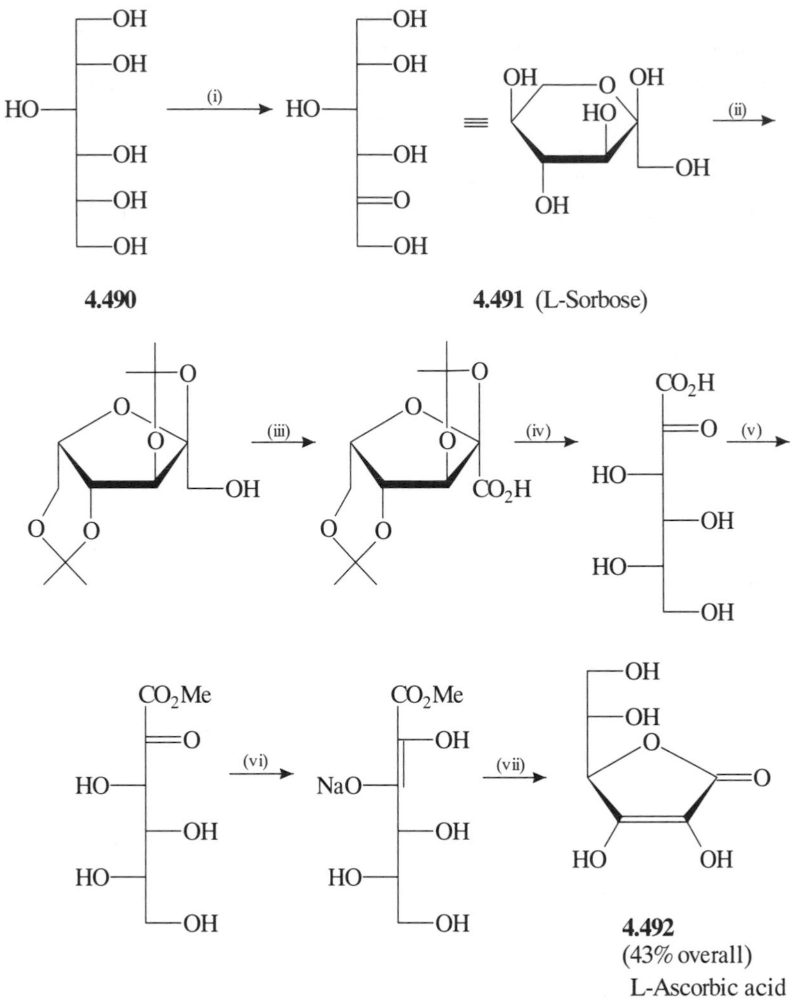

4.490   **4.491** (L-Sorbose)

4.492
(43% overall)
L-Ascorbic acid

(i) *A. suboxydans*; (ii) Me$_2$CO, H$^+$; (iii) KMnO$_4$; (iv) H$_3$O$^+$; (v) MeOH, H$^+$;
MeONa; (vii) HCl

Scheme 4.143

several ways,[291] and the other is 5-acetamido-3,5-dideoxy-D-*glycero*-D-*galacto*-2-
nonulosonic acid (*N*-acetylneuraminic acid) **8.22**, which is a sialic acid found
in many animal and bacterial polysaccharides (Section 8.1.4).[292] It is usually
obtained from natural sources since there is not a good large scale chemical
synthesis amongst the presently available methods.[293] However, chemoenzymic
methods are currently being explored to overcome this deficiency,[294] as are tin-
and indium-mediated allylations of unprotected carbohydrates.[295] For example,
Whitesides' group has chain extended *N*-acetyl-D-mannosamine **4.495** on a
preparatively useful scale by treating it with ethyl α-(bromomethyl)acrylate in

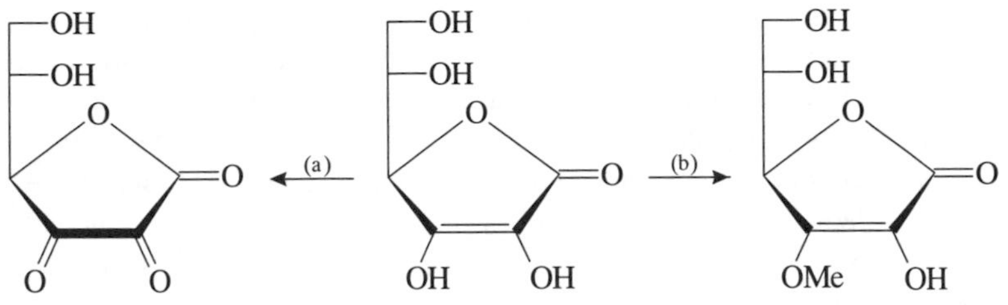

4.493   **4.494**

(a) air
(b) CH₂N₂

Scheme 4.144

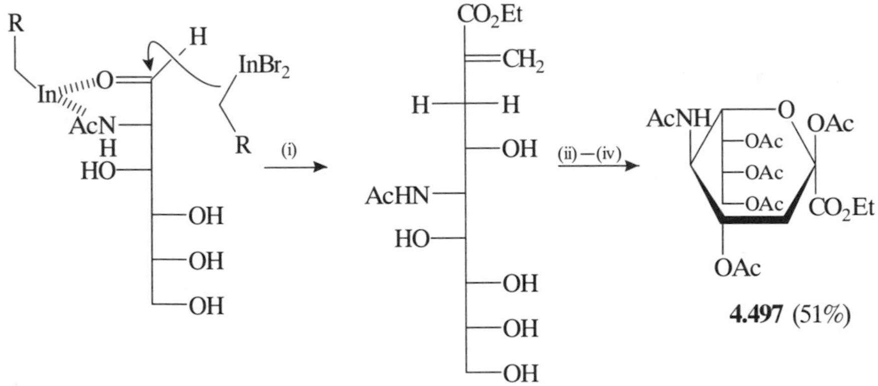

4.495
|
R = EtCO₂C=CH₂

4.496 (four parts)
+ C-4 epimer (one part) (90% overall)

4.497 (51%)

(i) ethyl α-(bromomethyl)acrylate, In, HCl, EtOH, 40 °C; (ii) O₃, MeOH, −78 °C;
(iii) H₂O₂, HCO₂H, MeOH, H₂O; (iv) Ac₂O, Py, DMAP, 22 °C

Scheme 4.145

the presence of indium under the acidic conditions outlined in Scheme 4.145.[296]
The addition of the organoindium reagent is stereoselective, the major product
isomer **4.496** being that predicted by the Cram chelated model (Section 3.1.4.a),
as illustrated in **4.495**. Ozonolysis of the isomeric enoate mixture followed
by oxidative ozonide decomposition and exhaustive acetylation gives the fully
protected neuraminic acid **4.497**.

4.9.3   URONIC ACIDS

The uronic acids are monosaccharide derivatives that contain aldehydo and
carboxylic acid chain-terminating functions, and they occur in nature as important
constituents of many polysaccharides, usually in the pyranose forms of hexose
derivatives (Section 8.1.3). The D-*gluco*-compound, which was the first to be

discovered, was isolated from urine (hence the name), in which it occurs as glycosides and glycosyl esters. Toxic substances in the body are commonly glycosylated by enzymic transfer of glucuronic acid and then excreted.

The uronic acids often crystallize from aqueous solution as their lactones, the ring size adopted being dependent upon the relative configurations at the asymmetric centres. When 6→3-lactones are formed the sizes of the sugar rings are also configurationally dependent. D-Glucuronic acid, for example, is usually isolated as its furanurono-6→3-lactone **4.503** as is D-mannuronic acid, but D-galacturonic acid gives the pyranurono-6→3-lactone **4.498** since the furanoid forms would have *trans*-fused five-membered ring systems (cf. 3,6-anhydrohexosides, Section 4.6.3).

| **4.498** | **4.499** | **4.500** X = H |
| | | **4.501** X = Br |

(a)   Syntheses

If a uronoside with a simple aglycon is required, selective oxidation of the primary hydroxyl group of a hexopyranoside in aqueous solution with oxygen in the presence of platinum can be employed (Section 4.9.1.a.iii). Methyl α-D-glucopyranoside, for example, gives the corresponding pyranosiduronic acid **4.499** in 87% yield when so treated;[297] otherwise, sodium hypochlorite in the presence of catalytic amounts of the 2,2,6,6-tetramethyl-1-piperidinyloxy free radical can be used for selective reactions of this type.[298] When, as is usually the case, greater flexibility is called for, glycosides selectively blocked at the secondary hydroxyl groups must be used, and then the primary position is often oxidized with a reagent compatible with organic solvents.[299,300] Jones's reagent is frequently used, as illustrated by Scheme 4.146, in the synthesis of the glucopyranosiduronate **4.502**, which has protecting groups at C-1, C-6 and the secondary positions that can subsequently be selectively removed.[299] Efficient direct conversion of primary alcohols in otherwise protected compounds to methyl carboxylate groups can be effected by the use of calcium hypochlorite(I) in acetonitrile containing small proportions of acetic acid (to generate HOCl) and methanol.[301]

Glucopyranosiduronic acid derivatives can be synthesized from methyl 2,3,4-tri-O-acetyl-α-D-glucopyranosyl bromide uronate **4.504** which is available from the abundant 'D-glucurone' **4.503**, as illustrated in Scheme 4.147.[302]

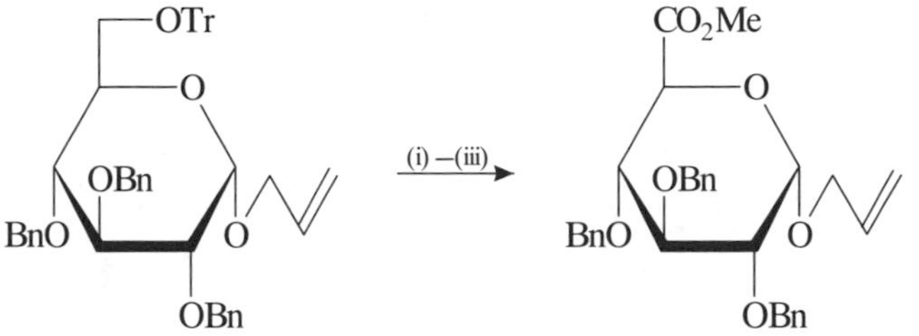

4.502 (54%)

(i) HCl, Me$_2$CO; (ii) CrO$_3$, H$_2$SO$_4$, Me$_2$CO, $-10°$, 3 h; (iii) CH$_2$N$_2$, CH$_2$Cl$_2$

Scheme 4.146

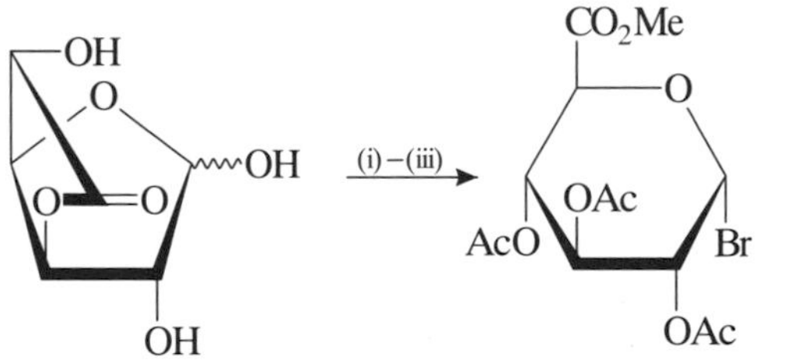

4.503   **4.504** (71%)

(i) MeONa, MeOH, 22 °C, 0.5 h; (ii) Ac$_2$O, HClO$_4$, 40 °C; (iii) AcOH, HBr

Scheme 4.147

(b)   Reactions

As would be expected, the carboxylate residue in these carbohydrate compounds increases the range of reactions that they undergo.[303] Two recently introduced transformations will be mentioned here.

Uronate esters are susceptible to radical-induced bromination at the carbon atoms adjacent to their carbonyl groups (see Section 4.5.1.d).[121] Thus, when methyl tetra-O-acetyl-β-D-glucopyranuronate **4.500** is treated with bromine in refluxing carbon tetrachloride over a 275 W heat lamp, the 5-bromoester **4.501** is obtained as a stable, sublimable product in 68% yield.

Branched-chain compounds, of value in the synthesis of polyether antibiotics, have been formed by Claisen–Ireland rearrangements on uronic acid allyl esters. For example, when benzyl 2,3-O-isopropylidene-α-D-lyxofuranosiduronic acid (E)-crotyl ester **4.505** is treated as indicated in Scheme 4.148, the enol ether form **4.506** of the ester is trapped and on warming undergoes a [3,3]-sigmatropic rearrangement to give, after desilylation and esterification, an isomeric mixture of branched-chain uronates **4.507**.[304]

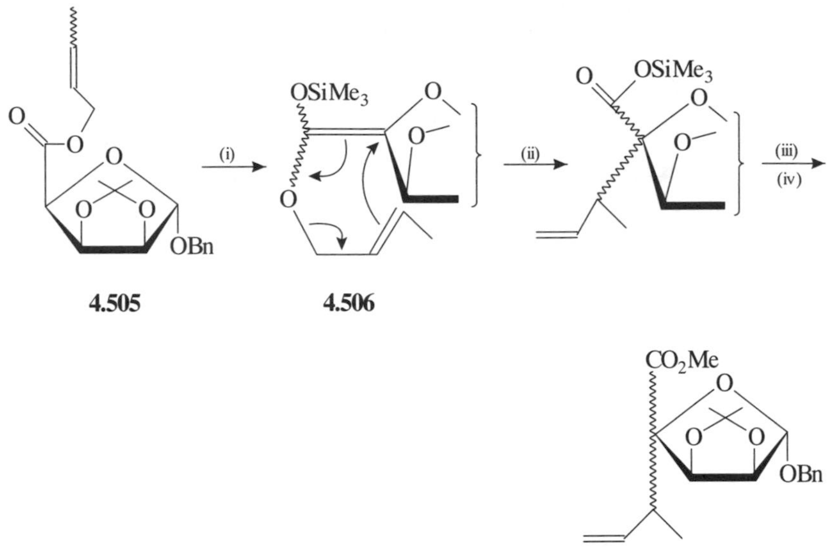

4.505   **4.506**

4.507 (80%)

(i) i-Pr$_2$ NLi, Me$_3$SiCl, HMPIT, THF, $-100\,^{\circ}$C, 8 min; (ii) $-78\,^{\circ}$C for 8 min, 24 $^{\circ}$C
for 2 h; (iii) n-Bu$_4$NF, THF, 30 min; (iv) CH$_2$N$_2$, Et$_2$O

Scheme 4.148

4.10   UNSATURATED SUGARS[161,177,305,306]

A wide range of unsaturated monosaccharide derivatives are known and many
examples, each having a multiple bond at one of all of the possible sites, have
been described. Together they represent the most versatile and synthetically useful
category of derivatives, as is illustrated in many places in this book (see especially
this Chapter and Chapters 3,5 and 7). Methods for naming unsaturated derivatives
are described in Appendix 2.

Introduction of a double bond into sugars can result in the formation of alkene,
enol or enediol types of functionality (Schemes 4.149a, 4.149b, and 4.149c, respectively), the double bonds of which may be *endo* or *exo* with respect to furanoid or
pyranoid rings, and in addition acyclic compounds or parts of cyclic molecules
may be involved. Basically, such introductions can be effected by elimination
procedures or by incorporating unsaturation during C–C bond-forming processes
such as the Wittig reaction, and initial products can take part in rearrangements to
provide further unsaturated derivatives. Alkyne groups may also be incorporated
into acyclic parts of sugar molecules.

Of particular significance are 'glycals', which are furanoid or pyranoid compounds having double bonds between C-1 and C-2 that make them vinyl ethers
and therefore subject to many reactions which often show high regio- and stereoselectivity. Compounds with 'isolated' double bonds between, for example, C-2 and

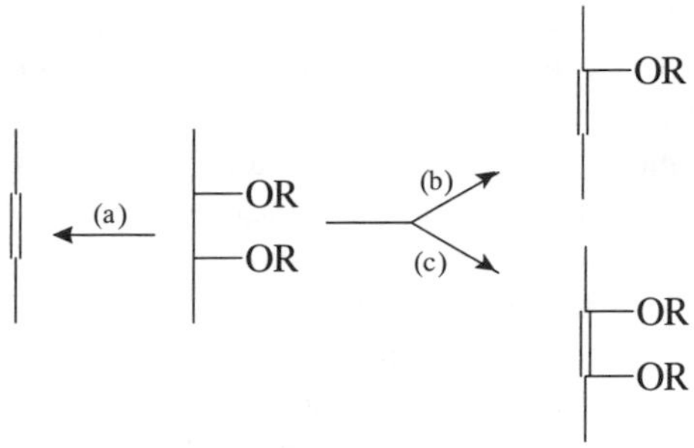

Scheme 4.149

C-3 or C-3 and C-4 of pyranoid derivatives exhibit normal alkene chemistry, but, in that they commonly have oxygen-bonded groups at allylic positions, they often take part in reactions which involve allylic displacements and in rearrangement processes. Because their double bonds have the same relationship to the ring oxygen atoms, C-4 alkenes of pyranose derivatives and glycals have similar chemistry, but hexose compounds with double bonds at C-5 show unique chemical reactivity. The glycals, 'isolated alkenes' and 5-enes will be dealt with in turn.

Unsaturated furanoid and pyranoid compounds exhibit parallel chemistry, but the former tend to be much more reactive and prone to further elimination to give furans.

4.10.1 GLYCALS

(a) Syntheses

Acylated glycals are traditionally prepared by the action of zinc and acetic acid on acylated glycosyl halides, e.g. **4.508**. These conditions are typical of dissolving metal reductions, and the elimination may occur following reduction of C-1 carbocations by the metal, but it is not altogether unlikely that organometallic intermediates could be involved. By this reaction, 3,4,6-tri-O-acetyl-D-glucal **4.509** was first synthesized in Fischer's laboratory. Base-catalysed deacetylation yields

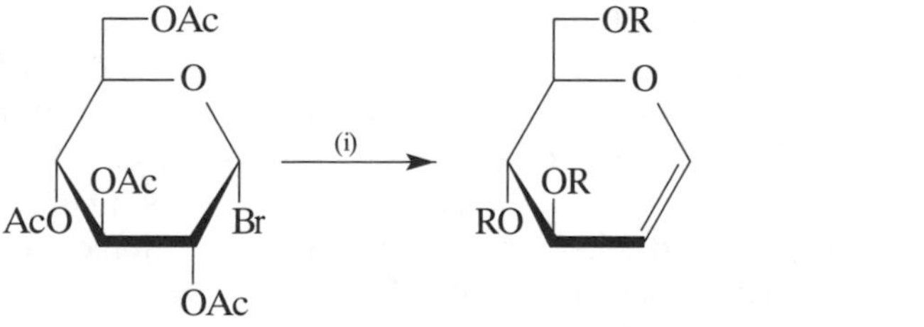

4.508   **4.509** R = Ac (70%)
                                       4.510 R = H   (ii)

(i) Zn, HOAc; (ii) MeO⁻, MeOH

Scheme 4.150

the parent D-glucal **4.510** (Scheme 4.150) which was given its anomalous, trivial name from the incorrect observation that it shows aldehydic properties (see Section 4.10.1.b). The zinc-based method has its limitations in that it can be somewhat cumbersome to apply, and the conditions are too agressive for use in making furanoid compounds. For the latter category furanosyl halide derivatives are best taken as starting materials, and for compounds with good leaving groups at C-2 eliminations can be effected by use of, for example, sodium iodide in acetone. Much use, however, has been made of glycofuranosyl halides having *O*-substituents which permit the use of basic or other reagents with which ester groups are incompatible. In particular, isopropylidene acetals at C-2 and C-3 have proved useful, the D-mannofuranosyl compound **4.511** affording glycal **4.512** in high yield by use of zinc/silver graphite, as illustrated in Scheme 4.151.[307] Analogous eliminations have been carried out by use of lithium in liquid ammonia[308] or sodium

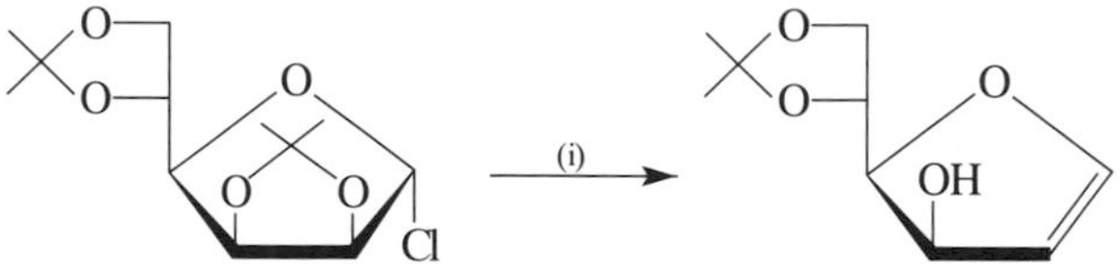

4.511   **4.512** (81%)

(i) Zn/silver graphite, THF, $-20\,^\circ$C

Scheme 4.151

naphthalenide,[309] and some of these reagents have also been used with appreciable success in making pyranoid glycals. Thus, tri-*O*-acetyl-D-glucal can be made in near quantitative yield from the traditional bromide **4.508** by use of the zinc/silver graphite reagent,[307] and 3-*O*-benzyl-4,6-*O*-benzylidene-D-glucal is produced from phenyl 2,3-di-*O*-benzyl-4,6-*O*-benzylidene-1-thio-β-D-glucopyranoside in similar yield on treatment with lithium naphthalenide in THF at $-78\,^\circ$C.[310] The latter reaction is presumably initiated by transfer of two electrons from two molecular equivalents of the naphthalenide to the sulphur atom in the thioglycoside, followed by fragmentation into thiophenate and a glycopyranosyl anion which subsequently eliminates benzyl oxide anion.

An ingenious method of producing D-allal derivatives, which involves stereospecific oxygenation at C-3 of the tri-*O*-acetyl-D-glucal-derived sulphoxide **4.513**, depends most probably on a [2,3]-sigmatropic rearrangement, as illustrated in Scheme 4.152. The final product is formed by base-catalysed acetyl migration from O-4 to O-3.[311] More commonly, Claisen rearrangements applied, for example, to vinyl hex-2-enopyranosides have, by thermal [3,3]-sigmatropic processes, given access to glycals having branch points at C-3.[312]

Glycals may also be obtained by deoxygenation of 1,2-anhydroaldose analogues[313] or by radical-induced elimination of 1,2-thiocarbonates.[314]

A series of glycals are known which have substituents at C-2; the most common derivatives are the so-called '2-hydroxyglycals',[315] the parent substances being enols and hence not known in this tautomeric form. However, elimination of the

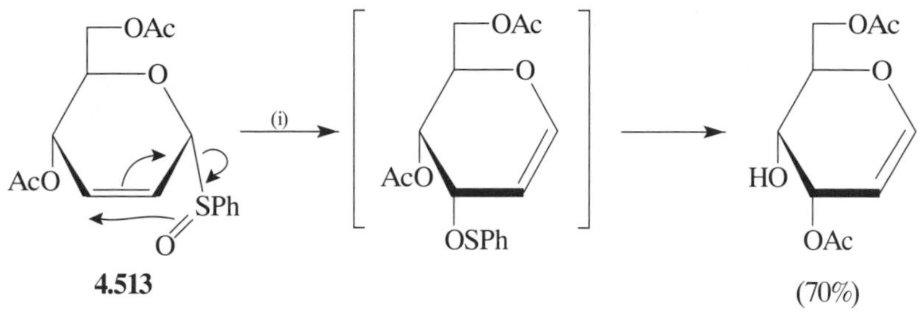

4.513   (70%)

(i) piperidine, 20 °C

Scheme 4.152

elements of hydrogen halide from acylglycosyl halides gives stable *O*-acylated 2-hydroxyglycals. They may be made by use of acetonitrile as solvent and diethylamine in the presence of tetrabutylammonium bromide,[316] by conversion of the bromides to iodides prior to reaction with the base[317] or by use of such bases as DBU in DMF (see Section 4.5.2.a),[318] and are of considerable use as precursors of other unsaturated compounds, notably enones (see below).

(b) Reactions

(i) Addition reactions

Most straightforward alkene addition reactions can be applied to glycals and their *O*-substituted derivatives, and because of the vinyl ether character of all these compounds and the presence of bulky substituents at the C-3 allylic centres, considerable regio- and stereo-selectivities are commonly shown.[161,167,305,306] Electrophilic additions are initiated at C-2; nucleophiles then add at C-1 and often subsequently are themselves nucleophilically displaced in glycosyl transfer reactions. Glycal derivatives are thus important precursors of glycosylating reagents (see below); addition of acetyl hypofluorite, for example, to the furanoid glycal **4.514** gives the glycosyl acetate **4.515** (Scheme 4.153), from which 2′-fluorinated D-*arabino*-nucleosides can be made.[319]

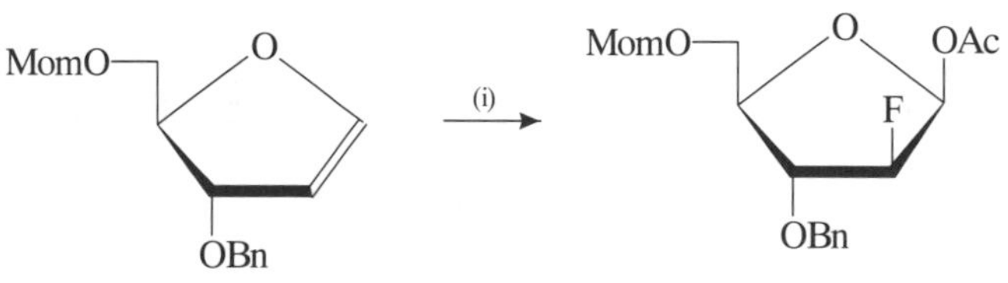

4.514   **4.515** (51%) +debenzylated
                                                      product (12%)

(i) AcOF, CH$_2$Cl$_2$, C$_6$H$_{14}$, 20 °C

Scheme 4.153

Hydrogenation can be effected, usually with mild catalysts to avoid hydrogenolytic loss of the allylic groups, and peracid hydroxylation affords a means

Table 4.1 Use of glycal derivatives in the synthesis of glycosides

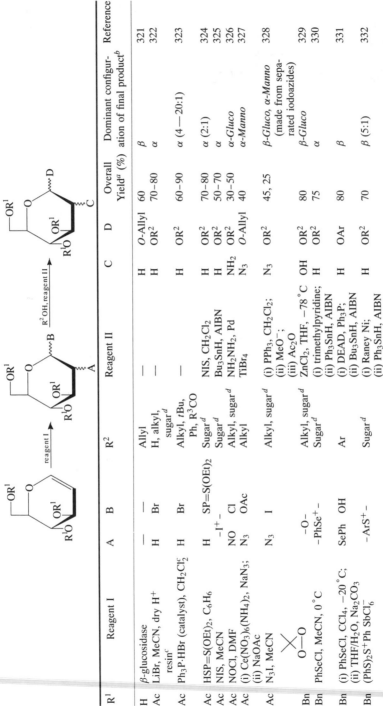

R¹	Reagent I	A	B	R²	Reagent II	C	D	Overall Yield^a (%)	Dominant configuration of final product^b	Reference
H	β-glucosidase	—	—	Allyl	—	H	O-Allyl	60	β	321
Ac	LiBr, MeCN, dry H⁺ resin^c	H	Br	H, alkyl, sugar^d	—	H	OR²	70–80	α	322
Ac	Ph₃P·HBr (catalyst), CH₂Cl₂	H	Br	Alkyl, tBu, Ph, R³CO	—	H	OR²	60–90	α (4—20:1)	323
Ac	HSP=S(OEt)₂, C₆H₆	H	SP=S(OEt)₂	Sugar^d	NIS, CH₂Cl₂	H	OR²	70–80	α (2:1)	324
Ac	NIS, MeCN	-I⁺-		Sugar^d	Bu₃SnH, AIBN	H	OR²	50–70	α	325
Ac	NOCl, DMF	NO	Cl	Alkyl, sugar^d	NH₂NH₂, Pd	NH₂	OR²	30–50	α-Gluco	326
Ac	(i) Ce(NO₃)₆(NH₄)₂, NaN₃; (ii) NaOAc	N₃	OAc	Alkyl	TiBr₄	N₃	O-Allyl	40	α-Manno	327
Ac	N₃I, MeCN	N₃	I	Alkyl, sugar^d	(i) PPh₃, CH₂Cl₂; (ii) MeO⁻; (iii) Ac₂O	N₃	OR²	45, 25	β-Gluco, α-Manno (made from separated iodoazides)	328
Bn	[epoxide O—O]	-O-		Alkyl, sugar^d	ZnCl₂, THF, −78 °C	OH	OR²	80	β-Gluco	329
Bn	PhSeCl, MeCN, 0 °C	-PhSe⁺-		Sugar^d	(i) trimethylpyridine; (ii) Ph₃SnH, AIBN	H	OR²	75	α	330
Bn	(i) PhSeCl, CCl₄, −20 °C; (ii) THF/H₂O, Na₂CO₃	SePh	OH	Ar	(i) DEAD, Ph₃P; (ii) Bu₃SnH, AIBN	H	OAr	80	β	331
Bn	(PhS)₂S⁺Ph SbCl₆⁻	-ArS⁺-		Sugar^d	(i) Raney Ni; (ii) Ph₃SnH, AIBN	H	OR²	70	β (5:1)	332

^{a} Approximate.

^{b} Where no ratios are noted the yields are of isolated main products.

^{c} HBr may not formally add to the double bond but be catalytic for the addition of alcohol.

^{d} Designates carbohydrate derivatives fully protected except at one hydroxyl group.

of reverting to aldoses, the relatively inaccessible D-talose (*O*-2, and *O*-3 *cis*), for example, being obtained in modest yield from its *C*-2 epimer, D-galactose, via D-galactal.[320] Glycals with substituents on O-3, on the other hand, epoxidize mainly to give O-2–O-3 *trans*-related products. Additions such as alkoxymercuration and hydroformylation proceed, as expected, to give mainly 2-mercurated glycosides and 2-deoxy-1-*C*-formyl compounds, respectively.

Of major importance are several addition reactions which result in products having electrophilic centres at C-1, i.e. compounds that can be used as glycosylating agents. Aldosides can be produced by this approach, and recently developed methods which allow good stereoselectivity have, in particular, made glycals very significant starting materials for the preparation of important glycosides of 2-amino sugars and especially 2-deoxy sugars (see Sections 4.3.1.e and 4.2.1.f. respectively). Commonly, acylated glycals are used as starting materials, but with some reagents benzylated (or silylated) compounds should be employed as illustrated in Table 4.1, which gives examples of the applicability of these compounds in various glycoside syntheses. The use of glycal esters in the synthesis of 2,3-unsaturated glycosides is referred to in Section 4.10.1.b.ii.

Esters of 2-hydroxyglycals are suitable sources of several pyranoid enone derivatives. (See elsewhere[233] for a general review of these compounds.) For example, chlorination of the benzoate **4.516**, followed by hydrolysis, gives a good yield of **4.519** by way of the benzoxonium ion **4.517** and the derived 1-*O*-benzoyl-2-ulose **4.518**, which undergoes β-elimination of benzoic acid (Scheme 4.154a).[333]

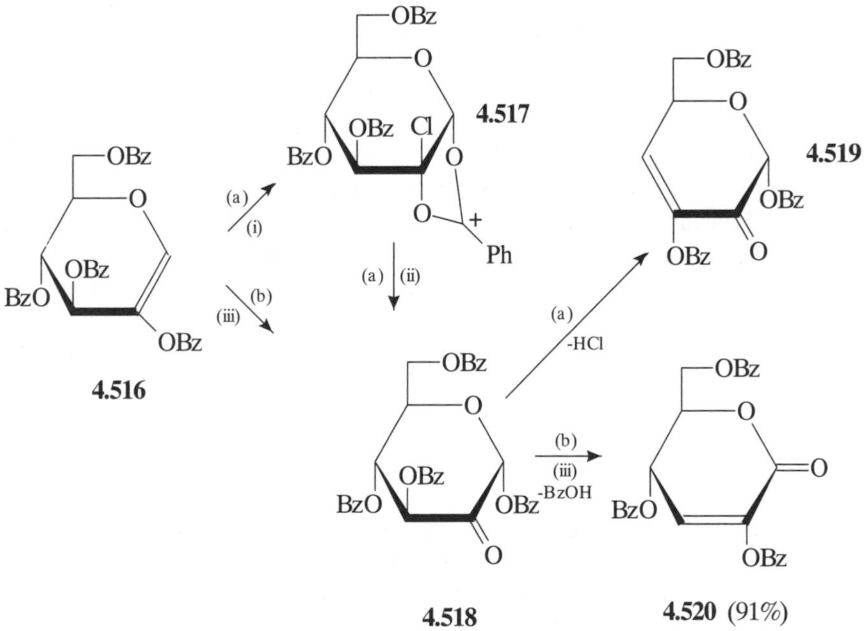

(i) Cl$_2$, PhMe, −30 °C; (ii) H$_2$O; (iii) MCPBA, BF$_3$·OEt$_2$, CH$_2$Cl$_2$

Scheme 4.154

X   Y   R
4.522 O CMe$_2$ Ac

4.521   4.523 CO NTs Tbdms   **4.524**

Otherwise, **4.516**, on treatment with *m*-chloroperbenzoic acid, affords the same 2-ulose derivative **4.518**,[334,335] but with this reagent together with boron trifluoride etherate, benzoyl migration from *O*-1 to *O*-2 in the enol form of **4.518** precedes β-elimination and leads to the unsaturated lactone derivative **4.520** almost quantitatively (Scheme 4.154b).[335]

These esters may also be used to produce useful products by direct additions. The 2,6-anhydroheptonic acid derivative **4.521**, for example, can be made in 55% yield by photochemical reaction of the tetraacetate analogue of **4.516** with formamide in the presence of acetone as initiator.[336]

Of particular interest for synthetic purposes are cycloaddition reactions undergone by glycals and their derivatives: very efficient epoxidation of tri-*O*-benzyl-D-glucal is referred to in Table 4.1 and Scheme 3.29, and analogous dichlorocarbene addition to the trimethyl ether affords the α-*gluco*-dichlorocyclopropano adduct in high yield.[337] Photochemical addition of ethyne to a glycal derivative is described in Section 7.2.2, the oxetane **4.522** is available by similar addition of acetone in quantitative yield,[338] and the β-lactam **4.523** is formed specifically and in 90% yield when the silylated glucal precursor is treated in chloroform with tosyl isocyanate (three molecular equivalents) at room temperature for 50 h.[339]

Several [4+2]-cycloaddition processes resulting in fusion of cyclohexane rings to pyranoid compounds have been recorded,[340] the epimers **4.524** being obtained in 65% yield when the corresponding glucal derivative is heated (110 °C, 10 days) with 7-cyanocyclobutabenzene, the reaction proceeding by way of the *o*-xylylene diene formed by initial thermal rearrangement.[341]

(ii) Rearrangement reactions

The most common rearrangement reaction of acylated glycals, and to a lesser extent their *O*-alkylated analogues, results in the introduction of a nucleophilic group at C-1 and the replacement of the double bond to C-2. To assume that S_N' reactions are involved or that all examples occur by the same mechanism is to oversimplify the situation, however, because, for example, tri-*O*-acetyl-D-glucal **4.509** is unstable under many of the conditions used for the displacements and isomerizes in the presence of acid catalysts or thermally to the 2-enose derivative **4.525** (Scheme 4.155).[342] Therefore, whether **4.509** or **4.525** undergoes nucleophilic attack may depend upon the reaction conditions, and these may also determine whether isolated products are formed under kinetic control or after equilibration.

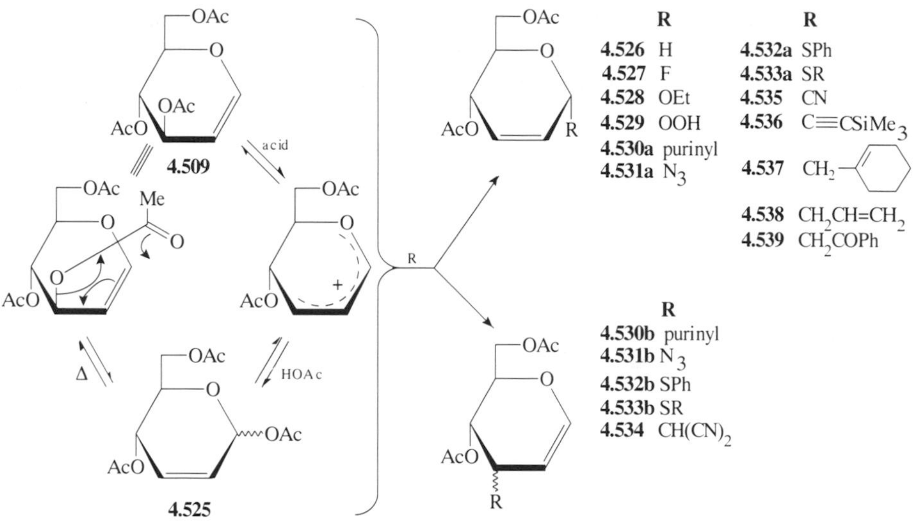
OAc
O
OAc
AcO
4.509
acid
OAc
Me
O
O
AcO
OAc
O
+
AcO
Δ
OAc
O
OAc
AcO
4.525
HOAc
R
OAc
O
AcO
R
OAc
O
AcO
R
R
4.526 H
4.527 F
4.528 OEt
4.529 OOH
4.530a purinyl
4.531a N_3
R
4.532a SPh
4.533a SR
4.535 CN
4.536 C≡CSiMe$_3$
4.537 CH$_2$
4.538 CH$_2$CH=CH$_2$
4.539 CH$_2$COPh
R
4.530b purinyl
4.531b N_3
4.532b SPh
4.533b SR
4.534 CH(CN)$_2$

Scheme 4.155

Notwithstanding these complications, the reactions illustrated in Scheme 4.155 have considerable synthetic utility and offer means of introducing hydrogen and fluorine and many groups bonded to the sugar rings *via*, for example, oxygen, nitrogen, sulphur or carbon.

Hydrogen-linked products With triethylsilane in dichloromethane containing an equivalent of boron trifluoride etherate, **4.509** gives the anhydrohex-2-enitol **4.526** in 95% yield.[343]

Fluorine-linked products A yield of the fluoride **4.527** of about 80% and an α/β ratio of 9:1 are obtained when tri-*O*-acetyl-D-glucal is treated with pyridinium poly(hydrogen fluoride) in dichloromethane at 0 °C.[344]

Oxygen-linked products The primary use to which this rearrangement reaction can be put is in the synthesis of 2,3-unsaturated *O*-glycosides[345] such as the ethyl compound **4.528**, which, because it crystallizes directly (70% isolated) following reaction of **4.509** with ethanol in benzene containing a catalytic amount of boron trifluoride etherate, has become a readily available carbohydrate derivative.[346] Stoichiometric proportions of complex alcohols can also be used; thus, for example, compound **4.540** is produced in 75% yield from the corresponding 2,5-dideoxystreptamine diol, and from it aminoglycoside antibiotic analogues (Section 8.2.1.c) can be prepared.[347] Alcohols react exclusively at C-1, and with tri-O-acetyl-D-glucal give glycosidic products with α/β ratios of approximately 8:1.

An interesting development of the reaction involves the use of 3-O-n-pentenoylglycal esters and their activation as leaving groups with iodonium dicollidine perchlorate. In this way, 1,2:5,6-di-O-isopropylidene-α-D-glucose reacts with 4,6-O-isopropylidene-3-O-pent-4-enoyl-D-glucal to give the corresponding 2,3-unsaturated 1,3-linked disaccharide **4.541** in 65% yield (α/β ratio of 6.7:1).[348] Thus, even in the presence of the electron-rich glycal double bond, it is the unactivated alkene function in the substituent group that leads to reaction following electrophilic addition.

4.540   **4.541**

When tri-O-acetyl-D-glucal **4.509** is heated in refluxing aqueous dioxane, 'normal' allylic displacement occurs and compound **4.542** is produced. However, under light, this product isomerises to the (E)-alkene **4.543** with a free aldehydic group which may have been responsible for the reducing properties attributed to glucal and to its anomalous naming. Treated with dilute sulphuric acid in dioxane in the presence of mercury(II) sulphate, the same starting material is again converted quantitatively to **4.543** (Scheme 4.156).[306]

In dioxane containing catalytic sulphuric acid and 85% hydrogen peroxide, tri-O-acetyl-D-glucal gives unsaturated glycosyl hydroperoxides, the α-compound **4.529** (Scheme 4.155) being obtainable in 72% yield.[349]

Nitrogen-linked products   Nitrogen nucleophiles in acidic conditions can give 2,3-unsaturated products in good yield (Scheme 4.155); for example, **4.509** condensed with silylated purines in dichloromethane containing trityl perchlorate readily affords the unsaturated nucleosides **4.530a**, but with poor stereoselectivity.[350] This represents, however, an oversimplified representation of the reaction, because when products of this type are heated strongly or subjected to stronger acidic conditions they rearrange to the 3-substituted isomers **4.530b**.[351] It is not uncommon, therefore,

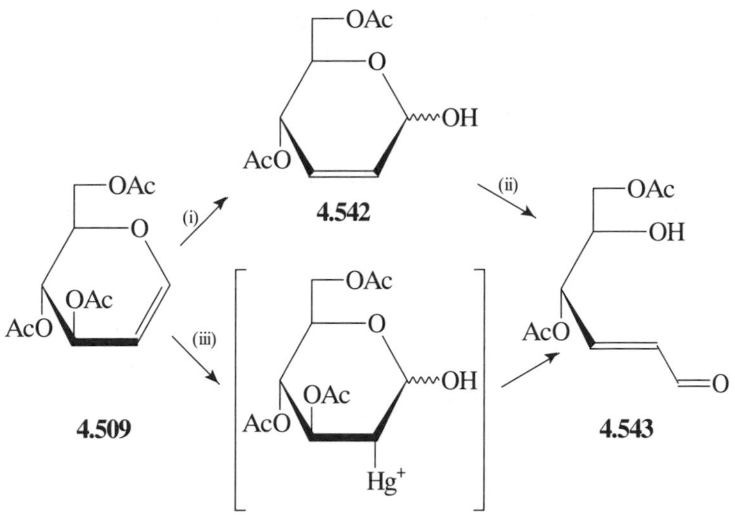

(i) Δ, H₂O; (ii) hv; (iii) Hg²⁺, H₂O

Scheme 4.156

for such products to be reported from this type of reaction. Likewise, tri-*O*-acetyl-
D-glucal, on treatment with sodium azide in acetonitrile in the presence of an
excess of boron trifluoride, affords mixed products **4.531a** and **4.531b** with the
3-azidoglycals **4.531b** dominating.[352] The possible sigmatropic rearrangement of
allylic azides further complicates the interpretation of the observations.

Sulphur-linked products   Thiophenol reacts with tri-*O*-acetyl-D-glucal in benzene
in the presence of catalytic amounts of boron trifluoride etherate to give initial products from which the α-glycoside **4.532a** can be isolated in 70% yield together with
9% of the β-anomer and small amounts of the 3-thioglycals **4.532b**.[353] Alkane thiols
with tin(IV) chloride as catalyst likewise give thioglycosides **4.533a** under kinetic
control, but under equilibrium conditions the 3-thioalkylglycals **4.533b** dominate.[354]

Carbon-linked products   Reaction of glycal esters with various reagents offering
nucleophilic carbon centres in the presence of Lewis acids affords important means
of access to 2,3-unsaturated *C*-glycosides (Scheme 4.155).[355] It is very unusual for
the products to be of the *C*-3-substituted glycal type, but an exception exists in the
case of the sodio derivative of methyl dicyanoacetate which, in acetonitrile with
4.509 and boron trifluoride etherate, gives **4.534** as the main products (31 and 27%
of the D-*ribo*-compound and D-*arabino*-compound isolated, respectively) following
ester cleavage and decarboxylation of the initally formed branched compounds.[356]

   Normally reaction occurs at C-1 and thus **4.509** with diethylaluminium cyanide,
which acts both as a nucleophile and a Lewis acid, gives the α-*C*-glycoside
analogue **4.535** in 90% yield.[357] Other reagents used and products formed by this
approach are, for example, bis(trimethylsilyl)ethyne, CH₂Cl₂, −20 °C, TiCl₄ to
give **4.536** (75%);[358] methylenecyclohexane, hexane/ethyl acetate, 20 °C, SnBr₄ to

give **4.537** (94%);[359] allyltrimethylsilane, CH_2Cl_2, $-78\,^\circ C$, $TiCl_4$ to give **4.538** (85%, α/β 16:1);[360] α-(trimethylsilyloxy)styrene, $-40\,^\circ C$, CH_2Cl_2, $BF_3\cdot OEt_2$, to give **4.539** (99%, α/β 4:1).[361]

With appropriate unsaturated substituents at C-3, glycal rearrangements of the type discussed in this section occur intramolecularly: for example, the Claisen rearrangement of 3-*O*-vinyl derivatives gives access to formylmethyl *C*-glycosides of 2-enes, and, as noted above, 3-azido-3-deoxy derivatives of glycals equilibrate with the 1-azido-2-enes.[306,357]

An alternative means of converting furanoid or pyranoid glycals to 2,3-unsaturated *C*-glycosides with or without the retention of the oxygen substituents at C-3 involves the use of palladium reagents and π-complexes derived from them.[362]

Glycal derivatives with various substituents (notably acyloxy groups) at C-2 undergo many of the above reactions.

(iii) Substitution reactions at C-1

Reaction of the *O*-silylated or *O*-benzylated glycals **4.544** with *t*-butyllithium in THF gives the lithiated intermediates **4.545** which, with tributyltin chloride at $-78\,^\circ C$, are converted into the stannylated glycals **4.546** (71%).[363] Compounds of this type can be converted in good yield into *C*-arylglycals **4.547** by application of palladium-catalysed coupling either directly[364] or *via* the iodoanalogues **4.548**, which are made by treatment of the tin compounds with iodine (Scheme 4.157).[365] Sulphones, e.g. **4.549**, in this series, made by oxidation of the corresponding thio-glycosides followed by base-catalysed elimination, may also be used as sources of the stannylated compounds (e.g. **4.546**), the reaction involving application of tri-*n*-butyltin hydride and a radical initiator (cf. Section 3.1.2.c.iv).[364]

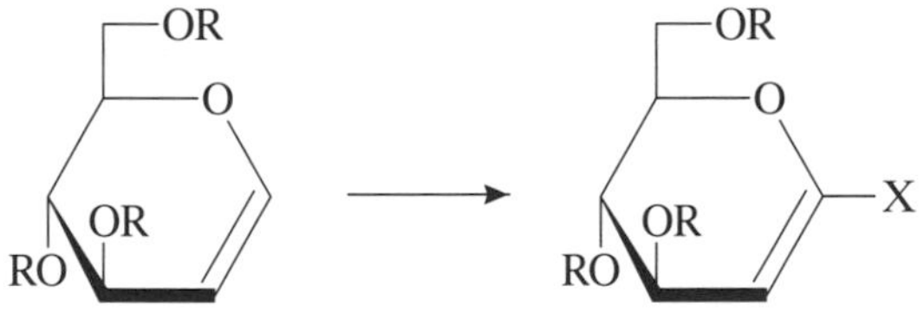

$R = SiPh_2Bu\text{-}t, Si(Pr\text{-}i)_3, Bn$

	X
4.545	Li
4.546	SnBu_3
4.547	Ar
4.548	I
4.549	SO_2Ph

Scheme 4.157

4.10.2 'ISOLATED' ALKENES

This section, which is necessarily highly selective, refers to monosaccharide derivatives which contain double bonds other than those of the glycal type or at C-5 and C-6 of hexopyranose compounds. Typically, therefore, 2-enes of furanoid and pyranoid derivatives, 3-enes of the latter type, 5-enes of hexofuranose sugars and many alkenes derived from acyclic monosaccharide derivatives show chemistry of the type illustrated in this section.

(a) Syntheses

Monosaccharide compounds having one unprotected hydroxyl group are commonly used for alkene synthesis following conversions to sulphonate esters; for example, the unsaturated nucleoside derivative **4.551** is obtained from the mesylate **4.550** (Scheme 4.158). Otherwise, provided the hydroxyl group has an unoxygenated neighbouring carbon atom, the pyrolytic Chugaev reaction of the derived xanthate esters may be used.[167,177,306] Eliminations from tertiary alcohol centres may be effected simply by treatment with thionyl chloride in pyridine.[366]

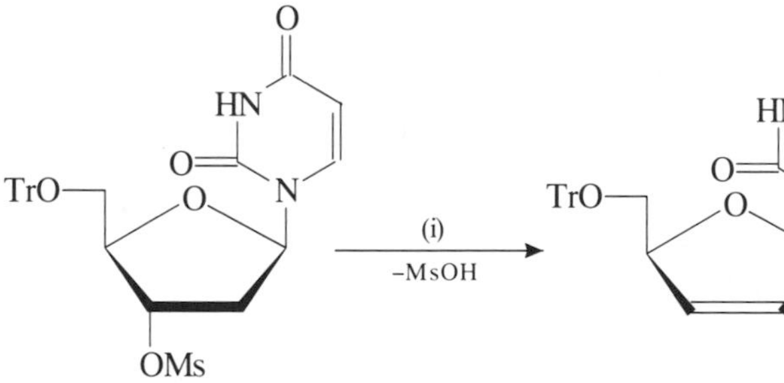

(i) *t*-BuOK, DMSO

Scheme 4.158

Many examples are known of the formation of enes from α-diols, one of the most-used methods for the conversion being the Tipson–Cohen reaction in which derived α-disulphonates are treated with sodium iodide and zinc in refluxing DMF. By this process methyl 4,6-*O*-benzylidene-β-D-glucopyranoside disulphonates **4.552** are converted in high yield to the corresponding alkenes **4.556**, whereas the α-anomers **4.553** react less readily to give only modest yields of alkenes **4.557**, showing that the reaction is subject to subtle structural factors (Scheme 4.159).[367] However, *cis*- and *trans*-diols can both be converted to enes by this procedure. A further reaction which also shows this versatility and has the advantage of being applicable to underivatized diols involves the use of triiodoimidazole, imidazole and triphenylphosphine in refluxing toluene. In this way, **4.554** gives 74% of alkene **4.556** directly by a process in which the hydroxyl groups are activated as described in Section 4.5.1.e.i.[368]

Radical reactions applied to various thioesters of α-diols have been used with marked success for effecting eliminations, Barton and colleagues having contributed notably in this area by developing procedures to the point of very high efficiency. Use of diphenylsilane and AIBN in boiling dioxane for 3 h converts the dixanthate **4.555** to the corresponding alkene **4.557** in 97% yield, and similar success is achieved in converting the ribonucleoside derivative **4.561** into alkene **4.562** (Scheme 4.160).[369]

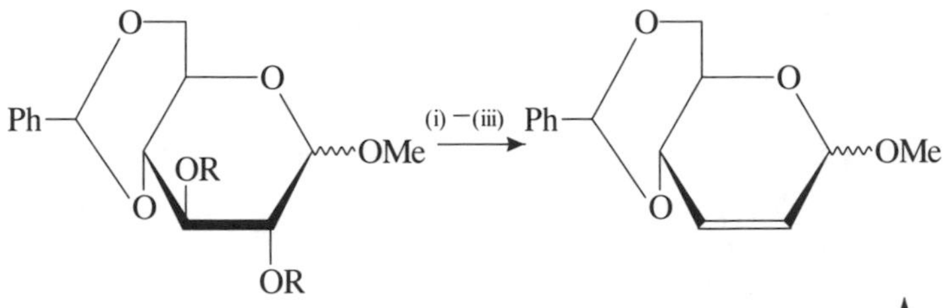

4.552 R = Ms, Ts; β-anomer
4.553 R = Ms, Ts; α-anomer
4.554 R = H; β-anomer
4.555 R = CSSMe; α-anomer

4.556 β-anomer
4.557 α-anomer (iv)–(vi)

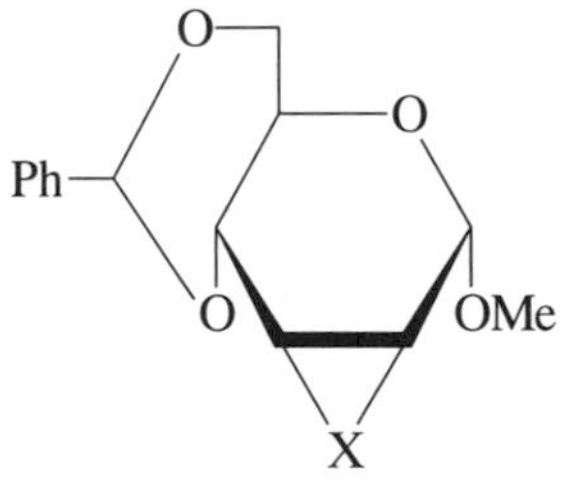

 (i) **4.552**, R = Ms, NaI, Zn, DMF,
    reflux 20 min (81% **4.556**);

    4.552, R = Ts, NaI, Zn, DMF,
    reflux 5 min (85% **4.556**);
    4.553, R = Ms, NaI., Zn, DMF,
    reflux 44h (13% **4.557**);
    4.553, R = Ts, NaI, Zn, DMF,
    reflux 12h (55% **4.557**);

 (ii) **4.554**, triiodoimidazole, imidazole, TPP,
    PhMe, reflux 3h (74% **4.556**);

(iii) **4.555**, Ph$_2$SiH$_2$, AIBN, dioxane, argon,
    reflux 3h (97% **4.557**);

4.558 X = O
4.559 X = S
4.560 X = NH

(iv) X = O, MeI, Me$_2$CO, reflux, 1 h, TsCl, Py, reflux, 20 min (90% **4.557**);
(v) X = S, P(OEt)$_3$, 155 °C, 0.5 h (70% **4.557**); (vi) X = NH, HONO, NaOH
(80% **4.557**)

Scheme 4.159

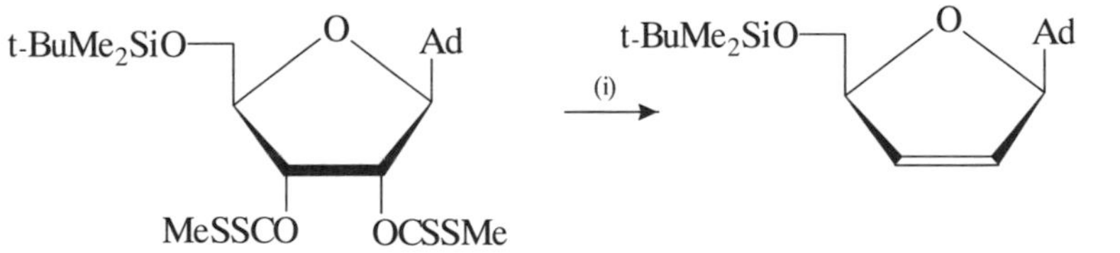

4.561

4.562 (97%)

(i) Ph$_2$SiH$_2$, AIBN, PhMe, 80–110 °C, 1.5 h

Scheme 4.160

Carbohydrate epoxides (e.g. **4.558**) and the corresponding episulphides (e.g. **4.559**) also serve as potential sources of alkenes, as do the less common epimines (aziridines; e.g. **4.560**) (Scheme 4.159).[161,167,306]

Elimination reactions occur much more readily in compounds having leaving groups β to electron-withdrawing functions, and compounds **4.563**–**4.566** represent many unsaturated derivatives produced in this way.

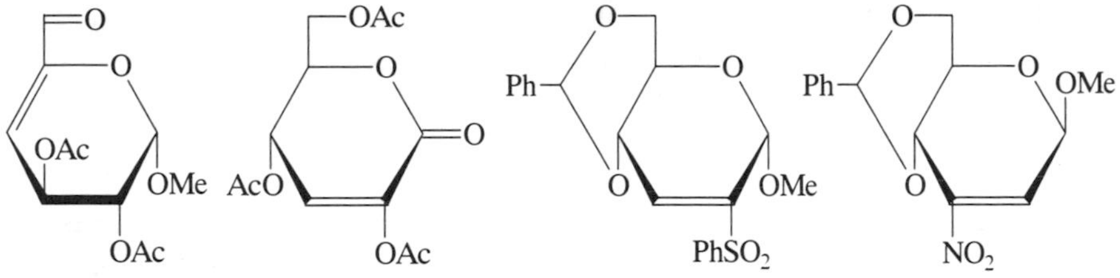

4.563   **4.564**   **4.565**   **4.566**

In Scheme 4.155 several examples of 2,3-unsaturated hexopyranose derivatives made from tri-*O*-acetyl-D-glucal by allylic rearrangement were given; related routes to many monosaccharide alkenes involve intramolecular (sigmatropic) allylic rearrangements, and have the advantage of giving products regio- and stereo-specifically.

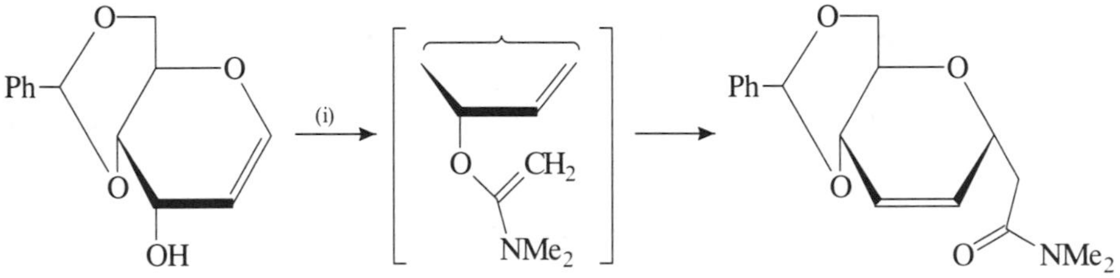

4.567   **4.568**   **4.569**

(i) MeC(OMe)$_2$NMe$_2$, xylene, reflux, 2h

Scheme 4.161

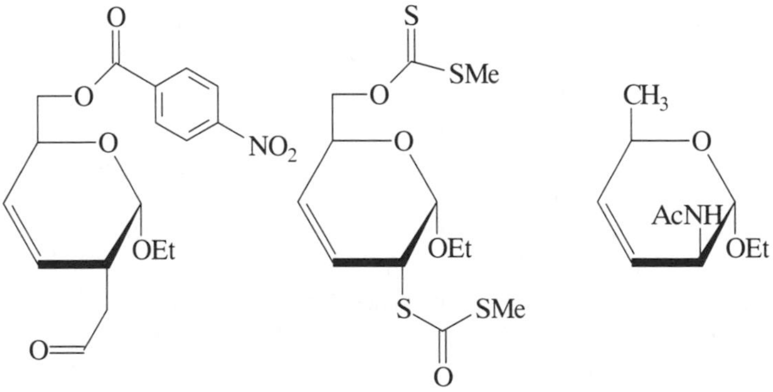

4.570   **4.571**   **4.572**

Thus, the allal acetal **4.567**, under the conditions of the Eschenmoser modification of the Claisen rearrangement, gives the 2-ene **4.569** in excellent yield by way of the intermediate **4.568** (Scheme 4.161).[357] Other compounds which have been made by related processes are **4.570**, **4.571** (respectively derived from the 4-vinyl ether[370] and 4-xanthate ester[371] of the corresponding D-*erythro*-2-enosides) and **4.572** (from either the corresponding 2-azido or 2-isothiocyanato derivative, made from either the 4-azido- or 4-thiocyanato-2-enoside with the D-*threo*-structure).[372]

(b) Reactions

Alkenes in 'isolated' (i.e. central) parts of cyclic monosaccharide derivatives and in acyclic compounds undergo a host of addition and substitution reactions, enolic compounds differing as expected from simple alkenes and being sources of ulose derivatives.

(i) Addition reactions

Unsubstituted alkenes undergo normal additions, often with high regio- and stereo-selectivity. Methyl 4,6-*O*-benzylidene-2,3-dideoxy-α-D-*erythro*-hex-2-eno-pyranoside **4.557** thus yields 70% of the 2,3-dibromo-D-*altro*-product with bromine, and acetyl hypobromite gives the 2-*O*-acetyl-3-bromo-3-deoxy adduct in high yield with the same configuration, showing that the reactive bromonium ion intermediate is formed on the lower (α) face of the ring.[306] Similarly, the same compound undergoes cyclopropanation with diiodomethane and a zinc–copper couple in refluxing ether (1 h) to give the α-product in 62% yield, which also illustrates that, in such alkenes, O-1 can stabilize intermediates and transition states of the kind involved in these addition reactions.[373] On the other hand, *cis*-hydroxylation with hydrogen peroxide and osmium tetroxide or with neutral permanganate, and *cis*-oxyamination with osmium tetroxide and Chloramine T result mainly in the products of addition to the more exposed β-face of the double bond.[306] In the last of these cases there is no strong regiochemical bias and both hydroxyamines are obtained (see Section 4.3.1.e.iii). Epoxidation with benzonitrile and hydrogen peroxide of the above alkene again leads to preferential addition at the β-face of the double bond, but with 2-enopyranosides having free hydroxyl groups at C-4 the oxygen atom is introduced in the *cis*-relationship to the hydroxyl functions as a consequence of hydrogen-bonding stabilization of the transition states.

Cycloaddition processes offer particularly powerful means of amplifying the structure of unsaturated monosaccharides. For example, condensation of the enone **4.573**, available from the 4,6-diol, with butadiene in dichloromethane at low temperature in the presence of aluminium chloride gives the Diels–Alder adduct **4.574** in 81% yield (Scheme 4.162).[374] Another cycloaddition reaction, using the 1,3-dipolar nitrile oxide **4.575** with the D-glucose-derived alkene **4.576**, gives 65% of the adduct **4.577** from which the 6-deoxyoctofuranos-7-ulose **4.578** can be prepared as illustrated in Scheme 4.163. Reduction of the keto group and removal of the protecting groups gives, in 65% yield, the corresponding 6-deoxyoctoses as a 1:2, mixture of C-7 epimers.[375]

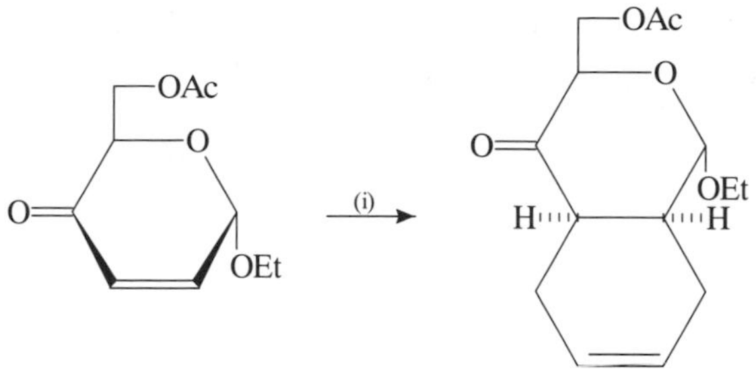

4.573   **4.574** (81%)

(i) CH$_2$=CHCH=CH$_2$, AlCl$_3$, CH$_2$Cl$_2$, $-40\,^\circ$C, 2.5 h

Scheme 4.162

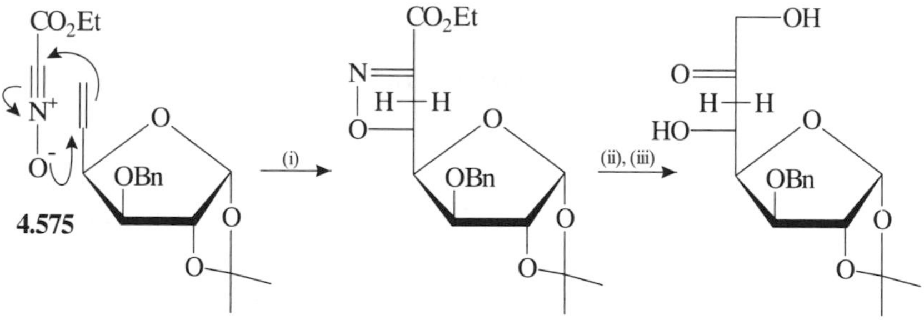

4.575

4.576   **4.577** (65%)   **4.578** (78%)

(i) $^-$O-$\overset{+}{N}$≡CCO$_2$Et, Et$_3$N, Et$_2$O; (ii) NaBH$_4$, THF; (iii) Pd/C, H$_2$, B(OH)$_3$, H$_3$O$^+$,
MeOH, 20 °C, 18 h

Scheme 4.163

As expected, addition reactions applied to conjugated enones and unsaturated
nitro compounds and sulphones show high regioselectivities.

(ii) Allylic substitution reactions

The presence of the double bonds in unsaturated derivatives makes substituents
at the allylic positions particularly susceptible to nucleophilic displacement. Most
often, direct substitution with inversion of configuration occurs, as illustrated in
Scheme 4.164, which indicates the potential value of reactive compounds such as
mesylate **4.579** in the synthesis of derivatives like azides **4.580** and **4.582** which can
serve as precursors of amino sugars. The allylic iodide **4.581** acts as an intermediate
with suitable reactivity for the preparation of the latter azide.[376]

For the introduction of functional groups by processes that involve allylic rearrangement it is best to use thermal intramolecular reactions of isomers analogous
to those indicated for the conversion of 3-O-substituted glycals into 1-substituted
2-enose derivatives (Section 4.10.1.b.ii). An example involving such rearrangement

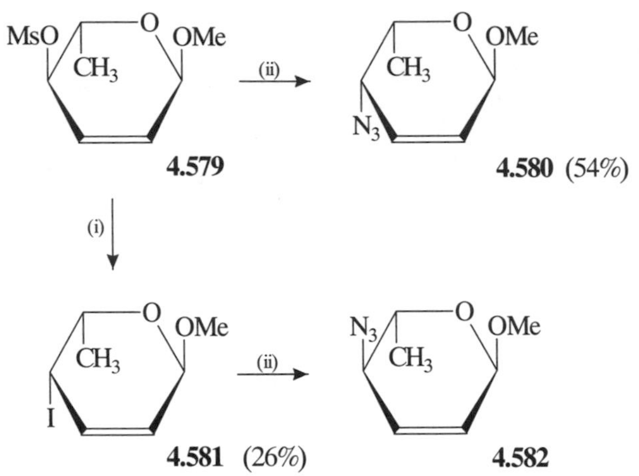

MsO—O OMe   (ii) →   O OMe
     CH₃   CH₃
                                N₃

4.579   **4.580** (54%)

(i) ↓

O OMe   (ii) →   N₃—O OMe
 CH₃   CH₃
I

4.581 (26%)   **4.582**

(i) NaI, Me₂CO, 22 °C; (ii) NaN₃, Me₂CO

Scheme 4.164

of a 4-*O*-substituted hex-2-enose compound to a 2-*N*-3-enose isomer is illustrated in Section 7.2.1, and advantage can be taken of the propensity of allylic azides, thiocyanates, xanthate esters and vinyl ethers to undergo isomerizations with allylic rearrangements (cf. Sections 4.10.2.a and 4.10.2.b) to obtain very many modified unsaturated compounds.

4.10.3   6-DEOXYHEX-5-ENOPYRANOSE DERIVATIVES

This group of unsaturated derivatives is selected for specific attention because of two unique reactions they undergo which give good access to functionalized cyclopentane and cyclohexane derivatives, respectively. Both are shown in outline in Scheme 4.165. The first utilizes acyclic 5,6-dideoxyhex-5-enose compounds containing the molecular feature **4.584** (see Scheme 4.91 for a specific synthesis), which undergo 1,3-dipolar cycloadditions *via* nitrones on treatment with *N*-alkylhydroxylamines to give bicyclic isoxazolidines **4.585**.[377]

The second depends upon the formation of the reactive mercury-containing intermediates containing the feature **4.587** which are made by treatment of 6-deoxyhex-5-enes **4.586** with mercury(II) salts in aqueous acetone, and which under the conditions of their synthesis cyclize to β-hydroxycyclohexanones **4.588** as indicated.[378] A notable feature of the reaction is the high selectivity shown, with main products having the *trans*-relationship between the newly formed hydroxyl groups and the substituents on the carbon atoms identified in **4.588**. This transformation was discovered during the authors' researches, is referred to as the Ferrier reaction, and is used as an illustration on the cover of this book.

Both reactions depend on 6-deoxy-6-halohexopyranosyl starting materials **4.583**, and the uses of each and further discussion are given in Sections 7.2.2.b and 7.2.2.c, respectively.[306,340]

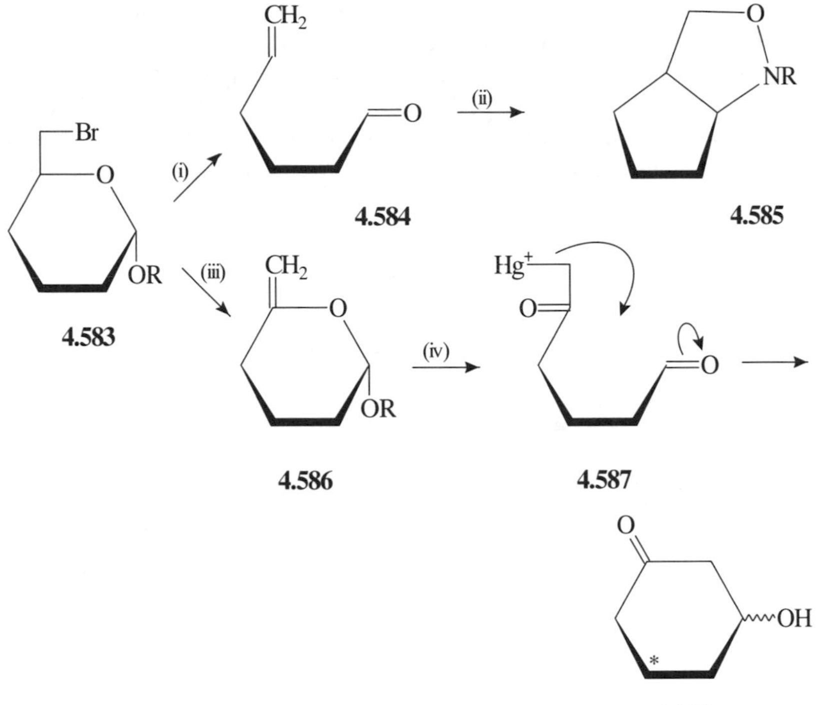

4.584

4.585

4.583

4.586

4.587

4.588

(i) Zn, EtOH; (ii) RNHOH; (iii) AgF, Py; (iv) HgX$_2$, H$_2$O
Scheme 4.165

4.10.4   PRODUCTS DERIVED BY WITTIG-LIKE PROCESSES

Several examples are given elsewhere in this text of the Wittig reaction being used to prepare alkenes from *aldehydo*-sugar derivatives (Section 3.1.4.g) and sugar-based ketones (Section 4.8.2.b); from these, extended chain and branched-chain compounds are obtainable. Tebbe and Tebbe-like methylenating agents may

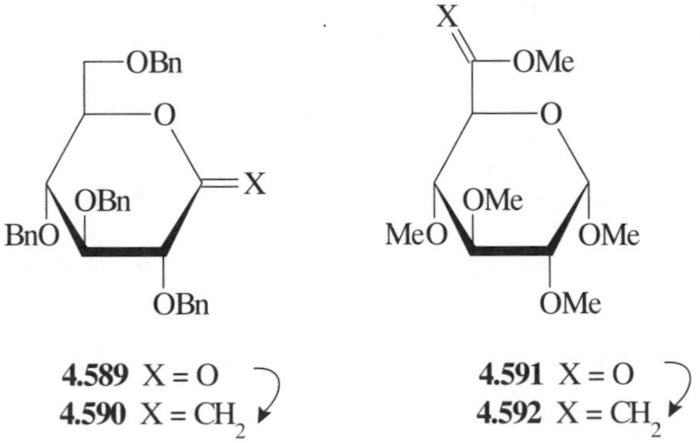

4.589 X = O
4.590 X = CH$_2$

4.591 X = O
4.592 X = CH$_2$

be used with sugar acid lactones and even with esters, and thus, for example, tetra-*O*-benzyl-D-glucono-δ-lactone **4.589** affords the *exo*-alkene **4.590** (Tebbe reagent in toluene, THF and catalytic amounts of pyridine at −40°C) in 88% yield, and the uronic acid ester **4.591** can be converted into the 6-deoxyhept-6-enose derivative **4.592** in 86% yield.[379] The related conversion of carbohydrate acetates into propenyl ethers[380] is referred to in Sections 3.1.1.a.ii and 3.1.1.b.vi.

4.11 REFERENCES

1. *Chem. Rev.*, 1969, **69**, 1.
2. *Adv. Carbohydr. Chem. Biochem.*, 1969, **24**, 139.
3. *Synthesis*, 1981, 1.
4. *Adv. Carbohydr. Chem.*, 1953, **8**, 107.
5. *J. Chem. Soc.*, 1965, 3496.
6. *Chem. Rev.*, 1969, **69**, 407.
7. *J. Org. Chem.*, 1959, **24**, 1618.
8. *J. Am. Chem. Soc.*, 1965, **87**, 4579.
9. *J. Chem. Soc.*, 1965, 2236.
10. *J. Org. Chem.*, 1974, **39**, 3223.
11. *Carbohydr. Res.*, 1980, **78**, 173.
12. *Tetrahedron Lett.*, 1968, 2271.
13. *Methods Carbohydr. Chem.*, 1972, **6**, 218.
14. *Methods Carbohydr. Chem.*, 1972, **6**, 215; *J. Org. Chem.*, 1969, **34**, 3819.
15. *J. Org. Chem.*, 1980, **45**, 4387.
16. *Can. J. Chem.*, 1987, **65**, 412.
17. *Carbohydr. Res.*, 1979, **77**, 262.
18. *J. Chem. Soc., Chem. Commun.*, 1984, 1530.
19. *Tetrahedron*, 1987, **43**, 971.
20. *Adv. Carbohydr. Chem.*, 1967, **22**, 109; *MTP International Rev. Sci. Org. Chem.* (eds D. H. Hey and G. O. Aspinall), Series 1, Vol. 7, Butterworths, London, 1973, p. 31; *MTP International Rev. Sci. Org. Chem.* (eds D. H. Hey and G. O. Aspinall), Series 2, Vol. 7, Butterworths, London, 1976, p. 35.
21. *Chem. Ber.*, 1965, **98**, 93.
22. *J. Chem. Soc.*, 1963, 5295.
23. *Tetrahedron*, 1963, **19**, 1711.
24. *J. Org. Chem.*, 1961, **26**, 537.
25. *J. Org. Chem.*, 1966, **31**, 226.
26. *J. Am. Chem. Soc.*, 1964, **86**, 5352.
27. *Adv. Carbohydr. Chem. Biochem.*, 1970, **25**, 109.
28. *J. Chem. Soc., Chem. Commun.*, 1973, 505.
29. *J. Chem. Soc.*, 1954, 2443.
30. *Carbohydr. Res.*, 1966–1967, **3**, 333.
31. *J. Chem. Soc.*, 1962, 4770.
32. *The Carbohydrates* (eds W. Pigman and D. Horton), Vol. IB, Academic Press, New York, 1980, p. 761.
33. *Adv. Carbohydr. Chem.*, 1966, **21**, 143; *Adv. Carbohydr. Chem. Biochem.*, 1971, **26**, 279.
34. *J. Chem. Soc., Perkin Trans. 1*, 1988, 999.
35. *J. Am. Chem. Soc.*, 1948, **70**, 3955.
36. *J. Chem. Soc.*, 1963, 4701.
37. *Adv. Carbohydr. Chem.*, 1967, **22**, 177.

38. *Carbohydr. Res.*, 1970, **14**, 255; *Carbohydr. Res.*, 1970, **15**, 397.
39. *Synthesis*, 1987, 665.
40. *Bull. Chem. Soc. Jpn.*, 1972, **45**, 567, 3614.
41. *Adv. Carbohydr. Chem. Biochem.*, 1981, **38**, 105.
42. *J. Chem. Soc., Perkin Trans. 1*, 1975, 1773.
43. *Carbohydr. Res.*, 1989, **188**, 228.
44. *J. Chem. Soc., Perkin Trans. 1*, 1975, 1574; *Tetrahedron*, 1983, **39**, 2609.
45. *J. Org. Chem.*, 1986, **51**, 2613.
46. *J. Am. Chem. Soc.*, 1983, **105**, 4059.
47. *Tetrahedron Lett.*, 1989, **30**, 2619.
48. *J. Org. Chem.*, 1981, **46**, 4843.
49. *J. Chem. Soc., Perkin Trans. 1*, 1977, 1718.
50. *J. Chem. Soc., Chem. Commun.*, 1977, 927; *Tetrahedron*, 1989, **45**, 4989.
51. *Chem. Ber.*, 1980, **113**, 3067.
52. *Tetrahedron Lett.*, 1982, **23**, 281; *J. Org. Chem.*, 1984, **49**, 176.
53. *Tetrahedron Lett.*, 1979, **20**, 2805.
54. *J. Chem. Soc., Perkin Trans. 1*, 1977, 1564.
55. *Chem. Ber.*, 1976, **109**, 2537.
56. *Chem. Pharm. Bull.*, 1979, **27**, 2838; *Carbohydr. Res.*, 1979, **76**, 79.
57. *Carbohydr. Res.*, 1978, **65**, 35.
58. *Top. Curr. Chem.*, 1990, **154**, 285.
59. *J. Org. Chem.*, 1983, **48**, 5093.
60. *The Amino Sugars* (ed. R. W. Jeanloz), Vol. IA, Academic Press, New York, 1969, p. 1; *The Carbohydrates* (eds W. Pigman and D. Horton), Vol. IB, Academic Press, New York, 1980, p. 644.
61. *Synthesis*, 1977, 45; *Tetrahedron Lett.*, 1987, **28**, 5941; *Carbohydr. Res.*, 1990, **197**, 318.
62. *Methods Carbohydr. Chem.*, 1962, **1**, 242.
63. *Carbohydr. Res.*, 1970, **14**, 231.
64. *Carbohydr. Res.*, 1986, **153**, 150.
65. *J. Am. Chem. Soc.*, 1987, **109**, 2821.
66. *Chem. Ber.*, 1974, **107**, 1188.
67. *J. Org. Chem.*, 1990, **55**, 5700; *Collect. Czech. Chem. Commun.*, 1983, **48**, 2386.
68. *Acta Chem. Scand.*, 1988, **42B**, 605.
69. *J. Am. Chem. Soc.*, 1986, **108**, 2466.
70. *J. Org. Chem.*, 1982, **47**, 2691; *J. Chem. Soc., Chem. Commun.*, 1987, 1073.
71. *Adv. Carbohydr. Chem. Biochem.*, 1969, **24**, 67; *Methods Carbohydr. Chem.*, 1972, **6**, 245.
72. *Angew. Chem., Int. Ed. Engl.*, 1967, **6**, 568.
73. *J. Org. Chem.*, 1965, **30**, 2735.
74. *Top. Curr. Chem.*, 1970, **14**, 556; *Can. J. Chem.*, 1967, **45**, 991; *Can. J. Chem.*, 1968, **46**, 2511.
75. *J. Chem. Soc.*, 1961, 4166.
76. *Can. J. Chem.*, 1968, **46**, 401, 405, 413.
77. *Can. J. Chem.*, 1979, **57**, 1244.
78. *J. Chem. Soc., Perkin Trans. 1*, 1981, 2070.
79. *Bull. Chem. Soc. Jpn.*, 1986, **59**, 3131.
80. *Carbohydr. Res.*, 1985, **136**, 153.
81. *Carbohydr. Res.*, 1979, **68**, 257.
82. *Liebigs Ann. Chem.*, 1986, 564.
83. *Methods Carbohydr. Chem.*, 1976, **7**, 29.
84. *Carbohydr. Res.*, 1986, **150**, 111.
85. *Carbohydr. Res.*, 1986, **146**, 193.

86. *Carbohydr. Res.*, 1985, **142**, 141.
87. *Can. J. Chem.*, 1973, **51**, 33.
88. *Tetrahedron Lett.*, 1967, 1549; *Methods Carbohydr. Chem.*, 1972, **6**, 208.
89. *Glycoconjugate J.*, 1984, **4**, 313.
90. *Am. Chem. Soc. Symp. Ser.*, 1976, **39**, 90.
91. *J. Carbohydr. Chem.*, 1988, **7**, 701.
92. *J. Carbohydr. Chem.*, 1988, **7**, 359.
93. *Liebigs Ann. Chem.*, 1985, 775.
94. *Adv. Carbohydr. Chem. Biochem.*, 1975, **31**, 9.
95. *Acta Chem. Scand.*, 1983, **37B**, 72.
96. *Carbohydr. Res.*, 1984, **129**, 267.
97. *J. Chem. Soc.*, 1963, 5288.
98. *Carbohydr. Res.*, 1973, **29**, 135.
99. *Methods Carbohydr. Chem.*, 1972, **6**, 282.
100. *J. Am. Chem. Soc.*, 1975, **97**, 4063.
101. *Carbohydr. Res.*, 1982, **103**, 286.
102. *Carbohydr. Res.*, 1986, **150**, C7.
103. *Chem. Rev.*, 1992, **92**, 1167.
104. *Adv. Carbohydr. Chem.*, 1968, **23**, 115.
105. *Carbohydr. Res.*, 1979, **68**, 391.
106. *J. Org. Chem.*, 1987, **52**, 3337.
107. *Adv. Carbohydr. Chem. Biochem.*, 1984, **42**, 135.
108. *The Carbohydrates*, (eds W. Pigman and D. Horton), Vol. IB, Academic Press, New York, 1980, p. 799.
109. *Experientia*, 1986, **43**, 1301.
110. *J. Chem. Soc., Chem. Commun.*, 1987, 1200.
111. *Angew. Chem., Int. Ed. Engl.*, 1991, **30**, 1387; *Acc. Chem. Res.*, 1991, **24**, 235.
112. *J. Org. Chem.*, 1975, **40**, 1337.
113. *J. Org. Chem.*, 1989, **54**, 1884.
114. *Chem. Ber.*, 1927, **60**, 232; *Collect. Czech. Chem. Commun.*, 1956, **21**, 1003.
115. *Carbohydr. Res.*, 1990, **208**, 287.
116. *Can. J. Chem.*, 1961, **39**, 1765.
117. *Methods Carbohydr. Chem.*, 1972, **6**, 286.
118. *J. Chem. Soc.*, 1965, 6666.
119. *J. Org. Chem.*, 1970, **35**, 606.
120. *Tetrahedron Lett.*, 1970, 2869.
121. *Adv. Carbohydr. Chem. Biochem.*, 1991, **49**, 37.
122. *Chem. Pharm. Bull.*, 1987, **35**, 2148; *Tetrahedron Lett.*, 1989, **30**, 4505.
123. *J. Am. Chem. Soc.*, 1961, **83**, 3827.
124. *J. Chem. Soc., Chem. Commun.*, 1967, 881.
125. *J. Chem. Soc. C*, 1966, 2366.
126. *J. Chem. Soc., Chem. Commun.*, 1970, 1038.
127. *J. Chem. Soc., Perkin Trans. 1*, 1973, 1400.
128. *J. Org. Chem.*, 1981, **46**, 3193.
129. *Adv. Carbohydr. Chem. Biochem.*, 1973, **28**, 225.
130. *Adv. Carbohydr. Chem. Biochem.*, 1981, **38**, 195; *Adv. Carbohydr. Chem. Biochem.*, 1990, **48**, 91.
131. *J. Org. Chem.*, 1966, **31**, 2817.
132. *Chem. Ber.*, 1941, **74**, 1807.
133. *Can. J. Chem.*, 1957, **35**, 955.
134. *Can. J. Chem.*, 1987, **65**, 412.
135. *Carbohydr. Res.*, 1967, **5**, 292.
136. *J. Org. Chem.*, 1982, **47**, 1506.

137. *J. Org. Chem.*, 1981, **46**, 1790.
138. *Carbohydr. Res.*, 1966, **2**, 181.
139. *J. Org. Chem.*, 1969, **34**, 2163.
140. *Adv. Carbohydr. Chem. Biochem.*, 1981, **39**, 71.
141. *J. Org. Chem.*, 1969, **34**, 1035; *Org. Synth.*, 1985, **65**, 243.
142. *J. Chem. Soc., Chem. Commun.*, 1988, **94**, 272.
143. *Carbohydr. Res.*, 1978, **61**, 511; *Methods Carbohydr. Chem.*, 1980, **8**, 227.
144. *J. Org. Chem.*, 1972, **37**, 2289.
145. *Carbohydr. Res.*, 1971, **16**, 375.
146. *Liebigs Ann. Chem.*, 1982, 1245.
147. *Helv. Chim. Acta*, 1980, **63**, 327.
148. *J. Chem. Soc., Perkin Trans. 1*, 1980, 2866; *J. Chem. Soc., Perkin Trans. 1*, 1982, 681; *Pure Appl. Chem.*, 1984, **56**, 845.
149. *Carbohydr. Res.*, 1972, **24**, 45; *Methods Carbohydr. Chem.*, 1976, **7**, 49.
150. *Tetrahedron*, 1963, **19**, 973.
151. *J. Org. Chem.*, 1970, **35**, 2319.
152. *Methods Carbohydr. Chem.*, 1972, **6**, 190.
153. *J. Chem. Soc., Perkin Trans. 1*, 1977, 1715.
154. *Carbohydr. Res.*, 1981, **94**, C14; *J. Org. Chem.*, 1983, **48**, 393, 4734.
155. *J. Org. Chem.*, 1992, **57**, 7001.
156. *Tetrahedron*, 1982, **38**, 2547.
157. *Tetrahedron Lett.*, 1986, **27**, 3219.
158. *J. Org. Chem.*, 1979, **44**, 1434.
159. *Carbohydr. Res.*, 1977, **58**, 139; *Bull. Chem. Soc. Jpn.*, 1984, **57**, 2535.
160. *Chem. Lett.*, 1988, 1703; *Carbohydr. Res.*, 1983, **116**, 227.
161. *Adv. Carbohydr. Chem.*, 1965, **20**, 67.
162. *Helv. Chim. Acta*, 1979, **62**, 1990, 2400, 2411; *Helv. Chim. Acta*, 1984, **67**, 1328.
163. *J. Org. Chem.*, 1989, **54**, 2307.
164. *Carbohydr. Res.*, 1977, **54**, 85.
165. *Angew. Chem., Int. Ed. Engl.*, 1989, **28**, 969.
166. *Tetrahedron Lett.*, 1988, **29**, 3691.
167. *Tetrahedron Lett.*, 1989, **30**, 57.
168. *J. Med. Chem.*, 1986, **29**, 151.
169. *J. Chem. Soc.*, 1946, 625.
170. *Methods Carbohydr. Chem.*, 1963, **2**, 188.
171. *Carbohydr. Res.*, 1986, **158**, 245.
172. *Aust. J. Chem.*, 1980, **33**, 1021; *J. Chem. Soc., Perkin Trans. 1*, 1974, 650.
173. *Carbohydr. Res.*, 1988, **183**, 135.
174. *Collect. Czech. Chem. Commun.*, 1966, **31**, 315; *J. Chem. Soc. B*, 1969, 377.
175. *Synth. Commun.*, 1982, **12**, 931.
176. *Aust. J. Chem.*, 1981, **34**, 1997; *J. Chem. Soc., Perkin Trans. 1*, 1981, 2328.
177. *Adv. Carbohydr. Chem. Biochem.*, 1969, **24**, 199.
178. *Carbohydr. Res.*, 1965, **1**, 242.
179. *J. Biol. Chem.*, 1933, **102**, 331.
180. *Tetrahedron Lett.*, 1988, **29**, 1449.
181. *Tetrahedron Lett.*, 1964, 2013.
182. *Adv. Carbohydr. Chem. Biochem.*, 1970, **25**, 181.
183. *Methods Carbohydr. Chem.*, 1963, **2**, 172.
184. *J. Carbohydr. Chem.*, 1987, **6**, 423.
185. *J. Chem. Soc.*, 1938, 1088.
186. *J. Chem. Soc.*, 1940, 620.
187. *Carbohydr. Res.*, 1971, **17**, 57; *Carbohydr. Res.*, 1971, **20**, 305.
188. *J. Chem. Soc. C*, 1969, 2263.

189. *Chem. Ber.*, 1980, **113**, 3919.
190. *Adv. Carbohydr. Chem.*, 1950, **5**, 191.
191. *Adv. Carbohydr. Chem. Biochem.*, 1970, **25**, 229.
192. *Can. J. Chem.*, 1951, **29**, 1079.
193. *J. Am. Chem. Soc.*, 1951, **73**, 3742.
194. *Adv. Carbohydr. Chem. Biochem.*, 1991, **49**, 93.
195. *J. Chem. Soc.*, 1947, 1403.
196. *Can. J. Chem.*, 1974, **52**, 3367.
197. *Adv. Carbohydr. Chem.*, 1955, **10**, 1.
198. *Chem. Ber.*, 1963, **96**, 3195.
199. *Can. J. Chem.*, 1960, **38**, 136.
200. *J. Chem. Soc.*, 1965, 1616.
201. *J. Chem. Soc.*, 1950, 371.
202. *Top. Curr. Chem.*, 1970, **14**, 367.
203. *J. Am. Chem. Soc.*, 1966, **88**, 2073.
204. *J. Chem. Soc. C*, 1967, 372.
205. *Adv. Carbohydr. Chem. Biochem.*, 1984, **42**, 69.
206. *Adv. Carbohydr. Chem. Biochem.*, 1975, **31**, 135.
207. *J. Chem. Soc.*, 1962, 3544.
208. *Progress in Stereochemistry*, (eds B. J. Aylett and M. M. Harris), Vol. 4, Butterworths, London, 1969, p. 43.
209. *Can. J. Chem.*, 1958, **36**, 480.
210. *Carbohydr. Res.*, 1971, **20**, 251.
211. *Chem. Ber.*, 1974, **107**, 2992.
212. *Radicals in Organic Synthesis: Formation of Carbon–Carbon Bonds*, Pergamon Press, Oxford, 1986.
213. *Angew. Chem., Int. Ed. Engl.*, 1984, **23**, 69.
214. *Synthesis*, 1988, 417, 489.
215. *Tetrahedron*, 1985, **41**, 4079.
216. B. Giese *Radicals in Organic Synthesis: Formation of Carbon–Carbon Bonds*, Pergamon Press, Oxford, 1986, p. 62.
217. *Tetrahedron*, 1990, **46**, 1.
218. *J. Org. Chem.*, 1969, **34**, 1029.
219. *Carbohydr. Res.*, 1982, **100**, C27.
220. *Can. J. Chem.*, 1975, **53**, 2017.
221. *Carbohydr. Res.*, 1970, **15**, 1.
222. *Tetrahedron Lett.*, 1982, **23**, 1763.
223. *J. Org. Chem.*, 1984, **49**, 2347; *J. Org. Chem.*, 1989, **54**, 5350.
224. *Tetrahedron*, 1984, **40**, 2083.
225. *Tetrahedron Lett.*, 1984, **25**, 4579.
226. *Can. J. Chem.*, 1978, **56**, 1836; *J. Chem. Soc., Perkin Trans. 1*, 1986, 61; *Helv. Chim. Acta*, 1972, **55**, 467; *Chem. Lett.*, 1973, 789.
227. *J. Am. Chem. Soc.*, 1972, **94**, 1353.
228. *J. Am. Chem. Soc.*, 1979, **101**, 5839.
229. *Synthesis*, 1971, 70.
230. A.H. Haines *Methods for the Oxidation of Organic Compounds*, Academic Press, London, 1988.
231. *The Carbohydrates*, (eds W. Pigman and D. Horton), Vol. IB, Academic Press, New York, 1980, pp. 1013, 1101.
232. *Synthesis*, 1981, 165; *Org. React.*, 1990, **39**, 297.
233. *Chem. Rev.*, 1982, **82**, 287.
234. *J. Org. Chem.*, 1965, **30**, 2304.
235. *Methods Carbohydr. Chem.*, 1972, **6**, 315.

236. *Carbohydr. Res.*, 1967, **3**, 318.
237. *Chem. Lett.*, 1977, 1327.
238. *Oxidation in Organic Chemistry* (ed. W. S. Trahanovsky), Vol. B, Academic Press, New York, 1973, p. 177.
239. *J. Chem. Soc. C*, 1966, 1131.
240. *Carbohydr. Res.*, 1969, **11**, 199.
241. *Carbohydr. Res.*, 1969, **10**, 456.
242. P.M. Collins Unpublished work.
243. *J. Chem. Soc., Chem. Commun.*, 1987, 1625.
244. *Carbohydr. Res.*, 1980, **84**, 35.
245. *Tetrahedron Lett.*, 1968, 3363; *J. Org. Chem.*, 1970, **35**, 4000.
246. *Carbohydr. Res.*, 1978, **67**, 267.
247. *Carbohydr. Res.*, 1973, **26**, 441.
248. *Synthesis*, 1982, 245.
249. *J. Chem. Soc., Perkin Trans. 1*, 1982, 1967.
250. *Adv. Carbohydr. Chem. Biochem.*, 1984, **42**, 227.
251. *Carbohydr. Res.*, 1984, **129**, C1.
252. *Carbohydr. Res.*, 1970, **14**, 405; *Adv. Carbohydr. Chem. Biochem.*, 1981, **38**, 105.
253. *Carbohydr. Res.*, 1986, **155**, 141.
254. *Carbohydr. Res.*, 1975, **44**, 227.
255. *Adv. Carbohydr. Chem.*, 1962, **17**, 169; *Methods Carbohydr. Chem.*, 1972, **6**, 342.
256. *J. Chem. Soc.*, 1965, 2292.
257. *Can. J. Chem.*, 1965, **43**, 955.
258. *J. Org. Chem.*, 1962, **27**, 1800.
259. *Carbohydr. Res.*, 1973, **30**, 386.
260. *Carbohydr. Res.*, 1970, **15**, 215.
261. *J. Chem. Soc. C*, 1966, 398.
262. *J. Chem. Soc., Chem. Commun.*, 1986, 775.
263. *J. Chem. Soc., Chem. Commun.*, 1967, 726.
264. *J. Chem. Soc., Perkin Trans. 1*, 1973, 632.
265. *J. Org. Chem.*, 1981, **46**, 3764.
266. *J. Chem. Soc.*, 1965, 3448.
267. *Chem. Ber.*, 1980, **113**, 489.
268. *Can. J. Chem.*, 1969, **47**, 3989.
269. J. March *Advanced Organic Chemistry*, 4th Edn, Wiley, New York, 1991, p. 367.
270. *J. Chem. Soc., Chem. Commun.*, 1983, 1324.
271. W. Caruthers *Some Modern Methods of Organic Synthesis*, 3rd Edn, Cambridge University Press, Cambridge, 1986, p. 12.
272. *J. Chem. Soc., Chem. Commun.*, 1984, 60.
273. *J. Chem. Soc., Chem. Commun.*, 1983, 141.
274. *Tetrahedron*, 1984, **40**, 1289.
275. *Liebigs Ann. Chem.*, 1987, 759.
276. *Carbohydr. Res.*, 1972, **21**, 166.
277. *Methods Carbohydr. Chem.*, 1972, **6**, 129.
278. *J. Chem. Soc., Perkin Trans. 1*, 1972, 1670.
279. *J. Chem. Soc. C*, 1971, 1960, 1965; *J. Chem. Soc., Perkin Trans. 1*, 1980, 779.
280. *J. Chem. Res. (S)*, 1979, 266.
281. *J. Chem. Soc., Perkin Trans. 1*, 1980, 277.
282. *J. Chem. Res. (S)*, 1978, 446.
283. *J. Chem. Soc., Chem. Commun.*, 1974, 292.
284. *J. Chem. Soc., Perkin Trans. 1*, 1985, 575.
285. *J. Chem. Soc., Chem. Commun.*, 1985, 1038.
286. *J. Chem. Soc., Perkin Trans. 1*, 1977, 2423.

287. *Adv. Carbohydr. Chem. Biochem.*, 1980, **37**, 79.
288. *Helv. Chim. Acta*, 1934, **17**, 510.
289. *J. Biol. Chem.*, 1944, **152**, 585.
290. *Adv. Carbohydr. Chem. Biochem.*, 1981, **38**, 323.
291. *J. Am. Chem. Soc.*, 1993, **115**, 413 and references therein.
292. *Adv. Carbohydr. Chem. Biochem.*, 1982, **40**, 131.
293. *Synthesis*, 1991, 583.
294. *J. Am. Chem. Soc.*, 1988, **110**, 6481.
295. *J. Org. Chem.*, 1993, **58**, 5500.
296. *J. Org. Chem.*, 1993, **58**, 7937.
297. *Adv. Carbohydr. Chem.*, 1962, **17**, 169.
298. *Tetrahedron Lett.*, 1993, **34**, 1181.
299. *Recl. Trav. Chim. Pays-Bas*, 1985, 259.
300. *Carbohydr. Res.*, 1988, **173**, 306; *Carbohydr. Res.*, 1987, **159**, 229.
301. *Tetrahedron Lett.*, 1993, **34**, 2741.
302. *J. Am. Chem. Soc.*, 1955, **77**, 3310.
303. *Adv. Carbohydr. Chem. Biochem.*, 1979, **36**, 57.
304. *J. Am. Chem. Soc.*, 1985, **107**, 3279.
305. *The Carbohydrates* (eds W. Pigman and D. Horton), Vol. IB, Academic Press, New York, 1980, p. 843.
306. *Rodd's Chemistry of Carbon Compounds* (ed. M.F. Ansell) Vol. IF,G Supplement, Elsevier, Amsterdam, 1983.
307. *J. Chem. Soc., Chem. Commun.*, 1986, 1149.
308. *J. Org. Chem.*, 1980, **45**, 48.
309. *J. Chem. Soc., Perkin Trans. 1*, 1978, 595.
310. *Tetrahedron Lett.*, 1989, **30**, 2537; *Carbohydr. Res.*, 1989, **188**, 81.
311. *J. Org. Chem.*, 1990, **55**, 1979.
312. *Chem. Ber.*, 1978, **111**, 1632.
313. *Tetrahedron*, 1981, **37**, 2989.
314. *Aust. J. Chem.*, 1989, **42**, 2127.
315. *Adv. Carbohydr. Chem.*, 1954, **9**, 97.
316. *Can. J. Chem.*, 1965, **43**, 94.
317. *J. Chem. Soc. C*, 1966, 2339.
318. *Carbohydr. Res.*, 1972, **25**, 242.
319. *Carbohydr. Res.*, 1987, **162**, 13.
320. *J. Am. Chem. Soc.*, 1957, **79**, 3234.
321. *Tetrahedron Lett.*, 1991, **32**, 6125.
322. *J. Org. Chem.*, 1991, **56**, 5468.
323. *J. Org. Chem.*, 1990, **55**, 5812.
324. *Synthesis*, 1992, 1133.
325. *Synthesis*, 1978, 696.
326. *Can. J. Chem.*, 1973, **51**, 7, 33.
327. *Carbohydr. Res.*, 1984, **133**, C1.
328. *Carbohydr. Res.*, 1989, **193**, 61.
329. *J. Am. Chem. Soc.*, 1989, **111**, 6661.
330. *J. Chem. Soc., Chem. Commun.*, 1981, 572.
331. *J. Org. Chem.*, 1991, **56**, 5740.
332. *J. Org. Chem.*, 1992, **57**, 2084.
333. *Pure Appl. Chem.*, 1978, **50**, 1343.
334. *Chem. Ber.*, 1980, **113**, 489.
335. *Liebigs Ann. Chem.*, 1989, 1153.
336. *Can. J. Chem.*, 1976, **54**, 91.
337. *Carbohydr. Res.*, 1967, **4**, 239.

338. *Bull. Chem. Soc. Jpn.*, 1972, **45**, 3496.
339. *J. Org. Chem.*, 1986, **51**, 2395.
340. *Chem. Rev.*, 1993, **93**, 2779.
341. *J. Org. Chem.*, 1983, **48**, 3269.
342. *J. Chem. Soc. C*, 1969, 581.
343. *Carbohydr. Res.*, 1984, **128**, C9.
344. *Tetrahedron Lett.*, 1988, **29**, 1363.
345. *J. Chem. Soc. C*, 1969, 570.
346. *Methods Carbohydr. Chem.*, 1972, **6**, 307.
347. *Carbohydr. Res.*, 1979, **68**, 43.
348. *J. Chem. Soc., Chem. Commun.*, 1992, 94.
349. *Leibigs Ann. Chem.*, 1987, 637.
350. *Carbohydr. Res.*, 1988, **176**, 219.
351. *J. Chem. Soc. C*, 1971, 553.
352. *Carbohydr. Res.*, 1980, **82**, 207.
353. *Carbohydr. Res.*, 1991, **216**, 93.
354. *Tetrahedron*, 1980, **36**, 287; *J. Org. Chem.*, 1988, **53**, 845.
355. *Tetrahedron*, 1992, **48**, 8545.
356. *Chem. Ber.*, 1976, **109**, 3262.
357. *J. Org. Chem.*, 1984, **49**, 518.
358. *Carbohydr. Res.*, 1987, **171**, 193.
359. *J. Chem. Soc., Perkin Trans. 1*, 1990, 1995.
360. *J. Org. Chem.*, 1982, **47**, 3803.
361. *J. Org. Chem.*, 1984, **49**, 522.
362. *Acc. Chem. Res.*, 1990, **23**, 201.
363. *J. Org. Chem.*, 1991, **56**, 1944.
364. *Carbohydr. Res.*, 1992, **228**, 103.
365. *J. Org. Chem.*, 1991, **56**, 4821.
366. *Carbohydr. Res.*, 1973, **31**, 331.
367. *J. Chem. Soc., Perkin Trans. 1*, 1977, 1981.
368. *Synthesis*, 1979, 813.
369. *Tetrahedron Lett.*, 1991, **32**, 2569.
370. *J. Chem. Soc., Perkin Trans. 1*, 1973, 1791.
371. *Carbohydr. Res.*, 1977, **58**, 481.
372. *J. Chem. Soc. C*, 1971, 1907.
373. *Can. J. Chem.*, 1972, **50**, 2909.
374. *J. Am. Chem. Soc.*, 1983, **105**, 5874.
375. *J. Chem. Soc., Chem. Commun.*, 1991, 132.
376. *J. Chem. Soc., Perkin Trans. 1*, 1972, 2977; *J. Chem. Soc., Perkin Trans. 1*, 1973, 1295.
377. *Helv. Chim. Acta*, 1979, **62**, 1990, 2400, 2411.
378. *J. Chem. Soc., Perkin Trans. 1*, 1979, 1455.
379. *Carbohydr. Res.*, 1990, **205**, 428.
380. *J. Am. Chem. Soc.*, 1992, **114**, 6354.

5 Reactions and Products of Reactions of the Hydroxyl Groups

The hydroxyl groups of carbohydrates show all the chemical properties associated with simple alcohols, and undergo substitution of the hydrogen atoms to give ethers, esters and acetals as would be expected. Their effect on each other is to increase their mutual acidity and thus increase their activity in some substitution reactions. Cyclic derivatives, most often acetals, can be formed from two or three hydroxyl groups within the same molecule; the periodate ion may, for example, complex with an α-diol system to give a cyclic ionic intermediate which undergoes further reaction with cleavage of the carbon–carbon bond. In this chapter ethers, esters and acetals of individual hydroxyl groups will be treated in addition to cyclic species derived from diols and triols. Sugar anhydrides, i.e. compounds formally derived by elimination of water from two hydroxyl groups within the same molecule, are intramolecular ethers. However, they are produced by intramolecular attack by oxygen nucleophiles at carbon atoms and are treated in Chapter 3 if that carbon is the anomeric centre or in Chapter 4 if at other positions. As would be expected for compounds having several dissimilar hydroxyl groups, not all such groups in monosaccharides exhibit the same reactivities, and the question of relative reactivity is of vital importance because syntheses of partially and specifically substituted derivatives are frequently carried out. The important issue of the relative reactivities of hydroxyl groups is therefore considered at the outset.

Very often, partially substituted derivatives are made with the aid of *protecting (blocking) groups*.[1] Such groups are initially selectively introduced at particular hydroxyl sites, remaining hydroxyl groups are then manipulated, often also by substitution, and then the protecting functions are removed. Frequently, glycosides are used as derivatives that not only protect the anomeric hydroxyl group but also hold the sugar in a defined ring size. Most often in this context they are made by oxygen-nucleophilic substitution reactions at the anomeric carbon centres and are, accordingly, treated in Chapter 3; on occasion, however, they may be made by *O*-alkylation of the anomeric hydroxyl groups as is noted below.

The importance of the selective introduction of protecting groups in synthetic carbohydrate chemistry cannot be overemphasized.

5.1   RELATIVE REACTIVITIES OF HYDROXYL GROUPS[2]

Because so many variables are involved it is not possible to order the reactivities of the various hydroxyl groups in polyhydroxyl compounds towards different

O-substituting reagents. There are, however, some generalizations which provide useful indications of the selective reactivities of alcohol groups under particular circumstances.

The selective enzymic acylation and deacylation of carbohydrates is referred to in Section 5.3.3.

5.1.1   SELECTIVE *O*-SUBSTITUTION AT THE ANOMERIC HYDROXYL GROUP

Such substitution can give useful and discrete glycosidic products, but frequently reactions are carried out under conditions in which reversible sugar ring opening may occur, and in these circumstances products of different ring sizes and anomeric configurations are formed. In consequence, reactions of this kind may have limited value; nevertheless, as will be seen, they can be useful in particular circumstances.

Aldoses are slightly more acidic than the corresponding alditols, D-glucose and D-glucitol having, for example, pK_a values at 18 °C of 12.43 and 13.57, respectively.[3] This is a result of the greater ease of dissociation of the anomeric hydroxyl group arising from the indirect stabilization by the ring oxygen atom of the glycosyloxy anion. In consequence of this, anomeric hydroxyl groups can be *O*-substituted selectively in basic conditions,[2] but it should be recognized that this selectivity is usually not pronounced, and the reactions involved are not favoured in the same way as are acid-catalysed reactions at the anomeric carbon atoms (Chapter 3).

Treatment of D-glucose by slow addition of dimethyl sulphate and aqueous alkali at pH 9 (to limit base-catalysed reactions of the sugar; Section 3.1.7) and 25–30 °C affords mainly methyl β-D-glucopyranoside which can be isolated directly in 35% yield following acetylation and deacetylation.[4] The measured slightly higher acidity of β-D-glucopyranose relative to the α-anomer, which has been ascribed to a larger entropy of ionization,[3] could be partially responsible for the β-selectivity. In addition, substitution at equatorially oriented hydroxyl groups favours β-substitution; furthermore, it has been concluded that oxyanions derived from equatorial anomeric hydroxyl groups are enhanced selectively as nucleophiles because of repulsions with the ring oxygen atom orbitals (see **5.1**).[5] Recognizing this, Schmidt has coined the term *kinetic anomeric effect* (cf. Section 2.3.1.b).[6]

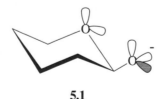

5.1

More effective selectivity is observed in the case of 2-acetamido-4,6-*O*-benzylidene-2-deoxy-D-glucose **5.3** from which, with dimethyl sulphate in aqueous sodium hydroxide, the analogous β-glycoside **5.2** precipitates in 70% yield. Surprisingly, when the solvent is changed to DMSO the α-glycoside **5.4** is

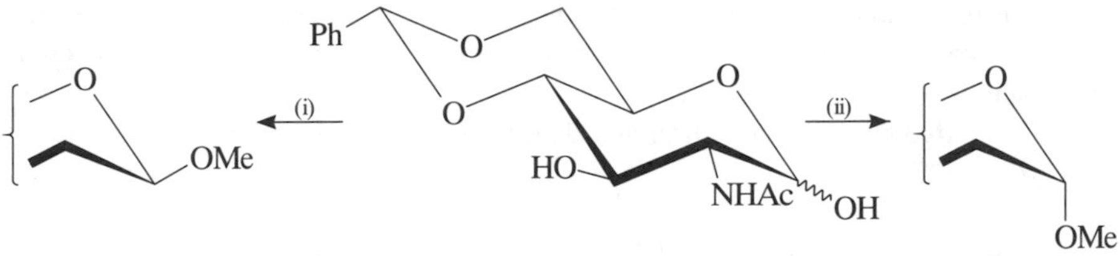

5.2 (70%)   **5.3**   **5.4** (86%)

(i) Me$_2$SO$_4$, NaOH (aqueous); (ii) Me$_2$SO$_4$, DMSO
Scheme 5.1

obtained in 86% yield on pouring the reaction mixture into an ice/water mixture (Scheme 5.1).[7] The work of Schmidt's group reveals that the answer to this remarkable observation lies in the formation of specific metal ion complexes in aprotic solvents such as DMSO.[5] When they treated 2,3-O-isopropylidene-D-ribose **5.5** with potassium t-butoxide in THF the complex **5.6** was formed, as evidenced by its reaction with triflate esters (ROTf) to give β-ribosides (including disaccharides) **5.7** in high yield. Alternatively, when the 5-trityl ether **5.8**, with the bulky trityl group impeding complexation on the β-side of the molecule, is treated with sodium hydride in dioxane, complex **5.9** is formed which gives high yields of the alternative α-linked compounds **5.10** (Scheme 5.2).

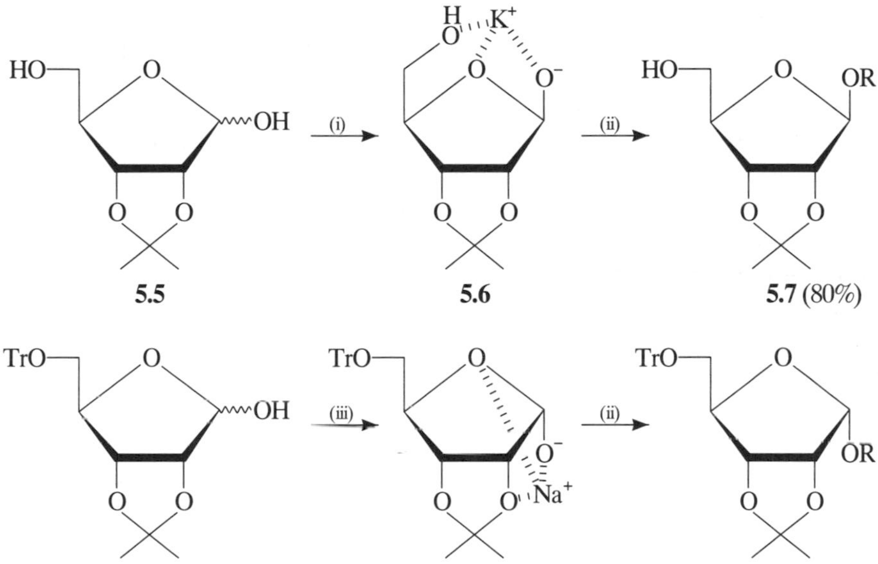

5.5   **5.6**   **5.7** (80%)

5.8   **5.9**   **5.10** (80%)

(i)KOBu-t, THF, 20 °C; (ii) ROTf; (iii) NaH, dioxane, 20 °C
Scheme 5.2

It seems safe to conclude that in the above-mentioned reaction of 2-acetamido-2-O-benzylidene-2-deoxy-D-glucose **5.3** in DMSO a sodium ion complex akin to

5.9 activates the α-anomer towards methylation. The potential of this approach in the wider syntheses of glycosides has been exploited by the use of the complex formed on treatment of 2,3,4,6-tetra-O-benzyl-D-glucose with sodium ions, which leads to an efficient synthesis of β-glucopyranosides.[5]

Selective acylations can be effected at the anomeric position, the α-L-fucose ether **5.11** giving a 54% yield of the C-1 equatorial β-glycoside **5.12** on treatment with p-nitrobenzoyl chloride in pyridine (Scheme 5.3).[8] Not all such substitution reactions, however, favour the anomeric position; selective tritylations and tosylations in pyridine occur at primary positions rather than at anomeric centres.

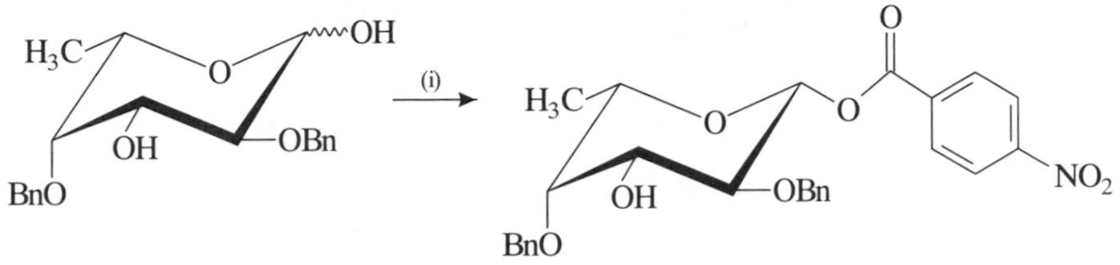

5.11   **5.12** (54%)

(i) p-O$_2$NC$_6$H$_4$COCl, Py, CH$_2$Cl$_2$, 20 °C, 20h

Scheme 5.3

5.1.2   SELECTIVE O-SUBSTITUTION AT PRIMARY HYDROXYL GROUPS

If anomeric hydroxyl groups are disregarded (see above), it is generally safe to assume that primary alcohol groups are appreciably more reactive than are secondary groups, and therefore that selective etherifications and esterifications can be effected at the former. This is particularly so if space-demanding reagents are involved, and pivaloyl (trimethylacetyl) chloride, t-butyldimethylsilyl chloride, triphenylmethyl (trityl) chloride and p-toluenesulphonyl (tosyl) chloride are frequently used when preferential substitution of primary hydroxyl groups is required. It has to be appreciated, however, that these reagents do react at secondary sites, and therefore primary-selective substitution reactions are conducted with just more than stoichiometric proportions of reagents and at appropriately low temperatures.

5.1.3   SELECTIVE O-SUBSTITUTION AT SECONDARY HYDROXYL GROUPS

Consistent with well-known generalizations from cyclohexane chemistry, equatorial secondary hydroxyl groups of pyranoid compounds tend to be substituted more readily than are otherwise comparable axial groups as is illustrated in Scheme 5.4, which involves two D-mannopyranosyl compounds in the $^4C_1$ and $^1C_4$ conformations, respectively, the hydroxyl groups at C-2, C-3 thus being axial, equatorial and equatorial, axial, respectively.[9]

(i) TsCl (one equivalent), Py, $-5\,°$C

Scheme 5.4

Further generalizations are difficult to make, although it is frequently noted that the hydroxyl group at C-2 of glucopyranosides is the most reactive of the secondaries while that at C-4 is least so. Methyl α-D-glucopyranoside, therefore, on treatment with two equivalents of acylating halide in pyridine gives mainly 2,6-disubstituted compounds, while the main products of trisubstitution are the 2,3,6-triesters. In the case of β-anomers, however, there is less or even reversed selectivity, and competitive experiments have shown that this is because with such compounds the hydroxyl group at C-2 is not activated as is the corresponding group of the α-anomers. The source of this activation appears to be associated with the *cis*-relationship between the anomeric oxygen atom and O-2, with the former conceivably acting as an intramolecular base catalyst which helps to remove the proton from its thereby-activated neighbouring group. A similar influence is present during the sulphonylation of 1,4:3,6-dianhydro-D-glucitol (**5.13**); with tosyl chloride in pyridine it is the sterically protected *endo*-hydroxyl group at C-5 which is selectively esterified.[2]

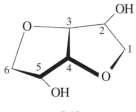

5.13

Selectivities of benzoylations can be enhanced by the use of *N*-benzoylimidazole as the esterifying reagent. Thus, from methyl 4,6-*O*-benzylidene-α-D-glucopyran-oside the 2-ester is produced almost exclusively with this reagent, as is the 3-ester from the β-galactoside analogue in which the reactive equatorial hydroxyl group

also has an axial neighbouring oxygen atom.[10] In methyl 4,6-*O*-benzylidene-α-
D-altropyranoside, in which both hydroxyl groups are axial, it is O-2 which is
selectively reactive.[11]

To complicate the matter still further, different selectivities are found under
different reaction conditions. Compound **5.13**, for example, is preferentially
substituted with acetic anhydride in pyridine at the C-2 *exo*-hydroxyl group. To
account for this, it has been suggested that in this case the reaction is dependent on
attack of the dissociated alcohol oxyanion on the anhydride, and the dissociation of
the *endo*-hydroxyl group is impeded by O-1. On the other hand, when acid chlorides
are used the undissociated alcohols are the nucleophiles, and now spatially adjacent
oxygen atoms enhance their nucleophilicity.[2] Perhaps such reasoning explains why
carboxylic acid anhydrides react preferentially at position 3 of glucopyranosides,
whereas the corresponding chlorides give mainly 2-esters.

Overall, as indicated above, there appears to be a tendency for the hydroxyl
group at C-2 to be the most reactive of the secondary groups on glycopyranoside
rings (the inductive effect of the anomeric centre being responsible). As instanced
above, however, it is not invariably so, and other factors can overcome the influence. As already noted (Scheme 5.4), when the hydroxyl group at C-2 is axial,
the steric impeding factor dominates, and in the methyl mannopyranoside example
given, and with methyl 6-*O*-trityl-α-D-mannoside, selective tosylation with tosyl
chloride occurs at the equatorial O-3.

Etherification of various hydroxyl groups in carbohydrates is similarly dependent
on many factors, but again the broad generalization is that primary groups and then
hydroxyl groups at C-2 tend usually to react most favourably.[2]

Because of the complexity of the problem it has been possible only to give
indications of the relative reactivities of different hydroxyl groups.[2] Adding to the
difficulties which preclude the establishment of useful reactivity factors are such
variables as temperature, solvent, reagent concentration and changes in reactivity
of particular hydroxyl groups with substitution changes of neighbouring groups.

5.1.4   MODIFICATION OF THE SELECTIVE *O*-SUBSTITUTION OF
HYDROXYL GROUPS BY USE OF ORGANOTIN DERIVATIVES

Formation of trialkylstannyl ethers from single hydroxyl groups or cyclic stannylene
derivatives from diols enhances the nucleophilicity of the oxygen atoms involved
and thus activates the hydroxyl groups towards electrophilic reagents, especially
acylating agents.[12] Selective stannylation or stannylene formation can consequently
alter relative reactivities within polyhydroxyl compounds, and practical use can be
made of this finding.

Stannyl derivatives are usually made by use of hexabutyldistannoxane in benzene solution with azeotropic removal of water. Stoichiometric proportions are used
according to the number of hydroxyl groups it is intended to substitute

$$ROH + \tfrac{1}{2}(Bu_3Sn)_2O \longrightarrow ROSnBu_3 + \tfrac{1}{2}H_2O$$

and the products are usually used without isolation.

Compounds containing primary hydroxyl groups are preferentially stannylated at these positions, the inherently high reactivity of which being thereby increased. Reactive secondary groups likewise are activated so that, for example, distannylation of methyl α-D-glucopyranoside followed by treatment with three equivalents of benzoyl chloride (no base is required) leads to the 2,6-dibenzoate in 82% yield together with some of the 2,3,6-triester. Similarly, the α-D-mannopyranoside and β-D-galactopyranoside give high yields of the 3,6-diesters by way of the intermediates **5.14** and **5.15**. The α-D-galactopyranoside, however, affords a mixture of several products in which considerable substitution at O-2 occurs, suggesting that a tin derivative at this position, stabilized by the axial O-1, competes with that at O-3.

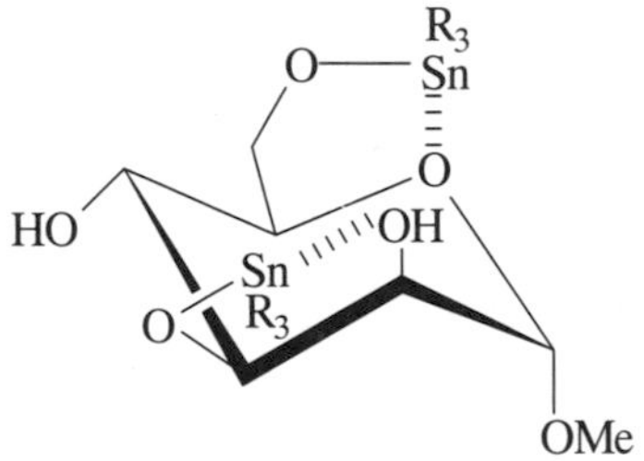

R = Bu   **5.14** 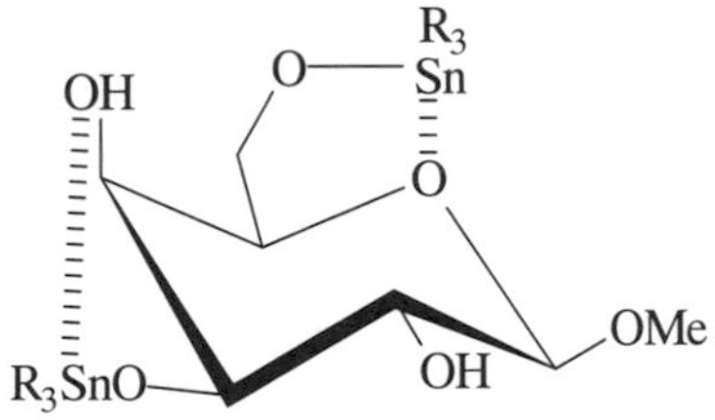   **5.15**

The situation relating to α-diols and their cyclic stannylene derivatives (made by use of polymeric dibutyltin oxide) is interesting because often one of the oxygen

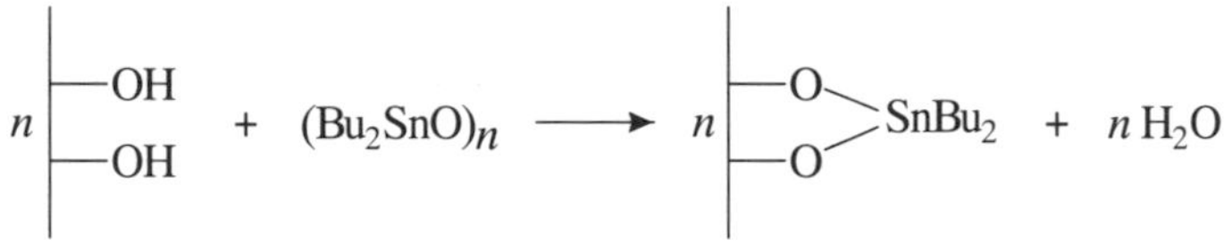

atoms is selectively activated for further substitution by properties of these derivatives rather than of the initial hydroxyl groups. In non-polar solvents the stannylenes exist as dimers illustrated by **5.16**, which shows that the oxygen atoms of the five-membered rings are chemically distinct, those at positions A being bonded to two tin atoms while those at sites B have one tin-bonded neighbour. The latter are more accessible sterically as well as more nucleophilically enhanced by electronic effects and are thus more reactive. X-Ray analysis of the α-glucopyranoside stannylene dimer of **5.17** shows that O-2 is in the favoured apical position and the compound gives the 2-benzoate in near quantitative yield. That electronic factors determine which oxygen is apical is suggested by the selective benzoylation at the axial 2-position of the corresponding stannylenated mannoside, but the situation is not clear-cut; benzylation of this compound affords the 3-ether in 85% yield, and generally in the case of *cis*-diols on pyranoid rings the equatorial hydroxyl groups are selectively activated.[12]

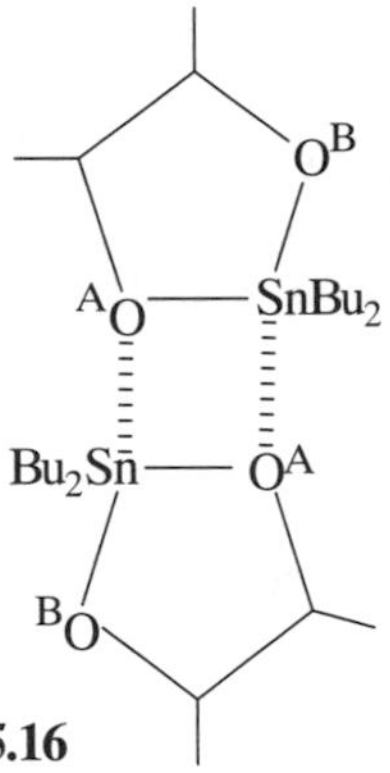

5.16

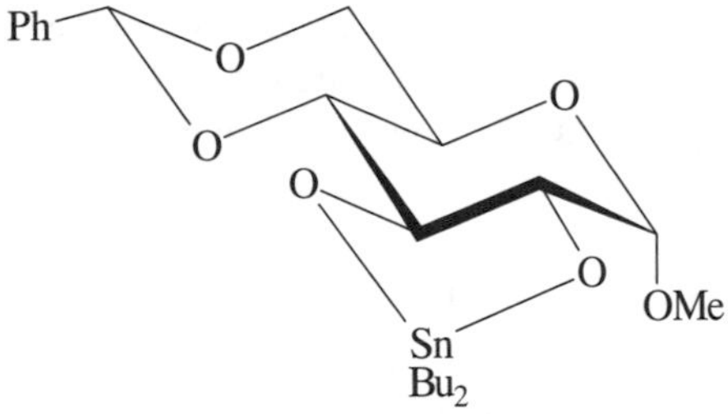

5.17

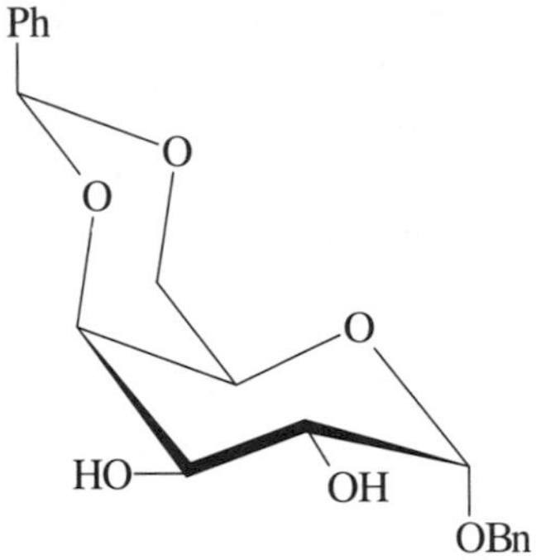

5.18

Benzyl 4,6-O-benzylidene-α-D-galactopyranoside **5.18**, following stannylenation, reacts at O-3 to give an almost quantitative yield of the monobenzoate.
Clearly, a factor is operating in this case which is difficult to rationalize since
there is considerable symmetry in the C-1–C-4 section in the molecule and the
inductive influence of the anomeric centre would have been expected to favour
substitution at O-2. Presumably the key lies in the structure of the dimer.

Selective stannylenation of methyl β-D-glucopyranoside takes place at the 4,6-
diol positions and allows the synthesis of the 6-benzoate in 80% yield. It is therefore
very surprising that similar treatment of the α-anomer gives 80–90% of methyl 2-
O-benzoyl-α-D-glucopyranoside, thereby suggesting that a 2,3-O-stannylene intermediate is preferentially involved.[13]

The use of stannylene derivatives in methyl ether synthesis and in the selective
formation of carbonyl compounds is referred to in Sections 5.2.1 and 4.9.1.a.ii,
respectively.

5.2 ETHERS

In this section derivatives produced by direct substitution of the hydrogen atoms
of hydroxyl groups at positions other than the anomeric centre are considered.

Substitution at the anomeric centre gives rise to glycosides which are acetals rather than ethers and is dealt with in Sections 3.1.1 and 5.1.1.

The most commonly used ethers are described, but methoxymethyl (Mom) 'ethers' (MeOCH$_2$OR) and tetrahydropyranyl 'ethers', which are frequently employed in syntheses, are referred to in Section 5.4.2 as they are acetals rather than true ethers.

5.2.1 METHYL ETHERS[14]

Partially methylated sugars, e.g. 3-*O*-methyl-D-galactose, 6-deoxy-2-*O*-methyl-D-allose (javose) and 6-deoxy-2,3-di-*O*-methyl-D-galactose, are known to occur in natural products, but the main interest in such compounds has evolved from the classical methylation procedure adopted for the structural analysis of higher saccharides (Section 8.1.3.a). By this procedure the free hydroxyl groups in polysaccharides are converted to methyl ethers and then the glycosidic bonds linking the units are cleaved specifically by acid hydrolysis to give partially methylated sugars, the structures and proportions of which reveal the main architectural features of the polymers. Partially methylated sugars have therefore been studied in great detail. In the early days of structural work these derivatives had the advantage of having relatively high volatility and they were separated by fractional distillation of their glycosides. Nowadays, their volatility makes them susceptible to investigation by gas–liquid chromatographic and mass spectrometric techniques which are applied with great advantage in polysaccharide analysis.

A variety of methylation procedures have been developed, a versatile and efficient method being the *Kuhn modification* of the traditional methyl iodide/silver oxide treatment (Purdie method). In this, the reaction is carried out in DMF; which is a better solvent for hydroxylic compounds than is methyl iodide. Silver oxide can be replaced as acid acceptor by barium or strontium oxides or hydroxides, and complete methylation of polyhydroxyl-compounds is usually readily effected. Free sugars give mixtures of methylated methyl glycosides which often contain high proportions of furanosides, specific glycosides being methylated directly to give single, fully substituted products.

Alternatively, methylation can be efficiently carried out with methyl bromide in DMF containing suspended sodium hydride, and other variations of the same general procedure involve the use of dimethyl sulphate or methyl iodide in DMSO containing powdered sodium hydroxide or sodium hydride.

Particularly favoured is the Hakomori procedure which involves the use of the dimethylsulphinyl anion as base,[15] this being prepared by dissolving sodium hydride in DMSO, and the carbohydrate is added in this solvent prior to the addition of methyl iodide. Otherwise, methyl triflate can be used (as a source of methyl groups) in an aprotic solvent in the presence of a hindered base, e.g. 2,6-di-*t*-butylpyridine.[16]

A different feature of DMSO chemistry is involved when, with acetic anhydride and acetic acid, it is used to convert alcohols (primary, secondary or tertiary)

$$\text{ROH} \xrightarrow{\text{(i)}} \text{ROCH}_2\text{SMe} \xrightarrow{\text{(ii)}} \text{ROMe}$$
$$(>70\%) \qquad\qquad (>80\%)$$

(i) DMSO, Ac$_2$O, AcOH; (ii) NiCl$_2$, H$_3$BO$_3$, NaBH$_4$

Scheme 5.5

into methylthiomethyl ethers which can then be reduced to methyl analogues (Scheme 5.5).[17]

Traditionally, the Haworth procedure, which utilizes dimethyl sulphate in strong aqueous alkali, was used in polysaccharide work, but this has the disadvantage that many derivatives are unstable under these conditions and it has been superseded largely by the above methods involving the use of aprotic solvents. The harshness of the conditions may otherwise be avoided by use of methyl halide in phase transfer conditions with aqueous alkali and tetrabutylammonium bromide as catalyst.[18]

For methylation of derivatives which are particularly sensitive to base (for example those containing acetyl esters which may cleave or migrate), diazomethane/boron trifluoride etherate can be used.[19]

Partially methylated compounds are normally prepared either by alkylation of protected derivatives followed by removal of the protecting groups (e.g. Scheme 5.6) or they can be obtained by nucleophilic displacement of, for example, sulphonate esters (Sections 4.1.2.a and 5.3.6) or by methoxide ion attack on an epoxide ring (Sections 4.1.3 and 4.6.1).

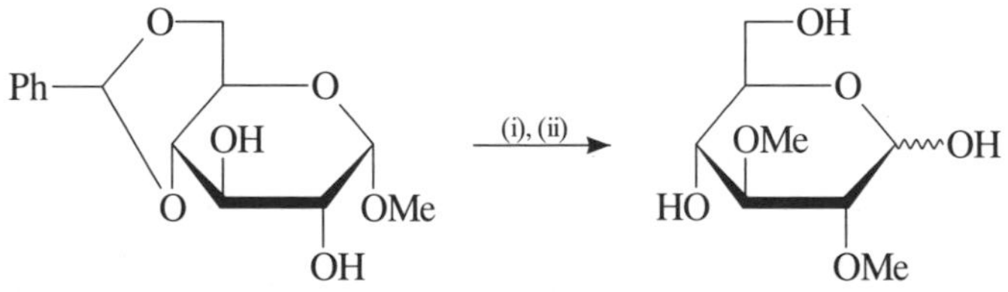

(i) MeI, DMF, Ag$_2$O; (ii) H$_3$O$^+$

Scheme 5.6

Attempts to effect good selectivity in the partial substitution of compounds containing more than one hydroxyl group did not meet with success until stannylene derivatives (Section 5.1.4) were applied.[12] With methyl iodide in warm DMF the 2,3-O-dibutylstannylene derivative of methyl 4,6-O-benzylidene-α-D-mannopyranoside gives the 3-methyl ether exclusively. A variant uses tin(II) chloride and diazomethane in methanol/dichloromethane with which, for example, methyl 4,6-O-benzylidene-α-D-glucopyranoside is substituted almost specifically at O-3.[20]

Methyl ethers are stable to most acidic and basic reagents and are therefore usually retained once introduced into a carbohydrate. They can, however, be removed with Fenton's reagent (hydrogen peroxide in the presence of iron(II) ions), or better, since the demethylated products can undergo further oxidation with this reagent, with boron trichloride at low temperatures. Under these latter conditions glycosidic, ether, ester and acetal bonds are cleaved and the resulting sugars (if aldoses) are stable. A noteworthy feature of both these reactions is that they give products having the same configuration as the initial ethers, which indicates that the methyl–oxygen bonds rather than the sugar–oxygen bonds undergo fission. This suggests that bimolecular attack by water on the complexes formed occurs more favourably at the methyl carbon atoms (Scheme 5.7a) than at the sterically more hindered sugar carbon atoms (Scheme 5.7b).

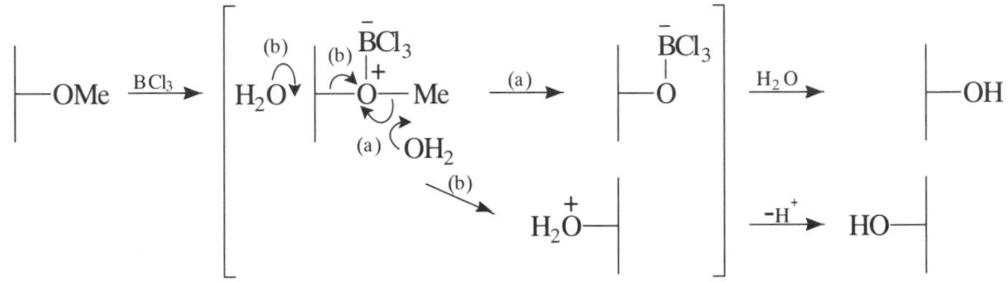

Scheme 5.7

A further method of cleaving methyl ethers involves oxidation with chromium trioxide in acetic acid followed by alkaline hydrolysis of the derived formate esters.[21] Ester groups tend to survive the oxidation step but are then cleaved with the formates; glycosidic and other acetal bonds are not stable. A mixture of tetrabutylammonium iodide and boron trifluoride etherate also cleaves methyl ethers while leaving benzoates unaltered,[22] and lithium diethylamide similarly removes these ethers and shows some selectivity for groups at primary sites.[23]

Several instances are known in which methoxy groups attached to anomeric or acetal centres migrate during displacement reactions, and because of the stabilized carbocations they generate when they leave this is not unexpected. Treatment of the ribose dimethyl acetal derivative **5.19** with tetra-N-butylammonium benzoate in N-methylpyrrolidone thus gives the 4-O-methyl-L-lyxose derivative **5.21** following attack of the nucleophile on the oxonium ion intermediate **5.20** (Scheme 5.8).[24] Much more surprising is the observation that methyl ether (rather than acetal)

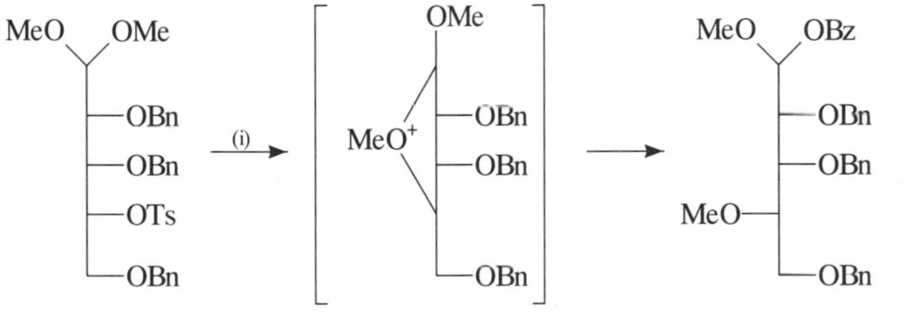

5.19   **5.20**   **5.21** (71%)

(i) Bu₄NOBz, N-methylpyrrolidone, 110 °C, 16 h

Scheme 5.8

groups can be labile under such conditions. Solvolysis of methyl 6-*O*-methane-sulphonyl-2,3-di-*O*-methyl-β-D-galactopyranoside **5.22** in aqueous methanol in the presence of sodium acetate gives the illustrated products presumably by way of the oxonium ion **5.23** (Scheme 5.9).[25] Such reactions are, however, unusual and methyl ethers of sugars can normally be assumed to be chemically stable derivatives.

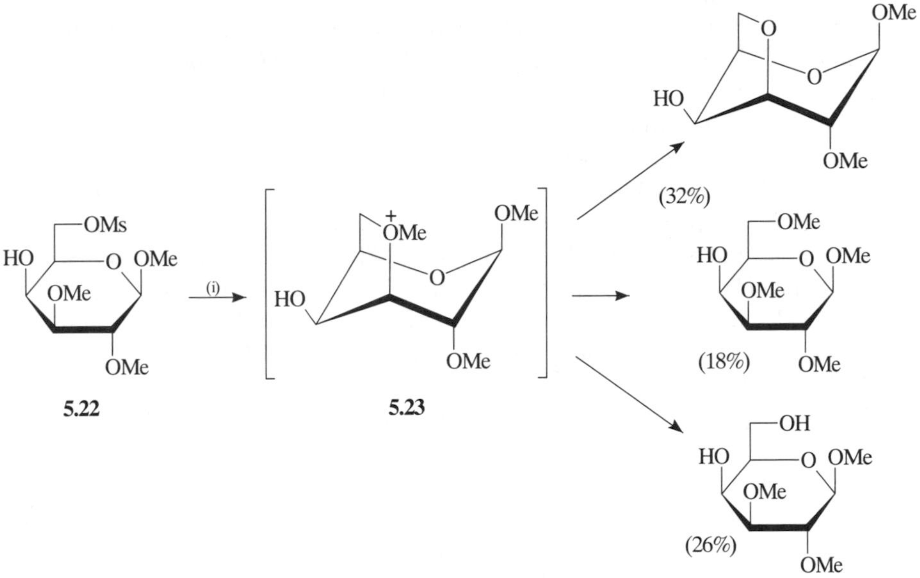

(i) MeOH, H_2O, NaOAc, reflux, 96 h

Scheme 5.9

5.2.2   BENZYL ETHERS[26]

The characteristics of benzyl ethers which make them of importance in carbohydrate chemistry are their relative stability under acidic and basic conditions and their susceptibility to cleavage to give the alcohols and toluene by hydrogenolysis, usually over palladium catalysts. Benzyl protecting groups can therefore be removed under conditions in which most other groups are stable, and they are thus of great value in syntheses. Azido groups can be selectively reduced in their presence. (See Scheme 4.37).

Most frequently, hydrogenolysis is carried out using hydrogen gas with a palladium catalyst absorbed on charcoal the hydroxide giving a very active non-pyrophoric product,[27] but a very convenient modification applies the concept of hydrogen transfer and uses easily dehydrogenated donors such as cyclohexene, cyclohexadiene, isopropanol or ammonium formate as the source of hydrogen together with the same type of catalyst.[27] Vigilance has to be observed when using this technique, however, since the benzyl ethers themselves may act as hydrogen donors and thereby be oxidized to benzoyl esters which are stable.[28] Presumably air is required for this alternative reaction.

Otherwise, benzyl ethers may be cleaved by use of dissolving sodium in ethanol or sodium in liquid ammonia.[29] The lability of the ether bond under all these conditions follows from the stability of benzyl radicals or related species formed as intermediates, and for the same reason trityl ethers (Section 5.2.3) and benzylidene acetals (Section 5.4.4) are similarly cleaved. Other debenzylation procedures include photochemical bromination at the benzylic centres by use of NBS with spontaneous hydrolysis of the very reactive products,[30] oxidation with chromium trioxide in acetic acid to give base-labile benzoates,[21] or with ruthenium tetroxide[31] or ozone,[32] and electrolytic oxidation.[33]

Adaptation of the Kuhn methylation procedure provides a means of preparing benzyl ethers, e.g. benzyl bromide is used in DMF in the presence of silver oxide and/or barium oxide. Otherwise, benzyl chloride can be employed in DMF or DMSO in the presence of powdered sodium hydroxide or sodium hydride, and this procedure has the advantage of avoiding the highly lachrymatory bromide. Milder methods include the use of the phase transfer technique, benzyl triflate together with a hindered base such as 2,6-di-t-butylpyridine,[16] benzyl trichloroacetimidate with catalytic triflic acid,[34] and phenyldiazomethane (EXPLOSIVE POTENTIAL) with catalytic fluoroboric acid.[35]

Normally, partial benzylation of polyhydroxyl compounds is not very selective, but benzyl bromide in the presence of tetrabutylammonium hydroxide in dichloromethane/aqueous sodium hydroxide permits good conversion of methyl 4,6-O-benzylidene-α- and -β-D-glucopyranoside into the 2- and 3-ether in the ratio 2.5:1 in each case.[36] When, however, the benzylation of the β-compound is preceded by stannylenation, this ratio is reversed (Section 5.1.4).[37] Methyl α-L-rhamnopyranoside, after conversion to the 2,3-stannylene compound, reacts in DMF with benzyl bromide to give the equatorial 3-ether which then affords largely the 2,3-diether on application of the above-mentioned phase transfer technique.[38] An alternative way of making monobenzyl ethers involves selective cleavage of benzylidene acetals, either 4- or 6-ethers being obtainable from hexopyranoside 4,6-acetals (Section 5.4.4).

The ethers are stable towards basic reagents and are tolerably stable in acid solutions, so that glycosidic bonds or acetals can be hydrolysed in their presence. They are, however, unstable in strong acid and are cleaved, for example, by acetolysis (acetic acid, acetic anhydride and sulphuric acid). Under conditions in which sulphonyloxy groups are displaced, benzyl ethers can participate, as, for example, in the production of the 3,6-anhydro derivative **5.25** from the 5,6-disulphonate **5.24** (Scheme 5.10).

p-Methoxybenzyl ethers have the advantage of being cleavable electro-oxidatively under conditions in which benzyl ethers are stable. In addition, they may be selectively removed in the presence of benzyl ethers by 2,3-dichloro-5,6-dicyano-1,4-benzoquinone (DDQ) oxidation, and 3,4-dimethoxybenzyl ethers are so much more susceptible that they may be removed with this reagent under conditions in which methoxybenzyl groups are stable. Conversely, benzyl groups can be cleaved by hydrogenolysis over Raney nickel while the substituted ethers remain unreacted,

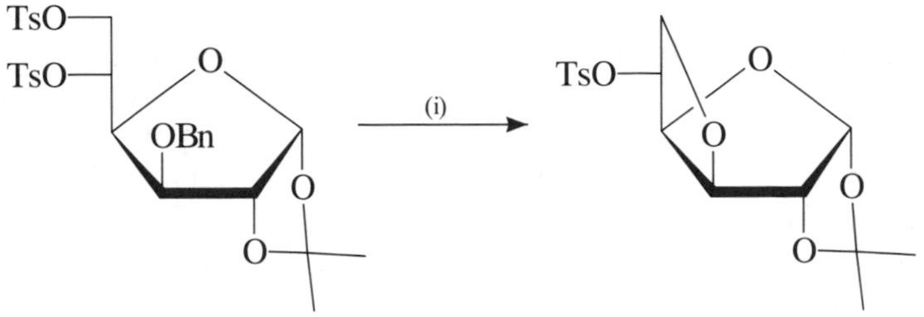

5.24   **5.25**   (45%)

(i) MeOH, H_2O, NaOAc, reflux, 115 h

Scheme 5.10

and thus this set of derivatives represents a very useful group for the synthesis of partially protected compounds by selective deprotection procedures.[1] Under the conditions of DDQ oxidation the following groups are stable: alkenes, epoxides, ketones, acetals, tosylates, silyl ethers and isopropylidene acetals.[39]

5.2.3   TRIPHENYLMETHYL (TRITYL) ETHERS[40]

The resonance stabilization of the triphenylmethyl carbocation causes trityl chloride to ionize readily and to resemble an acyl chloride in its reaction towards alcohols in pyridine solution. Substitution occurs readily, and because of the bulk of the reagent marked selectivity is shown for primary hydroxyl groups. Aldohexoses give predominantly 6-ethers; fructose gives the 1,6-diether, and the aldopentoses also react mainly in the furanose form to give 5-substituted derivatives. Reaction is slow with secondary groups and high selectivity is exhibited for equatorial hydroxyl groups in the presence of axial alcohol functions on six-membered rings. For reactions of hindered alcohols trityl perchlorate or tetrafluoroborate can be used together with bases like 2,4,6-tri-t-butylpyridine in dichloromethane. With such reagents, 2,3,5-triethers of ribofuranosyl derivatives can be prepared. The primary purpose to which trityl ethers are put, however, is to the selective substitution of primary hydroxyl groups, and 5′-ethers of pentofuranosyl nucleosides are frequently prepared as the first step in nucleotide syntheses.

Trityl chloride in which one of the phenyl groups is part of a polymer chain can be used to bond primary positions of carbohydrates to the resin, and thus permit solid phase reactions at the unprotected sites. By this method, for example, methyl 2,3,4-tri-O-benzoyl-α-D-glucopyranoside can be made in 86% yield.[41]

Trityl ethers are stable under basic conditions, so that alkylations and acylations can be effected in their presence, but they are highly susceptible to acid-catalysed hydrolysis and to hydrogenolysis. Usually a reagent such as hydrogen bromide in acetic anhydride is used for their removal, but following the observation that these groups can be cleaved during adsorption chromatography, a method based on percolation down a column of silica gel has been developed.[42]

These ethers have a strong propensity to give trityl carbocations, which enhances
the nucleophilicity of the oxygen atoms and renders them more reactive with electrophiles such as acylating or glycosylating agents than is the case with unprotected
hydroxyl groups. This has been utilized to facilitate the syntheses of disaccharides involving primary sites, especially with orthoesters and related compounds as
glycosylating agents.[43] (See also Section 3.2.2.)

Methoxy-substituted trityl ethers were developed as more acid-sensitive groups,
the tris-(p-methoxyphenyl)methyl ethers being so sensitive that they are seldom
used. On occasion, (p-methoxyphenyl)diphenylmethyl compounds have suitably
balanced properties for specific syntheses, and are used as protecting groups at
primary alcohol sites. Under acidic conditions they cleave about 10 times faster
than do trityl analogues.[1] Unlike trityl compounds, they also cleave with sodium
naphthalenide in HMPIT.[44]

5.2.4 SILYL ETHERS[45]

Ethers of this set have become some of the most commonly used in carbohydrate
chemistry, appropriate members showing combined properties which make them
ideal protecting groups.[46]

Treatment of polyhydroxyl compounds with trimethylsilyl chloride or
hexamethyldisilazane (Me$_3$SiNHSiMe$_3$) in pyridine solution converts them to
the pertrimethylsilyl ethers which are well suited to examination by gas–liquid
chromatography,[47] and to purification by vaccum distillation. They have therefore
been adapted for use in analysis of sugar mixtures and also in mass spectrometry
studies. The five products obtained by trimethylsilylation of D-fructose were thus
separated and shown to be the two pyranose, the two furanose and the acyclic
perethers.

The ethers are usually heavy oils but some, for example the tetrasubstituted
derivative of α-D-xylopyranose (obtained by gas chromatographic separation from
its isomers),[48] are solids. They are stable under normal storage conditions and
can be hydrolysed to the initial carbohydrate compounds by heating in aqueous
alcohol, the hydrolyses being catalysed by acids and bases. Since the latter allow
selective hydrolyses at primary positions these derivatives offer alternative means
of obtaining primary-substituted compounds. o-Nitrophenyl β-D-galactopyranoside
6-phosphate, required as an analytical substrate for a galactosidase enzyme, for
example, has been made from the glycoside by trimethylsilylation at the free
hydroxyl groups (O-2, O-3, O-4 and O-6), selective solvolysis (potassium carbonate
in methanol) of the C-6 ether, phosphorylation and deprotection.[49] In the light of
the availability of more suitable derivatives it is now, however, considered that
the trimethylsilyl ethers are too hydrolytically sensitive for general synthetic use,
a conclusion that has led to the introduction of analogues having more-hindered
silicon centres. In particular, the t-butyldimethylsilyl (Tbdms) ether group was
developed by Corey[50] and has become one of the most useful of the protecting
groups available for use with carbohydrates. Imidazole is usually used to catalyse

the reaction of the silyl chloride with the alcohol in DMF, the bulky nature of the reagent ensuring that it reacts selectively with sterically accessible hydroxyl groups. This selectivity is, however, enhanced in the absence of the catalyst, methyl α-D-glucopyranoside **5.26** with the silyl chloride in pyridine giving the crystalline 6-substituted compound in effectively quantitative yield following an acetylation–deacetylation isolation procedure.[51] The 2,6-diether can then be made by use of imidazole/DMF with two equivalents of the silyl chloride (Scheme 5.11),[52] and in the ribonucleoside series the 5'-position, O-2' and O-3' are substituted in that order.

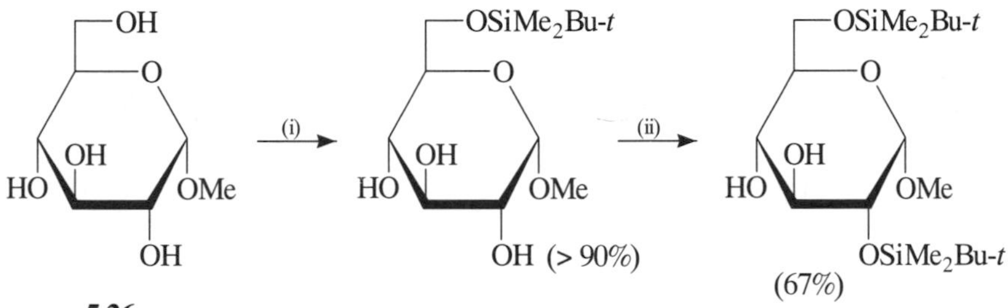

5.26

(i) t-BuMe$_2$SiCl, Py; (ii) t-BuMe$_2$SiCl, DMF, imidazole
Scheme 5.11

The group is approximately 1000 times more stable to acid hydrolysis than is the trimethylsilyl ether group, but it can be cleaved under forcing acidic conditions. It is effectively base-stable. Normally it is removed by use of the specific tetrabutylammonium fluoride in, for example, THF, but these conditions are sufficiently basic to cause the migration of susceptible ester groups. Fully silylated ribonucleosides cleave preferentially in acidic conditions at the primary position allowing other substituents to be introduced at this site, but, surprisingly, under the basic tetrabutylammonium fluoride conditions, 2',5'-diethers can give rise to the nucleoside 5'-ethers, suggesting selective reaction at the secondary site. This, however, may be an invalid conclusion because silyl ethers have been found to migrate within partially substituted carbohydrate derivatives. Thus, migration (usually in basic conditions) has been observed between O-2' and O-3' in 2',5'-disubstituted nucleosides and, on treatment with triphenylphosphine and diethyl azodicarboxylate, the 3,6-bis-(Tbdms) derivative of methyl β-D-glucopyranoside isomerizes into the 4,6- and 2,6-substituted compounds. Sodium hydride in HMPIT,[53] ammonium fluoride in methanol at $60\,^{\circ}\mathrm{C}$[54] and silicon tetrafluoride in dichloromethane[55] are other reagents applicable to the cleavage of these and other silyl ethers.

t-Butyldiphenylsilyl (Tbdps) derivatives are another 100 times more stable than the Tbdms analogues to hydrolysis in acid and will resist conditions that cleave these latter ethers and also trityl and tetrahydropyranyl groups and acetals. They are somewhat less liable to migrate under basic circumstances than the other silyl ethers, but nevertheless at room temperature saturated methanolic potassium carbonate induces rapid migration: methyl 4,6-*O*-benzylidene-2-*O*-*t*-butyldiphenylsilyl-α-D-glucopyranoside, which is initially formed in a 94:6 ratio with the 3-substituted isomer by silylation of the glycoside diol, gives a 1:2 mixture at equilibrium.[56] Fluoride ions still bring about cleavage of these ethers which are also labile towards the other non-acidic reagents that cleave Tbdms ethers.

1,3-Dichloro-1,1,3,3-tetraisopropyldisiloxane has proved to be of particular value in ribonucleoside chemistry for the selective substitution of the 3′,5′-hydroxyl group pairs and then for the specific release of each of these groups (Scheme 5.12).[57]

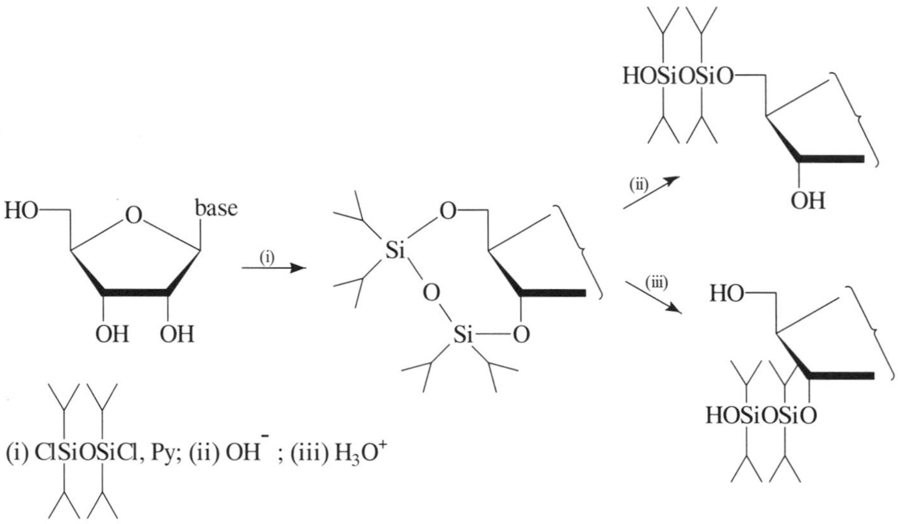
HO
O
base
OH OH
(i)
Si
O
Si
O
O
(ii)
(iii)
HOSiOSiO
OH
HO
HOSiOSiO
(i) ClSiOSiCl, Py; (ii) OH⁻ ; (iii) H3O+

Scheme 5.12

These cyclic Tips ethers can be cleaved, like the other silyl compounds, with fluoride ions.

5.2.5 ALKENYL ETHERS

Vinyl ethers can be prepared from base-stable carbohydrate derivatives using ethyne and potassium hydroxide at elevated temperatures and pressures or by vinyl exchange from, for example, ethyl vinyl ether by use of mercury(II) acetate. Yields obtained by this milder method are usually about 50%, partly because the reactions involved are reversible

$$\mathrm{ROH} + \mathrm{CH_2{=}CHOEt} + \mathrm{Hg(OAc)_2} \rightleftharpoons \mathrm{AcOHgCH_2CH(OR)OEt} + \mathrm{HOAc} \rightleftharpoons \mathrm{CH_2{=}CHOR} + \mathrm{EtOH} + \mathrm{Hg(OAc)_2}$$

The products give ethyl ethers on hydrogenation and cleave hydrolytically in the presence of acid or mercury(II) salts. With acid catalysts they add alcohols to

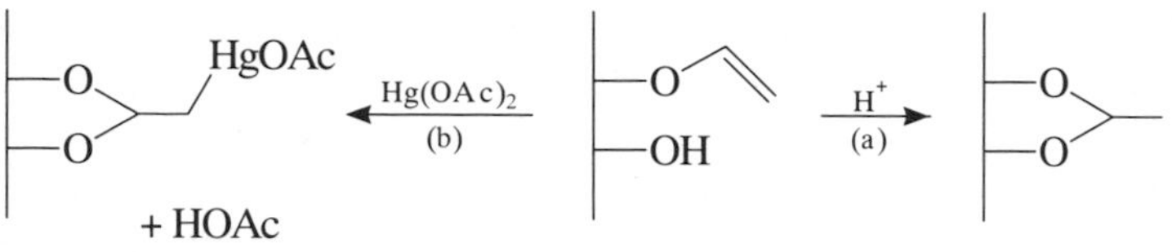

Scheme 5.13

give acetals which, if these alcohols are intramolecular, are cyclic ethylidene acetals (Scheme 5.13a). Analogous additions occur with mercury(II) salts (Scheme 5.13b). In unsaturated sugar derivatives vinyl ethers at allylic sites can permit specific thermal Claisen rearrangements to give useful methods of preparing branched-chain derivatives (Sections 4.8.2.b, and 4.10.2).

Monosaccharide isopropenyl ethers, made from acetates by use of the Tebbe methylenating reagent, react with sugar derivatives having free hydroxyl groups at the anomeric positions and the adducts collapse to form disaccharides (Section 3.1.1.a.ii).

Allyl ethers of carbohydrates have been used extensively for the ingenious reason that, whereas they themselves are stable to acidic conditions, they can be isomerized to prop-1-enyl ethers, and these are readily susceptible to acid-catalysed hydrolysis or transvinylation onto water in the presence of mercury salts, as would be expected for enol ethers. By application of the observation that the 2-allyl group of methyl 2,3-di-*O*-allyl-4,6-*O*-benzylidene-α-D-glucopyranoside **5.27** rearranges preferentially, a method has been developed for the synthesis of 2-*O*-substituted glucoses, as illustrated in Scheme 5.14.[58]

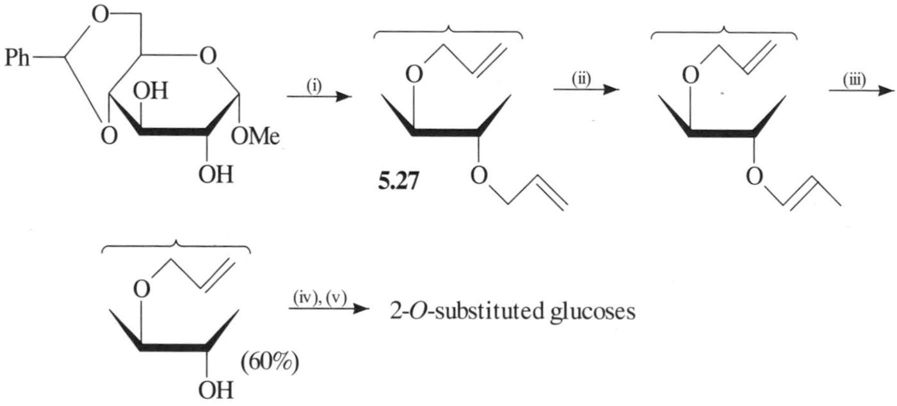

(i) CH_2=CHCH$_2$Br, NaH; (ii) *t*-BuOK, DMSO, 40 °C, 0.5 h; (iii) HgCl$_2$, H$_2$O; (iv) O-2 substitution; (v) removal of protecting groups

Scheme 5.14

Appreciable changes occur in the reactivities of allyl ethers following the introduction of methyl groups into the allyl substituents. Whereas potassium *t*-butoxide in DMSO leads to isomerization of allyl ethers, it causes cleavage of but-2-enyl analogues (ROCH$_2$CH=CHMe), which means that these can be used in distinguishable ways within the same molecules.[59] Prenyl (3-methylbut-2-enyl) ethers, on the other hand, are relatively stable to reactions with chlorotris(triphenylphosphine)rhodium(I), whereas with this reagent allyl and but-2-enyl analogues isomerize to the 1-enyl compounds at different rates.[60] It is possible, therefore, to utilize the three types of ethers selectively in the presence of each other.

Allyl ethers and their analogues are made from the relevant halides using the methods noted for benzyl compounds (Section 5.2.2). Otherwise, allyl ethyl carbonate in the presence of palladium(0) catalysts offers a means of synthesis under neutral conditions.[61]

5.2.6   OTHER ETHERS

2,2,2-Trichloroethyl ethers and *p*-chlorobenzyl ethers have the advantages of being removable by use of zinc in acetic acid/sodium acetate[62] and of showing very good crystallizing characteristics, respectively. The complex ether **5.28** condensed with catechol in butanol in the presence of sodium hydroxide powder gives 65% of the crown ether **5.29** which shows enantioselective complexing abilities for several *α*-amino acids.[63]

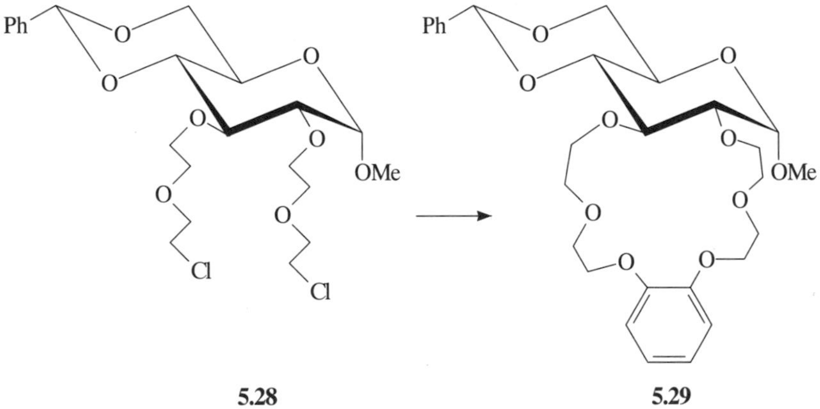

5.28   5.29

Methoxymethyl and tetrahydropyranyl 'ethers' are in fact acetals and are noted in Section 5.4.2.

5.3   ESTERS[64]

Carbohydrates can be condensed through their hydroxyl groups with a variety of both organic and inorganic acid derivatives to give esters. In biochemistry the sugar phosphates are of pre-eminent significance (Sections 2.5.1, and 2.5.2), but sulphated and acetylated carbohydrates and other esters are frequently found in natural materials. Synthetic acetates, xanthates and nitrates of cellulose are, additionally, of particular commercial significance, and a wide variety of esters are of appreciable value in synthetic chemistry.

Although the majority of the esterification reactions of sugars proceed by direct attack of electrophilic reagents at hydroxyl oxygen atoms regardless of whether or not the anomeric centre is involved, the reactivity of the products (as with the analogous ethers) is very dependent upon this distinction. Glycosyl esters, particularly under acidic conditions, can readily cleave by glycosyl C–O bond fission to give stabilized carbocations, and are in consequence appreciably more reactive than are esters attached to other positions (Section 3.2).

Acetates and benzoates are the best-known carboxylate esters which are frequently used for the preparation of suitable crystalline derivatives of hydroxylic compounds. *p*-Nitrobenzoates and other substituted aromatic esters, however, can be more suitable for this purpose. The basic chemistry of all the carboxylate esters is similar, but often they vary in their reactivity, and the exchange of ester groups at one site in a molecule can affect the reactivities of other centres. Benzoylated glycosyl halides are thus frequently less reactive than acetylated analogues, presumably because of the steric protection and conformational inertia provided by the bulky aromatic rings. Ester groups have a general deactivating influence on glycosyl donors and also on neighbouring hydroxyl groups as glycosyl acceptors, facts that have been put to significant use in di- and oligo-saccharide syntheses (Sections 6.3 and 6.4).

5.3.1   ACETATES

Much of the material in this Section applies to acyl esters generally, as well as to acetates.

Carbohydrate acetates have been encountered in monomeric and polymeric natural materials, but they are best known as synthetic products which often show favourable crystalline properties, and from which the parent alcoholic compounds can be recovered under basic or acidic conditions. Generally, however, deacetylation is effected by the convenient *Zemplén procedure*, in which sodium methoxide is employed in catalytic amounts in methanolic solution. As is shown in Scheme 5.15, the reagent used in the first step of this reaction is regenerated in the last.

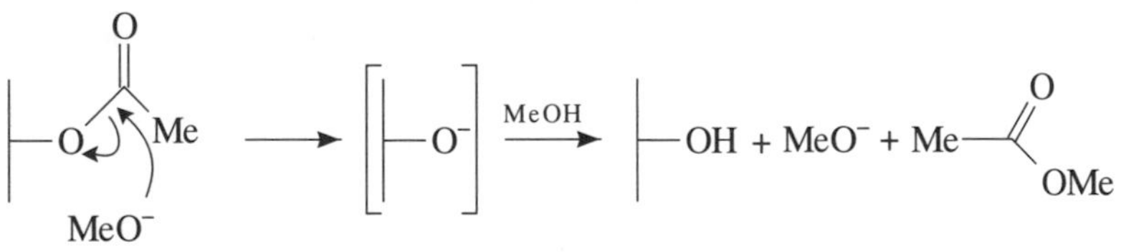

Scheme 5.15

Other deacetylation methods are given in Section 5.3.1.b. As well as being useful as manageable derivatives, sugar acetates are highly suited to examination by n.m.r. spectroscopy and mass spectrometry, and are capable of purification and analysis by all forms of chromatography including gas–liquid methods.

(a) Synthesis

Complete acetylation of unsubstituted sugars produces mixtures of isomeric products the proportions of which depend upon the conditions used and upon the sugar, and to some degree the esterifying conditions can be selected to favour a required isomer. Acetic anhydride is almost always the reagent used, and the catalysts most frequently employed are zinc chloride (or other similar Lewis acids), sodium acetate and pyridine. Since acidic catalysts also cause anomerization of the glycosyl acetates (Section 3.2.1.a), thermodynamically favoured products predominate when these are used, and since the anomeric effect (Section 2.3.1.b) is of pre-eminent importance under these conditions, the pyranose peracetates having axial anomeric ester groups are generally formed in largest proportions. Therefore, for example, acetylation of glucose, mannose, galactose or xylose in the presence of acidic catalysts provides a means of obtaining good yields of esters of the α-pyranoid sugars. D-Ribose, on the other hand, gives predominantly the β-pyranose form because the α-anomer is subject to strong 1,3-diaxial interactions in both of the chair conformations.

With cold pyridine as solvent and reaction catalyst, the esterification reaction proceeds faster than mutarotation of the unreacted sugar, and consequently the main product is the perester of the modification of the sugar used. On the other hand, in hot acetic anhydride in the presence of sodium acetate, equilibration of the anomeric forms of the free sugars occurs faster than the esterification reaction, and the equatorial anomeric hydroxyl groups are substituted preferentially (Section 5.1.3). Glucose, therefore, under these conditions can be converted into the crystalline β-pentaacetate in good yield. Scheme 5.16 illustrates the main products of acetylation of D-glucose under these different conditions.

Sugars which form relatively stable furanose rings give more complex mixtures of acetates. Galactose, for example, in the presence of sodium acetate or pyridine at elevated temperatures gives appreciable proportions of furanose acetates, and the β-anomer can be isolated in fair yield. More generally, however, furanosyl acetates (which are of value as glycosylating reagents, especially useful in nucleoside synthesis) are prepared from precursors which contain the furanosyl ring. Thus, methyl D-ribofuranoside **5.30**, prepared by mild methanolysis of the

5.30   5.31

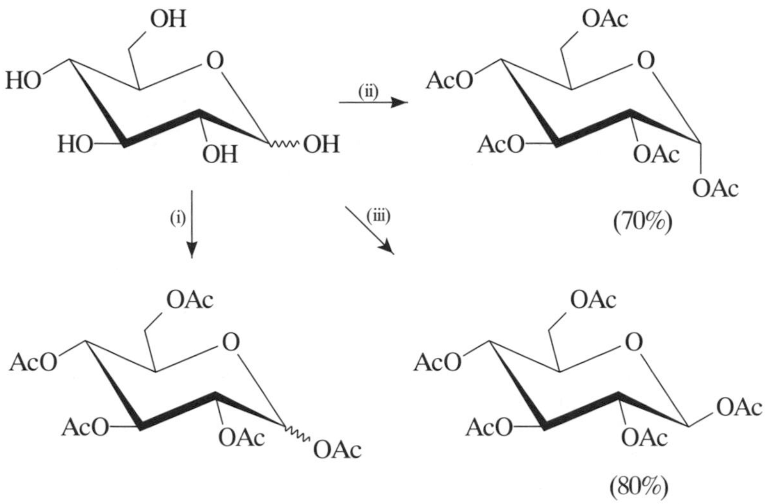

(i) Ac₂O, Py; (ii) Ac₂O, ZnCl₂, 100 °C, 1 h; (iii) Ac₂O, NaOAc, 100 °C, 1 h

Scheme 5.16

free sugar (Section 3.1.1.a), on treatment with acetic acid, acetic anhydride and sulphuric acid undergoes acetylation of the free hydroxylic groups and acetolysis of the C-1–O-1 bond to give the furanosyl acetates from which the crystalline β-anomer **5.31** can be isolated in 50% yield. In related fashion, aldono-γ-lactones on acetylation followed by reduction of the lactone group and further acetylation give related compounds, and thiofuranosides (Section 3.1.2) can likewise be used to provide 1,2-*trans*-related glycofuranosyl acetates, the exchange at the anomeric centre being effected with mercury(II) acetate in acetic acid.

Acyclic sugar acetates (e.g. **5.32**, R=Ac) which possess true aldehydic properties are obtainable from aldose dialkyl dithioacetals (Section 3.1.2) by acetylation and deacetalation by use of, for example, mercury catalysts. Acetylation of the related aldehyde **5.32** (R=H), which exists at least partially in the cyclic form, gives further isomers of D-galactose pentaacetate, in this case the seven-membered septanose derivatives **5.33** (cf. Section 2.4.1).

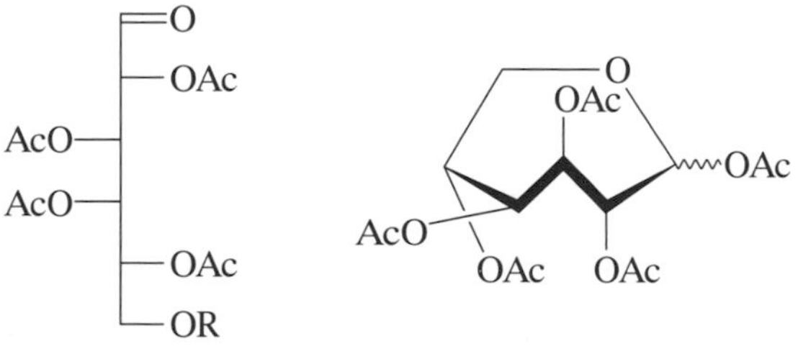

5.32   **5.33**

Acetylation of ketoses can lead to appreciable proportions of the esters of the acyclic keto forms, the L-sorbose keto pentaacetate, for example, being obtainable in 70% yield by use of acetic anhydride and zinc chloride.[65]

On occasion, particularly unreactive hydroxyl groups, for example those in tertiary positions, can be acylated, and in such circumstances 4-(dimethylamino)pyridine is often used as promoter.[66] Otherwise, *t*-butylammonium fluoride in THF enhances the reactivity of acid anhydrides.[67] *N*-Acylimidazoles tend to react with polyhydroxyl systems with enhanced selectivity (Section 5.1.3).

The Mitsunobu method (Sections 4.1.2 and 4.1.2.c) is highly applicable to the formation of ester groups from alcohols, but since the key step involves a reaction at carbon rather than at a hydroxyl group it is treated in Section 4.1.2.c.

Acetolyses reactions by which glycosidic bonds are cleaved to give glycosyl acetates are dealt with in Section 3.1.1.b.vii.

(b) Reactions

In addition to sodium methoxide (see above), barium hydroxide, ammonia, dimethylamine or potassium cyanide in methanol[68] or triethylamine in aqueous ethanol can be used in base-catalysed deacetylations, but these milder reagents also may be unsuitable when other base-labile groups or free carbonyl groups are present. In such circumstances, esters can be deacetylated in methanolic solution with hydrochloric acid as catalyst.[69] Selective enzymic deacetylations are noted in Section 5.3.3.

Selective deacetylation can be achieved at the anomeric centre under a range of conditions,[70] and furthermore the reaction of penta-*O*-acetyl-α- or

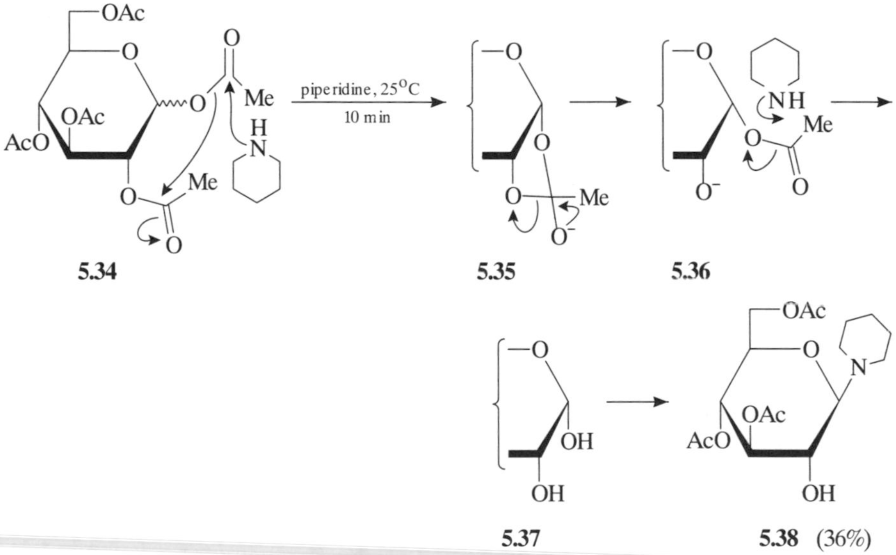

5.34   5.35   5.36

5.37   5.38 (36%)

Scheme 5.17

-β-D-glucopyranose **5.34** with three equivalents of piperidine gives crystalline N-(3,4,6-tri-O-acetyl-β-D-glucopyranosyl)piperidine **5.38** in about 30% yield in each case.[71] Since this compound has an unprotected hydroxyl group at C-2, it can be used to prepare 2-O-substituted glucose derivatives. It is produced itself, presumably, by initial nucleophilic attack at the anomeric ester carbonyl group (Scheme 5.17) with the formation of the orthoacid anion **5.35** which ring opens to give the 1,3,4,6-tetraacetate (from **5.36**) prior to cleavage at C-1 to the diol **5.37** and the formation of the glycosylamine (Section 3.1.3.a).

While the intermediate **5.35** is shown as giving the 1-ester **5.36**, it could ring open to give the isomer with the ester group at C-2 and the hydroxyl group at the anomeric position. In basic conditions either the 1-acetate **5.36** or the isomeric 2-ester would form **5.35**, which means that the ester group can migrate from one hydroxyl group to another. In practice, such acyl migration is frequently observed,[2] 1,3,4,6-tetra-O-acetyl-α- or -β-D-glucopyranose **5.39**, for example, reacting with methyl iodide and silver oxide to give methyl 2,3,4,6-tetra-O-acetyl-β-D-gluco-pyranoside **5.40**, the migration occurring away from the anomeric position and the free sugar methylating in the equatorial β-orientation (Scheme 5.18).[72]

5.39   **5.40** (81% from α-anomer,
                                                           51% from β-anomer)

(i) MeI, Ag$_2$O; α-anomer, 20 °C, 44 h; β-anomer, reflux, 4 h
Scheme 5.18

In appropriate instances ester groups tend to migrate towards primary hydroxyl groups, the methylation of triacetate **5.41** (α-anomer) giving the 2-methyl ether **5.42** as the main product under the above conditions.[73] The 4-substituted isomer **5.43** is produced, however, with Kuhn reagents applied to the β-anomer (Scheme 5.19),[74] which suggests that the ester groups move sequentially, compound **5.42** being formed following migration of acetates from positions 4 to 6, 3 to 4 and 2 to 3. The possibility, however, of a migration from O-2 directly to O-4 in this case cannot be ruled out.

When migrations of ester groups involve centres with similar electronic and steric characteristics, as for example O-2 and O-3 of β-ribofuranosyl derivatives, mixed monoesters result. In amino sugar chemistry, acetyl migrations from hydroxyl to amino groups are very common under basic conditions (Section 4.3.2.a).

An unexpected feature of the chemistry of acetates, which they share with the pivaloates (t-BuCO$_2$R), is their cleavage by alkyl-oxygen fission (to give deoxy

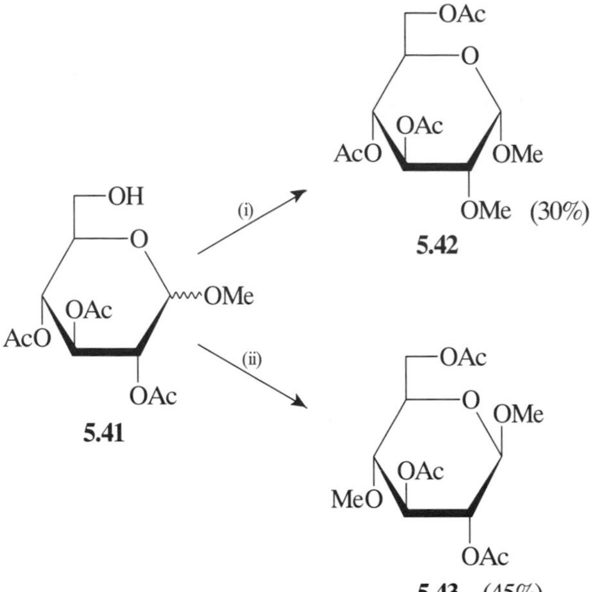

5.42 (30%)

5.41

5.43 (45%)

(i) MeI, Ag$_2$O, reflux, 48 h; (ii) MeI, Ag$_2$O, DMF, 20 °C, 6.5 h
Scheme 5.19

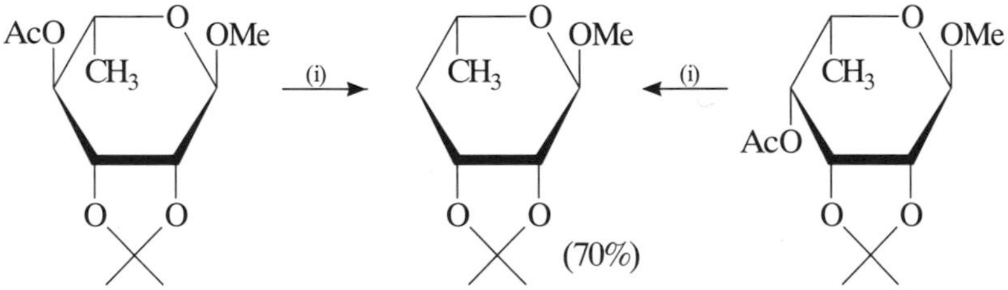

(70%)

(i) $h\nu$ (254 nm), H$_2$O/HMPIT (5:95), 60 h
Scheme 5.20

compounds, Section 4.2.1.d) on u.v. photolysis in aqueous HMPIT, as is illustrated in Scheme 5.20.[75] In this way primary and secondary hydroxylated centres may be deoxygenated in moderate yield.

Other important features of the chemistry of acetates (and related carboxylates) are their participation in displacement reactions (Section 4.1.2.b.ii) and their conversion to orthoesters (Section 3.3.2). Reactions exclusive to glycosyl esters are described in Chapter 3.

5.3.2   OTHER CARBOXYLATE ESTERS

A wide selection of other carboxylate esters of carbohydrates are known, the most common being the benzoates which are normally prepared by use of benzoyl chloride in pyridine. They can also be formed selectively from benzylidene acetals (Section 5.4.4), and can be introduced by attack of the benzoate ion at carbon atoms bearing good leaving groups (Section 4.1).

Benzoylated derivatives of many kinds are significantly less reactive than are their acetylated analogues; for example, benzoylated glycosyl halides hydrolyse more slowly than do the acetates and are in consequence more bench-stable. Nevertheless, benzoyl groups are far from being inert during reactions; they readily participate as neighbouring groups in nucleophilic displacements, for example. Mesitoates (2,4,6-trimethylbenzoates) are more stable, but can be cleaved reductively with aluminium hydride.[76]

Ingenious use of benzoates and substituted benzoates in circular dichroism studies has led Nakanishi to methods for determining absolute configurations of alcohols, and in extensions of the work the natures of glycosidic linkages in oligosaccharides have been investigated.[77]

Acetates bearing halogen substituents are useful, particularly because of their relative ease of solvolysis. Thus, chloroacetyl groups can be selectively cleaved with hydrazine acetate or with thiourea in methanol even in the presence of anomeric acetyl groups.[78] Similarly, trichloroacetates can be preferentially hydrolysed with pyridine as catalyst,[79] and trifluoroacetates are so unstable that, with methanol in inert solvents, they cleave sometimes selectively in a manner that can be put to use synthetically. For example, the ester group at O-3 of methyl 4,6-O-benzylidene-2,3-di-O-(trifluoroacetyl)-α-D-glucopyranoside, on treatment of the compound with methanol in carbon tetrachloride, can be selectively removed to allow the formation of 3-substituted derivatives.[80]

Levulinoates ($MeCOCH_2CH_2CO_2R$), which are removable with hydrazine under conditions in which even acetates and benzoates are stable, can be used as protecting groups resistant to migration in partially substituted sugars undergoing oxidation under acid conditions,[81] and the photolability of pyruvates ($MeCOCO_2R$) is of value in the preparation of keto compounds (Section 4.9.1.a.ii). Pivaloates may be converted to deoxy sugar analogues (see acetates above).

Long-chain fatty acid esters of sugars have surface-active and liquid crystal properties and many have been made for exploitation of these characteristics. Others are of potential value in the food industry. For example, sucrose esters have been made in near quantitative yield by treating the octaacetate and fatty acid methyl esters under transesterifying conditions with sodium metal. The products have properties akin to those of salad oils.[82]

5.3.3   ENZYMIC ACYLATIONS AND DEACYLATIONS[83]

Lipases, enzymes which catalyse the formation and hydrolysis of esters, have recently been used in carbohydrate chemistry and have shown considerable selectivity. Further advantages are that they can be applied in organic solvents and on large scales.

By use of the lipase of pig pancreas, acyl groups have been transferred in pyridine solution from trichloroethyl carboxylates to free sugars with the consequential formation of esters at the primary sites. For example, 6-O-acetyl- and -butanoyl-D-glucose can be obtained in 76 and 50% yields, respectively, by this method, while D-fructose is esterified at each of its primary positions to give the 1-acetate and 6-acetate in 43 and 17% yields, respectively.[84] With the same enzyme in tetrahydrofuran, 6-O-butanoyl-D-glucose can be converted into the 2,6-diester in 50% yield. Alternatively, with an enzyme from *Chromobacterium viscosum*, 80% of the 3,6-disubstituted isomer can be made from the same starting material.[85]

Enzymes, being chiral and therefore enantioselective, can show extreme changes in selectivity with enantiomeric carbohydrate substrates. For example, a lipase from *Pseudomonas fluorescens* is almost completely selective in catalysing reactions with methyl 6-deoxy-α-D- and -L-galactopyranoside at O-2 and O-4, respectively.[86]

Acting in the reverse sense enzymes show selective hydrolytic activity, the pig pancreatic enzyme preferentially removing the axial ester group at O-2 from 2,3,4-tri-O-acetyl-1,6-anhydro-β-D-galactose to give the 3,4-diacetate.[87]

5.3.4   CARBOXYLATE ORTHOESTERS

The best-known and most useful class of carbohydrate orthoesters are the bicyclic 1,2-glycopyranosyl derivatives which may be derived from acylated glycopyranosyl halides and used in the synthesis of glycosides. These are treated in Section 3.2.2.

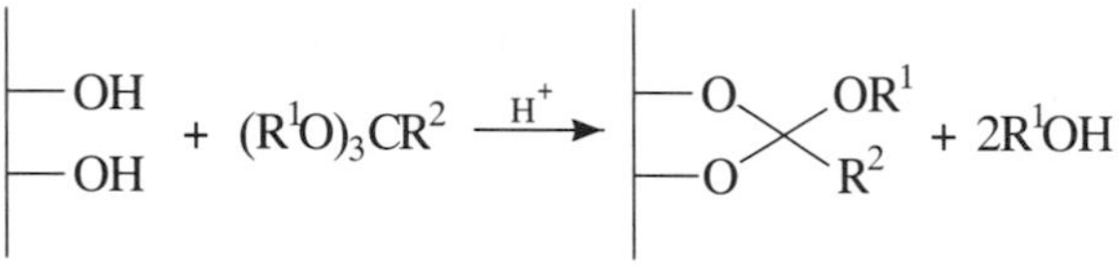

Scheme 5.21

Orthoesters at other sites are less well known but can be prepared by (i) acid-catalysed condensation between *cis-vic*-diols and trialkyl orthoesters (Scheme 5.21) or (ii) by alcoholysis of cyclic acyloxonium ions formed, for example, by treatment of epoxides having neighbouring acyl ester groups with Lewis acids. An example is given in Scheme 5.22.[88] Intermediates such as **5.44** react with diborane to give ethylidene acetals, the diastereoisomers having the methyl group *exo* being favoured.

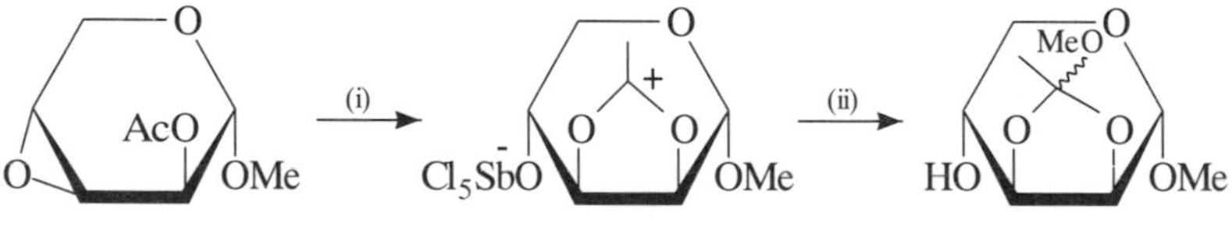

5.44   (80%, *endo/exo*-methoxy 1:4.5)

(i) SbCl$_5$, CCl$_4$, CH$_2$Cl$_2$; (ii) NaOMe, MeOH

Scheme 5.22

With suitable triols the former method results in tridentate products in which three hydroxyl groups are simultaneously protected, such compounds also being available by other methods. For example, in weakly acidic conditions α-D-ribo-pyranose benzyl 1,2-orthobenzoate gives the 1,2,4-triorthobenzoate **5.45** in 38% yield,[89] and on treatment of 1,2-*O*-isopropylidene-α-D-glucofuranose with tosyl chloride in pyridine containing DMF compound **5.46** is produced (70%), the ortho-formyl component being derived from the solvent.[90]

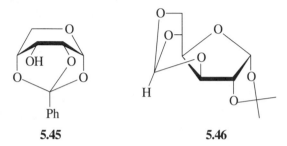

5.45   **5.46**

Orthoesters are devoid of carbonyl functions and therefore are stable to base, but are highly labile in acid media because of their acetal-like character. Under such conditions, acyloxonium ions are formed initially and these may hydrolyse or otherwise they may be intercepted by, for example, reduction, and thereby converted into the 2,3-*O*-ethylidene compounds.[88] Under stereoelectronic control, five-membered cyclic orthoesters fused to pyranoid rings hydrolyse to give mainly axially oriented monoesters so that 1,2-substituted α-D-glucopyranose and β-D-mannopyranose compounds usually afford mainly 1- and 2-*O*-acyl derivatives, respectively.[91] Acid-catalysed hydrolysis of O-2′, O-3′ orthoesters of ribofuranose derivatives normally gives mixtures of 2′- and 3′-esters because of the lack of strong stereoelectronic control in such cases.

5.3.5   CARBONATES AND SUBSTITUTED CARBONATES[92]

(a)   Carbonates

Carbohydrate carbonates have been known for many years but mainly as products prepared as synthetic intermediates; however, the anomeric glycosides **5.47**, which

5.47

contain a cyclic carbonate associated with the branching group of the branched-chain sugar aldgarose, have been isolated from the methanolysis products of an antibiotic substance elaborated by a species of *Streptomyces* (Section 4.8.2.a).

Reagents most frequently used in the preparation of carbohydrate carbonates are phosgene, alkyl chloroformic esters and diaryl carbonates, but a different approach to the synthesis of cyclic carbonate derivatives involves treatment of primary sulphonyl esters which have neighbouring hydroxyl groups with potassium hydrogen carbonate in dimethyl sulphoxide at room temperature. Compound **5.49** is thus prepared in high yield from the 6-sulphonate **5.48** (Scheme 5.23).[93]

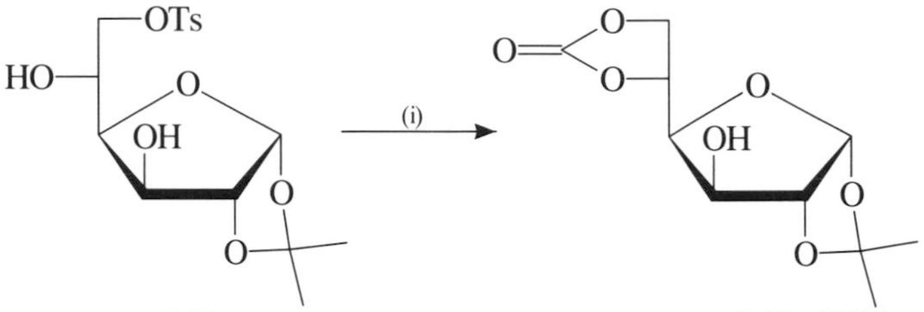

5.48   **5.49** (85%)

(i) KHCO$_3$, DMSO, 20 °C, 72 h

Scheme 5.23

Phosgene (COCl$_2$), usually used in pyridine, tends to give cyclic carbonates having five-membered rings; D-glucose and D-mannose are thus converted into the 1,2:5,6-and 2,3:5,6-furanoid esters **5.50** and **5.51**, respectively, and alkylation of the latter offers a suitable means of preparing α-D-mannofuranosides. With monohydroxyl compounds, chloroformate esters (ROCOCl) are produced which are themselves carbonating reagents and which give mixed carbonates with other available alcohols or else react with a further molecule of sugar to give bridged intermolecular esters (ROCO$_2$R).

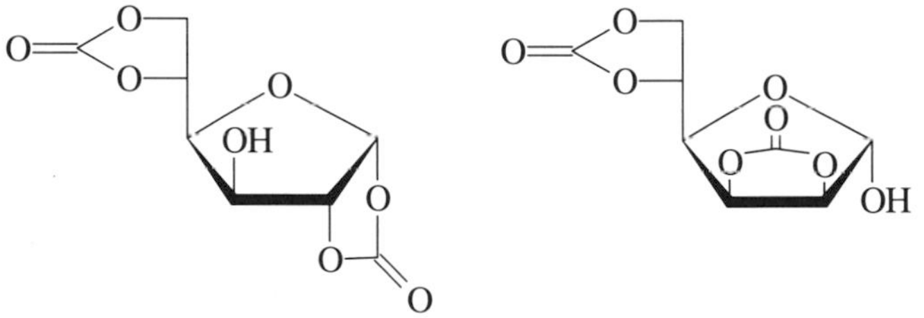

5.50   **5.51**

Reaction of hydroxyl groups with chloroformic esters in organic bases usually occurs in simple fashion to give mixed esters, e.g. **5.52**, which can be employed in the synthesis of 3-substituted xylose derivatives, and the specificity of the primary hydroxyl substitution reaction can be increased by using bulkier esters such as isobutyl chloroformate. In aqueous alkaline solution, intramolecular transesterification occurs when possible and cyclic carbonates result, so that, for example, 1,2-*O*-isopropylidene-α-D-glucofuranose **5.53** on treatment with benzyl chloroformate in alkaline aqueous dioxane gives the 3-*O*-benzyloxycarbonyl 5,6-carbonate **5.54**, which on hydrogenolysis loses the substituent at C-3. 1,2-*O*-Isopropylidene-α-D-glucofuranose 5,6-carbonate **5.49** can be obtained alternatively from the 1,2-*O*-isopropylidene acetal by reaction with diethyl carbonate in the presence of sodium methoxide, or by base-catalysed cyclization of a 6-alkoxy- or 6-aryloxy-carbonyl compound or also in a direct esterification with phosgene.

5.52   **5.53**   **5.54**

Being esters, carbonates are readily cleaved with bases, but they are usefully stable in acid. Furthermore, since cyclic esters are formed preferentially from *vicinal* diols they can be used for protecting such diols in a complementary fashion to the acid-labile isopropylidene acetals (Section 5.4.3). The synthesis of α-mannofuranosides mentioned above illustrates the value of carbonates in preparative work.

2,2,2-Tribromoethyl chloroformate exhibits marked specificity for primary hydroxyl groups and gives esters **5.55** which can be reconverted into the parent alcohols under neutral conditions with a zinc–copper couple. In related fashion, *p*-nitrophenyl chloroformate yields derivatives **5.56** which cleave on treatment with imidazole. Under these conditions acetates and most other carboxylates are stable. These reagents, therefore, are of appreciable value for protecting purposes in synthetic work.

$$ROCO_2CH_2CBr_3 \qquad\qquad ROCO_2C_6H_4NO_2-p$$

5.55   **5.56**

(b)   Thiocarbonates

Replacement of oxygen by sulphur in carbonate esters gives thiocarbonates of which there are seven different types.[†] Xanthates, cyclic thiocarbonates and (benzylthio)carbonyl derivatives have particular significance in carbohydrate chemistry.

[†] If R^1 represents a carbohydrate moiety and R^2 represents a carbohydrate or non-carbohydrate moiety these are $R^1SC(O)OR^2$, $R^1OC(S)OR^2$, $R^1OC(O)SR^2$, $R^1SC(S)OR^2$, $R^1SC(O)SR^2$, $R^1OC(S)SR^2$ and $R^1SC(S)SR^2$.

Xanthate esters, which are the basis of the viscose rayon synthesis from cellulose, are well known in the monosaccharide series, and methyl xanthates (ROC(S)SMe) are prepared by treatment of sodium derivatives of hydroxyl compounds with carbon disulphide followed by methyl iodide (Scheme 5.24).[92] Because they readily react with tributyltin hydride in the presence of an initiator by a free radical mechanism, they afford very useful access to deoxy compounds (Section 4.2.1.c) and to carbohydrate carbon radicals useful for synthetic purposes (Section 4.8.2.b). On pyrolysis, monosaccharide xanthates **5.57** normally isomerize to S-[(alkylthio)carbonyl]thio derivatives **5.58** which can be desulphurized reductively to give a further (less convenient) route to deoxy derivatives (Scheme 5.24). If, however, a carbohydrate xanthate ester function has a neighbouring unoxygenated carbon atom as in **5.59**, thermal elimination does occur in the normal Chugaev manner.[94] Alkene derivatives (Section 4.10.2.a) may also be produced from 1,2-dixanthates (e.g. **5.61**) by treatment with tributyltin hydride[95] or from 1,2-cis-diols, via thiocarbonates (e.g. **5.60**), by heating in trimethyl phosphite (Scheme 5.25).[96]

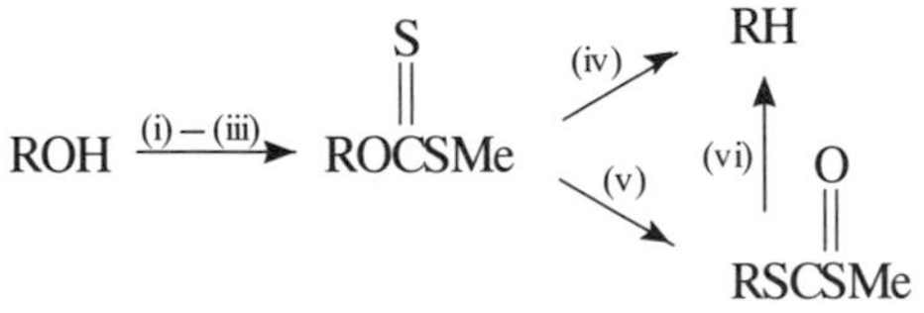

$$ROH \xrightarrow{\text{(i)}-\text{(iii)}} \underset{\substack{\| \\ S}}{ROCSMe}$$

(iv) → RH

(v) ↘   (vi) ↑ $\underset{\substack{\| \\ O}}{RSCSMe}$

5.57   **5.58**

(i) NaH; (ii) CS$_2$; (iii) MeI; (iv) Bu$_3$SnH, AIBN; (v) heat; (vi) H$_2$/Ni

Scheme 5.24

Phenoxy(thiocarbonyl) esters, prepared by use of the corresponding chloride (PhOC=SCl), are also useful precursors of carbohydrate carbon radicals and deoxy sugars, both of which they afford on treatment with tributyltin hydride and a radical initiator.[97]

Cyclic thiocarbonates derived from terminal 1,2-diols can be used to give monoalcohols which are either the primary compounds (radical ring opening by treatment with tributyltin hydride followed by alkaline hydrolysis of the derived primary ester intermediates) or the secondary (treatment with methyl iodide to give primary iodo, secondary (methylthio)carbonates which can be reductively converted into terminal deoxy derivatives).[98] (See also Scheme 4.29, Section 4.2.1.c.)

Introduction of (benzylthio)carbonyl esters into carbohydrate derivatives can be effected by the use of (benzylthio)carbonyl chloride (PhCH$_2$SCOCl), and they offer a means of protecting hydroxyl groups in acidic media. 3-O-(Benzylthio)carbonyl-β-D-glucose has been prepared, for example, by removal of the acetals from the 1,2:5,6-di-O-isopropylidene derivative. The 3-ester may be O-substituted and then de-esterified by mild alkali treatment or with hydrogen peroxide or with hydrogen over Raney nickel to give products with the 3-hydroxy group free.[99]

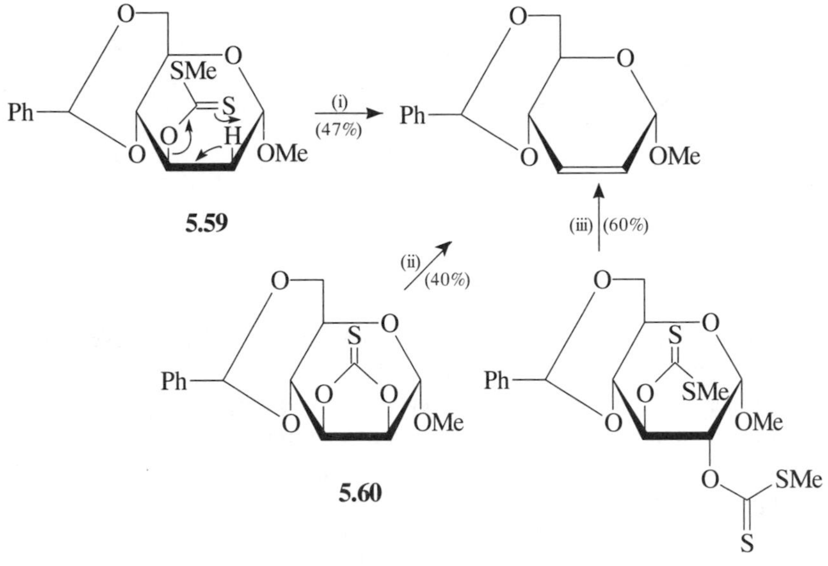

5.59

5.60

5.61

(i) heat; (ii) P(OMe)₃; (iii) Bu₃SnH, AIBN

Scheme 5.25

(c)   Carbamates (urethanes)

Phenycarbamates also have the advantage of being relatively stable, particularly in acidic media, and can be introduced under mild conditions. They have thus been found suitable for the determination of the sites of substitution of partially acylated carbohydrates for which direct methylation is invalidated by the propensity of the acyl groups to migrate. Esterification of the free hydroxyl groups is brought about by heating in benzene or pyridine with phenyl isocyanate (Scheme 5.26), and the products are deacylated with acid. Methylation, followed by reduction of the carbamate groups with lithium aluminium hydride, gives stable methylated products having methyl groups at the sites of initial acetylation.

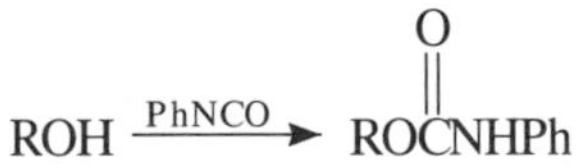

$$\text{ROH} \xrightarrow{\text{PhNCO}} \text{ROC(=O)NHPh}$$

Scheme 5.26

5.3.6   SULPHONATES[100,101]

The very great importance of sulphonate esters in carbohydrate chemistry stems from the excellent 'leaving properties' of sulphonyloxy groups in nucleophilic

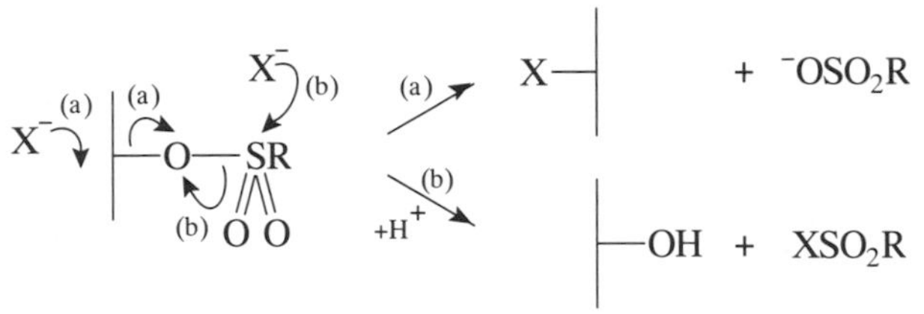

5.62

Scheme 5.27

displacement reactions (Section 4.1.2), i.e. from the propensity of these derivatives to react by alkyl–oxygen fission (Scheme 5.27a) rather than by sulphur–oxygen fission (Scheme 5.27b). Most other esters react by processes equivalent to route (b), and consequently do not offer a means of carrying out chemical operations at the carbon atoms of sugar chains. A wide range of reactions have been effected with sulphonates (Chapter 4), and many substituted products (Scheme 5.27a) or unsaturated compounds formed by elimination of sulphonic acid are known. Inter-molecular nucleophiles offer a way of introducing a variety of substituents onto a sugar chain (Section 4.1.2); intramolecular hydroxyl nucleophiles give rise to anhydro derivatives (Section 4.6), and other carbon atoms of the carbohydrate can act as nucleophiles to yield rearranged products with modified ring systems (Section 4.7). In the majority of cases such reactions also provide a method for inverting configurations at asymmetric carbon atoms bearing the sulphonate groups.

Methanesulphonates (mesylates; **5.62**, R = Me) and p-toluenesulphonates (tosyl-ates; **5.62**, R = C_6H_4Me-p) have been used very frequently, and these often undergo required displacement (or elimination) reactions particularly in dipolar aprotic solvents, especially HMPT, DMSO or DMF. There is, however, espe-cially for displacement reactions which are inherently difficult because of restricted access by nucleophiles, appreciable advantage in using esters containing strongly electron-withdrawing groups. p-Nitrobenzenesulphonates **5.62** (R = $C_6H_4NO_2$-p) have been used, but very much more effective are (especially) the trifluoromethane-sulphonates (triflates; **5.62**, R = CF_3)[102] and also the imidazolylsulphonates (imi-dazylates; **5.62**, R = $C_3H_2N_2$).[103] Examples of both have been found to undergo efficient displacement when analogous tosylates are unreactive, as in the case of the 3-esters **5.63** of 1,2:5,6-di-O-isopropylidene-D-glucose with which nucleophiles attacking C-3 have to enter the *endo*-space between the sugar and the O-1 and O-2 acetal rings, thereby suffering electrostatic repulsion with the unshared electron orbitals of O-1 and O-2. Charged nucleophiles are particularly impeded, and only rarely have they been found to react with the tosylate **5.63** (R = p-tosyl). Iodide is unreactive, but with the triflate[101,102] and imidazylate[103] it gives the *allo*-iodides in 84 and 71% yields, respectively (see Sections 4.1.1, 4.1.2 and 4.5).

Several other structural features of carbohydrate compounds inhibit displace-ments of sulphonyloxy groups, and these have been rationalized in terms relating to impedance to the approach of the nucleophiles. Thus, all compounds containing

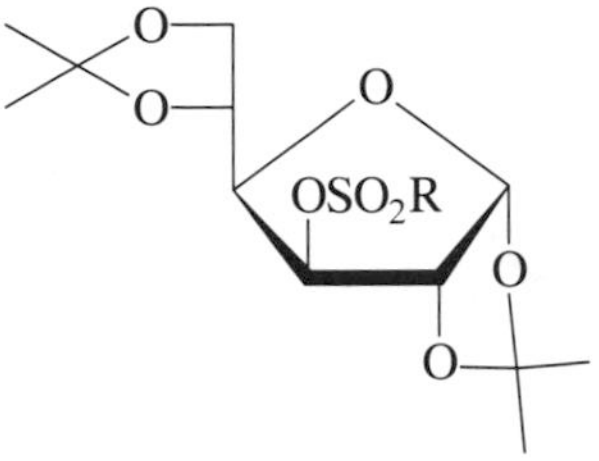

5.63 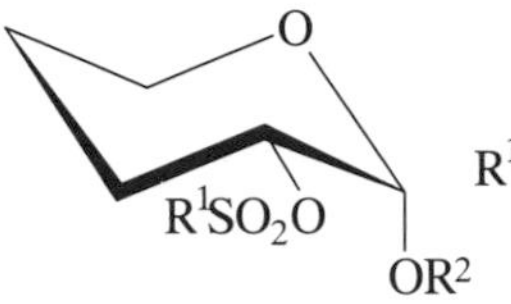   **5.64** 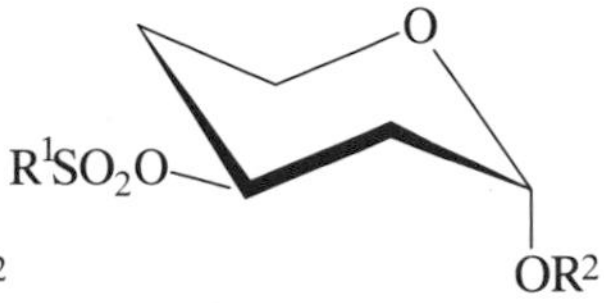   **5.65**

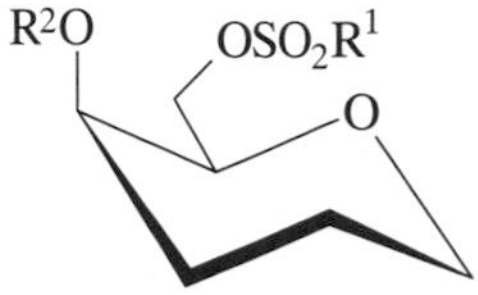

5.66

structural units **5.64**–**5.66** are relatively inert to attack by nucleophiles since, in the establishment of S_N2 transition states, inhibiting interactions develop involving, in the respective cases, the ring oxygen atoms and the axial O-1, O-1 and O-4.[104] Many further aspects of the use of sulphonates in displacement and elimination reactions are discussed in Chapter 4, especially Sections 4.1.1 and 4.1.2.

Sulphonates are usually prepared by use of the sulphonyl chloride in cold pyridine, but for triflates the anhydride is the reagent of choice; occasionally, *N*-sulphonylimidazoles have been employed, and for the very reactive imidazylates the esters are best made from the carbohydrate chlorosulphates. *p*-Toluenesulphonyl (tosyl) chloride in pyridine shows marked selectivity for primary hydroxyl groups and thus, for example, aldohexosides give 6-tosylates as the initial products. Under more forcing conditions, however, secondary hydroxyl groups are substituted and readily displaced sulphonates can react with the halide ions in the systems to yield chlorinated products. Thus, esterification of methyl α-D-glucopyranoside with an excess of the reagent in hot pyridine yields compounds **5.67** and **5.68** (Scheme 5.28).[105] Such displacements would also be expected with glycosyl tosylates, and indeed reaction of D-glucose with an excess of tosyl chloride in

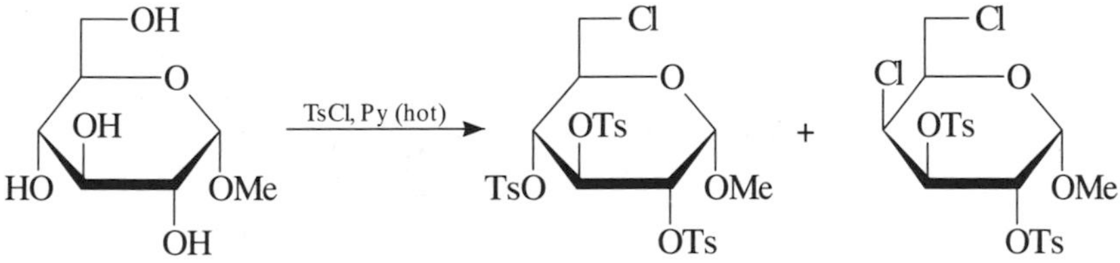

5.67   **5.68**

Scheme 5.28

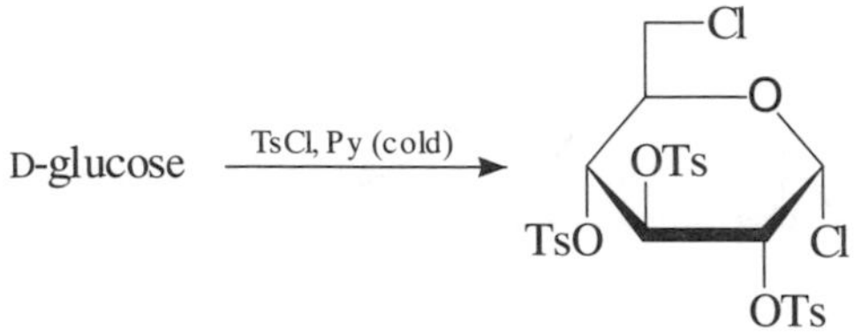

D-glucose $\xrightarrow{\text{TsCl, Py (cold)}}$

5.69

Scheme 5.29

the cold yields the glycosyl chloride **5.69**, (Scheme 5.29). With limiting amounts of the reagent, position 6 of the sugar is first esterified followed by position 2, the anomeric hydroxyl group appearing unexpectedly (Section 5.1.1) to be the least reactive. Esterification of free sugars in pyridine can also lead to glycosylpyridinium salt formation either by direct displacement of glycosyl sulphonyloxy groups by the solvent or through the intermediacy of the chlorides, and conceivably their hydrolysis on work-up to give products unsubstituted at O-1 accounts for the apparent low reactivity of the anomeric hydroxyl group.

Treatment of hexopyranosides with mesyl chloride in DMF has led to an efficient method for replacing the primary hydroxyl groups by chlorine. However, this reaction probably does not proceed by way of 6-sulphonate esters (Section 4.5.1.a).

Sulphonates are stable towards mild acids and towards mildly basic conditions such as are used in methyl iodide methylations (Section 5.2.1), but are cleaved with aqueous alkali. Frequently under these conditions anhydro derivatives arise as a consequence of the presence of suitably disposed intramolecular nucleophiles (Section 4.6), but in the absence of such groups hydrolysis, with retention of configuration (i.e. with the unusual S–O bond fission), can occur. More usually, however, sulphonates are reconverted to the parent alcohols by treatment with sodium amalgam or sodium in liquid ammonia or with Raney nickel, but again anhydride formation, if it is possible, can interfere with such desulphonylations. Tosylates may be cleaved photolytically in methanol containing triethylamine in which benzoates and benzyl ethers are stable.[106] Conversely, they resist basic conditions which cleave benzoates and hydrogenolysis which removes benzyl ethers, and therefore tosylates, benzoates and benzyl ethers are a trio of protecting groups which may be selectively removed in the presence of each other in all possible ways. Removal of tosylates can also be accomplished by electron transfer from photoactivated electron-rich aromatic compounds.[107]

A further method which removes sulphonate esters involves the use of lithium aluminium hydride reductions. However, such reductions can proceed by attack at carbon (Scheme 5.27a) or sulphur (Scheme 5.27b) ($X = H^-$), and hence the parent alcohol or the deoxy analogue can be produced. Usually primary esters give deoxy compounds, whereas secondary derivatives react to give alcohols as is illustrated by the preparation from the ditosylate **5.70** of the dideoxymonohydroxy compound **5.71** which is the α-glycoside of the antibiotic sugar chromose A (Scheme 5.30).[108]

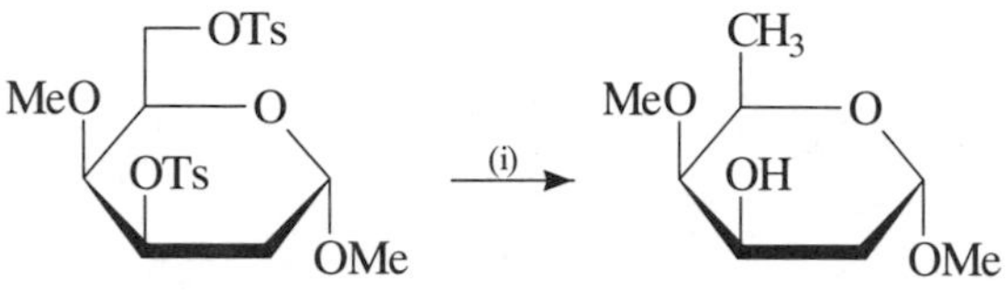

5.70   **5.71** (35%)

(i) LiAlH$_4$, Et$_2$O, benzene, reflux, 48 h

Scheme 5.30

The reaction is frequently used to give terminal deoxy compounds from primary esters, but it is less reliable as a means of de-esterifying secondary analogues. The delicate balance between the possible reduction processes is illustrated by the finding that treatment of the dimesylate **5.73** with lithium aluminium hydride in anhydrous tetrahydrofuran gives the diol **5.74** together with its 3,6-anhydro derivative, which must be formed by initial de-esterification of the secondary group,[109] whereas the expected alcohol **5.72** is obtained as the main product when the reaction is carried out in ether/benzene (Scheme 5.31). An alternative means of reductively removing oxygen from primary positions of sugar derivatives involves displacements of sulphonyloxy groups (which can be introduced specifically) using sodium iodide in acetone solutions and reduction of the iodo products (Section 4.2.1.b).

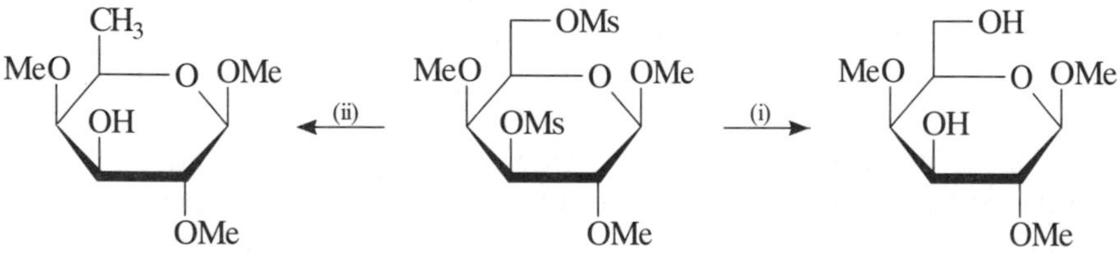

5.72 (60%)   **5.73**   **5.74** (44%)

(i) LiAlH$_4$, THF, 50 °C, 24 h; (ii) LiAlH$_4$, Et$_2$O, benzene, reflux, 72 h

Scheme 5.31

While sulphonates are not normally considered to be subject to intramolecular migration, such isomerization has been observed with the anhydronucleoside triflate **5.75** which, at room temperature in pyridine, HMPIT or DMSO, rearranges to the 3′-substituted ester.[110]

An alternative type of sulphonic acid derivative in which a sugar carbon atom is attached directly to sulphur has attracted attention since 1′-(6-deoxy-α-D-glucopyranosyl)glycerol 6-sulphonic acid **5.76** was found as a constituent of the sulpholipid occurring in photosynthetic plant tissue. The sugar component has been synthesized by tosyloxy displacement with sulphite from a 6-sulphonylated glucose derivative, and also by radical addition of bisulphite to the double bond of compound **5.77**.[111]

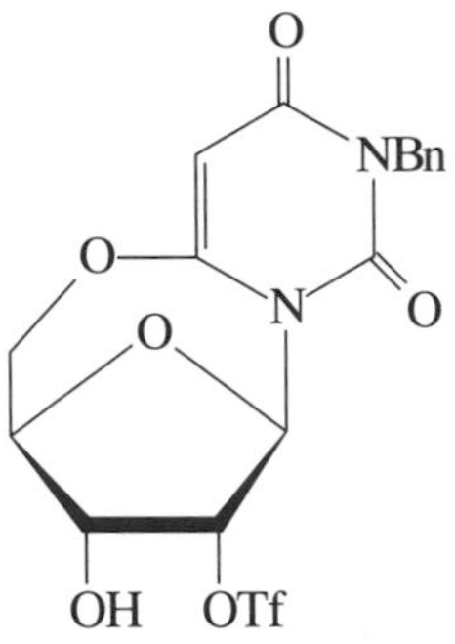

5.75 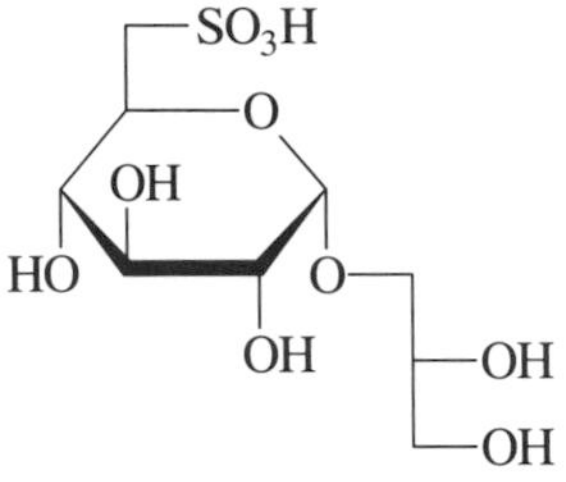   **5.76** 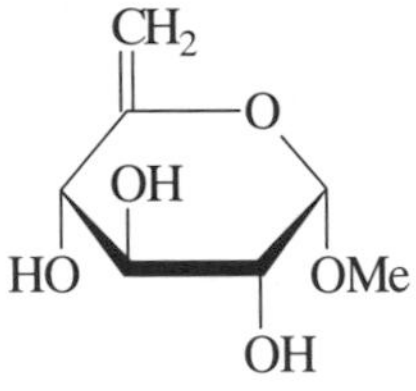   **5.77**

5.3.7 SULPHATES AND CHLOROSULPHATES

(a) Sulphates[112]

Particular interest in the chemistry of sugar sulphates stems from the occurrence of such esters in the seaweed polysaccharides carrageenan and agar (Section 8.1.3.b.vii) and in various animal mucopolysaccharides, notably heparin (Section 8.1.4.b).

Specific monosaccharide sulphates are best prepared by esterification of suitably protected derivatives with either chlorosulphonic acid ($ClSO_3H$) in pyridine or with the sulphur trioxide–pyridine complex. D-Glucose 3-sulphate **5.78** and 6-sulphate can thus be obtained from the 1,2:5,6-di-O-isopropylidene acetal (Scheme 5.32) and the 1,2,3,4-tetraacetate, respectively. The esters are therefore stable under the mildly acidic and basic conditions required to remove the protecting groups from the first products of reaction, but they are unstable in strong acid or alkali. In strong acid, direct hydrolysis occurs with sulphur–oxygen bond fission and at a rate comparable with that of the hydrolysis of glycopyranosidic bonds, but in alkali carbon–oxygen bond fission usually occurs and sugar anhydrides are formed where possible, in the same way as with sulphonates. Methyl α-D-galactopyranoside 6-sulphate thus is readily converted into the 3,6-anhydro

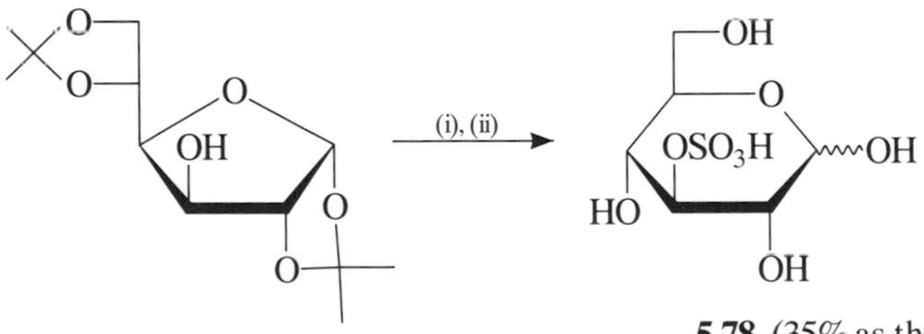

5.78 (35% as the Ba^{2+} salt)

(i) SO_3, Py; (ii) H_3O^+

Scheme 5.32

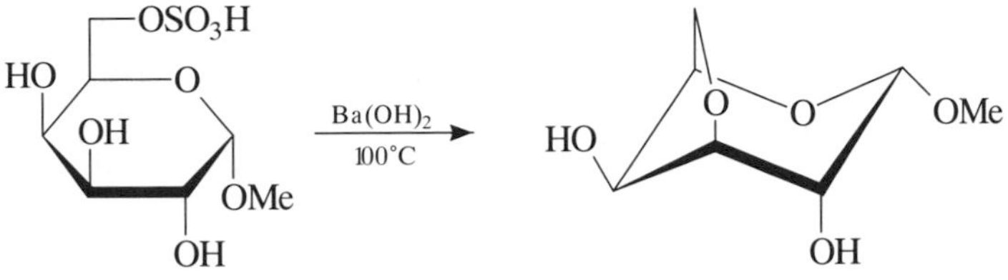

Scheme 5.33

compound **5.79** (Scheme 5.33), but the stability of the ester is appreciably enhanced when the C-3 hydroxyl group is protected by alkali-stable substituents.

When, on treatment with alkali, displacements occur to give epoxides (Section 4.6.1), these products react with methoxide ions to give methylated compounds, the identification of which locates the positions of the original sulphate residues. In this way a sulphated 1,4-linked D-xylose-containing polymer, isolated from an algal source, gives 2-*O*-methyl-D-xylose, which reveals that the epoxide has structure **5.80** and therefore that the sulphate was also located at the 2-position (Scheme 5.34).[113]

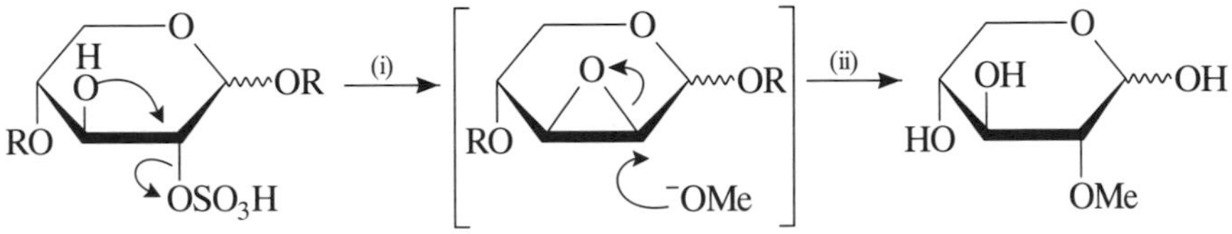

5.80

(i) MeO$^-$; (ii) H$_3$O$^+$ R = polymer chain
Scheme 5.34

Frequently, sugar sulphates, which are highly acidic, are handled and stored in the form of their calcium, barium, cyclohexylamine or alkaloid (e.g. brucine) salts, and they are best purified by chromatographic procedures.

N-Sulphation of amino groups can be carried out selectivity at pH 9 in aqueous solution by use of the sulphur trioxide–trimethylamine complex.[114]

(b)   Chlorosulphates, cyclic sulphates and sulphites

Sulphuryl chloride (SO$_2$Cl$_2$) offers a means of synthesizing cyclic sulphates from suitable diols: methyl 4,6-*O*-benzylidene-α-D-glucopyranoside gives, for example, the 2,3-cyclic ester **5.81** when pyridine is used in excess, but the di(chlorosulphate) **5.82** is obtained when small amounts of pyridine are used (Scheme 5.35), presumably because, under these conditions in which the reagent concentrations are high, intermolecular esterification of the first-formed monoesters is favoured relative to an

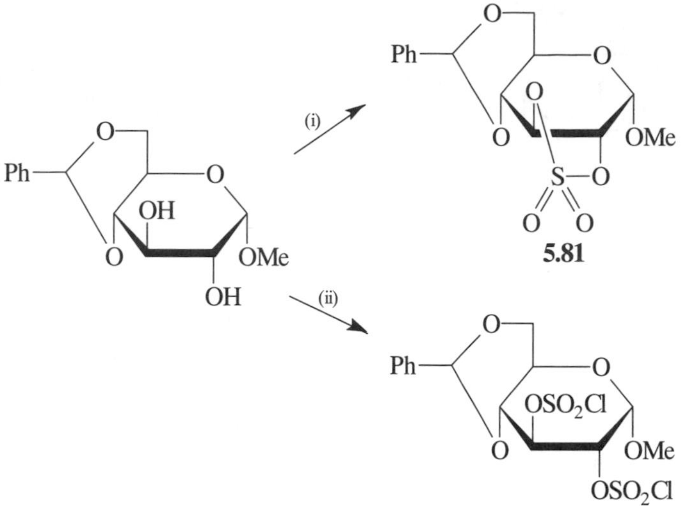

5.81

5.82 (94%)

(i) SO_2Cl_2, Py (excess); (ii) SO_2Cl_2, Py, $CHCl_3$

Scheme 5.35

intramolecular reaction.[112] It is noteworthy, however, that the diester **5.82** cyclizes to **5.81** in pyridine even at $0\,°C$.[115]

Polyhydroxyl compounds can react further in a manner reminiscent of their behaviour with sulphonyl chlorides (Section 5.3.6), i.e. not only does esterification occur, but labile esters can be nucleophilically displaced by chloride ions. Depending upon the conditions used, methyl α-D-glucopyranoside yields the tetra-ester **5.83** or the dichloro derivative **5.84** (Scheme 5.36),[116] whereas from D-glucose itself the α-glycosyl chloride **5.85** is obtained following reaction at $20\,°C$ with the reagents given in Scheme 5.36.[117] D-Xylose likewise gives the pentose analogue of **5.85** when so treated, but β-D-xylopyranosyl chloride 2,3,4-tri(chlorosulphate) can be isolated in good yield when the reaction is carried out at $-70\,°C$. This latter

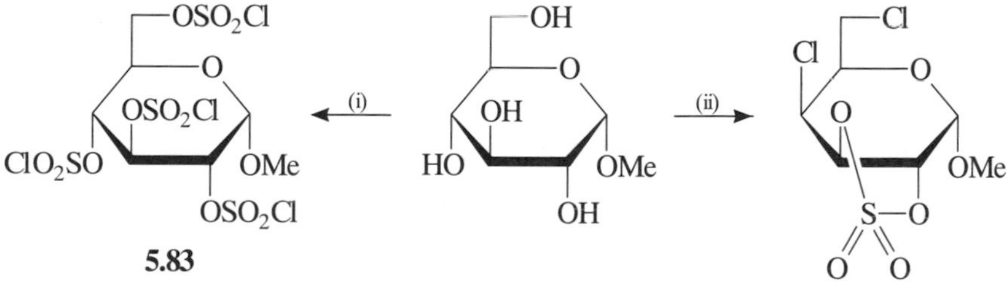

5.83

5.84 (30%)

(i) SO_2Cl_2, Py, $CHCl_3$, $-70\,°C$; (ii) SO_2Cl_2, Py, $CHCl_3$, $5\,°C$, 2 h

Scheme 5.36

product is useful as a glycosylating agent for the preparation of α-D-xylopyranosides since the chlorosulphate groups do not participate in the displacements at C-1 and may be removed on treatment in aqueous methanol with sodium iodide in the presence of barium carbonate.

Reaction of sucrose and sulphuryl chloride gives initially the octa(chlorosulphate) but then, successively, chlorine is introduced at C-6′, C-6, C-4 and C-1′. Following de-esterification, a series of chlorodeoxy disaccharides can be obtained, some of which have remarkable sweetness.[118] The commercial 'sucralose' **5.86**, (R = OH) is 650 times sweeter than sucrose, and when the hydroxyl group at C-4′ is replaced by a halogen the sweetness increases significantly and with the size of the new halogen atom. For **5.86** with R = F, Cl, Br and I the sweetness factors are 1000, 2000, 3000, and 7500, respectively.[119]

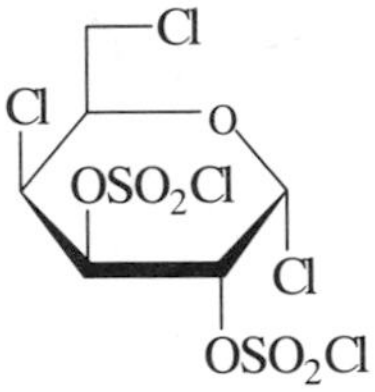

5.85 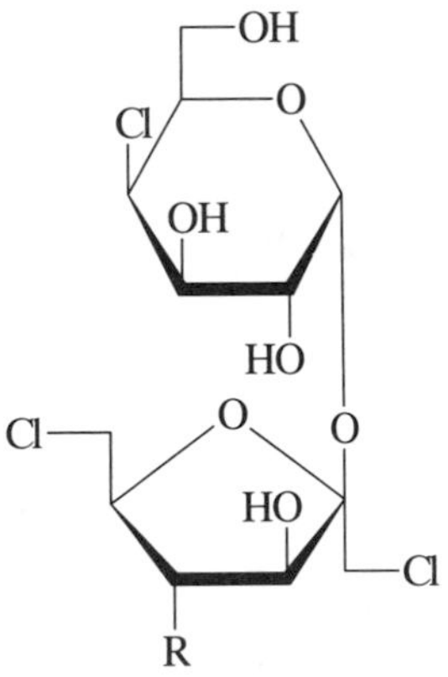   **5.86**

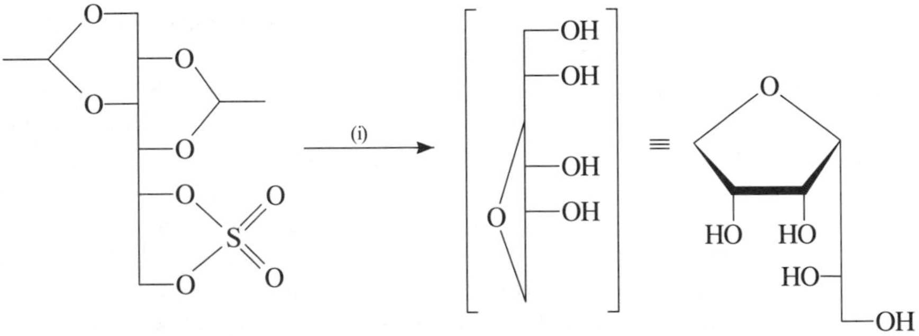

(i)

5.87   **5.88** (72%)

(i) 0.015 mol l^{-1} H$_2$SO$_4$, 8 h, 100 °C

Scheme 5.37

Cyclic sulphates may also be prepared from the corresponding sulphites (obtained from diols with thionyl chloride, SOCl$_2$) by oxidation with permanganate or ruthenium(III) chloride and periodate. Unlike simple sulphates they show considerable tendency to cleave by carbon–oxygen bond fission in acid solution: 1,3:2,4-di-*O*-ethylidene-D-glucitol 5,6-sulphate **5.87** thus gives 3,6-anhydro-D-glucitol **5.88** in high yield (Scheme 5.37).[120] Cyclic sulphates and also cyclic sulphites of 1,2-diols can be used as epoxide analogues in ring-opening reactions, as illustrated in Scheme 5.38, which further exemplifies their leaving group tendencies.[121–123]

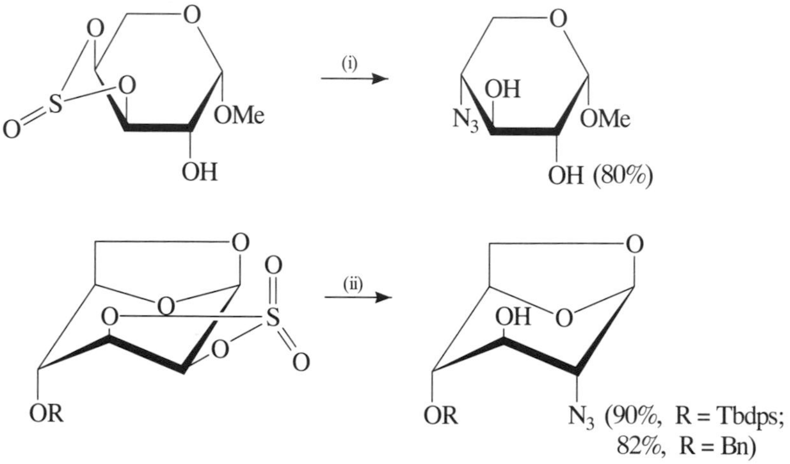

(i) NaN₃, HMPIT, 110 °C;[122] (ii) LiN₃, DMF, 80 °C[123]

Scheme 5.38

5.3.8   PHOSPHATES[124]

Interest in carbohydrate phosphates derives mainly from their unique significance in biochemistry where they play vital roles in such fundamental processes as the biosynthesis and metabolism of sugars (Section 2.5.1). In addition, phosphates of D-ribose and its 2-deoxy derivative form fundamental components of the nucleic acids and of various coenzymes (Section 8.2.2).

Both chemical and enzymic methods are available for the synthesis of specific phosphates. Chemically, glycosyl phosphates are prepared either from glycosyl halides or compounds having an unsubstituted anomeric centre, and phosphate esters at other positions are formed by use of specifically substituted sugar derivatives. Biochemically, they are produced by the action of appropriate enzymes on provided substrates. Contrary to what could be written in the first edition, such biochemical procedures are suited to the preparation of specific phosphates in bulk, since in the interval, and thanks largely to the work of Whitesides it has proved possible to isolate the relevant enzymes and apply them in the laboratory (Section 2.5.2). It emerges that they can also be utilized to make wide ranges of natural compounds and their analogues.

Chemical synthetic methods remain important. Treatment of acetylated glycosyl bromides (Section 3.3.2) with, for example, silver dibenzyl phosphate gives glycosyl phosphates after catalytic removal of the protecting groups, as is illustrated

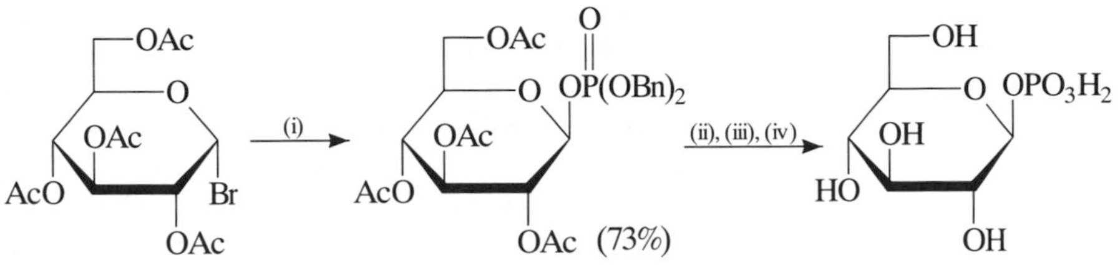

(i) $(BnO)_2\overset{\overset{\text{O}}{\|}}{P}OAg$, benzene, 55 °C, 2 h; (ii) H_2, Pd; (iii) KOH; (iv) H^+

Scheme 5.39

in Scheme 5.39 for β-D-glucopyranosyl phosphate **5.89**. Alternatively, these compounds are obtainable by treatment of aldose peracetates with anhydrous phosphoric acid in a reaction which resembles those leading to acetylated glycosyl halides (Section 3.2.1.b) and acetylated aryl glycosides (Section 3.2.1.c.i). The α-anomer of compound **5.89** can be prepared in this way from penta-*O*-acetyl-β-D-glucopyranose, and the corresponding 1,6-diphosphate in similar fashion from the 1,2,3,4-tetraacetate 6-phosphate. Otherwise, glycosyl phosphates may be made by use of glycosyl trichloroacetimidates[5] or glycose 1,2-orthoesters.[125]

Esterification at sites other than the anomeric is usually carried out on specifically substituted compounds, and with such phosphorylating agents as dibenzyl or diphenyl phosphorochloridate [$(PhO)_2POCl$] in pyridine solution. D-Galactose 2-phosphate **5.90** is thus synthesized as shown in Scheme 5.40.[126] Phosphate ester groups can also be introduced by nucleophilic displacement methods involving

(i) $(PhO)_2POCl$, Py, 20 °C, four days; (ii) H_2, Pt; (iii) EtO^-

Scheme 5.40

reagents such as lithium diphenyl phosphate applied to such electrophiles as methyl 2,3-anhydro-4,6-O-benzylidene-α-D-allopyranoside or 1,2,3,4-tetra-O-acetyl-6-O-tosyl-α-D-glucose.

Appreciable interest has recently been taken in phosphate esters of inositols, especially D-*myo* inositol 1,4,5-trisphosphate **5.91** which is a vital second messenger for calcium mobilization in animal cells, and several more recent phosphorylating methods have been applied to its preparation (Section 8.3). Tetrabenzyl pyrophosphate [(BnO)$_2$P(O)OP(O)(OBn)$_2$] has been used following conversion of the free hydroxyl groups of a tri-O-substituted compound to lithio derivatives by treatment with lithium diisopropylamide.[127] Otherwise, dibenzyl phosphorofluoridate [(BnO)$_2$POF], employed in acetonitrile with caesium fluoride as promoter or in THF with butyllithium, has been successful.[128] A quite different approach depends upon the initial preparation of phosphite esters which are oxidized to phosphates with, for example, t-butyl hydroperoxide, this method having the advantage that it can be used to prepare diesters with two sugar units linked by a phosphate bridge.[129]

$$\text{OH} \quad \text{OH}$$
$$\text{OPO}_3\text{H}_2$$
$$^4\text{OPO}_3\text{H}_2 \quad ^1$$
$$\text{H}_2\text{O}_3\text{PO}$$
$$5$$
$$\text{OH}$$

5.91

Monophosphate esters, which are strongly acidic, are usually isolated and stored as metal or cyclohexylammonium salts, and are best purified by ion exchange chromatography. Glycosyl phosphates are labile to acid and are relatively stable to bases but, in common with other phosphates, will hydrolyse with strong alkali. D-Glucose 6-phosphate is 60% hydrolysed by 0.2 mol l^{-1} alkali in three minutes at 100 °C. Aldose 3-phosphates which have the ester group in the β-relationship to the carbonyl group, however, degrade rapidly by an elimination mechanism to yield metasaccharinic acids (Section 3.1.7.b). Aldose 2-phosphates are labile in acid media (probably hydrolysing as enol esters), while other phosphates are more stable towards acids.

Ester migration by way of cyclic diesters also occurs in acidic and particularly basic media (cf. acetyl migrations; Section 5.3.1.b), and provides a further means of preparing specific esters. 1,2-O-Isopropylidene-D-xylofuranose 3-phosphate **5.92** can be cyclized to give the 3,5-diester **5.93** which hydrolyses to a mixture of the 3-phosphate and 5-phosphate (Scheme 5.41). Cyclization of other suitable monophosphates can also be effected with dicyclohexylcarbodiimide, glucose 6-phosphate and ribofuranoside 2-phosphates giving, for example, 4,6- and 2,3-cyclic esters, respectively. Alternatively, cyclic diesters can be prepared by direct substitution of suitable diols with phenyl phosphorodichloridate (PhOPOCl$_2$).

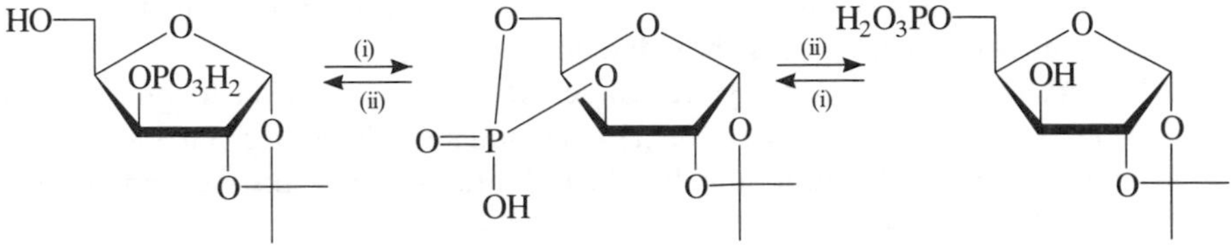

5.92   **5.93**

(i) Py, DCC; (ii) 1 mol l^{-1} KOH, 100 °C

Scheme 5.41

5.3.9   NITRATES[130]

A wide variety of carbohydrate derivatives containing nitrate ester groups are known, varying from monoesters to peresters, but as yet few, if any, have been found in natural products.

Glycosyl nitrates are obtainable most readily by reaction of acylglycosyl halides or peresters (usually acetates or benzoates) with nitric acid in chloroform solution or, in the case of the halides, with silver nitrate. Nitration at sites other than the anomeric can also be brought about with nitric acid in chloroform, but cold acetic anhydride is sometimes preferred as solvent since, under these conditions, acetal rings are not cleaved: the dinitrate **5.94** is prepared directly from the corresponding diol (Scheme 5.42). As an alternative reagent, dinitrogen pentoxide in chloroform solution can be employed. A specific route to primary esters involves the displacement of iodide with silver nitrate from iodomethyl groups which are available from primary sulphonates (Section 4.5.1.a).

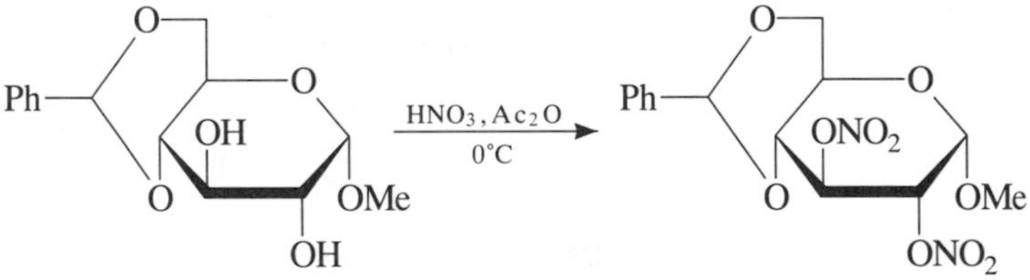

5.94 (60%)

Scheme 5.42

Sugar mononitrates and dinitrates are often stable crystalline compounds, *but more highly substituted derivatives may be heat or shock sensitive.* The ester groups, except when attached to glycosyl centres, are tolerably stable to acid, so that acetals can be removed in their presence; they are also unaffected by acylation and mild alkylation procedures. With strong alkali their reactions are complex, varying between those of sulphonates and carboxylates, so that both carbon–oxygen and nitrogen–oxygen bond fission take place. In addition, however, degradation to carbonyl compounds can occur (by β-elimination of nitrous acid), and these undergo subsequent reactions. Removal of nitrate esters is best carried out reductively with lithium aluminium hydride or with hydrazine.

Although secondary nitrates suffer direct de-esterifications on treatment with iodide ion, primary esters behave as sulphonates (Section 5.3.6) and give iodomethyl compounds, and terminal vicinal dinitrates (again like the sulphonates; Section 4.5.1) undergo eliminations to give alkenic products.

5.3.10   BORATES, BORONATES[131] AND BORINATES

In aqueous solution various diols can react with boric acid to give ionic esters (e.g. **5.95**) which have been profitably utilized in carbohydrate chemistry. By investigating the electrical conductivities of aqueous glucose solutions to which boric acid was added, Böeseken was first able to assign the anomeric configurations to α- and β-glucopyranose (Section 2.1.4). The same type of ions can be used in electrophoretic examinations of sugars during which the ions migrate across potential gradients at rates dependent upon the degree of complexing. Valuable analytical techniques have been developed based on this principle, which can also be used as a means for assigning configurations to diols. In this way D-mycarose **5.96**, the enantiomer of a naturally occurring branched-chain sugar, is determined to have a *cis*-related C-3–C-4 diol, since its methyl glycopyranoside migrates appreciably on electrophoretograms whilst the corresponding derivative of the C-3 epimer is immobile.

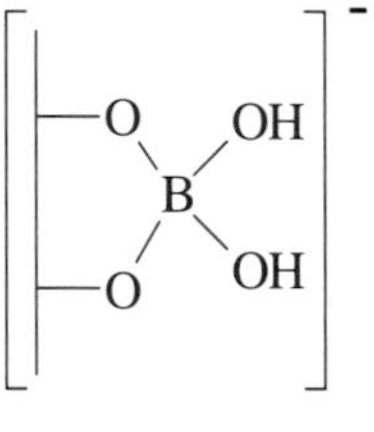

5.95 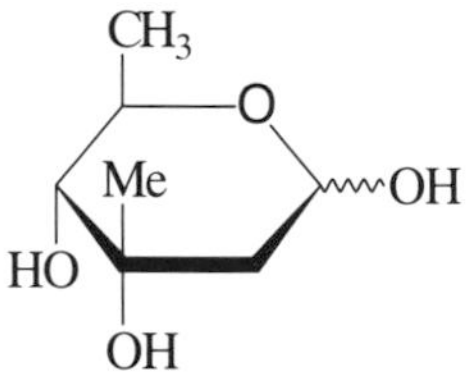   **5.96**

Isolated borate esters of carbohydrates are not well known, very probably because the syntheses of discrete derivatives are complicated by the formation of dimeric species **5.97** or, in the case of polyhydroxyl compounds, by various triesters **5.98**.

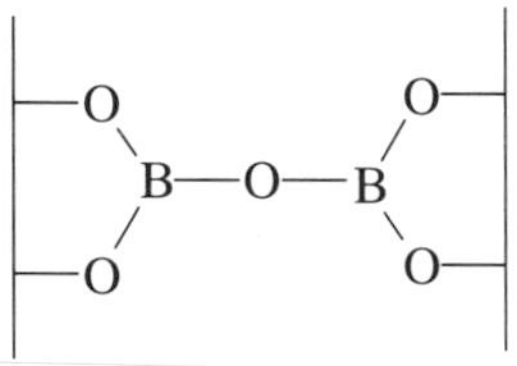

5.97 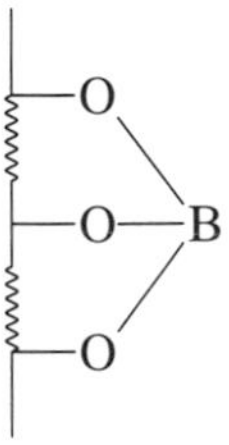   **5.98** 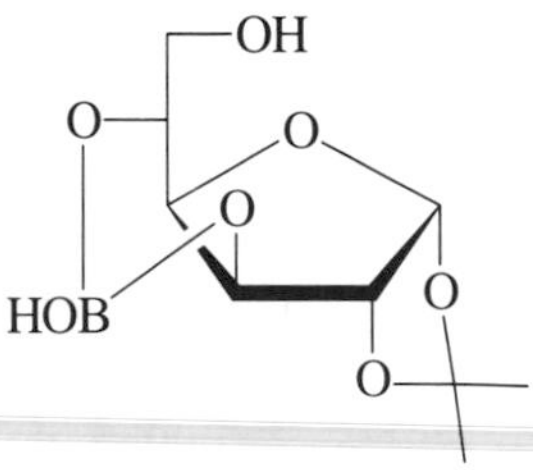   **5.99**

The potential value of such esters is, however, well illustrated by the finding that treatment of D-glucose with boric acid in acetone in the presence of sulphuric acid yields a crystalline derivative **5.99** which is a suitable starting material for the preparation of 6-substituted D-glucoses. Treatment of the products with methanol liberates the borate protecting group, and distillation removes the acid by-product as its volatile trimethyl ester.

A much more satisfactory reagent for the synthesis of boric acid derivatives of carbohydrates is phenylboronic acid [PhB(OH)$_2$], which normally gives simple cyclic esters from diols. It has the advantages of forming, under mild conditions, esters which are stable during esterifications, glycosylations or dimethyl sulphoxide oxidations of unprotected hydroxyl groups, and of being removable under very mild conditions. Formation of the esters is usually carried out by heating the carbohydrate with the reagent or its cyclic trimeric anhydride in a solvent such as benzene and removing the water formed by azeotropic distillation; conversely, the esters are cleaved by adding water to their solutions in organic solvents. The liberated free acid can then be removed by use of anion exchange resins or by distillation in the presence of propane-1,3-diol, with which it forms a volatile cyclic ester. From methyl α-D-glucopyranoside a 4,6-boronate is obtainable in excellent yield from which 2,3-substituted methyl glucosides can be prepared. In the pentopyranoside series, arabinosides and lyxosides give 3,4- and 2,3-cyclic esters, respectively, indicating that 1,2-*cis*-diol systems are preferred by the reagent, but with ribosides and xylosides 2,4-cyclic esters are formed. In this latter respect the reagent appears to have unique selectivity, and from the 2,4-esters the rare 3-substituted glycosides can be synthesized. The use of the phenylboronate protecting group in synthetic work is illustrated in Schemes 5.43 and 5.44.[131]

Phenylboronic acid has also found use as a test reagent for *cis-cis*-1,2,3-triols on pyranoid rings, since compounds possessing this structural feature show enhanced paper chromatographic mobility when the compound is added to solvents. This results from stabilization of the esters formed from 1,3-related diaxial diols by specific coordination from the intermediate equatorial hydroxyl groups, as shown in **5.100**.

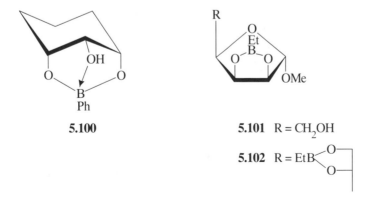

5.100

5.101 R = CH$_2$OH

5.102 R = EtB

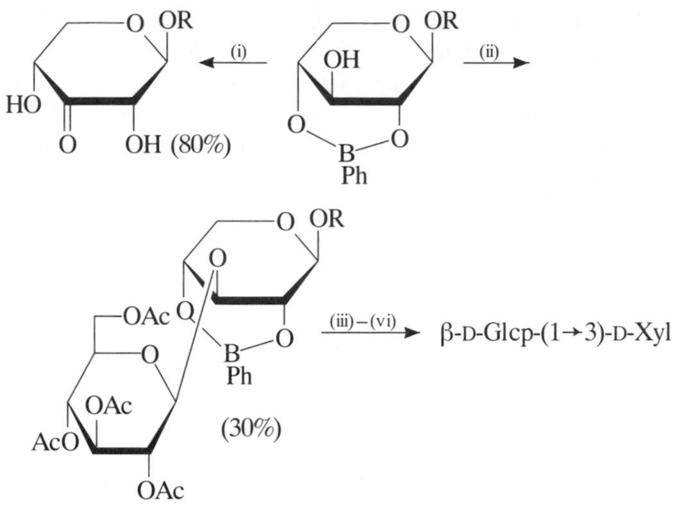

(i) DMSO, Ac$_2$O, 40 °C, 75 min, resin; (ii) tetra-acetylglucosyl bromide (R = Me); (iii) HOCH$_2$CH$_2$CH$_2$OH; (iv) MeO$^-$; (v) resin; (vi) H$_2$,Pd (R = Bn)

Scheme 5.43

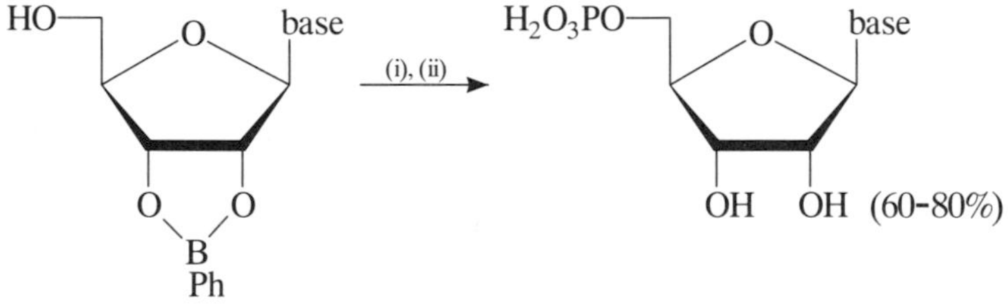

(i) Morpholinophosphodichloride, diphenyl phosphate, dioxane, lutidine, 20 °C, 48 h; (ii) deboronation

Scheme 5.44

Ethylboronates have also found specific uses in the field, D-lyxose and D-mannose, for example, on heating under reflux with dimethyl ethylboronate in the presence of an acid resin, giving the methyl furanoside boronates **5.101** and **5.102**, respectively, from which the unsubstituted glycosides can be obtained in excellent yield.[132]

Borinates (ROBR$_2$) have usually been encountered following treatment of hydroxylated carbohydrates with diethylborinic acid or similar reagents, a process that can lead to useful hydroborative reduction of C–O bonds. Thus, phenyl β-D-galactopyranoside, on heating at 120 °C with triethylborane and hexaethyldiborane in the presence of an activating mixed borate/mesylate anhydride, gives, via the perester **5.103**, the 1-phenylgalactitol **5.104** which can be isolated as its acetate in excellent yield following removal of the diethylborinate esters from each of the hydroxyl groups.[133]

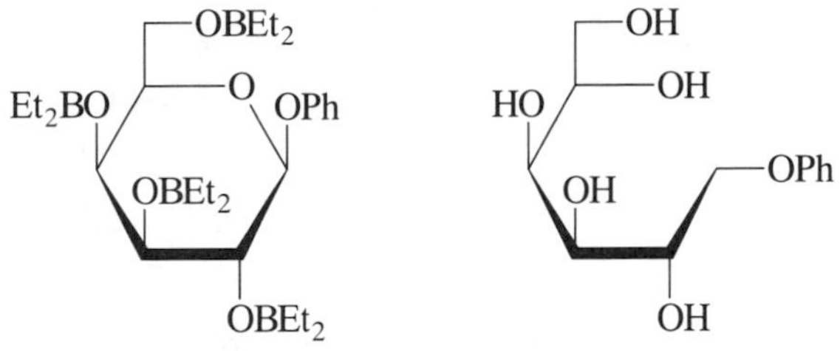

$$\text{5.103} \qquad \text{5.104}$$

5.4   ACETALS

5.4.1   GENERAL

Acetals are normally formed as the products of acid-catalysed condensation of one carbonyl and two alcohol groups

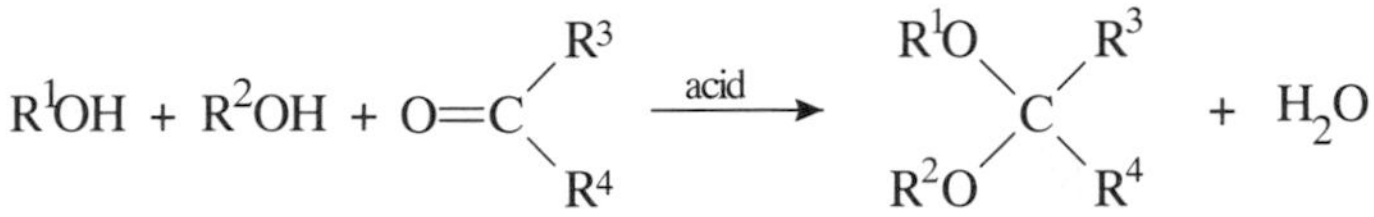

$$R^1OH + R^2OH + O{=}C{<}^{R^3}_{R^4} \xrightarrow{\text{acid}} {}^{R^1O}_{R^2O}{>}C{<}^{R^3}_{R^4} + H_2O$$

In carbohydrate chemistry the carbonyl groups of free sugars form such derivatives, and when the sugars are treated with simple alcohols in the presence of acids, glycosides, which are mixed acetals, are the products (Scheme 5.45a, cf. Section 3.1.1). If, in acyclic sugar derivatives, there is no suitable hydroxyl group available within the molecule, reaction occurs wholly intermolecularly to give acyclic acetals (Scheme 5.45b). (In the reactions of free sugars with thiols in the presence of acid catalysts the dominant products isolated are the dithioacetals; see Section 3.1.2.a.)

In reciprocal manner, single hydroxyl groups of two carbohydrate molecules may react with aldehydes or ketones (Scheme 5.45c), but products of this type are unusual. Nevertheless, impressive instances of the use of compounds consisting of two monosaccharides joined together by way of oxygen atoms linked as parts of acetals in the intramolecular synthesis of disaccharides have been reported (Sections 3.1.1.a.ii and 6.5; see also compound **5.109**). More common are the mixed acetals obtained by treatment of sugar derivatives containing single alcohol groups with either enol ethers (Scheme 5.45d) or simple acetals (Scheme 5.45e). Otherwise, and importantly, suitable diols within monosaccharides react with aldehydes or ketones to give cyclic acetals (Scheme 5.45f), such derivatives being very common and useful compounds in carbohydrate synthesis since they afford a means of protecting specific pairs of hydroxyl groups.

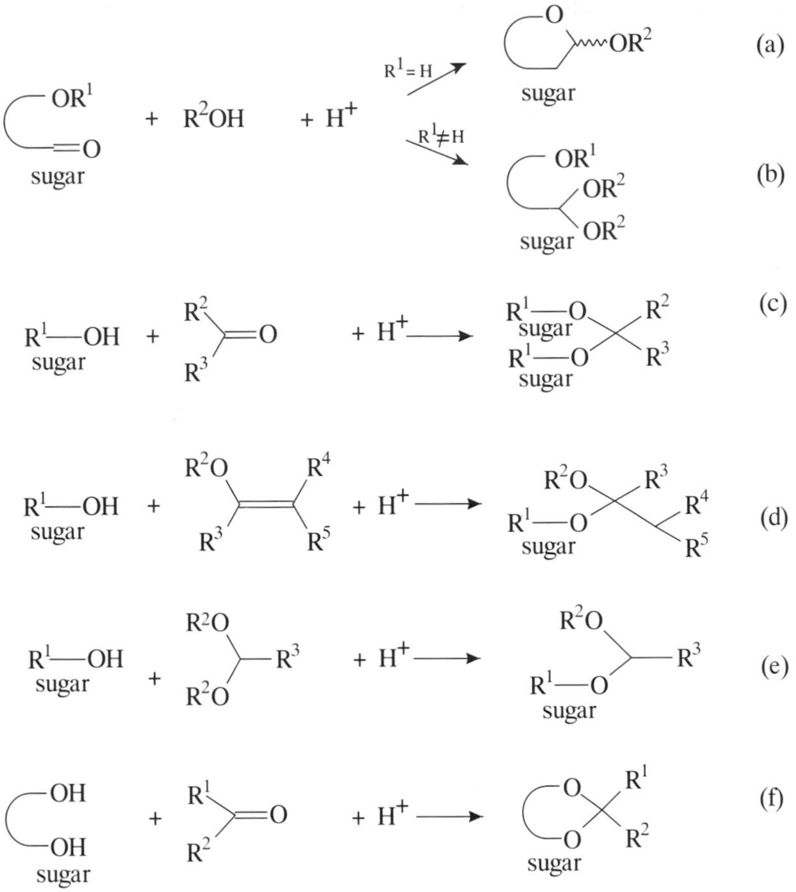

Scheme 5.45

5.4.2   ACETALS FROM SINGLE HYDROXYL GROUPS OF CARBOHYDRATES

Methoxymethyl (Mom) derivatives are commonly used to protect single hydroxyl groups of carbohydrates, the best method of synthesis being by use of di-methoxymethane (Scheme 5.46a),[134] which provides an example of the reaction in Scheme 5.45e. Otherwise, they may be made under basic conditions using chloromethyl methyl ether (Scheme 5.46b).[131] They are stable to many reagents often used in carbohydrate synthesis and are readily removed by acid treatment.

Although tetrahydropyranyl derivatives suffer the serious disadvantage that they give mixtures of diastereoisomeric acetals with asymmetric alcohols of the type normally encountered with carbohydrates, they are in common use, being easily prepared by the enol ether method (Scheme 5.45d) using dihydropyran (Scheme 5.47). Again, they are cleaved in the presence of acid catalysts.

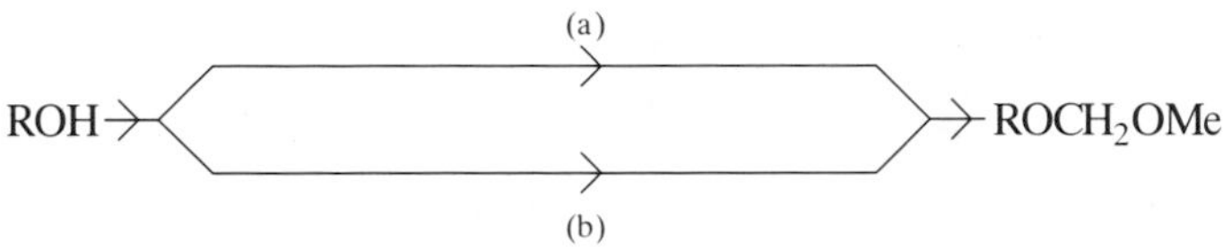

(a) $CH_2(OMe)_2$, P_2O_5, $CHCl_3$
(b) $ClCH_2OMe$, NaH

Scheme 5.46

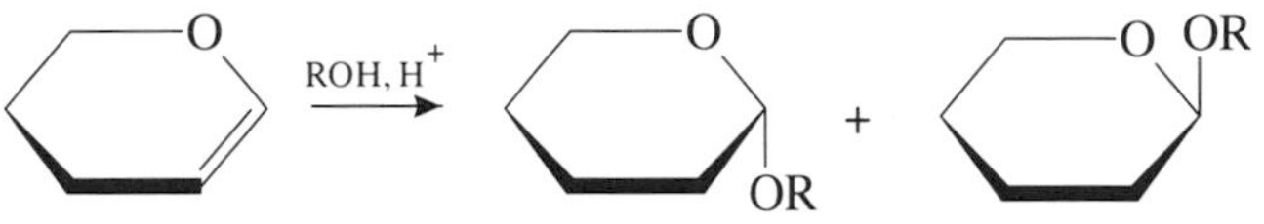

Scheme 5.47

Hexopyranosides form acetals preferentially at the primary position and this affords means of obtaining 4,6-substituted derivatives. With ethyl vinyl ether, for example, methyl α-D-glucopyranoside in DMSO solution containing small amounts of boron trifluoride etherate gives initially compound **5.105** which, on further reaction, with acid, cyclizes to the ethylidene acetal **5.106** together with some of the starting material (Scheme 5.48).[135]

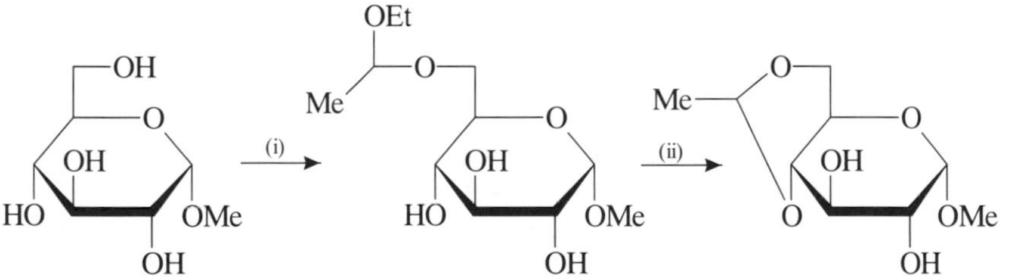

(i) CH_2=CHOEt, DMSO, $BF_3 \cdot OEt_2$, 40 °C, 2 h; (ii) TsOH, $CHCl_3$, 20 °C, 15 min

Scheme 5.48

5.4.3   ACETALS FROM CARBOHYDRATE DIOLS[136]

The cyclic compounds possessing the structural feature shown in Scheme 5.45f are all 'acetals', but when ketones are the non-carbohydrate reactants the term 'ketal' is occasionally used to describe the products. Diols within free ketoses[137] or aldoses[138] or their derivatives, notably glycosides, react with a range of aldehydes or ketones in this way. Products of this type, formed between the keto group of pyruvic acid and sugar diols, are found in polysaccharides of Gram-negative bacteria (Section 8.1.3.d) and some seaweed polysaccharides (Section 8.1.3.b.vii). Otherwise, acetals derived from diols of monosaccharides are not common in natural products.

Acetals can often be formed from specific diols within polyhydroxy compounds in good yield and with high specificity, and since they are stable to hydroxyl substitution reaction conditions and are readily cleaved with dilute acid, they are of the greatest value for the selective protection of pairs of hydroxyl groups in synthetic work. Acetaldehyde, benzaldehyde and acetone are most frequently employed to give ethylidene, benzylidene and isopropylidene derivatives, respectively, but many other carbonyl compounds have also been used. Usually the carbohydrate is taken with a large excess of the aldehyde or ketone, and acidic catalysts such as concentrated sulphuric acid or Lewis acids (for example, zinc chloride) are used, and since water is liberated during the condensation a desiccant such as anhydrous copper sulphate is frequently added to favour the formation of the products. Wide variations of these types of conditions have been applied, and a modification which is useful for effecting energetically unfavoured condensations involves acetal exchange with, for example, 2,2-diethoxypropane and removal of the liberated ethanol by distillation. Procedures for preparing cyclic acetals from vinyl ethers and from orthoesters are referred to in Sections 5.4.2 and 5.3.4, respectively.

Since the condensation reactions are usually carried out under forcing conditions, and since acetal isomerization is usually possible under the influence of the acidic catalysts, highly substituted and thermodynamically favoured products are usually obtained, and whereas *ketones* tend to give *five-membered cyclic products, aldehydes* favour the formation of *six-membered rings*. It should be noted, however, that six-membered ketals and five-membered cyclic products from aldehydes can form when more suitable alternatives are not possible, and also that isolated compounds are not the exclusive products of many condensations. Thermochemical studies reveal that the 1,3-dioxolane ring **5.107** is less stable than the 1,3-dioxane ring **5.108**, and consequently carbonyl derivatives would be expected to favour the formation of the latter. When ketones form such rings, however, one of the large groups at the acetal centre must occupy an axial position in either chair form of the ring, and this destabilizes such a system relative to alternative five-membered analogues. Under acidic conditions, aldehydes give only the diastereoisomers with the larger groups equatorial at the acetal centres of 1,3-dioxane rings, but when fused in five-membered 1,3-dioxolane rings, such carbonyl compounds can give both possible diastereoisomers.

5.107   **5.108**

Acetalation of diols under kinetic control can be effected by use of geminal dihalides such as benzal bromide (PhCHBr$_2$) or dichloromethane in basic conditions, and in this way products are available which are difficult to obtain in acid conditions.[139] For example, benzal bromide in pyridine can allow the formation of benzylidene acetals of appropriate diols which contain acid-fragile groups such as trityl ethers or other acetals. Otherwise, phase transfer procedures can be employed, and in this way the *trans*-fused 2,3-methylene acetal of methyl 4,6-*O*-benzylidene-α-D-glucopyranoside can be made in 63% yield using dibromomethane and aqueous sodium hydroxide with tetrabutylammonium bromide as phase transfer catalyst. These conditions give the unusual diribosid-5-yl acetal **5.109** in 34% yield from methyl β-D-ribofuranoside.[140] Under such basic conditions aldehydes (other than formaldehyde) and unsymmetrical ketones can be expected to give mixtures of epimers at the acetal centres.

A feature of cyclic acetals fused to glycopyranosylating agents is that they diminish their reactivities (see Section 6.4.1).

(a)   Acetals and ketals of glycosides[136,138]

Methyl hexopyranosides react with aldehydes in acidic conditions and give mainly 4,6-acetals; from methyl α-D-glucopyranoside and -galactopyranoside the acetals **5.110** and **5.111**, which respectively have ring structures resembling *trans*- and *cis*-decalin, are readily obtainable in high yield, the phenyl groups occupying equatorial

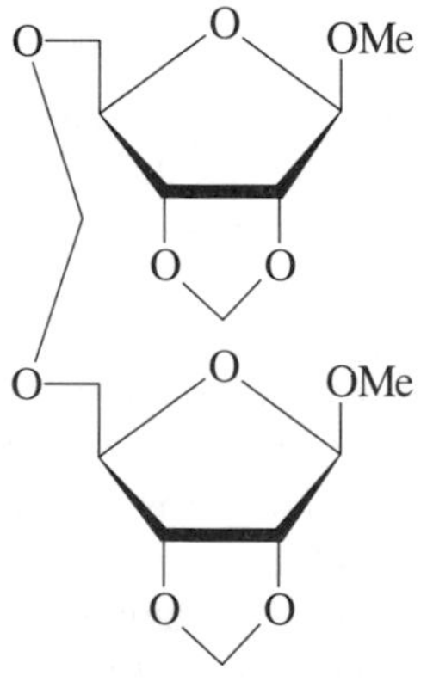

5.109 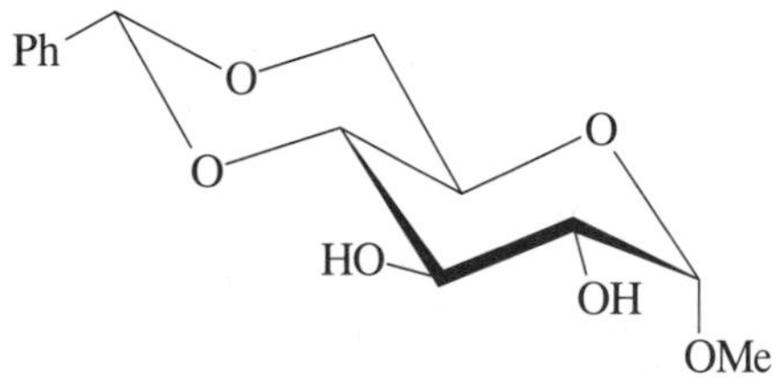   **5.110**

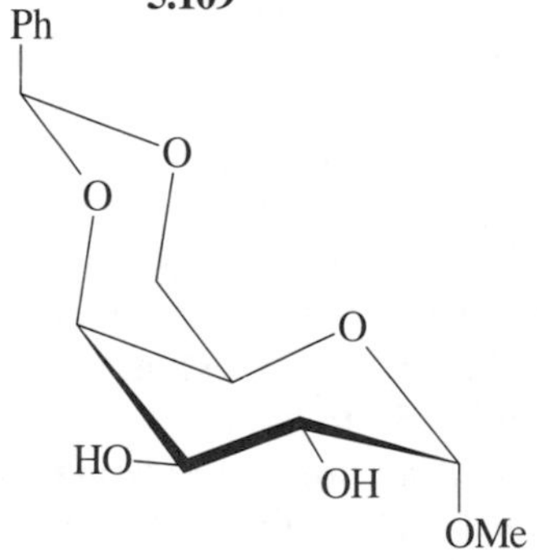

5.111a 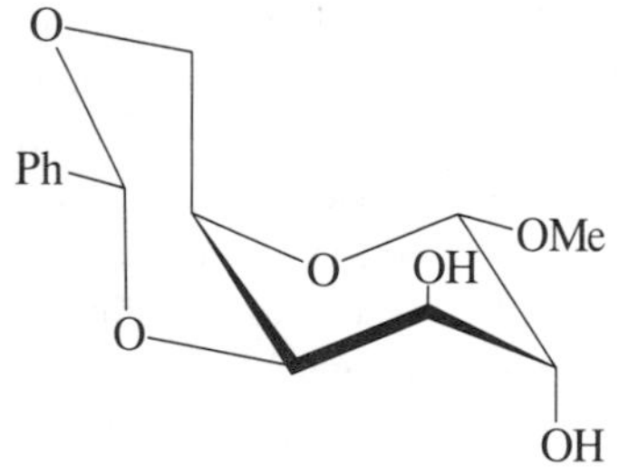   **5.111b**

positions. In the latter case, because of the flexibility of the *cis*-decalin ring system, two chair conformations **5.111a** and **5.111b** are possible and the former is favoured. With the α-mannoside analogue **5.112** reaction can occur further at the 2,3-diol site, and two 2,3:4,6-diacetals **5.113** are produced which differ in their stereochemistry at the five-membered ring acetal carbon atom (Scheme 5.49). In such a case the energy difference between the two diastereoisomers is less than when one has an axial and the other an equatorial phenyl group, and both are formed in appreciable proportions.

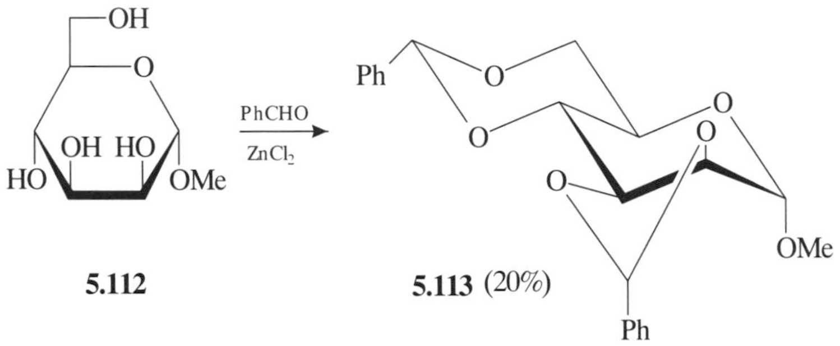

5.112   **5.113** (20%)

Scheme 5.49

Treatment of methyl 2,3-di-*O*-methyl-α-D-glucopyranoside with benzal bromide in the presence of potassium *t*-butoxide gives two acetals which differ in stereochemistry at the new chiral benzylic position and which, on treatment with acid, afford the dimethyl ether of **5.110** exclusively.

With ketones, pyranosides give five-membered ketal products involving 1,2-*cis*-diols where possible. In the presence of protonic acids, lyxosides and mannosides therefore give 2,3-cyclic products, whereas 3,4-substituted compounds are obtained from arabinosides and galactosides. However, when zinc chloride is used as catalyst, the galactosides also give appreciable proportions of 4,6-acetals. Under forcing acetal exchange conditions other types of rings can be formed, an example being the preparation in high yield of the 4,6-*O*-isopropylidene derivative **5.114** of methyl α-D-glucopyranoside, which is accompanied by a small amount of the 2,3:4,6-diacetal **5.115** (Scheme 5.50). The same glycoside on heating with 1,1-dimethoxy-cyclohexane in DMF in the presence of tosic acid gives the analogous cyclohexyli-dene diacetal in very high yield, showing that energetically unfavoured rings can be formed when the correct forcing conditions are applied. Methyl 4,6-*O*-benezylidene-α-D-glucopyranoside, on treatment with dichloromethane and sodium hydride in DMF, gives a 50% yield of the fully substituted 2,3-*O*-methylene compound.

Selective protection of *vic-trans*-diequatorial diols in the presence of other diols can be effected by use of the dihydropyran dimer **5.116** which gives the product **5.117** in high yield when heated with methyl α-D-galactopyranoside in chloroform in the presence of camphorsulphonic acid as catalyst. Selective protection of O-6 of the product then allows glycosylation at O-4, and use of the *O*-benzylated *S*-ethyl thioglycoside analogue of **5.117** permits galactosylation of alcohols, including sugar

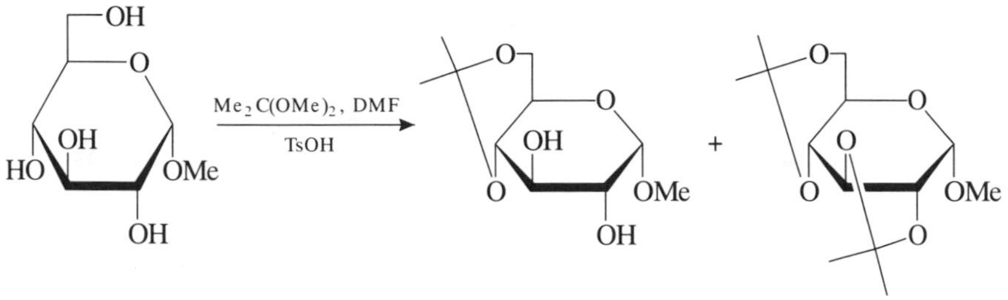

5.114 (80%)   **5.115** (2%)

Scheme 5.50

alcohols.[141] This novel means of protecting diols therefore shows considerable potential as an aid in oligosaccharide synthesis.

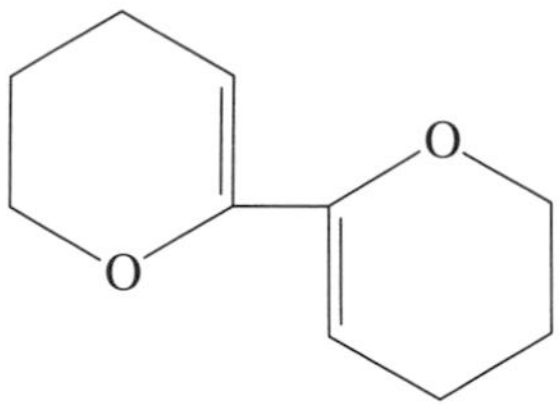

5.116 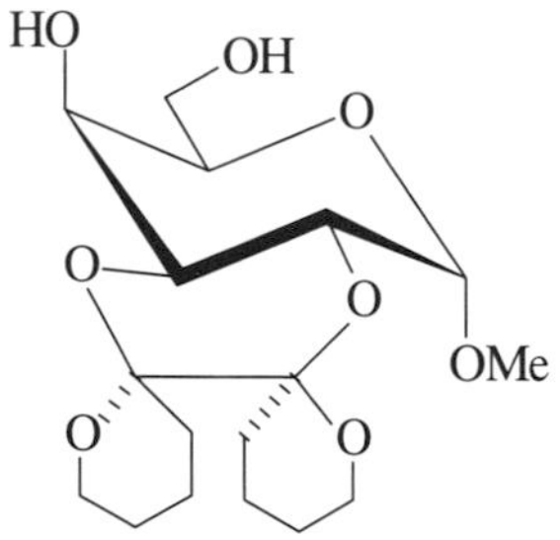   **5.117**

(b)   Acetals and ketals of free sugars[136-138]

Because of the possible variations in ring size and anomeric configuration of free sugars, their reactions with aldehydes and ketones are frequently complex, the nature of the products being dependent upon many factors. Nevertheless, some important acetals can be prepared simply and in high yield. For example, 4,6-*O*-ethylidene-D-glucose is obtainable with 90% efficiency.

Benzaldehyde also tends to give 4,6-acetals from hexoses on condensation in the presence of zinc chloride, and the D-glucose derivative (obtainable in 40% yield) reacts further to give some of the 1,2:4,6-diacetal **5.118** (Scheme 5.51). In the presence of acetic acid, however, the reaction proceeds differently to give some of the 1,2:3,5-furanoid compound **5.119**, perhaps because isomerizations are favoured by protic acids. D-Xylose, likewise, undergoes 1,2:3,5-substitution with benzaldehyde in the presence of protic acids, and in this case two products, differing only in their stereochemistries at the benzylidene acetal centres of the five-membered rings, are formed (Scheme 5.52). When this last reaction is repeated in the presence of primary alcohols, the conformationally stable 2,4:3,5-di-*O*-benzylidene derivatives of the corresponding dialkyl acetals (e.g. **5.120**) are obtained, and from these the acylic dialkyl acetals can be prepared by hydrogenolytic cleavage of the benzylidene rings.[142]

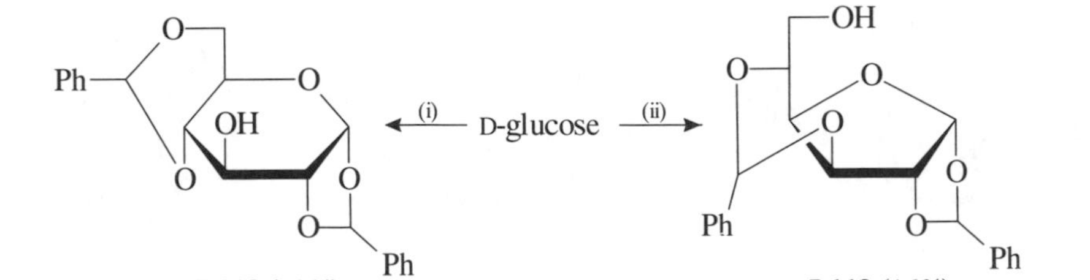

5.118 (10%)   **5.119** (16%)

(i) PhCHO, ZnCl$_2$; (ii) PhCHO, ZnCl$_2$, HOAc
Scheme 5.51

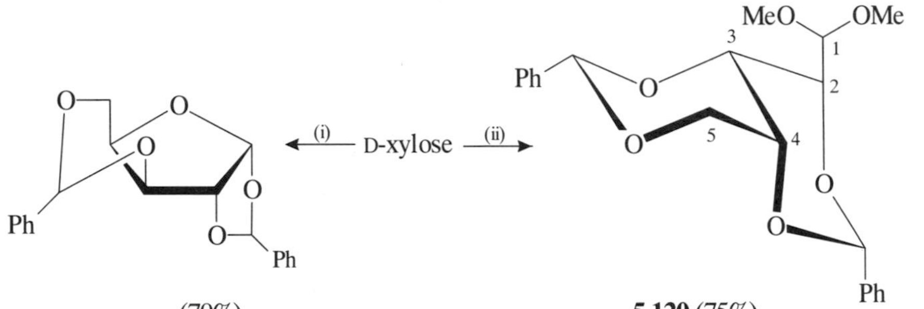

(70%)   **5.120** (75%)

(i) PhCHO, HCl; (ii) PhCHO, MeOH, HCl
Scheme 5.52

Notwithstanding the expectation that complex mixtures of products would be expected on treatment of sugars with mixtures of different carbonyl compounds, D-ribose with acetone and formaldehyde (one equivalent) in the presence of sulphuric acid gives 82% of the crystalline mixed acetal **5.121**.[143]

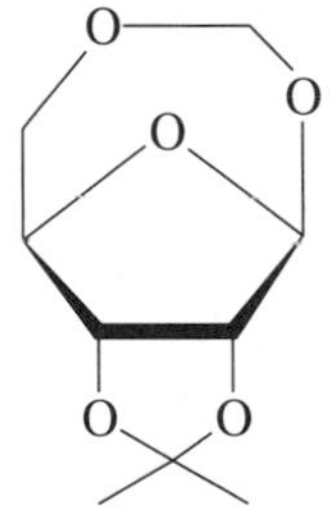

5.121

The situation is apparently somewhat simpler with ketones themselves, and in many cases good yields of pure products are obtained. D-Glucose and D-galactose, for example, yield the important 1,2:5,6-diacetal and 1,2:3,4-diacetal **5.122** and **5.123**, respectively, which in turn serve as ideal substances from which 3- and

6-substituted glucoses and galactoses can be prepared. Formation of the former compound is favoured because in the furanoid form glucose offers two suitable *vicinal* diols for condensation, and fusion of a five-membered acetal to a furanoid ring causes less strain than similar fusion to a pyranoid ring. With the latter sugar, however, the second factor is offset by the existence in the 1,2:5,6-di-*O*-isopropylidene derivative **5.124** of a strong interaction between the *endo*-5,6-acetal and the inside of the 'V'-shaped bicyclic ring system. This compound, however, is formed in small proportions together with the main product, and it has been elegantly characterized by separation by gas–liquid chromatography and the observation that its mass spectrum resembles closely that of the *gluco*- analogue **5.122**.

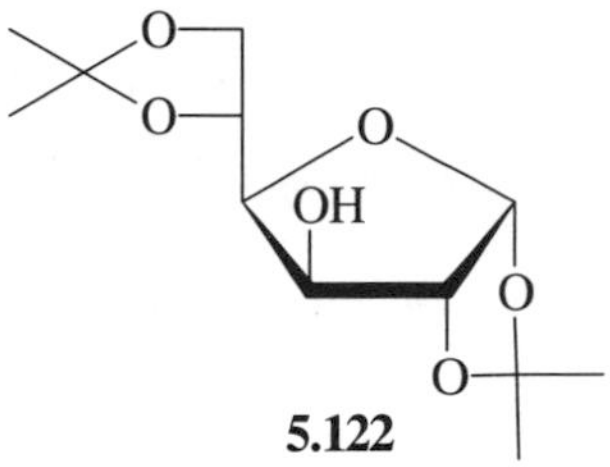

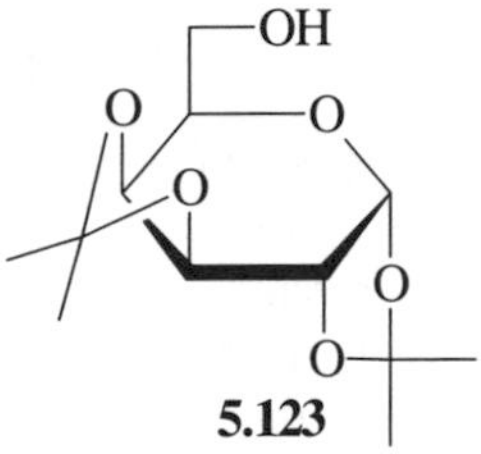

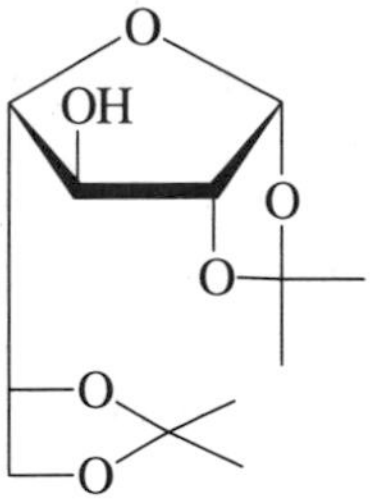

Likewise, in the D-glucose case there are several products formed in the normal acetone/sulphuric acid reaction apart from the diacetal **5.122**, which is normally obtained in a yield somewhat greater than 60%. Careful fractionation of the other products gives the septanose derivatives **5.125** and **5.126**.[144] Alteration in conditions can give rise to other products, as in the case when acetone is used in DMF in the presence of copper(II) sulphate and the 4,6-acetal and 5,6-acetals can be isolated.[145]

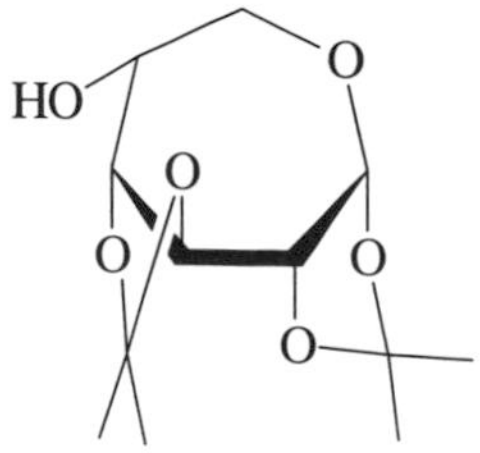

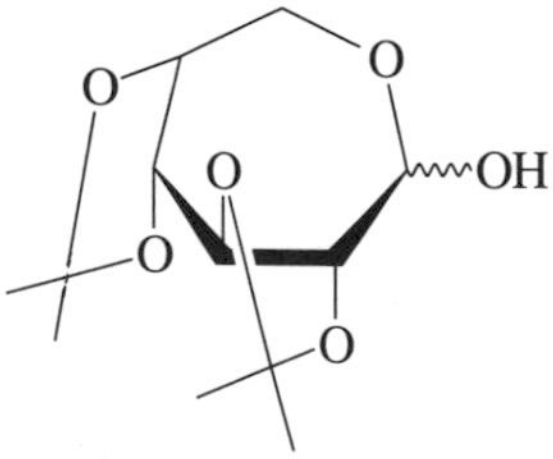

D-Xylose condenses smoothly with acetone and catalytic amounts of sulphuric acid to give the 1,2:3,5-diacetal **5.127** which is useful in the synthesis of partially substituted xyloses since it can be selectively hydrolysed to give the 1,2-derivative **5.128**, illustrating the relative stability of the five-membered acetal ring. Despite its glycosidic character, the same 1,2-ring system of 1,2:5,6-di-*O*-isopropylidene-*α*-D-glucofuranose **5.122** is also the more stable, and the compound can be smoothly hydrolysed under controlled conditions to the 1,2-acetal **5.129** which serves as a suitable starting material for the synthesis of many glucofuranose derivatives.

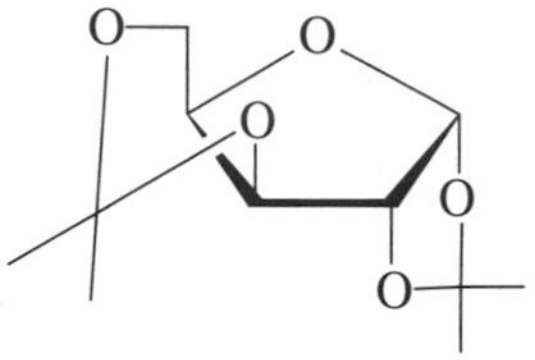

5.127 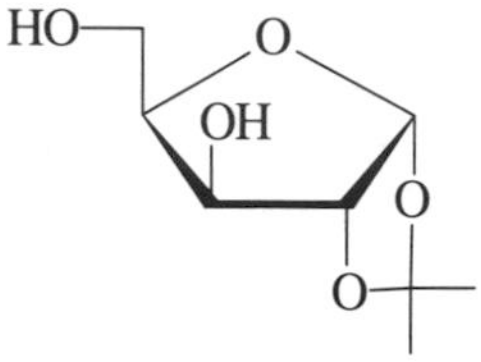   **5.128** 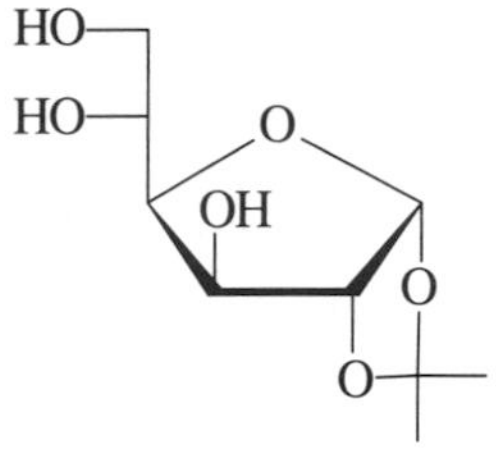   **5.129**

Analogous treatment of other sugars can be extremely complex. From the reaction of D-ribose in the presence of strong protic acids, for example, compounds **5.130**–**5.134** are formed, i.e. intramolecular and intermolecular anhydrides as well as acetals are produced. As might be readily predicted from the earlier considerations, the 2,3-acetal **5.130** is the predominant product since, especially in the *β*-modification, it has minimum *endo*-interactions.

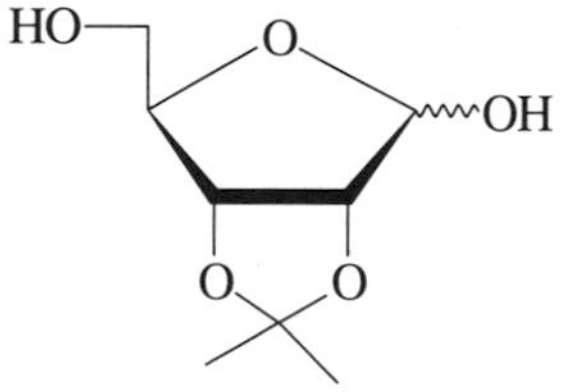

5.130 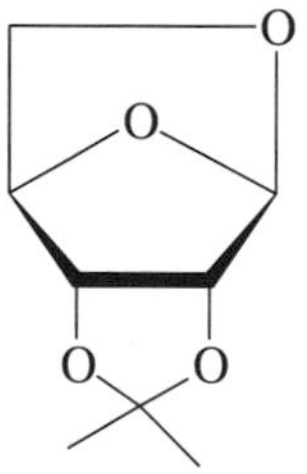   **5.131**

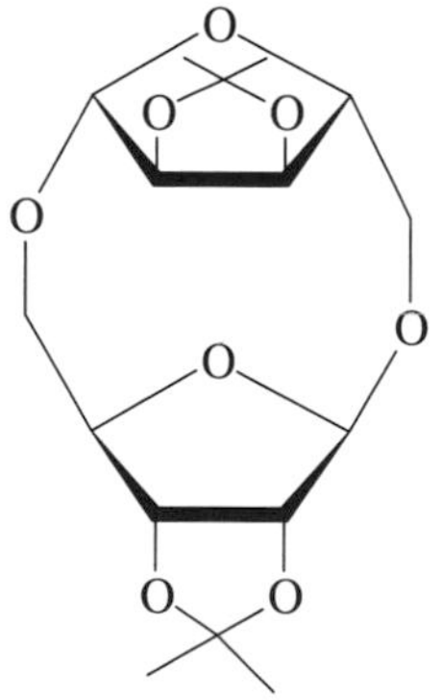

5.132 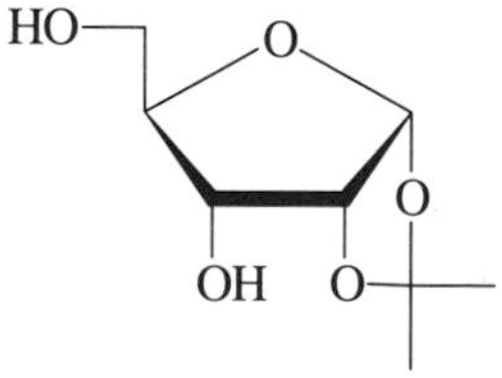   **5.133** 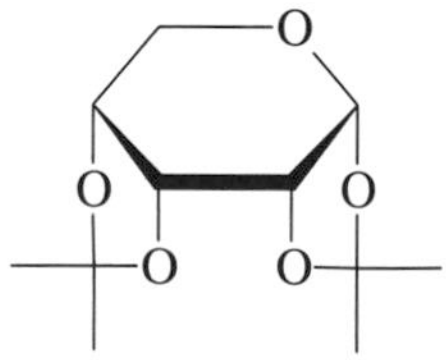   **5.134**

Although the acetonations of the pentuloses must undoubtedly be complex, they can be used as a means of isolating the free sugars from mixtures produced by pyridine-catalysed isomerizations of D-xylose and D-arabinose (Section 3.1.7.a). Crystalline 2,3-*O*-isopropylidene-β-D-*threo*-pentulofuranose **5.135** and 1,2:3,4-di-*O*-isopropylidene-β-D-*erythro*-pentulofuranose **5.136** can be isolated in the respective cases.[146]

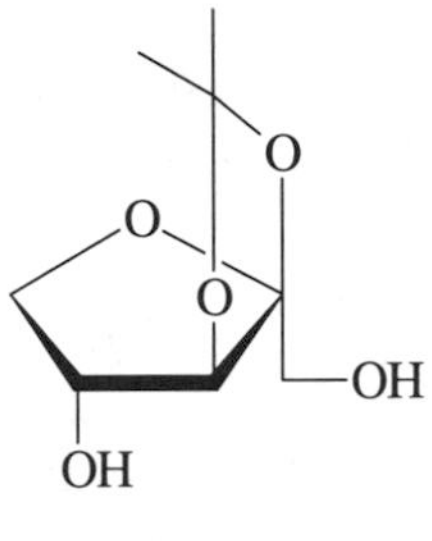

5.135 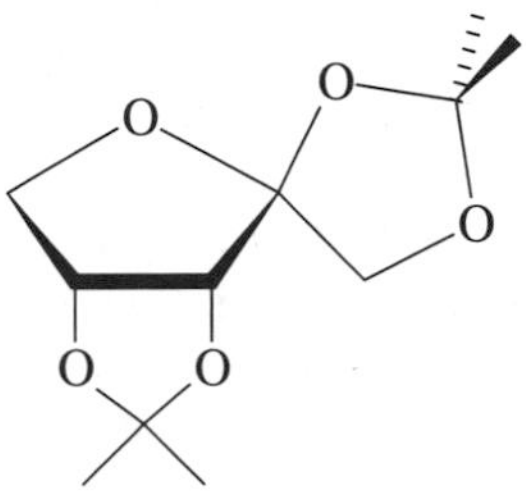   **5.136**

(c)   Acetals and ketals of alditols and other acyclic derivatives[136]

Carbonyl compounds condense with acyclic polyhydroxyl derivatives to yield a variety of products with different structures and with different degrees of substitution. Furthermore, this difficult subject is complicated by the fact that the acetals present at any stage in a condensation reaction may be either kinetically or thermodynamically controlled products. Thus, for example, the first product of condensation of D-glucitol with acetaldehyde in aqueous acid is the 2,3-cyclic acetal, but this then isomerizes to the 2,4-substituted compound. Nevertheless, the following generalizations have been developed relating to the thermodynamic products formed on condensation of acyclic polyhydroxyl compounds with aldehydes:[147]

(i)   Six-membered rings formed from secondary *erythro*-diols are most favoured, and are preferred to six-membered rings involving a primary hydroxyl group. (Such products allow the 1,3-dioxolane ring to adopt a chair conformation with the maximum number of large substituents equatorial; see Scheme 5.53)

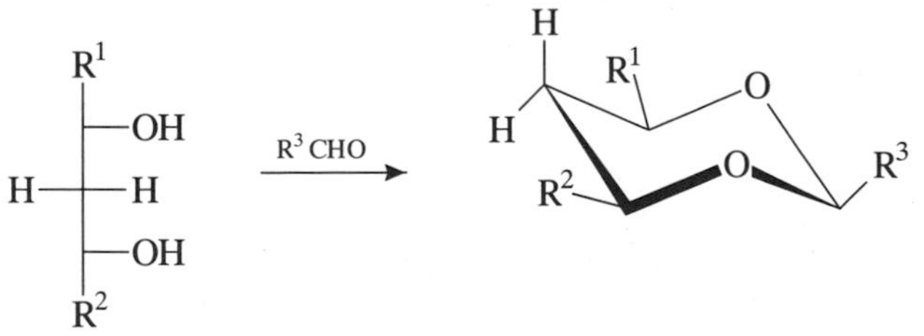

Scheme 5.53

(ii)   Less favoured are five-membered rings formed from terminal or *threo*-diols
       (conformationally *gauche*) and six- or seven-membered rings obtained by
       condensations from *threo*-diols. The relative stabilities of these ring systems
       vary with the aldehydes used.

(iii) Other ring systems are highly unfavoured.

The formation of the 2,4-acetal **5.137** from D-glucitol under thermodynamic
control is thus in keeping with these generalizations, and further substitution (using
benzaldehyde, for example) yields the 1,3:2,4-diacetal **5.138** followed by the fully
substituted 1,3:2,4:5,6-triacetal **5.139** (Scheme 5.54). Conversely, graded hydrolysis of the final product gives first the above diacetal followed by the 2,4-substituted
compound. Again, in keeping with these generalizations D-xylose, D-lyxose and D-
ribose dimethyl acetals **5.140a**–**5.142a** produce mainly the 2,4:3,5-diacetals, the
conformations of which are as shown (**5.140b**–**5.142b**), but D-arabinose dimethyl
acetal **5.143a** gives the 2,3:4,5-isomeric product **5.143c**.[148] In the last two cases
the generalizations are insufficiently precise to predict the products, but the observations are readily interpreted in conformational terms (Scheme 5.55).

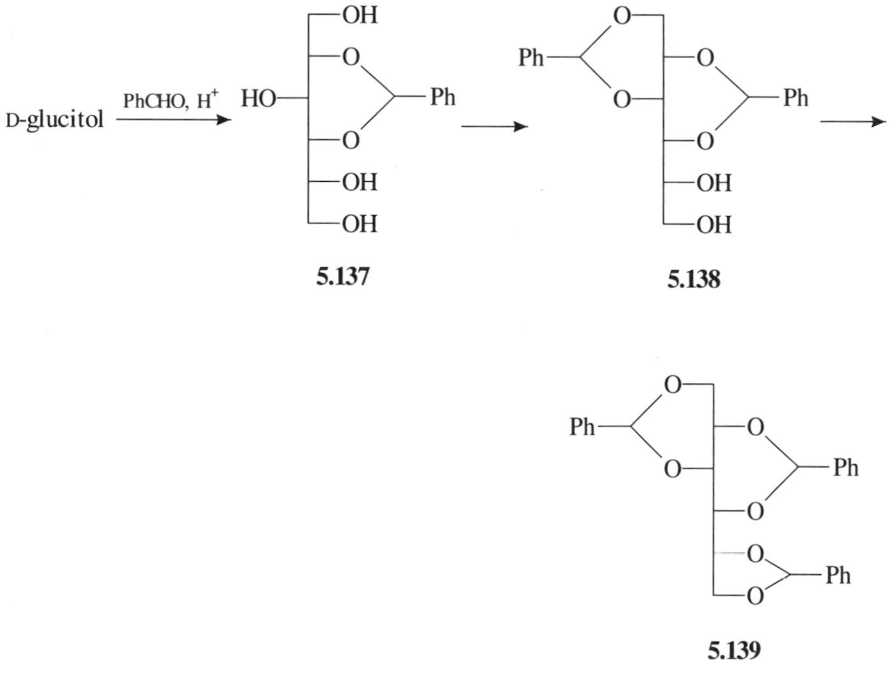

5.137   **5.138**

5.139

Scheme 5.54

Whereas the 2,4:3,5-diacetals formed from the first three pentose derivatives can
adopt ring conformations devoid of strong interactions as shown, the arabinose
analogue **5.143b** is held in the *trans*-decalin type of ring system in which the
C-1 group is necessarily axial, and the molecule is consequently strained. The

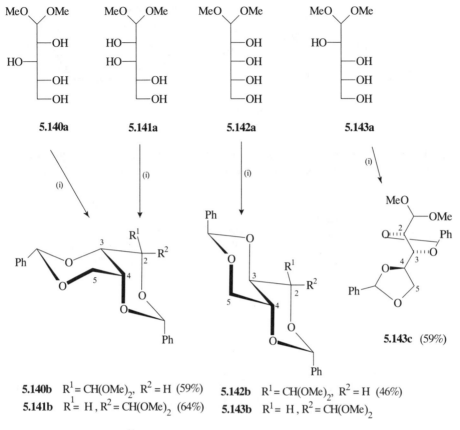

5.140a   **5.141a**   **5.142a**   **5.143a**

5.143c (59%)

5.140b   R¹ = CH(OMe)₂, R² = H (59%)   **5.142b**   R¹ = CH(OMe)₂, R² = H (46%)
5.141b   R¹ = H , R² = CH(OMe)₂ (64%)   **5.143b**   R¹ = H , R² = CH(OMe)₂

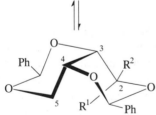

(i) PhCHO, HCl

Scheme 5.55

isomer containing five-membered rings **5.143c** has least possible interactions for such a bicyclic system and therefore it is preferred. It is noteworthy that under the conditions used the dimethyl acetal groups remain unreacted in all these cases.

With ketones, five-membered rings are favoured, and more stable products are obtained from *threo-* rather than *erythro-α*-diols since in the former case the ring substituents are *trans*-related (Scheme 5.56). The product of acetonation of D-xylose diethyl dithioacetal, therefore, is the 2,3:4,5-di-*O*-isopropylidene derivative

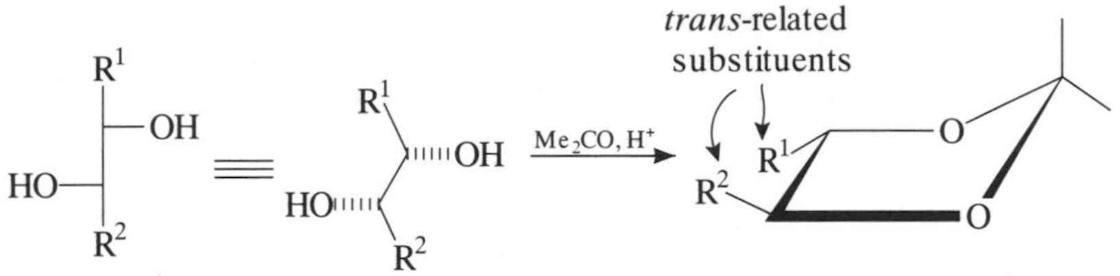

Scheme 5.56

5.144, but the finding of the 3,4-acetal and 2,4-acetal amongst the products of partial substitution reveals that the reaction is more complex than might appear.

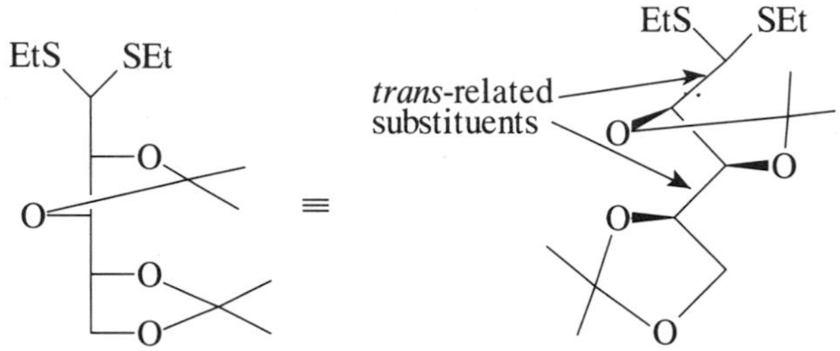

5.144

D-Glucitol and D-mannitol yield 1,2:3,4:5,6-tri-O-isopropylidene derivatives **5.145**, and **5.146**, whereas D-galactitol, which has an *erythro*-3,4-diol, gives only the 2,3:4,5-diacetal and 2,3:5,6-diacetals **5.147**, and **5.148**.

5.4.4   REACTIONS OF CYCLIC ACETALS[149]

All acetal rings are susceptible to acid-catalysed hydrolysis, the rates of the cleavages being dependent upon the stabilities of the ring systems and upon the nature of the substituents at the acetal centres. The importance of the steric factor is illustrated by the observation that rings which are difficult to form, as for example from 1,3-*threo*-diols on acyclic molecules, are readily cleaved. The significance of the electronic factor is shown by the finding that whereas fluorinated isopropylidene acetals are exceptionally stable to acids, the electron-releasing p-methoxy group in anisylidene acetals renders them significantly more labile than benzylidene analogues; this can have value in their application as protecting groups. A useful reagent for the selective removal of isopropylidene and benzylidene acetals from compounds containing halo, benzoyl, sulphonyl and amino groups is 90% aqueous trifluoroacetic acid.

Acetals are stable under a wide range of conditions used for the substitution of hydroxyl groups, but they may migrate, particularly in acid conditions. Thus, for example, 4,5-O-isopropylidene-D-xylose dimethyl dithioacetal isomerizes in acid to the 3,4-substituted compound which gives the 2,3:4,5-diacetal on

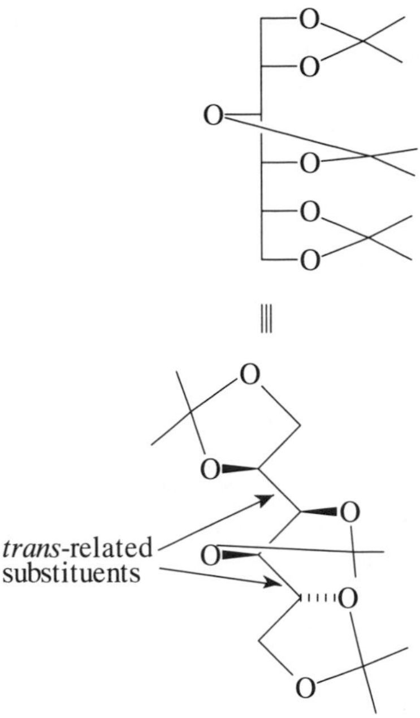

5.145

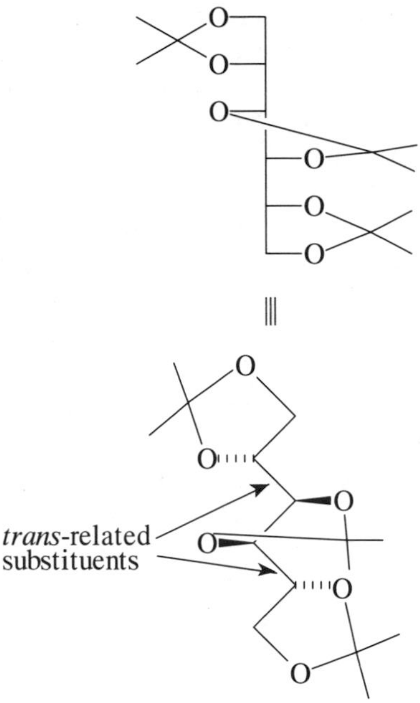

5.146

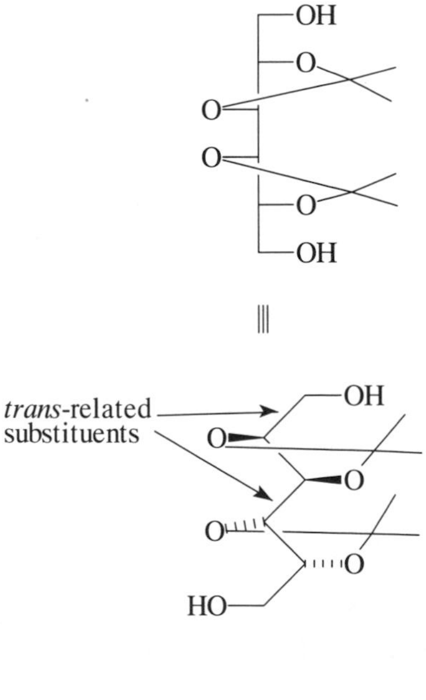

5.147

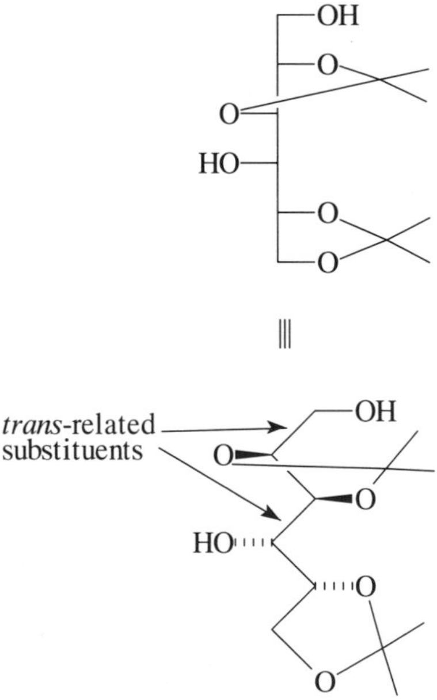

5.148

treatment with acetone in the presence of zinc chloride. Less expected are the migrations that occur when 1,2:5,6-di-O-isopropylidene-α-D-glucofuranose is treated with various reagents based on triphenylphosphine and halogen donors, which often give 6-deoxy-6-halo-1,2:3,5-di-O-isopropylidene-D-glucoses rather than the 3-deoxy-3-halo derivatives (Section 4.5.1.e).

Benzylidene rings can be cleaved hydrogenolytically (cf. Section 5.2.2), five-membered systems being more vulnerable than six-membered analogues, and thus, for example, hydrogenolysis using hydrogen transfer from ammonium formate allows the formation of methyl 4,6-O-benzylidene-α-D-mannopyranoside in 86% yield from the 2,3:4,6-di-O-benzylidene compound.[150]

The selective opening of benzylidene acetals is now an important route to differentially substituted compounds, reduction, for example, giving monobenzyl ethers. In the case of hexopyranoside 4,6-O-benzylidene compounds, reaction with aluminium chloride and lithium aluminium hydride[151] or diphenylboron bromide[152] results in excellent selectivity, and 4-benzyl ethers of the 6-alcohols are obtainable in high yield. On the other hand, if sodium cyanoborohydride is used in THF with dry HCl as catalyst,[153] ring opening occurs in the alternative direction to give the ethers at O-6. In this way, therefore, excellent methods are provided for preparing derivatives, including oligosaccharides substituted at O-4 and O-6, respectively. With the relatively labile p-methoxybenzylidene analogues, sodium cyanoborohydride can be used to cleave the acetal rings selectively in either direction, and even in the presence of benzyl ether groups (Scheme 5.57). With acid, protonation occurs at O-4 and leads to the 6-ether,[154] whereas trimethylsilyl chloride is more sterically demanding as an electrophile and ring opening occurs by cleavage of the O-6 acetal bond to give the 6-silyl ether which is hydrolysed during the isolation procedure.

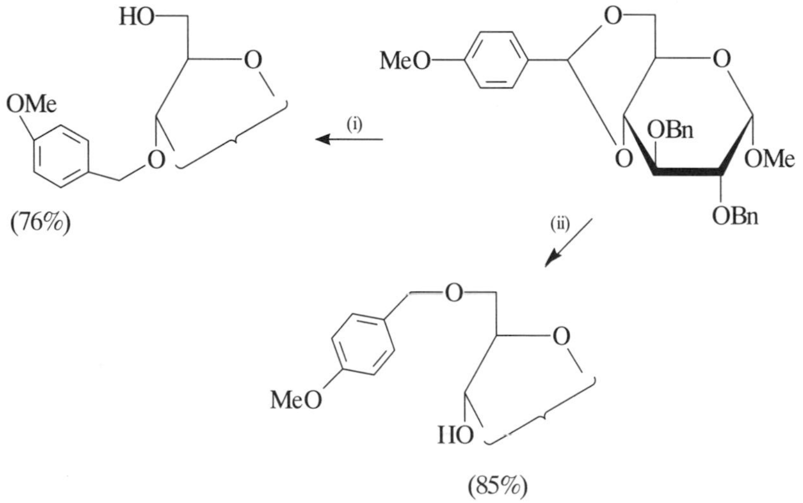

(76%)

(85%)

(i) NaCNBH$_3$, Me$_3$SiCl, MeCN; (ii) NaCNBH$_3$, DMF, CF$_3$CO$_2$H

Scheme 5.57

The direction of ring opening of five-membered benzylidene rings depends upon the stereochemistry at the acetal centre. Thus, with sodium cyanoborohydride and hydrogen chloride the (S)-isomer and (R)-isomer of methyl 4-O-benzyl-2,3-O-benzylidene-α-L-rhamnopyranoside, which differ in stereochemistry at the acetal centre of the five-membered ring, give the 3,4- and 2,4-dibenzyl products in 95 and 88% yield, respectively.[153]

Alternatively, oxidative ring opening of benzylidene acetals can be carried out with NBS to give 6-bromo-6-deoxy-4-benzoates of hexopyranosides with good regioselectivity (Section 4.5.1.c), and ozone,[155] t-butyl hydroperoxide[156] and trityl fluoroborate[157] all result in ring cleavage to give hydroxybenzoates, frequently, however, with no strong regioselectivity.

In a modified approach, O-nitrobenzylidene acetals are photochemically opened to produce hydroxyl O-nitrosobenzoates which, on oxidation, give the hydroxy O-nitrobenzoates in good yield and sometimes with high regioselectivity. While 4,6-acetals of hexopyranosides are converted into mixed products (glucoside and mannoside acetals give the 6-ester and 4-ester in the ratio 7:3, whereas this is reversed with the galactosides),[158] methyl 2-O-acetyl-3,4-O-(o-nitrobenzylidene)-α-L-fucopyranoside reacts with effective regiospecificity to give the 4-nitrobenzo-ate.[159] Glucosylation of the acetals **5.149** results in a disaccharide derivative which, when irradiated in methanol with a medium pressure arc lamp followed by oxidation with trifluoroperacetic acid, gives the corresponding nitrobenzoate that on further glucosylation affords mainly the derivative **5.150** of 2,3-di-O-glucopyranosyl-L-fucose (Scheme 5.58).[160]

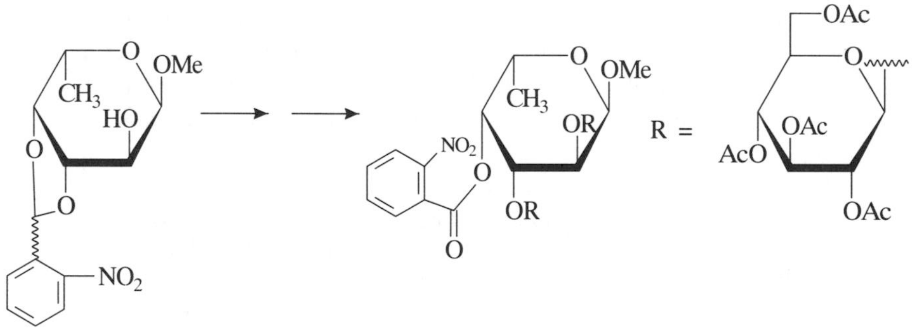

5.149   **5.150** (60%)

Scheme 5.58

On occasion, acetals can react with Grignard reagents. In benzene/ether solution, for example, methylmagnesium iodide reacts nucleophilically at the acetal centre of the 5,6-dioxolane ring of **5.151** to give the 5-ols **5.152** (Scheme 5.59). Under these conditions vulnerable isopropylidene acetals afford O-t-butyl monoalcohols.[161] Butyllithium can cause ring opening of some benzylidene acetals with loss of benzaldehyde and formation of lithium enolates. (Section 4.9.1.a.ii).

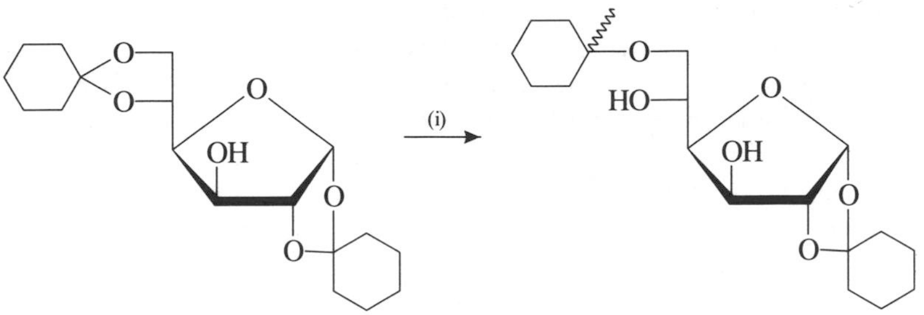

5.151   **5.152** (67%)

(i) MeMgI, benzene, ether, reflux, 1.25 h

Scheme 5.59

5.5   OXIDATIVE CLEAVAGE OF α-DIOLS BY WAY OF CYCLIC INTERMEDIATES[162]

5.5.1   GENERAL

Reaction of α-diols with periodic acid H_5IO_6 or its salts gives rise to cyclic ester intermediates which cleave in a two-electron oxidation process according to Scheme 5.60 to give two carbonyl products and iodate(V). Because of the wide range of its applications and the precision with which it can be used, this is one of the most valuable reactions in carbohydrate chemistry.[163] General examples of diol cleavages encountered with sugar derivatives are illustrated in Scheme 5.61. Since the variables in the oxidation reactions, i.e. the reduction of periodate and the formation of formic acid and formaldehyde or acetaldehyde, can all be determined accurately on a very small scale,[†] periodate oxidation is invaluable as an analytical technique, but in addition it has often been utilized in preparative work.

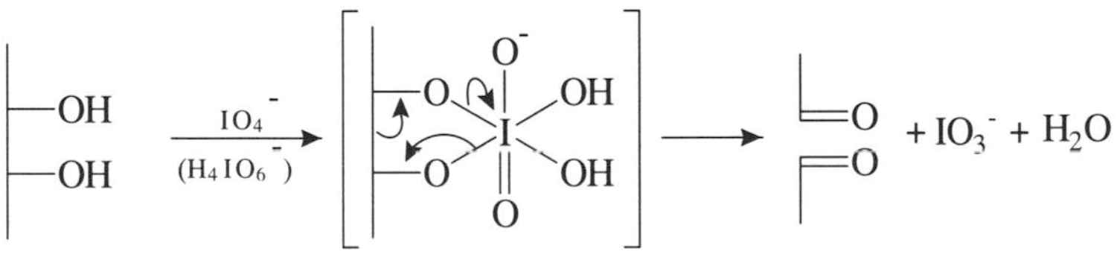

Scheme 5.60

Although aldoses are oxidized ultimately to give formaldehyde from the terminal hydroxymethyl groups, and formic acid from each of the other carbon atoms, interesting reaction intermediates have been recognized. Glucose, at slightly acid pH values (3–5), thus gives 2-*O*-formyl-D-glyceraldehyde **5.154** and the 2,3-diester **5.156** by way of 4-*O*-formyl-D-arabinose **5.153** and 3,4-di-*O*-formyl-D-erythrose

---

[†] Periodate concentrations can be followed using the 223 nm absorption of the ion, formaldehyde by a photometric method based on the colour it produces with chromotropic acid, and formic acid, similarly, using 2-thiobarbituric acid.

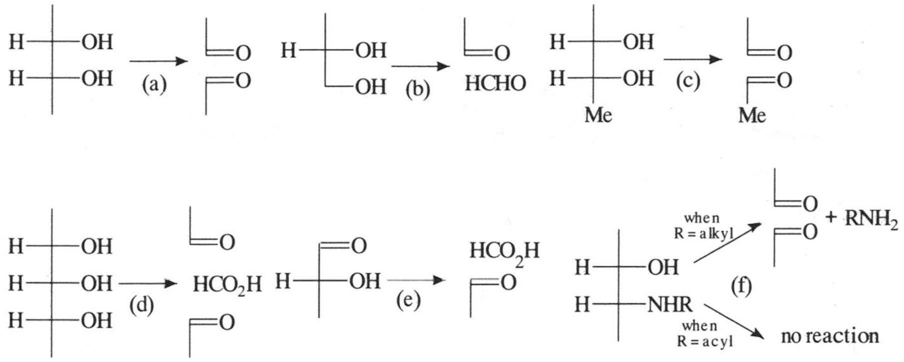

Scheme 5.61

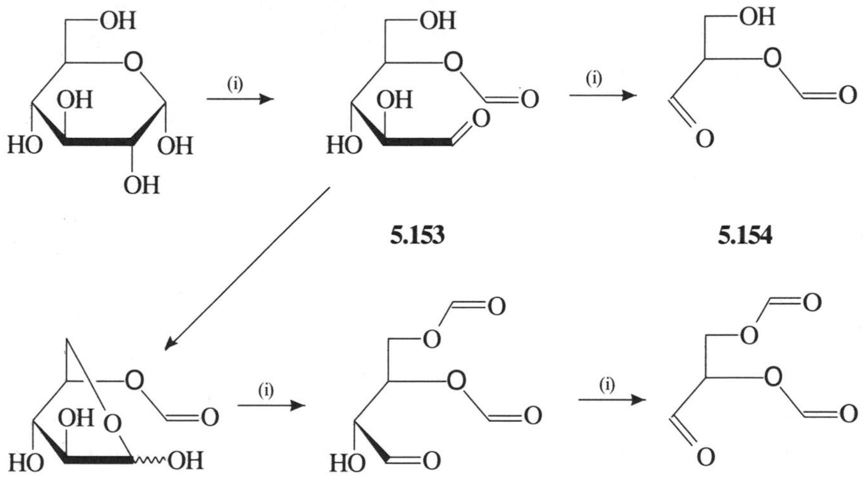

5.153   **5.154**

5.155   **5.156**

(i) IO_4^- (aq.) pH 3–5

Scheme 5.62

5.155 as shown in Scheme 5.62. This indicates that the sugar is oxidized initially at the 1,2-diol site of the α-pyranose modification. At other pH values, hydrolysis of these esters is rapid, and the oxidations proceed to completion. In a practical application of this reaction, 2-*O*-methyl-D-arabinose **5.159** can be synthesized from 3-*O*-methyl-D-glucose **5.157** — easily obtainable from the diacetal **5.122** — by way of the 4-*O*-formyl derivative **5.158** (Scheme 5.63).[164]

Steric factors very markedly control the rates at which α-diols undergo oxidative cleavage; *cis*-systems on furanoid and pyranoid rings are more readily brought into conformations suitable for complex formation, and consequently oxidize more readily than the related *trans*-systems. α-D-Glucopyranose therefore oxidizes more

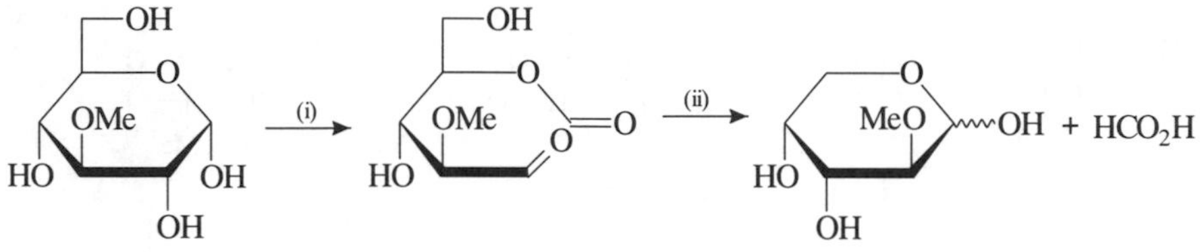

5.157   **5.158**   **5.159** (85%)

(i) IO_4^- (aqueous), pH 4; (ii) H_2O

Scheme 5.63

rapidly than the β-anomer (cf. the reaction of these isomers with borate ions; Section 5.3.10). When certain *trans*-diols are rigidly held by fused ring systems, their reactivity is suppressed appreciably so that, for example, 1,6-anhydro-β-D-glucofuranose **5.160** and methyl 4,6-*O*-benzylidene-α-D-altropyranoside **5.161** are not oxidized under conditions usually employed for such reactions (dilute aqueous solutions, pH 3–6, room temperature).

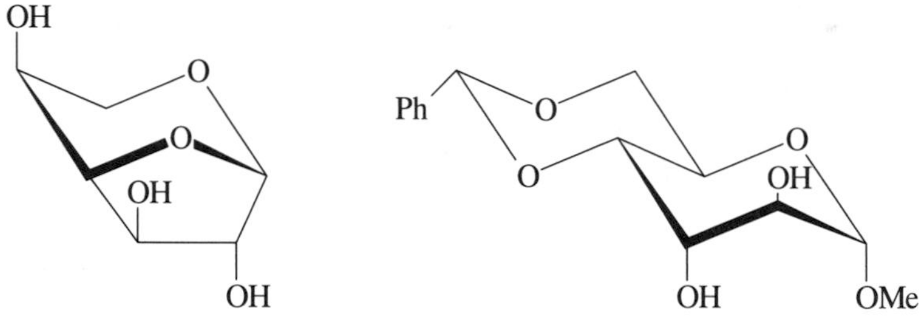

5.160   **5.161**

Further instances of α-diol-containing molecules which do not reduce the periodate ion smoothly under normal conditions are methyl β-D-ribopyranoside and 1,2-*O*-isopropylidene-α-D-glucofuranose which form stable tridentate complexes with the reagent. Such a reaction may also be used to diagnose the presence of 1,2,3-*cis-cis*-triols on pyranoid rings (cf. Section 5.3.10).

In acyclic compounds, α-diols with the *threo*-configuration oxidize more readily than the *erythro*-analogues (cf. Scheme 5.56).

5.5.2   EXAMPLES OF THE USE OF DIOL CLEAVAGE

The following examples illustrate the versatility of the diol cleavage reaction and its wide use in carbohydrate chemistry.

(a)   The determination of ring size and anomeric and absolute configurations of hexosides

On oxidation, aldohexopyranosides liberate formic acid (from C-3) while furanosides give formaldehyde (from C-6) (Schemes 5.64a and 5.64b), and the residual

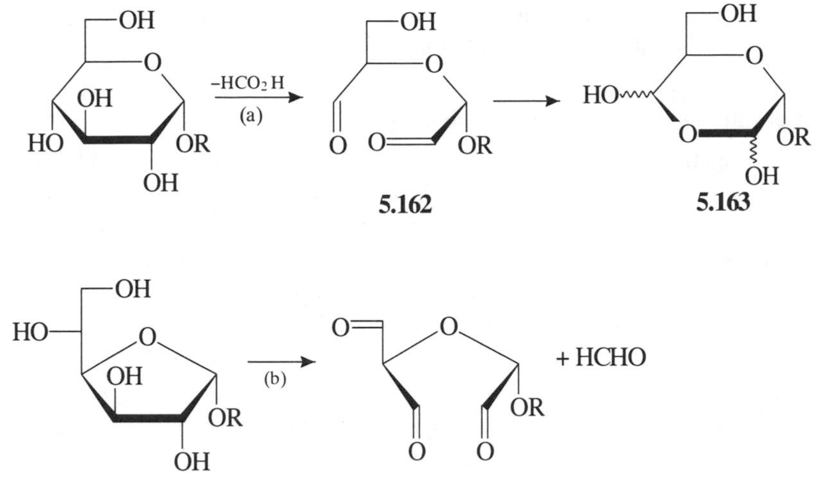

Scheme 5.64

fragments can be characterized by comparison with those similarly obtained from other glycosides. In this way, for example, all α-D-hexopyranosides (or α-D-pento-furanosides) with the same aglycon (R) yield the same product **5.162**, which exists in the hydrated cyclic form **5.163**, so an unknown compound can be degraded to yield the same fragment as is obtained from other related substances of known configurations at C-1 and C-5.

By use of this method, the anomeric configurations of naturally occurring ribo-furanosyl nucleosides (e.g. **5.166**, see also Section 8.2.2.a) have been assigned since they give the same oxidation products **5.165** as are obtained from corresponding synthetic pyranoid derivatives **5.164**, which are readily characterized by n.m.r. methods (Scheme 5.65).

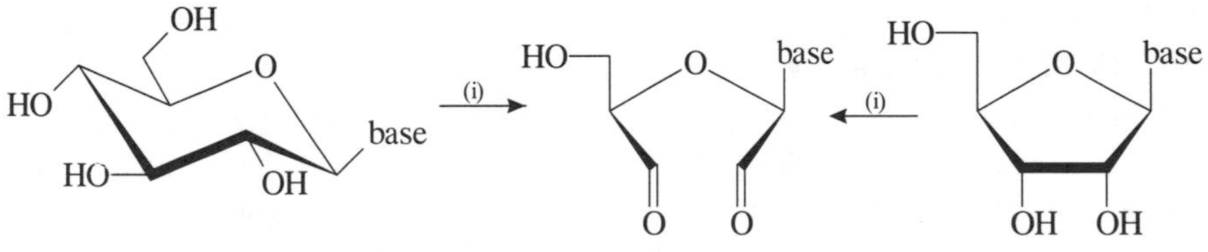

(i) IO$_4^-$ (aqueous)

Scheme 5.65

(b)   The analysis of mixtures

From the above example it follows that a mixture of glycosides of different ring sizes can be analysed by this method. Similarly, a mixture of 2-O-methylxylose and 3-O-methylxylose can be analysed by converting them to their pyranoid glycosides

(Section 3.1.1.a) and determining their abilities to reduce the periodate ion, such analyses being possible by spectrophotometry on the microscale.

(c)   Specific degradations of carbohydrates

In the investigation of the biosynthesis of ascorbic acid (Section 4.9.2) in strawberry plants, D-glucose- and D-galactose-1-^{14}C were found to be converted to the vitamin by different pathways. Whereas glucose is incorporated without rearrangement of the carbon skeleton, galactose gives a product with as much radiocarbon at C-6 as at C-1. For work of this type, methods for specifically fragmenting sugar derivatives are required, and amongst them periodate oxidation is most prominent. In the degradation of ascorbic acid the enediol is oxidized with sodium hypoidite to give oxalic acid and L-threonic acid which on further oxidation with periodate give fragments (Scheme 5.66) which may be independently subjected to radiochemical assay. From this experiment the ^{14}C contents of C-1 and C-2, C-3, C-4 and C-5 and C-6 can be determined.

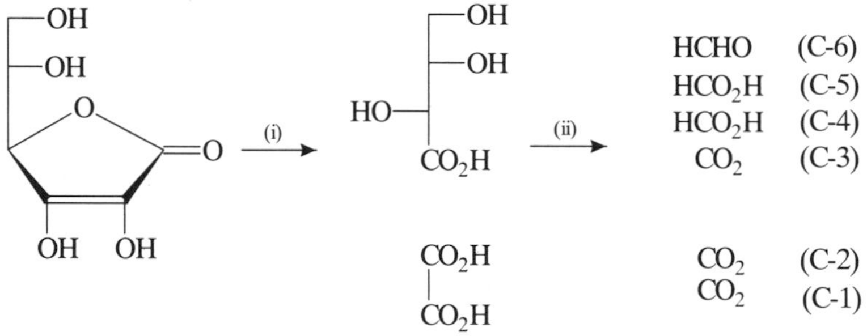

(i) NaOI; (ii) IO$_4^-$

Scheme 5.66

(d)   Configurational analysis

Addition of hydrogen cyanide to the pentuloses (cf. Section 3.1.4.d) followed by hydrolysis of the cyanohydrins gives rise to pairs of branched-chain aldonic acids, the lactones of which were configurationally characterized by observing their relative rates of reaction with periodate. Details are given in Section 4.8.1.

(e)   Synthesis of short-chain sugars

Several of the short-chain carbohydrates which are commonly used in chiral syntheses (Section 7.2) are best prepared from the more readily available higher sugars by oxidative degradation procedures. D-Erythrose, for example, is obtained most suitably by periodate oxidation of 4,6-O-ethylidene-D-glucose followed by hydrolysis of the acetal ring, and D-threose can similarly be prepared from 1,3-O-benzylidene-D-arabinitol.

Lead tetraacetate, which can also be used to cleave α-diols, tends to be more stereoselective than the periodate ion; it cannot be used, however, in aqueous solution.[165] In acetic acid, D-glucose is oxidized with this reagent in the α-furanose form to give 2,3-di-*O*-formyl-D-erythrose **5.167** (Scheme 5.67) from which, by hydrolysis, D-erythrose can also be prepared.

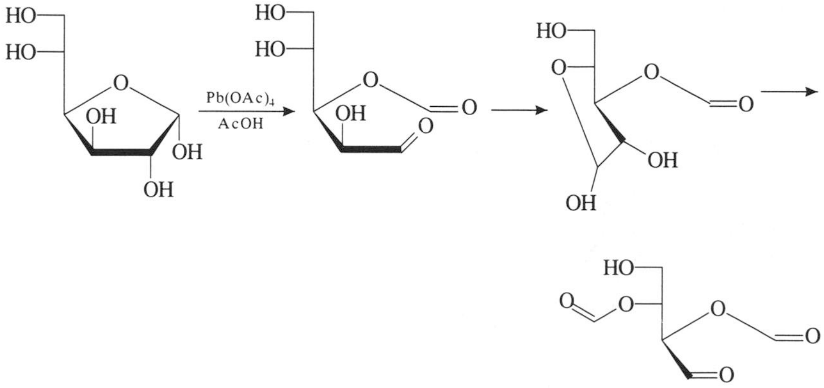

5.167

Scheme 5.67

(f)   Syntheses from periodate oxidation products

Although products of the oxidation of glycosides exist in cyclic modifications (cf. **5.163**) they react as dialdehydes and can be used in the syntheses of further cyclic products. Examples of the use of these derivatives given in Sections 4.3.1.c and 4.3.1.d illustrate their value in the synthesis of amino sugars.

5.6   REFERENCES

1. T. W. Greene and P. G. M. Wuts, *Protective Groups in Organic Synthesis*, 2nd Edn, Wiley-Interscience, New York, 1991.
2. *Adv. Carbohydr. Chem. Biochem.*, 1976, **33**, 11.
3. *Adv. Carbohydr. Chem.*, 1960, **15**, 11.
4. *Carbohydr. Res.*, 1970, **12**, 421.
5. *Angew. Chem., Int. Ed. Engl.*, 1986, **25**, 212.
6. *Tetrahedron Lett.*, 1984, **25**, 821.
7. *J. Am. Chem. Soc.*, 1960, **82**, 4608.
8. *Carbohydr. Res.*, 1973, **28**, 61.
9. *J. Chem. Soc.*, 1957, 2271.
10. *Carbohydr. Res.*, 1970, **12**, 463; *Carbohydr. Res.*, 1971, **16**, 495.
11. *Synthesis*, 1972, 83.
12. *Tetrahedron*, 1985, **41**, 643.
13. *J. Org. Chem.*, 1976, **41**, 1832.
14. *Adv. Carbohydr. Chem.*, 1950–55, **5**–**10**; 1958, **13**.
15. *J. Biochem.* (Tokyo), 1964, **55**, 205.

16. *Carbohydr. Res.*, 1975, **44**, C5.
17. *Aust. J. Chem.*, 1978, **31**, 1031.
18. *Carbohydr. Res.*, 1976, **48**, 271.
19. *Methods Carbohydr. Chem.*, 1972, **6**, 365.
20. *Chem. Pharm. Bull.*, 1970, **18**, 677.
21. *Carbohydr. Res.*, 1970, **12**, 147.
22. *Synthesis*, 1985, 274.
23. *Tetrahedron Lett.*, 1971, 1935.
24. *J. Chem. Soc. C*, 1967, 1182.
25. *Carbohydr. Res.*, 1969, **9**, 287.
26. *Adv. Carbohydr. Chem.*, 1957, **12**, 137.
27. *Tetrahedron Lett.*, 1967, 1663.
28. *Chem. Rev.*, 1985, **85**, 129; *Tetrahedron Lett.*, 1986, **27**, 2497.
29. *Carbohydr. Res.*, 1974, **34**, 208.
30. *J. Org. Chem.*, 1990, **55**, 378.
31. *Tetrahedron Lett.*, 1983, **24**, 3829.
32. *Synthesis*, 1985, 1123.
33. *J. Org. Chem.*, 1975, **40**, 1356.
34. *J. Chem. Soc., Perkin Trans. 1*, 1985, 2247.
35. *Tetrahedron Lett.*, 1989, **30**, 4759.
36. *Carbohydr. Res.*, 1976, **50**, C12.
37. *Carbohydr. Res.*, 1984, **133**, 147.
38. *Carbohydr. Res.*, 1980, **85**, 313.
39. *Tetrahedron*, 1986, **42**, 3021.
40. *Adv. Carbohydr. Chem.*, 1948, **3**, 79.
41. *Tetrahedron Lett.*, 1975, 3055.
42. *J. Org. Chem.*, 1967, **32**, 2544.
43. *Carbohydr. Res.*, 1979, **76**, 252; *Carbohydr. Res.*, 1980, **85**, 209.
44. *Tetrahedron Lett.*, 1975, 2081.
45. T.W. Greene and P.G.W. Wuts. *Protective Groups in Organic Synthesis*, 2nd Edn, Wiley-Interscience, New York, 1991, p. 68.
46. *Synthesis*, 1985, 817.
47. *Adv. Carbohydr. Chem. Biochem.*, 1973, **28**, 11.
48. *Tetrahedron*, 1962, **18**, 1143.
49. *Carbohydr. Res.*, 1968, **7**, 180.
50. *J. Am. Chem. Soc.*, 1972, **94**, 6190.
51. *Aust. J. Chem.*, 1977, **30**, 639.
52. *Helv. Chim. Acta*, 1978, **61**, 1832.
53. *Tetrahedron Lett.*, 1988, **29**, 6161.
54. *Tetrahedron Lett.*, 1992, **33**, 1177.
55. *Tetrahedron Lett.*, 1992, **33**, 2289.
56. *Angew. Chem., Int. Ed. Engl.*, 1990, **29**, 431.
57. *Tetrahedron Lett.*, 1988, **29**, 1561.
58. *J. Chem. Soc. C*, 1968, 1903.
59. *J. Chem. Soc., Perkin Trans. 1*, 1976, 1395.
60. *J. Chem. Soc., Perkin Trans. 1*, 1980, 738.
61. *Tetrahedron Lett.*, 1989, **30**, 4669.
62. *J. Am. Chem. Soc.*, 1975, **97**, 4069.
63. *J. Org. Chem.*, 1986, **51**, 1906.
64. *The Carbohydrates* (Eds W. Pigman and D. Horton), Vol. 1A, Academic Press, New York, 1972, p. 217.
65. *Ber. dt. Chem. Ges.*, 1933, **66**, 1251; *J. Am. Chem. Soc.*, 1937, **59**, 1467.
66. *Ber.*, 1977, **110**, 2911.

67. *Tetrahedron Lett.*, 1977, 1691.
68. *J. Org. Chem.*, 1986, **51**, 727; *J. Carbohydr. Chem.*, 1985, **4**, 215.
69. *Tetrahedron Lett.*, 1987, **28**, 2299.
70. *Tetrahedron Lett.*, 1993, **34**, 1359.
71. *J. Am. Chem. Soc.*, 1952, **74**, 1498.
72. *J. Org. Chem.*, 1959, **24**, 1388.
73. *J. Chem. Soc.*, 1931, 2858.
74. *Acta Chem. Scand.*, 1957, **11**, 1788.
75. *J. Chem. Soc., Chem. Commun.*, 1977, 927.
76. *J. Org. Chem.*, 1992, **57**, 3431.
77. *Pure Appl. Chem.*, 1984, **56**, 1031.
78. *Carbohydr. Res.*, 1970, **15**, 263.
79. *Carbohydr. Res.*, 1976, **46**, 201.
80. *J. Chem. Soc.*, 1953, 735.
81. *Carbohydr. Res.*, 1989, **189**, 135.
82. *J. Food Sci.*, 1990, **55**, 236.
83. *Synthesis*, 1991, 499.
84. *J. Am. Chem. Soc.*, 1986, **108**, 5638.
85. *J. Am. Chem. Soc.*, 1987, **109**, 3977.
86. *J. Org. Chem.*, 1990, **55**, 4187.
87. *Tetrahedron*, 1989, **45**, 7077.
88. *Tetrahedron*, 1967, **23**, 2315; *Carbohydr. Res.*, 1976, **49**, 289.
89. *J. Am. Chem. Soc.*, 1955, **77**, 5337.
90. *Proc. Chem. Soc.*, 1964, 399.
91. *Can. J. Chem.*, 1970, **48**, 1754; *Tetrahedron*, 1975, **31**, 2463.
92. *Adv. Carbohydr. Chem.*, 1960, **15**, 91.
93. *Chem. Ber.*, 1967, **100**, 3225.
94. *J. Chem. Soc.*, 1964, 5443.
95. *J. Chem. Soc., Perkin Trans. 1*, 1979, 2378.
96. *Carbohydr. Res.*, 1966, **2**, 349.
97. *J. Am. Chem. Soc.*, 1981, **103**, 932.
98. *J. Chem. Soc., Perkin Trans. 1*, 1975, 1773; *J. Chem. Soc., Perkin Trans. 1*, 1977, 1718.
99. *Can. J. Chem.*, 1964, **42**, 2560.
100. *Adv. Carbohydr. Chem.*, 1968, **23**, 233.
101. *Adv. Carbohydr. Chem. Biochem.*, 1969, **24**, 139.
102. *J. Org. Chem.*, 1980, **45**, 4387.
103. *Tetrahedron Lett.*, 1981, **22**, 3579.
104. *Carbohydr. Res.*, 1969, **10**, 395.
105. *Adv. Carbohydr. Chem.*, 1953, **8**, 107; *Carbohydr. Res.*, 1968, **6**, 503.
106. *J. Org. Chem.*, 1989, **54**, 3577.
107. *J. Org. Chem.*, 1988, **53**, 3386.
108. *Carbohydr. Res.*, 1965, **1**, 128.
109. *J. Org. Chem.*, 1967, **32**, 1643.
110. *J. Org. Chem.*, 1986, **51**, 1525.
111. *J. Am. Chem. Soc.*, 1964, **86**, 4469.
112. *Adv. Carbohydr. Chem.*, 1965, **20**, 183.
113. *J. Chem. Soc.*, 1963, 5459.
114. *Carbohydr. Res.*, 1988, **174**, 253.
115. *Can. J. Chem.*, 1963, **41**, 1151.
116. *Can. J. Chem.*, 1960, **38**, 1122.
117. *Can. J. Chem.*, 1962, **40**, 1408.
118. *Chem. Soc. Rev.*, 1985, **14**, 357; *Carbohydr. Res.*, 1987, **162**, 53.

119. *Carbohydrates as Organic Raw Materials* (Ed. F. W. Lichtenthaler), VCH, Wein-heim, 1991, p. 17; *Adv. Carbohydr. Chem. Biochem.*, 1987, **45**, 199.
120. *J. Chem. Soc.*, 1964, 2735.
121. *Synthesis*, 1992, 1035.
122. *Tetrahedron Lett.*, 1985, **26**, 6343.
123. *J. Carbohydr. Chem.*, 1992, **11**, 837.
124. *The Carbohydrates* (eds W. Pigman and D. Horton), Vol. IA, Academic Press, New York, 1972, p. 253; *Q. Rev. Chem. Soc.*, 1957, **11**, 61.
125. *Carbohydr. Res.*, 1978, **64**, 297.
126. *Biochem. J.*, 1968, **109**, 597.
127. *Tetrahedron*, 1990, **46**, 4995.
128. *Tetrahedron Lett.*, 1988, **29**, 5763.
129. *Tetrahedron Lett.*, 1987, **27**, 6271; *Biochem. J.*, 1987, **246**, 771.
130. *Adv. Carbohydr. Chem.*, 1957, **12**, 117.
131. *Adv. Carbohydr. Chem. Biochem.*, 1978, **35**, 31.
132. *Liebigs Ann. Chem.*, 1990, 807.
133. *Angew. Chem., Int. Ed. Engl.*, 1985, **24**, 519.
134. *Synthesis*, 1975, 276.
135. *J. Org. Chem.*, 1968, **33**, 1067.
136. J.F. Stoddart, *Stereochemistry of Carbohydrates*, Wiley-Interscience, New York, 1971; *Chem. Rev.*, 1979, **79**, 491.
137. *Adv. Carbohydr. Chem. Biochem.*, 1971, **26**, 197.
138. *Adv. Carbohydr. Chem. Biochem.*, 1977, **34**, 179.
139. *Acta Chem. Scand.*, 1972, **26**, 3895.
140. *Synthesis*, 1978, 48.
141. *Tetrahedron Lett.*, 1992, **33**, 4767.
142. *Carbohydr. Res.*, 1967, **5**, 132.
143. *Carbohydr. Res.*, 1988, **173**, 303.
144. *Aust. J. Chem.*, 1975, **28**, 525.
145. *Acta Chem. Scand.*, 1975, **29B**, 278.
146. *Carbohydr. Res.*, 1969, **10**, 549.
147. *J. Chem. Soc. B*, 1968, 827.
148. *Carbohydr. Res.*, 1966, **2**, 197.
149. *Adv. Carbohydr. Chem. Biochem.*, 1981, **39**, 71.
150. *Carbohydr. Res.*, 1985, **140**, C7.
151. *Adv. Carbohydr. Chem. Biochem.*, 1981, **39**, 71.
152. *Can. J. Chem.*, 1990, **68**, 897.
153. *Carbohydr. Res.*, 1982, **108**, 97.
154. *J. Chem. Soc., Perkin Trans. 1*, 1984, 2371.
155. *Can. J. Chem.*, 1975, **53**, 1204.
156. *Chem. Lett.*, 1988, 1699.
157. *Acta Chem. Scand.*, 1977, **31B**, 359.
158. *J. Chem. Soc., Perkin Trans. 1*, 1975, 1700.
159. *J. Chem. Soc., Perkin Trans. 1*, 1975, 1695.
160. *J. Chem. Soc., Perkin Trans. 1*, 1983, 921.
161. *Bull. Chem. Soc. Jpn.*, 1980, **53**, 230.
162. *The Carbohydrates* (eds W. Pigman and D. Horton), Vol. IB, Academic Press, New York, 1980, p. 1167.
163. *Adv. Carbohydr. Chem.*, 1956, **11**, 1; *Adv. Carbohydr. Chem.*, 1961, **16**, 105.
164. *J. Am. Chem. Soc.*, 1955, **77**, 4346.
165. *Adv. Carbohydr. Chem.*, 1959, **14**, 9.

6 The Chemical Synthesis of Oligosaccharides

6.1 BIOLOGICAL FUNCTIONS OF OLIGOSACCHARIDES; THE NEED FOR CHEMICAL SYNTHESIS

It is now established that carbohydrates play key roles in many complex biological processes that depend upon the specific molecular structures of oligosaccharides and upon specific interactions between them and other compounds, most frequently proteins.[1] Prior to the recognition of this it appeared that most oligomeric/polymeric carbohydrates played rather macro roles as, for example, structural or food storage materials, and that the other main polymeric compounds of nature, the nucleic acids and the proteins, were left to control the highly sophisticated functioning of organisms at the molecular level using chemical information coded into their detailed molecular structures. On the basis mainly of the discoveries since the 1970s it is now clear that nature also uses many oligosaccharides with fewer than 20 monosaccharide components as storehouses of biological information, their potential for this purpose being enhanced relative to that of the oligopeptides and oligonucleotides by the extensive variety of ways in which the constituent units may be joined together. Thus, for example, there are over 1000 ways of linking three of the same D-hexopyranose molecules by glycosidic bonds, and only one way of peptide bonding three molecules of the same α-amino acid; there are over 2×10^6 possible pentasaccharides comprising D-sugars![2] While this diversity makes major demands of the synthetic chemist, the recent identification of biologically based synthetic problems in the field has stimulated the development of specific and moderately efficient synthetic methods for the chemical synthesis of oligosaccharides. Many oligomers containing about five monosaccharide units have been made, and larger specific structures having more than 10 monomer units have been synthesized by purely chemical methods. In marked contrast, when the predecessor of this book was written in 1972 the chemical synthesis of a trisaccharide was a notable achievement.

Many of the roles played by oligosaccharides in natural processes are dependent on the mutual recognition between the oligosaccharides involved and antibody, lectin (plant proteins with specific carbohydrate-binding properties),[3] enzyme or other proteins, and in many cases the oligosaccharides are themselves bonded to proteins or lipids. Some of these roles are as blood group determinants;[4] as bacterial surface antigens which induce immune responses;[5] as mammalian cell surface compounds that bind viruses,[6] bacteria, lymphocytes and toxins; as constituents of the coat of the AIDS virus and implicated in the invasion of T-cells; and as elicitors that induce defence reactions in plants by activating the biosynthesis of phytoalexins.[7] (See also Sections 8.1.2 and 8.1.4.) Appreciation of these roles has led, for example, to the recognition that 70% of the antibodies produced by mammals immunized with foreign whole cells are directed against oligosaccharide components (the production of synthetic oligosaccharide antigens and vaccines thus becomes a possibility); metastasis in cancer involves changes in the cell-surface carbohydrates of tumour cells; and the states of arthritic conditions correlate with protein glycosylation.[1a]

With such biological importance now ascribed to specific, relatively small oligosaccharides it is not surprising that much recent effort has gone into their chemical synthesis, notably for the purpose of contributing to biochemical aspects of medical science.[2,8−10]

6.2   GENERAL FEATURES OF OLIGOSACCHARIDE SYNTHESIS[2,8−10]

There are so many variables involved in oligosaccharide synthesis that each target compound requires that a particular strategy be developed following consideration of the monomers to be bonded, the types of inter-unit linkages to be made and the sequence — in both the topological and chronological senses — in which they must be linked. As far as the types of bonds are concerned, they can be classified according to whether they involve the formation of equatorial or axial linkages from the glycosylating donor to glycosyl acceptor molecule (ROH) and whether the glycosylating agent requires an equatorial or axial group (X) at C-2, i.e. whether the products are, for example, of the β-D-glucopyranosyl/β-D-galactopyranosyl **6.1**, the α-D-mannopyranosyl **6.2**, the α-D-glucopyranosyl/α-D-galactopyranosyl **6.3** or the β-D-mannopyranosyl **6.4** types (cf. Section 3.3.2). Special methods have to be adopted if the C-2 substituent (X) is an amino or substituted amino group (Section 4.3.2.f),[11] or if 2-deoxy sugars (X=H) (see Section 6.4.2) or furanosyl donors are involved.

| **6.1** | **6.2** | **6.3** | **6.4** |

Normally all hydroxyl groups of the acceptor, other than those to be glycosylated, must be protected, and each succeeding step requires that the next hydroxyl group to be substituted be specifically revealed. Lastly, all the protecting groups must be removed to release the required oligosaccharide.

Traditionally, the chemical synthesis of inter-monosaccharide bonds has involved the intermolecular coupling of the components with the use of a mixture of rational and empirical features to provide optimal efficiencies, in particular those relating to the activation of the glycosylating agents and stereoselectivities of the couplings. An ideal alternative approach which, in principle, could overcome the activation and stereochemical issues involves holding the component sugar donor and acceptor units in appropriate orientations within the same molecules in such a way that they can be made to couple intramolecularly. Although this strategy is in its infancy, it appears to have such potential for oligosaccharide synthesis that first developments are highlighted separately in Section 6.5.

A very elegant approach to the preparation of linear oligosaccharides derived from one sugar (homo-oligosaccharides) involves the interative use of O-protected glycosylating agents which, when coupled to acceptors, give chain-extended oligomers that may then be specifically deprotected to give new acceptors for continuation of the chain from the non-reducing terminus (usually written at the left-hand end). Alternatively, the sequence can be built in the reverse way by using acceptors which, on glycosylation, give oligomers that may be converted into new glycosyl donors. In this case the growth is from the reducing end. These cases are illustrated in Scheme 6.1, in which M_D and M_A represent donating and accepting monosaccharide units, respectively, and M^R signifies the 'reducing' terminal units having anomeric centres unbonded to other sugars.

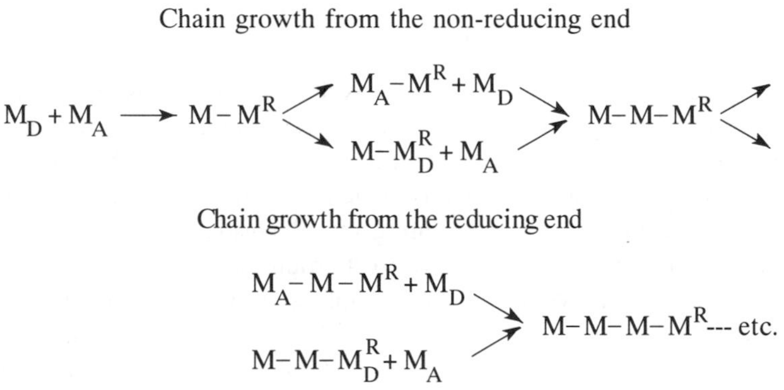

Chain growth from the non-reducing end

$$M_D + M_A \longrightarrow M - M^R \begin{cases} M_A - M^R + M_D \\ M - M_D^R + M_A \end{cases} M - M - M^R \begin{cases} \\ \end{cases}$$

Chain growth from the reducing end

$$\begin{cases} M_A - M - M^R + M_D \\ M - M - M_D^R + M_A \end{cases} M - M - M - M^R \text{--- etc.}$$

Scheme 6.1

Strategies of this kind cannot be applied in the majority of cases. Much more frequently, oligosaccharides are put together by the block coupling of small oligomers by which means the more complex intermediates involved in a synthesis are minimally exposed to potentially hazardous conditions.

While it is possible, in principle, to envisage the synthesis of oligosaccharides by use of specific, isolated enzymes, such approaches to complex hetero-oligosaccharides present formidable challenges, and the method is, as yet, of limited value.[12] Enzymes can, however, be used in specific circumstances: the terminal

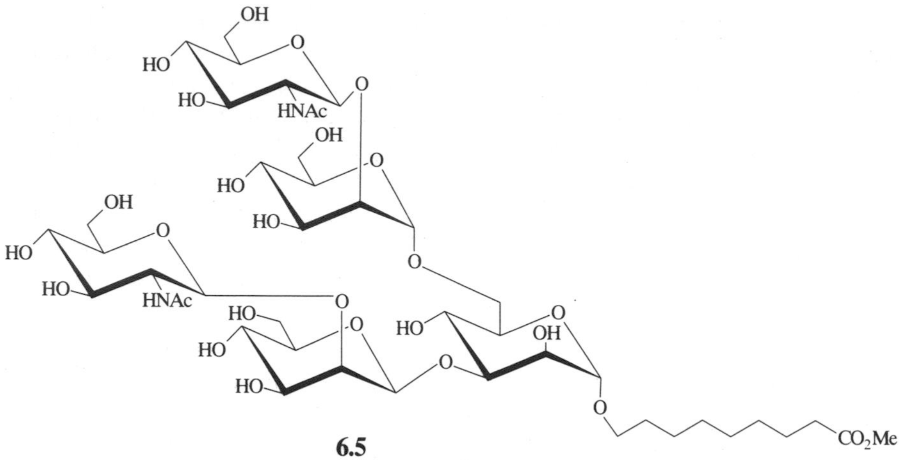

6.5

N-acetylglucosamine units of pentasaccharide **6.5** have been attached sequentially to a chemically synthesized mannotriose glycoside by use of transferase enzymes with UDP-*N*-acetylglucosamine as the glycosyl donor.[13]

Likewise, solid phase and automated procedures are not well advanced in this field, which again distinguishes the oligosaccharides from the oligopeptides and oligonucleotides.

6.3   GLYCOSYL ACCEPTORS

Normally, glycosyl acceptors have the hydroxyl group to be substituted specifically free, but, especially in the orthoester-based glycosidation method (Section 3.2.2), they can be advantageously activated by tritylation, the approach having the further advantage that water or simple alcohols, which are better acceptors than are carbohydrate alcohols, are not produced during the glycosidation step. Otherwise, the nucleophilicity of glycosyl acceptor hydroxyl groups may be enhanced by tributyl-stannylation.[14] Primary alcohols are appreciably more reactive than are secondary, the reactivities of the latter (cf. Section 5.1.3) being dependent on such factors as their axial or equatorial orientations, their positions in the acceptor molecules and, importantly, the substituents on other hydroxyl groups in the acceptors. Alkylated (e.g. benzylated) derivatives usually contain much more active acceptor hydroxyl groups than do their esterified analogues.[8a] An example is given in Section 3.3.2.a.i.

6.4   GLYCOSYL DONORS

6.4.1   GENERAL

Most control is achieved in glycosidic bond-forming reactions by the appropriate selection of the glycosylating reagents, which ideally should be shelf-stable and activatable, together with the associated reagents and solvents (Chapter 3). The main methods in use involve displacement of leaving groups from the anomeric centres of the donors: thioglycosides (Section 3.1.2.c), glycosyl sulphoxides (Section 3.1.2.c.iv), imidates (Section 3.2.3), 1,2-orthoesters (Section 3.2.2), 1,2-anhydrides (Section 3.1.1.c), bromides and chlorides (Section 3.3), fluorides (Section 3.3.5) and pentenyl glycosides (Section 3.1.1.b.vi). However, it is possible, in suitable circumstances, to couple glycosyloxy anions with carbohydrate triflates[2] and even anomeric hydroxyl groups with suitable acceptors (see also Section 3.1.1.a.ii) to produce inter-sugar linkages.

As a broad generalization the glycosyl halides, thioglycosides and glycosyl trichloroacetimidates have become established as the most suitable donors for oligosaccharide synthesis, and, in general, with D-glucose or D-galactose donors having O-2 substituents (usually esters) that may participate in the displacement of the C-1 leaving group, products of type **6.1** (X = OCOR) are formed. Conversely, α-glycosides **6.3** (e.g. X = O-benzyl) are formed when the groups at C-2 cannot participate. In the D-mannose and D-rhamnose (6-deoxy-D-mannose) series, α-products **6.2** normally predominate in all cases. To make β-D-mannopyranosyl linkages **6.4** special methods are required. It is possible in some cases, for example, to prepare analogous β-D-glucopyranosides and to invert the configuration at C-2 either by (ideally intramolecular) nucleophilic displacement reactions or by oxidation of hydroxyl groups to the ketones followed by reduction. As an alternative, however, O-etherified α-D-mannopyranosyl bromides can be induced, with silver silicate as activator, to take part in S_N2 displacements to give the required products.[8] (See Section 6.5 for intramolecular approaches to β-D-mannopyranosyl compounds.)

In the same way as alcohol groups in acceptors are more reactive in compounds with alkyl groups rather than acyl groups on other oxygen atoms, so O-alkyl glycosylated donors are more potent than are the acylated analogues. This marked influence identified by Fraser-Reid allows, for example, chemoselective reaction of an O-benzylated glycosylating agents in the presence of its O-acetylated analogue (Section 3.1.1.b.vi). The activities of glycosylating agents are also dependent upon the conformational mobilities of the pyranoid rings, with fused cyclic acetal substituents markedly diminishing their potencies.[15] This finding perhaps indicates that the lower reactivity of benzoylated glycosylating agents relative to their acetylated analogues can be attributed to relative conformational interia factors.

Because of the significance of 2-amino-2-deoxyhexoses in many biologically important oligosaccharides, their linking to other sugars has received special attention.[11] Fortunately, the main glycosylation methods involving the use of 1,2-oxazolines, 2-deoxy-2-phthalimido halides, thioglycosides, trichloroacetimidates or 2-azido-2-deoxy halides (Section 4.3.2) give rise to β-linked products in the *gluco*- and *galacto*-series. When α-linked components are required they can be made, for example, by using 2-azido-2-deoxy-β-D-glucopyranosyl chlorides or, alternatively, α-bromides under conditions in which the accessible and more reactive β-anomers take part in the displacements.[11]

6.4.2 2-DEOXYGLYCOSYL LINKING

Because of the absence of functional groups at C-2 to direct the stereochemistry of glycosylation processes, 2-deoxyglycosyl donors cannot be employed in the efficient synthesis of complex 2-deoxyglycosides of specified anomeric configuration. While, to a limited extent, direct additions to O-substituted glycals can be used (Section 4.10.1.b), several other efficient procedures have been developed which provide good stereocontrol in the glycosylation step. In the main, they are based on the use of reactive, electrophilic glycosylating species containing three-membered rings at C-1–C-2 and afford glycosides having functions at C-2 which can be reductively removed to leave the deoxy group. The principles involved in two approaches are outlined in Scheme 6.2.

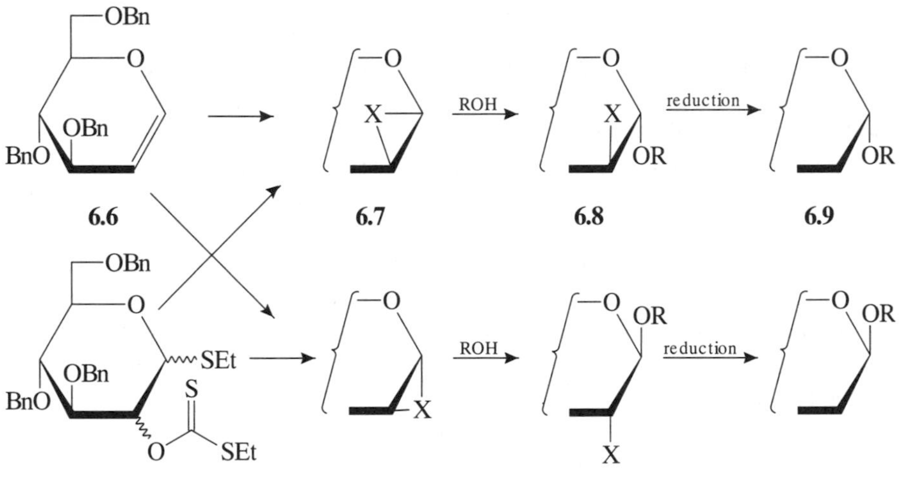

Scheme 6.2

Tri-O-benzyl-D-glucal **6.6** forms an iodonium ion **6.7** (X = I$^+$) on the β-face of the double bond and gives 2-iodo-α-D-glycosides **6.8** and hence access to 2-deoxy-α compounds **6.9** by reductive removal of the iodine.[16] In a related manner, S-ethyl 1-thio-β-D-glucopyranosides **6.10** (β-*gluco*-) carrying xanthate ester groups at C-2, on suitable electrophilic activation, generate episulphonium species **6.7** (X = EtS$^+$) and hence also give α-linked products **6.9** (Section 3.1.2.c.ii). An advantage of this strategy is that 2-deoxy-β-D-glycosides **6.13** can be made, again in good yield, starting from S-ethyl 1-thio-α-D-mannopyranoside 2-xanthates from which episulphonium **6.11** (X = EtS$^+$) ions and 2-ethylthio-β-D-glucopyranosides **6.12** (X = EtS) are derived, and hence the β-linked products **6.13**.[17] Otherwise, epoxidation of tri-O-benzyl-D-glucal with 3,3-dimethyldioxirane gives the α-D-*gluco* product **6.11** (X = O) which, condensed with glycosyl acceptors, leads to otherwise protected 2-hydroxy β-D-glucopyranosyl disaccharides **6.12** (X = OH) which can be deoxygenated by standard free radical procedures.[18]

Related chemistry utilizes the *O*-silylated 2-phenylthio-D-glucopyranosyl fluorides, which are made by treatment of *S*-phenyl 1-thioglucopyranosides specifically unprotected at O-2 with DAST in a process that involves migration of the phenylthio groups from C-1 to C-2 (cf. Section 4.5.1.e.ii). With acceptor alcohols the fluorides react in the presence of tin(IV) chloride to give 2-phenylthioglycosides in very high yields and with excellent selectivities. In ether the *β*-products **6.12** (X = SPh) are obtained (presumably by way of the intermediate having structural unit **6.11** X = PhS⁺); in dichloromethane *α*-linked glycosides are produced, and hence with desulphurization a route to the 2-deoxy *α*-compounds **6.9** is provided.[19] In the latter case, presumably, direct displacement of fluoride occurs.

6.5   THE INTRAMOLECULAR APPROACH

As indicated in Section 6.2, the intramolecular approach to the linking of sugar units appears to have good potential for the synthesis of specific sugar–sugar bonds of oligosaccharides. In particular, it offers novel means of addressing problems associated with the activation of glycosyl donors and with the control of the stereochemistry of the coupling processes. Only initial steps have been taken, but

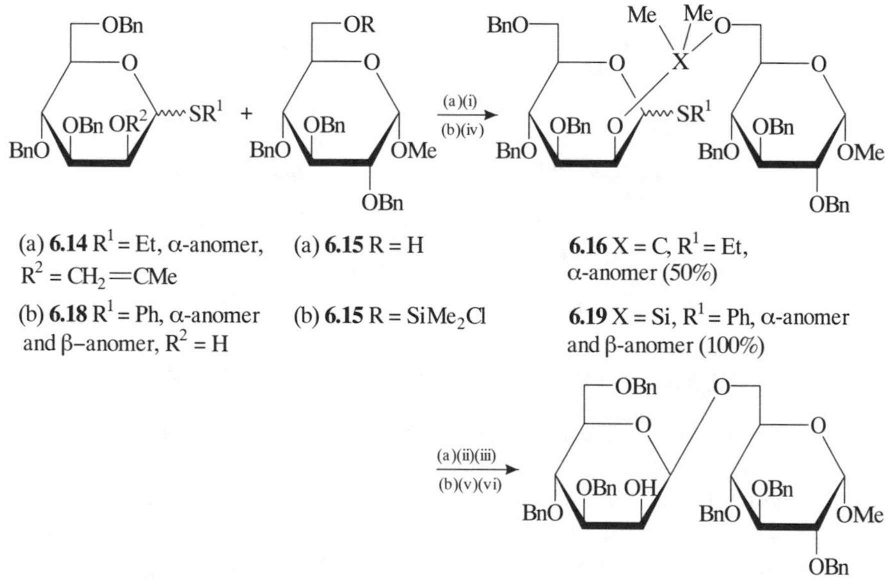

(a) **6.14** R¹ = Et, α-anomer,   (a) **6.15** R = H   **6.16** X = C, R¹ = Et,
R² = CH₂═CMe   α-anomer (50%)

(b) **6.18** R¹ = Ph, α-anomer   (b) **6.15** R = SiMe₂Cl   **6.19** X = Si, R¹ = Ph, α-anomer
and β–anomer, R² = H   and β–anomer (100%)

6.17 (a, 61%; b, 73% from α-anomer)

(a) (i) TsOH; (ii) NIS; (iii) H₂O
(b) (iv) imidazole; (v) MCPBA; (vi) Tf₂O, di-*t*-butylpyridine

Scheme 6.3

it seems safe to predict that they foreshadow a new dimension in oligosaccharide synthetic chemistry.

Two parallel approaches (Scheme 6.3) to the preparation of β-D-mannopyranosyl disaccharides utilize dimers **6.16** and **6.19** which, respectively, are made by acid-catalysed addition of the alcohol **6.15** (R = H) to the 2-O-isopropenyl 1-thiomannoside **6.14**[20] and by condensation of the chlorosilylether **6.15** (R = SiMe$_2$Cl) with the alcohol **6.18**.[21] While the latter reaction is nearly quantitative, the former gives a yield of about 50%. In both cases extrusion of the bridging groups follows the activation of the anomeric thio groups, and glycosylation of O-6 occurs from the β-direction to give **6.17**.

The expectation that this type of methodology will provide new means of making α-D-glucopyranosyl and -galactopyranosyl oligomers is encouraged by the conversion of the silicon-bridged compounds **6.20** (made using dichlorodimethylsilane) into alkyl α-glucopyranosides on activation with NIS (yields 60%)[22] and by the transient addition of 2,3,4,6-tetra-O-benzyl-D-galactose to the isopropenyl ether **6.21** in the presence of trimethylsilyl triflate, which subsequently also catalyses the formation of the α-(1→4)-linked galactosylglucose (CH$_2$Cl$_2$, −25 °C, 56%, α/β 5.4:1).[23] This latter reaction is described further in Section 3.1.1.a.ii where the reverse procedure (condensation of isopropenyl glycosides with monosaccharide alcohol acceptors) is also noted.

6.20   **6.21**

6.6   EXAMPLES OF OLIGOSACCHARIDE SYNTHESIS

The examples given in this section illustrate the complexity of the subject but do not include the largest or most complex compounds to have been synthesized. Although in the schemes the monosaccharide rings are drawn in the chair conformations expected to be the most stable, the represented relative orientations of the sugars bear no relationship to energy minima, i.e. to the preferred shapes of the oligomers (see Section 8.1.2.c.ii).

6.6.1   LINEAR HOMO-OLIGOSACCHARIDES

The rhynchosporosides are a family of β-(1→4)-linked D-glucopyranose oligosaccharides glycosidically bound to (R)- or (S)-propane-1,2-diol which cause scald disease in cereals. They have been synthesized by the sequential approach prior to final linking to the terminal monosaccharide in the glycosidic form. Thus, the cellobiose-based acetylated disaccharide *S*-phenyl thioglycoside **6.22** is converted to the corresponding glycosyl fluoride **6.23** (Section 3.1.2.c.ii) which is coupled with the acceptor **6.26** to give the trisaccharide *S*-phenyl thioglycoside **6.24**. The cycle is repeated, and the tetrameric fluoride **6.25** is condensed with the glycosidic acceptor **6.27**. Deprotection of the product gives **6.28**, which is (5*R*)-rhynchosporo-side (Scheme 6.4).[24] The yields are about 75% for each of the glycosidation steps and 85% for the thioglycoside to glycosyl fluoride conversions.

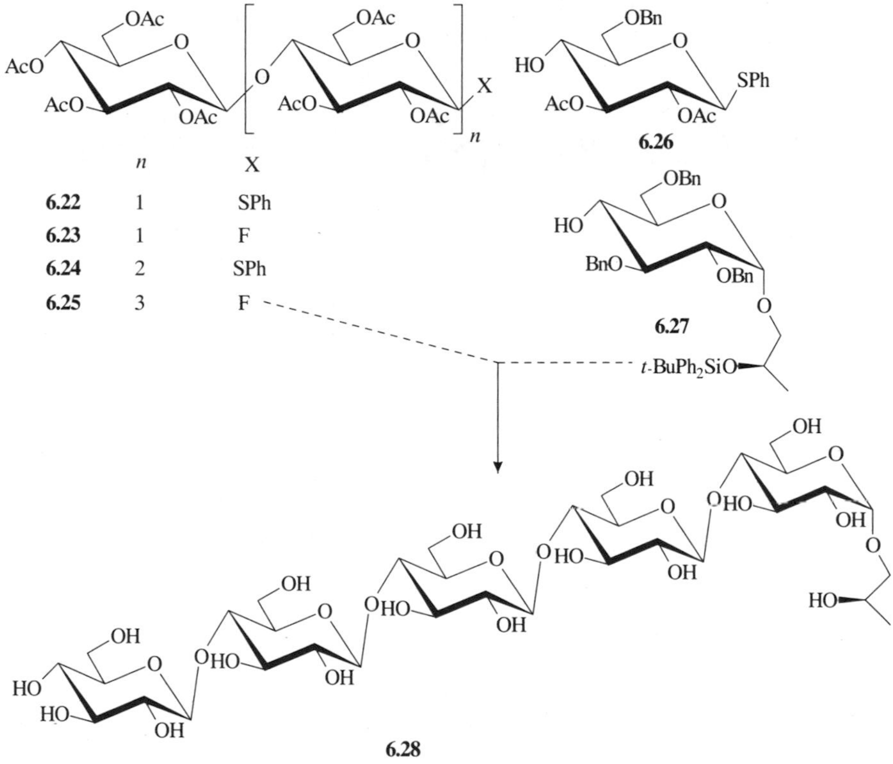

	n	X
6.22	1	SPh
6.23	1	F
6.24	2	SPh
6.25	3	F

Scheme 6.4

A related method, which results in the formation of α-(1→6)-linked D-glucose oligomers, depends on pent-4-enyl 6-*O*-*t*-butyldiphenylsilyl-D-glucopyranoside as the key building unit which, by benzylation, can be activated as a glycosylating agent (Section 3.1.1.b.vi) and, following debenzylation benzoylation and selective deprotection at *O*-6, can act as a glycosyl acceptor.[25]

6.6.2   LINEAR HETERO-OLIGOSACCHARIDES

As has been indicated in Section 3.1.1.b.vi, thioglycosides may be 'armed' or 'disarmed' as glycosyl donors and acceptors according to whether their protected hydroxyl groups are alkylated or acylated. It is therefore possible, by use of iodonium dicollidine triflate, to activate **6.29** in the presence of the thioglycosidic group of **6.30** and obtain 73% of the disaccharide **6.31**. Condensation of this, after replacement of the ester group by a methyl group, with the acceptor glycoside **6.32** in the presence of N-iodosuccinimide and triflic acid then gives access to 3-aminopropyl 3-O-[3-O-(2,3,4-tri-O-methyl-α-L-fucopyranosyl)-α-L-rhamnopyranosyl]-2-O-methyl-α-L-rhamnoside **6.33** (Scheme 6.5), which is a glycoside of a fragment of a glycolipid component of the cell wall of *Mycobacterium tuberculosis* made during studies aimed at the development of a vaccine against the organism.[26]

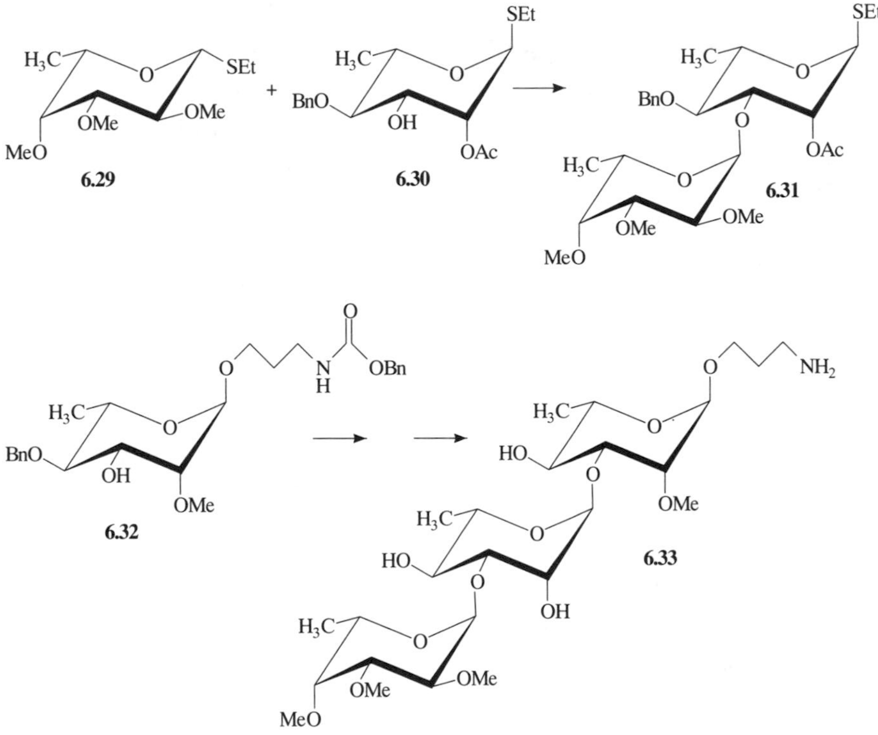

Scheme 6.5

Syntheses of higher hetero-oligosaccharides require the step by step development of each glycosidic linkage using appropriately substituted donors and acceptors. An example is given in Scheme 6.6, which illustrates the preparation of a pentaosyl glycolipid which occurs in human red blood cells and complexes with anti-Globoside I antibodies.[27] In this synthesis the amino sugar derivatives **6.34** and **6.35** are coupled to give the corresponding disaccharide in 94% yield using silver triflate as activator. This is then converted into the glycosyl donor **6.36** by hydrolysing the benzylidene acetal, acetylating, reducing the azide and converting the resulting amino group into the second phthalimido group. The allyl

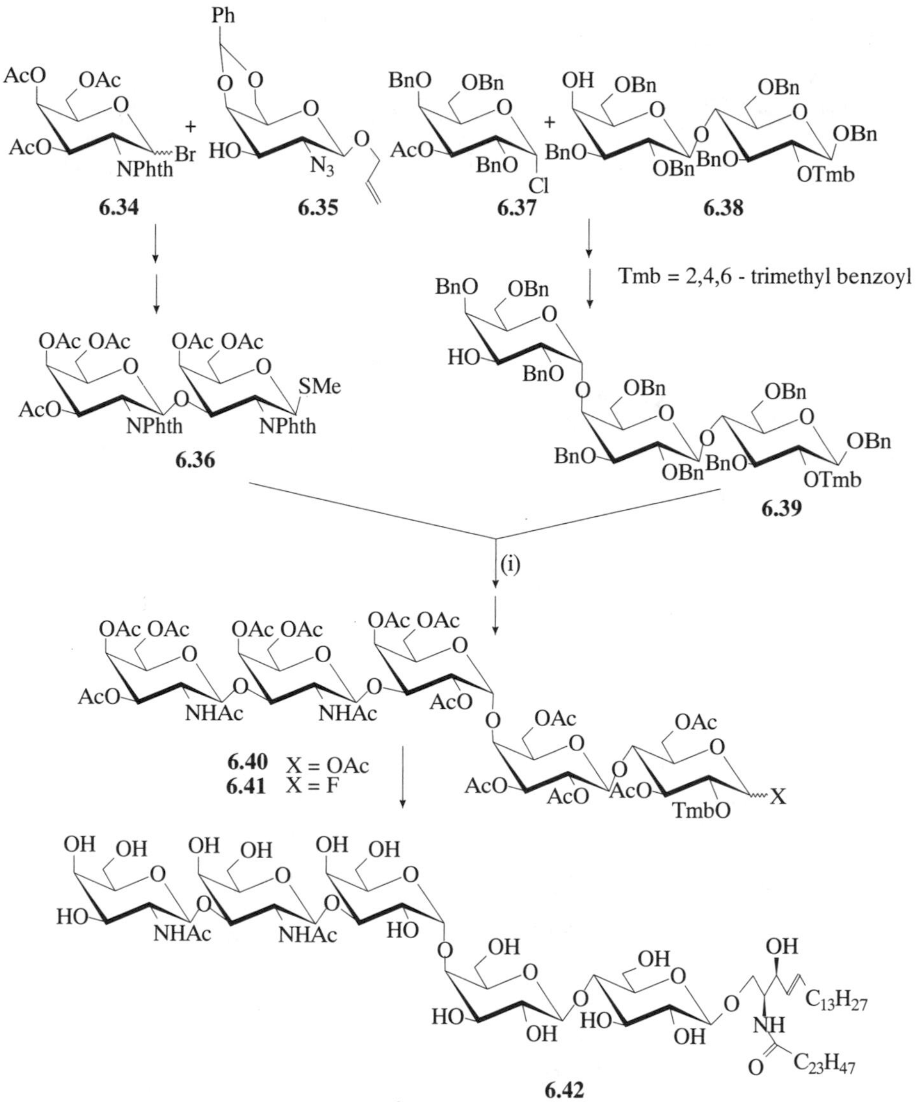

Tmb = 2,4,6 - trimethyl benzoyl

6.34   **6.35**   **6.37**   **6.38**

6.36

6.39

(i)

6.40 X = OAc
6.41 X = F

6.42

(i) (Bu₄N)₂CuBr₄, AgOTf, MeNO₂, 20 °C

Scheme 6.6

aglycon is then removed by use of a rhodium catalyst, and from the free sugar the glycosylating disaccharide thioglycoside **6.36** is made.

The acceptor trisaccharide **6.39** is prepared by silver triflate promoted condensation of the active glycosyl chloride **6.37** with the lactose-derived **6.38** in 1,2-dichloroethane at −20 °C, the α-linkage being established with good efficiency. Coupling of **6.36** with **6.39** gives 65% (+30% recovered trisaccharide) of the required pentamer (α/β ratio of 1:8). From this the perester **6.40** is made and hence

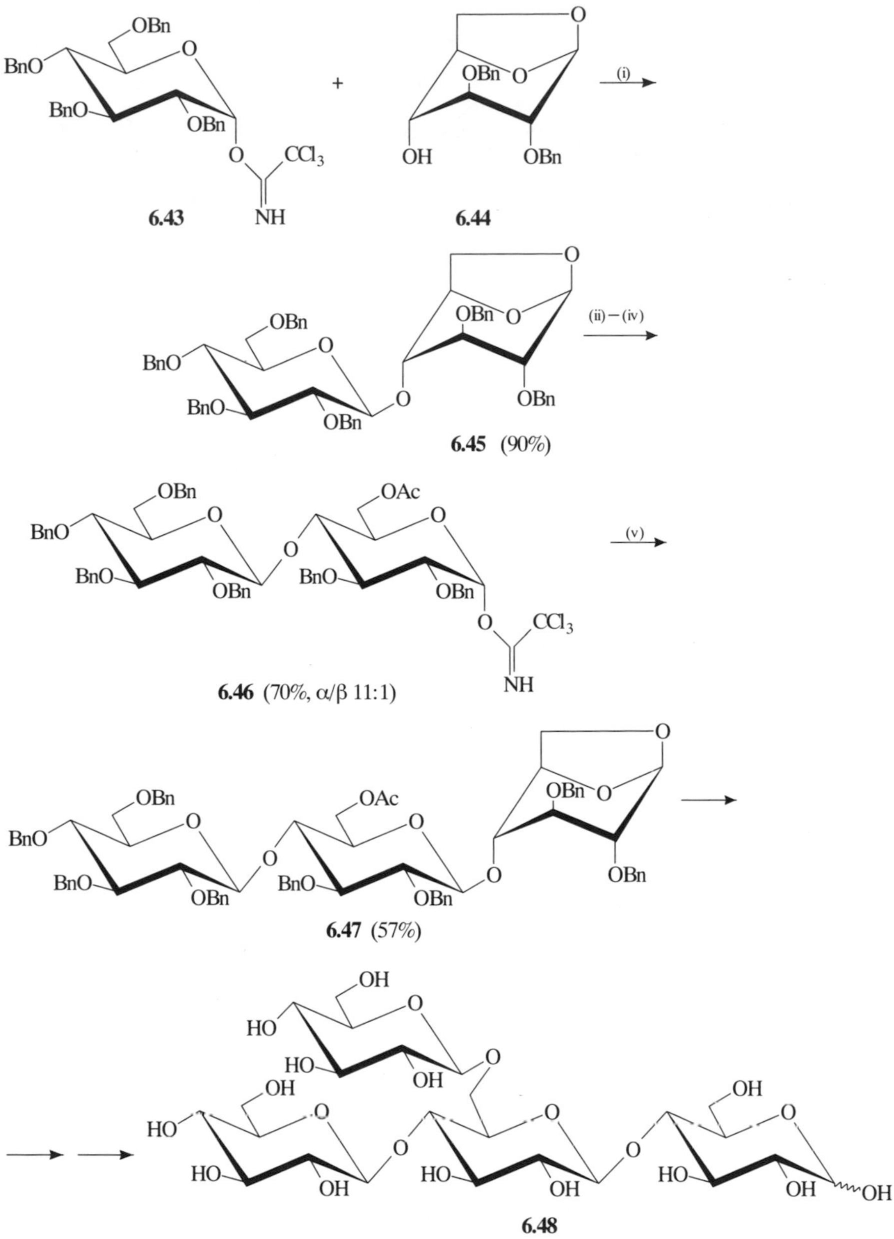

(i) BF$_3$·OEt$_2$, CH$_2$Cl$_2$, −38 °C, 2 h; (ii) CF$_3$CO$_2$H, AcOH/Ac$_2$O, 20 °C, 2 h; (iii) BnNH$_2$, Et$_2$O, 20 °C, 1.5 h; (iv) Cl$_3$CCN, NaH, CH$_2$Cl$_2$, 20 °C, 3 h; (v) **6.44**, BF$_3$·OEt$_2$, CH$_2$Cl$_2$, −40 °C, 2 h

Scheme 6.7

the ceramide glycoside **6.42** by use of the fluoride **6.41** followed by deprotection. The yield of product is 44% based on **6.41**.

6.6.3   BRANCHED HOMO-OLIGOSACCHARIDES

The example illustrated in Scheme 6.7 reveals the value of the glycosyl trichloroacetimidates (Section 3.2.3) as glycosylating agents, all of the reactions proceeding in high yield with the anomeric ratios being more than 5:1 in favour of the β-linkages. Condensation of the glycosylating agent **6.43** with the 1,6-anhydride acceptor **6.44** gives the disaccharide derivative **6.45** which is ring opened by acetolysis of the anhydride ring and is converted into the new, reactive imidate **6.46** and hence the trisaccharide derivative **6.47**, again by using **6.44**. Release of the acetylated hydroxyl group and subsequent glycosylation with **6.43** afford, after anhydride ring opening and deblocking, the branched tetrasaccharide **6.48** in about 60% yield from **6.46**.[28]

For the synthesis of the heptasaccharide **6.49**, which is bioactive as an elicitor of the biosynthesis of phytoalexins in plants, the trisaccharide unit which appears twice in the sequence is used as the glycosylating agent in the form of a S-ethyl β-thioglycoside perester. This is condensed with 1,2,3,4-tetra-O-benzyl-β-D-glucose in ether using methyl triflate as activator to give a derivative of the tetramer of the reducing end of **6.49** (65% yield). Release of the appropriate primary hydroxyl group (previously acetylated while other hydroxyl groups were benzoylated or

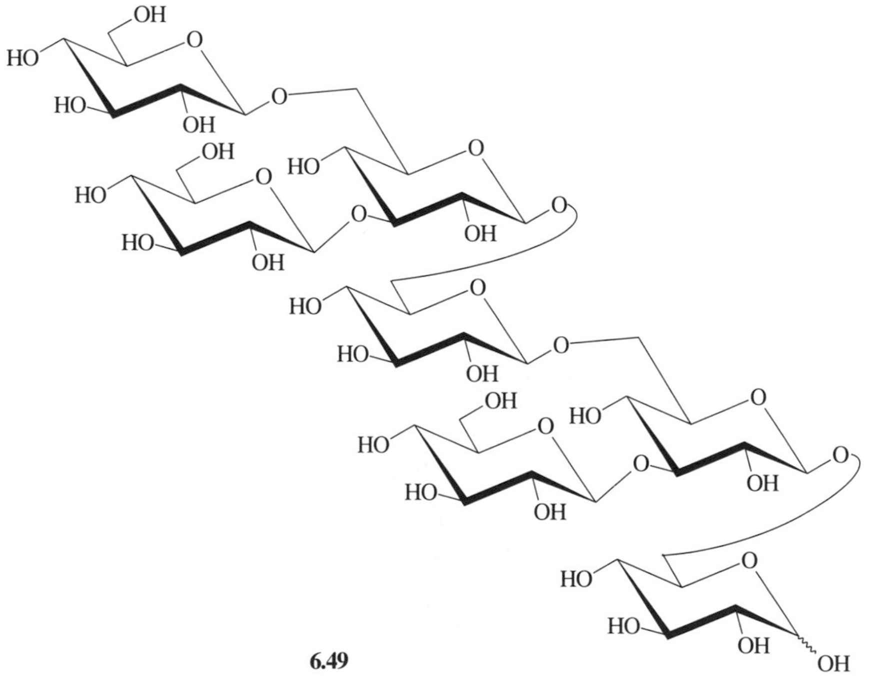

6.49

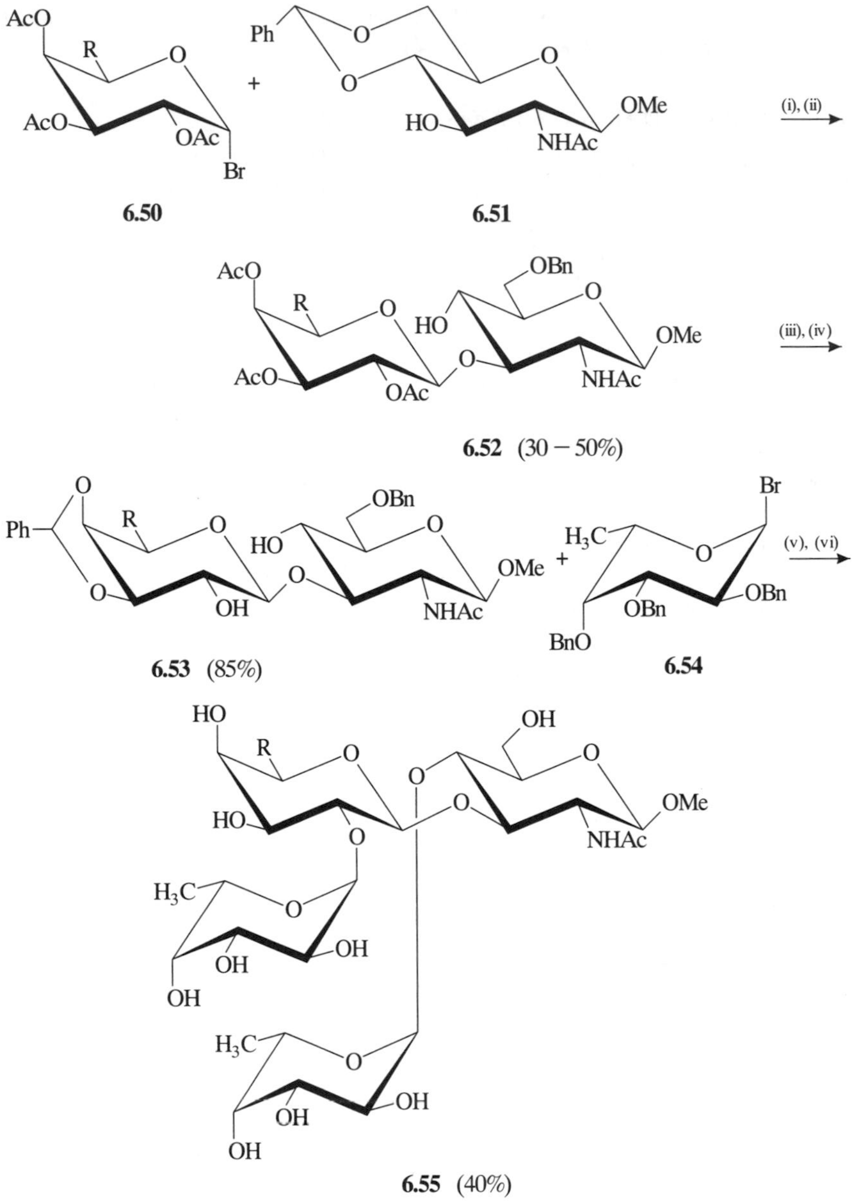

6.50   **6.51**

6.52 $(30-50\%)$

6.53 (85%) **6.54**

6.55 (40%)

(i) $Hg(CN)_2$, toluene, $MeNO_2$, $20\,^{\circ}C$, 20 min; (ii) $NaCNBH_3$, HCl, Et_2O; (iii) NaOMe, MeOH; (iv) $PhCH(OMe)_2$, H^+, DMF, MeCN; (v) Et_4NBr, DMF, CH_2Cl_2; (vi) H_2, Pd/C

Scheme 6.8

benzylated) followed by a second use of the trisaccharide S-ethyl thioglycoside gives the protected heptasaccharide (36%) from which the required product **6.49** is obtainable in excellent yield.[7]

6.6.4   BRANCHED HETERO-OLIGOSACCHARIDES

The ability to synthesize natural oligosaccharides implies the ability to prepare analogues, and from a comparison of the biochemical characteristics of the natural with those of the analogues comes a powerful means of correlating chemical structure with biochemical activity. In the course of Lemieux's major pioneering studies of the human blood group substances, the interactions between the Lewis b blood group determinant **6.55** (R = CH$_2$OH) and a monoclonal antibody were probed by use of a set of synthetic analogues having the hydroxymethyl group at C-5 of the D-galactose component replaced by hydrogen, methyl, fluoromethyl and chloromethyl groups. From the study it was concluded that this particular hydroxyl group is involved in non-polar interactions at the combining site with the antibody, probably intramolecularly hydrogen bonded with the adjacent axial hydroxyl group of the galactose unit.[29]

Scheme 6.8 illustrates how the analogues are produced. Variously modified glycosyl bromides **6.50** are condensed with the amino sugar derivative **6.51** and the disaccharides formed are treated with sodium cyanoborohydride to give the mono-alcohols **6.52**. De-O-acetylation and benzylidenation at the cis-α-diols gives **6.53** which, on fucosylation with **6.54** and deprotection, affords the required specifically modified branched-chain oligosaccharide glycosides **6.55** having R = H, Me, CH$_2$F and CH$_2$Cl.

An indication of the complexity of oligosaccharides of this type that can now be synthesized is given by structure **6.56**, which represents the hexasaccharide present as an inositol-containing glycolipid on the surface of *Trypanosoma brucei*, a parasitic protozoan. Not only has the compound been synthesized, but it has been linked to the inositol and the phospholipids with which it occurs in nature in a major synthetic achievement by T. Ogawa's group.[30]

The type of abbreviation employed in **6.56** is commonly used to represent complex structures of this kind. At the left-hand end of the formula the symbolism indicates α-D-mannopyranose linked glycosidically to O-2 of the neighbouring mannopyranose unit. Glucose, mannose and galactose are given the normal abbreviations Glc, Man and Gal, respectively. Glc$p$NH$_2$ signifies 2-amino-2-deoxy-D-glucopyranose.

$$\alpha\text{-D-Man}p\text{-}(1\rightarrow2)\text{-}\alpha\text{-D-Man}p\text{-}(1\rightarrow6)\text{-}\alpha\text{-D-Man}p\text{-}(1\rightarrow4)\text{-}\alpha\text{-D-Glc}p\text{NH}_2$$

$$\uparrow^3_1$$

$$\alpha\text{-D-Gal}p\text{-}(1\rightarrow6)\text{-}\alpha\text{-D-Gal}\,p$$

6.56

6.7 REFERENCES

1. (a) *Annu. Rev. Biochem.*, 1988, **57**, 785; (b) *Chem. Soc. Rev.*, 1989, **18**, 347; (c) *Chem. Br.*, 1990, **26**, 679.
2. *Angew. Chem., Int. Ed. Engl.*, 1986, **25**, 212.
3. *Science*, 1989, **246**, 227.
4. *Chem. Soc. Rev.*, 1978, **7**, 423.
5. *Chem. Br.*, 1990, **26**, 669.
6. *J. Am. Chem. Soc.*, 1991, **113**, 5865.
7. *Angew. Chem., Int. Ed. Engl.*, 1991, **30**, 1681; *Acc. Chem. Res.*, 1992, **25**, 77.
8. (a) *Angew. Chem., Int. Ed. Engl.*, 1982, **21**, 155; *Angew. Chem., Int. Ed. Engl.*, 1990, **29**, 823; (b) *Chem. Soc. Rev.*, 1984, **13**, 15; (c) *Pure Appl. Chem.*, 1991, **63**, 519; (d) *Acc. Chem. Res.*, 1992, **25**, 575; (e) *Chem. Rev.*, 1993, **93**, 1503; (f) *Synthetic Oligosaccharides* (ed. P. Kovac), *Am. Chem. Soc. Symp. Ser.* 560, American Chemical Society, Washington, 1994.
9. *Carbohydrate Chemistry* (ed. J. F. Kennedy), Oxford University Press, 1988, p. 500.
10. *Rodd's Chemistry of Carbon Compounds* (ed. M. Sainsbury), Vol. 1E, F, G 2nd Supplement, Elsevier, Amsterdam, 1993, p. 437.
11. *Chem. Rev.*, 1992, **92**, 1167.
12. *Angew. Chem., Int. Ed. Engl.*, 1985, **24**, 617.
13. *Carbohydr. Res.*, 1991, **210**, 145.
14. *Carbohydr. Res.*, 1976, **51**, C13.
15. *J. Am. Chem. Soc.*, 1991, **113**, 1434.
16. *J. Am. Chem. Soc.*, 1989, **111**, 6656.
17. *Tetrahedron Lett.*, 1992, **33**, 2063.
18. *J. Org. Chem.*, 1991, **56**, 5448.
19. *J. Am. Chem. Soc.*, 1986, **108**, 2466.
20. *J. Am. Chem. Soc.*, 1991, **113**, 9376.
21. *J. Am. Chem. Soc.*, 1992, **114**, 1087.
22. *J. Chem. Soc., Chem. Commun.*, 1992, 913.
23. *J. Am. Chem. Soc.*, 1992, **114**, 6354.
24. *J. Am. Chem. Soc.*, 1985, **107**, 5556.
25. *J. Am. Chem. Soc.*, 1990, **112**, 5665.
26. *J. Carbohydr. Chem.*, 1990, **9**, 783.
27. *Tetrahedron Lett.*, 1989, **30**, 5619.
28. *Angew. Chem., Int. Ed. Engl.*, 1982, **21**, 72.
29. *Can. J. Chem.*, 1988, **66**, 3083.
30. *Tetrahedron Lett.*, 1991, **32**, 671.

7 Synthesis of Enantiomerically Pure Non-carbohydrate Compounds by Use of Monosaccharides

There is great demand for sources of asymmetric compounds in enantiomerically discrete forms for all manner of purposes including those related to the control of biological processes, notably many in the field of human health. The world's enormous modern pharmaceutical industry utilizes very many such products and its initial failure to recognize the need to provide thalidomide free of the teratogenic optical isomer stands as a permanent and potent reminder of the biological dissimilarity of enantiomers. Natural products, biosynthesized in the chiral environments of the cells of plants, animals and micro-organisms with the aid of asymmetric catalysts (enzymes), have traditionally offered an abundant source of pure enantiomers; chemical synthetic methods are now also of major importance.

Vast efforts have gone into the development of a range of chemical methods for obtaining 'optically pure' compounds including the use of reagents and reactants bound to asymmetric compounds as 'auxiliaries'. In the latter case the chiral components are temporarily attached to the reactants which therefore acquire asymmetric environments and develop induced asymmetry during reaction. Chiral reagents cause selective formation of specific enantiomers from prochiral groups. Monosaccharides and their derivatives may serve as sources of both chiral reagents and reactants and, furthermore, are very frequently used as inexpensive and versatile starting materials which, by specific chemical manipulations, may be converted into an extensive range of single enantiomers of non-carbohydrate products.

7.1 CARBOHYDRATES AS CHIRAL AUXILIARIES

Whereas the creation of chiral centres from single prochiral carbon atoms, in the absence of any asymmetric influence such as solvent, catalyst or reagent, gives racemic products (as, for example, in the reduction of unsymmetrical ketones to secondary alcohols), enantiomers are formed in unequal proportions if the environments are asymmetric. It is therefore possible to induce asymmetry by carrying out appropriate reactions in the presence of asymmetric compounds such as carbohydrates. Most often they are employed as chiral auxiliaries by bonding to reactants, thereby providing localized asymmetric environments, and they are removed once new chiral centres are established.[1] Also, however, they can be used to render reagents asymmetric and can thus be effective without bonding to the substrates.

7.1.1　CARBOHYDRATE-BOUND REAGENTS

Treatment of 1,2:5,6-di-O-isopropylidene-α-D-glucofuranose with 9-borabicyclo-[3.3.1]nonane gives a borinic acid ester ($R^1R^2BOR^3$) which, with potassium hydride, is converted into the complex borohydride **7.1**. With this unsymmetrical reagent aryl ketones are reduced in near quantitative yield and with high selectivity, the (R)-alcohol **7.2** being obtained in this way in 93% yield and in > 97% enantiomeric excess (ee), as illustrated in Scheme 7.1.[2] With less sterically demanding ketones the reagent, however, shows much less selectivity, 2-butanone giving butan-2-ol with only 3% enantiomeric excess. Such enantioefficiency can easily be exceeded simply by use of complexes derived from carbohydrate alcohols and sodium borohydride or lithium aluminium hydride, and these can also reduce oximes to asymmetric amines with enantioselectivities of up to 50%.[3]

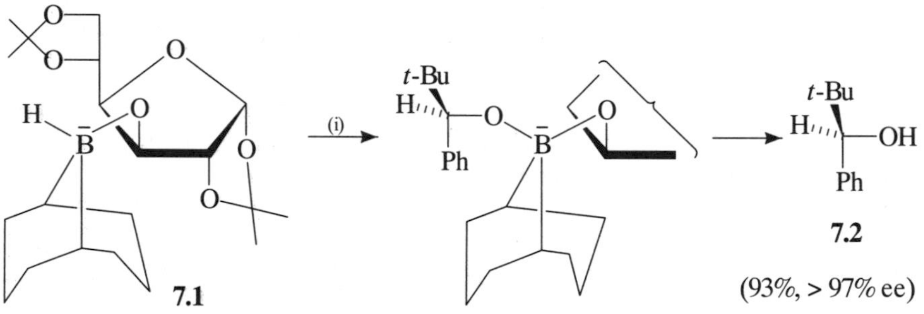

(i) PhCOBu-t, THF, $-78\,^\circ$C

Scheme 7.1

Asymmetric C-alkylations at activated carbon centres can also be effected. Thus, C-methylation of the Schiff base **7.3** in the enolate form **7.4** gives the (S)-ester **7.6** in 40% enantiomeric excess on treatment with the carbohydrate methyl sulphate **7.5** (Scheme 7.2).[4]

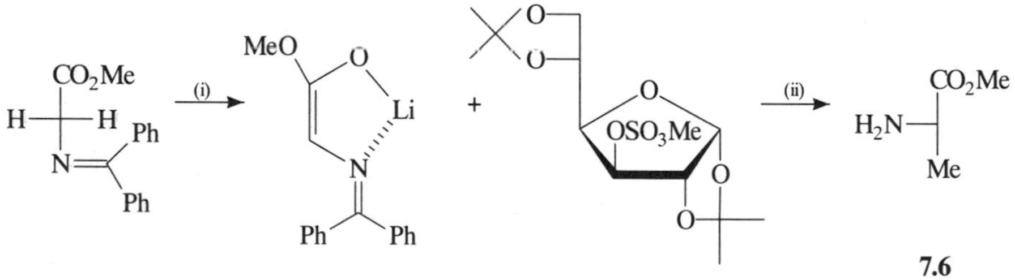

(i) LDA, THF, HMPIT, $-40\,^\circ$C; (ii) H_3O^+

Scheme 7.2

7.1.2   CARBOHYDRATE-BOUND REACTANTS

A particularly simple example of asymmetric induction by this approach is provided
by the Claisen rearrangement of **7.7** in the presence of sodium borohydride, which
gives the diastereoisomers **7.8** comprising 3:2 proportions of the glycosides of the
(R)-alcohol **7.9** and (S)-alcohol **7.10** (Scheme 7.3). Separation of the acetates of
the first products followed by hydrolysis gives the enantiomeric alcohols.[5]

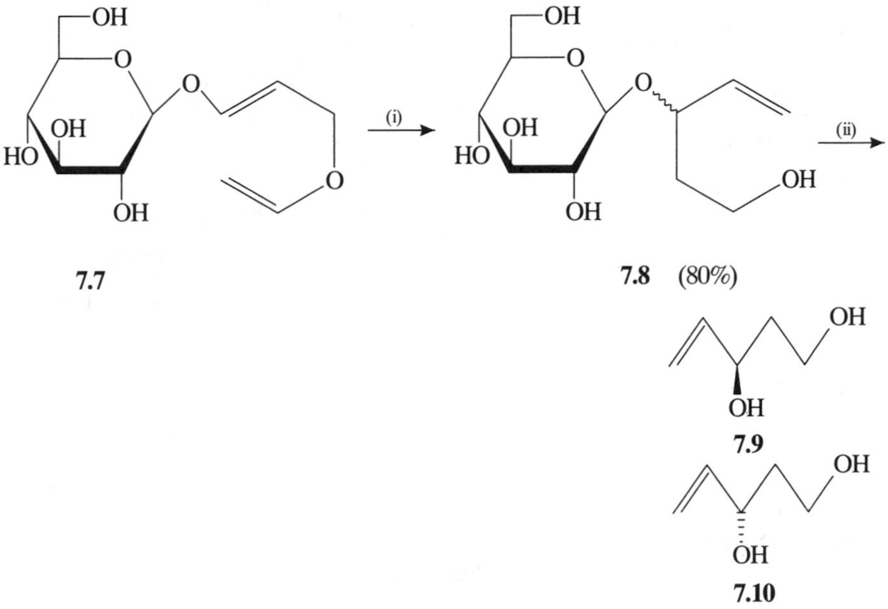

7.7

7.8   (80%)

7.9

7.10

(i) H_2O, $NaBH_4$, 60 °C, 4 h; (ii) H_3O^+

Scheme 7.3

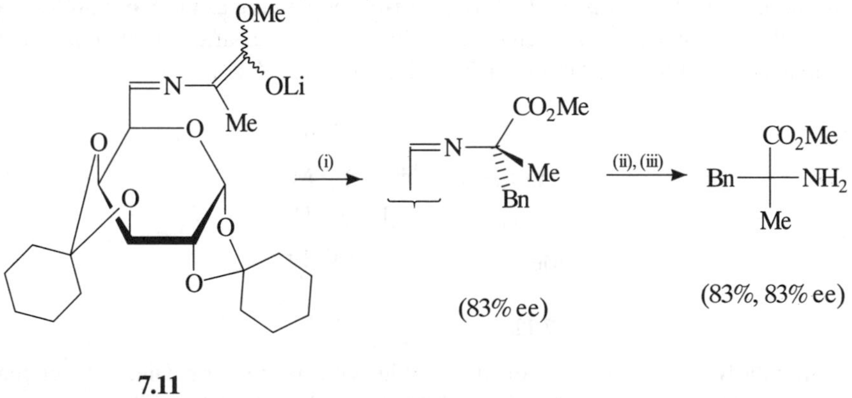

(83% ee)

(83%, 83% ee)

7.11

(i) BnBr, THF, HMPIT, −78 °C, 2 h; (ii) MeOH, HCl, H_2O; (iii) NH_3

Scheme 7.4

A broader approach to optically active α-alkyl-α-amino acids relies on the alkylation of carbohydrate-bound Schiff base enolates, e.g. **7.11** (Scheme 7.4),[6] and α-amino acids themselves are obtainable via cyanide addition (Strecker synthesis) to glycosylamine Schiff bases (Scheme 7.5).[7]

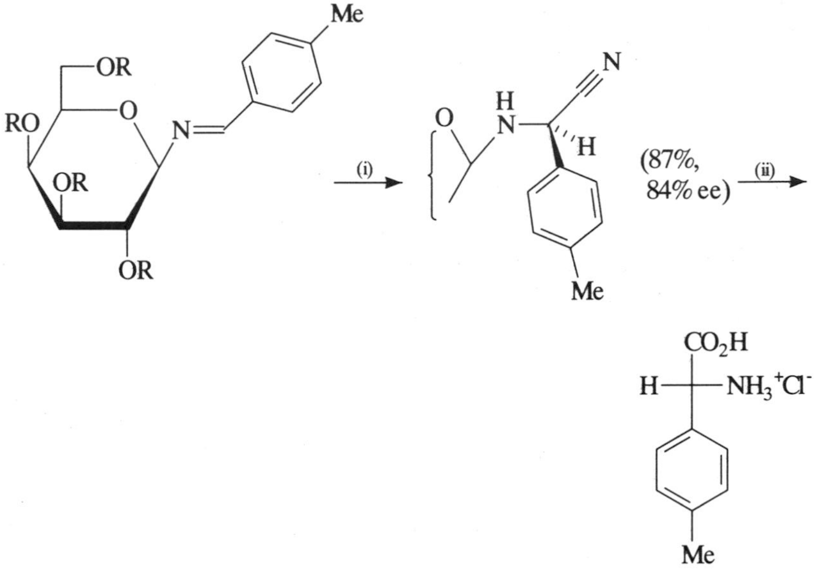

R = COBu-*t*
(i) Me$_3$SiCN, SnCl$_4$, THF, $-78\,^{\circ}$C, 8 h; (ii) HCl
Scheme 7.5

Simple carbohydrate ester groups can be transferred under basic conditions as carbon nucleophiles to alkylate carbonyl compounds and thereby generate asymmetric products. For example, the *C*-carboxymethyl group (CH$_2$CO$_2$H) can be introduced into ethyl pyruvate **7.12** using diisopropylamine as base and 3-*O*-acetyl-1,2:5,6-di-*O*-isopropylidene-D-glucose as source to give, after ethylation, diethyl citramalate with the (*S*)-enantiomer **7.13** dominant (29% ee).[8]

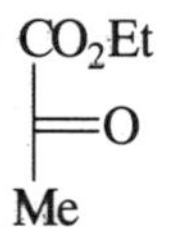

7.12 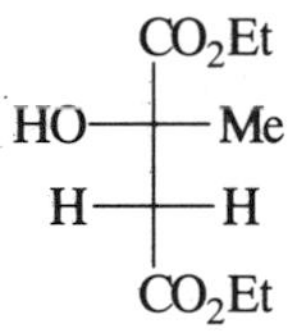   7.13

Particularly striking results are obtainable with asymmetric Diels–Alder reactions, the carbohydrate auxiliary being bound to either the dienophile or the diene.

Reaction of the furanose acrylate **7.14** with titanium(IV) chloride gives the complex **7.15** which, with cyclopentadiene, affords the norbornyl carboxylic acid ester **7.16** in 73% yield and with diastereoisomeric selectivity of > 98% in favour of the illustrated (1*R*, 2*R*)-compound, as revealed by the alcohol **7.17** formed by reductive de-esterification (Scheme 7.6).[9]

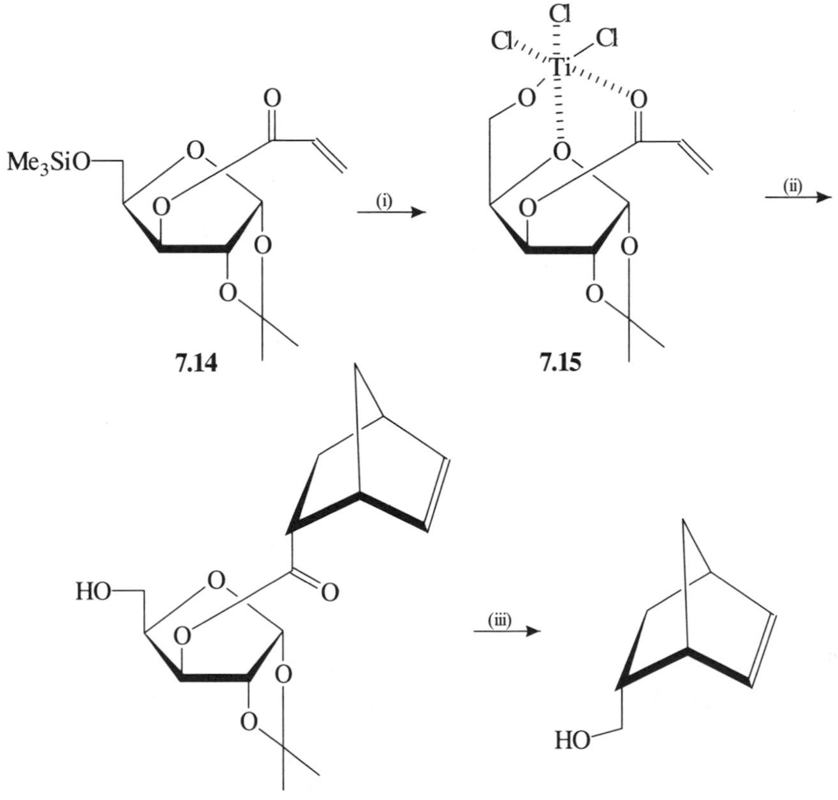

7.14

7.15

7.16 (73%)

7.17 (98% ee)

(i) $TiCl_4$, CH_2Cl_2, -78°C; (ii) ⬠ ⟋⟋ ,-78°C, 12h; (iii) $NaBH_4$

Scheme 7.6

When used in the reverse manner the carbohydrate is bound to the diene as in **7.19,** which reacts without catalyst with the epoxide dienophile **7.18** at 5°C in benzene to give **7.20**, containing five asymmetric centres in the aglycon, in 74% yield (Scheme 7.7). From it, the anthracyclinone (+)-4-demethoxydaunomycinone **7.21** can be made.[10]

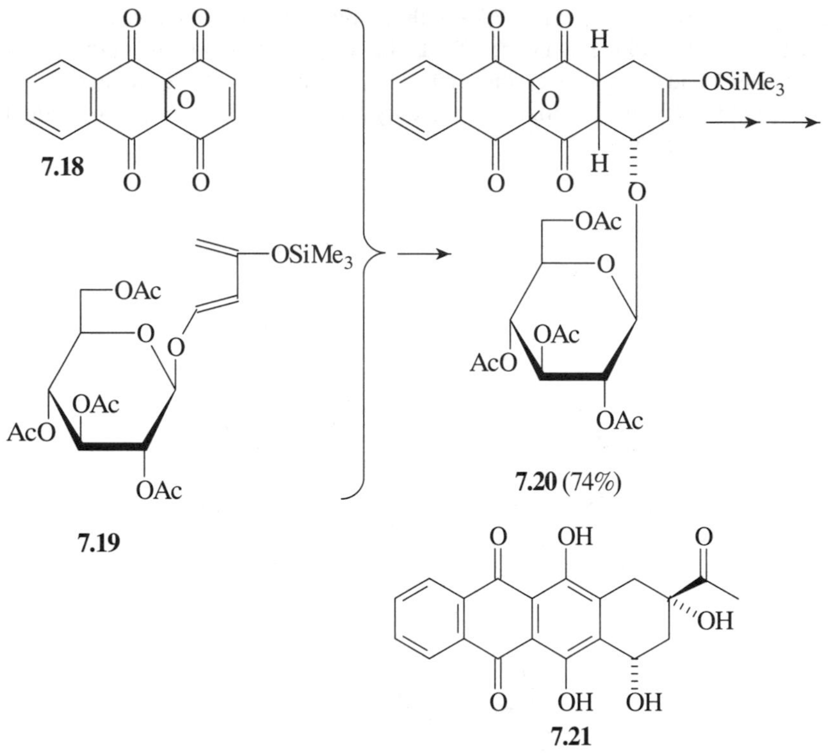

7.18

7.19

7.20 (74%)

7.21

Scheme 7.7

7.2 THE CONVERSIONS OF CARBOHYDRATES INTO ENANTIOMERICALLY PURE NON-CARBOHYDRATE COMPOUNDS

This now very extensive subject has developed rapidly since the 1970s and represents a commonly selected means of making a very wide range of compounds, or parts of compounds, which are normally natural product in origin.[11-14] Target substances, which have clear similarity to monosaccharides, for example polyhydroxylated cyclo-pentanes or -hexanes, are usually formed by conversions applied monosaccharides. In other cases extended-chain or branched-chain compounds elaborated from sugars are used. On the other hand, when the intended products contain only a few asymmetric centres, it is common practice to use as starting materials fragments obtainable from monosaccharides; e.g. 2,3-*O*-isopropylidene-D-glyceraldehyde **7.22** made by periodate oxidation of 1,2:5,6-di-*O*-isopropylidene-D-mannitol, or other sugar derivatives from which all but a very few asymmetric centres have been removed. For example, the enone **7.26** (Scheme 7.8), easily made from a readily available unsaturated carbohydrate via the oxime **7.25**,[15] can be converted into the α-keto oxime **7.27** which, on selective reduction, gives mixed

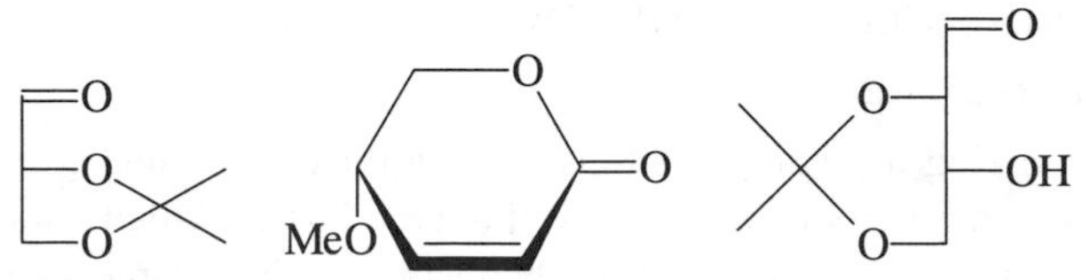

7.22   **7.23**   **7.24**

epimeric ketoamines the dimerization of which results in (*S, S*)-palythazine **7.28**, a natural product made by the salt water invertebrate *Palythoa tuberculosa*.[16] Compounds **7.23** and **7.24** are other examples of simple asymmetric compounds available from sugars.

To be effective, this approach ideally requires that the carbohydrate starting materials be available easily and cheaply, that the synthetic steps be efficient and not too numerous, that simple and inexpensive reagents be involved, that the reactions show appropriate and good selectivity, and that no difficult separations are involved.[17] Finally, the overall operations should be applicable on reasonably large scales. While these are demanding criteria, they serve as challenges which improvements in synthetic methodology are meeting continuously more effectively.

Most of the conversions that follow are rather straightforward examples that illustrate the extensive range of enantiomerically pure products obtainable from readily available carbohydrate compounds. Emphasis is placed on the initial reactions by which other types of compounds are made; subsequent details relating to transitions to specific target compounds are often glossed over. Particular attention is given to the important production of functionalized cyclopentanes and cyclohexanes (Sections 7.2.2.b and 7.2.2.c) which require specific conversion processes; the other sections cover such ranges of compounds that the choice of illustrations is highly selective.

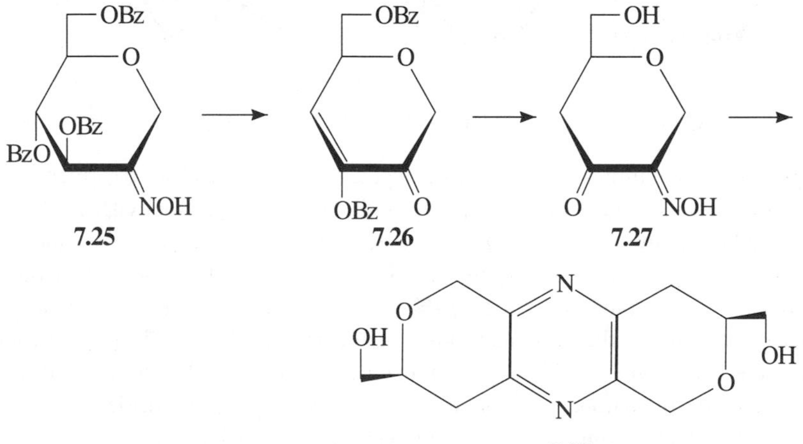

7.25   **7.26**   **7.27**

7.28

Scheme 7.8

7.2.1 HETEROCYCLIC COMPOUNDS

(a) Oxygen heterocycles

With nature utilizing a plethora of bioactive chiral compounds containing highly substituted oxygen heterocyclic systems, this area has provided some of the most demanding targets for synthetic organic chemists, and these targets have often been reached by use of carbohydrate starting materials. Palytoxin, with its 65 asymmetric centres and eight rings represented by the seven structural features **7.29**–**7.35**, is the pre-eminent example, its synthesis having been accomplished by Kishi's group in an extraordinary display of the power of modern methodology.[18] The rings are *C*-glycosidic in character, and not surprisingly, therefore, carbohydrates were selected as the starting materials for several of them.

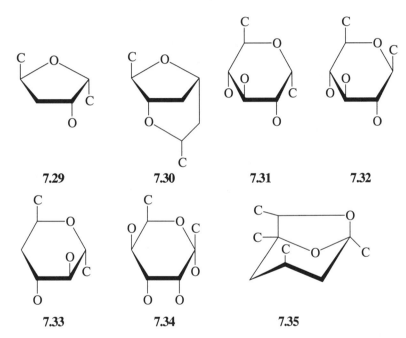

Only a few examples of syntheses of specific compounds can be given. Simple synthons are available from 2,3-*O*-isopropylidene-D-glyceraldehyde **7.22**; reduction, tosylation and hydrolysis of the acetal affords the specifically substituted glycerol **7.36** from which various chiral epoxides can be made (Scheme 7.9). Treatment of the epichlorohydrin (14% from D-mannitol) with sodium dimethyl malonate then gives the (−)-enantiomer of the lactone **7.37** and hence a route to enantiomerically pure, highly functionalized cyclopropanes.[19] The product can be accounted for by malonate carbanions first displacing the chloride ion and then opening the epoxide ring to give a hydroxyl diester that lactonizes.

Thermal [3,3]-sigmatropic rearrangement of the trichloroacetimidate **7.38** affords the *N*-trichloroacetamido compound **7.39**, the derived lactone **7.40** (Scheme 7.10)

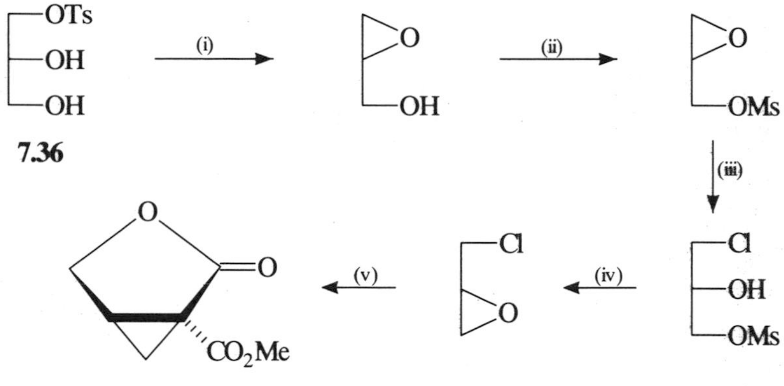

7.36

7.37 (15%)

(i) Na, MeOH; (ii) MsCl, Et$_3$N, toluene; (iii) HCl, MeOH, 20 °C;
(iv) Na, (HOCH$_2$)$_2$, 20 °C; (v) NaHC(CO$_2$Me)$_2$, MeOH, reflux

Scheme 7.9

being a versatile precursor of, for example, variously hydroxylated (S)-α-amino acids.[20]

An example of a compound having two elaborate carbon-bonded substituents on a tetrahydropyranyl ring is the antibiotic pseudomonic acid C **7.45,** an antimicrobial agent against Gram-positive bacteria used in dermatology. It has been made from the iodide **7.41** which was obtained from the commercially available L-lyxose. Radical coupling to the allylic sulphone **7.42** with sulphonyl radical displacement gives the branched-chain **7.43** with good stereoselectivity. A Wittig reaction applied to the free sugar derived from **7.43** then affords the C-glycoside **7.44**; subsequently, with an Emmons phosphonate reagent followed by deprotection the natural product **7.45** is produced.[21]

Other relevant compounds prepared from carbohydrates are **7.46**, the enantiomer of the natural (+)-cryptosporin, made by condensation of a sulphonyllactone with

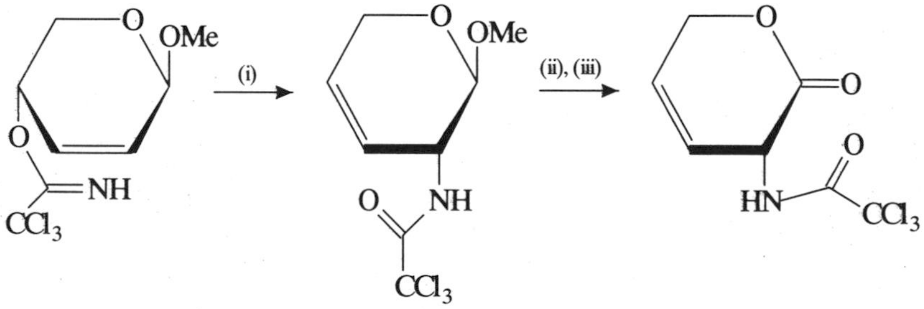

7.38   **7.39** (80%)   **7.40** (77%)

(i) reflux, xylene; (ii) HCl (aqueous); (iii) DMSO, Ac$_2$O

Scheme 7.10

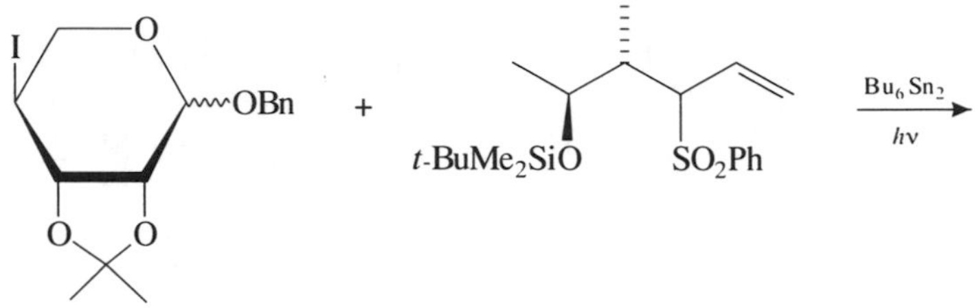

7.41   **7.42**

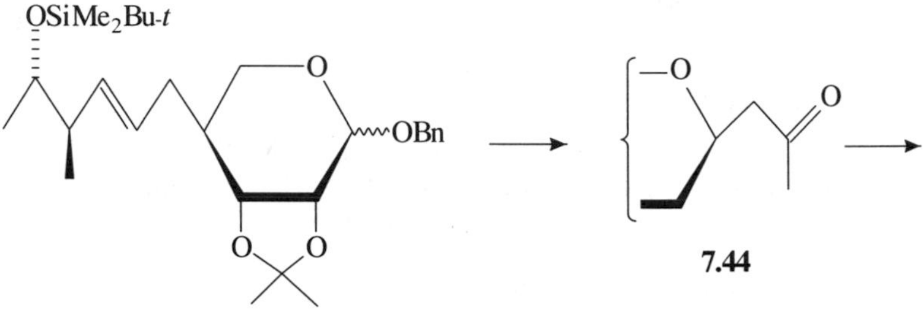

7.43 (74%)   **7.44**

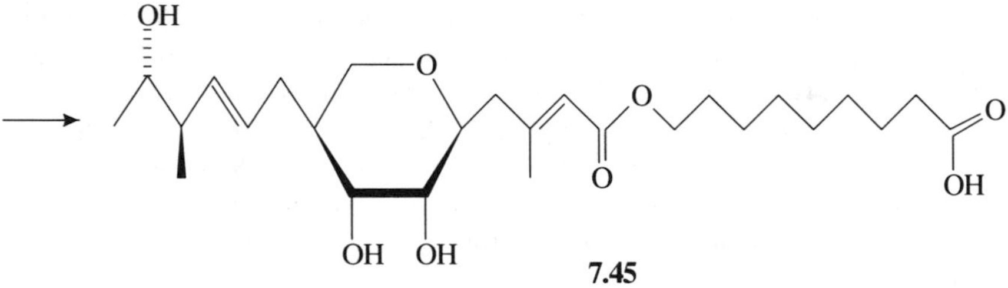

7.45

Scheme 7.11

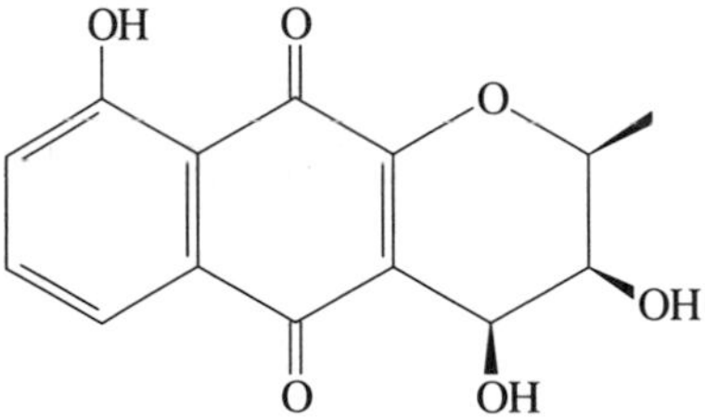

(-)-Cryptosporin **7.46** 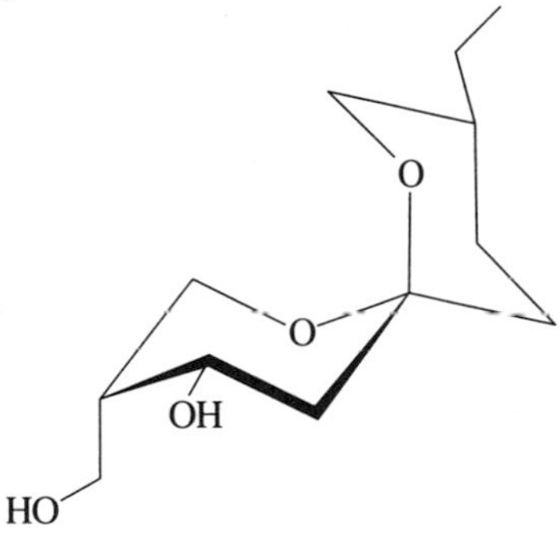   Talaromycin A **7.47**

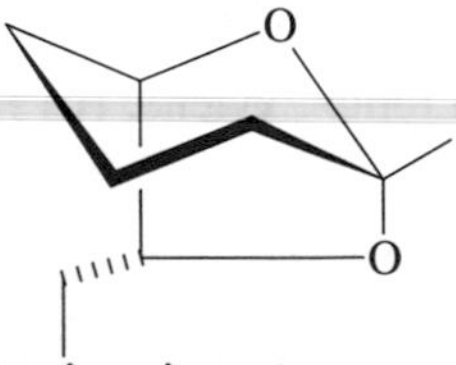

(+)-*exo*-Brevicomin **7.48**

a 1-nitro-L-fucal derivative;[22] the spiroketal talaromycin A **7.47**, produced by C-1 extension of 2,3:4,5-di-O-isopropylidene-D-fructose;[23] and the Western pine beetle pheromone (+)-*exo*-brevicomin **7.48**, which is the anhydride of the hemiacetal form of a 1,3,4,5,8,9-hexadeoxynonulose and can be synthesized by Wittig extension from a readily available 5,6-dideoxy-hex-6-enose compound.[24]

(b) Nitrogen heterocycles

Interest has been shown in saturated, hydroxylated six-membered nitrogen heterocycles because of their potent biological activity.[25] Particularly, some polyhydroxypiperidines are excellent and specific inhibitors of the stereochemically corresponding glycosidase enzymes, the functionings of which control many biological processes. Deoxynojirimycin (1,5-imino-D-glucitol) is thus a glucosidase inhibitor, the isomer deoxymannojirimycin **7.50** inhibits the functioning of mannosidases, and stereoisomers likewise operate on respective enzymes. Consequently, these compounds have important implications for anticancer and antiviral chemotherapy through their impact on the biosynthesis of relevant oligosaccharides (Section 6.1).[26] Sugars serve as very suitable starting materials for their synthesis, the preparation of **7.50** (35% from 1,2:5,6-di-O-isopropylidene-α-D-glucofuranose) being outlined in Scheme 7.12. The key step involves intramolecular displacement of the triflate ester in **7.49** by the amino group formed by reduction of the azide at C-6 by the unusual use of triphenylphosphine followed by aqueous potassium carbonate.[27]

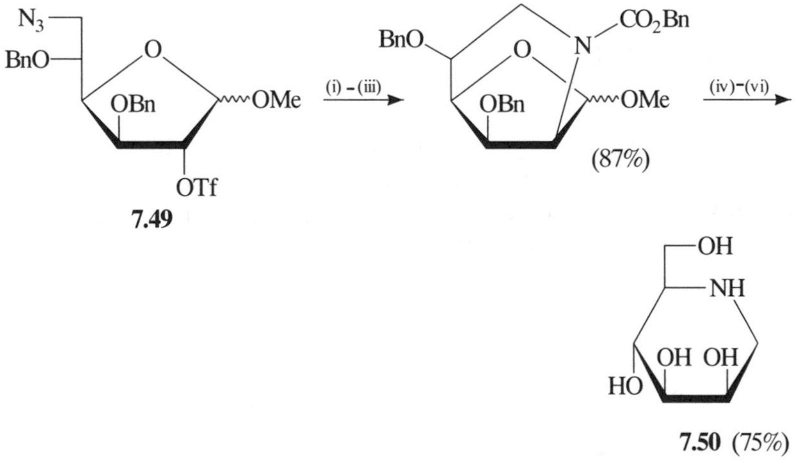

7.49

7.50 (75%)

(i) Ph$_3$P, CH$_2$Cl$_2$ reflux, 2 h; (ii) KOH, H$_2$O; (iii) ClCO$_2$Bn; (iv) H$_3$O$^+$;
(v) NaBH$_4$; (vi) H$_2$, Pd/C

Scheme 7.12

Not surprisingly, the indolizidine plant alkaloid (+)-castanospermine **7.52** also inhibits glycosidases and, in addition, it shows anticancer, antiviral and anti-AIDS activity; it therefore is of major interest. A relatively efficient synthesis is available from 1,2-*O*-isopropylidene-α-D-glucofuranurono-6,3-lactone **7.51** (Scheme 7.13).[28] Likewise, the related alkaloid swainsonine **7.53** shows glycosidase inhibitory and other biochemical activities and has been prepared from carbohydrate precursors.[29]

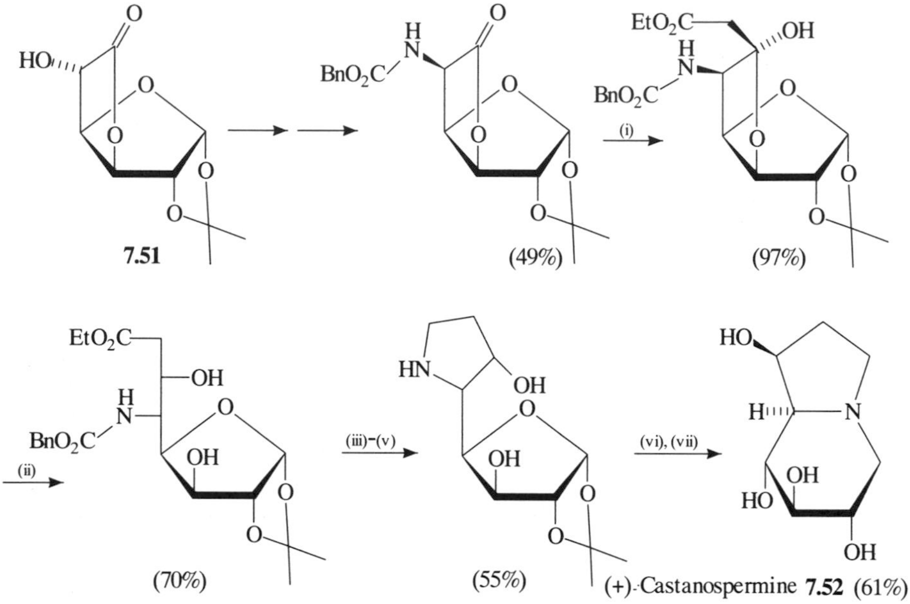

7.51   (49%)   (97%)

(70%)   (55%)   (+)-Castanospermine **7.52** (61%)

(i) *i*-Pr$_2$NLi, MeCO$_2$Et; (ii) H$_2$, PtO$_2$; (iii) HCO$_2$H; (iv) resin (OH$^-$); (v) LiAlH$_4$; (vi) CF$_3$CO$_2$H; (vii) H$_2$, Pt/C

Scheme 7.13

Access to nitrogen heterocycles of interest in the area of β-lactam antibiotics can be gained by addition of trichloroacetyl isocyanate to glycals, compound **7.55** being available in this way from the corresponding D-allal derivative in 72% yield. From this the triol **7.56** can be made and hence, by periodate oxidation of the C-3,

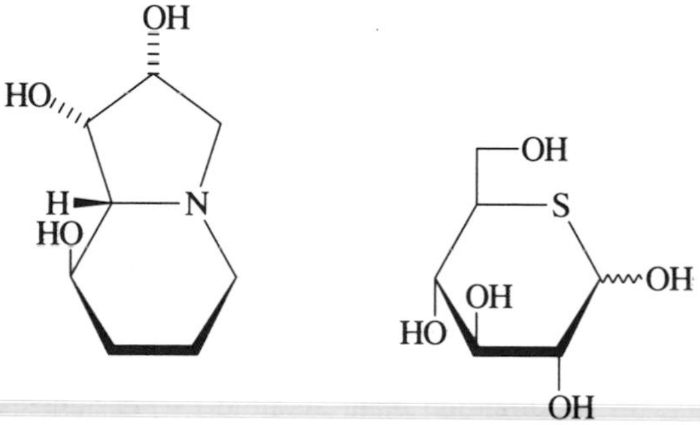

Swainsonine **7.53**   5-Thioglucose **7.54**

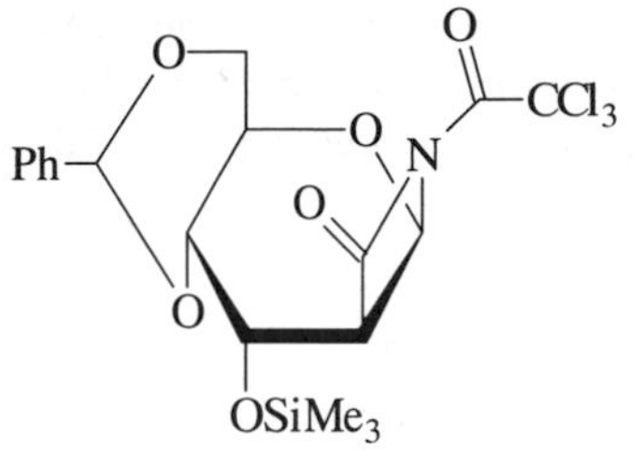

7.55 (81%) 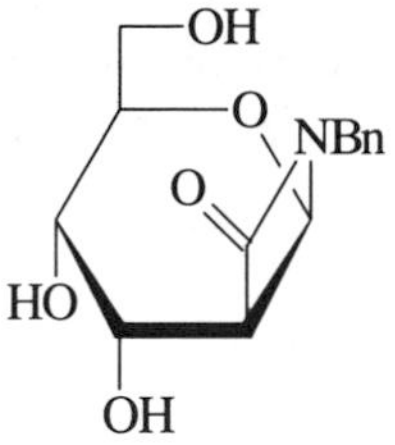   **7.56** 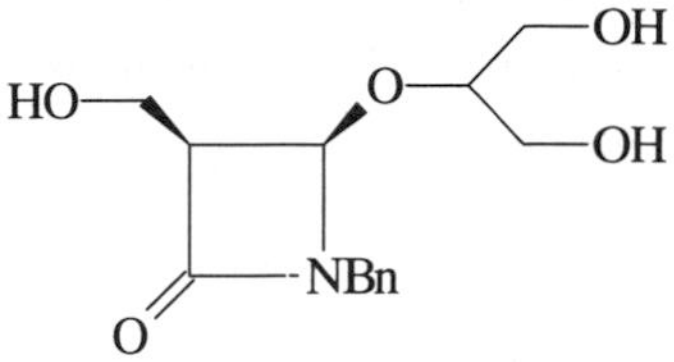   **7.57**

C-4 diol and sodium borohydride reduction, the azetidinone **7.57** with the appropriate configuration for use in the synthesis of 1-oxaphenams and 1-oxacephenes is obtainable.[30]

(c) Sulphur heterocycles

Although it has been reported as a reversible sterilant of male rats, 5-thio-D-glucose **7.54** has not retained the attention given to several 'nitrogen in the ring' analogues (see above). It can be made in 85% yield from the episulphide **4.197** by treatment with potassium acetate in acetic acid/acetic anhydride followed by deprotection.[31]

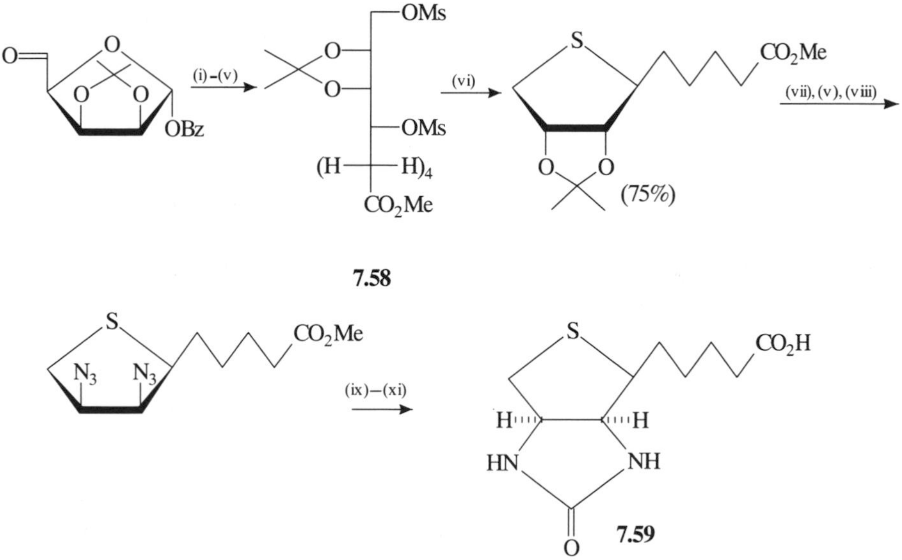

7.58

7.59

(i) $Ph_3P=\!\!\diagup\!\!\diagdown CO_2Me$ (ii) H_2, Pd/C; (iii) NaOMe; (iv) NaBH$_4$; (v) MsCl;

(vi) Na$_2$S, HMPIT; (vii) HCO$_2$H; (viii) NaN$_3$; (ix) PtO$_2$, MeOH, Ac$_2$O;

(x) Ba(OH)$_2$; (xi) COCl$_2$

Scheme 7.14

Somewhat further removed from parent sugars, the vitamin B complex constituent (+)-biotin **7.59** has been prepared from D-mannose, the critical step being
treatment of the derived dimesylate **7.58** with sodium sulphide in the aprotic HMPIT
(Scheme 7.14).[32]

7.2.2 CARBOCYCLIC COMPOUNDS

Although carbocyclic compounds can be made from many types of organic precursors, carbohydrates can be particularly suitable, especially for enantiomerically pure
products rich in oxygen-bonded substituents. Several efficient synthetic routes are
now available for the preparation from sugar derivatives of the most important
cyclic systems, the cyclopentanes and cyclohexanes,[33] but rings of other sizes can
also be produced.

(a) Cyclopropane, cyclobutane and cycloheptane derivatives

Commonly the Simmons–Smith reaction with diiodomethane and zinc–copper
couple is used to prepare cyclopropane compounds from carbohydrate alkenes,
the stereochemistry of the carbene addition being particularly subject to the influence of allylic groups.[34] Otherwise they can be made from epoxides, as illustrated
by the conversion of the D-*allo*-compound **7.60** to the cyclopropane **7.61** on heating
with a phosphonate (Scheme 7.15). From this product the *gem*-dimethyl analogue
and hence (−)-chrysanthemic acid **7.62** (and its enantiomer) have been prepared
(See also Scheme 7.9).[35]

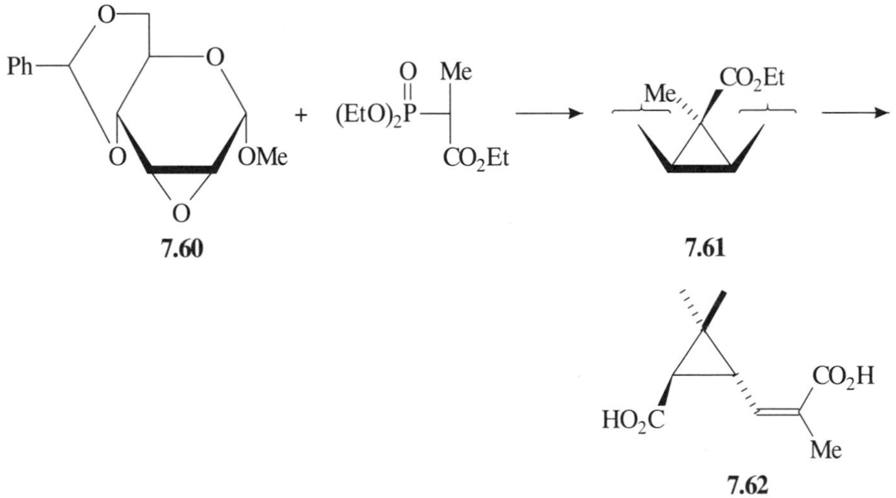

Scheme 7.15

When carbanions are established in a β-relationship to leaving groups, as at
C-1 of dithioacetal **7.63**, and provided elimination or alternative nucleophilic

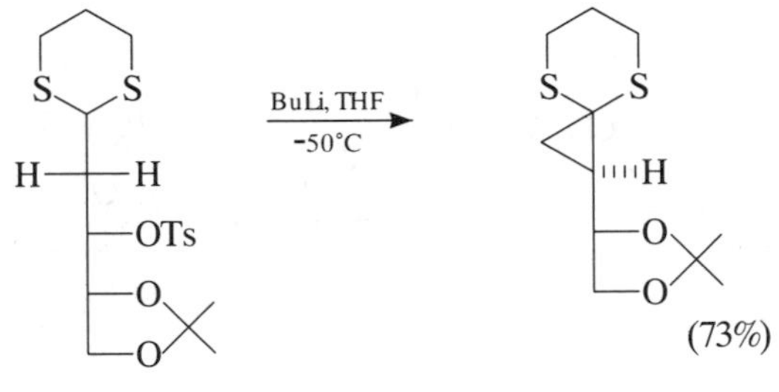

7.63   (73%)

Scheme 7.16

displacement cannot occur, cyclopropane derivatives are produced in good yield
(Scheme 7.16).[36]

Cyclobutanes are not common products of intramolecular nucleophilic displacement reactions, the 3,5-di-O-methyl-4-tosylate analogue of compound **7.63** giving
only 18% of the four-membered cyclic product with n-butyllithium.[36] Photochemical [2+2]-cycloaddition reactions, however, are more satisfactory, compound **7.64**
being obtained in 50% yield by addition of ethyne to a glycal-based enone. Conversion to **7.65** by deacetoxylation at C-4, methyllithium addition at the carbonyl centre
and formic acid catalysed rearrangement results in the 2-oxabicyclo[3.2.1]octane
system akin to the B and C rings of the trichothecene sesquiterpenes.[37] In similar
fashion, (−)-grandisol **7.66**, a boll-weevil pheromone, and its enantiomer can be
made from the product **7.67** of ethene addition to the appropriate enone.[38]

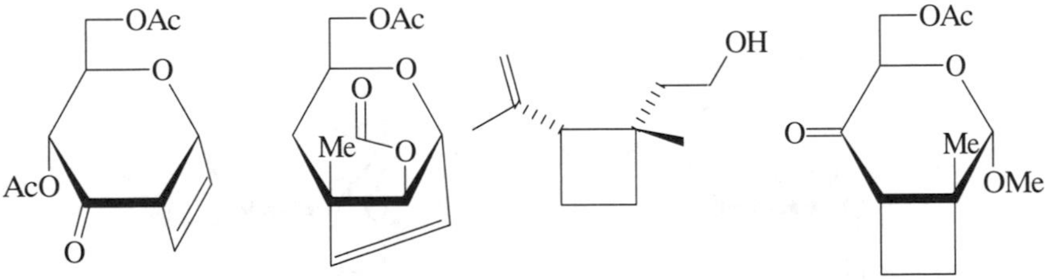

7.64   **7.65**   **7.66**   **7.67**

Although the conversion is uncommon, carbohydrate derivatives can be used to
provide functionalized cycloheptanes, the D-mannose-based diene **7.68**, on treatment
with a palladium(0) catalyst, giving **7.70** by way of the enolate π-complex **7.69**
(Scheme 7.17).[39] In this case steric factors inhibit cyclization to the corresponding
cyclopentanone, and analogous inhibition leads to an excellent yield of a further
cycloheptane derivative in a radical cyclization noted below under 'cyclohexane
derivatives'.

(b)   Cyclopentane derivatives

Although they have only been developed in the main since 1980, there are
now several good methods available for the conversion of carbohydrates into

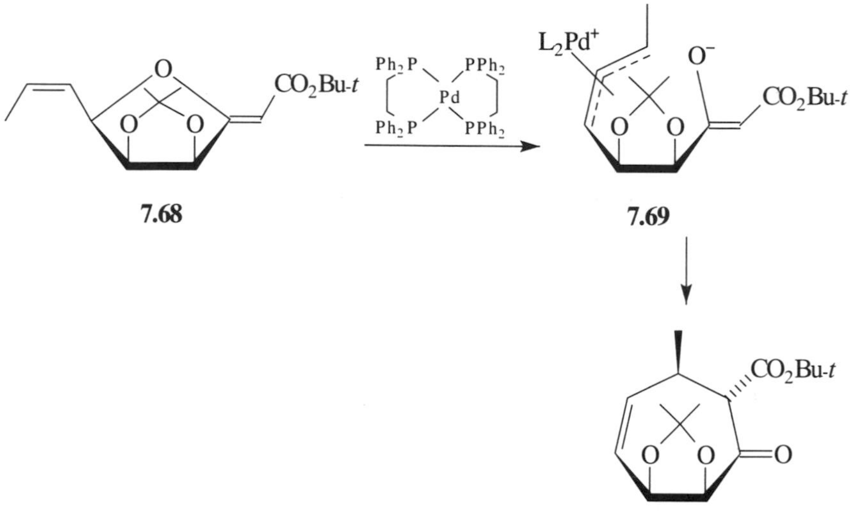

7.68   **7.69**

7.70 (64%)

Scheme 7.17

cyclopentane compounds. Straightforward intramolecular displacements can be used as in the case of the 2-deoxy-D-ribose dithioacetal **7.71**, from which, with n-butyllithium, a sulphur-stabilized carbanion is produced to displace the tosyloxy group and afford the cyclopentane derivative **7.72** (Scheme 7.18).[36]

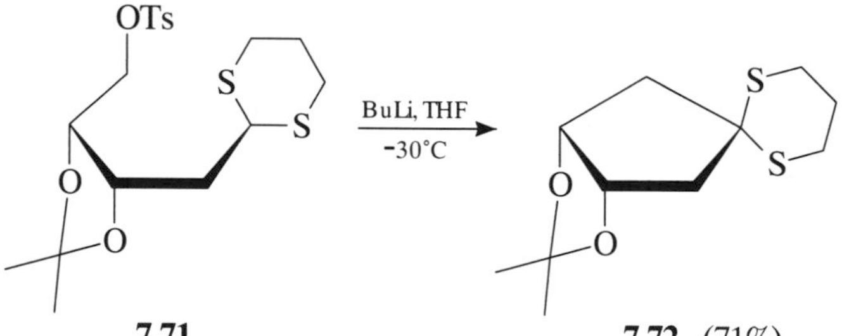

7.71   **7.72** (71%)

Scheme 7.18

In a very sophisticated synthesis of prostaglandin $F_{2\alpha}$ **7.74**, a similar (but nitrile-activated) carbanion cyclization is applied to the D-glucose-based **7.73** (Scheme 7.19, glucose carbon atoms identified by number).[40]

1,4-Dicarbonyl compounds can be derived from sugars by, for example, oxidation at C-4 of aldose derivatives, and potentially are sources of cyclopentanes by aldol attack of C-5 carbanions at C-1. Such processes are, however, kinetically unfavoured, and although in some instances cyclopentanes have been made in this way, difficulties in producing products in good yields are normally encountered.

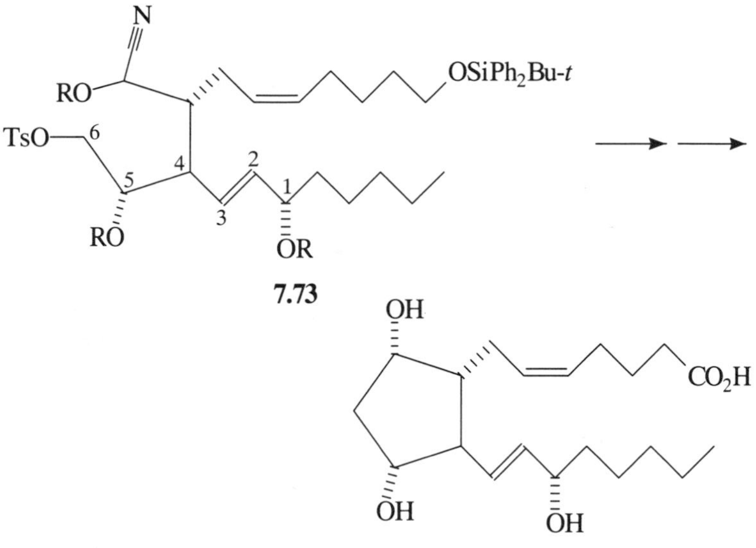

7.73

R = CHMe(OEt)   **7.74**

Scheme 7.19

A method which circumvents the problem utilizes the D-ribose-derived enollactone **7.75** which, by treatment with a complex hydride followed by quenching in aqueous ammonium chloride, is converted into **7.76** and hence, by a β-elimination induced by use of mesyl chloride, into the enone **7.77** (Scheme 7.20).[41] Mannostatin A **7.78**, a natural product which inhibits the action of mannosidases, has been prepared by use of the cyclohexylidene analogue of **7.75**.[42]

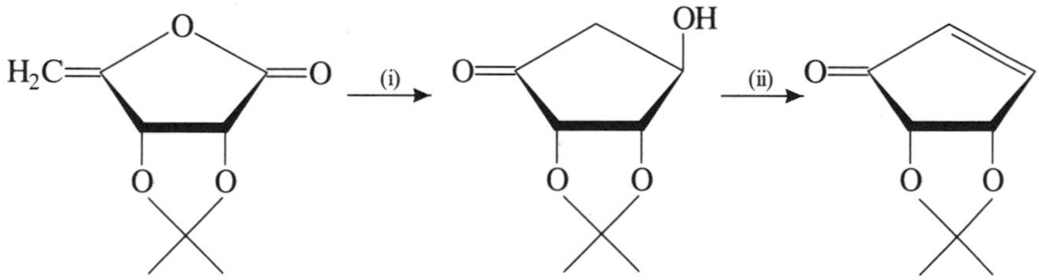

7.75   **7.76**   **7.77** (74% overall)

(i) Li(t-BuO)₃AlH; (ii) MsCl, Py

Scheme 7.20

Free radical intramolecular cyclization processes are extremely suitable for cyclopentane syntheses, and as appropriate radicals can be prepared from unsaturated carbohydrate derivatives, this approach is a very good way of producing highly functionalized cyclopentanes. For example, alcohol **7.79**, obtained by a Wittig reaction applied to 2,3,4,6-tetra-O-benzyl-D-glucose, when converted to an

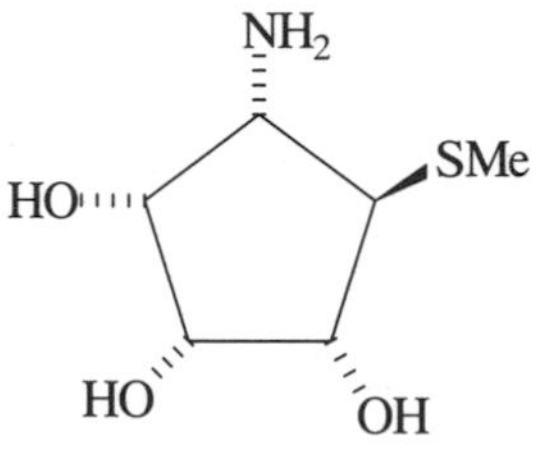

Mannostatin A **7.78**

appropriate thioester and treated with tributyltin hydride and a radical initiator produces radical **7.80** which is trapped intramolecularly to give **7.81** together with small amounts of two stereoisomers (Scheme 7.21).[43]

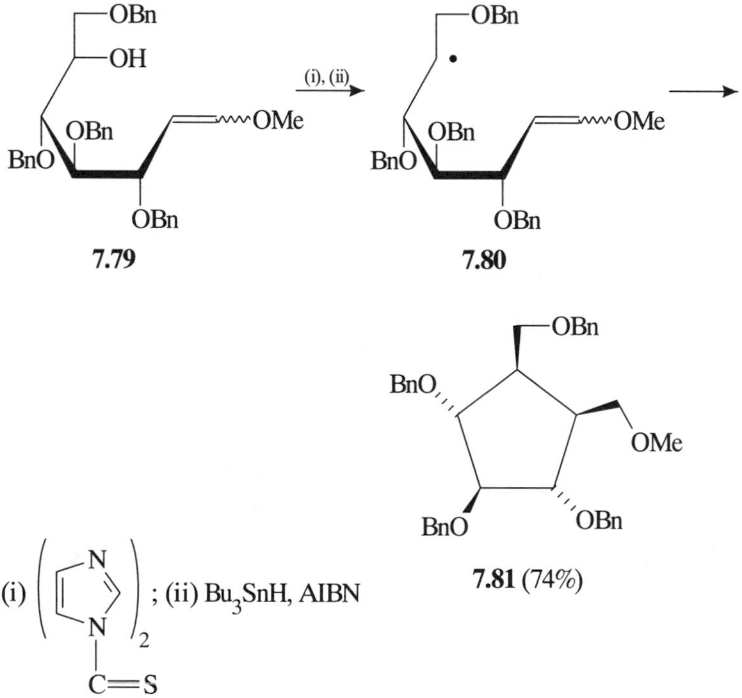

(i) $\left(\begin{array}{c} \text{N} \\ \text{N} \end{array} \right)_2$; (ii) Bu$_3$SnH, AIBN   **7.81** (74%)

C=S

Scheme 7.21

Treatment of 6-deoxy-6-halohexopyranosides with zinc in aqueous ethanol causes ring opening and the production of enals; for example, the relevant methyl 6-deoxy-6-iodo-α-D-glucopyranoside yields **7.82** with high efficiency (see Section 4.5.2.a). On reaction with N-methylhydroxylamine this is converted into nitrone **7.83** which immediately undergoes a 1,3-dipolar cycloaddition reaction to give the isoxazolidine **7.84** from which the prostaglandin precursor **7.85** can be made.[44] Otherwise, **7.82** can be chain extended by a Wittig synthesis to the dienone **7.86** which undergoes photocycloaddition to **7.87** from which the other prostaglandin precursor **7.88** is obtainable (Scheme 7.22).[45]

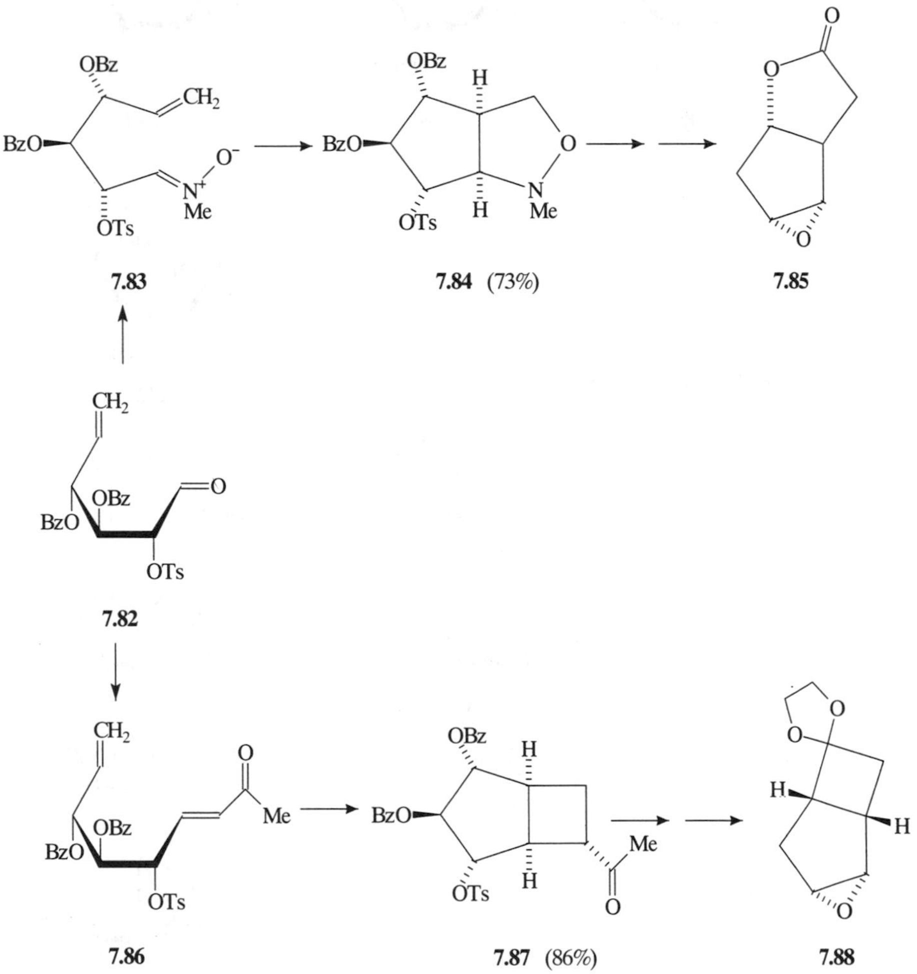

7.83   **7.84** (73%)   **7.85**

7.82

7.86   **7.87** (86%)   **7.88**

Scheme 7.22

(c)   Cyclohexane derivatives

While 4-ketoaldehydes do not aldol cyclize readily to cyclopentanones, 5-keto-hexoses provide an excellent route to cyclohexanones.[46] Acid catalysed removal of the isopropylidene ring from **7.89** (R = H) and treatment of the product **7.90** (R = H) with base therefore gives access to the inosose **7.91** (plus stereoisomers), and then by reduction the epimeric inositols **7.92** (Scheme 7.23).[47] An analogous conversion of the corresponding 6-phosphate **7.90** (R = PO$_3$H$_2$) mimics the key

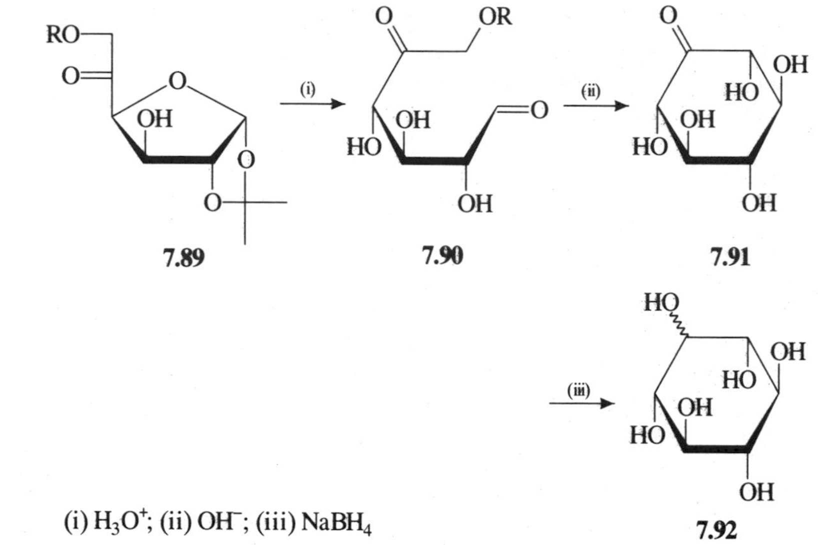

7.89   **7.90**   **7.91**

(i) H$_3$O$^+$; (ii) OH$^-$; (iii) NaBH$_4$ **7.92**

Scheme 7.23

7.93   **7.94**   **7.95** (73% from **7.93**)

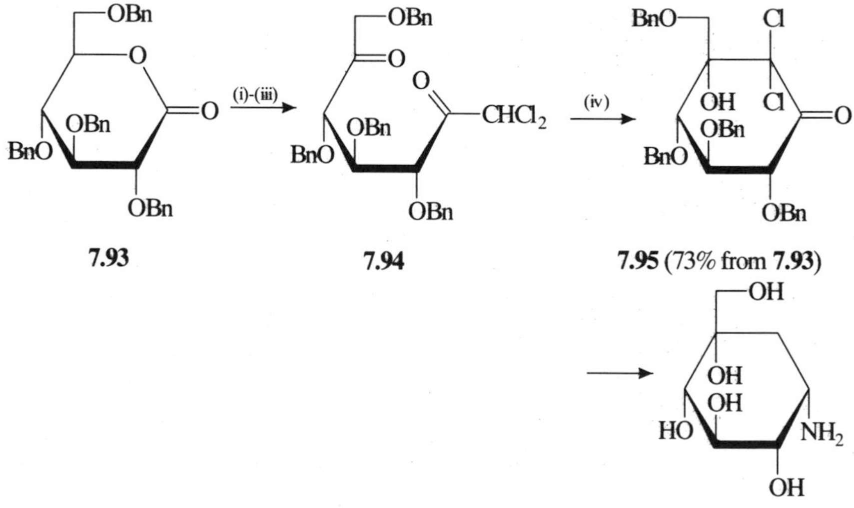

Valiolamine **7.96**

(i) LDA, CH$_2$Cl$_2$; (ii) NaBH$_4$; (iii) (CF$_3$CO)$_2$O, DMSO; (iv) Et$_3$N
Scheme 7.24

step in the natural route to *myo*-inositol from D-glucose.[48] (Inositols are described in Section 8.3.)

A more recent adaptation of the process involves addition of a dichloromethyl carbanion to D-gluconolactone ether **7.93** followed by reduction (of the derived ketone at the former C-1) to give a 2,6-dihydroxyl compound which is oxidized to the corresponding diketone **7.94**. Base-catalysed cyclization gives the tertiary alcohol **7.95** from which the naturally occurring *carba*-aminosugar valiolamine **7.96** can be made (Scheme 7.24).[49]

Frequent use has also been made of ring closures onto aldehydic groups using carbanions stabilized by groups other than ketones; indeed, the first carbohydrate to inositol interconversion was effected by H. O. L. Fischer (the son of Emil Fischer) who, in 1948 with Grosheintz, produced isomeric deoxynitroinositols **7.98** from the 6-deoxy-6-nitrohexoses **7.97** by alkaline treatment (Scheme 7.25).[50]

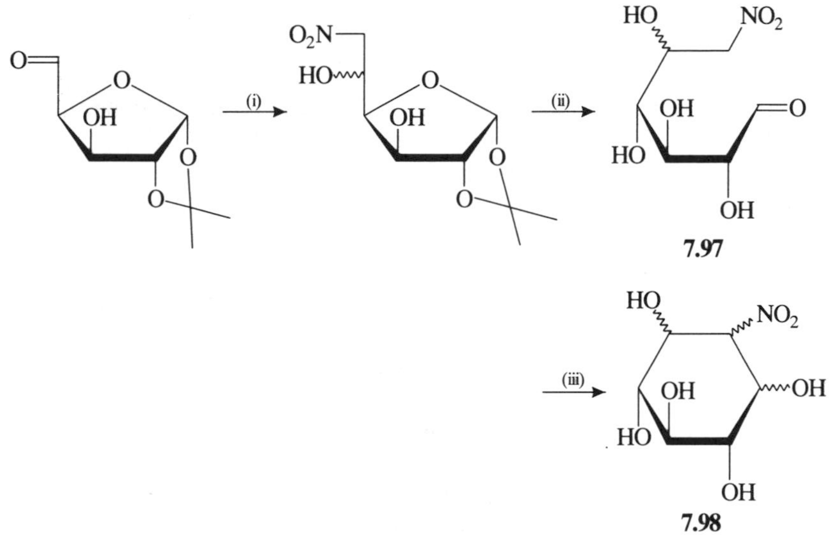

7.97

7.98

(i) MeNO₂, NaOEt; (ii) H₃O⁺; (iii) Ba(OH)₂

Scheme 7.25

More relevant compounds, e.g. streptamine **7.99**[51] and validamine **7.100** (Section 8.3),[52] have been made using this procedure in work related to the aminoglycoside antibiotics. Commonly, this approach can be abbreviated by use of pentose 1,5-dialdehydes and nitromethane in basic conditions which initially give 6-nitrohexoses (e.g. **7.97**) and then ring-closed products.[53]

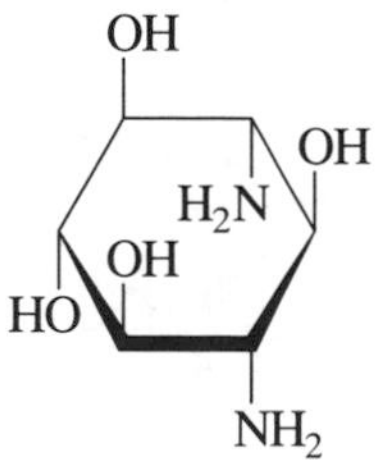

7.99 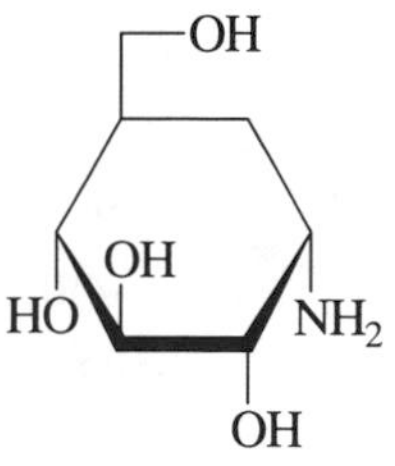   **7.100**

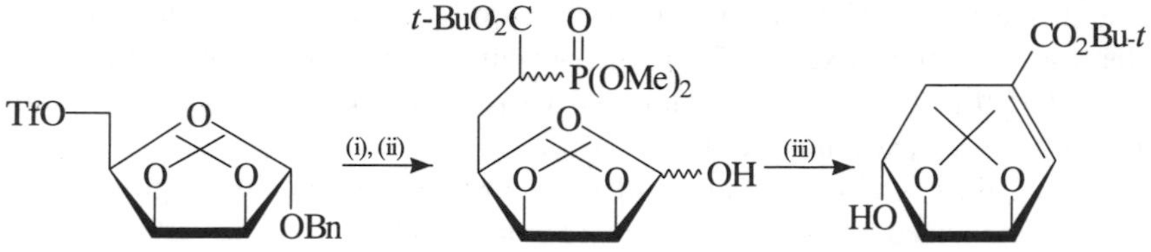

7.101 (73%)

(i) $(MeO)_2\overset{\overset{\displaystyle O}{\|}}{P}CH_2CO_2Bu\text{-}t$, NaH; (ii) H_2, Pd/C; (iii) NaH, THF

Scheme 7.26

An efficient cyclization which uses a phosphorus-stabilized carbanion and results in the shikimic acid derivative **7.101** is illustrated in Scheme 7.26.[54]

A very simple, convenient and efficient method for preparing cyclohexanone derivatives utilizes mercurial intermediates made by mercuration of hex-5-enopyranose or hex-5-enopyranoside compounds (e.g. **7.102**) with mercury(II) salts in inert organic solvents in the presence of some water. Regiospecific hydroxymercuration occurs to give adducts e.g. **7.103** which, being hemiacetals, can spontaneously ring open to 5-ketohexoses mercurated (and thus further activated) at C-6 (e.g. **7.104**). Under the conditions of their formation these undergo intramolecular aldol reactions to result in deoxyinosose products (e.g. **7.105**, Scheme 7.27) in high yield.[33,55] Whereas, originally, this reaction was conducted in refluxing aqueous acetone using molar proportions of mercury(II) acetate or chloride, it has subsequently been found that 0.05 molar equivalents of mercury(II) trifluoroacetate used in this solvent at room temperature are sufficient to allow the reaction to proceed.[56] Other mercury(II) salts also react in catalytic quantities, but more slowly.

A notable and convenient feature of this reaction is the high stereoselectivity shown at the alcohol asymmetric centre generated during the ring closure step, the hydroxyl groups of the isolated products being *trans*-related to the substituents in the β-positions (C-3 of the starting materials). While this could result from complexing, for example, of the mercury ion and O-3 in intermediate **7.104**,

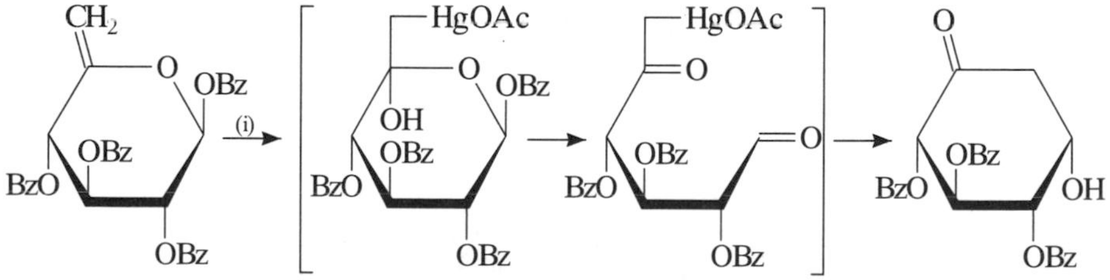

(i) $Hg(OAc)_2$, Me_2CO, H_2O

Scheme 7.27

it appears more probable that it results from the complexing indicated in the intermediate/transition state **7.106**, which clearly would be destablized were the stereochemistry at C-3 inverted. This reaction is illustrated on the cover of the book and was discovered during the authors' researches.

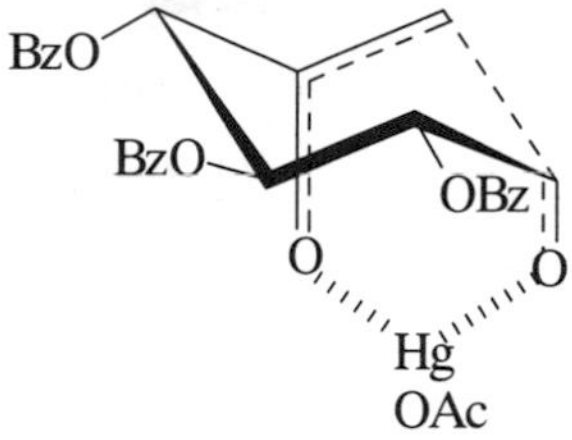

7.106 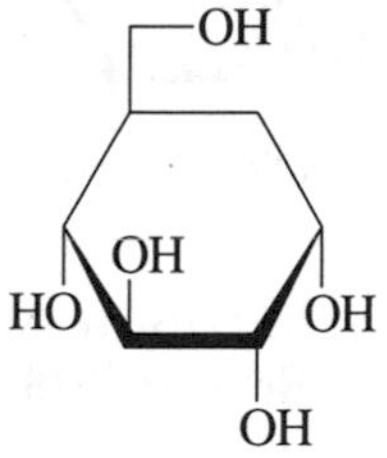   **7.107** 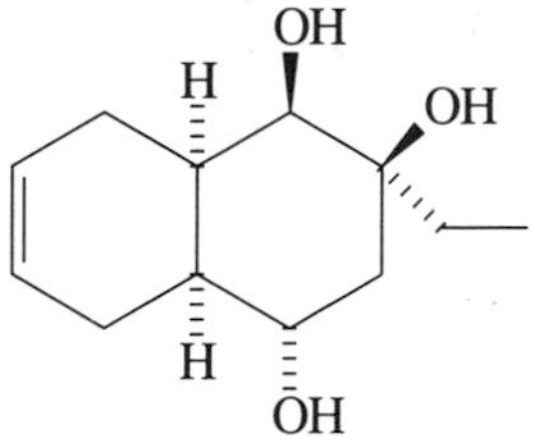   **7.108**

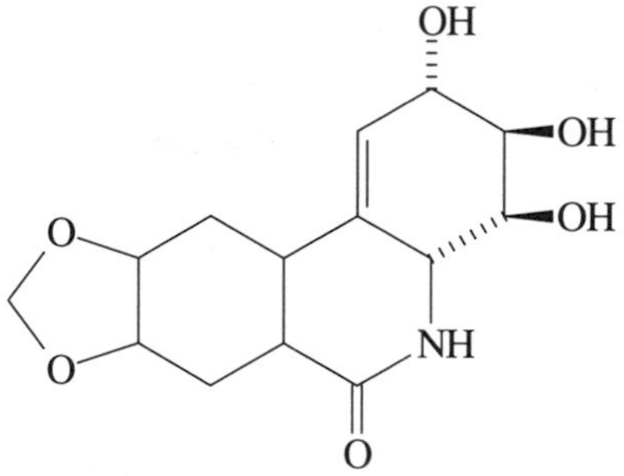

7.109

Many applications have resulted in the production of cyclitols and aminocyclitols of relevance in aminoglycoside antibiotic work, another has given access to *carba-α-D-*glucopyranose **7.107**,[57] while others have led to target compounds outside carbohydrate chemistry; **7.108** and **7.109** are examples which are, respectively, a compound related to the AB ring system of β-rhodomycinone[58] and the alkaloid (+)-lycoricidine are examples.[59] Of particular significance is the applicability of the method to the synthesis of enantiomerically pure inositol derivatives either by hydroxylation of the alkenes readily obtainable by carrying out β-elimination of water from the ketonic products (e.g. **7.105**)[60] or by conducting the cyclizations with alkenes such as **7.110** bearing oxygen-bonded substituents at C-6. In this way compound **7.110**, prepared from the 6-aldehyde, can be converted stereospecifically to the inositol derivative **7.111** (Scheme 7.28).[61] Such approaches are of relevance to the synthesis of specific inositol phosphates, notably *myo-inositol* 2,4-di-phosphate.[62]

Free radical cyclizations can be effective in producing cyclohexane derivatives; for example, **7.112**, derived from the corresponding D-*allo*-iodide by treatment with tributyltin hydride and a radical initiator, gives **7.113** in 87% yield.[63] When, however, six-membered ring formation is sterically impeded, other products can

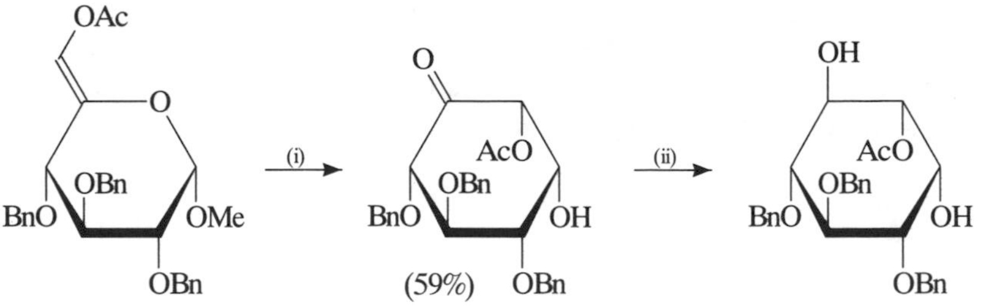

7.110   **7.111** (100%)

(i) $Hg(O_2CCF_3)_2$, Me_2CO, H_2O; (ii) $NaBH(OAc)_3$

Scheme 7.28

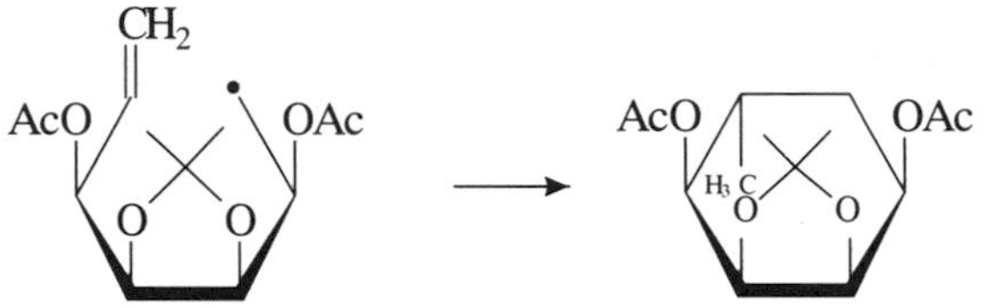

7.112   **7.113** (87%)

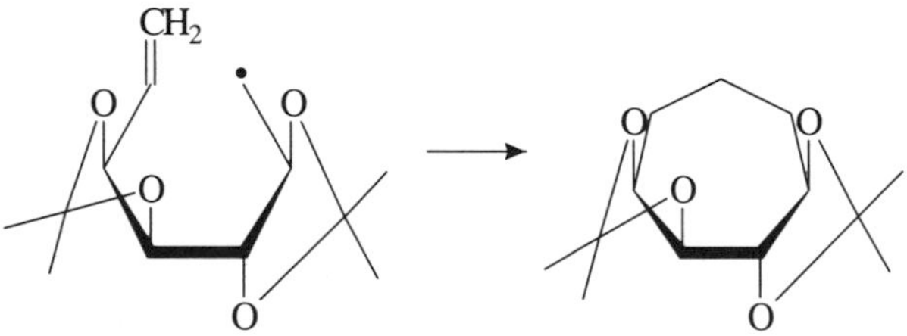

7.114   **7.115** (81%)

Scheme 7.29

ensue, the D-*gulo*-heptitol radical **7.114** giving the cycloheptane derivative **7.115** (81% yield) by an unusual *endo*-closure (Scheme 7.29).

Applied to the branched-chain ester **7.116**, radical cyclization gives access to the bicyclic lactone **7.117**, which is a key intermediate in a synthesis of (+)-phyllantocin, an antileukaemic plant product.[64]

Cycloaddition processes, particularly the Diels–Alder reaction, which may be conducted thermally or in the presence of Lewis acid catalysts, are of considerable value for making functionalized cyclohexane derivatives from carbohydrates. For example, the addition of the highly reactive *o*-xylylene **7.118**, made from *o*-di(bromomethyl)benzene, to the D-glucose-based enone dienophile **7.119** gives the tricyclic product **7.120** related to cryptosporin **7.46** (Scheme 7.30).[65] In reciprocal fashion the carbohydrate diene **7.121** undergoes [4+2]-cycloaddition with

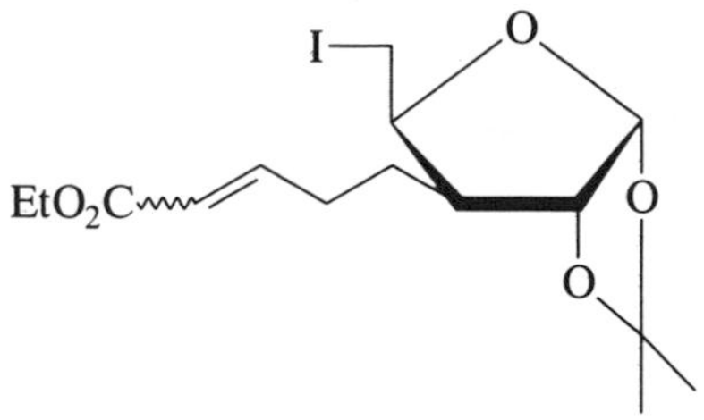

7.116 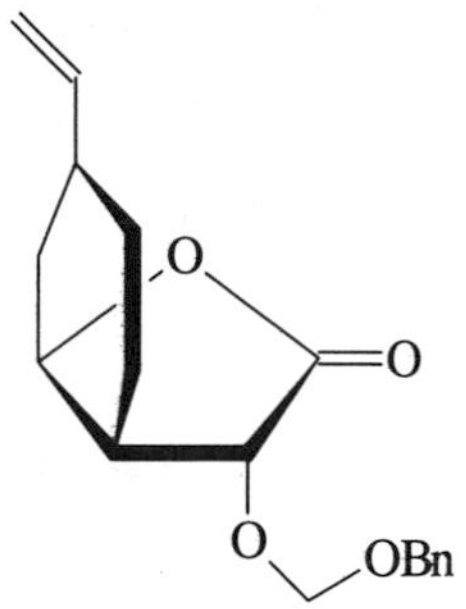   **7.117**

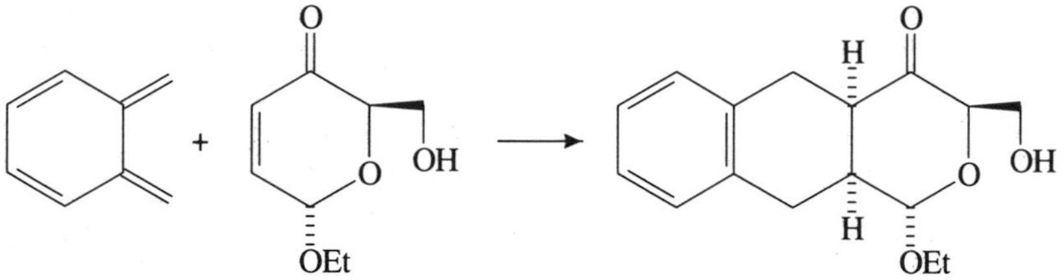

7.118   **7.119**   **7.120** (70%)

Scheme 7.30

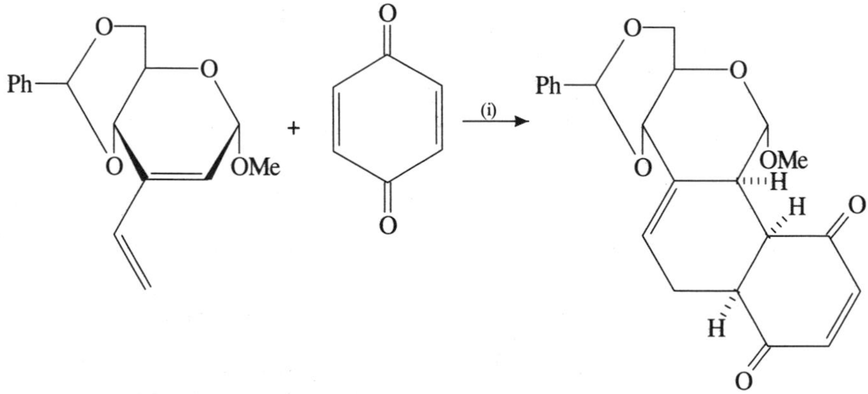

7.121   **7.122** (95%)

(i) MeCN, reflux, 4 h

Scheme 7.31

benzoquinone to yield adduct **7.122** almost quantitatively (Scheme 7.31),[66] and to complete the various possible approaches the nonitol-based triene **7.123** rearranges intramolecularly to give 81% of **7.124** on heating at 160 °C in toluene (Scheme 7.32).[67] Given the possibilities for rapidly elaborating structures by these approaches, it is not surprising that they have been used to prepare an extensive range of complex non-carbohydrate products.[68]

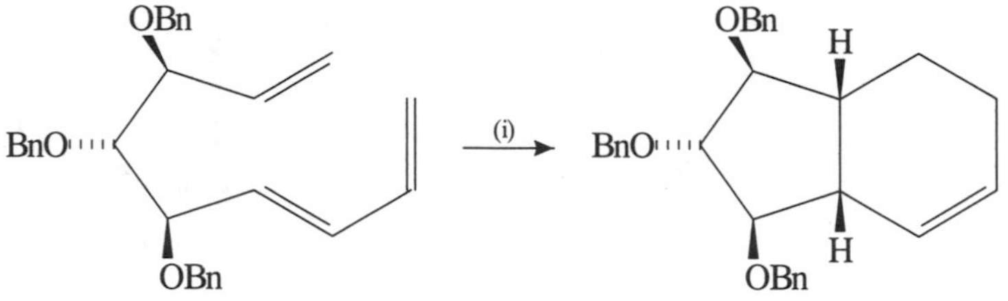

7.123   **7.124** (81%)

(i) toluene, 160 °C, 4h

Scheme 7.32

7.2.3 ACYCLIC COMPOUNDS

Research has revealed how important are many chiral acyclic compounds in biological processes, the arachidonic acid derived leukotrienes being notable in this regard.[69] Several of these and many other acyclic compounds have been synthesized from carbohydrates.

For the synthesis of leukotriene B_4 **7.127**, D-mannitol is ingeniously used to make, in duplicate, the chiral aldehydes **7.125** and **7.126** which are then linked by way of a four-carbon conjugating bridge using Wittig-like reactions (Scheme 7.33).[70]

Chain extensions from C-1 and C-5 of D-ribose lead to the synthesis of (11R, 12S,13S)-trihydroxy-(9Z,15Z)-octadecadienoic acid **7.128** which is implicated in rice blast disease.[71]

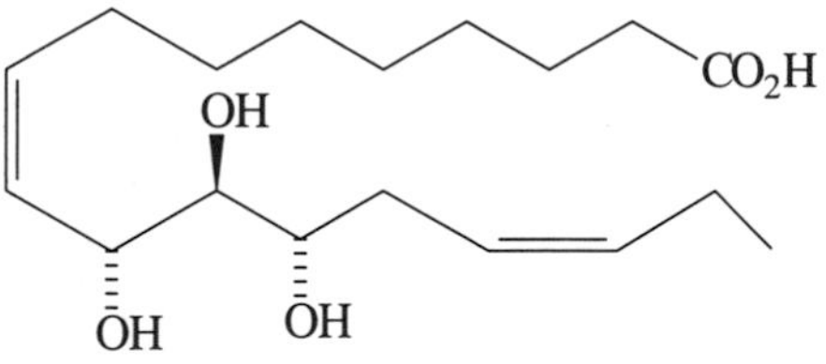

7.128 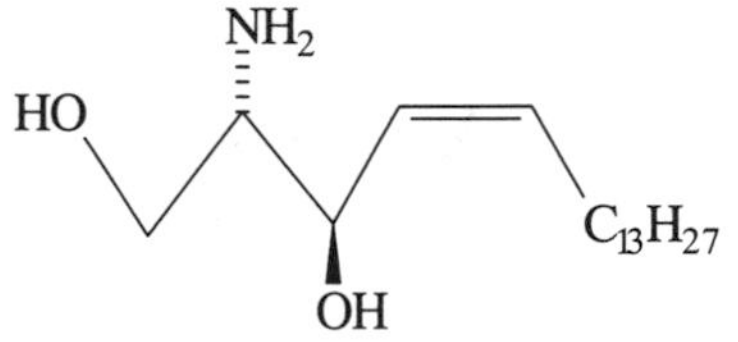   **7.129**

Inversion of configuration at C-3 of a derivative of 2-amino-2-deoxy-D-glucose and periodate cleavage of the C-4–C-5 bond gives a suitable 3-amino-3-deoxy-L-erythrose compound for elaboration to the *erythro*-sphingenine **7.129**, a ceramide component of biologically important glycosphingolipids.[72]

Good opportunities are provided by sugars for the preparation of unusual amino acids, compound **7.131**, a component of the immunosuppressive peptide cyclosporin, having been produced by way of the D-glucose-derived **7.130**.[73] Also from glucose, statine **7.133**, a γ-amino acid component of renin inhibitors, has been made via the 5-azido-3-deoxy derivative **7.132**.[74] In both these cases, C-1 was removed by periodate cleavage *en route* to the products.

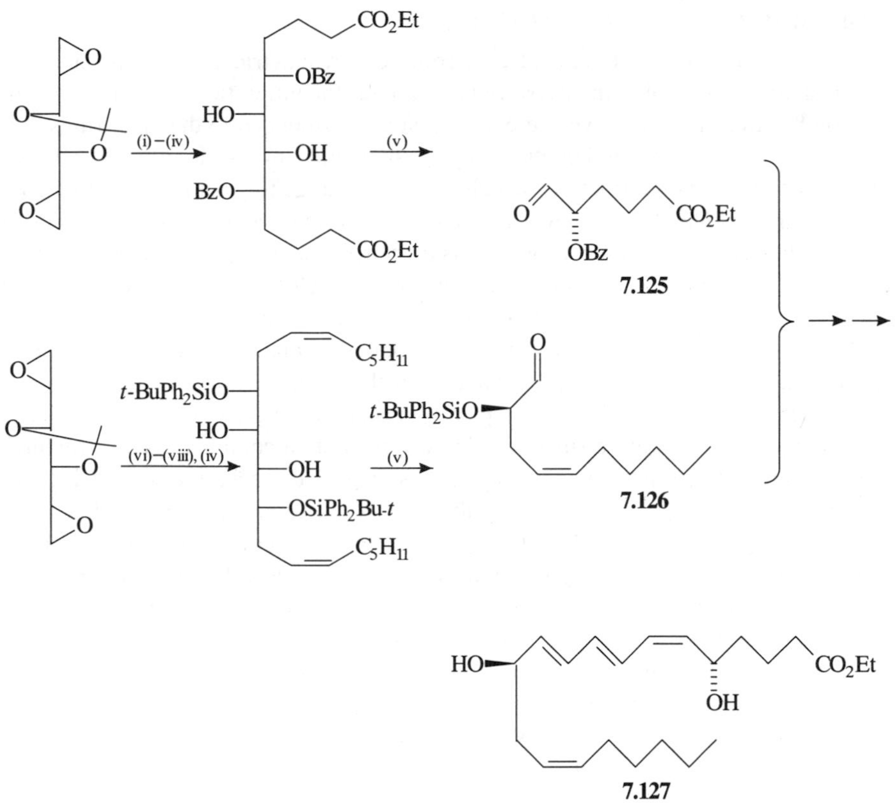

7.125

7.126

7.127

(i) LiC≡CCO$_2$Et, BF$_3$; (ii) BzCl; (iii) H$_2$, Pt/C; (iv) H$_3$O$^+$; (v) Pb(OAc)$_4$;
(vi) LiC≡CC$_5$H$_{11}$; (vii) t-BuPh$_2$SiCl; (viii) H$_2$, Pd/BaSO$_4$

Scheme 7.33

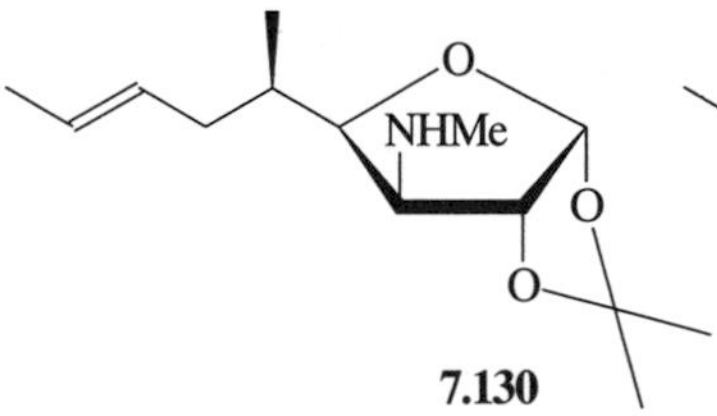

7.130

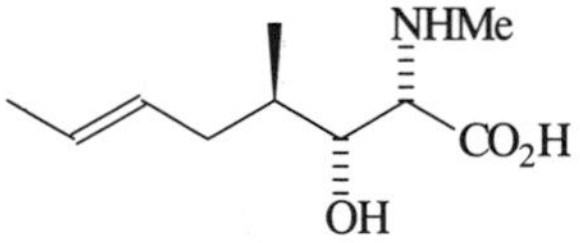

7.131

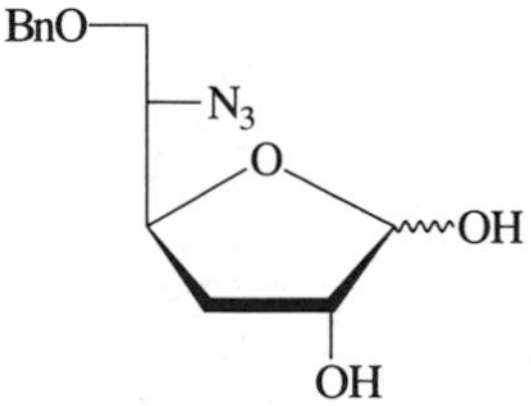

7.132

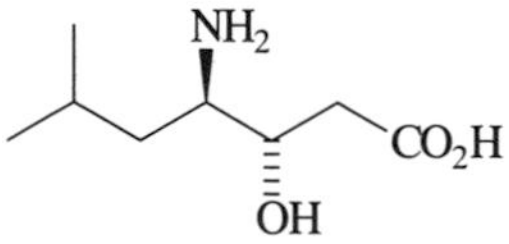

Statine

7.133

7.2.4 MACROCYCLIC COMPOUNDS

Nature has provided a vast range of very complex asymmetric macrocyclic substances, most of which have been discovered in searches for antibiotics; some are noted in Section 8.2.1.c. Many are glycosidic having sugars O-bonded to the large rings, but carbohydrates have impacted in other ways in synthetic work by acting as precursors for both complex cyclic structures which are parts of large rings or highly substituted acyclic components of these rings. The synthesis of non-cyclic carbohydrate derivatives is therefore important here as well as for purposes noted above (Section 7.2.3). Because of the complexity of this topic only two examples are considered, and many outstanding instances of the art of synthesis have to be disregarded.

The avermectins are a set of similar compounds isolated by the Merck, Sharp and Dohme Company from a *Streptomyces* broth which are potently active against nematodes and arthropods and consequently used extensively in animal husbandry. Belonging to the category of macrocycles which contain complex cyclic structures as parts of their large rings, the avermectins also have a disaccharide comprising two 2,6-dideoxyhexose units linked glycosidically as indicated in **7.134**, which gives the structure of avermectin A_{1a}. The synthesis of the whole compound has been reported from Danishefsky's group who prepared the disaccharide, the spiroketal and the oxahydrindene components from different monosaccharide precursors.[75] Only the lightest outlines are given of the methods used for the aglycon components (Scheme 7.34). The illustrated products **7.135** and **7.136** are linked by a six-carbon bridge, the oxahydrindene and the macrocyclic lactone rings are closed, and the disaccharide glycosylation is finally carried out to complete a monumental synthesis typical of the best work in the field.

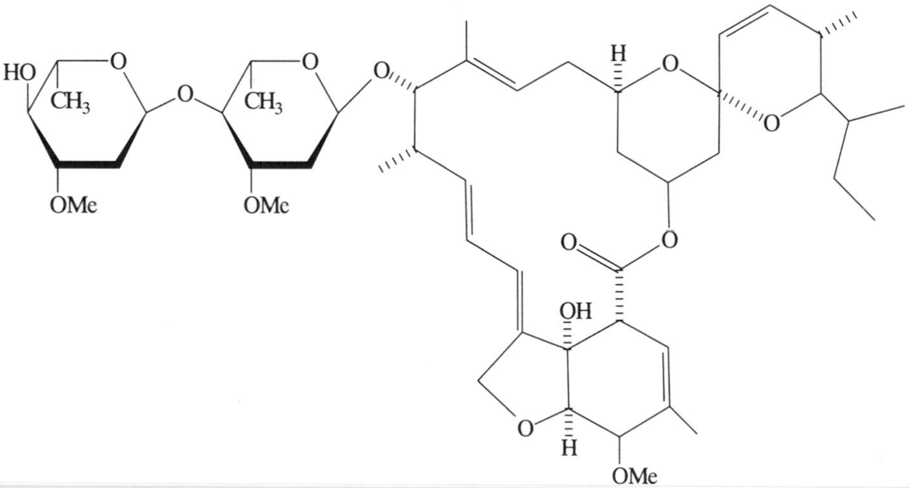

Avermectin A_{1a} **7.134**

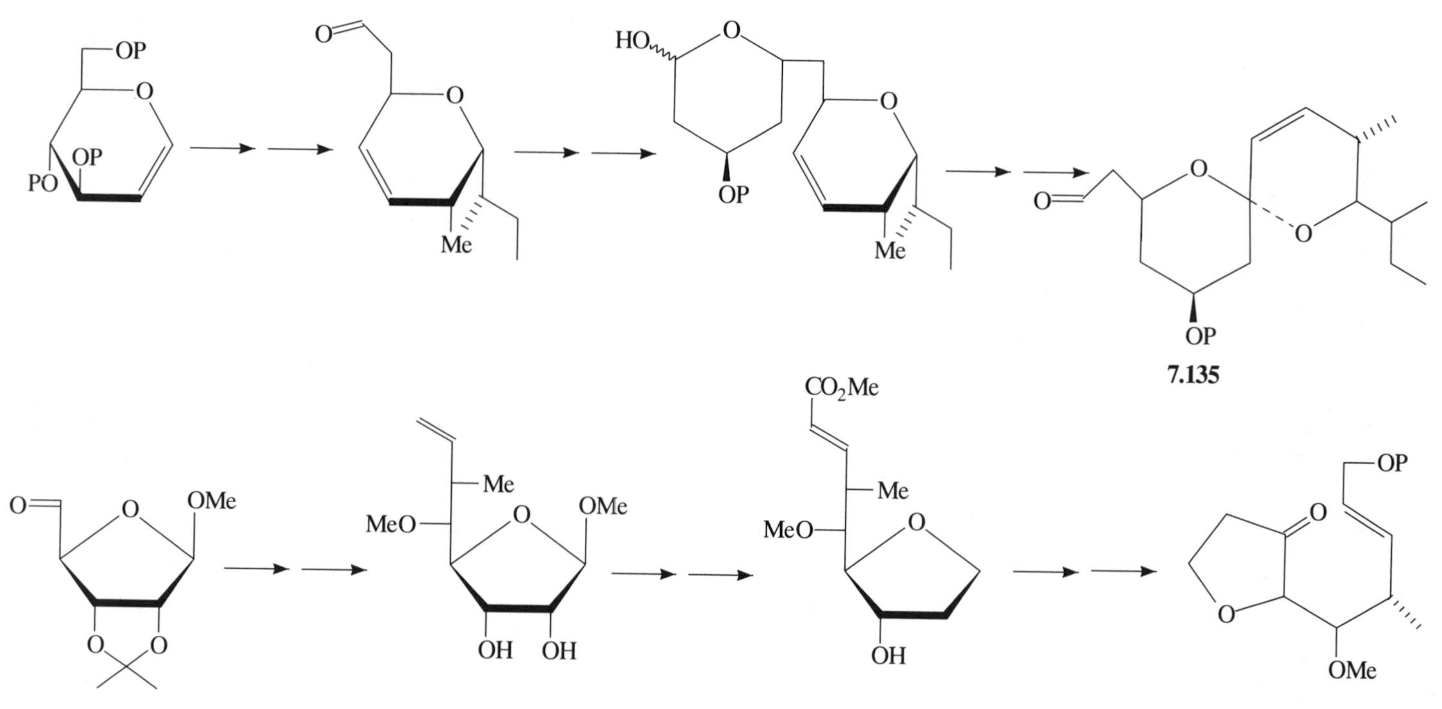

7.135

7.136

Scheme 7.34

P = protecting group

The synthesis of rifamycin W **7.137** by Tatsuta's group represents one requiring a very complex acyclic unit, the ansa chain **7.139**, to form a major part of the macrocyclic lactam structure.[76] This chain is elaborated from the doubly branched 1,6-anhydro-D-allopyranose derivative **7.138** as indicated very briefly in Scheme 7.35. The synthesis was completed by aldol coupling of the aldehydo group with a propanoyl group on a suitable nitronaphthalene derivative, chain extending from the ester end and finally lactam ring closing after reduction of the nitro group.

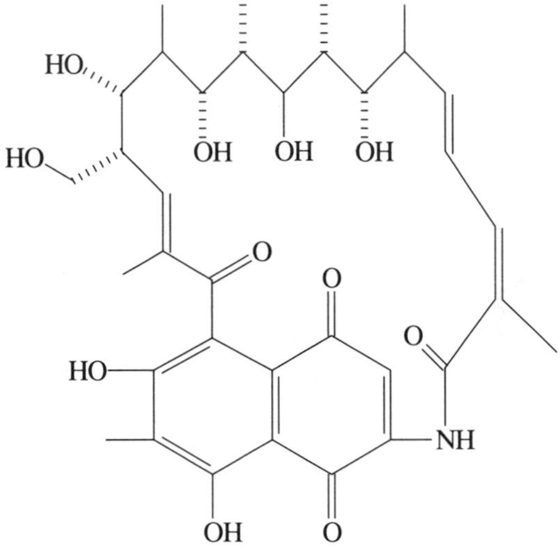

7.137

7.3 REFERENCES

1. *Rodd's Chemistry of Carbon Compounds*, (ed. M. Sainsbury), Vol. IE,F,G 2nd Supplement, Elsevier, Amsterdam, 1993, p. 273; *Angew. Chem., Int. Ed. Engl.*, 1993, **32**, 336.
2. *J. Org. Chem.*, 1986, **51**, 1934.
3. *J. Chem. Soc., Perkin Trans. 1*, 1974, 1902.
4. *Tetrahedron Lett.*, 1987, **28**, 3801; *Bull. Chem. Soc. Jpn.*, 1989, **62**, 3026.
5. *Tetrahedron Lett.*, 1990, **31**, 4147.
6. *Synthesis*, 1983, 789.
7. *Angew. Chem., Int. Ed. Engl.*, 1987, **26**, 557. See *Tetrahedron*, 1991, **47**, 6079 for a relevant review of α-amino acid synthesis.
8. *Acta Chem. Scand.*, 1981, **35B**, 273.
9. *Angew. Chem., Int. Ed. Engl.*, 1987, **26**, 267.
10. *Tetrahedron*, 1984, **40**, 4657.
11. S. Hanessian, *Total Synthesis of Natural Products: The Chiron Approach*, Pergamon Press, New York, 1983.
12. *Prog. Chem. Org. Nat. Prod.*, 1980, **39**, 1.
13. *Tetrahedron*, 1984, **40**, 3161.

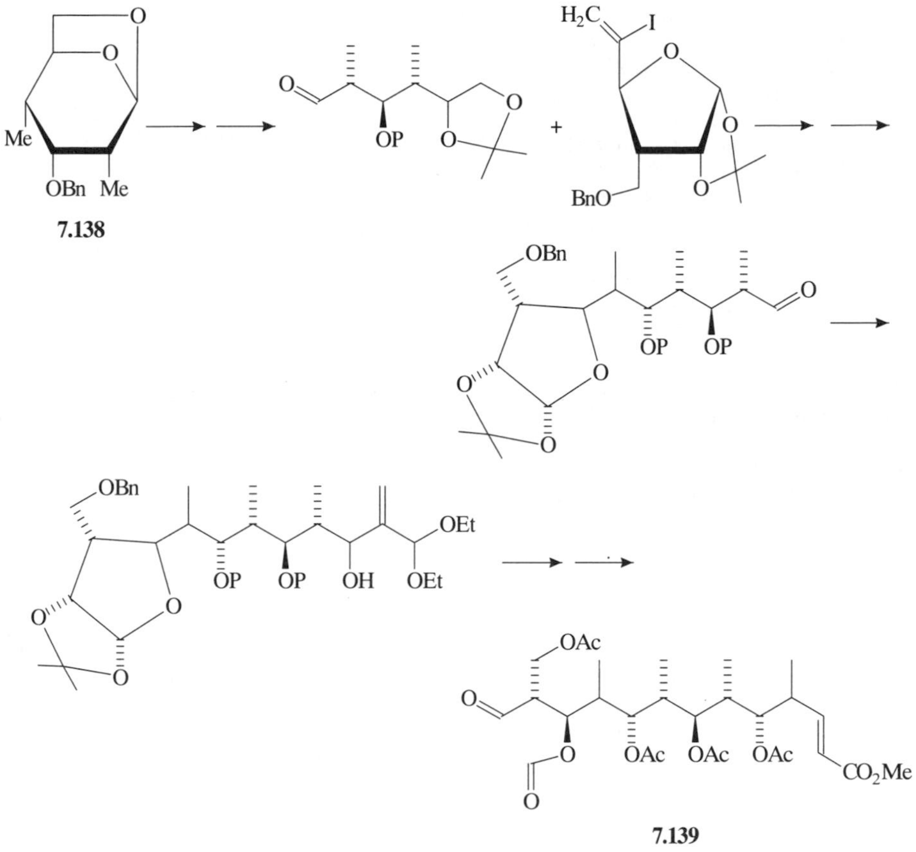

7.138

7.139

P = protecting group

Scheme 7.35

14. *Rodd's Chemistry of Carbon Compounds* (ed. M. Saimsbury), Vol. 1E, F, G 2nd Supplement, Elserier, Amsterdam, 1993, p. 315.
15. *Angew. Chem., Int. Ed. Engl.*, 1987, **26**, 1271.
16. *Angew. Chem., Int. Ed. Engl.*, 1982, **21**, 141.
17. *Modern Synthetic Methods* (ed. R. Scheffold), Vol. 6, VCH, Weinheim, 1993, p. 273.
18. *J. Am. Chem. Soc.*, 1989, **111**, 7525, 7530.
19. *Helv. Chim. Acta*, 1989, **72**, 1301.
20. *Tetrahedron Lett.*, 1989, **30**, 1993.
21. *J. Org. Chem.*, 1989, **54**, 5845.
22. *Helv. Chim. Acta*, 1989, **72**, 1649.
23. *Carbohydr. Res.*, 1990, **205**, 293.
24. *J. Chem. Soc., Perkin Trans. 1*, 1983, 1645.
25. G.W.J. Fleet, *R. Soc. Chem. Spec. Publ.*, 1988, 65; *Chem. Br.*, 1987, **23**, 842.
26. *Chem. Br.*, 1989, **25**, 287.
27. *Tetrahedron*, 1989, **45**, 327.
28. *Tetrahedron Lett.*, 1990, **31**, 4321.
29. *J. Am. Chem. Soc.*, 1989, **111**, 2580; *J. Am. Chem. Soc.*, 1990, **112**, 8100.
30. *Tetrahedron*, 1989, **45**, 7195.

31. *Methods Carbohydr. Chem.*, 1972, **6**, 286.
32. *Tetrahedron Lett.*, 1975, **16**, 2765.
33. *Chem. Rev.*, 1993, **93**, 2779.
34. *Acc. Chem. Res.*, 1975, **8**, 192.
35. *Tetrahedron*, 1984, **40**, 1279.
36. *J. Org. Chem.*, 1991, **56**, 6038.
37. *Tetrahedron Lett.*, 1986, **27**, 1777.
38. *Carbohydr. Res.*, 1982, **108**, 41.
39. *J. Am. Chem. Soc.*, 1981, **103**, 7559.
40. *J. Am. Chem. Soc.*, 1978, **100**, 8272.
41. *Tetrahedron Lett.*, 1988, **29**, 5521.
42. *J. Org. Chem.*, 1991, **56**, 4096.
43. *Acc. Chem. Res.*, 1991, **24**, 139.
44. *J. Chem. Soc., Perkin Trans. 1*, 1983, 1629.
45. *J. Chem. Soc., Perkin Trans. 1*, 1983, 1635, 1641.
46. *Tetrahedron*, 1982, **38**, 2939.
47. *J. Org. Chem.*, 1969, **34**, 1386.
48. *J. Am. Chem. Soc.*, 1975, **97**, 6810.
49. *J. Org. Chem.*, 1992, **57**, 3642.
50. *J. Am. Chem. Soc.*, 1948, **70**, 1479.
51. *J. Org. Chem.*, 1974, **39**, 812.
52. *Chem. Pharm. Bull.*, 1988, **36**, 4236.
53. *Adv. Carbohydr. Chem.*, 1969, **24**, 67.
54. *J. Chem. Soc., Perkin Trans. 1*, 1984, 905.
55. *J. Chem. Soc., Perkin Trans. 1*, 1979, 1455; *J. Chem. Soc., Perkin Trans. 1*, 1985, 2413.
56. *Bull. Chem. Soc. Jpn.*, 1991, **64**, 2118.
57. *J. Chem. Soc., Chem. Commun.*, 1987, 1008.
58. *Liebigs Ann. Chem.*, 1986, 551.
59. *Tetrahedron Lett.*, 1981, **32**, 4525.
60. *Chem. Lett.*, 1991, 1473.
61. *J. Am. Chem. Soc.*, 1991, **113**, 9883.
62. *J. Am. Chem. Soc.*, 1991, **113**, 9885.
63. *Carbohydr. Res.*, 1992, **226**, 57.
64. *J. Chem. Soc., Chem. Commun.*, 1989, 1160.
65. *J. Chem. Soc., Chem. Commun.*, 1984, 911.
66. *Tetrahedron*, 1988, **44**, 3355.
67. *Tetrahedron*, 1989, **45**, 2793.
68. *Am. Chem. Soc. Symp. Ser.*, 1992, **494**, 112.
69. *Nat. Prod. Res.*, 1988, **5**, 1.
70. *Tetrahedron Lett.*, 1986, **27**, 4161.
71. *Tetrahedron Lett.*, 1990, **31**, 4349.
72. *Carbohydr. Res.*, 1989, **194**, 125.
73. *Tetrahedron Lett.*, 1989, **30**, 6769.
74. *Chem. Lett.*, 1989, 687.
75. *J. Am. Chem. Soc.*, 1989, **111**, 2967.
76. *Tetrahedron*, 1990, **46**, 4629.

8 Natural Products Related to and Containing Monosaccharides

Sugars are found in nature in widely different forms, in chemical association with a vast number of compounds, including other sugars, and in materials which perform a range of functions varying from structural to specifically functional in many biochemical ways. Attachment of monosaccharide units to other molecules is usually through the anomeric centre because biochemical glycosylation is a favoured process (Section 2.5.1), but other hydroxyl groups frequently carry substituents, usually esters. In this chapter an indication is given of the main chemical features of the best-known classes of natural products which contain sugars. Cyclitols, which are closely related to the monosaccharides and biosynthesized from them, are described briefly at the end.

8.1 COMPLEX SACCHARIDES

8.1.1 DISACCHARIDES

One monosaccharide unit can be linked glycosidically to another either by way of the anomeric centre of the latter, in which case a *non-reducing disaccharide* results, or through one of the other oxygen atoms, in which case each resulting dimer has an unsubstituted anomeric centre and is, consequently, a *reducing disaccharide*. In nature both types of compound are abundant.

(a) Non-reducing disaccharides

The best-known and most significant member of this class is sucrose (cane sugar or beet sugar) **8.1**, which is the most important low molecular weight carbohydrate of animal diets. World production of the compound in highly pure form for human consumption is around 10^8 tonnes per annum, two thirds coming from the cane industry and one third from beet. Since it possesses a furanosyl glycosidic bond it is very sensitive to acid hydrolysis (Section 3.1.1.b.ii) during which it undergoes the 'inversion' process, so called because the optical rotations of solutions change from *dextro* to *levo* on liberation of the constituent free sugars D-glucose and D-fructose.

Methylation analysis, which involves complete methylation (dimethyl sulphate and aqueous alkali) followed by hydrolysis, gives 2,3,4,6-tetra-*O*-methyl-D-glucose **8.2** and 1,3,4,6-tetra-*O*-methyl-D-fructose **8.3** (Scheme 8.1a), establishing the pyranoid nature of the glucose ring and the furanoid character of the fructose

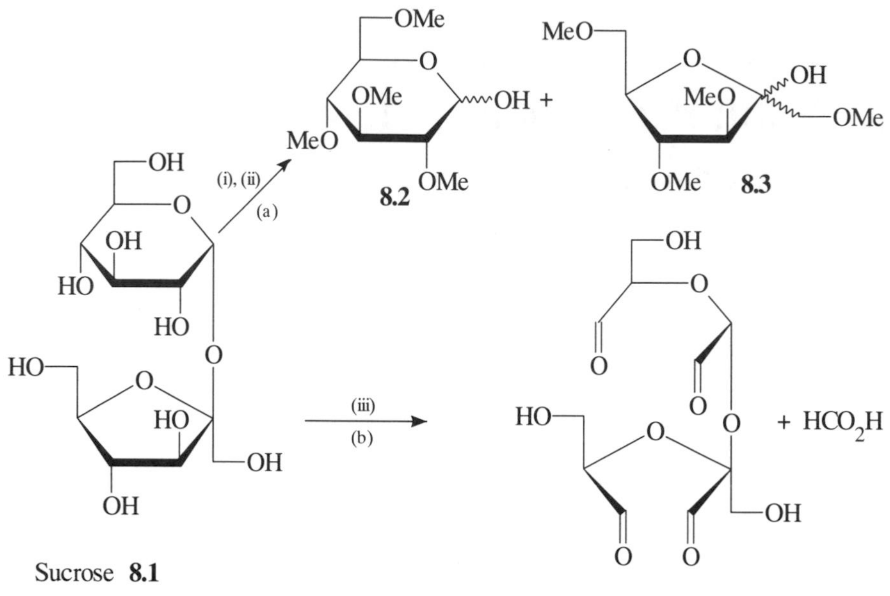

Sucrose **8.1**

(i) Me$_2$SO$_4$, NaOH; (ii) H$_3$O$^+$; (iii) IO$_4^-$

Scheme 8.1

moiety. In agreement with this, sucrose reduces three moles of periodate ion per mole (Section 5.5) and gives one mole of formic acid derived from C-3 of the glucose unit as shown (Scheme 8.1b). There remains the question of the anomeric configurations of the two glycosidic bonds. First evidence came from enzymic studies which illustrated nicely the value of enzymology in configurational analysis of carbohydrates. Enzymes are proteins which catalyse biochemical reactions in a manner which is usually characterized by high specificity, and since sucrose is hydrolysed by an α-glucosidase from yeast (an enzyme known to catalyse the cleavage of α- and not β-D-glucosidic bonds) it is assigned an α-D-glucosidic structure. Similarly, it is hydrolysed in the presence of an enzyme specific for β-D-fructosides and is consequently a β-D-fructoside. It is therefore α-D-glucopyranosyl β-D-fructofuranoside. Confirmation of this assignment comes from application of Hudson's isorotation rule (Appendix 4), and the structure is put beyond question by an X-ray diffraction analysis of a sodium bromide complex. N.m.r. spectroscopy can be applied to the assignment of such configurations and to the determination of the preferred solution conformation of the disaccharide (Section 8.1.2.c). The chemical synthesis of sucrose remained as a challenge for many years and was finally achieved by Lemieux and Huber by the condensation of 3,4,6-tri-O-acetyl-1,2-anhydro-α-D-glucose **3.80** and 1,3,4,6-tetra-O-acetyl-D-fructose (Section 3.1.1.c).[1]

Trehalose **8.4** is another well-known non-reducing disaccharide which is found as a storage carbohydrate in certain plants, algae, fungi and yeasts, and is the main carbohydrate component of the blood of a variety of insects. Methylation of **8.4** followed by hydrolysis gives only 2,3,4,6-tetra-O-methyl-D-glucose, and

Trehalose **8.4**

periodate oxidation occurs with the reduction of four equivalents of reagent and produces two of formic acid; the compound, therefore, is a D-glucopyranosyl D-glucopyranoside. Three isomers are possible — the α,α-, the α,β- and the β,β-linked compounds — for which the calculated specific optical rotations are $+180$, $+71$ and -58, respectively (Appendix 4). The natural product has an $[\alpha]_D$ value of $+178$ and can, in consequence, be assigned the α,α-structure, a conclusion that is readily confirmed by ^{1}H n.m.r. spectroscopy. The α,α-compound and α,β-compound (also known as *neotrehalose*) can be obtained synthetically by heating 2,3,4,6-tetra-*O*-acetyl-D-glucose (mixed anomers) with 3,4,6-tri-*O*-acetyl-1,2-anhydro-D-glucose at $100\,^\circ$C,[2] and the β,β-isomer (isotrehalose) is produced, together with *neotrehalose*, by use of acylated glycosyl halides.[3]

(b)   Reducing disaccharides

Altogether there are 20 ways of glycosidically linked two of the same hexose molecules to give reducing disaccharides (α-linked and β-linked furanosyl and pyranosyl forms bonded to O-2–O-6), so the total possible number of such compounds is very large. Many occur naturally or have been obtained by chemical or enzymic synthetic methods, or by partial hydrolysis of higher molecular weight materials. Of these, maltose and cellobiose will be mentioned as representatives of homodisaccharides which are composed of two units of the same sugar; lactose will be discussed as a typical heterodisaccharide.

Maltose, produced by enzymic hydrolysis of starch, is 4-*O*-α-D-glucopyranosyl-D-glucose **8.5a**, which behaves as a normal reducing sugar, showing mutarotation and existing finally in aqueous solution as an equilibrated mixture of the α-form and β-form. On complete methylation (best carried out by alkylation of the methyl glycoside, which is prepared by way of the acetobromo sugar; see Section 3.3.2.a), followed by acid-catalysed hydrolysis of both glycosidic bonds of the product, 2,3,4,6-tetra-*O*-methyl-D-glucose is obtained from the non-reducing unit and the 2,3,6-trimethyl ether from the reducing moiety. Since these two products could be derived from either a 4- or 5-linked dimer (the latter having the reducing moiety in the furanoid form), a further experiment must be conducted to distinguish between these possibilities. Methylation of the aldonic acid obtained by bromine water oxidation of the disaccharide (Section 3.1.6.a), followed by hydrolysis, gives the tetramethylglucose as before, together with 2,3,5,6-tetra-*O*-methyl-D-gluconic acid, showing that the oxygen atom at C-5 is initially involved in pyranose ring formation. This establishes that position 4 is involved in the inter-unit linkage. The specific optical rotation of the sugar (+130, equilibrated in water) is consistent only with an α-linked dimer (Appendix 4), a conclusion confirmed by the finding that maltose is hydrolysed in the presence of α-glucosidases, and by the observation in the n.m.r. spectrum of the compound of a narrow doublet ($J_{1,2} = 3.4$ Hz) for the resonance of the anomeric proton of the non-reducing unit.

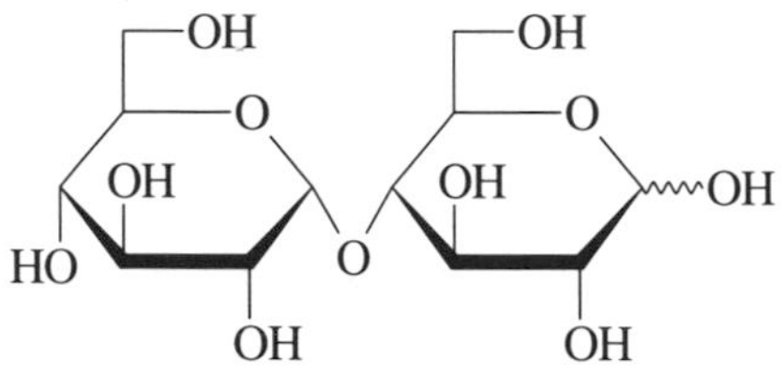

Maltose **8.5a** 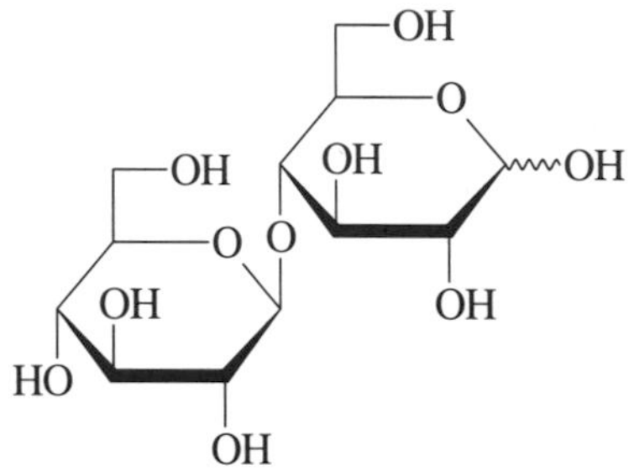   Cellobiose **8.6a**

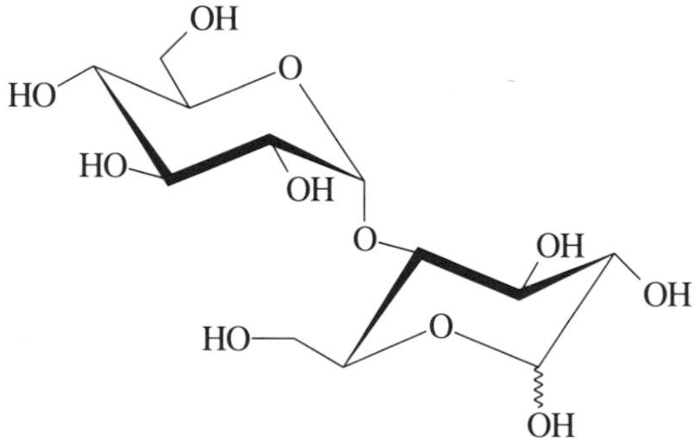

8.5b

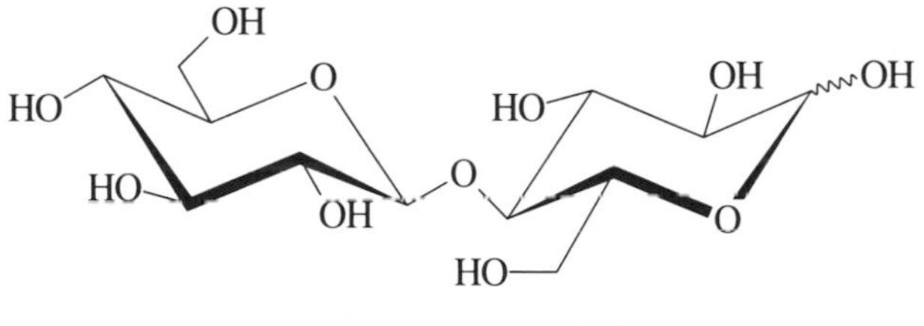

8.6b

Cellobiose **8.6a**, the dimer upon which cellulose is built, on the other hand, is also a 4-O-D-glucopyranosyl-D-glucose, but its specific rotation (+35, equilibrated in water), its cleavage by β-glucosidases and its n.m.r. spectrum (which shows a doublet for the resonance of the anomeric proton of the non-reducing unit with $J_{1,2} = 7.0$ Hz) establish the β-character of the glycosidic linkage. Maltose and cellobiose are therefore structurally very similar but, as is seen from their

preferred conformations **8.5b** and **8.6b**, respectively (Section 8.1.2.c),[4] they lead
to the homopolymers amylose and cellulose (Section 8.1.3.b) with very different
three-dimensional structures, the former being spiral and the latter linear.

Structural analysis of heterodisaccharides, which are composed of two different
monosaccharides, employs the same procedures as have been illustrated for the
glucobioses, but in addition it must be determined which of the units possesses the
reducing end-group. Lactose, or milk sugar, 4-O-β-D-galactopyranosyl-D-glucose
8.7, is present (5%) in the milk of mammals, and is shown to have a reducing
glucose moiety by the characterization of D-galactose and D-gluconic acid as the
hydrolysis products of lactobionic acid, the aldonic acid formed by bromine water
oxidation. The β-inter-unit bond again may be characterized by enzymic methods,
and is further indicated by the relatively low specific rotation of the disaccharide
(+55, equilibrated in water) and by the ^{1}H n.m.r. $J_{1,2}$ value of the non-reducing
moiety.

Lactose **8.7**

Since the traditional methods for determining disaccharide structure described
above are somewhat cumbersome and, in particular, are difficult to apply to small
quantities of materials, other methods have been developed which are based mainly
on oxidative techniques. Mild oxidation with lead tetracetate (Section 5.5.2) of
3-, 4- or 6-linked hexose dimers gives formate esters (e.g. **8.8** from a 3-linked
compound) which stabilize the products towards further oxidation. Hydrolysis
of these gives pentoses (as illustrated in Scheme 8.2), tetroses and glyceraldehyde, respectively, which reveal the position of the dimer linkage. A related
technique involves reducing the dimers to the glycosylpolyols and treating the
products with very dilute sodium periodate solution which causes specific oxidation of the more reactive acyclic moieties. The consumption of four molecular
equivalents of oxidant (followed by a further two when a higher concentration of
reagent is used) is thus diagnostic of a 6-linked hexosylhexose. The analysis of the
unsaturated disaccharide derivative **8.9** was aided by this method, since hydroxylation of the double bond followed by removal of the protecting groups gives a
mannosyl-D-galactose **8.10**, shown by this test to have a 6-link. Oxidation of the
disaccharide itself gives a product **8.11** having the same optical rotation as that
obtained from 6-O-α-D-glucopyranosyl-D-glucose and different from that derived

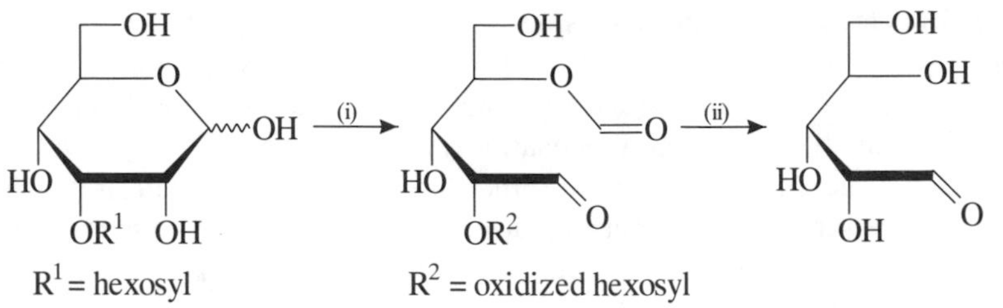

R^1 = hexosyl   R^2 = oxidized hexosyl

8.8

(i) Pb(OAc)$_4$; (ii) H$_3$O$^+$

Scheme 8.2

from the β-isomer. Consequently, the α-glycosidic configurations of the mannosyl-galactose and the unsaturated derivative are established. N.m.r. methods are also available for complete structural analysis of disaccharides (Section 8.1.2.c).

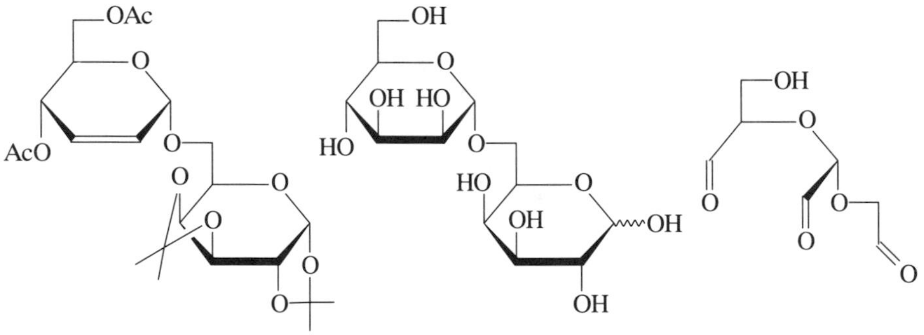

8.9   **8.10**   **8.11**

(c)   Disaccharide synthesis

The following glycosylating reagents, fully protected at all hydroxyl groups, are most commonly used for the chemical synthesis of disaccharides: glycosyl chlorides or bromides (Section 3.3.2.a.ii), glycosyl fluorides (Section 3.3.5), 1,2-orthoesters (Section 3.2.2), glycosyl trichloroacetimidates (Section 3.2.3.b), glycosyl acetates (Section 3.2.1.c), thioglycosides (Section 3.1.2.c.ii), glycosyl sulphoxides (Section 3.1.2.c.iv), pent-4-enyl glycosides (Section 3.1.1.b.vi), 1,2-anhydropyranoses (Section 3.1.1.c) and nucleoside diphosphate sugars with transferase enzymes (Section 3.1.1.b.v). General matters relating to the synthesis of inter-monosaccharide bonds are dealt with in Chapter 6. In addition, some disaccharides can be obtained by modification of available compounds: 4-O-β-D-galactopyranosyl-D-fructose (lactulose) is thus made by mild alkaline treatment of lactose (Section 3.1.7.a), and anomerization of 6-O-β-D-glucopyranosyl-D-glucose (gentiobiose) octaacetate with titanium(IV) chloride gives the thermodynamically more stable α-linked (isomaltose) derivative. Hexosylpentoses can be prepared by application of appropriate degradations of reducing hexosylhexoses (e.g. Section 3.1.6.a.iv).

(d) Naturally occurring disaccharides

Some of the other more common reducing disaccharides available from natural sources are known by trivial names as follows: gentiobiose (6-*O*-β-glucopyranosyl-D-glucose, conventionally abbreviated to β-D-Glc*p*-(1→6)-D-Glc), isomaltose (α-D-Glc*p*-(1→6)-D-Glc), kojibiose (α-D-Glc*p*-(1→2)-D-Glc), lactulose (β-D-Gal*p*-(1→4)-D-Fru), laminaribiose (β-D-Glc*p*-(1→3)-D-Glc), leucrose (α-D-Glc*p*-(1→5)-D-Fru), maltulose (α-D-Glc*p*-(1→4)-D-Fru), melibiose (α-D-Gal*p*-(1→6)-D-Glc), nigerose (α-D-Glc*p*-(1→3)-D-Glc), planteobiose (α-D-Gal*p*-(1→6)-D-Fru), rutinose (α-L-Rha-*p*(1→6)-D-Glc), sophorose (β-D-Glc*p*-(1→2)-D-Glc) and turanose (α-D-Glc*p*-(1→3)-D-Fru).

8.1.2   OLIGOSACCHARIDES[5]

(a) Discrete compounds

Raffinose is the most abundant oligosaccharide found in plants with the exception of sucrose. It is the non-reducing trisaccharide *O*-α-D-galactopyranosyl-(1→6)-*O*-α-D-glucopyranosyl-(1→2)-β-D-fructofuranoside (α-D-Gal*p*-(1→6)-α-D-Glc*p*-(1→2)-β-D-Fru*f*), a galactosyl derivative of sucrose, which on mild acidic hydrolysis gives 6-*O*-α-D-galactopyranosyl-D-glucose (melibiose) and D-fructose, and with an α-galactosidase gives galactose and sucrose. The non-reducing tetrasaccharide stachyose (α-D-Gal*p*-(1→6)-α-D-Gal*p*-(1→6)-α-D-Glc*p*-(1→2)-β-D-Fru*f*) is also found in plants and is a further member of the series of galactosyl sucroses; on enzymolysis it gives galactose, raffinose and sucrose, which establishes the sequence of the constituent sugars.

In the same way as the above series of oligomers based on sucrose is found in plants, a series built on lactose exists in human milk. D-Galactose, L-fucose and 2-acetamido-2-deoxy-D-glucose are contained in such oligosaccharides, one of which has structure **8.12**.

$$\alpha\text{-L-Fuc}p$$
$$\downarrow 1$$
$$\downarrow 4$$

α-L-Fuc*p*-(1→2)-β-D-Gal*p*-(1→3)-β-D-Glc*p*NAc-(1→3)-β-D-Gal*p*-(1→4)-D-Glc*p*

Fuc*p* = fucopyranose, Gal*p* = galactopyranose, Glc*p*NAc = 2-acetamido-2-deoxy-D-glucopyranose, Glc*p* = glucopyranose

8.12

Whereas these compounds probably have food storage as a main role, it is becoming increasingly apparent that nature uses specific oligosaccharides in much more sophisticated ways for control of biological processes, and such is the significance of the recognition of this that the term 'glycobiology' has been coined to describe this new part of molecular biology. Oligosaccharides, whether independent or as parts of larger molecules, have become identified as information-storing compounds which are involved in specific associations with other biochemical species (usually proteins) to trigger biological processes. Their immense potential as data stores and some of their roles in biology are noted later in this chapter. The heptaglucose compound **6.49** is an example of a bioactive oligosaccharide produced by partial enzymic hydrolysis of the cell wall of the pathogen *P. megasperma* which acts as an elicitor of the defence response in plants which host the fungus, for example soyabean. The specificity of the interaction of the oligosaccharide with the host protein, which results in the production of the antibiotic defence substances phytoalexins, can be judged from the finding that removal of the 1,6-linked non-reducing glucose unit diminishes the bioelicitor activity by a factor of 4000. Similarly, some plants, prone to fungal attack, recognize specific oligosaccharide sections of chitin and chitosan, which are major polysaccharides of fungal cell walls, and respond defensively to them.

Albersheim, who has pioneered this work,[6] has also recognized that specific oligosaccharide components of plant cell walls can have biochemical activity which results in regulatory responses. Parts of α-$(1{\rightarrow}4)$-linked D-galacturonic acid polysaccharides, which are the pectic components of plant cell walls, have regulatory effects on plant growth and development, while others induce defence responses, the dodecamer being most active as an elicitor of the defence responses. Similarly, xylose- and glucose-containing oligosaccharides of plant hemicelluloses, which occur naturally and are apparently formed by partial cleavage of the polymers, inhibit artificially induced plant growth.

Free oligosaccharides with biological regulatory activities have been named 'oligosaccharins', and there is evidence that their importance in animal as well as plant biochemistry is much greater than is currently recognized.

Of well-established and particular significance are the *cyclodextrins*, a set of cyclic α-$(1{\rightarrow}4)$-linked D-glucopyranose oligomers obtainable enzymically by the partial hydrolysis of starch.[7] Although members having up to 12 glucose units are known, the so-called α-, β- and γ-cyclodextrins with six, seven and eight sugars, respectively, are most common and are commercially available.

They were first encountered in the late 1800s, and Schardinger in 1904 described their production by the action of an amylase of *Bacillus macerans* on starch. In consequence, they have otherwise been known as 'Schardinger dextrins' or 'cycloamyloses'.[8] Apparently the enzyme can excise small sections from the tubularly structured amylose (Section 8.1.3.b.ii) and concurrently link the terminal units of the excised parts to give products, for example α-cyclodextrin **8.13a**, having toroidal structures with hydrophilic exteriors and insides lined largely by H-3 and H-5 of the glucose rings and the inter-unit oxygen atoms which together generate hydrophobic environments. The diameters of these tubular cavities are 0.57, 0.78 and 0.95 nm for the α-, β- and γ-compounds, respectively, being suitable, therefore, to 'host' small hydrophobic organic compounds, particularly benzene

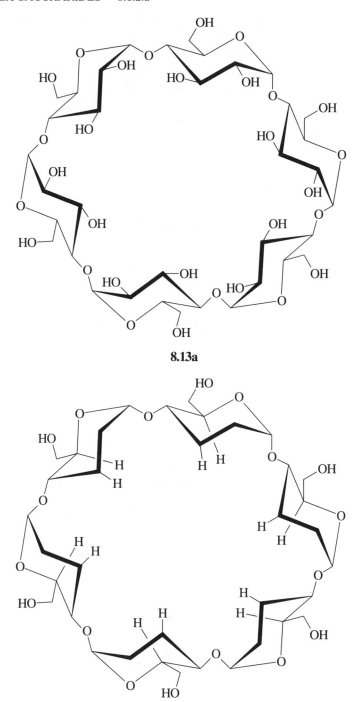

8.13a

8.13b

derivatives, and in this way they act in a manner reminiscent of amylose itself. They provide the best example of relatively small and available organic compounds which have the ability to complex with others. An impression of a cross-section of α-cyclodextrin is given in **8.13b**, which identifies the hydrophobic interior. The secondary hydroxyl groups are in an annulus above the sugar rings while the primaries are coplanar below this ring, as indicated in **8.13c**. (For representational purposes the 'tub' in this last diagram is approximately half as deep as it is calculated to be.)

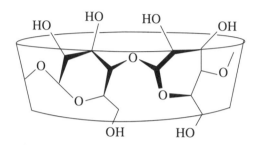

With this ability to bind other compounds in environments in which the 'guests' are held in close proximity to specific hydroxyl groups they have been perceived as simple models of enzymes and have been found to exhibit catalytic activity. At pH 10.6, for example, the base-catalysed hydrolysis rate of phenyl acetate is enhanced 9.7 times by α-cyclodextrin, but for *p*-tolyl acetate the factor is only 3.8, presumably because the additional methyl group renders the transition state less favourable in a small way. With the *meta*-isomer, however, the enhancement is 39, indicating that in this case the methyl group apparently permits complexation but disturbs the symmetry of the fit of the aromatic ring in the cavity, allowing a nucleophilic catalytic hydroxyl group better access to the carbonyl carbon atom. A further increase to 226 for *m-t*-butylphenyl acetate strongly supports this conclusion, Scheme 8.3 illustrating the first steps in the process. The hydroxyl groups involved in the catalysis are those at C-2 and C-3 of the glucose units.[9]

Scheme 8.3

For better catalysis and closer mimicking of enzymic function, appropriate catalytically active groups may be chemically introduced into the cyclodextrin structures. Thus, the products obtained by introducing two imidazole groups, which play key roles in the action of some hydrolytic enzymes, can be regioselectively catalytic. For example, the cyclic phosphate **8.15** is cleaved to the monophosphate **8.14** by a di-6-*O*-imidazolylcyclodextrin, whereas an analogue which has the imidazole groups bound to the cyclodextrin by means of a two-atom linking arm shows different selectivity, producing the isomeric product **8.16** (Scheme 8.4). Without the catalytic groups poor selectivity is shown, but in their presence optimum catalysis occurs at pH 6, indicating that one of the imidazoles acts as a base while the other in its protonated form serves as an acid catalyst.[10] Related regioselectivity is shown in some cases by the cyclodextrins themselves, ribonucleotide $2',3'$-cyclic phosphates, for example, undergoing preferential cleavage of the P–O-$2'$ bond in the presence of the α-compound, and alternatively giving $2'$-monophosphates with the β-compound and γ-compound.[11]

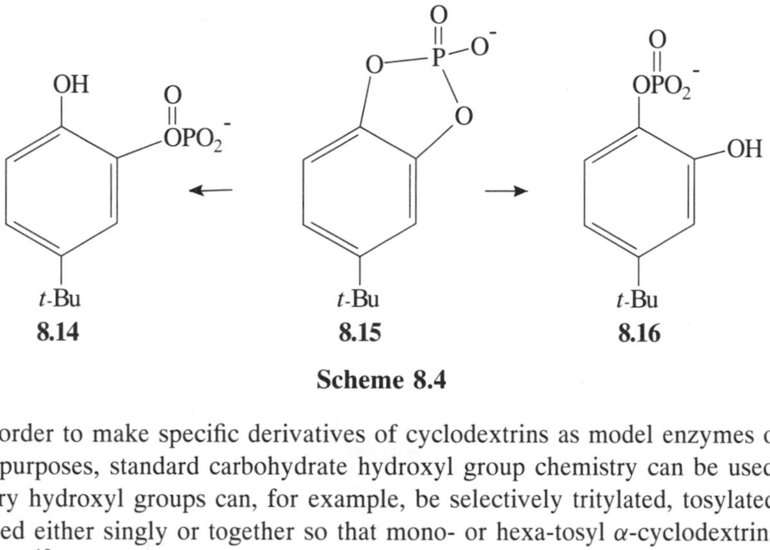

Scheme 8.4

In order to make specific derivatives of cyclodextrins as model enzymes or for other purposes, standard carbohydrate hydroxyl group chemistry can be used; the primary hydroxyl groups can, for example, be selectively tritylated, tosylated and silylated either singly or together so that mono- or hexa-tosyl α-cyclodextrins can be made.[12] From the latter, surprisingly, the hexa-3,6-anhydro derivative, with all its rings conformationally inverted, can be made in good yield by treatment with aqueous alkali.[13]

α-Cyclodextrin[14] and its α-D-mannose analogue[15] have been prepared synthetically, and further analogues with other inter-unit linkages may also be produced, as evidenced by the chemical synthesis of the β-(1→3)-linked D-glucopyranose hexamer.[16]

Not only have the cyclodextrins been of interest in synthetic chemistry, but many uses have been found in analytical[17] and in industrial[9,18] work. Since several physical properties of compounds are altered on complexing within cyclodextrins, these changes can be utilized for analytical purposes — pK_a values, diffusion properties, electrochemical and spectral properties can be used in this way. Their use as n.m.r. shift reagents is a case in point, and they have found many applications in chromatography, including use as enantioselective adsorbents.

In the chemical industry they are used to stabilize food, toiletry and pharmaceutical products, and in the last of these to solubilize medicinal compounds and thereby increase their bioavailability. Moreover, they can be applied to improve the selectivity of reactions and in separation and purification processes on an industrial scale.

(b)   Components of complex molecules

Of very great significance in biology is the phenomenon of cell recognition, by which oligosaccharides bonded to large molecules and cellular structures act as surface 'markers' to enable cells to interact appropriately with others (see Section 8.1.4).[19] They must, for example, be able to recognize cells from their own organism as being distinct from those of others. In the simplest case, red blood cells carry carbohydrate blood group antigens which determine blood group types in humans and require that type A people, for example, who have the tetrasaccharide **8.17** as the key antigen linked by lipid components to the surfaces of the red blood cells, are not transfused with type B blood the antigen of which is **8.18**. Should this happen the 'foreign' blood with antigen **8.18** will be recognized by antibodies present in the type A blood person and clumping and precipitation will occur.

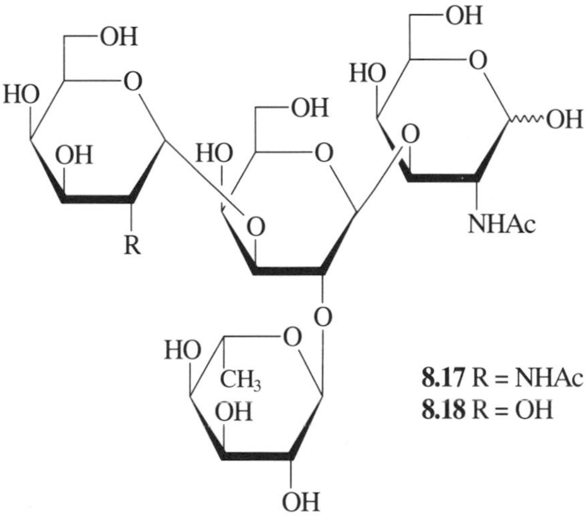

8.17 R = NHAc
8.18 R = OH

These determinants are also present elsewhere in the body attached to proteins and are then referred to as 'blood group substances'. For example, they occur attached to a soluble glycoprotein of ovarian cyst fluid.

In related fashion, bacterial polysaccharides of both Gram-positive and Gram-negative organisms can show specific antigenic activity which is characteristic of small oligosaccharide repeat units. For example, the structural and antigenic

8.19

units of *Salmonella* serogroups A, B and D are the tetrasaccharide **8.19** and the analogues which are epimers at C-4 and at C-2 and C-4 of the dideoxyhexose sugar, respectively.[20] Knowledge of this has led to the synthesis of these compounds which can then be attached to large, non-carbohydrate carriers to act as antigens for the production of monoclonal antibodies to be used in immunological research and potentially in medicine as components of vaccines.

(c)   Structural and conformational analysis of oligosaccharides

(i)   Structural analysis

Traditionally, methylation analysis as described for disaccharides (Sections 8.1.1.a and 8.1.1.b) and polysaccharides (Sections 8.1.3.a and 8.1.3.b) was used for establishing the primary structures of oligosaccharides, and while it provides outlines of main structural features, the method is largely unsatisfactory in respect of the sequencing of constituent sugars. N.m.r. spectroscopic and mass spectrometric methods can now be used to make good this deficiency, both having the advantage of being applicable to small samples. Considerable amounts of both ^{1}H and ^{13}C n.m.r. chemical shift data are available in databases,[21,22] and computer programs have been complied which allow structural determination of unknown oligosaccharides by comparative methods.[23] ^{1}H-based n.m.r. techniques are extremely powerful,[24] but empirical methods can be applied using ^{13}C chemical shifts of reference compounds and known effects caused by glycosylation.[25] For direct structural analysis by high field n.m.r. methods several techniques, e.g. ^{1}H–^{1}H and ^{1}H–^{13}C correlation spectroscopy (COSY) and multistep relayed correlations (RECSY), ^{1}H-detected ^{13}C spectroscopy and Hartmann–Hahn spectroscopy (HOHAHA), are applicable.

Mass spectrometry coupled with gas chromatography is commonly used for the analysis of the products of methylation/hydrolysis of oligosaccharides, but technological advances in the field, particularly in the development of high resolution instruments, mild ionization techniques such as fast-atom bombardment (f.a.b), laser desorption and electrospray and high performance data systems, have made mass spectrometry potent for direct oligosaccharide structural analysis. Derivatized samples are usually employed, but recent advances make the method applicable to underivatized compounds in some cases. Ions of mass up to 4000 Da can be handled by the f.a.b. technique, but electrospray methods, amazingly, allow examination of ions with masses of up to 100 000 Da. The method is best applied to molecular weight determination and for sequence analysis since, during fragmentation, glycosidic bonds are preferentially cleaved, but techniques for linkage determination are also available.[26]

(ii) Conformational analysis[27]

Because of the importance of oligosaccharides in biochemical processes, many of which follow from mutual molecular recognition between the carbohydrates and proteins, the question of their preferred conformations in aqueous solution and the distortability from them has taken on major significance. As frequently happens in science, methods of addressing these questions have developed concurrently with the recognition of their importance.

As indicated in Section 2.3.4, a conformation of a disaccharide is described by the ring shapes of the constituent sugars and the torsion angles ϕ and ψ defining the rotamer states about each of the two C–O bonds involved in the glycosidic linkage. For oligosaccharides, therefore, conformational analysis requires that the ring conformations and two torsion angles be determined for each sugar and each interunit bond, respectively. N.m.r. spectroscopy with instruments operating at 300 MHz or greater and computations of the relative energies of all possible conformations and identification of those representing energy minima[28] are two methods which can be applied complementarily with considerable power.

^{1}H n.m.r. methods can be used to determine ring conformations, $^3J_{H,H}$ values giving the steric relationships of pairs of vicinal C–H bonds. Chemical shifts, on the other hand, provide data relevant to the steric relationships of the rings connected by glycosidic bonds since, for example, protons are deshielded by oxygen atoms of neighbouring sugars which approach within 0.27 nm. Clearly, for this application, basic data on chemical shifts of relevant protons on appropriate monosaccharide compounds are required. Measurement of ^{1}H–^{1}H nuclear Overhauser enhancements provides a further powerful means of interrelating the steric relationships between the monosaccharide units of an oligomer. For interpretation of any of the data provided by the above techniques full assignments of all resonances in the spectra have to be carried out, and two-dimensional correlation methods are most powerful for this purpose.

Likewise, for application of ^{13}C n.m.r. methods, full spectral assignments must be completed, usually by use of ^{1}H–^{13}C correlations or chemical shifts of model compounds. $^4J_{C,H}$ values can then give data on the critical torsion angles involved in the glycosidic linkages, and chemical shifts can be informative in indicating centres of molecular crowding. Relaxation rate measurements are most useful for studying the dynamics of changes in molecular conformations.

For the purposes of calculating relative energies of oligosaccharides *ab initio* methods and the force field approach are normally too expensive of computer time, and 'hard sphere' methods, which consider only the non-bonded interactions between the sugar units, are normally adopted. When taken together quantitatively with the *exo*-anomeric effect (Section 2.3.4), which is significant in determining the rotamer states about the glycosidic C–O bonds (ϕ), this 'hard sphere *exo*-anomeric effect' (HSEA) method[29] has given results which correlate well with those obtained by n.m.r. procedures, and is of major significance.

By use of ^{1}H and ^{13}C n.m.r. methods, which include J, δ, T_1 and nOe measurements, sucrose has been found to adopt conformation **8.20** in aqueous and DMSO solution, this also being the pronounced energy minimum determined by the HSEA approach. The results of the latter analysis are illustrated in Figure 8.1 from which ψ and ϕ values of approximately $+79°$ and $-28°$ describe the calculated energy minimum, and the shape of the energy surface reveals a sharp minimum and thus point $\times$ (Figure 8.1) represents a strong conformational preference.[27] This being so, it is not surprising that in the crystalline state sucrose also adopts conformation **8.20**; however, there is now an additional intramolecular hydrogen bond between the hydroxyl group at C-6 of the fructose moiety and the ring oxygen atom of the glucosyl unit.[30]

Calculations and n.m.r. parameters show that α-maltose and β-cellobiose exist in conformations **8.5b** and **8.6b**, respectively,[4] and reveal that α- and β-(1→4)-linked D-glucose oligomers would be helical and linear, respectively, in their conformational minima, just as is found for amylose and cellulose (Section 8.1.3.b).

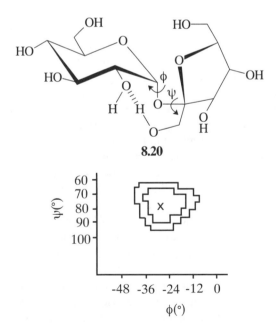

Figure 8.1 Energy surface for rotations about the glycosidic bonds of sucrose. Isocontour diagrams for 4.2 and 8.4 kJ mol^{-1}. The symbol $\times$ represents the energy minimum (reproduced with permission from *Pure Appl. Chem.*, 1983, **55**, 609)

Very impressively, both the n.m.r. and the HSEA approaches can be applied to complex oligosaccharides, and in consequence, for example, the preferred conformations of the blood group determining oligosaccharides (Section 8.1.2.b) are known,[19b] and hence their specific interactions with proteins, notably antibodies and lectins, have been studied in detail,[31] the bases of the affinities being associations between complementary lipophilic surfaces. Applied to the branched octasaccharide **8.21**, a sequence found in glycoproteins (Section 8.1.4.a), the methods reveal that whereas the mean planes of the monosaccharide rings lie in very irregular relationships to each other, the molecule is largely linear with the 1,6-linked trisaccharide component folded over relative to its representation in **8.21** and lying close to the reducing terminal N-acetylglucosamine unit.[32]

β-D-Gal$p$-(1→4)-β-D-GalNAcp-(1→2)-α-D-Man$p$ $\underset{1}{}$

6

β-D-Man$p$-(1→4)-D-GlcNAc

3

β-D-Gal$p$-(1→4)-β-D-GlcNAcp-(1→2)-α-D-Man$p$ $\underset{1}{}$

8.21

8.1.3   POLYSACCHARIDES[33]

(a)   General[34]

Polysaccharides are elaborated by plants, animals and micro-organisms to serve a host of functions, many of which are well defined while others are less clear. Some, such as cellulose, are structural materials; some, like starch and glycogen, serve as food storage compounds; and others, for example, the polysaccharides of bacterial capsules, confer virulence and immunological specificity.

A large variation occurs in the complexity of the polysaccharides; the homopolysaccharides, being built from one sugar, are described by use of the suffix 'an'. Thus, polyglucoses are 'glucosans'. On the other hand, heteropolysaccharides may contain up to six constituent monosaccharides. Inter-unit connections are by way of glycosidic bonds, and therefore in linear compounds there are single reducing and non-reducing ends to the chains. In branched-chain polysaccharides the number of non-reducing ends equals the number of branch points and there is one reducing end. The degree of polymerization, i.e. the number of monosaccharide units in a polysaccharide, can vary from tens to several thousands.

A variety of simple sugars frequently occur in the polysaccharides, notably D-glucose, D-mannose, D- and L-galactose, D-fructose, D-xylose and L-arabinose, and often, especially in the more complex polymers, the parent monosaccharides occur in modified form. 6-Deoxy-L-galactose (fucose) and -L-mannose (rhamnose), D-glucuronic, D-galacturonic and L-iduronic acid, and 2-amino-2-deoxy-D-glucose and -galactose (and their N-acetates) are often found, and sometimes sugars can carry methyl ether or acetate or sulphate ester groups. Bacterial polymers, in particular, frequently contain sugars which are modified by having unusual stereochemistry and functionality. In addition, *N*-acetylneuraminic acid **8.22** (Neu5Ac) is widely found in bacterial and animal sources, and muramic acid (2-amino-2-deoxy-3-*O*-D-lactyl-D-glucose) **8.23** and 'keto-D-*manno*-deoxyoctulosonic acid' (Kdo) **8.24** also occur in bacterial polysaccharides. In many instances, heteropolysaccharides are based on oligosaccharide repeat moieties which may contain two monosaccharide units or several.

8.22   **8.23**   **8.24**

From the list of constituent sugars it is apparent that most belong to the D-series and are linked in the pyranoid ring form. However, arabinose, galactose and fructose are sometimes present as furanosides, and the first of these is also exceptional in being found most frequently in the L-configuration. This fact is accounted for by the biosynthetic decarboxylation of D-galacturonic acid which gives rise to L-arabinose (Section 2.5.1).

Polysaccharides are usually isolated from their natural source material by use of inert solvents such as water or DMSO, and when dilute acid or alkali must be used doubt always remains regarding the structural relationships between the polysaccharides in the isolated and natural states. Initial extracts are often dialysed or treated by ion exchange chromatography or gel filtration to remove ionic and other low molecular weight impurities prior to the precipitation of the polysaccharides with water-miscible solvents such as alcohol or acetone. Otherwise, they may be isolated by freeze drying or by complexing with metal salts or with quaternary ammonium salts such as cetyltrimethylammonium bromide (Cetavlon). The purification procedures used are frequently repeated until the products are homogeneous, as indicated by as many physical tests (chromatography, electrophoresis, and particularly gel electrophoresis) as possible. Homogeneity, however, may take on a particular significance in this field because of microheterogeneity by which some structural variations can occur within samples, thus causing any elucidated molecular structure to imply 'average structure' (see Section 8.1.4).

For the determination of primary structures of polysaccharides it is necessary to identify the component sugars and their ratios, the manner in which they are linked and their sequence. This last requirement, in appropriate cases, means the recognition of repeating oligosaccharide sequences. For monosaccharide analysis polymers are normally hydrolysed with acid and the resulting sugars are characterized and quantitatively assayed by chromatographic methods. High performance liquid chromatography and gas chromatography of volatile derivatives are best for

480

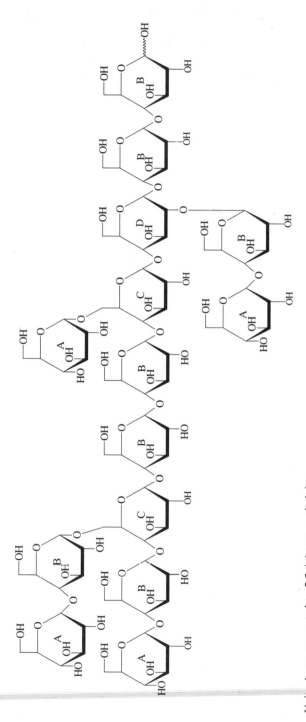

Scheme 8.5

Units A are converted to 2,3,4,6-tetramethylglucose
Units B to 2,3,6-trimethylglucose
Units C to 2,3-dimethylglucose
Units D to 3,6-dimethylglucose

this purpose, and when the latter is coupled with mass spectrometry a very powerful characterizing tool is available. Otherwise, enzymes can be used in particular cases, D-glucose, for example, being specifically determinable by use of D-glucose oxidase, the principle being applicable with immobilized enzyme and electrochemical detection, i.e. D-glucose-specific electrodes can be used. Various classes of sugars can sometimes be assigned by a range of colorimetric methods (Section 3.1.8.c).

Linkage analysis is traditionally carried out by the classical methylation procedure whereby the unsubstituted hydroxyl groups in polymers are fully methylated and the inter-unit bonds are then cleaved hydrolytically to give sets of partially methylated monosaccharides. The free hydroxyl groups, which always include the anomeric groups, indicate the positions which were bonded to other monosaccharides in the polymers. Methylations can be carried out by the use of dimethyl sulphate and aqueous sodium hydroxide (Haworth conditions), often followed by treatment with methyl iodide and silver oxide (Purdie conditions), but commonly the Hakomori method (dimethylsulphinyl sodium in DMSO) is employed (see Section 5.2.1). If neutral conditions are required methyl triflate may be used in trimethyl phosphate.[35] Identification of the methylated monosaccharides is best done by gas chromatography/mass spectrometry of derivatives such as the partially methylated alditol acetates which, being specific compounds, avoid the problem raised by the anomeric mixtures formed, for example, on trimethylsilylation of partially methylated free sugars. An example of the results of methylation analysis applied to a branched glucan is given in Scheme 8.5. Should confirmation of sugar ring sizes be required (the finding of 2,3,6-tri-O-methyl-D-glucose, for example, leaving the furanoid/pyranoid question regarding the relevant units unanswered), methylated polysaccharides may be cleaved by use of triethylsilane and boron trifluoride to give methylated 1,4- or 1,5-anhydroalditols which are stable, of defined ring size and anomerically discrete (Scheme 8.6).[36] Polymers bearing acetyl groups can be similarly probed for the positions of these esters since they are stable under neutral methylation (see above) and reductive cleavage conditions.[37]

With the positions of inter-unit linkages known, it is then necessary, if appropriate, to determine the sugar sequences in the chains, and the chemical approach to this requirement involves partial hydrolysis to determine the nature of the constituent oligosaccharides. Acid-catalysed hydrolysis is of key importance and can be used to remove selectively parts of the molecules which are bonded via relatively weak glycosidic links. In similar fashion, but with different selectivities, acetolysis by use of acetic anhydride and sulphuric acid can also be used. Very importantly, specific enzymes can sometimes be employed which either cleave chains at specific, central monosaccharide sites, e.g. α-D-glucopyranose (*endo*-enzymes), to give sequences of oligosaccharides or at terminal, non-reducing positions (*exo*-enzymes) to give monosaccharides and disaccharides.[38]

Uronic acid containing polysaccharides are subject to selective partial cleavage because of the special chemical features of the uronic acid moieties. Firstly, it is usually possible to identify the sugar to which the acid is attached because the linking bonds are particularly resistant to acid hydrolysis (Section 3.1.1.b.i)

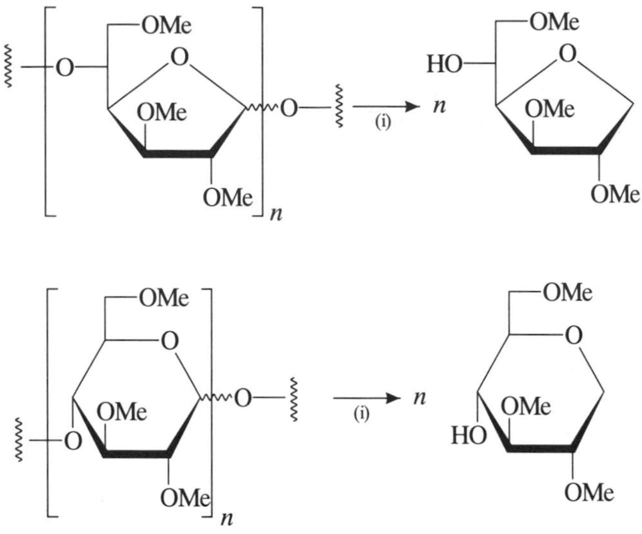

(i) Et$_3$SiH, BF$_3$

Scheme 8.6

and aldobiouronic acids, e.g. **8.25**, are consequently obtainable following such treatment. Secondly, the glycosidic bonds on either side of a hexuronic acid moiety may be specifically cleaved by chemical methods. Scheme 8.7 illustrates that the derived methyl ester **8.26** (R = CO$_2$Me), on treatment with sodium methoxide, releases oligomer-1 by a β-elimination process, while Hofmann rearrangement of the amide can give oligomer-2 by way of the unstable intermediate **8.26** (R = NH$_2$). Otherwise, glycosiduronic acid linkages of methylated polymers may be specifically cleaved by decarboxylation by use of lead tetraacetate or anodic oxidation in acetic acid, both methods leading to replacement of the carboxyl by an acetoxy group. A further, analogous method applicable to methylated polysaccharides involves specific reduction of the uronic acid residues with lithium aluminium hydride and use of the derived primary hydroxyl groups to produce acid-sensitive 5,6-alkenes by way of 6-deoxy-6-iodides (Section 4.10.3).

With many polymers having equatorially oriented 2-amino-2-deoxyhexopyrano-sidic bonds, nitrous acid effects selective deamination and 2,5-anhydrohexose forma-tion (see Section 4.3.2.c); *N*-acetyl analogues must be *N*-deacetylated by use of

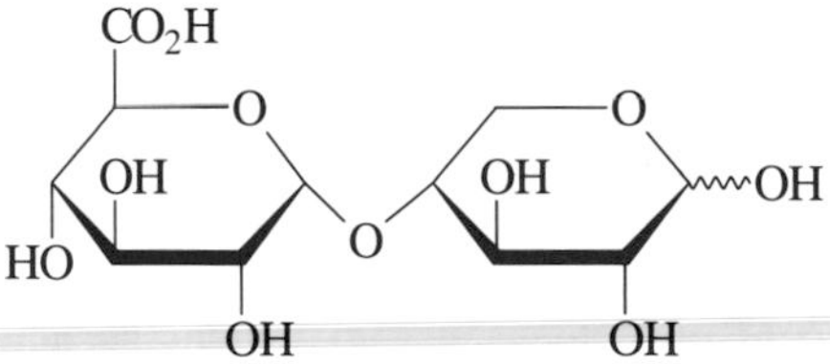

8.25

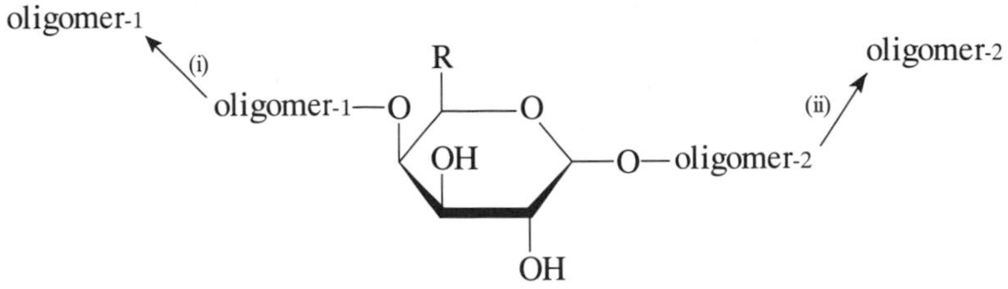

8.26

(i) NaOMe(R = CO$_2$Me); (ii) NaOBr (R = CONH$_2$)
Scheme 8.7

hydrazine or sodium hydroxide in DMSO prior to deamination, but N-sulphated compounds may be deaminated without desulphation and this reaction has been valuable in studies of heparin (Section 8.1.4.b) (Scheme 8.8).[39]

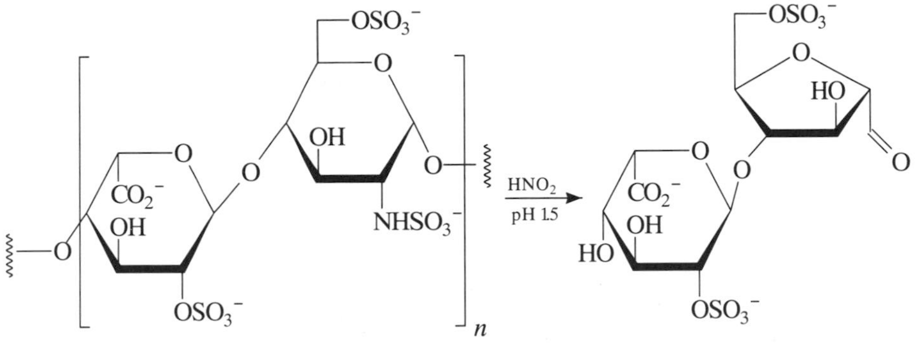

Scheme 8.8

While it is often difficult to devise specific degradation methods for polymers which contain many similar units, oxidations of available 1,2-diols in the constituent sugars with the periodate ion can sometimes be used to complement studies based on the methylation and partial hydrolysis approaches. The glucan callose, which is found on the sieve plates of the phloem of the grape vine during winter dormancy, is tedious to isolate even in small quantities and was analysed in this way. It consumes periodate only to the extent of 0.05 molar equivalents per glucose residue, which is consistent with the presence of 1,3-linked glucopyranose units since only monomers thus linked are devoid of oxidizable 1,2-diol systems. Alternatively, the degree of branching in 1,3- and 1,4-linked glucans and other similarly bonded hexosans can be assayed by determining the number of equivalents of formic acid liberated on oxidation; each is diagnostic of a non-reducing end-group since these alone have *vicinal* triols.

Other applications of periodate oxidation which are of appreciable value involve the examination of residues formed after oxidation of the polymers and removal of the oxidized fragments. Thus, a linear glucan, containing both 1,3- and 1,4-linkages, on oxidation followed by reduction of the derived aldehydic groups and very mild acid hydrolysis gives oligosaccharide derivatives as shown in Scheme 8.9.

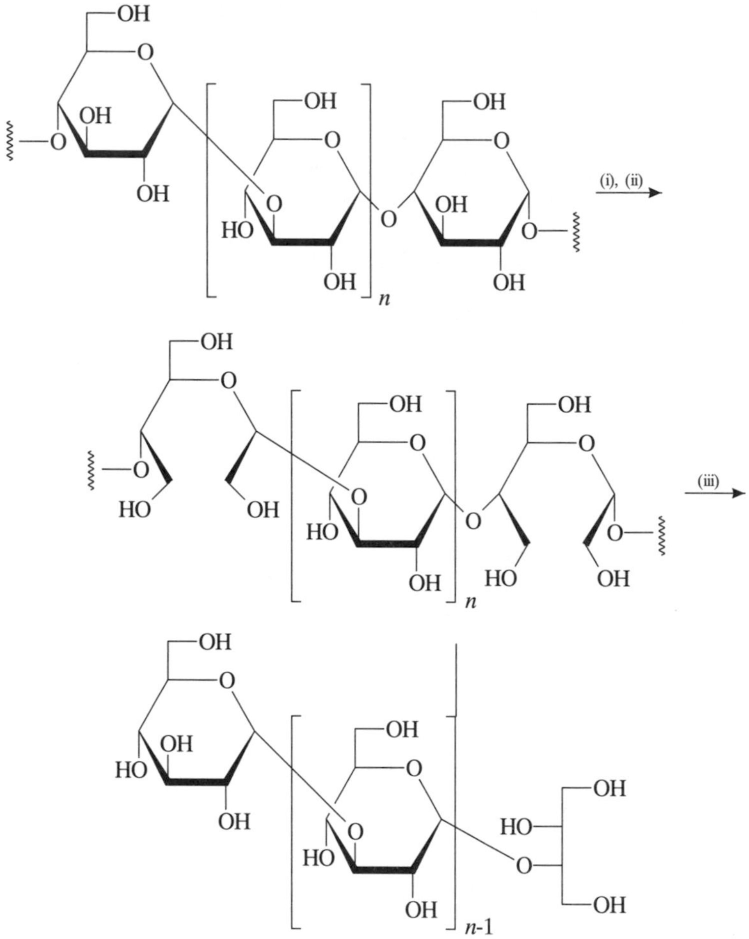

(i) IO_4^-; (ii) $NaBH_4$; (iii) H_3O^+

Scheme 8.9

Analysis of the degradation fragments therefore provides a means of determining whether the 1,3-linked residues appear together in the polymer or whether they alternate with the 1,4-linkages or if there is a random distribution. These procedures, devised by Smith and his colleagues, are termed the 'Smith degradation'.

There remains, as far as the primary structure of the polysaccharide is concerned, the issue of the anomeric configuration of the glycosidic bonds which is of major significance since it governs to a very large degree the overall shape of polysaccharide molecules and their resultant physical and chemical properties (cf. amylose and cellulose). The question of the secondary, tertiary and quaternary structures of these polymers is outside the scope of this book, but the importance of these in biology, technology and commerce should not be overlooked.

Knowledge relating to the ability of specific enzymes to cleave inter-unit bonds can give valuable data on the anomeric configurations. Otherwise, this information may be obtained by n.m.r. methods or by consideration of measured optical rotations, α-D- and β-L-linkages leading to much more dextro rotations than β-D- and α-L-analogues (Appendix 4).

More reliably, the $J_{1,2}$ values in ^{1}H n.m.r. spectra may be used in cases of polymers with regular structures. Anomeric configurations involving sugars with H-1 and H-2 axial, as in β-D-glucopyranosides, give characteristically large values ($\sim$ 8 Hz), while the values for axial–equatorial (e.g. α-D-glucopyranosides or β-D-mannopyranosides) and equatorial–equatorial (e.g. α-D-mannopyranoside) relationships are approximately 3 and 1 Hz, respectively. ^{1}H–^{13}C coupling constants and both ^{1}H and ^{13}C chemical shifts may also be used to assign anomeric configurations.

Given the molecular complexity of polysaccharides and the possibility that chemically identical monosaccharide units within them could be magnetically distinguishable, n.m.r. spectroscopy can be remarkably useful as a tool in structural analysis, and with its many techniques is becoming increasingly important in studies of molecular conformation and dynamics.[40] In practice, while appreciable line broadening often occurs, good quality spectra can frequently be obtained by use of high field instruments, by measuring spectra of heated solutions to eliminate localized geometry-related factors and to move the proton signal of the HO^{2}H formed by deuterium exchange of the hydroxyl groups of the polymers, and by use of a battery of n.m.r. techniques. Application, for example, of convolution difference processing of the free induction decay signal obtained from beef lung heparin in ^{2}H$_2$O at 90 °C, gives the ^{1}H spectrum shown in Figure 8.2.[41] It is essentially that of the constituent repeating disaccharide shown in Figure 8.2, and features of the conformations of the monosaccharide rings can be identified from it. While the amino sugar ring (A) is α-linked and in the $^4C_1$ conformation (small $J_{1,2}$, large $J_{2,3}$, $J_{3,4}$ and $J_{4,5}$ values), the coupling constants associated with the vic-trans-proton pairs of the L-iduronic acid ring (I) (H-1, H-2; H-2, H-3; H-3, H-4) are noticeably smaller and reveal mixed conformations with significant proportions of the $^1C_4$ chair.[42]

The method is highly suited to the structural analysis of other polymers composed of repeat units. The capsular polysaccharide of *Streptococcus pneumoniae* Type 17A, for example, has been determined to have the complicated octasaccharide

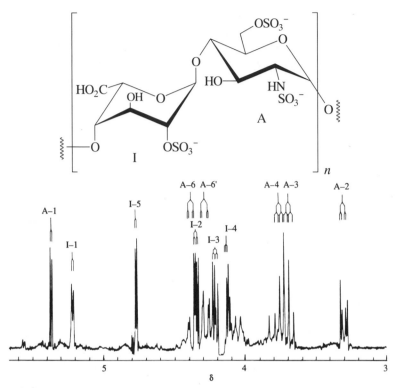

Figure 8.2 ^{1}H n.m.r. spectrum (270 MHz) of beef lung heparin (largely composed of the illustrated repeat unit) in D_2O at 90 °C after convolution difference processing (reprinted with permission from *Macromolecules*, **12**, 1001–1007. Copyright (1979) American Chemical Society)

8.27 as the repeat unit following the application of ^{1}H n.m.r. spectroscopy and methylation and specific degradation analytical methods.[43]

$$\beta\text{-D-Gal}p$$
$$\downarrow 1$$
$$\downarrow 4$$
$$\rightarrow 3)\text{-}\beta\text{-D-Glc}p\text{-}(1\rightarrow 3)\text{-}\alpha\text{-D-Gal}p\text{-}(1\rightarrow 3)\text{-}\beta\text{-L-Rha}p\text{-}(1\rightarrow 4)\text{-}\alpha\text{-L-Rha}p\text{-}(1\rightarrow 4)$$
$$2|$$
$$\text{OAc}$$
$$\beta\text{-D-GlcA}p\text{-}(1\rightarrow 3)\text{-}\beta\text{-D-Gal}f\text{-}(1\rightarrow$$
$$\uparrow 2$$
$$|1$$
$$\alpha\text{-D-Glc}p$$

8.27

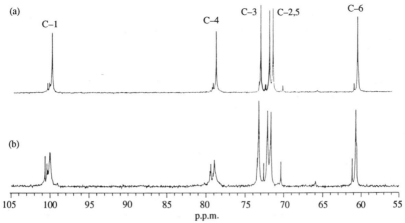

Figure 8.3 ^{13}C n.m.r. spectra (75 MHz) of high amylose corn starch (a) and waxy rice starch (b) in DMSO-d$_6$ at 80 °C (reproduced with permission from *Carbohydr. Res.*, 1987, **160**, 58)

^{13}C n.m.r. spectra likewise can also be simple and highly informative.[44] In Figure 8.3 the spectrum of a high amylose corn starch can be compared with that of waxy rice, which is essentially amylopectin (Section 8.1.3.b.ii).[45] Both are α-(1→4)-linked D-glucopyranose polymers, the former being largely linear and giving essentially a simple spectrum while the spectrum of the latter is complicated by the heterogeneity introduced by branching of the chains through C-6, and a consequent relatively high number of non-reducing terminal groups and branch point groups. The resonances at δ 61.0 and 70.1 are assignable to C-6 and C-4, respectively, of the terminal, non-reducing moieties. These spectra clearly show how powerful ^{13}C n.m.r. spectroscopy can be in studying polymers of these types.

Up until the early 1980s 2000 Da represented the approximate upper limit for mass spectral analysis of carbohydrate derivatives; 10 years later sector-type instruments are operating up to 15 000 Da, and the introduction of electrospray and laser ionization techniques has since made the examination of intact biopolymers of molecular weight up to 100 000 Da possible. It appears that structural polysaccharide chemistry is soon to have available to it mass spectrometry as a potent tool for analyses which could include linkage as well as sequence and molecular weight determinations.

No structural assessment of a polysaccharide is complete without a determination of molecular size, and although chemical methods, for example those which can identify reducing terminal units, can assist with smaller polysaccharides, larger compounds require application of physical methods such as osmometry, ultracentrifugation, light scattering as well as mass spectrometry (see above). Sometimes indications of molecular weight are obtained during isolation procedures which involve, for example, gel filtration using calibrated supports.

(b)   Polysaccharides of plants[46]

Plants have a central role in carbohydrate biochemistry because it is largely within them that sugar production via photosynthesis takes place, and they use the products most directly for structural, food storage and functional purposes. In relative terms plants tend to produce polysaccharides which are structurally the simplest, and cellulose, and starch to a lesser extent, have particular significance because of the massive quantities in which they are biosynthesized.

(i)   Cellulose[47]

Cellulose is an unbranched glucan composed of β-(1→4)-linked monomer units whose characteristics demonstrate most markedly how polymeric features affect physical and chemical properties. It is the most abundant organic material found in the plant kingdom, forming the principal constituent of the cell walls of higher members and providing them with their structural strength. Cotton wool is almost pure cellulose, but in wood, the other chief source of the polymer, it is found in close association with other polysaccharides (mainly hemicelluloses) and lignin. Its biological function depends upon its chemical inertness as well as on its physical strength. It is completely insoluble in water and in other common solvents; it does, however, dissolve in cuprammonium solution (Schweizer's reagent) and in 1,2-diaminoethane and can be methylated in strong alkaline solution with dimethyl sulphate, and acetylated using acetic anhydride and a variety of acid catalysts. Cellulose is very much more resistant to acid hydrolysis than are other polysaccharides, but with 70% sulphuric acid it can be converted into D-glucose in good yield. However, it is more satisfactorily degraded by acetolysis with acetic anhydride, acetic acid and sulphuric acid, with which it gives cellobiose octaacetate (see Section 3.1.1.b.vii).

Physical measurements indicate a degree of polymerization of about 5000 and the β-glycosidic linkages confer on the molecule a linear structure as indicated by formula **8.28**. These long molecules held in this linear conformation aggregate to give fibrils bound together by a multiplicity of hydrogen bonds and also by van der Waals attractive forces, and it is this aggregation which confers the insolubility, chemical inertness and physical strength. Such is the regularity of the fibril structures that portions diffract X-rays; that is, they show crystalline properties, so the resultant picture of cellulose is of bundles of parallel molecules which possess regions of high order (the crystalline regions) alternating with regions where intermolecular association is less well ordered (the amorphous regions). Cellulose fibres are then built up of these fibrils wound spirally in opposite directions around a central axis to give a product of high physical strength.

Cellulose **8.28**

Enzymes of the human alimentary tract do not catalyse the hydrolysis of cellulose, but a wide range of cellulases occur in various animals, plants and microorganisms; these, like acids, attack first the amorphous regions of the fibres.

Apart from its value in the unmodified state, cellulose can be converted to a series of O-substituted derivatives which are of appreciable commercial significance. Cellulose nitrate, with a high degree of substitution, is used as gun cotton, while a product with a degree of substitution of 2.3 nitrate groups per anhydroglucose unit is known as 'collodion'. Lower esters are plasticized with camphor to give celluloid. Cellulose acetates in fully and partially substituted forms are used as textile fibres, and other carboxylate esters and alkyl ethers are manufactured for use as plastics. Most significantly, viscose rayon is prepared from sodium cellulose xanthate (ROC(S)SNa) (Section 5.3.5.b.) by treatment with dilute sulphuric acid which regenerates the parent material in a form suitable for use as a textile fibre.

(ii)   Starch[48]

Like cellulose, starch is a glucan found extensively in the plant kingdom, but it has vastly different physical and chemical properties, and instead of serving a structural function it is elaborated as an energy reserve and is stored in the roots, tubers, fruit and seeds in the form of insoluble granules which are also sufficiently ordered to enable them to diffract X-rays. As the principal carbohydrate component of animal diets, it has great commercial significance, but in addition starch and its derivatives have other important economic values, for example in the textile industry.

Starch is generally believed to be a mixture of two structurally different α-glucans, *amylose* and *amylopectin*. These can be separated by fractionation procedures involving either preferential extraction of the amylose from the granules by a leaching procedure with warm water or dimethyl sulphoxide or, more commonly, dissolution of the granules in warm water which causes them to rupture and the contents to disperse, followed by preferential precipitation of the amylose component with a complexing agent such as butanol, thymol or pyridine. High purity polysaccharides are obtainable by repetition of these precipitation procedures or by use of concanavalin A in affinity column chromatography. Most starches contain $20 \pm 5\%$ of amylose, but materials with appreciably more and others with very little (see Figure 8.3) are also known.

The amylose component is essentially a linear α-$(1\rightarrow4)$-linked glucan ($[\alpha]_D = +220$) which, in solution, readily adopts a helical structure as a consequence of the steric relationship of adjoining units imposed by the α-bonds (see maltose **8.5b**). In this form it is therefore tubular in shape (**8.29**) and in cross-section it resembles the cyclodextrins (Section 8.1.2.a). Like the cyclodextrins, it readily forms inclusion complexes with hydrophobic molecules, probably as a result of *hydrophobic bonding* with the interior of the helix. Usually the helix is believed to contain six anhydroglucose residues per turn, but to some extent the polymer can adapt to accommodate various compounds with which it complexes; the helix which encloses n-butanol, for example, is tighter than that which reacts with t-butanol. It is this ability to complex which results in the preferential precipitation of amylose

Amylose **8.29**

from aqueous solution on the addition of the fractionating agents mentioned above, and it also accounts for the interaction of starch with iodine which gives the well-known deep-blue colour. In this complex a chromophoric polyiodine anion is stabilized by inclusion within the six-unit helix of the amylose component, and the colour is discharged on warming because of the uncoiling which takes place at elevated temperatures. This specific iodine-binding power of amylose is the basis of the potentiometric titration method used to determine the amylose content of starch samples.

Methylation analysis leads to the conclusion that amylose is essentially linear since no dimethylglucoses are obtained, and the proportion of the tetramethyl ether indicates a chain length of about 300. However, physical methods such as osmometry, viscometry, light scattering and ultracentrifugation indicate values higher by a factor of two or three.

Enzymic methods have been applied more successfully to starch than to any other polysaccharide, and have confirmed an essentially linear structure for amylose, but have also indicated the presence of occasional anomalous linkages. Thus, β-amylase, an enzyme which causes cleavage of successive maltose units from the non-reducing ends of α-(1→4)-glucans, but which is blocked by a branch point or other structural irregularity, does not hydrolyse some amyloses completely. Finally, amylose shows the property of *retrogradation*, i.e. spontaneous precipitation from aqueous solution owing to aggregation of the polymer chains.

Amylopectin is also an α-linked glucan ($[\alpha]_D = +150$), and since 2,3,6-tri-O-methyl-D-glucose is the main product of methylation and hydrolysis, it too is composed largely of 1,4-linkages. However, methylation also reveals that there is a branch point through position 6 every 25 units, and periodate studies are consistent with this, since one mole of formic acid derived from non-reducing end-groups is liberated by the same number of anhydroglucoses. The polymer is therefore branched, and enzyme studies have indicated a structure in which highly branched clusters form parts of larger aggregates.

Although amylopectin is appreciably larger than amylose since it contains many thousands of monomer units, it produces only a dull-red colour with iodine, indicating that it cannot coil effectively to give the helices required for the formation of the inclusion complexes. The branched structure accounts for this, and also for the fact that amylopectin molecules cannot aggregate in solution and so no retrogradation is observed. β-Amylase in this case removes the exterior chains, leaving a degraded polymer known as a β-limit dextrin with stubs containing either two or three units (depending upon whether the degraded chains initially contained even or odd numbers of sugar units — Scheme 8.10). About 55% of the polymer weight appears as maltose during this process and this establishes that the average 'repeat unit' of 25 anhydroglucose residues contains 14 which are removable with the enzyme, and therefore on average 16.5 between the non-reducing group and the nearest 1,6-linkage.

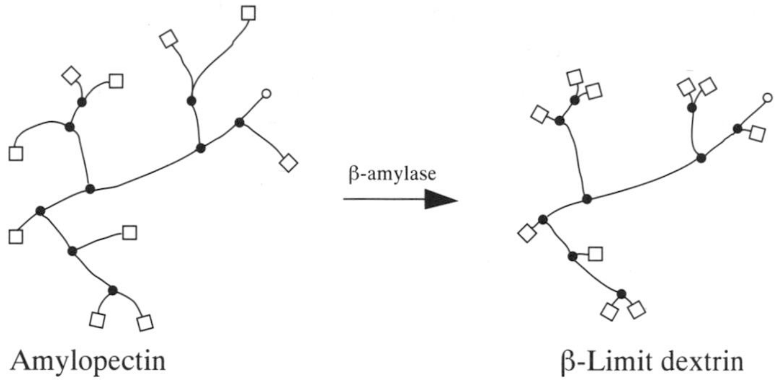

Amylopectin β-Limit dextrin

⌣ Represents an α–(1→4)-linked D-glucopyranose chain;
o Represents the reducing centre;
• Represents a 1,6-branch point;
□ Represents a non-reducing end-group

Scheme 8.10

(iii) Hemicelluloses[49]

These are polymers of relatively small size (a few hundred sugar units) which are found in close association with cellulose and lignin in land plant cell walls. Most contain a small number of different monosaccharides, their nature being dependent on the source of the polymer. Xylans are the most prevalent, occurring in most land plants and comprising β-(1→4)-linked sugars in a cellulose-like backbone, soft wood xylan having α-L-arabinofuranose and 4-O-methyl-α-D-glucopyranosyluronic acid units bonded to O-2 or O-3. L-Arabinose is also found attached to β-(1→3)-linked galactans in some softwoods.

Ivory nuts are the common source of β-(1→4)-linked D-mannans which contain small amounts of D-galactose while other, related polysaccharides, are genuine galactomannans containing the two sugars in equal proportions, the galactose being

α-(1→6)-linked to the mannan backbones. The galactose/mannose ratios can be manipulated and the properties of the products thus altered to produce useful sizes and gelling agents for foods. D-Glucomannans occur in hardwoods and softwoods as do arabinogalactans which have β-(1→3)-bonded galactan main chains with multiple D-galactose and L-arabinofuranose substituents.

It is a characteristic of the hemicelluloses that they frequently consist mainly of long chains made of one sugar carrying occasional or frequent substituents, usually other sugars or short oligosaccharides. In this way the chains cannot aggregate as they do in cellulose, and thus these polymers often have high solubility in water.

(iv)   Gums

Several plants produce hard gums, either spontaneously or in response to injury (including cutting), that have found considerable use in industry as, for example, thickening and stabilizing agents and as adhesives. They tend to have relatively complex structures commonly containing five or six different monosaccharides within their molecules, *gum arabic*, for example, having chains of β-(1→3)-linked galactose units to which are bonded other chains containing D-galactose, L-arabinofuranose, L-rhamnose and D-glucuronic acid groups. Other *Acacia* gums have different types and degrees of substitution. *Gum ghatti, gum tragacanth* and *guar gum* are other members of the group, their variations in structure being exemplified by the fact that the last is a β-(1→4)-linked mannan having α-D-galactopyranosyl moieties bonded to O-6 of every second main chain sugar unit.

(v)   Pectins

Pectins are made up primarily of α-(1→4)-linked D-galacturonic acid monomers or their methyl esters and are found in land plant cell walls. The *pectinic acids*, which comprise the methyl esters, are very water soluble and are common gelling agents used in preservatives and jellies. Their molecular architecture normally contains proportions of such monosaccharides as L-arabinose, D-xylose, L-fucose and L-rhamnose, the last of these sometimes within the uronic acid main chains. *Pectic acids* are non-methylated analogues of the pectinic acids.

(vi)   Other food storage polysaccharides

Some plants choose to store carbohydrates in the form of homopolysaccharides other than starch, some lichens, for example, producing the β-(1→6)-linked glucan *pustulan*, and Iceland moss storing the β-(1→3)-β-(1→4)-linked glucan *lichenan* as well as the α,α-linked *isolichenan*. Others use D-fructose for storage purposes, the β-(2→1)-linked furanose polymer being *inulin* found in dahlia and Jerusalem artichoke tubers, while *levan* is a β-(2→6)-joined short polymer found in grasses. Both polymers are unusual in having the 'reducing ends' protected by glycosidically linked α-D-glucopyranose units, i.e. they terminate with sucrose molecules, and inulin is unique in having the fructofuranose rings as substituents on the (O-1, C-1, C-2)$_n$ polymer chain of **8.30**.

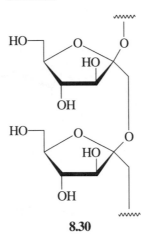

8.30

(vii)   Seaweed polysaccharides[50]

Several of these have commercial value such as the cell wall materials *agar, carrageenan* and *alginic acid* and the food storage material *laminaran*. *Agar* is found in the red algae *Rhodophyceae* and is the name given to a mixture of three polysaccharides: *agarose*, which is made up of the repeat unit **8.31** comprising very largely unsulphated D-galactose and 3,6-anhydro-L-galactose, *agaropectin*, which also contains D-glucopyranosyluronic acid and, unusually, pyruvic acid acetal groups, the latter spanning the O-4, O-6 diol of the D-galactose moieties, and some sulphate ester groups, and a sulphated galactan which does not contain the 3,6-anhydro-L-galactose groups. Agar is a prized gelling agent, 1.5% solutions of which set at about 35 °C and do not melt below 60 °C. It is used extensively in the food and pharmaceutical industries and in biological laboratory work, for example in gel electrophoresis and for culturing micro-organisms.

Carrageenans, a set of sulphated polysaccharides, are also found in certain red algae families and consist of polymers of D-galactopyranose bonded through, alternatively, α-(1→3)- and β-(1→4)-linkages, i.e. they are based on the repeat unit **8.32**, but there are important additional features in the various members. In particular, some have the 4-substituted galactose molecule in the 3,6-anhydro form, and all are highly sulphated polymers and are available commercially in the form of metal ion salts. Again, the gelling characteristics make the carrageenans of value in the food industry, in particular for the enhancement of the texture of various dairy products, and in the pharmaceutical industry for the making of stable suspensions of insoluble compounds and to provide texture to products such as toothpastes.

Alginic acid is the most important polysaccharide of brown seaweeds of the *Phaeophyceae* family, again being used for food thickening and in pharmaceutical and cosmetic products. Chemically it is very unusual, being made up of β-(1→4)-bonded D-mannuronic and α-(1→4)-bonded L-guluronic acid units which are C-5 epimers of each other (the epimerization taking place after the biosynthetic poly-merization), and which occur in blocks rather than alternately as in the cases of agarose and the carrageenans. In the presence of calcium ions very strong gels are

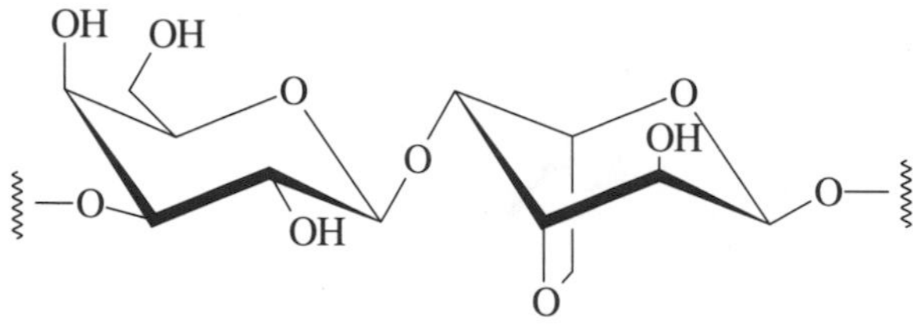

8.31

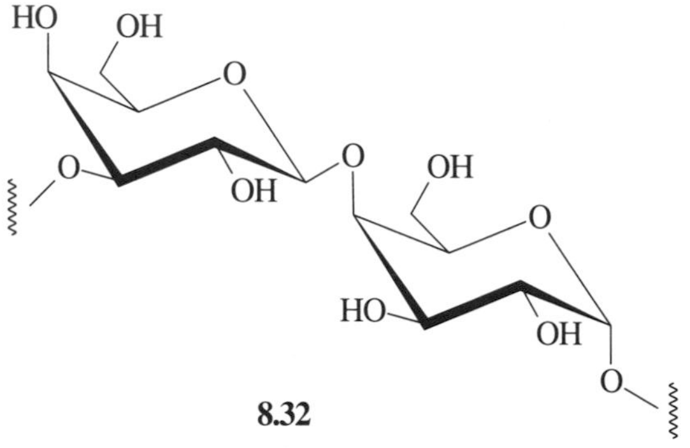

8.32

formed following the pairing of the linear L-guluronic acid sections by coordination around the metal ions.

Elaborated by members of the *Laminaria* species, *laminarin* is a β-(1→3)-bonded D-glucopyranose polymer with branching through O-6 and, in some cases, having D-mannitol units terminating the chains.

(c) Polysaccharides of animals[51]

Only a few homopolysaccharides are found in animal sources, the majority being complex heteropolysaccharides often having simple repeat units and containing amino sugar components.

The simplest homopolymer is *glycogen*, which is a reserve carbohydrate found ubiquitously in the muscles and livers of mammals, fish and insects and called upon as a glucose source when energy requirements demand. It is a glucan of very high molecular weight (degree of polymerization about 10^5–10^7), having mainly α-(1→4)-linked units with branching through position 6 approximately every 10 units; it is, therefore, similar to amylopectin but more highly branched. Several human diseases which can be traced to genetically determined enzymic irregularities in glycogen metabolism have been recognized, some leading to the accumulation of the polysaccharide in specific tissues or organs. Snails, as well as producing glycogen, also synthesize a galactan which has a highly branched structure, linkages in this case being through positions 3 and 6.

Chitin is a fibrous β-(1→4)-linked 2-acetamido-2-deoxy-D-glucose polymer, the N-acetylation being about 20% incomplete, of which the shells of arthropods including insects and shellfish are composed. Fungi also utilize the compound for structural purposes. Its inertness derives from the same factors that affect cellulose, of which chitin is the 2-acetamido-2-deoxy analogue. Alkali-catalysed deacetylation affords *chitosan* which is soluble in dilute acids, whereas the prime characteristic of the parent compound is insolubility in most solvents, *N*, *N*-dimethylacetamide being an exception. Commercial uses of chitin derivatives include as films, membranes, fibres and surface coatings.

Very important animal polymers contain protein as well as complex carbohydrate components. These glycoproteins and proteoglycans are dealt with in Section 8.1.4.

(d)   Polysaccharides of micro-organisms[52]

As well as elaborating polysaccharides as key constituents of their cell walls and, where appropriate, capsules (see below), many bacteria synthesize extracellular polymers which, in general terms, are similar to some plant products; structurally they tend to be relatively simple, and there are, for example, bacterial analogues of *alginic acid* and *levan*. Some are of considerable commercial importance: *dextrans* are elaborated by members of the *Leuconostoc* group and consist mainly of α-$(1\rightarrow 6)$-linked α-D-glucopyranose units with a small proportion of other linkages and branch points, the natures and proportions of which depend upon the strain of the organism. They are used as blood plasma substitutes and, on cross linking with epichlorohydrin, afford Sephadex gels employed in the fractionation of biological macromolecules. Dextran sulphates can be used as anticoagulants and in ulcer treatment. Several *Streptococci* produce different α-glucans.

β-Glucans produced by *Alcaligenes faecalis* are insoluble in cold water but give elastic, firm gels on warming. These *curdlans*, containing β-$(1\rightarrow 3)$-linked D-glucose units with some C-6 branching, have value as gelling agents and food additives. Further β-linked glucose-based polymers, produced by *Xanthamonas* strains, are the *xanthan gums* which are also used in the food industry, but in addition applications have been found in textile and paper manufacture. In the oil industry they are useful in aqueous solution as drilling additives for the improved displacement of oil and increase in recovery from wells. The main chains are 1,4-linked and bear trisaccharide branched chains comprising two units of D-mannose and one of D-glucuronic acid.

Very importantly, bacteria also synthesise carbohydrate polymers for production of their cell walls and capsules, many showing high immunological specificity and being typical of narrow ranges of bacteria. Commonly this specificity is dependent on the oligosaccharide repeat units present in the polymers which contain some of a range of unusual sugars, for example mono- and di-deoxy, aminodeoxy and diamino sugars, aminouronic acids, and mono-, di- and even tri-*O*-methyl derivatives, altogether over 100 such modified sugars having been identified.

The cell walls of Gram-negative (those not reacting to Gram stain) and Gram-positive bacteria are morphologically complex and quite different in chemical terms, the latter having mainly a peptidoglycan as the structural cell wall polymer with bonded heteropolysaccharides and teichoic acids which are polymers of sugar or alditol phosphates. Together these complexes represent some

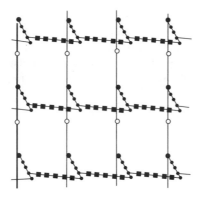

○ = 2-Acetamido-2-deoxy-D-glucopyranose
● = 2-Acetamido-2-deoxy-3-O-lactyl-D-glucopyranose
· = Amino acid
■ = Glycine

Figure 8.4 Cell wall structure of Gram-positive bacteria

of the most complicated molecular structures known. The peptidoglycans are typically made up of β-(1→4)-glycosidically linked chains of amino sugars, for example 2-acetamido-2-deoxy-D-glucose alternating with muramic acid (a 3-O-lactyl derivative) crosslinked with polypeptide chains which may vary appreciably in structure. A typical sheet is represented in Figure 8.4 and several matrices are crosslinked to give the main fabric of the cell walls. The unique structure presents a selective target for attack by bacteriocidal agents; indeed, β-lactam antibiotics prevent bacterial growth by interfering with the crosslinking of the polypeptide chains.[53] Also, lysozyme, which is able to hydrolyse the glycosidic links in the peptidoglycan chains (see Section 3.1.1.b.iv), confers bacteriocidal activity to tears, nasal mucus and saliva by its presence in these bodily fluids.[54]

Teichoic acids are phosphate-containing polymers with the phosphates acting in diester linkages to join polyhydroxyl species such as glycerol (1,2- or 1,3-linked), ribitol (1,5-linked) or 3-O-β-D-glucopyranosylglycerol (1,6′-linked) which may have sugars or alanine bonded to unsubstituted hydroxyl groups, a simple general structure being given by **8.33**. Sometimes related substances, which are antigenically active, are elaborated extracellularly together with capsular polysaccharides. The heteropolysaccharides of the cell walls typically have small oligosaccharide repeat units (often tetrasaccharides) and vary appreciably in structure.

$$\text{H}_2\text{O}_3\text{PO—alditol—O—} \left[\begin{array}{c} \text{sugar} \\ | \\ \overset{\text{O}}{\overset{\|}{\text{P}}}\text{—O—alditol—O—} \\ | \\ \text{OH} \quad\quad \text{D-alanine} \end{array} \right]_n \begin{array}{c} \text{sugar} \\ | \\ \overset{\text{O}}{\overset{\|}{\text{P}}}\text{—O—alditol} \\ | \\ \text{OH} \quad\quad \text{D-alanine} \end{array}$$

sugar atop alditol, D-alanine below — Teichoic acids **8.33**

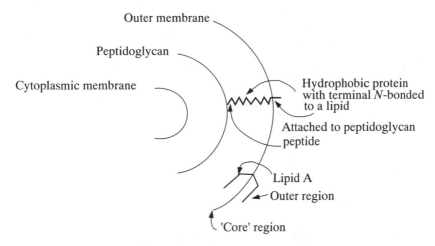

Figure 8.5 Cell wall structure of Gram-negative bacteria

Gram-negative bacteria, on the other hand, have thin (perhaps monolayer) sheets of peptidoglycan between cytoplasmic membranes and outer membranes which appear to be non-covalently attached to the peptidoglycan by a hydrophobic protein consisting of 57 amino acids (see the cartoon of the cell wall, Figure 8.5).[55] This protein is bonded by its terminal carboxyl group to an amino group of a diamino acid in the peptidoglycan which is not involved in crosslinking to the pentaglycine peptide (cf. Figure 8.4). The nitrogen terminus of the protein is attached to a lipid, which is embedded in the outer membrane and anchored there by hydrophobic forces.[56]

The outer membrane contains very complex lipopolysaccharides of which there are at least three main bonded components: the 'lipid' A component, the 'core' (backbone) region and the 'outer' region. Lipid A does not vary greatly between species and consists of the β-(1→6)-linked 2-amino-2-deoxy-D-glucose dimer carrying two phosphate esters and heavily substituted with long-chain fatty acid groups, as illustrated by the R^1 and R^2 substituents in **8.34.** It is joined to the core region by way of a 3-deoxy-D-*manno*-octulosonic acid (Kdo, for ketodeoxyoctulosonic acid) bridge which is often composed of three Kdo glycosidically (2→7) linked units, heptoses (L-*glycero*-D-*manno*-heptose) and hexoses, as illustrated in **8.34** for the salmonella complex. Considerable variation is, however, found in this region. To it are also bound the antigenically active outer oligo- and poly-saccharide chains which, for example, allow classification of Gram-negative bacteria according to serotypes, each one bearing its own oligosaccharide-based antigen. Structural variation of these is also appreciable (see Section 8.1.2.b).

Gram-negative bacteria commonly produce polysaccharide capsules, in which case these take over the immunological characteristics. Normally they are acidic in character, containing uronic or ulosonic acids, pyruvic acid ($MeCOCO_2H$) as cyclic acetal substituents on sugar residues, or phosphate groups. They are further characterized by having structures based on repeating oligosaccharides and extreme

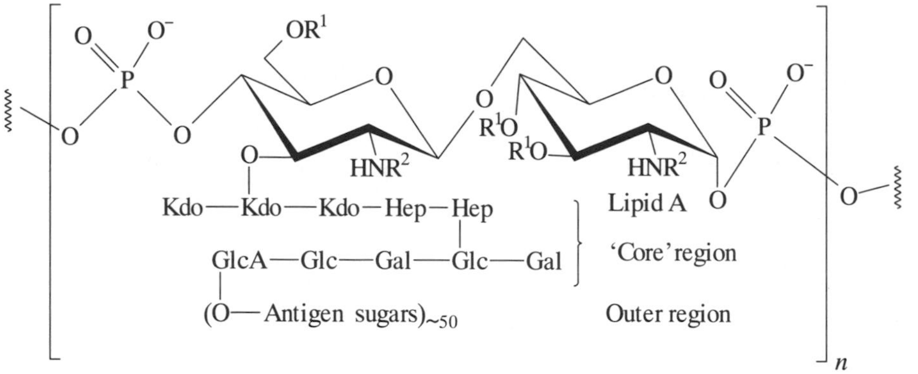

$$R^1 = \text{long chain fatty acid} \quad R^2 = COCH_2CH(OH)(CH_2)_{10}Me$$

8.34

variation is found within the units. N.m.r. spectroscopy has provided a powerful means of determining these structures (Section 8.1.3.a).

Other micro-organisms, for example yeasts and fungi, have polysaccharide profiles of their own.

8.1.4 GLYCOPROTEINS, PROTEOGLYCANS AND GLYCOLIPIDS

As well as being found in nature linked to other sugar units as in oligo- and poly-saccharides, carbohydrates form important constituents of the glycoproteins, proteoglycans and glycolipids in which they are covalently bound to proteins and glycerides or other hydrophobic entities. These sets of compounds are extensively varied in structure and occur naturally in plants, animals and micro-organisms, but many have particular importance in human biology and medicine.

(a) Glycoproteins

Very many proteins of plants, animals and micro-organisms qualify to belong to this category by having oligosaccharides as parts of their structures in variable proportions from 0.5 up to 85%. They function in ways ranging from structural (for example, collagen) to hormonal (thyroglobulin), enzymic (prothrombin and taka-amylase, the human intestinal maltose- and sucrose-hydrolysing enzyme), protective (gastric mucins, immunoglobulins and blood group substances), transport (58 of the 60 plasma proteins including α_1-acid glycoprotein and ceruloplasmin)[57] and food reserve (milk proteins and all of the hen egg white proteins except lysozyme, of which ovalbumin is a well-known example). The number of oligosaccharides attached to protein chains varies enormously; for example, there is only one glycosylation site in ovalbumin, whereas there are about 200 in sheep submaxillary mucin. Furthermore, different structural types of oligosaccharides can also be attached to one protein chain. The most common monosaccharide constituents are D-glucose, D-galactose, D-mannose and L-fucose which are often found together with the amino sugars 2-acetamido-2-deoxy-D-glucose and -D-galactose and N-acetylneuraminic acid.[58] The saccharide chains are either N-glycosidically linked to the amido group of L-asparagine, e.g. **8.35**, or *O*-glycosidically linked to the hydroxyl groups of L-serine or L-threonine, e.g. **8.36**. The two groups are chemically distinguishable since, in the latter case, the attached oligosaccharides can be readily removed from the protein chain with very mild base (1 mol l^{-1} sodium carbonate at 20 °C), whereas those attached by amide links **8.35** are stable under these conditions. The remarkable ease with which β-elimination occurs for the *O*-bonded compounds has been attributed to an increase in the functioning of the sugars as leaving groups by complexation with cations.[59]

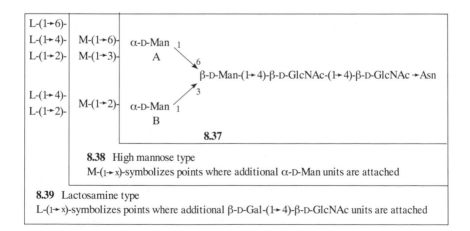

8.35   **8.36**

Some generalizations can be made about the *N*-glycosidically linked glycoproteins; for instance, they normally possess the core pentasaccharide **8.37**.[60] One type, the high mannose variety, contains further mannose units usually attached at *O*-3 and *O*-6 of sugar A and at *O*-2 of sugar B, as illustrated in **8.38** by the M-$(1{\rightarrow}x)$-symbols. Some are further elaborated by having additional *N*-acetylglucosamine and galactose molecules at various sites, thereby giving rise to a wide range of oligosaccharides, an example being the octasaccharide **8.41** found in ovalbumin.

L-(1→6)-			
L-(1→4)-	M-(1→6)-	α-D-Man ₁	
L-(1→2)-	M-(1→3)-	A	
			β-D-Man-(1→4)-β-D-GlcNAc-(1→4)-β-D-GlcNAc →Asn
L-(1→4)-			
L-(1→2)-	M-(1→2)-	α-D-Man ₁	
		B	

8.37

8.38 High mannose type
M-(1→x)-symbolizes points where additional α-D-Man units are attached

8.39 Lactosamine type
L-(1→x)-symbolizes points where additional β-D-Gal-(1→4)-β-D-GlcNAc units are attached

The other type is elaborated by attachment of N-acetyllactosamine units to O-2, O-4 and/or O-6 of mannose A and positions 2 and 4 in mannose B, as illustrated in **8.39** by the L-$(1\rightarrow x)$-symbols. These N-linked glycoproteins are consequently referred to as the N-acetyllactosamine type; they are often further substituted with fucose units or several N-acetylneuraminic acid groups as terminal residues, yielding oligosaccharides as elaborate as **8.40**, which occurs in α_1-acid glycoprotein.

α-Neu5Ac-(2→ 3/6)-β-D-Gal-(1→ 4)-β-GlcNAc

α-Neu5Ac-(2→3/6)-β-D-Gal-(1→ 4)-β-GlcNAc

α-D-Man

α-Neu5Ac-(2→ 3/6)-β-D-Gal-(1→ 4)-β-GlcNAc

β-D-Man-(1→4)

α-Neu5Ac-(2→ 3/6)-β-D-Gal-(1→ 4)-β-GlcNAc

α-D-Man

β-D-GlcNAc-(1→4)-GlcNAc

8.40

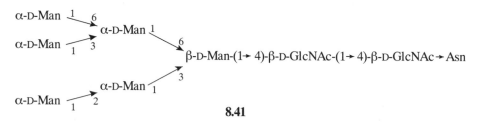

8.41

A purified glycoprotein, even one that has only one glycosylation site, is usually a mixture comprising identical protein chains with different, although related, oligosaccharides attached. This is called *microheterogeneity* or *site-heterogeneity*, and each individual molecule is called a *glycoform*.[61] Thus, purified ovalbumin, for example, which is a high mannose type of glycoprotein, has about 10 glycoforms with structures created as described above, **8.41** being one example. As would be anticipated, when there are several glycosylation sites on a protein chain the number of glycoforms increases dramatically.

Fewer generalizations can be made about the structures of O-linked glycoproteins: β-D-Gal-$(1\rightarrow 3)$-α-GalNAc-$(1\rightarrow O)$-Ser (or Thr) is, however, found at the points of attachment of several examples, but many other linking units occur.

Workers in the field of glycobiology have convinced the scientific community of the significant role played by the oligosaccharides of these complex macromolecules in the biological functions of living organisms. In consequence, the need for specific oligosaccharides for biochemical purposes has become intense, and this has acted as a stimulus to the development of improved glycosidation methods. These points are discussed further in Section 6.1, where the biological importance of specific oligosaccharides and their syntheses are introduced. The carbohydrate components of the glycoproteins are vital; in some cases the oligosaccharides influence the physicochemical properties of proteins. For example, the carbohydrates of the so-called antifreeze glycoprotein **8.42**, present in the sera of an arctic fish, are

$$\left[\begin{array}{c} \text{———— Ala-Ala-Thr} \\ \beta\text{-D-Gal-}(1\rightarrow3)\text{-}\alpha\text{-D-GalNAc-1} \end{array} \right]_n$$

8.42

responsible for lowering its freezing point to $-2\,°C$. In other cases they play a vital role in cell surface recognition phenomena. Thus, they are implicated in intercellular recognition, intercellular adhesion and cell contact inhibition; they serve as cell surface antigens, receptor sites for viruses, proteins and hormones; their structures and, consequently, functions are modified in virus-infected and cancerous cells; and catabolism of circulating proteins by different organs is regulated by their presence. It is this role of the oligosaccharides in molecular recognition processes, which commonly involve glycoproteins, that is a central topic in glycobiology.

The circulating lifetimes of plasma glycoproteins are dependent upon their oligosaccharide structures.[53,62] For example, ceruloplasmin (used to carry copper) suffers prompt removal from circulation by the liver when the sialic acid residues, which are attached either $2\rightarrow6$ or $2\rightarrow3$ to most of the terminal galactose units, are removed by treatment with neuraminidase. A glycoprotein receptor in the liver specifically binds the branched oligosaccharides terminating in galactoses that are present in the asialoceruloplasmin, one glycoform of which is depicted in **8.43**.

β-Gal-(1→4)-β-GlcNAc-(1→2)-α-Man ⟍1
                                        ⟍6
                                            β-D-Man-(1→4)-β-D-GlcNAc-(1→4)-β-D-GlcNAc
β-Gal-(1→4)-β-GlcNAc ¹ ⟍4   ⟋3
                            α-Man ¹
β-Gal-(1→4)-β-GlcNAc ⟋²
                    ¹

8.43

The inflammatory response that occurs on injury or infection leads to the rapid appearance on the tissue membranes of a special adhesive glycoprotein called E-selectin which binds white blood cells (leukocytes), and by this means the white blood cells are recruited to help in the normal repair of an injury. The mechanism of the adhesion process involves the specific binding of the E-selectin to sialyl-Lewis^x **8.44**, a tetrasaccharide found on the surface of white blood cells.[63] Again, it is found that terminal sialic acid is essential since sialidase-treated white cells do not bind. On some occasions too many leukocytes are recruited to a traumatized site, thereby normal tissue can be destroyed, as is observed in septic shock, chronic inflammatory diseases such as rheumatoid arthritis and in reperfusion injury following heart attacks and strokes. In addition, sialyl-Lewis^x is also found on some lung cancer and colon cancer cells, suggesting that it could be implicated in the metastasis of human cancers. Consequently, sialyl-Lewis^x is being viewed as a potential sugar-based antiinflammatory and antitumour drug since it could be used to block the E-selectin binding sites, thereby preventing leukocytes from binding.

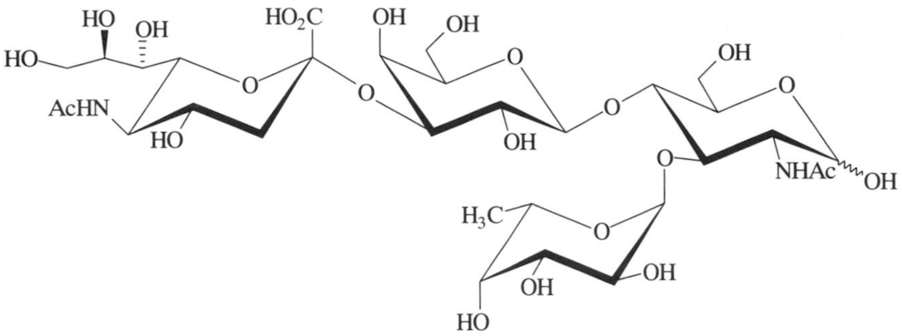

Sialyl-Lewis^X **8.44**

Detailed attention has been paid to the synthesis and conformational analysis of oligosaccharides that occur in glycoproteins.[64]

(b)   Proteoglycans

These are a group of glycoproteins which contain a very high proportion of polysac-charide, so much so that at one time the protein was thought to be a contaminating impurity, and this led to methods of purification which cleaved the proteins and consequently the products were classified as polysaccharides. As a family they are referred to as glycosaminoglycans and are high molecular weight polymers composed of disaccharide repeat units comprising one 2-amino-2-deoxy sugar with, most often, a uronic acid partner.[65] There are five recognized types of glycosamino-glycans: chondroitin sulphate, dermatan sulphate, keratan sulphate, heparin and hyaluronic acid, the first four usually being *O*-glycosidically attached to the protein chain through the linkage trisaccharide indicated in **8.45**. *Chondroitin sulphate*, probably the most abundant glycosaminoglycan in the human body, occurs in many mammalian tissues and in highest concentration in cartilage. It is present, in its non-sulphated form, in placenta, where it decreases with term. It is structurally a polymeric form of **8.46** in which the O-4 or O-6 of the amino sugar unit is sulphated to varying proportions, and 100 repeat units may be present. *Dermatan sulphate* is located in fibrous connective tissue, e.g. skin, tendons and joint capsules, and is structurally related to chondroitin sulphate, but for the most part it has the C-5 epimeric uronic acid L-iduronic acid instead of D-glucuronic acid. The repeat unit is mainly **8.47**, but sulphate may also arise at O-2 of the uronic acid moiety.

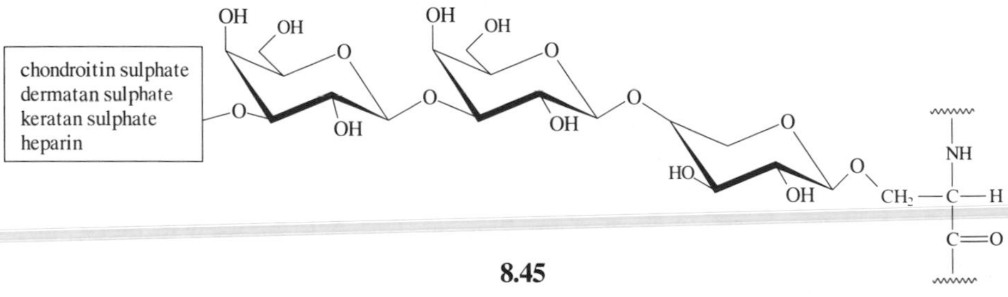

chondroitin sulphate
dermatan sulphate
keratan sulphate
heparin

8.45

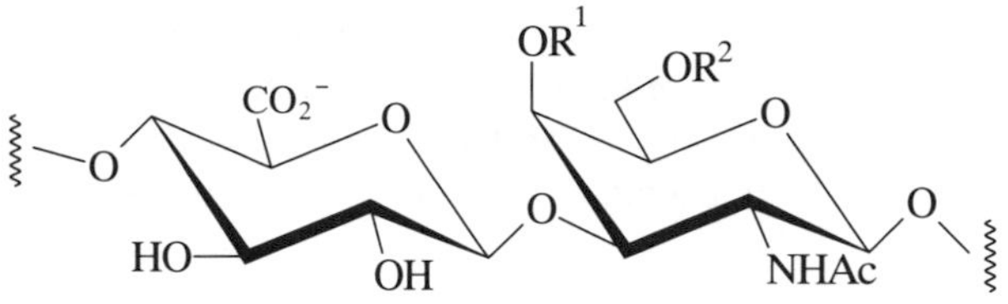

R^1, R^2 = H, SO_3^- Repeat unit of chondroitin sulphate

8.46

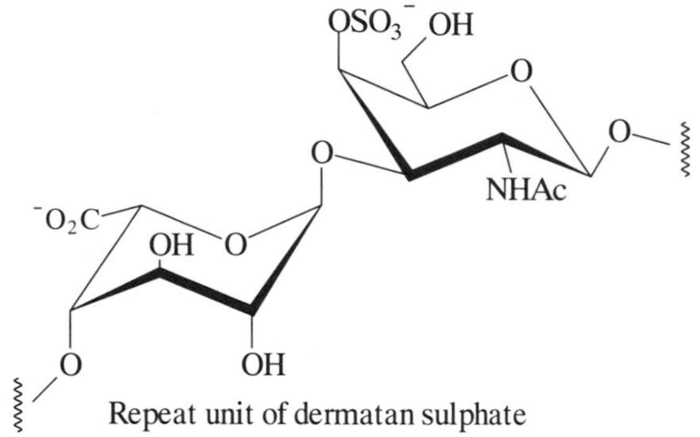

Repeat unit of dermatan sulphate

8.47

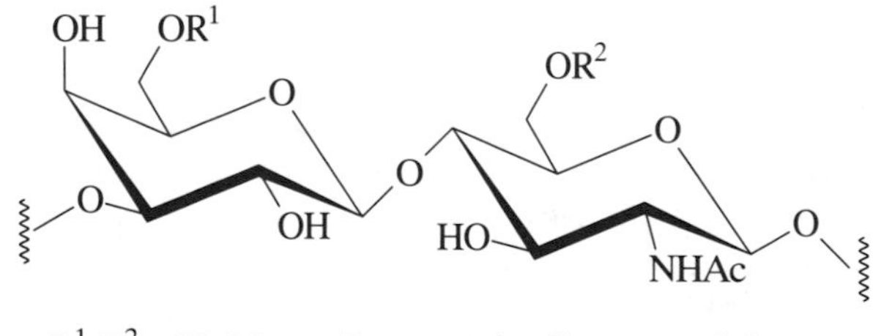

R^1, R^2 = H, SO_3^- Repeat unit of keratan sulphate

8.48

Keratan sulphate is found in the cornea, in embryonic liver and lung tissue and in slightly modified form in skeletal tissue such as cartilage and bone. Its chemical structure is odd in being non-acidic and is based on *N*-acetyllactosamine (repeat unit **8.48**), and either sugar unit may carry a sulphate ester group at O-6. As with the previous polymer, linkages are alternately 1→3 and 1→4, but the degree of polymerization is about 10% less. *Heparin* is a useful anticoagulant because of its specific affinity for the protease antithrombin which inhibits proteolytic enzymes involved in blood clotting. As such, it is used intravenously after surgery and in the treatment of arterial and venous thrombosis. It occurs naturally in the liver and heart, but the major source for clinical use is lung and mast cells. It is mainly a polymer of the L-iduronic acid disaccharide repeat unit **8.49**, but with blocks of the polymer containing analogues in which the α-L-iduronic acid is replaced with

β-D-glucuronic acid. The average sulphate group content is two to three groups per disaccharide unit but other analogues, given the name *heparan sulphate*, with one ester group per disaccharide, are known to occur as cell surface components. The pentasaccharide **8.50** is the smallest molecule to exhibit anticoagulant properties and several derivatives and analogues of it have been synthesised.[66]

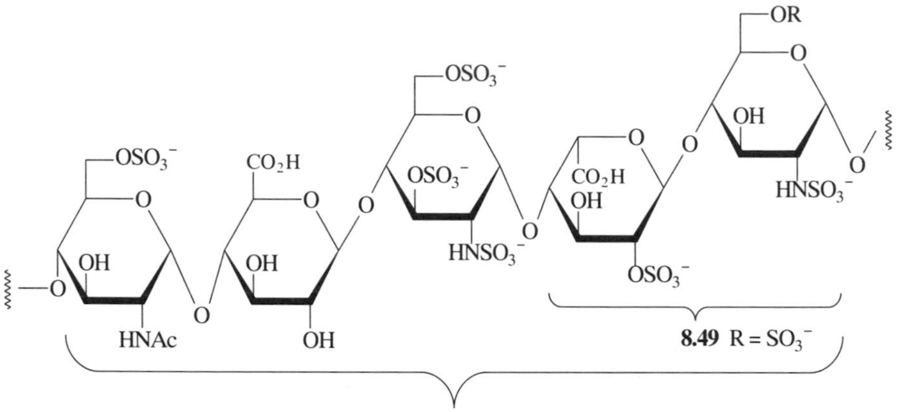

8.50 R = H

Hyaluronic acid is widely distributed in the bodies of mammals, in the skin, arterial walls, cartilage, eye and umbilical cord. At low concentration it produces viscous aqueous solutions which function as lubricants and shock absorbers in joints, which when afflicted with arthritis contain depolymerized hyaluronic acid. It is obtainable from cockscomb, although biotechnologists have now introduced a bacterial fermentation method of preparation to meet the needs of a variety of medical applications such as the treatment of glaucoma and joint degeneration.[67]

It is an unusual glycosaminoglycan since proteins are tethered to it by non-covalent means. Structurally it is a polymeric form of disaccharide unit **8.51** with the highest molecular weight of any glycosaminoglycan; in some cases, there may be as many as 10 000 repeat units. Although it may not be a true proteoglycan, its treatment in this section is justified since it is always found in close association with other proteoglycans; nowhere is this more clearly illustrated than in cartilage, where it is present as the neutral organizing molecule for several proteoglycan chains which are non-covalently anchored to it.[68]

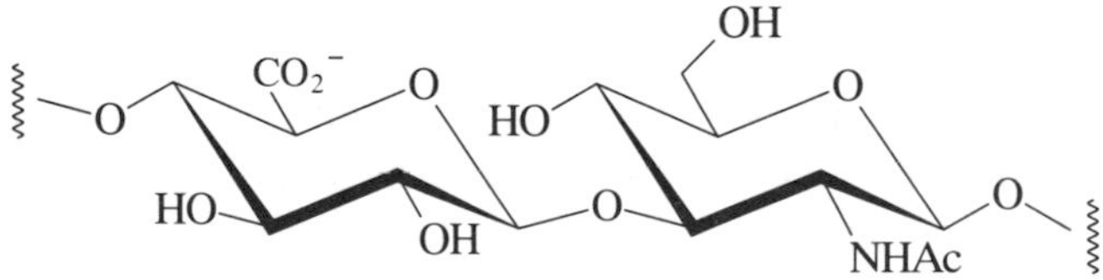

Repeat unit of hyaluronic acid

8.51

(c) Glycolipids[69]

All such compounds are amphipathic in having hydrophilic mono- or oligo-saccharides glycosidically bonded to hydrophobic entities, often long-chain alkanols or amines or glycerides, steroids or terpenes. The sugars found are effectively the same as those occurring in glycoproteins.

In the animal kingdom the major compounds of this class are a set of *glycosphingolipids* having different mono- or oligo-saccharides bonded to O-1 of a ceramide, for example **8.52**, which is a long-chain fatty acid amide of the base sphingosine. Monoglycosylceramides are called *cerebrosides*, the name indicating that they are found in and available from the brain, and a β-D-*galacto*-compound is the major member of the cells of the mammalian central nervous system, while a β-D-glucosyl analogue predominates in other cells. The main glycolipid of human erythrocytes has the trisaccharide α-D-Gal*p*- (1→4)-β-D-Gal*p*-(1→4)-β-D-Glc as the constituent carbohydrate, and other members of the family carry oligosaccharides which are the blood group determinants (Section 8.1.2.b). Imbalance in the glycolipids brought on by irregularities relating to enzymes of carbohydrate transfer cause specific human disease conditions. Many glycolipids containing *N*-acetylneuraminic acid within their oligosaccharides are major constituents of the human brain and are named *gangliosides*.

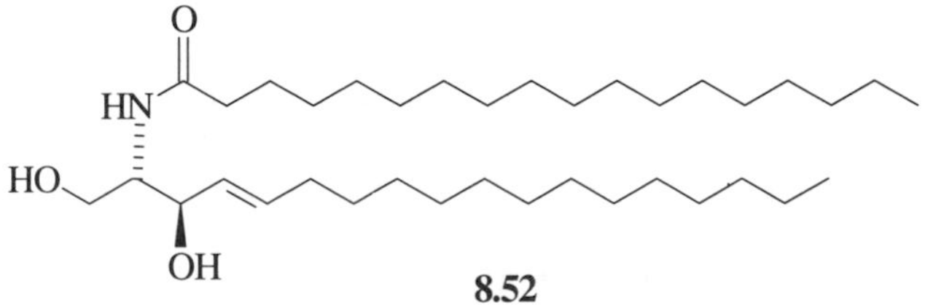

8.52

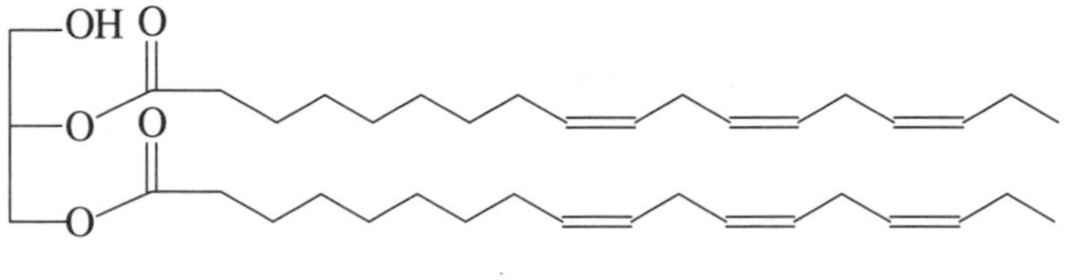

8.53

Plants, in which glycolipids are widely used, particularly in the leaves, predominantly synthesize these compounds by glycosylation of 1,2-deacylglycerol (for example **8.53**, the α-linolenate), usually with D-galactose or a dimer of this sugar. Otherwise, sugars may be bonded to steroids or terpenes either glycosidically or by esterification at O-6 of D-glucose when the steroid or terpene has a carboxylic acid group. Complex glycolipids containing glycosylated inositols linked through phosphate bridges to ceramides are also found.

Some bacteria produce galactosyldiacylglycerols of the kind found in plants, but others elaborate quite different structures. Thus, the 'cord-factors' of *Mycobacteria* are 6,6'-diesters of α,α-trehalose, the *M. tuberculosis* compound having acids of about 88 carbon atoms in length and bearing a few hydroxyl and methoxy groups. Unusually, some bacterial glycolipids are glycosylated glycerols having long-chain ethers, for example phytanyl $C_{20}H_{37}$, instead of esters. Otherwise, oligosaccharides may be linked to inositol and then, via a phosphate linkage, to diacylglycerols.

Reference is made in Section 8.1.3.d to the lipid A glycolipids of Gram-negative bacteria which are fully acylated derivatives of the β-(1→6)-linked 2-amino-2-deoxy-D-glucose dimer bearing phosphate groups at O-1 and O-4'.

8.2 OTHER GLYCOSIDES

As well as occurring in nature bound to other carbohydrates and to protein and lipid compounds (Section 8.1), monosaccharides are found glycosidically attached to an immense range of other substances. Although the parts played by carbohydrates in the biochemistry and biology of the products are extensive, a common factor is the hydrophilic feature they introduce which confers water solubility, selective polarity and, in appropriate situations, amphipathic character. Notwithstanding this generalization, however, the interactions between carbohydrates and other species are often hydrophobic in character.

In this section attention is mainly paid to the vast range of natural compounds which have sugars O-linked to non-carbohydrates, but reference is also made to S-bonded and N-bonded compounds, the latter group including the nucleosides. Lastly, a subsection is provided on the so-called 'C-glycosides' which are chemically somewhat anomalous in being devoid of acetal (or equivalent) functionality at the 'anomeric centre'. The nature of these topics is so extensive that only highly selective illustrations can be provided.

8.2.1   OTHER O-GLYCOSIDES[70]

Again, such compounds are found in the plant, animal and microbiological worlds, the first and last of these, in particular, providing a plethora of natural products having widely differing carbohydrate and aglycon components.

(a) Plant O-glycosides

Plants provide a vast range of O-glycosides with some, for example flavanoid compounds, being widely distributed in the botanical kingdom, while others are typical of particular species. D-Glucose, D-mannose, D-galactose, L-rhamnose, D- and L-fucose, L-arabinose, D-xylose and hexuronic acids are the most common sugars present, often being bound in hetero-oligosaccharide form, but the branched apiose (3-C-hydroxymethyl-D-*glycero*-tetrose), D-allose and an extensive range of deoxy- and dideoxy-hexoses, often carrying O-methyl groups, are also to be found in plant products. Many of the last range of sugars have been given trivial names, amongst the most common being D-cymarose (2,6-dideoxy-3-*O*-methyl-D-*ribo*-hexose), D-digitoxose (2,6-dideoxy-D-*ribo*-hexose), L-oleandrose (2,6-dideoxy-3-*O*-methyl-L-*arabino*-hexose) and D-thevetose (6-deoxy-4-*O*-methyl-D-glucose).

Amongst relatively simple compounds are glycosylated derivatives of alditols, inositols and a range of phenols, arbutin **8.54** being an example of the last group, found in the leaves of *Rosaceae* and yielding hydroquinone on enzymolysis. The strong reducing properties of this product have biochemical consequences, as does the hydrogen cyanide produced on enzymic hydrolysis of the *Rosaceae* seed component amygdalin **8.55**. Such compounds belong to a set known as the cyanogenic glycosides which are found in, for example, apricot and peach stones, lima beans and sorghum, amygdalin giving gentiobiose and mandelonitrile on hydrolysis, and since the latter is the cyanohydrin of benzaldehyde, hydrogen cyanide is liberated spontaneously. As 'laetrile' amygdalin has gained some contentious prominence as a cancer chemotherapeutic agent.

Arbutin **8.54**   Amygdalin **8.55**

More complex compounds are represented by the cardiac glycosides, the best known being digitoxin, a trisaccharide glycoside comprising three D-digitoxose molecules bonded as R in the steroid-like compound **8.56**, which is found in the *Digitalis* family. Members of these cardiac glycosides bear the name because of their digitalis-like effect on heart muscle. A related set, found also in other sources, are the saponins which often have complex oligosaccharides bonded to sapogenins, e.g. **8.57**. Members of this group, which includes digitonin, are toxic to fish and have in consequence been used as selective poisons. Nature also provides related plant products based on triterpene and alkaloid derivatives closely related to **8.57**. The list of plant-derived glycosides extends to glycosylated flavonols and related compounds, hydroxyindoles, carotenoids and many more. Some of these products, on enzymolysis, release the aglycons for functional use by the plant and presumably are produced in the glycosidic form for storage or transport reasons.

(b) Animal *O*-glycosides

In animals, glycosides play their major role in assisting with the solubilization of both foreign compounds and endogenous substances, and in this way glycosylation is a key step in detoxification processes. Natural steroids, for example, may

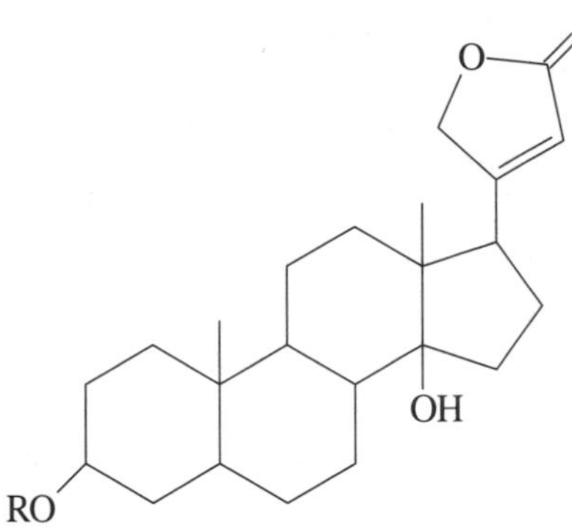

Digitoxin **8.56**

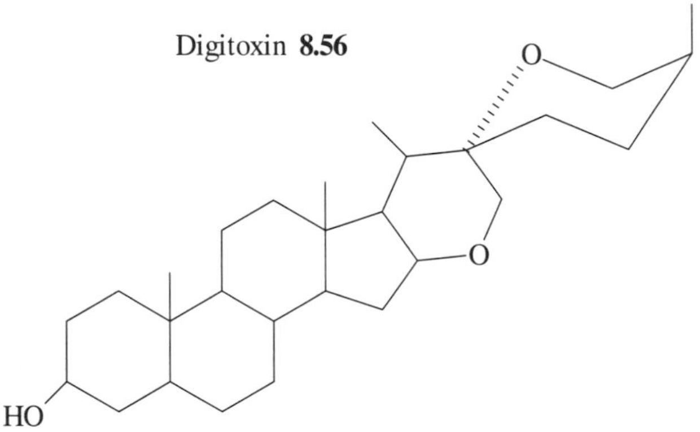

8.57

be excreted in the urine conjugated (a biochemical use of this word not to be confused with the normal chemical use) with D-glucose or its 2-acetamido-2-deoxy derivative, and bilirubin, a breakdown product of haem, is excreted via the bile as a derivative containing two glycosidically bonded D-glucuronic acid groups. Indeed this sugar acid provides the main carbohydrate used in conjugation and is involved in the processes employed by animals to rid themselves of very many foreign compounds. These may range from such disparate materials as trichloroethanol to morphine, while compounds devoid of hydroxyl groups sometimes may be enzymically hydroxylated prior to glycosylation. Carboxylic acids can be derivatized in this way (to give glycosyl esters), as can amines and thiols (to give N- and S-linked products, respectively). In the case of the antiinflammatory compound phenylbutazone the carbon centre between the carbonyl groups is activated to the extent that it acts nucleophilically in the glycosylation step to give the uronic acid C-glycoside **8.58** as the main metabolite in humans. In contrast, detoxification in the dog occurs mainly by prior hydroxylation followed by normal O-glycosylation. One consequence is that the drug is cleared many times faster from the blood plasma of dogs than of humans.

Transportation of some compounds in their glycosylated forms has been implicated in aspects of carcinogenesis. The connection between some aromatic amines

8.58

such as benzidine and bladder cancer, for example, is believed to relate to the retention of glycosylated derivatives in the bladder for appreciable periods of time.

(c)   Microbial O-glycosides[71]

Glycosides from microbiological sources are of particular significance because of the large number that possess antibiotic activity, with many important pharmaceutical compounds belonging to this category. Commonly, *Streptomyces* serve as the source of antibiotics whose trivial names conventionally have the suffix '*mycin*', whereas antibiotics derived from other genera have names ending in '*micin*'.

A main feature of the antibiotic glycosides is the diversity of structures exemplified not just by the aglycons but by the carbohydrate components, many having branched-chain, alkylamino, diamino and other mixed functions and unusual configurations not found outside microbiological carbohydrate products. In Appendix 5 the trivial names of the most common microbial glycoside sugars are listed together with their formal chemical names. Most are derived from antibiotic substances, and their discovery (mainly in the 1960s and 1970s) and hence the requirement for them served as an important stimulus to the development of synthetic methods in monosaccharide chemistry.

A few examples illustrate the range of biological actions of glycosidic antibiotics and the diversity of their chemical structures. The erythromycins are a family belonging to the 'macrolide' group which are based on a macrocyclic lactone as the aglycon and inhibit protein synthesis at the ribosomal level. They have been used extensively against staphylococcal, pneumonococcal, streptococcal and mycoplasmal infections. Erythromycin A has structure **8.59** with the amino sugar D-desosamine and the branched-chain L-cladinose as the constitutent monosaccharides. One of the best-known antibiotics, streptomycin **8.60**, used in the treatment of tuberculosis, also interferes with protein biosynthesis by binding to bacterial ribosomes. Structurally it belongs to the extensive aminoglycoside family of antibiotics which contain amino sugars in association with aminocyclitols, in this case the diaminocyclitol streptidine to which is bonded the unique branched-chain formylpentose L-streptose and then the very unusual 2-deoxy-2-methylamino-L-glucose. Other members of the family are the neomycins, kanamycins and gentamicines.

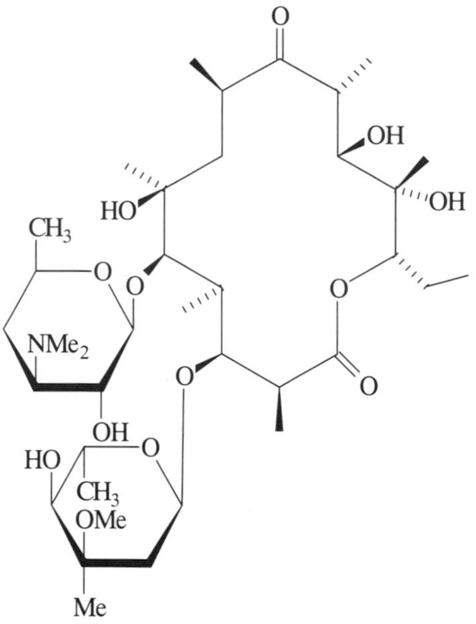

Erythromycin A **8.59**

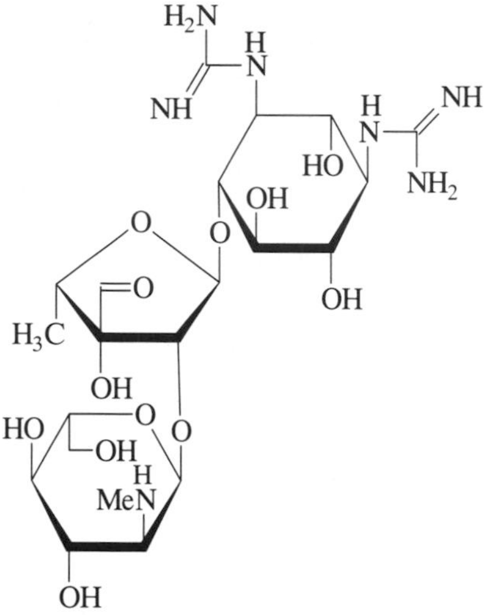

Streptomycin **8.60**

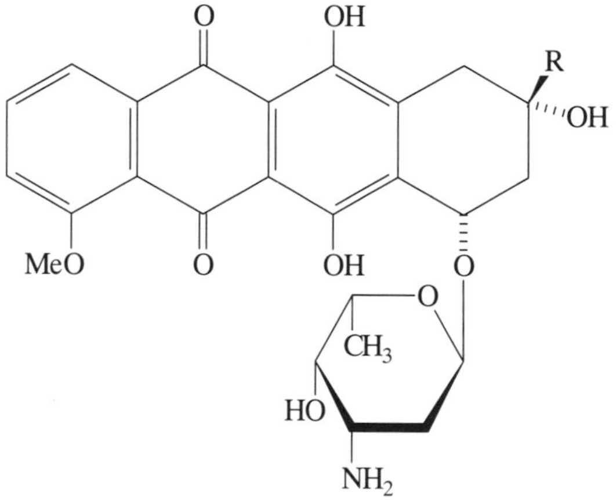

Daunomycin **8.61** R = COMe
Adriamycin **8.62** R = COCH$_2$OH

The anthracyclines show good anti-Gram-positive activity but chemically they are more valuable as anticancer compounds consequent upon their reacting in an intercalating manner with DNA to inhibit the synthesis of RNA. Daunomycin **8.61** and adriamycin **8.62**, which contain L-daunosamine, are used against leukaemias and soft-tissue sarcomas, but some of the family which contain trisaccharide components are less cardiotoxic and more useful for the treatment of solid tumours. Other antibiotics are mentioned in Section 8.2.2.a.

8.2.2 *N*-GLYCOSIDES

Compounds of this category are glycosylamines (Section 3.1.3.a) and occupy several vitally significant niches in biochemistry and biology. As seen earlier, for example, key linkages between carbohydrate and protein components of glycoproteins involve 2-acetamido-2-deoxy-D-glucose bonded to the amido group of L-asparagine **8.35**. Some *N*-glycosides occur naturally as microbiological metabolites, the antibiotic streptothricin F, which is active against both Gram-negative and Gram-positive organisms, being the glycosylamine **8.63** derived from the rare sugar 2-amino-2-deoxy-D-gulose.

Streptothricin F **8.63**

(a) Nucleosides[72]

This category of glycosylamines is much the most important in nature because as compounds consisting of D-ribose or 2-deoxy-D-ribose linked to purine or pyrimidine bases they represent key components of the nucleotides and the nucleic acids. Furthermore, many natural antibiotics and synthetic pharmaceuticals with various therapeutic activities belong to the nucleoside group.

Hydrolysis by selective chemical or enzymic methods of ribonucleic acid (RNA) produces mainly adenosine **8.64** (R = OH), guanosine **8.65** (R = OH), cytidine **8.66** (R^1 = OH) and uridine **8.67** (R^1 = OH, R^2 = H), whereas in deoxyribonucleic acid (DNA) deoxyadenosine **8.64** (R = H), deoxyguanosine **8.65** (R = H), deoxycytidine **8.66** (R^1 = H) and thymidine **8.67** (R^1 = H, R^2 = Me) are the main constituent nucleosides. All these compounds therefore consist of the β-D-furanosyl modifications of ribose or 2-deoxyribose attached to N-9 of the purine bases or N-3 of the pyrimidines.

While the famous double helix of DNA derives its structure and function from the well-known, specific base pairing between the adenine and thymine **8.68** and the guanine and cytosine **8.69** components of the nucleosides, and while the genetic code relates triplets of these bases to specific amino acids, the nucleic acids also contain minor amounts of other nucleosides. Thus, DNA contains small proportions of 5-methyl-2′-deoxycytidine **8.66** (R^1 = H, R^2 = Me) and RNA has more than 50 recognized minor nucleoside components, many of which are methyl derivatives

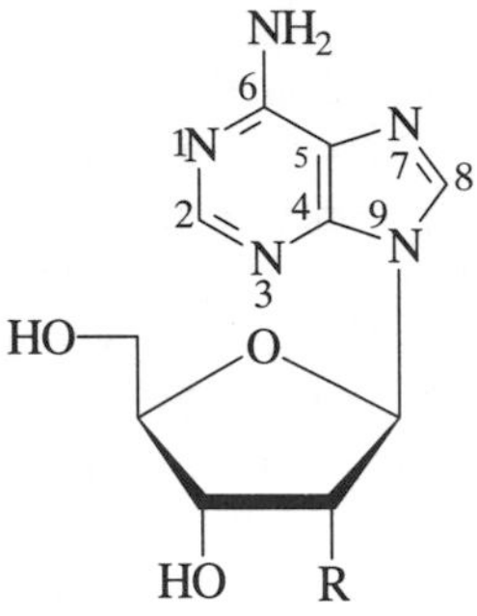

Adenosine R = OH
Deoxyadenosine R = H

8.64

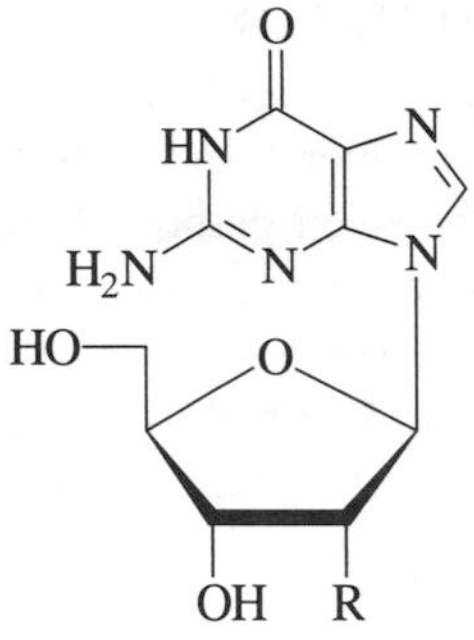

Guanosine R = OH
Deoxyguanosine R = H

8.65

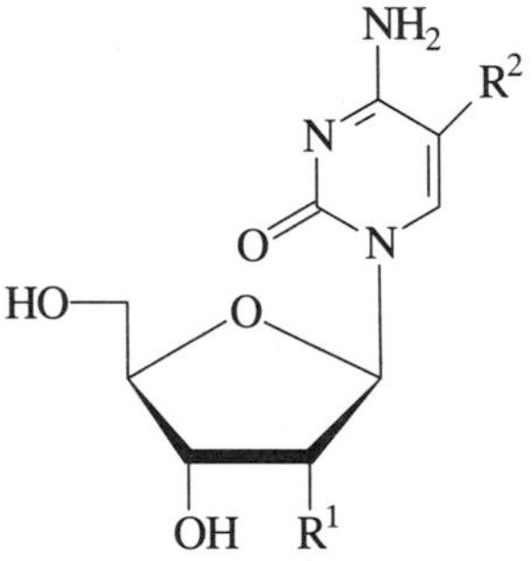

Cytidine R^1 = OH, R^2 = H
Deoxycytidine R^1 = H, R^2 = H

8.66

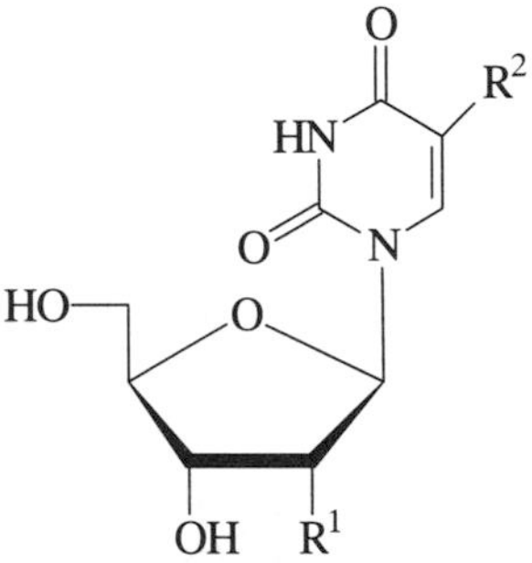

Uridine R^1 = OH, R^2 = H
Thymidine R^1 = H, R^2 = Me

8.67

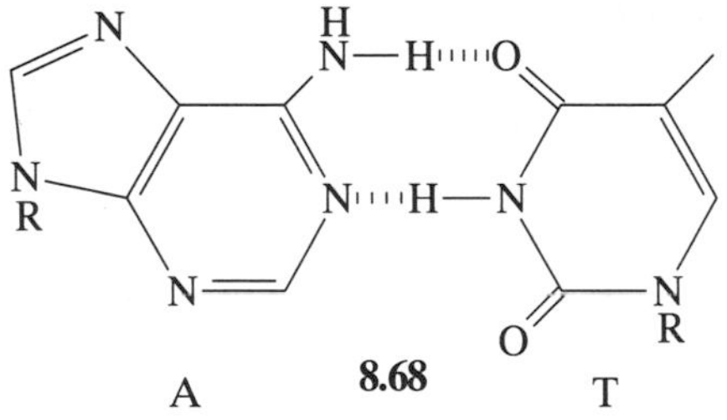

A   **8.68**   T

G   C

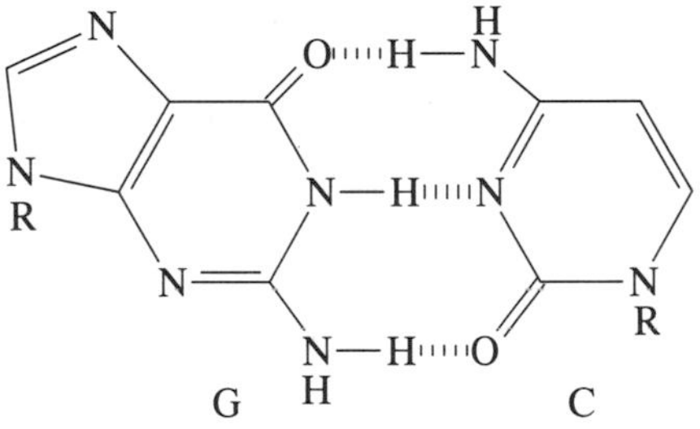

8.69

R =

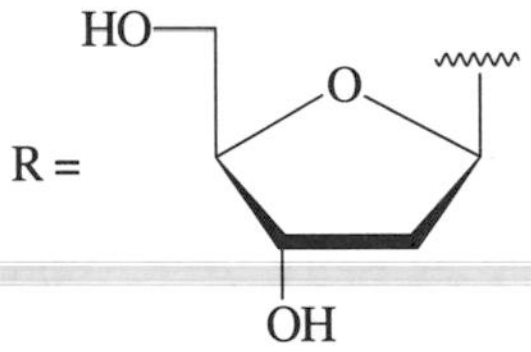

of the major compounds, but the most prevalent is the C-linked analogue pseudouridine (Section 8.2.4).

As well as occuring in nucleic acids, nucleosides occur naturally in an extensive range of modified compounds formed, mainly, as microbial metabolites. Many have antibiotic or other biological activities and together they exemplify a wide range of structures. Examples are oxetanocin **8.70**, cordycepin **8.71**, aristeromycin **8.72**, nucleocidin **8.73**, tubercidin **8.74** and several C-linked analogues (Section 8.2.4).

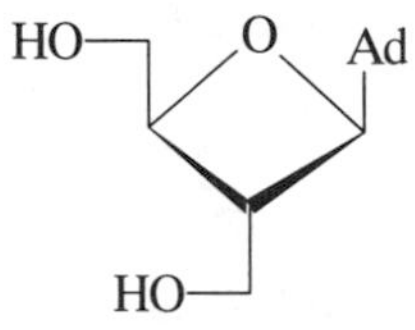

Oxetanocin **8.70** 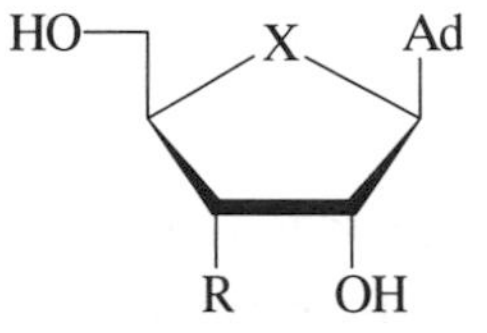   Cordycepin   **8.71** X = O, R = H   Nucleocidin **8.73**
                          Aristeromycin **8.72** X = CH$_2$, R = OH

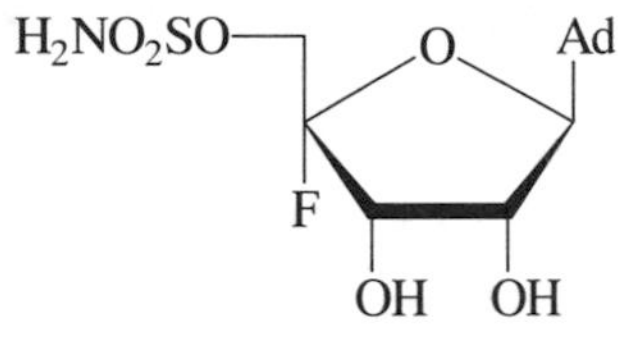

Ad = Adenin-9-yl

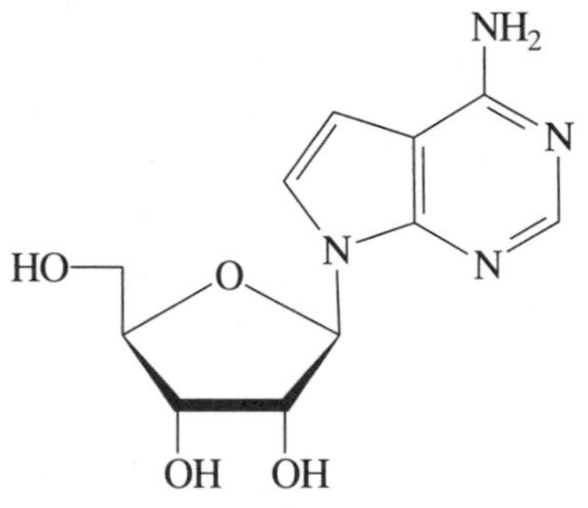

Tubercidin **8.74**

Since nucleosides are involved in the most basic of biological processes, synthetic analogues of the natural compounds often exhibit potent biological activity, many being of great pharmaceutical value.[73] Amongst these are ara-C, β-D-arabinofuranosylcytidine **8.75**, effective against myelocytic leukaemia; acyclovir **8.76**, a potent antiherpes agent; and AZT, 3'-azido-3'-deoxythymidine **8.77**, the most effective anti-HIV compound.

(b)   Nucleotides and nucleic acids[74]

Nucleotides are phosphate esters of the nucleosides and represent the monomeric units of the nucleic acids, being obtained by alkaline or enzymic hydrolysis of the ribonucleic acids or enzymic depolymerization of the deoxyribonucleic acids. Such phosphates may be synthesized from suitably protected nucleosides (Section 5.3.8), and specific products have been used to show that alkaline hydrolysis of RNA gives both 2'- and 3'-nucleoside phosphates, and that enzymic degradation of DNA with snake venom diesterase and deoxyribonuclease gives the 5'-esters.

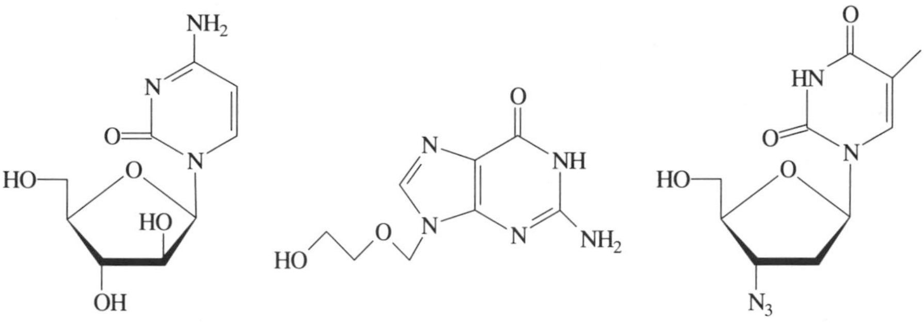

Ara-C **8.75**   Acyclovir **8.76**   AZT **8.77**

Apart from their role as nucleic acid components, the nucleotides play other important functions in biochemistry, some of the most important nucleotide coenzymes being adenosine 5'-triphosphate (ATP), nicotinamide adenine dinucleotide (NAD) and uridine diphosphate glucose (UDPG).

RNA and DNA have occupied a central position in the science of molecular biology which has developed so rapidly and with such consequences in recent years. While the former controls cellular protein synthesis the latter is responsible for the transmission of genetic information between parent and offspring.

Ribonucleic acid **8.78a** comprises a series of ribonucleosides linked by phosphodiester bridges between positions 3' and 5'. The polymer is susceptible to alkaline hydrolysis and gives 2'- and 3'-phosphates of the nucleosides by virtue of neighbouring group participation by the ionized OH group at C-2' (Scheme 8.11).

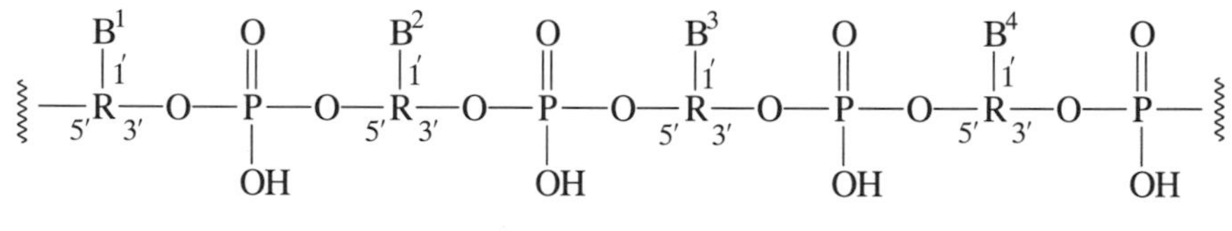

RNA **8.78a** R = β-D-ribofuranose, B = adenine, guanine, cytosine and uracil
DNA **8.78b** R = 2-deoxy-β-D-*erythro*-pentofuranose, B = adenine, guanine, cytosine and thymine

Deoxyribonucleic acid, alternatively, has the primary structure **8.78b**, is stable to alkali, and with acid loses the purine bases preferentially. The initial ratio of purines to pyrimidines is unity, which led ultimately to the famous Watson and Crick double-helical model for the polymer and opened the way to the most fundamental developments in basic biology.

The coverage given here to this class of compounds in no way reflects their relative significance. Brief mention has only been made of the central role occupied by ribose and 2-deoxy-D-*erythro*-pentose in their molecular architecture.

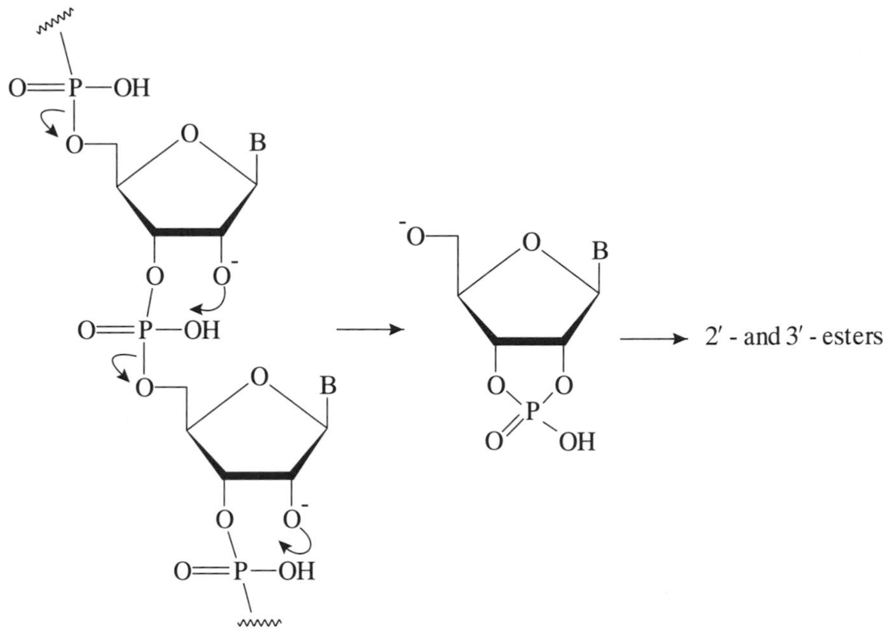

Scheme 8.11

8.2.3   *S*-GLYCOSIDES[75]

The occurrence of thioglycosides in natural products is surprisingly uncommon, the best-known compounds being the glucosinolates more than 80 of which occur as the mustard oil glucosides **8.79** (R very variable) of the *Cruciferae, Capparidaceae* and *Resedaceae* families of plants.[76] Sinigrin **8.79** (R = allyl), found in the seed of black mustard, gives D-glucose on enzymic cleavage with myrosinase, which occurs in the plants with the glucosinolates, and mainly 1-thio-D-glucose with base. However, treatment with methanolic potassium methoxide converts it in low yield to merosinigrin **8.80**[77] In the case of the compound **8.79** (R = (*R*)-2-hydroxybut-3-enyl), enzymolysis gives goitrin **8.81**, a potent goitrogen which constitutes a major problem in the economic use of rape-seed oil.[78]

Lincomycin **8.82** is a member of a set of related thioglycoside antibiotics which contain an *N*-methylproline unit linked by an amide bond to *S*-alkyl 6-amino-6,8-dideoxy-1-thio-α-D-*erythro*-D-*galacto*-octopyranosides.[79] It is used against infections caused by Gram-positive bacteria and anaerobic organisms.

8.2.4   *C*-GLYCOSIDES[80]

A series of sugar derivatives having the carbohydrate attached through the anomeric carbon centres to various aromatic moieties have been isolated from plant and

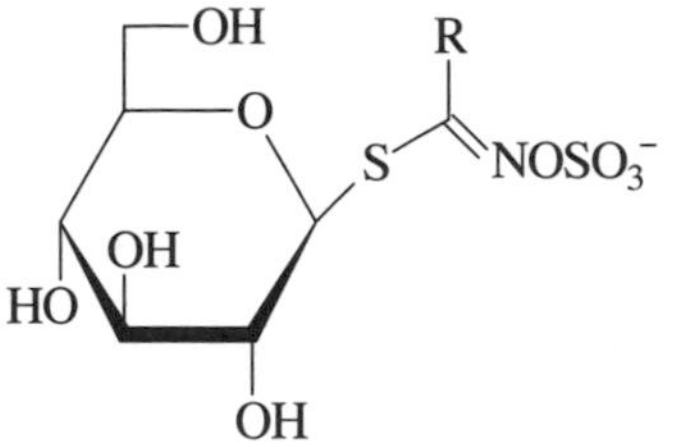

Sinigrin **8.79**

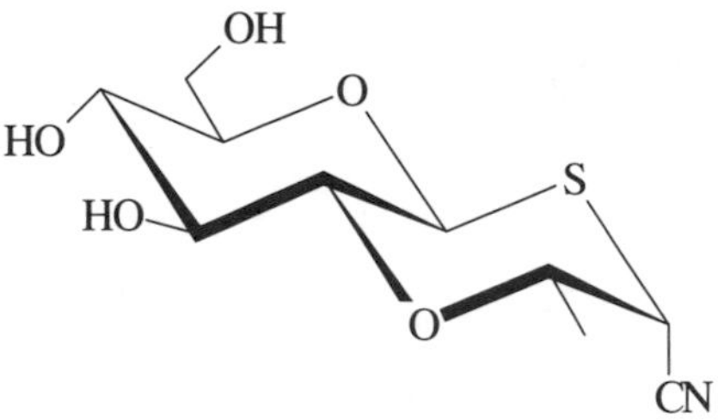

8.80

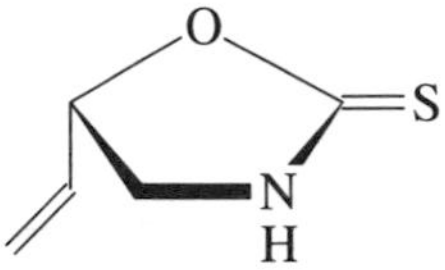

Goitrin **8.81**

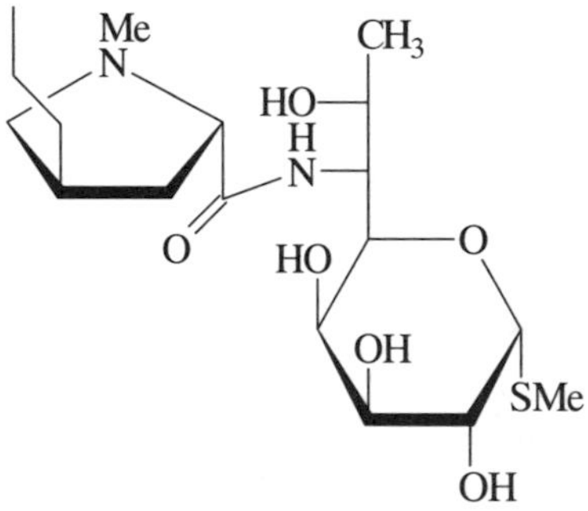

Lincomycin **8.82**

microbial sources. They are termed 'C-glycosides' despite their non-glycosidic nature, i.e. the absence in their structures of acetal functions and their consequent stability to acid. Several are anthracene or flavone derivatives, amongst the first recognized being barbaloin **8.83** and vitexin **8.84** which occur in the juices of the aloe plant and in New Zealand puriri wood, respectively. While they do not give glucose on acidic hydrolysis, they can be oxidized with hot aqueous iron(III) chloride to D-arabinose.

Micro-organisms have more recently provided a prolific source of a large range of such polyketide compounds, many of which have significant biological activity.[81] Two examples are vineomycinone B2 **8.85**, a member of the antitumour antibiotics the vineomycins,[82] and the further antitumour compound gilvocarcin V **8.86**.[83]

C-Nucleosides are an important class of C-glycosides which, because of their pharmacological activity, have become a significant group of natural products.[80]

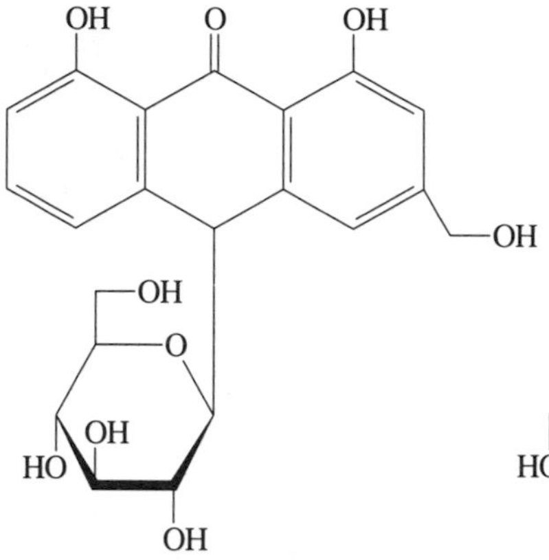

Barbaloin **8.83**

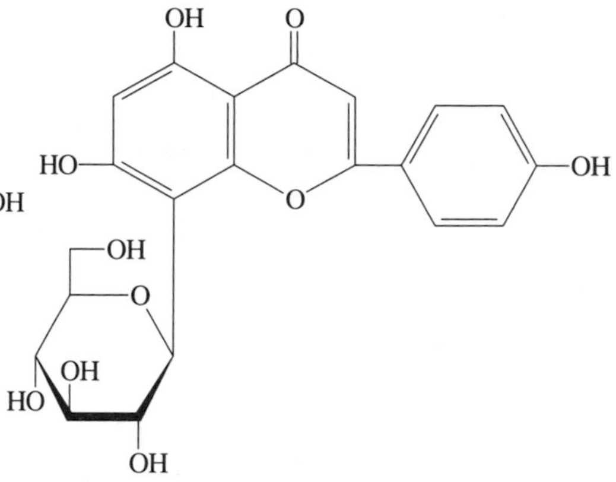

Vitexin **8.84**

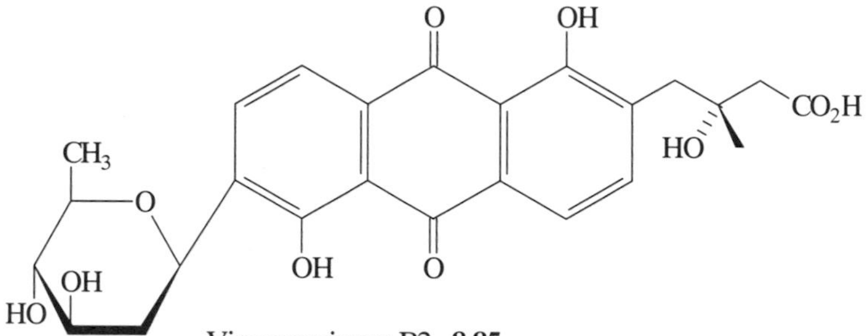

Vineomycinone B2 **8.85**

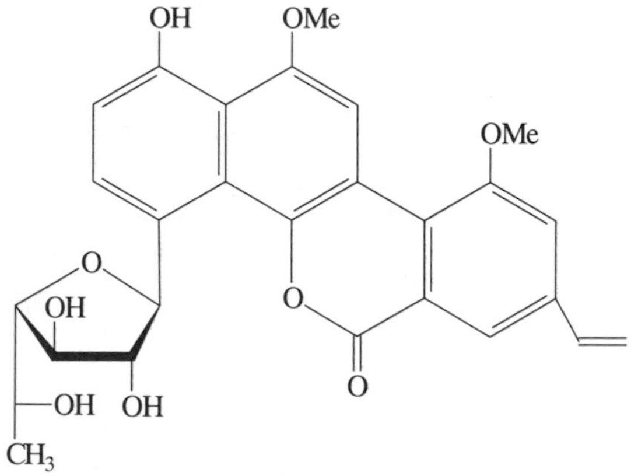

Gilvocarcin V **8.86**

The first to be discovered (1959) was pseudouridine **8.87**, found as one of the minor nucleosides of transfer RNA (Section 8.2.2.b). Others with valuable biological properties to have been isolated from fermentation broths are showdomycin **8.88** (antitumour and antibacterial), formycin **8.89** (antitumour) and pyrazomycin **8.90** (antiviral).

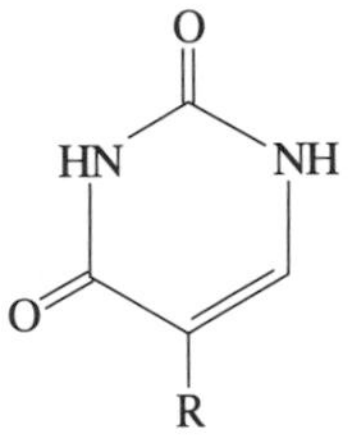

Pseudouridine **8.87** 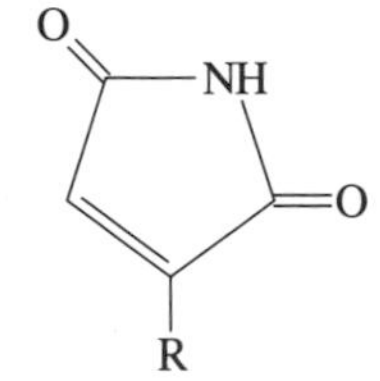   Showdomycin **8.88** 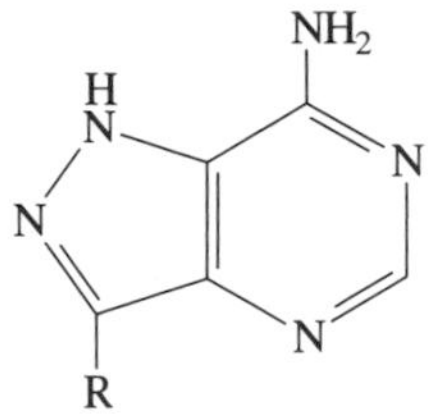   Formycin **8.89** 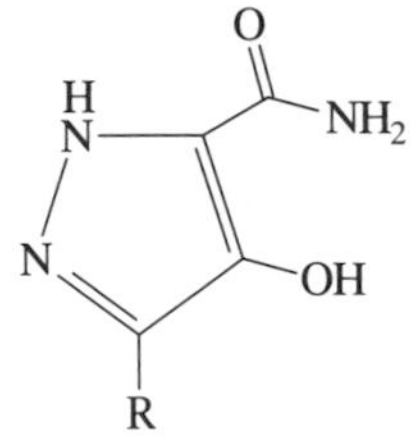   Pyrazomycin **8.90**

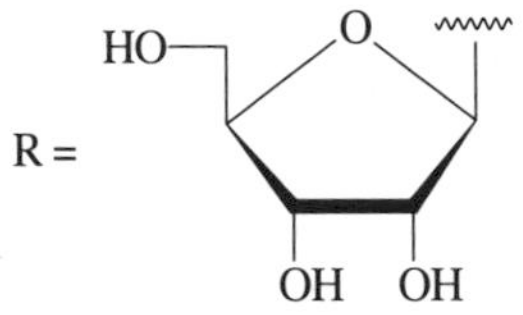

Compounds with tetrahydro-furanyl or -pyranyl rings and C-linked substituents at both positions adjacent to the ring oxygen atoms, and therefore extended C-glycosides, form parts of many complex natural products. Other natural products, for example the antibiotic papulacandin B[84] which contains the structural unit **8.91**, have apparent C-substituents at the anomeric centre, but since they also retain the second oxygen atom at that centre they are more appropriately considered to be ketose derivatives rather than C-glycosides.

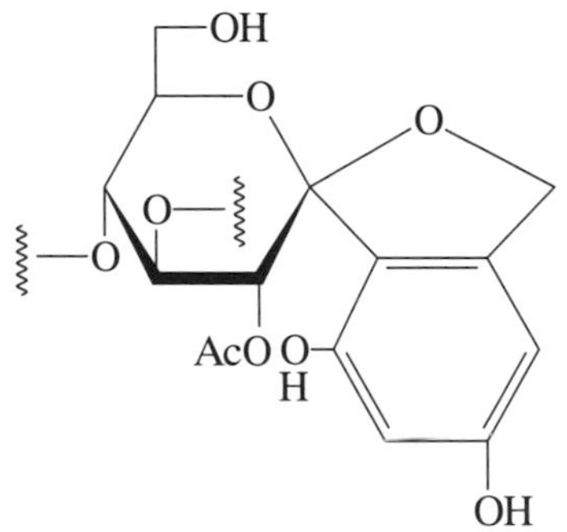

8.91

8.3   CYCLITOLS

Of the polyhydroxycycloalkanes (cyclitols), hexahydroxycyclohexanes or *inositols* are best known.[85] All nine possible isomers (Scheme 8.12) have been prepared, and it should be noted that all but the *chiro*-inositols are *meso* and therefore optically inactive compounds. Because of their obvious similarity to pyranose sugars they are discussed very briefly.

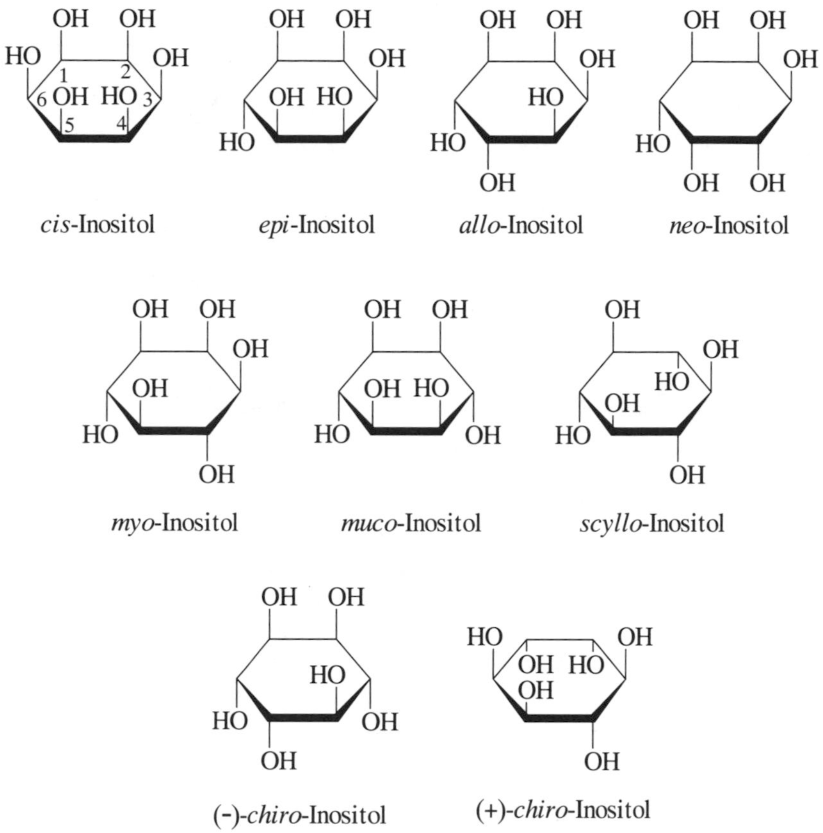

OH   OH
HO   OH
  1   2
6 OH HO 3
  5   4

cis-Inositol

OH   OH
        OH
OH HO
HO

epi-Inositol

OH   OH
        OH
     HO
HO
     OH

allo-Inositol

OH   OH
        OH
HO
OH   OH

neo-Inositol

OH   OH
        OH
  OH
HO
  OH

myo-Inositol

OH   OH
        OH
OH HO
HO   OH

muco-Inositol

OH
        OH
     HO
  OH
HO
  OH

scyllo-Inositol

OH   OH
     HO
HO   OH
     OH

(-)-*chiro*-Inositol

HO   OH
  OH HO
  OH
     OH

(+)-*chiro*-Inositol

Scheme 8.12   The inositols (each ring is numbered as is indicated for the *cis*-isomer)

myo-Inositol, often referred to simply as inositol (or misleadingly as *meso*-inositol) occurs widely in nature in the free form and as its derivatives, particularly phosphate esters. Commercially it is obtained from corn-steep liquors. *scyllo*-Inositol is found in plants (as monomethyl ethers), as are both forms of *chiro*-inositol, but the others must be prepared by chemical methods, such as by nucleophilic displacements at specific carbon atoms or reductions of specifically oxidized keto products. Generally this involves isomerization of available compounds (as in Scheme 8.13), but less common members have also been obtained by chromatographic separation of the hydrogenation products of hexahydroxybenzene or tetrahydroxybenzoquinone. By these reductive reactions *myo*-inositol is produced predominantly.

More specific routes to members of the inositol group are available from monosaccharides (Section 7.2.2.c), and a further approach involves the use of cyclohexa-3,5-diene-1,2-diol, which is a microbial oxidation product of benzene. Selective hydroxylation processes have been used to obtain several members of the family.[86]

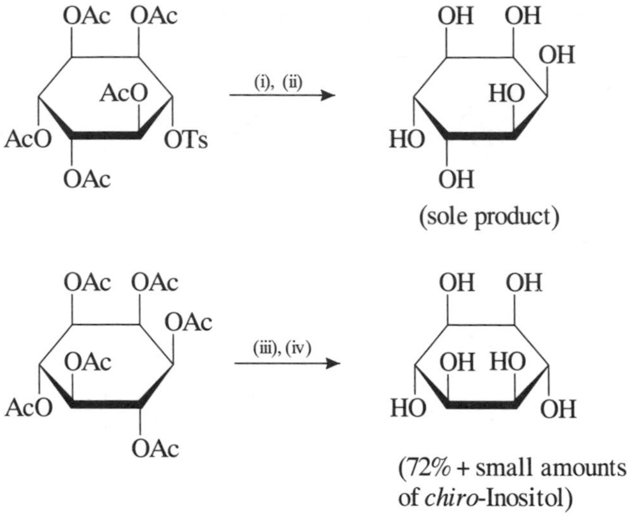

OAc OAc

AcO

AcO   OTs

OAc

— (i), (ii) →

OH   OH

OH

HO

HO

HO

OH

(sole product)

OAc OAc

OAc

OAc

AcO

OAc

— (iii), (iv) →

OH   OH

OH HO

HO   OH

(72% + small amounts
of *chiro*-Inositol)

(i) moist DMF, heat; (ii) MeO⁻; (iii) HF; (iv) H₂O

Scheme 8.13

Important compounds related to the inositols are the *inososes* or monoke-toinositol derivatives which occur as biosynthetic intermediates *en route* from D-glucose to inositols, the pentahydroxycyclohexanes (quercitols) and the 3,4,5,6-tetrahydroxycyclohexenes (conduritols), both of which occur in plants, and the inosamines in which one or two hydroxyl groups of the inositols are replaced by amino groups. These last compounds are important constituents of the aminogly-coside antibiotics (Section 8.2.1.c).

Closely related to the inositols are the *carba*-sugars, previously known as *pseudo*-(ψ)-sugars,[87] which have methylene groups instead of the ring oxygen atoms; one, the α-D-galactopyranose analogue **8.92**, has been found in *Streptomyces* fermenta-tion broths and shown to inhibit the growth of *Klebsiella* bacteria. The analogues validamine **8.93** and valienamine **8.94** are, respectively, key *N*-linked components of the antibiotic validamycin B and acarbose, a potent α-glucosidase-inhibiting compound with the cyclitol being a component of a modified α-D-glucopyranose-based tetrasaccharide.

Several methyl derivatives of inositols, both *O*-methyl, e.g. D-(+)-pinitol **8.95** and L-(−)-quebrachitol **8.96**, and *C*-methyl e.g. (−)-laminitol **8.97**, occur in nature, as do glycosylated derivatives, notably the widely distributed galactinol 1L-1-*O*-α-D-galactopyranosyl-*myo*-inositol.

Of great biological significance are the phosphates of *myo*-inositol, particularly D-*myo*-inositol 1,4,5-triphosphate **8.98** which acts as a second messenger in cells to stimulate the release of calcium from storage sites; it occurs in conjunction with mono-, di- and tetra-esters, and the recognition of its importance has recently caused great activity in biochemistry and synthetic chemistry (see Section 5.3.8).[88] *myo*-Inositol hexaphosphate is phytic acid, found in plants and in soil.

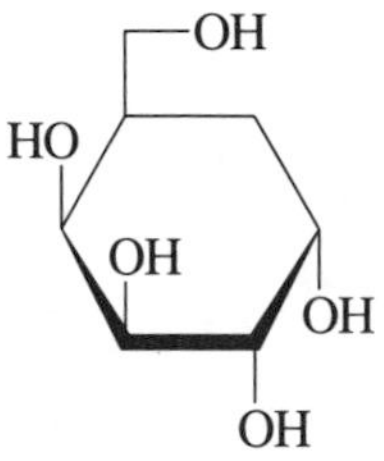

Carba-α-D-
galactopyranose
8.92

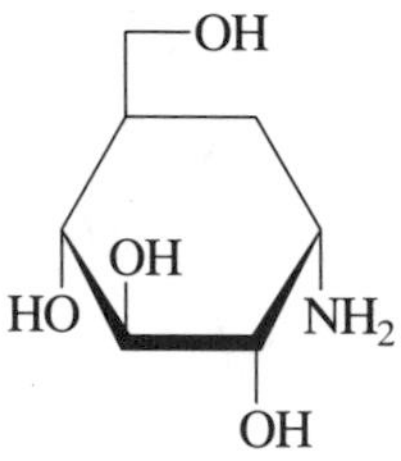

Validamine **8.93**

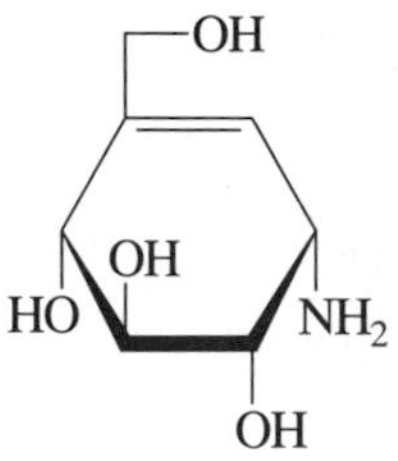

Valienamine **8.94**

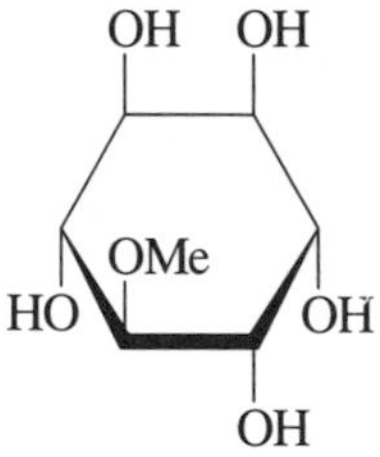

D-(+)-Pinitol **8.95**

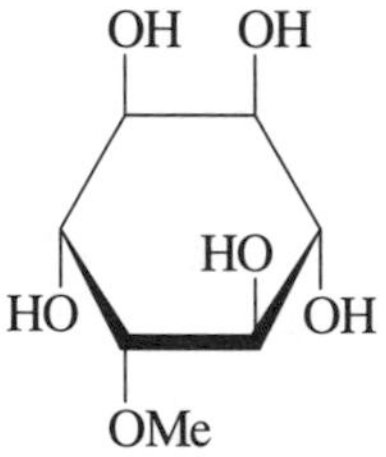

L-(−)-Quebrachitol **8.96**

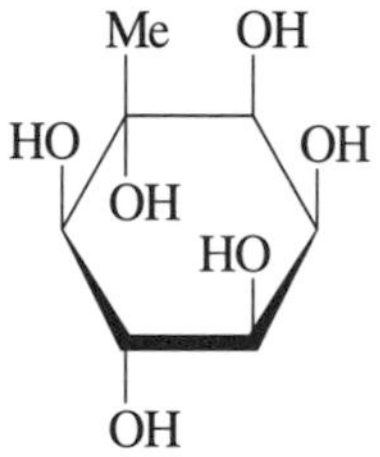

(−)-Laminitol **8.97**

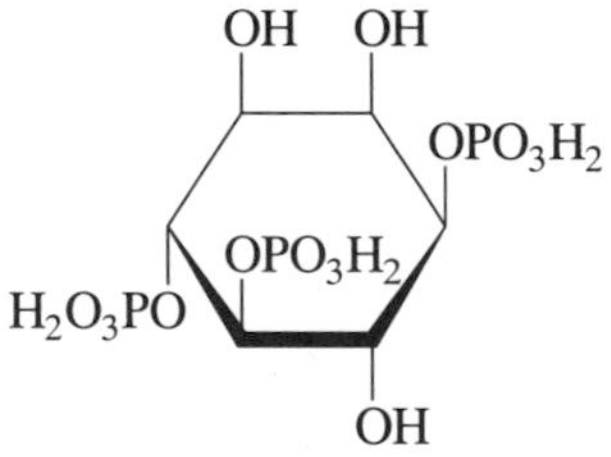

8.98

In a sequence of historical importance, which mimics the biochemical conversion of D-glucose to inositols by involving C-6 carbanion attack at C-1, *scyllo*-inosamine **8.99** and isomers were obtained by reduction of the initial products **8.100** formed on treatment of 6-deoxy-6-nitro-D-glucose and/or -L-idose **8.101** with mild alkali (Scheme 8.14). Reaction of the penta-*O*-acetate of the inosamine **8.99** with nitrous acid followed by deacetylation gave *myo*-inositol in the first chemical synthesis of the compound (Section 7.2.2.b).[89]

Polyhydroxycyclopentanes do not appear to parallel the inositols in their importance as natural products. Nevertheless, many related compounds have been found; two nucleoside analogues aristeromycin **8.72** and naplanocin A **8.102** contain *carba*-β-D-ribofuranose and a dehydro analogue, respectively.

Several natural products which contain polyfunctionalized cyclopentane rings, e.g. mannostatin A **8.103** (Section 7.2.2.b), have been found to inhibit glycosidase enzymes.[90]

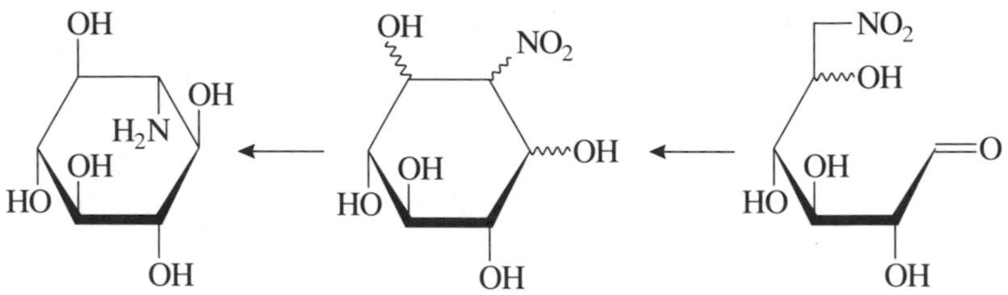

scyllo-Inosamine **8.99**   **8.100**   **8.101**

Scheme 8.14

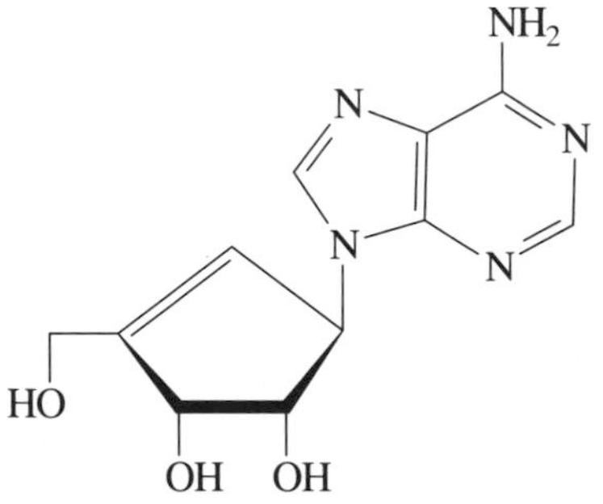

Naplanocin A **8.102** 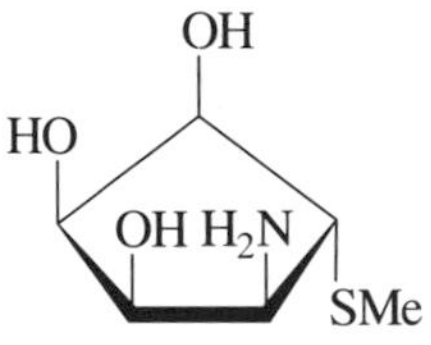   Mannostatin A **8.103**

8.4   REFERENCES

1. *J. Am. Chem. Soc.*, 1956, **78**, 4117.
2. *Can. J. Chem.*, 1954, **32**, 340.
3. *Adv. Carbohydr. Chem.*, 1963, **18**, 201.
4. *J. Chem. Soc., Perkin Trans. 1*, 1988, 889.
5. *The Carbohydrates* (eds W. Pigman and D. Horton), Vol. IIA, Academic Press, New York, 1970, p. 69; *Rodds Chemistry of Carbon Compounds* (ed. S. Coffey), Vol. 1F, Elsevier, Amsterdam, 1967; *Rodd's Chemistry of Carbon Compounds* (ed. M. Ansell), Vol. 1F, G Supplement, Elsevier, Amsterdam, 1983.
6. *Acc. Chem. Res.*, 1992, **25**, 77.
7. *Carbohydr. Res.*, 1989, **192**.
8. *Adv. Carbohydr. Chem.*, 1957, **12**, 190.
9. *Angew. Chem., Int. Ed. Engl.*, 1980, **19**, 344.
10. *Chem. Br.*, 1983, **19**, 126.
11. *Carbohydr. Res.*, 1989, **192**, 97.
12. *Tetrahedron*, 1983, **39**, 1417.
13. *J. Org. Chem.*, 1991, **56**, 7274.
14. *Carbohydr. Res.*, 1989, **164**, 277.
15. *Carbohydr. Res.*, 1989, **192**, 131.
16. *Tetrahedron Lett.*, 1990, **31**, 4517.
17. *Chem. Rev.*, 1992, **92**, 1457.
18. *Chem. Br.*, 1987, **23**, 455.
19. (a) *Chem. Soc. Rev.*, 1978, **7**, 423; (b) *Chem. Soc. Rev.*, 1989, **18**, 347.

20. *The Polysaccharides* (ed. G. O. Aspinall), Academic Press, New York: Vol. 1, 1982; Vol. 2, 1983, p. 287; Vol. 3, 1985.
21. *Carbohydr. Res.*, 1990, **205**, 19.
22. *Adv. Carbohydr. Chem. Biochem.*, 1984, **42**, 193.
23. *Carbohydr. Res.*, 1984, **133**, 173.
24. *Adv. Carbohydr. Chem. Biochem.*, 1983, **41**, 209; *Methods Enzymol.*, 1989, **179**, 122.
25. *Carbohydr. Res.*, 1988, **175**, 59.
26. *Adv. Carbohydr. Chem. Biochem.*, 1987, **45**, 19; *Methods Biochem. Anal.*, 1990, **34**, 91.
27. *Pure Appl. Chem.*, 1983, **55**, 605; *Am. Chem. Soc. Symp. Ser.*, 1990, **430**, 162.
28. *Pure Appl. Chem.*, 1989, **61**, 1201.
29. *Can. J. Chem.*, 1980, **58**, 631; *Can. J. Chem.*, 1982, **60**, 44.
30. *Carbohydrates as Organic Raw Materials* (ed. F.W. Lichtenthaler), VCH, Weinheim, 1990, p. 6.
31. *Can. J. Chem.*, 1992, **70**, 241.
32. *Liebigs Ann. Chem.*, 1985, 489.
33. *The Polysaccharides* (ed. G. O. Aspinall), Vol. 3, Academic Press, New York, 1985.
34. *The Polysaccharides* (ed. G. O. Aspinall), Vol. 1, Academic Press, New York, 1982.
35. *Carbohydr. Res.*, 1980, **78**, 372.
36. *J. Am. Chem. Soc.*, 1982, **104**, 3539.
37. *Carbohydr. Res.*, 1993, **241**, 321.
38. *The Polysaccharides* (ed. G. O. Aspinall), Vol. 3, Academic Press, New York, 1985, p. 1.
39. *The Polysaccharides* (ed. G. O. Aspinall), Vol. 1, Academic Press, New York, 1982, p. 118; *Adv. Carbohydr. Chem. Biochem.*, 1985, **43**, 51.
40. *The Polysaccharides* (ed. G. O. Aspinall), Vol. 1, Academic Press, New York, 1982, p. 135; *Adv. Carbohydr. Chem. Biochem.*, 1983, **41**, 209.
41. *Macromolecules*, 1979, **12**, 1001; *The Polysaccharides* (ed. G. O. Aspinall), Vol. 1, Academic Press, New York, 1982, p. 139.
42. *Carbohydr. Res.*, 1990, **195**, 169.
43. *Carbohydr. Res.*, 1983, **118**, 157.
44. *Adv. Carbohydr. Chem. Biochem.*, 1981, **38**, 13.
45. *Carbohydr. Res.*, 1987, **160**, 57.
46. *Carbohydrate Chemistry* (ed. J. F. Kennedy), Oxford University Press, Oxford, 1988, p. 220.
47. *The Polysaccharides* (ed. G. O. Aspinall), Vol. 2, Academic Press, New York, 1983, p. 12.
48. *The Polysaccharides* (ed. G. O. Aspinall), Vol. 3, Academic Press, New York, 1985, p. 210.
49. *The Polysaccharides* (ed. G. O. Aspinall), Vol. 2, Academic Press, New York, 1983, p. 98.
50. *The Polysaccharides* (ed. G. O. Aspinall), Vol. 2, Academic Press, New York, 1983, p. 196.
51. *Carbohydrate Chemistry* (ed. J. F. Kennedy), Oxford University Press, Oxford, 1988, pp. 263, 303; *The Polysaccharides* (ed. G. O. Aspinall), Vol. 3, Academic Press, New York, pp. 284, 338.
52. *Carbohydrate Chemistry* (ed. J. F. Kennedy), Oxford University Press, Oxford, 1988, p. 245; *The Polysaccharides* (ed. G. O. Aspinall), Vol. 2, Academic Press, New York, 1983, p. 287.
53. *Annu. Rev. Biochem.*, 1983, **52**, 825.
54. *Chem. Rev.*, 1990, **90**, 1171.
55. N. Sharon, *Complex Carbohydrates*, Addison-Wesley, Reading MA, 1975.
56. *Biochem. Biophys. Acta*, 1975, **415**, 335.

57. *Angew. Chem., Int. Ed. Engl.*, 1973, **12**, 721.
58. *Adv. Carbohydr. Chem. Biochem.*, 1982, **40**, 131; *Trends Biochem. Sci.*, 1985, 357.
59. *Angew. Chem., Int. Ed. Engl.*, 1993, **32**, 336.
60. *Annu. Rev. Biochem.*, 1985, **54**, 631.
61. *Annu. Rev. Biochem.*, 1988, **57**, 785.
62. *Acc. Chem. Res.*, 1993, **26**, 319; *Trends Biochem. Sci.*, 1977, 76.
63. *Sci. Am.*, 1993, 74.
64. *Angew. Chem., Int. Ed. Engl.*, 1990, **29**, 823.
65. *Annu. Rev. Biochem.*, 1991, **60**, 443; *Annu. Rev. Biochem.*, 1986, **55**, 539.
66. *Carbohydr. Res.*, 1987, **167**, 67; *Tetrahedron Lett.*, 1988, **29**, 803.
67. *Chem. Br.*, 1986, **22**, 703.
68. *Sci. Am.*, 1984, 82.
69. *Carbohydrate Chemistry* (ed. J. F. Kennedy), Oxford University Press, Oxford, 1988, p. 196.
70. *The Carbohydrates* (eds W. Pigman and D. Horton), Vol. IIA, Academic Press, New York, 1970, p. 213.
71. *Carbohydrate Chemistry* (ed. J. F. Kennedy), Oxford University Press, Oxford, 1988, p. 73.
72. *Carbohydrate Chemistry* (ed. J. F. Kennedy), Oxford University Press, Oxford, 1988, p 134; *The Carbohydrates* (eds W. Pigman and D. Horton), Vol. IIA, Academic Press, New York, 1970, p. 1.
73. R.J. Suhadolnik, *Nucleoside Antibiotics*, Wiley-Interscience, New York, 1970.
74. See any modern biochemistry or molecular biology text, e.g. B. Lewin, *Genes V*, Oxford University Press, Oxford, 1994.
75. *Adv. Carbohydr. Chem.*, 1963, **18**, 123; *The Carbohydrates* (eds W. Pigman and D. Horton), Vol. IB, Academic Press, New York, 1980, p. 799.
76. *The Carbohydrates* (eds W. Pigman and D. Horton), Vol. IIA, Academic Press, New York, 1970, p. 230.
77. *Can. J. Chem.*, 1984, **62**, 1236.
78. *J. Chem. Soc., Perkin Trans. 1*, 1990, 1909.
79. *Carbohydrate Chemistry* (ed. J. F. Kennedy), Oxford University Press, Oxford, 1988, p. 106.
80. *Adv. Carbohydr. Chem. Biochem.* 1976, **33**, 111; *J. Chem. Soc. Perkin Trans. 1*, 1987, 2371; *J. Chem. Soc. Perkin Trans. 1*, 1990, 283; *J. Org. Chem.*, 1990, 55, 2572.
81. *Prog. Med. Chem.*, 1985, **22**, 1.
82. *J. Am. Chem. Soc.*, 1990, **112**, 8188.
83. *J. Am. Chem. Soc.*, 1991, **113**, 8516.
84. *Helv. Chem. Acta*, 1977, **60**, 578.
85. *The Carbohydrates* (eds W. Pigman and D. Horton), Vol. IA, Academic Press, New York, 1972, p. 519; *Biochem. J.*, 1976, **153**, 23.
86. *Synlett*, 1993, 672.
87. *Adv. Carbohydr. Chem. Biochem.*, 1990, **48**, 21.
88. *Chem. Soc. Rev.*, 1989, **18**, 83; D. C. Billington, *The Inositol Phosphates—Chemical Synthesis and Biological Significance*, VCH, Weinheim, 1993.
89. *Adv. Carbohydr. Chem. Biochem.*, 1969, **24**, 67.
90. *J. Am. Chem. Soc.*, 1993, **115**, 444.

Appendix 1: The Literature of Monosaccharide Chemistry

Until 1965 research work in the field was published in the general organic chemical and biochemical literature. That year, however, saw the introduction of the first specialized journal with the appearance of *Carbohydrate Research*, and such has been the rate of development that, in 1982, *the Journal of Carbohydrate Chemistry* was launched; at the time of writing the first publication committed to rapid communications, *Carbohydrate Letters*, is awaited. Relevant papers still also appear in all major journals dealing with organic chemistry and biochemistry.

Reviewers have served the subject particularly well with *Advances in Carbohydrate Chemistry*, first published in 1945 and to become *Advances in Carbohydrate Chemistry and Biochemistry* with Vol. 24 in 1969, having set a standard of excellence. With the publication of Vol. 50 'Advances' remains the source to which researchers seeking authoritative assessment of topics in monosaccharide chemistry should first turn. Relevant reviews are referred to throughout this book, as are many others taken from a wide range of sources.

Rodd's Chemistry of Carbon Compounds offers useful general surveys of the field (with references) in Vol. 1F, G (1967) and their Supplement (1983). In a 2nd Supplement to Vol. 1E, F, G (1993) selective attention has been given to the synthesis of monosaccharides and their use as chiral auxiliaries, templates and starting materials in syntheses of chiral non-carbohydrate compounds.

Chemical Abstracts covers all carbohydrate publications in Section 33, and further abstracts of all monosaccharide papers appear in the Royal Society of Chemistry's annual *Specialist Periodical Reports*, *Carbohydrate Chemistry*, which dates back to 1968. In these, papers are treated in groups so that, for example, all reports dealing with amino sugars in a particular year are grouped.

Major, more recent multiauthor texts covering monosaccharide and related chemistry are *The Carbohydrates* (eds. W. Pigman and D. Horton), 2nd Edn. Academic Press, New York, Vol. 1A, 1972, Vol. IB, 1980 and Vols IIA and IIB, 1970, and *Carbohydrate Chemistry* (ed. J. F. Kennedy), Oxford University Press, Oxford, 1988. Other books are H.S. El Khadem *Carbohydrate Chemistry, Monosaccharides and their Oligomers*, Academic Press, San Diego, 1988 and R.W. Binkley *Modern Carbohydrate, Chemistry*, Dekker, New York, 1988.

A listing of physical properties and references to the most common carbohydrate derivatives is to be found in *Carbohydrates — a Source Book* (ed. P. M. Collins), Chapman & Hall, London, 1987.

Appendix 2: Nomenclature in Monosaccharide Chemistry

In the same way as the systematic method for naming aliphatic and aromatic compounds utilizes certain parent names which can be compounded and adapted for use with complex compounds and their derivatives, so carbohydrate nomenclature depends upon adaptable names of parent compounds, in this case the simple aldoses and ketoses. Also, as in general organic chemistry, many trivial names, built out of tradition and usage, are common, and a mixed system prevails: 'glucosamine' for '2-amino-2-deoxy-D-glucose' and '2-deoxyribose' for '2-deoxy-D-*erthro*-pentose' are very familiar; 'L-rhamnose' is more prevalent than '6-deoxy-L-mannose', while, conversely, 'D-quinevose' is less often used than is '6-deoxyglucose'. As in aliphatic chemistry, accepted systematic nomenclature is always correct, but there can be excellent reasons for retaining familiarity with trivial names; many are in the organic chemical and biochemical literature and in, for example, suppliers' catalogues.

Internationally agreed Rules for Carbohydrate Nomenclature were published in 1952,[1] and in revised form in 1963.[2] Currently, the best authority is *Tentative Rules for Carbohydrate Nomenclature, Part I, 1969*, published jointly in several biochemical journals[3] by the IUPAC Commission on the Nomenclature of Organic Chemistry and the IUPAC–IUB Commission on Biochemical Nomenclature. While an extensive, revised form of this document which covers all aspects of the subject is to be published in 1995,[4] four specific sets of recommendations relevant to conformational nomenclature[5] and the naming of branched-chain[6] and unsaturated[7] monosaccharides and cyclitols[8] have been made by the IUPAC–IUB Commission.

In the following synopsis only an outline of the methods applied is given, and familiarity with Fischer projections and Haworth perspective formulae and the D,L and α,β conventions (Section 2.1.4) is assumed. Readers are advised to consult the literature[3,5–8] for further details.

A2.1 THE BASIS OF SYSTEMATIC NOMENCLATURE—ALDOSES AND KETOSES

All names are developed from those of the parent monosaccharides represented in the Fischer projections of the acyclic forms, and they may be trivial (for compounds up to hexoses) or systematic. Thus, the sugars up to the hexoses shown in Sections 2.2.1 and 2.2.2 provide the nomenclature base, the formal name for the most common member being 'D-glucose' and not 'D-*gluco*-hexose'. For higher sugars, however, systematic composite names made using the stereochemical descriptors (e.g. D-*erythro*-, L-*ribo*-) derived from the parent compounds are used.

For ketoses and their derivatives, trivial names are permitted up to the hexoses, but 'D-*erythro*-pentulose' is to be preferred to 'D-ribulose' on the grounds that the latter name contains redundancy since '*ribo*' implies three chiral centres while there are only two in the acyclic modification of the sugar (Section 2.2.2). 'D-Sedoheptulose' is exceptional in being an accepted trivial name for the biochemically important D-*altro*-hept-2-ulose.

'Ulose' thus implies 'ketose', and the position of the carbonyl group should be specified in systematic names: '*erythro*-pent-3-ulose' is thus the five-carbon *meso*-ketose.

The hemiacetal ring forms of all the above sugars have 'furanose' or 'pyranose' name endings to indicate ring size (Sections 2.2.1 and 2.2.2).

A2.2   DICARBONYL COMPOUNDS

Dialdehyde derivatives of aldoses are 'dialdoses', and compound **A2.1** could be 'D-*gluco*-hexodialdose' or 'L-*gulo*-hexodialdose' **A2.2**. On the grounds that '*gluco*-' precedes '*gulo*-' alphabetically it is given the former name. Compound **A2.3** is 'D-*threo*-hexo-2,5-diulose' and the aldosulose **A2.4**, being both an aldose and a ketose, is 'D-*glycero*-D-*manno*-octos-5-ulose'.

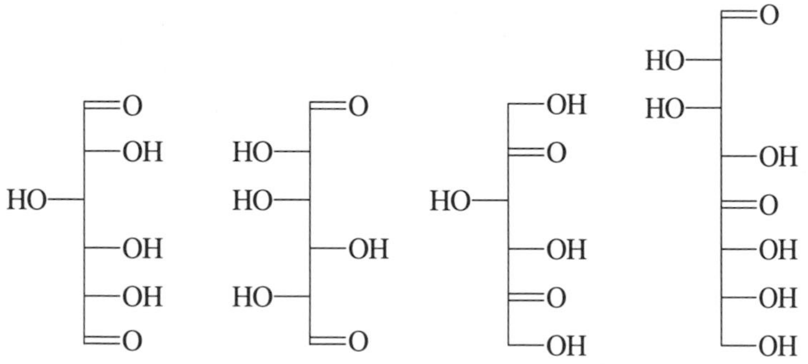

| **A2.1** | **A2.2** | **A2.3** | **A2.4** |

When such compounds or their derivatives have to be described in cyclic forms, clearly the carbonyl group involved in ring formation must be specified, and thus the methyl furanoside **A2.6**, formed from 'D-*arabino*-hexos-2-ulose' **A2.5** by ring closure involving O-5 and the ketonic group, is 'methyl α-D-*arabino*-hexos-2-ulo-2,5-furanoside'.

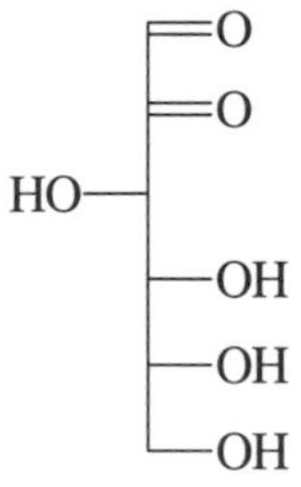

A2.5 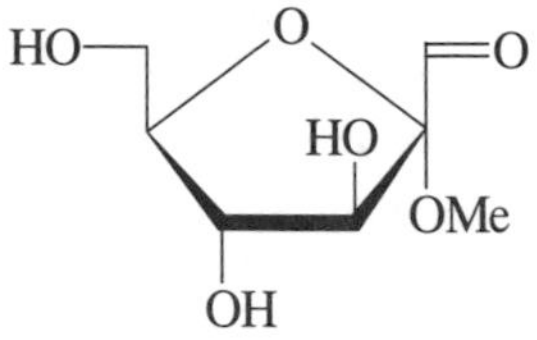   A2.6

A2.3 MONOSACCHARIDE ACIDS

Acids derived by oxidation of C-1 of aldoses are named by use of the suffix 'onic acid' instead of 'ose'. D-Glucose thus gives 'D-gluconic acid', a salt of which is 'sodium D-gluconate', and other derivatives are D-gluconamide, D-glucononitrile, D-gluconoyl chloride and D-glucono-1,5-lactone (or D-gluconic acid δ-lactone).

Aldonic acids with keto groups are 'ulosonic acids', the naturally occurring compound **A2.7** being 'Kdo' (Section 8.1.3.d), derived from the trivial 'ketodeoxy-octonic acid' or, systematically, '3-deoxy-D-*manno*-oct-2-ulosonic acid'. In the α-pyranose form **A2.8** it is '3-deoxy-α-D-*manno*-oct-2-ulopyranosonic acid'.

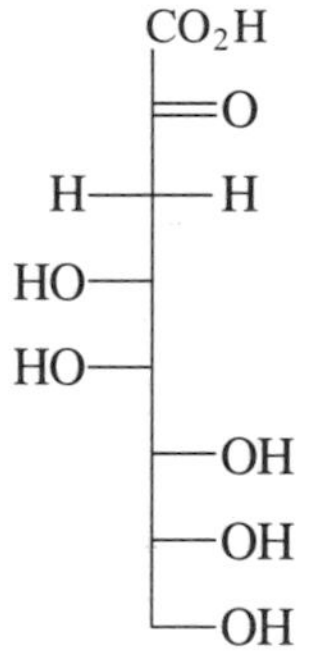

A2.7 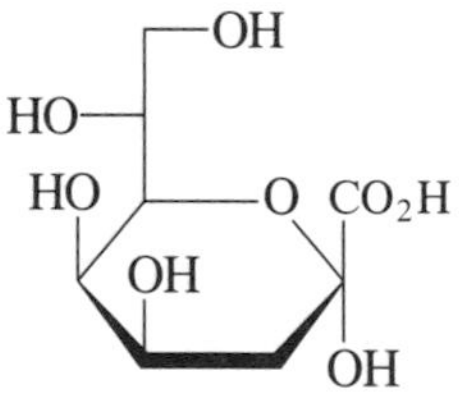   A2.8

Compounds derived from aldoses with carboxylic acid groups at the highest-numbered carbon atom are 'uronic acids', that from D-glucose being 'D-glucuronic acid' and leading to, for example, the ester 'methyl D-glucuronate' and the γ-lactone 'D-glucurono-6,3-lactone' which, in the furanose ring form **A2.9**, is 'D-glucofuranurono-6,3-lactone'.

Sugar dicarboxylic acids are 'aldaric acids' and give, for example, diesters, half-amides and dilactones, such derivatives of D-glucaric acid being 'diethyl D-glucarate', 'D-glucar-6-amic acid **A2.10**' and 'D-glucaro-1,4:6,3-lactone' **A2.11**.

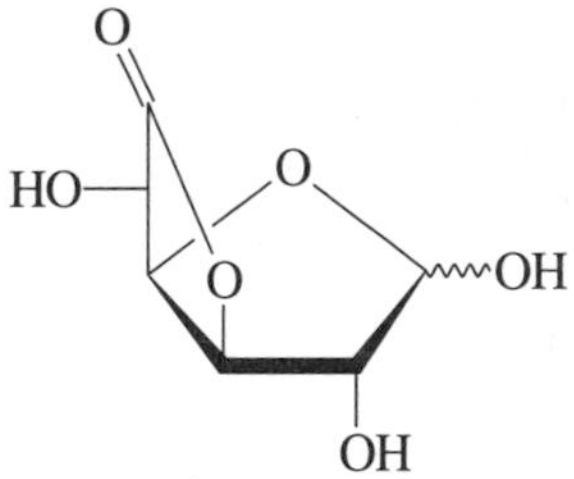

A2.9 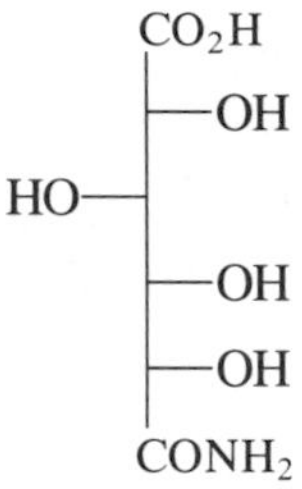   A2.10 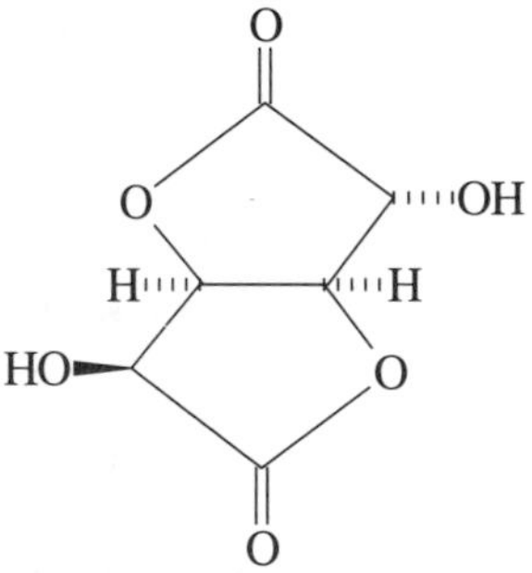   A2.11

A2.4   DERIVATIVES WITH REPLACED HYDROXYL GROUPS AND *C*-BONDED HYDROGEN ATOMS

Replacement of hydroxyl groups of parent sugars by hydrogen atoms gives 'deoxy sugars' (Section 4.2), many of which occur in natural products and are known by trivial names. When formal deoxygenation removes a chiral centre the systematic name changes to avoid redundancy, and thus the familiar '2-deoxyribose' becomes '2-deoxy-D-*erythro*-pentose'.

When hydroxyl groups are replaced by, for example, amino, thio or halo groups no such redundancy occurs and the names are developed by formal deoxygenation and replacement as appropriate. Thus, compounds **A2.12**–**A2.14** are '2-amino-2,6-dideoxy-D-galactopyranose' ('D-fucosamine'), '5-thio-D-glucopyranose' and '2-bromo-2-chloro-2-deoxy-D-glucose', respectively, the larger bromine in the last case taking precedence over chlorine for the purposes of assigning configuration. In the case of branched-chain derivatives, when the branching substituent replaces a hydrogen atom the residual hydroxyl group at the branching centre determines the configuration of the compound, but for compounds of the deoxy branched-chain type the branching carbon atom is used for stereochemical descriptive purposes. In this way, therefore, compounds **A2-15** and **A2-16** are, respectively, '2-*C*-methyl-D-glucopyranose' and '2-deoxy-2-*C*-methyl-D-mannopyranose'.

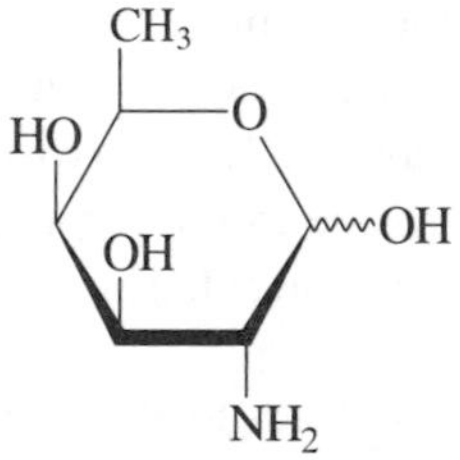

A2.12 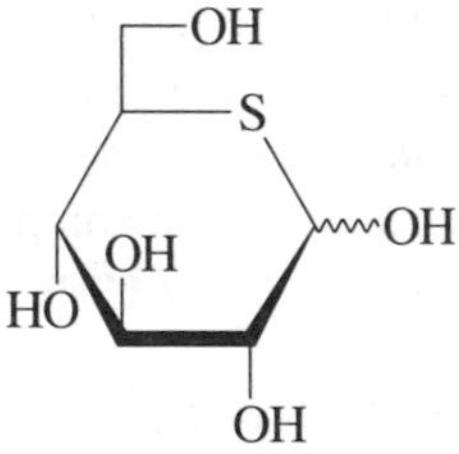   A2.13 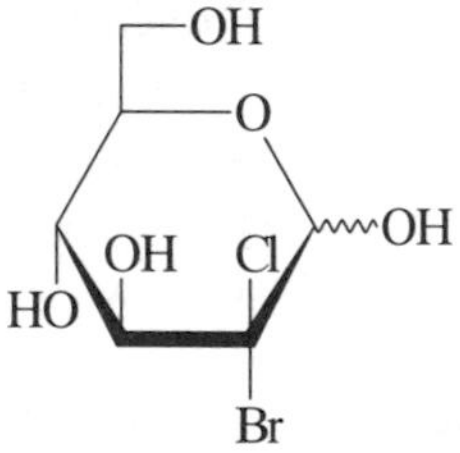   A2.14

On occasions, names are required for compounds in which non-carbohydrate portions are taken as parent and prefix names are required for the carbohydrate

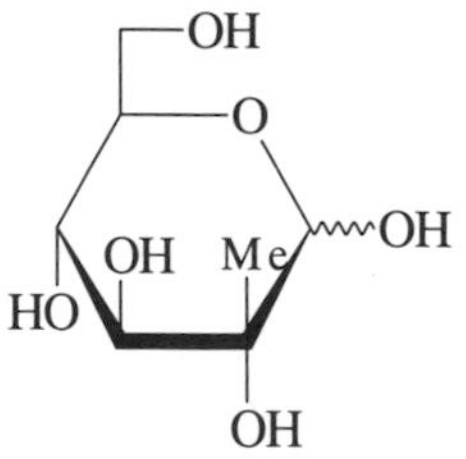

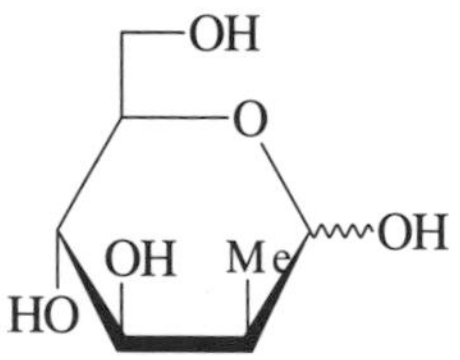

A2.15   A2.16

substituents. Frequently, bonding in such compounds is via the anomeric centre, the glycosyl halides, e.g. 'α-D-glucopyranosyl fluoride' **A2.17**, being the most common examples, but the same type of nomenclature can be used for the aromatic *C*-glycoside **A2.18**. There are occasions, however, when bonding through other carbon atoms has to be described, and for these the 'yl' suffix can also be used, compounds **A2.19** and **A2.20** being '*N*-(1-deoxy-D-fructopyranos-1-yl)glycine' and 'bis(5-deoxy-D-ribofuranos-5-yl) disulphide', respectively.

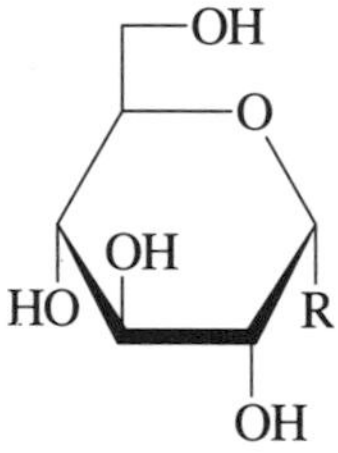

A2.17 R = F   A2.19 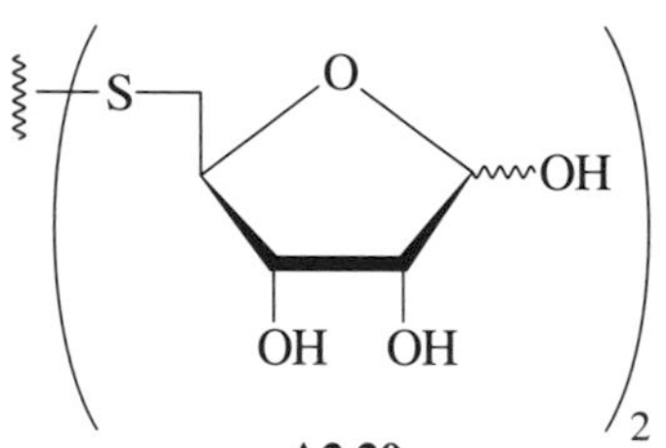   A2.20
A2.18 R = Ph

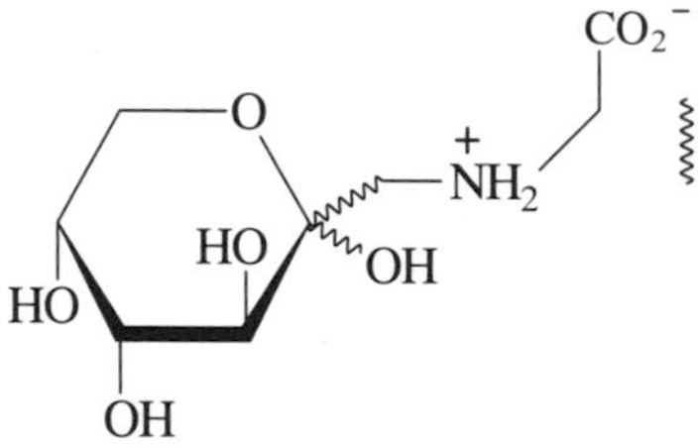

A2.5 DERIVATIVES WITH REPLACED *O*-BONDED HYDROGEN ATOMS

Glycosides are formed when alkyl (or equivalent) groups replace the hydrogen atoms of the anomeric hydroxyl groups, their normal nomenclature (e.g. methyl α-D-fructofuranoside) being illustrated extensively in Section 3.1.1. Otherwise, when it is desirable to use the aglycon as the parent component, the carbohydrate may be named as a glycosyloxy substituent as in '7-(β-D-glucopyranosyloxy)-8-hydroxy-coumarin' **A2.21**.

   O-Substitution at non-anomeric hydroxyl groups usually leads to ethers, esters or acetals, the names of which are composed by the use of appropriate *O*-alkyl, -acyl or -ylidene prefixes — hence, for example, '2,3,4,5,6-penta-*O*-methyl-D-galactose', '2,3,4-tri-*O*-acetyl-D-ribose' and '2,3:4,6-di-*O*-isopropylidene-D-mannose'. However, esters of inorganic acids are commonly named differently, as for example D-glucose 6-phosphate and D-galactose 6-sulphate.

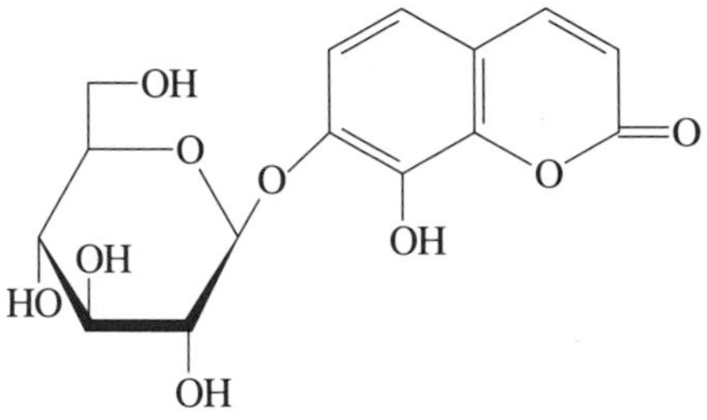

A2.21 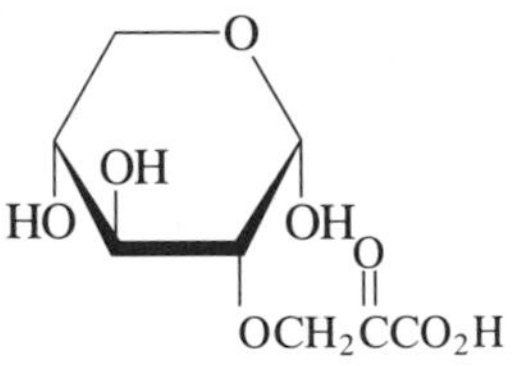   **A2.22** 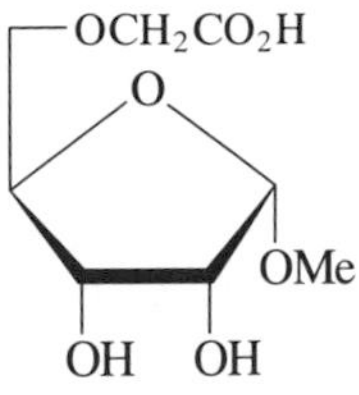   **A2.23**

When complex substituents are attached to carbohydrates they may take precedence for nomenclature purposes and 'glycos-X-*O*-yl' substituents may be used as in '(α-D-xylopyranos-2-*O*-yl)pyruvic acid' **A2.22** or '(methyl α-D-ribofuranosid-5-*O*-yl)acetic acid' **A2.23**.

Anhydro compounds, which are formal intramolecular anhydrides of *X*, *Y*-diols, are named by use of the '*X*, *Y*-anhydro' prefix in the name.

A2.6   UNSATURATED DERIVATIVES

Such monosaccharides having double or triple bonds between contiguous carbon atoms are named by use of '*X*-en' or '*X*-yn' within the name of the saturated analogue and are thus glycenose or glycynose derivatives. The position of unsaturation is defined by *X*, the lower-numbered carbon atom of the multiple bond. Compounds **A2.24**–**A2.28** are therefore respectively

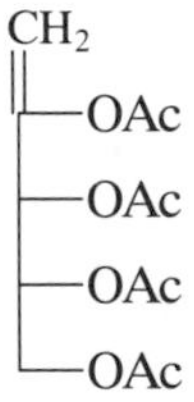

A2.24 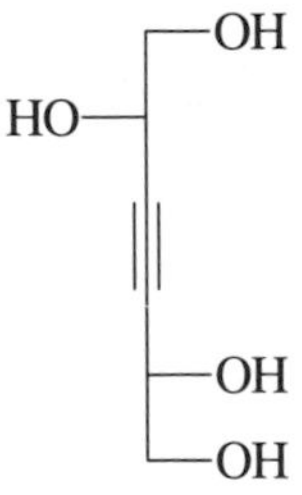   **A2.25** 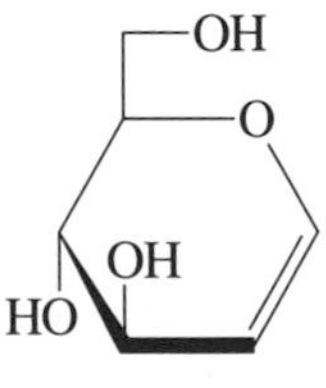   **A2.26** 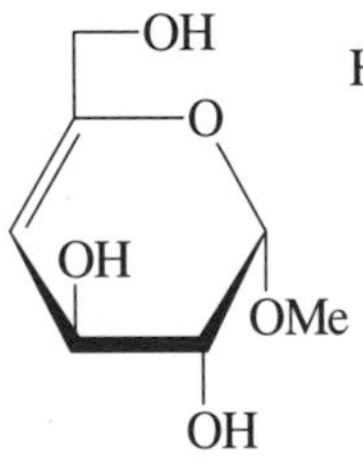   **A2.27** 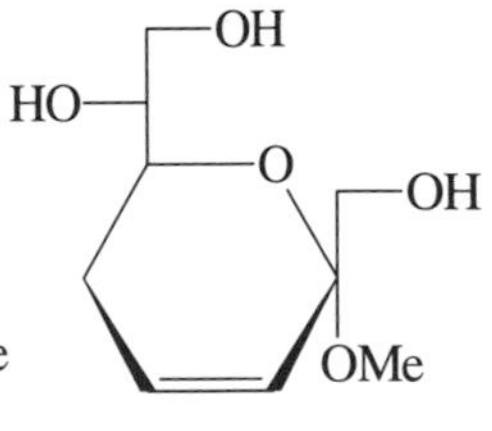   **A2.28**

'2,3,4,5-tetra-*O*-acetyl-1-deoxy-D-*erythro*-pent-1-enitol', '3,4-dideoxy-D-*threo*-hex-3-ynitol', '1,5-anhydro-2-deoxy-D-*arabino*-hex-1-enitol' (D-glucal for convenience), 'methyl 4-deoxy-β-L-*threo*-hex-4-enopyranoside' and 'methyl 3,4,5-trideoxy-α-D-*erythro*-oct-3-en-2-ulopyranoside'.

A2.7 REFERENCES

1. *J. Chem. Soc.*, 1952, 5108.
2. *J. Org. Chem.*, 1963, **28**, 281.
3. *Biochem. J.*, 1971, **125**, 673; *Eur. J. Biochem.*, 1971, **21**, 455; *J. Biol. Chem.*, 1972, **247**, 613.
4. Prof. D. Horton, Personal communication.
5. *Pure Appl. Chem.*, 1981, **53**, 1901.
6. *Pure Appl. Chem.*, 1982, **54**, 211.
7. *Pure Appl. Chem.*, 1982, **54**, 207.
8. *Pure Appl. Chem.*, 1974, **37**, 283.

Appendix 3: ^{1}H and ^{13}C Nuclear Magnetic Resonance Data for Monosaccharides and Their Peracetates

A3.1 DATA

This appendix lists some sources of n.m.r. data for a range of monosaccharides and their derivatives. The tables record specific chemical shifts (^{1}H and ^{13}C) and coupling constants ($^3J_{H,H}$ values) for many pyranoid aldopentoses and aldohexoses and their peracetates.

Table A3.1 Compounds covered in this appendix

Compounds	Reference
^{1}H data	
Aldopentoses, aldohexoses, methyl pentopyranosides, methyl hexopyranosides, methyl deoxyhexopyranosides, methyl 2-acetamido-2-deoxyhexopyranosides	1
Acetylated and benzoylated aldopentopyranoses	2
Acylated aldopentopyranosyl halides, glycosides, thioesters	3
Benzoylated aldohexopyranoses	4
^{13}C data	
Tetrofuranoses, aldopentoses and aldohexoses (furanose and pyranose forms), methyl tetrofuranosides, methyl pentosides and hexosides (furanose and pyranose forms), peracetylated pentoses and hexoses (furanose and pyranose forms), peracetylated methyl glycosides, glycosyl halides, etc.	1
Benzoylated aldohexopyranoses	4
Benzoylated glycosides and glycosyl halides, anhydro-hexoses and -heptoses, O-isopropylidene and O-benzylidene derivatives, deoxy-, aminodeoxy-, halo- and thio-derivatives, alditols and derivatives	5

More general information on the basic use of ^{1}H n.m.r. spectroscopy in monosaccharide chemistry can be found in the literature.[6-8]

Table A3.2 ¹H chemical shifts (δ values) and coupling constants (Hz, in parentheses) of aldopyranoses[a]

Compound	H-1	H-2	H-3	H-4	H-5	H-5′	H-6	H-6′
α-Ribose	4.75	3.71	3.83	3.77	3.82	3.50	—	—
	(3.0)	(3.0)			(5.3)	(2.6, 12.4)		
β-Ribose	4.81	3.41	3.98	3.77	3.72	3.57	—	—
	(6.5)	(3.3)	(3.2)		(4.4)	(8.8, 11.4)		
α-Arabinose	4.40	3.40	3.55	3.83	3.78	3.57	—	—
	(7.8)	(9.8)	(3.6)		(1.8)	(1.3, 13.0)		
β-Arabinose	5.12	3.70	3.77	3.89	3.54	3.91	—	—
	(3.6)	(9.3)	(9.8)		(2.5)	(1.7, 13.5)		
α-Xylose	5.09	3.42	3.48	3.52	3.58	3.57	—	—
	(3.6)	(9.0)	(9.0)		(7.5)	(7.5)		
β-Xylose	4.47	3.14	3.33	3.51	3.82	3.22	—	—
	(7.8)	(9.2)	(9.0)		(5.6)	(10.5, 11.4)		
α-Lyxose	4.89	3.69	3.78	3.73	3.71	3.58	—	—
	(4.9)	(3.6)	(7.8)		(3.8)	(7.2, 12.1)		
β-Lyxose	4.74	3.81	3.53	3.73	3.84	3.15	—	—
	(1.1)	(2.7)	(8.5)		(5.1)	(9.1, 11.7)		
α-Glucose	5.09	3.41	3.61	3.29	3.72	—	3.72	3.63
	(3.6)	(9.5)	(9.5)	(9.5)			(2.8)	(5.7, 12.8)
β-Glucose	4.51	3.13	3.37	3.30	3.35	—	3.75	3.60
	(7.8)	(9.5)	(9.5)	(9.5)			(2.8)	(5.7, 12.8)
α-Mannose	5.05	3.79	3.72	3.52	3.70	—	3.74	3.63
	(1.8)	(3.8)	(10.0)	(9.8)			(2.8)	(6.8, 12.2)
β-Mannose	4.77	3.85	3.53	3.44	3.25	—	3.74	3.60
	(1.5)	(3.8)	(10.0)	(9.8)			(2.8)	(6.8, 12.2)
α-Galactose	5.16	3.72	3.77	3.90	4.00	—	3.70	3.62
	(3.8)	(10.0)	(3.8)	(1.0)			(6.4)	(6.4)
β-Galactose	4.48	3.41	3.56	3.84	3.61	—	3.70	3.62
	(8.0)	(10.0)	(3.8)	(1.0)			(3.8)	(7.8)

[a]Measured at 400 MHz in D₂O at 23 °C.[1]

Table A3.3 ¹H chemical shifts (δ values) and coupling constants (Hz, in parentheses) of peracetylated pentopyranoses[a]

Peracetate of	H-1	H-2	H-3	H-4	H-5	H-5′
α-Ribose	3.92	4.85	4.42	4.92	5.99	6.26
	(3.6)	(3.3)	(3.2)		(9.3)	(4.7, 11.2)
β-Ribose	4.04	5.00	4.54	4.86	5.90	6.16
	(4.6)	(3.5)	(3.4)		(3.4)	(5.8, 12.4)
α-Arabinose	4.32	4.72	4.89	4.71	5.96	6.23
	(6.4)	(9.0)	(3.2)		(3.6)	(2.0, 13.0)
β-Arabinose	3.73	4.79	4.68	4.65	5.82	6.21
	(2.9)	(11.8)	(3.0)		(1.0)	(1.9, 13.2)
α-Xylose	4.29	5.01	4.56	4.99	6.10	6.30
	(3.5)	(9.8)	(9.6)		(5.5)	(11.6, 11.2)
β-Xylose	4.22	5.04	4.74	5.07	5.89	6.38
	(6.7)	(8.1)	(8.1)		(4.9)	(8.8, 11.8)
α-Lyxose	4.05	4.81	4.69	4.88	6.04	6.29
	(3.0)	(3.4)	(9.0)		(4.4)	(8.7, 11.6)
β-Lyxose	3.94	3.68, 3.78		5.01	5.82	6.38
	(2.5)				(3.3)	(5.4, 12.4)

[a]Measured at 100 MHz in (CD₃)₂CO at 31 °C.[2]

Table A3.4 ^{13}C chemical shifts (δ values) of aldopyranoses[a]

Compound	C-1	C-2	C-3	C-4	C-5	C-6
α-Ribose	94.3	70.8	70.1	68.1	63.8	—
β-Ribose	94.7	71.9	69.7	68.2	63.8	—
α-Arabinose	97.6	72.9	73.5	69.6	67.2	—
β-Arabinose	93.4	69.5	69.5	69.5	63.4	—
α-Xylose	93.1	72.5	73.9	70.4	61.9	—
β-Xylose	97.5	75.1	76.8	70.2	66.1	—
α-Lyxose	94.9	71.0	71.4	68.4	63.9	—
β-Lyxose	95.0	70.9	63.5	67.4	65.0	—
α-Allose	93.7	67.9	72.0	66.9	67.7	61.6
β-Allose	94.3	72.2	72.0	67.7	74.4	62.1
α-Altrose	94.7	71.2	71.1	66.0	72.0	61.6
β-Altrose	92.6	71.6	71.3	65.2	75.0	62.5
α-Glucose	92.9	72.5	73.8	70.6	72.3	61.6
β-Glucose	96.7	75.1	76.7	70.6	76.8	61.7
α-Mannose	95.0	71.7	71.3	68.0	73.4	62.1
β-Mannose	94.6	72.3	74.1	67.8	77.2	62.1
α-Gulose	93.6	65.5	71.6	70.2	67.2	61.7
β-Gulose	94.6	69.9	72.0	70.2	74.6	61.8
α-Idose	93.2	73.6[b]	72.7[b]	70.6[b]	73.6[b]	59.4
β-Idose	93.9	71.1[b]	68.8[b]	70.6[b]	75.6[b]	62.1
α-Galactose	93.2	69.4	70.2	70.3	71.4	62.2
β-Galactose	97.3	72.9	73.8	69.7	76.0	62.0
α-Talose	95.5	71.7	66.0	70.6	72.0	62.4
β-Talose	95.0	72.5[b]	69.6[b]	69.4	76.5	62.2

[a]Measured in D_2O. The review from which these data are taken[3] frequently cites several original sources.
[b]These assignments may be reversed.

Table A3.5 ^{13}C chemical shifts (δ values) of peracetylated aldopyranoses[a]

Peracetate of	C-1	C-2	C-3	C-4	C-5	C-6
α-Ribose	88.7	67.1	65.6	66.5	59.3	—
β-Ribose	90.7	67.1	66.0	66.0	62.5	—
α-Arabinose	92.2	68.2	69.9	67.3	63.8	—
β-Arabinose	90.4	67.3	68.7	66.9	62.9	—
α-Xylose	88.9	69.2	69.2	68.8	60.5	—
β-Xylose	91.7	69.3	70.8	68.1	62.5	—
α-Lyxose	90.7	68.2	68.2	66.6	61.9	—
β-Allose	90.1	68.2	68.2	65.8	71.2	61.9
α-Altrose	90.2	68.2	66.4	64.4	66.4	62.1
α-Glucose	89.2	69.4	70.0	68.1	70.0	61.1
β-Glucose	91.8	70.5	72.8	68.1	72.8	61.7
α-Mannose	90.4	68.6	68.2	65.4	70.5	62.0
β-Gulose	89.7	67.3[b]	67.1[b]	67.1[b]	71.1	61.3
α-Idose	90.4	65.9	66.2	65.9	66.2	61.8
α-Galactose	89.5	67.2	67.2	66.2	68.5	61.0
β-Galactose	91.8	67.8	70.6	66.8	71.5	61.0
α-Talose	91.4	65.2[b]	66.3[b]	65.3[b]	68.8[b]	61.5

[a]Measured in $CDCl_3$. The review from which these data were taken[3] cites different original sources.
[b]These assignments may be reversed.

A3.2 REFERENCES

1. *Annu. Rep. NMR Spectrosc.*, 1982, **13**, 1.
2. *J. Org. Chem.*, 1971, **36**, 2658.
3. *Carbohydrates in Solution*, ACS Washington, DC, 1973, p. 147.
4. *Carbohydr. Res.*, 1983, **124**, 177.
5. *Adv. Carbohydr. Chem. Biochem.*, 1983, **41**, 27.
6. *Annu. Rep. NMR Spectrosc.*, 1969, **2**, 35; *Annu. Rep. NMR Spectrosc.*, 1972, **5A**, 305.
7. *Chem. Rev.*, 1973, **73**, 669.
8. *Adv. Carbohydr. Chem. Biochem.*, 1972, **27**, 7; *Adv. Carbohydr. Chem. Biochem.*, 1974, **29**, 11.

Appendix 4: Polarimetry in Monosaccharide Chemistry

Unlike the more modern physical methods, notably n.m.r. spectroscopy and mass spectrometry, polarimetry has not developed 'everyday' character as a structure analytical tool, and its use in carbohydrate chemistry remains rather specialized. Its value, however, should not be overlooked; it is often used, for example, to allow simple distinction between enantiomers, but there are many more instances when it can be used profitably, often in conjunction with n.m.r. spectroscopy for, for example, configurational or even conformational assignment.[1] From the historical perspective it is particularly significant with Fischer having depended upon it for his configurational assignments to the sugars (Section 2.2), and Hudson having used it for the assignments of configurations to anomeric pairs of sugars and their derivatives, including glycosides (see below).

A4.1 BASIC ISSUES

An asymmetric compound causes rotation of the plane of polarization of plane-polarized light because the two circularly polarized components travel at different speeds through it or a solution of it. The optical activity of an asymmetric sample in solution is usually expressed as the *specific rotation* $[\alpha]$ measured at temperature T and wavelength λ with $[\alpha]_\lambda^T = 100\alpha/lc$, where α is the observed rotation (in degrees, the sign of α being important), l is the length of the sample tube (in dm) and c is the solution concentration (in g per 100 ml). The units for specific rotation are consequently complex (10^{-1} deg cm^2 g^{-1}), but have traditionally been (wrongly) recorded simply in degrees. To (partially) address this problem they are now sometimes recorded as unitless numbers. Alternatively, *molecular rotation* $[M]_\lambda^T$ or $[\phi]_\lambda^T$, often used for comparisons involving non-isomeric compounds, is ($[\alpha]_\lambda^T \times$ molecular weight)/100. Frequently however, the factor 1/100 has been omitted in carbohydrate chemistry.

Traditionally, measurements have been made at 589 nm (sodium D line) and specific rotations are recorded as $[\alpha]_D$ values, it being understood that room temperatures were involved. It is appropriate to consider what specifically is measured at 589 nm. When a chromophore in a compound acquires induced asymmetry from its molecular environment the associated electronic transition becomes 'optically active' and values of $[\phi]$ as a function of the wavelength of measurement give an optical rotatory dispersion plot with a Cotton effect near the centre of the absorption

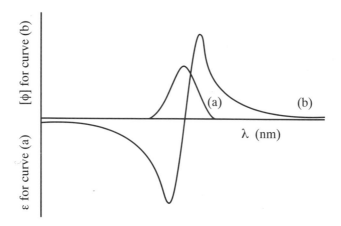

Figure A4.1 (a) Absorption spectrum of a typical optically active chromophore (ϵ versus λ) and (b) the rotatory dispersion ($[\alpha]$ versus λ; positive Cotton effect) associated with it

band (Figure A4.1). Away from the region of absorption the o.r.d. curve simplifies to a plain curve of the form $[\phi] = k/(\lambda^2 - \lambda_0^2)$ (Drude equation), where λ and λ_0 are the wavelengths of measurement and of the absorption maximum, respectively, and k is a constant, the rotation therefore diminishing with distance from the maximum. The observed rotation of a complex compound at any wavelength is thus the sum of the rotational components contributed by all the optically active chromophores present in the molecule, and the chromophore having its absorption maximum at highest wavelength contributes most to the rotation at wavelengths on the high wavelength side of the absorption bands.

For measurements on monosaccharides and their derivatives at 589 nm an important question is 'What are the highest wavelength chromophores in the molecule, and what are the intensities of their o.r.d. curves at this wavelength?' Normally there are no chromophores which absorb even in the accessible part of the ultraviolet (> 200 nm), the longest wavelength absorptions being associated with transitions of the unshared electrons of the oxygen atoms. Consequently, at 589 nm only relatively small residues of plain o.r.d. curves remain, the measured rotational values being the sum of many components. It follows that measurements are insensitive at this wavelength relative to values obtainable at lower wavelengths. The single optical rotation recorded for a particular compound combines a set of inextricable numbers which points clearly to the complex fundamental nature of the subject. Much effort has gone into the development of methods for interpreting rotational data in structural terms, and several empirical and more fundamental approaches have contributed as analytical techniques in monosaccharide chemistry.[1]

A4.2   EMPIRICAL METHODS OF CONFIGURATIONAL ANALYSIS BASED ON CONSIDERATION OF SINGLE ASYMMETRIC CENTRES

By far the best known of these empiricisms is *Hudson's isorotation rule*, which applies van't Hoff's 'principle of optical superposition' to the anomeric centres of monosaccharide and their cyclic derivatives.[2] This assumes that the optical rotation of glycosyl compounds is composed of two parts, A from the anomeric centre and B from the remainder of the molecule **A4.1**. By implication, therefore, structural changes in one position of a molecule do not appreciably alter the rotational contribution of the other. (This postulate has been found to be invalid in very general terms, but with restrictions it can still be applied successfully. Thus, the contribution A depends, for pyranoid compounds, upon whether the C-2 hydroxyl is equatorial or axial. B, while remaining essentially constant for free sugars, alkyl glycosides and disaccharides, varies considerably when the grouping X becomes highly polarizable as in aryl, benzyl and allyl glycosides or in glycosyl halides.)

A4.1

A is taken to be positive for α-D- and β-L-compounds and negative for β-D- and α-L-derivatives, so that the former are more dextrorotatory than the latter (for exceptions see below), and A and B can be evaluated from a knowledge of the rotations of anomeric pairs. The molecular rotations of methyl α- and β-D-xylopyranoside ($[\phi] = +252$ and -107, respectively), for example, can be expressed as $B + A$ and $B - A$, so that A and B are 72.5 and 179.5; A can then be used to calculate the rotations of other methyl glycosides (provided the C-2–OH bond remains equatorial); B can be applied to other xylopyranosyl compounds. Thus, the expected optical rotations of glycosyl derivatives can be calculated from information gained from related compounds, and this leads to the determination of the anomeric configurations of new glycosides and disaccharides. However, as might be expected from a consideration of the basis of polarimetric measurements (see above), the isorotation rules have been found to break down completely in several instances where chromophoric groups are close to the anomeric centre. For example, pyrimidine nucleosides such as thymidine and its anomer **A4.2**,[3] certain unsaturated compounds **A4.3**[4] and some amino sugar derivatives **A4.4**[5] have all been shown to be anomalous, the α-D-anomers being the less dextrorotatory members of anomeric pairs. In the second of these examples the Hudson empiricism contradicts a further observation[6] that cyclohexenols containing the unit **A4.5** are less dextrorotatory than their epimers.

A4.2   A4.3   A4.4

A4.5

Attempts to apply the principle of optical superposition to asymmetric centres other than the anomeric have not met with the same success. However, other generalizations have been noted which allow configurational assignments to be made at these other positions. It has been recognized, for example, that amides,[7] phenylhydrazides[8] and benzimidazoles[9] formed from aldonic acids are more dextrorotatory than the acids themselves when the configuration at C-2 is D. Also, the N-benzylphenylhydrazones derived from aldoses are laevorotatory when they have this configuration at C-2, and if C-1 and C-2 form part of a heterocyclic system, as in the phenylosotriazoles (obtained by oxidation of the phenylosazones), then the rotation is most strongly affected by the configuration at C-3.[10] In this case dextrorotation indicates the D-configuration at this centre and vice versa. The configuration at position 4 or 5 of an aldonic acid can be determined by *Hudson's lactone rule*,[11] which specifies that when lactonization is accompanied by a positive change in rotation, a centre with D-stereochemistry is involved. Such a method may depend on a knowledge of the lactone ring size, but since this is usually readily determined by infrared or n.m.r. measurements no restriction is placed on application of the principle.

A correlating generalization[12] has postulated that when any asymmetric carbon is viewed from the centre of the ring so that the bulkier group (B) is below the ring and the smaller (A) is above, if the larger adjacent ring group (L) is on the right of the smaller group (S), as in **A4.6**, this epimer is the more dextrorotatory. Relative 'sizes' of common ring groups are given as in Scheme A4.1

Although polarizability rather than the size of neighbouring groups is fundamentally more significant in determining their effect, these empirical rules are consistent with observations for many pairs of compounds. Therefore, the compounds

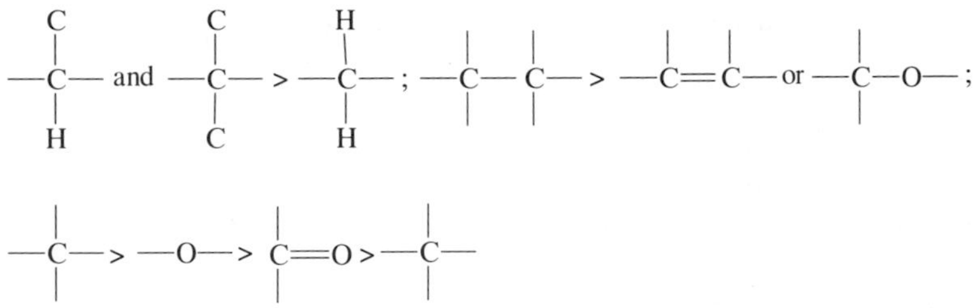

Scheme A4.1

having the structural features **A4.7** (α-D- or β-L-aldose), **A4.8** (e.g. the γ-lactone formed with a D-C-4 hydroxyl group), **A4.9** (a glycal with the D-configuration at C-3; C-4–C-5 >C-2=C-1) and **A4.10** (aldose, glycoside or lactone with the D-configuration at C-2) are all more dextrorotatory than their epimers. To obtain configurational information by application of these empirical rules it is necessary to know the optical rotation of both members of an epimeric pair, or that of a compound under consideration and of the deoxy derivative which lacks asymmetry at the centre concerned;[13] the approach thus has its limitations.

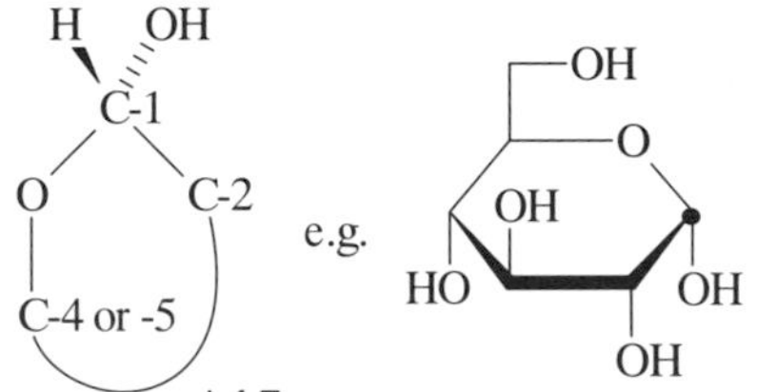

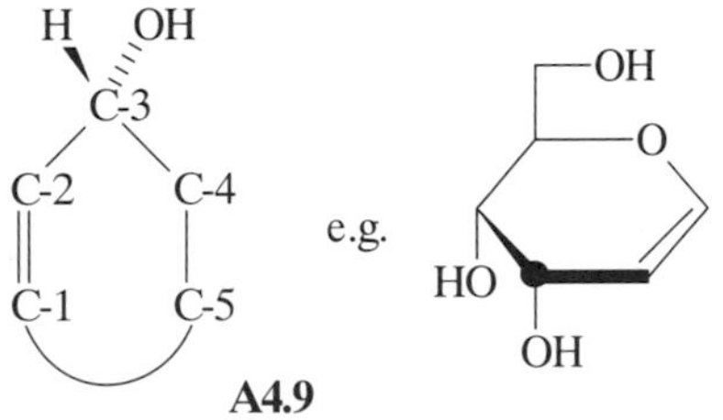

A4.3   EMPIRICAL METHODS OF CONFIGURATIONAL AND CONFORMATIONAL ANALYSIS BASED ON CONSIDERATION OF COMPLETE MOLECULES

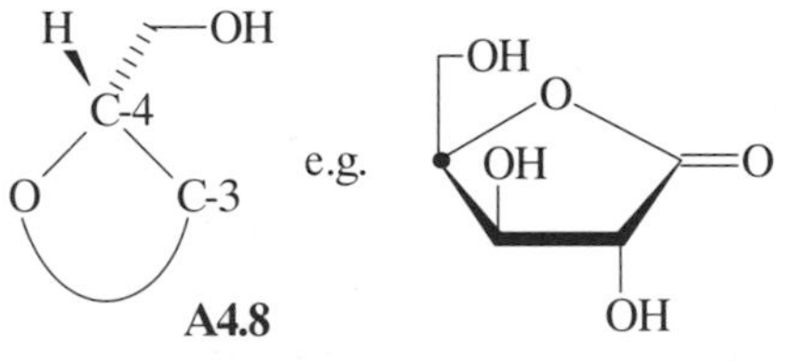

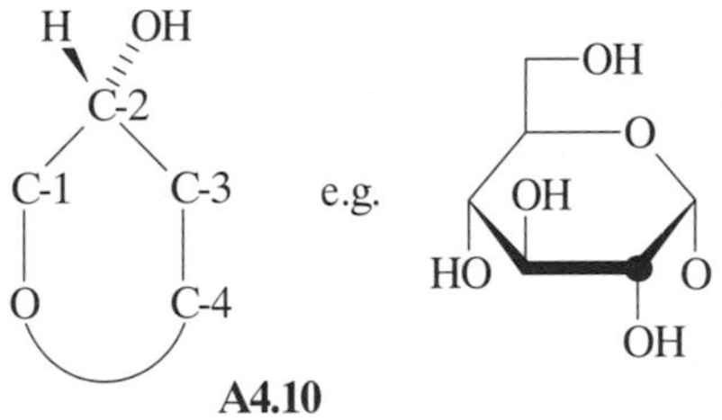

Early attempts to correlate optical rotations of carbohydrates with overall structures by application of the superposition principle did not meet with the success Hudson found in his analyses of the anomeric centre. The first significant progress was made by Whiffen[14] who analysed molecular rotation in terms of contributions from asymmetric bonds and established a set of six structural parameters, each assigned a rotational value, from which the molecular rotation of any pyranose sugar or glycoside in either regular chair conformation in aqueous solution could be calculated. Following this approach and their own observation that some conformationally labile monosaccharide derivatives adopt different conformations and have considerably different rotations in different solvents, Lemieux and Martin[15] simplified and extended the Whiffen approach[16] and that of Brewster.[17] They reduced the structural parameters to four which are based on pair-wise, vicinal interactions between *gauche*-related oxygen and carbon atoms spanned by either carbon–carbon or carbon–oxygen linkages, and they recognized that the rotamer states of the hydroxymethyl groups of hexopyranose compounds and the aglycons of glycosides make contributions to the optical rotations. For any entirely empirical method the Lemieux–Martin approach is remarkably useful.

A4.4   MORE FUNDAMENTAL APPROACHES

Apart from the above empirical approaches to the correlation of structure with the optical activity of sugars, semiempirical methods have been attempted; *ab initio* procedures are as yet, however, impractical. With the semiempirical approach, for example, polarizability theory has been used to establish the validity of the absolute configurations of the sugars that were assumed by Fischer (Section 2.1.4), Hudson's rules and the conformations adopted by sugars in solutions.[18] A deeper approach takes into account the electronic transitions on which optical activity depends and the effects of structure on them. It has been used with some success to calculate rotations of free sugars[19] and the solution conformations of disaccharides.[20]

A4.5   OPTICAL ROTATORY DISPERSION

Investigations of optical rotations as a function of wavelength of measurement can provide more meaningful information than measurements at single wavelengths, particularly with compounds having chromophores in the accessible part of the ultraviolet for which Cotton effects (Figure A4.1) are measurable. Most simple sugars and their glycosides give plain o.r.d. curves down to 200 nm,[21,22] that, for example, for methyl α-D-xylopyranoside becoming positive (increasing rapidly in the positive sense as 200 nm is approached) and that for the anomer being negative. For D-galactose the curve is exceptional, showing a maximum at 210 nm, but this is not the first extremum of a Cotton effect but simply an anomaly brought about by summing the o.r.d. curves of the two pyranoid anomers of the sugar present in equilibrium. By comparisions of the curves derived from pairs of compounds the contributions to the o.r.d. curves at long wavelengths and near 200 nm of specific structural features have been assessed.[22] For example, an α-hydroxyl group at the anomeric centre of a D-pyranose sugar contributes positively at both 200 nm and at long wavelengths (Hudson's rule), while a β-group makes negative contributions. At C-2, however, the effect of a D-substituent (e.g. in D-glucose) is positive at long wavelengths and negative at 200 nm; an L-substituent (e.g. D-mannose) makes the opposite contribution. These observations concur with the postulates of the various empirical methods relating to optical activity at long wavelengths (see above).

Monosaccharide derivatives with carbonyl chromophores, e.g. 2,3,4,5,6-penta-O-acetyl-D-glucose **A4.11**, give 'anomalous' o.r.d. curves with Cotton effects centred in the region of 290 nm, and for some rigid ketones the octant rule applies.[23] For several acyclic sugar derivatives containing chromophoric groups (e.g. 1-deoxy-1-nitroalditols **A4.12** and **A4.13**) the signs of the Cotton effects near 275 nm can be correlated with the configurations at C-2 (see Figure A4.2), but in the cases of the aldose diethyl dithioacetals other asymmetric centres impinge on the measured effects.[1]

A4.11

Fischer projection:
=O
—OAc
AcO—
—OAc
—OAc
—OAc

A4.12

Fischer projection:
—NO₂
—NHAc
HO—
HO—
—OH
—OH
—OH

A.4.13

Fischer projection:
—NO₂
AcHN—
HO—
HO—
—OH
—OH
—OH

A4.14

(pyranose ring with —OAc, O, AcO substituents, anomeric O—C₆H₄—NO₂)

While epimers **A4.3** are not just anomalous with respect to Hudson's rules at 589 nm but over the whole spectral range, the p-nitrophenyl glycosides **A4.14** give o.r.d. curves that cross at 290 nm near the centre of the Cotton effects, and they are anomalous only below this wavelength.[4] The situation with the nucleosides is complex, observed curves depending on secondary factors such as the relative orientations of the sugar and base rings, but many α- and β-D-glycosylpyrimidines give negative and positive curves, respectively, which is consistent with the anomalous $[\alpha]$ values of thymidine and its anomer **A4.2**. Purine-containing nucleosides behave in the opposite fashion.

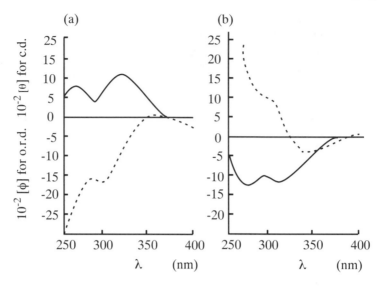

Figure A4.2 O.r.d. (broken lines) and c.d. (solid lines) curves of (a) 2-acetamido-1,2-di-deoxy-1-nitro-D-*glycero*-D-*galacto*-heptitol **A4.12** and (b) 2-acetamido-1,2-dideoxy-1-nitro-D-*glycero*-D-*talo*-heptitol **A4.13** (reproduced by permission of *Carbohydr. Res.*, 1967, **5**, p. 140)

A4.6 CIRCULAR DICHROISM

Although circular dichroism, which measures the difference in absorption between left- and right-circularly polarized light by an asymmetric compound, provides essentially the same information as does o.r.d., it can be applied more profitably since observed Cotton effects tend to be simpler and less subject to interference (Figure A4.2). Modern instruments permit measurements in the vacuum ultraviolet down to 165 nm and free sugars and glycosides have been found to display Cotton effects near 175 nm. The signs of the effects are positive and negative for α-D- and β-D-glycosides, respectively.[24] Carbohydrate acids can be examined by way of their Cotton effects near 215 nm and lower wavelengths, and 2-acetamido-2-deoxy sugars show effects near 210 and 190 nm.

For alditols having heterocyclic bases bonded to C-1 the signs of the Cotton effects observed relate to the dipole moment vectors of the heterocycles and the alditol-1-yl groups, i.e. they depend on the configuration at C-2.[25] Studies of nucleosides have led to rules correlating signs of Cotton effects with conformational features.[26]

Using Cotton effects related to $\pi-\pi^*$ intramolecular charge transfer transitions near 225 nm, which occur in dibenzoates of α-diols, Nakanishi has developed his 'exciton chirality method' as a means of determining the chirality of the diols and hence their configurations and conformations.[27] Extensions have led to methods for the structural analysis of complex carbohydrates.[28]

A4.7   REFERENCES

1. *The Carbohydrates* (eds W. Pigman and D. Horton), Vol. IB, Academic Press, New York, 1980, p. 1354.
2. *J. Am. Chem. Soc.*, 1909, **31**, 66; *J. Am. Chem. Soc.*, 1916, **38**, 1566.
3. *Chem. Ind.*, 1964, 2129.
4. *J. Chem. Soc.*, 1965, 2830.
5. *J. Org. Chem.*, 1967, **32**, 294.
6. *J. Chem. Soc.*, 1952, 4976.
7. *J. Am. Chem. Soc.*, 1918, **40**, 813.
8. *J. Am. Chem. Soc.*, 1917, **39**, 462.
9. *J. Am. Chem. Soc.*, 1942, **64**, 1612.
10. *J. Org. Chem.*, 1968, **33**, 2478.
11. *J. Am. Chem. Soc.*, 1910, **32**, 338.
12. *J. Org. Chem.*, 1958, **23**, 1425.
13. *J. Org. Chem.*, 1963, **28**, 428.
14. *Chem. Ind.*, 1956, 964.
15. *Can. J. Chem.*, 1968, **46**, 1453; *Can. J. Chem.*, 1969, **47**, 4427.
16. *Carbohydr. Res.*, 1970, **13**, 139.
17. *J. Am. Chem. Soc.*, 1959, **81**, 5483.
18. *Bull. Chem. Soc. Jpn.*, 1963, **36**, 473.
19. *Carbohydr. Res.*, 1987, **166**, 181.
20. *Carbohydr. Res.*, 1993, **239**, 1; *J. Am. Chem. Soc.*, 1990, **112**, 7406.
21. *J. Am. Chem. Soc.*, 1964, **86**, 3160.
22. *J. Am. Chem. Soc.*, 1965, **87**, 1765.
23. *Chem. Ber.*, 1967, **100**, 2317.
24. *Adv. Carbohydr. Chem. Biochem.*, 1987, **45**, 73.
25. *Carbohydr. Res.*, 1977, **59**, 11.
26. *J. Am. Chem. Soc.*, 1969, **91**, 831; *J.Am. Chem. Soc.*, 1971, **93**, 1600; *J. Am. Chem. Soc.*, 1972, **94**, 5487.
27. *Acc. Chem. Res.*, 1972, **5**, 257.
28. *Pure Appl. Chem.*, 1984, **56**, 1031; *Pure Appl. Chem.*,1989, **61**, 1193; *Carbohydr. Res.*, 1988, **176**, 175.

Appendix 5: Modified Sugars Found in (mainly) Microbiological Sources

Table A5.1

Trivial name	Formal name
Acosamine	3-Amino-2,3,6-trideoxy-L-*arabino*-hexose
Actinosamine	3-Amino-2,3,6-trideoxy-4-*O*-methyl-L-*arabino*-hexose
Aculose	2,3,6-Trideoxy-L-*glycero*-hexos-2-en-4-ulose
Aldgarose	4,6-Dideoxy-3-*C*-(1′-hydroxyethyl)-D-*ribo*-hexose 3,1′-carbonate
Amicetose	2,3,6-Trideoxy-D-*erythro*-hexose
Amosamine	4,6-Dideoxy-4-dimethylamino-D-glucose
Angolosamine	2,3,6-Trideoxy-3-dimethylamino-D-*xylo*-hexose
Arcanose	2,6-Dideoxy-3-*C*-3-*O*-dimethyl-L-*xylo*-hexose
Avidinosamine	3-Amino-2,3,6-trideoxy-3-*C*-methyl-L-*ribo*-hexose
Axenose	2,6-Dideoxy-3-*C*-methyl-L-*xylo*-hexose
Bacillosamine	2,4-Diamino-2,4,6-trideoxy-D-glucose
Bamosamine	4,6-Dideoxy-4-methylamino-D-glucose
Celestosamine	6-*O*-amino-6,8-dideoxy-7-*O*-methyl-D-*erythro*-D-*galacto*-octose
Chalcose	4,6-Dideoxy-3-*O*-methyl-D-*xylo*-hexose
Chromose A (olivinose)	2,6-Dideoxy-3-*O*-methyl-D-*lyxo*-hexose
Chromose B	3-*O*-Acetyl-2,6-dideoxy-3-*C*-methyl-L-*arabino*-hexose
Chromose C (olivose)	2,6-Dideoxy-D-*arabino*-hexose
Cinerulose	2,3,6-trideoxy-D- or L-*glycero*-hexos-4-ulose
Cladinose	2,6-Dideoxy-3-*C*-3-*O*-dimethyl-L-*ribo*-hexose
Curacose	6-Deoxy-4-*O*-methyl-D-galactose
Cymarose	2,6-Dideoxy-3-*O*-methyl-D-*ribo*-hexose
Daunosamine	3-Amino-2,3,6-trideoxy-L-*lyxo*-hexose
Deliconitrose	2,3,6-Trideoxy-3-*C*-methyl-3-nitro-D- or -L-*ribo*-hexose
Desosamine	3,4,6-Trideoxy-3-dimethylamino-D-*xylo*-hexose
Digitoxose	2,6-Dideoxy-D- or -L-*ribo*-hexose
Dihydrostreptose	5-Deoxy-3-*C*-hydroxymethyl-L-lyxose
Evermicose	2,6-Dideoxy-3-*C*-methyl-D-*arabino*-hexose
Evernitrose	2,3,6-Trideoxy-3-*C*-4-*O*-dimethyl-3-nitro-L-*arabino*-hexose
Forosamine	2,3,4,6-Tetradeoxy-4-dimethylamino-D-*erythro*-hexose
Garosamine	3-Deoxy-4-*C*-methyl-3-methylamino-L-arabinose
Gentosamine	3-Deoxy-3-methylamino-D-xylose
Holacosamine	4-Amino-2,4,6-trideoxy-3-*O*-methyl-D-*xylo*-hexose
Holosamine	4-Amino-2,4,6-trideoxy-3-*O*-methyl-D-*ribo*-hexose
Hydroxystreptose	3-*C*-Formyl-L-lyxose
Janose	4,6-Dideoxy-D-*lyxo*-hexose

Table A5.1 (*continued*)

Trivial name	Formal name
Kanosamine	3-Amino-3-deoxy-D-glucose
Kasugamine	2,4-Diamino-2,3,4,6-tetradeoxy-D-*arabino*-hexose
Kijanose	2,3,4,6-Tetradeoxy-3-*C*-methyl-4-(methoxycarbonyl)-amino-3-*C*-nitro-D-*xylo*-hexose
Labilose	6-Deoxy-2,4-di-*O*-methyl-D-galactose
Lincosamine	6-Amino-6,8-dideoxy-D-*erythro*-D-*galacto*-octose
Lividosamine	2-Amino-2,3-dideoxy-D-*ribo*-hexose
Megosamine	2,3,6-Trideoxy-3-dimethylamino-L-*ribo*-hexose
Mycaminose	3,6-Dideoxy-3-dimethylamino-D-glucose
Mycarose	2,6-Dideoxy-3-*C*-methyl-L-*ribo*-hexose
Mycinose	6-Deoxy-2,3-di-*O*-methyl-D-allose
Mycosamine	3-Amino-3,6-dideoxy-D-mannose
Nebrosamine	2,6-Diamino-2,3,6-trideoxy-D-*ribo*-hexose
Neosamine B (paromose)	2,6-Diamino-2,6-dideoxy-L-idose
Neosamine C	2,6-Diamino-2,6-dideoxy-D-glucose
Nogalose	6-Deoxy-tetra-3-*C*-2,3,4-*O*-methyl-L-mannose
Nojirimycin	5-Amino-5-deoxy-D-glucose
Noviose	6-Deoxy-5-*C*-4-*O*-dimethyl-L-*lyxo*-hexose
Oleandrose	2,6-Dideoxy-3-*O*-methyl-L-*arabino*-hexose
Oliose	2,6-Dideoxy-D-*lyxo*-hexose
Olivomycose	2,6-Dideoxy-3-*C*-methyl-L-*arabino*-hexose
Perosamine	4-Amino-4,6-dideoxy-D-mannose
Pillarose	2,3,6-Trideoxy-4-*C*-glycolyl-L-*threo*-hexose
Prumycin	4-*N*-(D-alanyl)amino-2-amino-2,4-dideoxy-L-arabinose
Purpurosamine A	2-Amino-2,3,4,6,7-pentadeoxy-6-methylamino-D-*ribo*-heptose
Purpurosamine B	2,6-Diamino-2,3,4,6,7-pentadeoxy-D-*ribo*-heptose
Purpurosamine C	2,6-Diamino-2,3,4,6-tetradeoxy-D-*erythro*-hexose
Quinovose A	2,6-Dideoxy-3-*C*-(1′-hydroxyethyl)-L-*xylo*-hexose
Quinovose B	3-*C*-Acetyl-2,6-dideoxy-L-*xylo*-hexose
Rhodinose	2,3,6-Trideoxy-L-*threo*-hexose
Rhodosamine	2,3,6-Trideoxy-3-dimethylamino-L-*lyxo*-hexose
Ristosamine	3-Amino-2,3,6-trideoxy-L-*ribo*-hexose
Rubranitrose	2,3,6-Trideoxy-3-*C*-4-*O*-dimethyl-3-*C*-nitro-D-*xylo*-hexose
Sibirosamine	4,6-Dideoxy-3-*C*-methyl-4-methylamino-D-mannose
Sisosamine	2,6-Diamino-2,3,4,6-tetradeoxy-D-*glycero*-hex-4-enose
Streptose	5-Deoxy-3-*C*-formyl-L-lyxose
Thomosamine	4-Amino-4,6-dideoxy-D-galactose
Tolyposamine	4-Amino-2,3,4,6-tetradeoxy-L-*erythro*-hexose
Vancosamine	3-Amino-2,3,6-trideoxy-3-*C*-methyl-L-*lyxo*-hexose
Vinelose	6-Deoxy-3-*C*-2-*O*-dimethyl-L-talose
Viosamine	4-Amino-4,6-dideoxy-D-glucose

General Index

Acarbose, 520
Acetals, 389
  of alditols, 399
  of free sugars, 395
  of glycosides, 393
Acetates
  general, 361
  photolysis, 365
  syntheses,
    by displacement reactions, 192, 199, 202
Acetic anhydride with DMSO, as oxidizing agent, 292
Acetobacter suboxydans, **49**, 299, 311
Acetolysis,
  of cellulose, 84
  of glycosides, 84
  of polysaccharides, 481
Acetonitrile as a carbon nucleophile, 303
Acetonitrilium ion, 161
Acetoxonium ion intermediates, 148, 150, 152, 168, 173
  to 1,2-orthoesters, 157, 176
3-Acetoxytetrahydropyran,
  conformation, 25
Acetyl hypobromite, 330
N-Acetylneuraminic acid, 48, 311, 479, 505
  in disaccharides, 100
Acosamine, 546
Acrolein, 53
Acrolein dimer, 54
Acrose, 53
Actinosamine, 546
Aculose, 546
Acyclic carbohydrates,
  alditols, 125
  aldose peracetates, 363
  conformations, 35
  dimethylacetals, 68
  dithioacetals, 94
  free sugars, 11, 43
Acyclic compounds, synthesis from carbohydrates, 456
Acyclovir, 513

Acylation,
  of amino sugars, 230
  enzymic, 367
N-Acylimidazoles, 364
Acyloxonium ion intermediates,
  (*see also* Acetoxonium) 247, 368
Addition reactions,
  to carbon double bonds, 84, 92, 214, 225, **319, 330**
Adenine, 511
Adenosine, 511
Adenosine 5'-triphosphate, 514
Adriamycin, 510
Agar, 493
Agaropectin, 493
Agarose, 493
Aglycon, defintion, 62
Agrobacterium, 80
AIDS virus, 416, 441, 513
Alanine, 496
Albersheim, P., 470
Alcaligenes faecalis, 495
Aldaric acids, 138
Aldehydo derivatives, 96, 110
Aldgamycin E, 285
Aldgarose, 281, 284, 303, 370, 546
Alditols,
  acetals, 399
  conformations, 35
  oxidation to ketoses, 49, 299, 311
  syntheses, 125
Aldobiouronic acids, 75, 482
Aldoheptoses, 19
Aldohexoses, 16
  chemical synthesis, 53
Aldol condensation, 123
Aldolase, 50
Aldonamides, 134
Aldonic acids, 126
  equilibrium with lactones, 128
  syntheses,
    from aldoses, 119, 126
    from β-glycosides, 87
Aldonolactones, 128

Selective Index of Compounds

The majority of derivatives of monosaccharides discussed in the text are not named in the General Index. This additional compound index refers to sources of data on the preparation of an arbitrarily selected set which are of importance in synthetic chemistry.